JACARANDA

CHEMISTRY 2

VCE UNITS 3 AND 4 | THIRD EDITION

JACARANDA CHEMISTRY 2

VCE UNITS 3 AND 4 | THIRD EDITION

ROBERT STOKES

ANGELA STUBBS

NEALE TAYLOR

BEN WILLIAMS

JASON BOURKE

MAIDA DERBOGOSIAN

CONTRIBUTING AUTHOR

Von Hayes

jacaranda
A Wiley Brand

Third edition published 2024 by
John Wiley & Sons Australia, Ltd
155 Cremorne Street, Cremorne, Vic 3121

First edition published 2017
Second edition published 2020

Typeset in 10.5/13 pt TimesLTStd

ISBN: 978-1-119-88614-3

Front cover image: © Valenty/Shutterstock

Illustrated by various artists, diacriTech and Wiley Composition
Services

Typeset in India by diacriTech

A catalogue record for this
book is available from the
National Library of Australia

The Publishers of this series acknowledge and pay their respects
to Aboriginal Peoples and Torres Strait Islander Peoples as the
traditional custodians of the land on which this resource was
produced.

This suite of resources may include references to (including
names, images, footage or voices of) people of Aboriginal
and/or Torres Strait Islander heritage who are deceased. These
images and references have been included to help Australian
students from all cultural backgrounds develop a better
understanding of Aboriginal and Torres Strait Islander Peoples'
history, culture and lived experience.

It is strongly recommended that teachers examine resources on
topics related to Aboriginal and/or Torres Strait Islander
Cultures and Peoples to assess their suitability for their own
specific class and school context. It is also recommended that
teachers know and follow the guidelines laid down by the
relevant educational authorities and local Elders or community
advisors regarding content about all First Nations Peoples.

The Publisher acknowledges ongoing discussions related to
gender-based population data. At the time of publishing, there
was insufficient data available to allow for the meaningful
analysis of trends and patterns to broaden our discussion of
demographics beyond male and female gender identification.

Printed in Singapore
M WEP224857 110923

Contents

UNIT 4 HOW ARE CARBON-BASED COMPOUNDS DESIGNED FOR PURPOSE? — 353

AREA OF STUDY 1 HOW ARE ORGANIC COMPOUNDS CATEGORISED AND SYNTHESISED?

12 Scientific investigations

Area of Study 3 Review

Practice past VCAA exam questions focused on key science skills.

APPENDIX Periodic table of the elements

About this resource

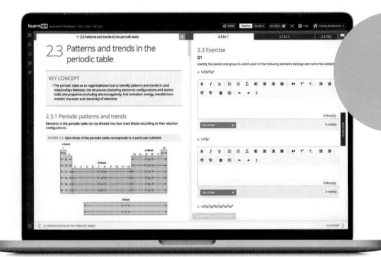

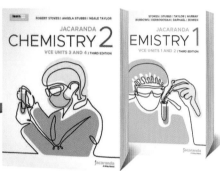

Aligned to the VCE Chemistry Study Design (2023–2027)

JACARANDA
CHEMISTRY 2 VCE UNITS 3 AND 4
THIRD EDITION

Developed by expert Victorian teachers for VCE students

Tried, tested and trusted. The NEW Jacaranda VCE Chemistry series continues to deliver curriculum-aligned material that caters to students of all abilities.

Completely aligned to the VCE Chemistry Study Design

Our expert author team of practising teachers and assessors ensures 100% coverage of the new VCE Chemistry Study Design (2023–2027).

Everything you need for your students to succeed, including:

- **NEW!** Access targeted questions sets including exam-style questions and all relevant past VCAA exam questions since 2013. Ensure assessment preparedness with practice SACs.

- **NEW!** Enhanced practical investigation support including practical investigation videos, and eLogbook with fully customisable practical investigations — including teacher advice and risk assessments.

- **NEW!** Teacher-led videos to unpack challenging concepts, VCAA exam questions, exam-style questions, investigations and sample problems to fill learning gaps after the COVID-19 disruptions.

Learn online with Australia's most

Everything you need for each of your lessons in one simple view

- Trusted, curriculum-aligned theory
- Engaging, rich multimedia
- All the teacher support resources you need
- Deep insights into progress
- Immediate feedback for students
- Create custom assignments in just a few clicks.

Practical teaching advice and ideas for each lesson provided in teachON

Each lesson linked to the Key Knowledge (and Key Science Skills) from the VCE Chemistry Study Design

Reading content and rich media including embedded videos and interactivities

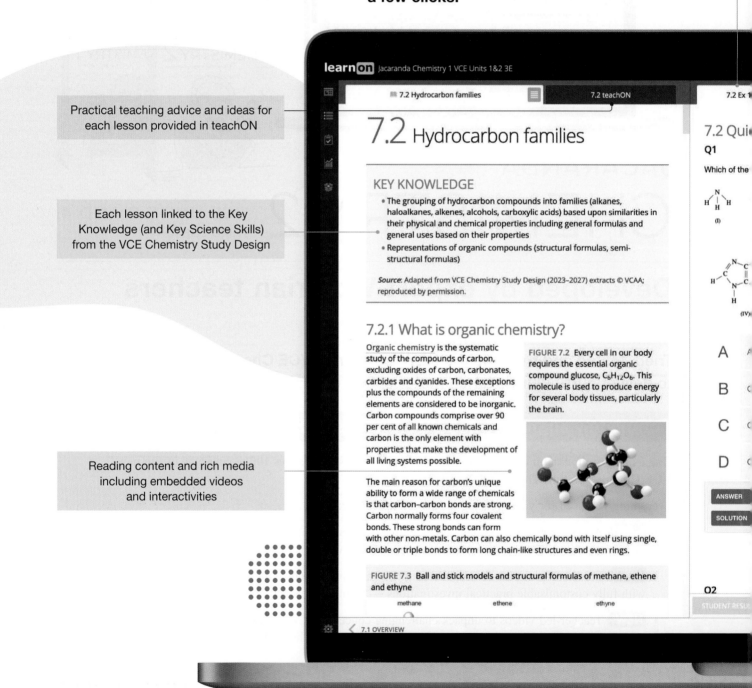

learn**on** Jacaranda Chemistry 1 VCE Units 1&2 3E

7.2 Hydrocarbon families | 7.2 teachON | 7.2 Ex 1

7.2 Hydrocarbon families

KEY KNOWLEDGE
- The grouping of hydrocarbon compounds into families (alkanes, haloalkanes, alkenes, alcohols, carboxylic acids) based upon similarities in their physical and chemical properties including general formulas and general uses based on their properties
- Representations of organic compounds (structural formulas, semi-structural formulas)

Source: Adapted from VCE Chemistry Study Design (2023–2027) extracts © VCAA; reproduced by permission.

7.2.1 What is organic chemistry?

Organic chemistry is the systematic study of the compounds of carbon, excluding oxides of carbon, carbonates, carbides and cyanides. These exceptions plus the compounds of the remaining elements are considered to be inorganic. Carbon compounds comprise over 90 per cent of all known chemicals and carbon is the only element with properties that make the development of all living systems possible.

FIGURE 7.2 Every cell in our body requires the essential organic compound glucose, $C_6H_{12}O_6$. This molecule is used to produce energy for several body tissues, particularly the brain.

The main reason for carbon's unique ability to form a wide range of chemicals is that carbon–carbon bonds are strong. Carbon normally forms four covalent bonds. These strong bonds can form with other non-metals. Carbon can also chemically bond with itself using single, double or triple bonds to form long chain-like structures and even rings.

FIGURE 7.3 Ball and stick models and structural formulas of methane, ethene and ethyne

methane   ethene   ethyne

7.1 OVERVIEW

7.2 Qui

Q1

Which of the

A

B

C

D

ANSWER

SOLUTION

Q2

STUDENT RESU

powerful learning tool, learnON

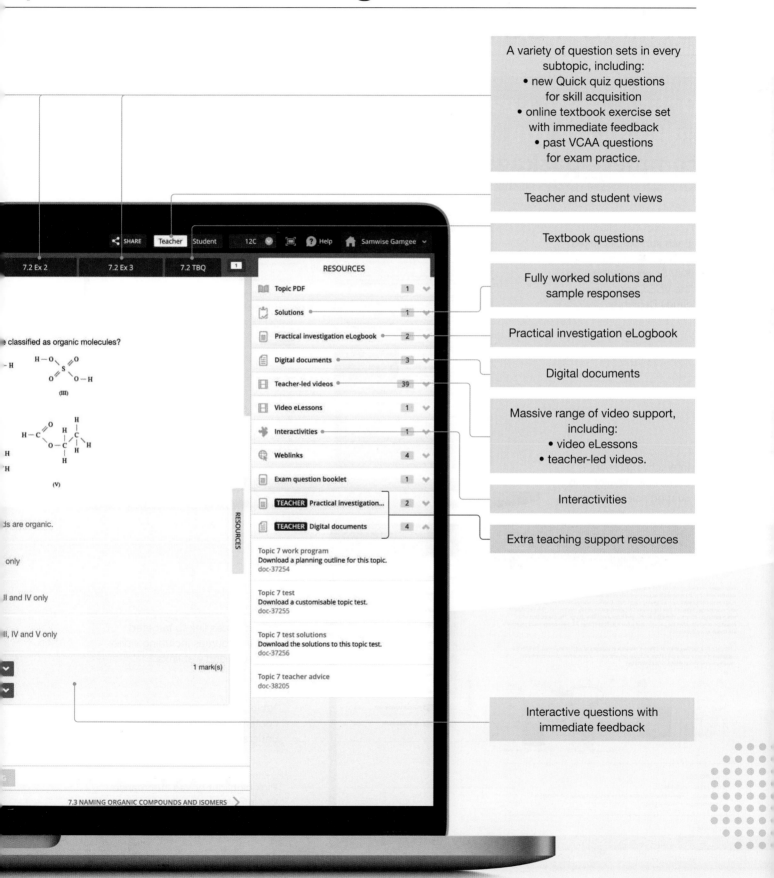

A variety of question sets in every subtopic, including:
- new Quick quiz questions for skill acquisition
- online textbook exercise set with immediate feedback
- past VCAA questions for exam practice.

Teacher and student views

Textbook questions

Fully worked solutions and sample responses

Practical investigation eLogbook

Digital documents

Massive range of video support, including:
- video eLessons
- teacher-led videos.

Interactivities

Extra teaching support resources

Interactive questions with immediate feedback

SHARE   Teacher   Student   12C   Help   Samwise Gamgee

7.2 Ex 2   7.2 Ex 3   7.2 TBQ   1

RESOURCES

Topic PDF   1
Solutions   1
Practical investigation eLogbook   2
Digital documents   3
Teacher-led videos   39
Video eLessons   1
Interactivities   1
Weblinks   4
Exam question booklet   1
TEACHER Practical investigation...   2
TEACHER Digital documents   4

Topic 7 work program
Download a planning outline for this topic.
doc-37254

Topic 7 test
Download a customisable topic test.
doc-37255

Topic 7 test solutions
Download the solutions to this topic test.
doc-37256

Topic 7 teacher advice
doc-38205

e classified as organic molecules?

(III)

(V)

ds are organic.

only

II and IV only

I, IV and V only

1 mark(s)

7.3 NAMING ORGANIC COMPOUNDS AND ISOMERS

Get the most from your online resources

Online, these new editions are the **complete package**

Trusted Jacaranda theory, plus tools to support teaching and make learning more engaging, personalised and visible.

Each subtopic is linked to Key Knowledge (and Key Science Skills) from the VCE Chemistry Study Design.

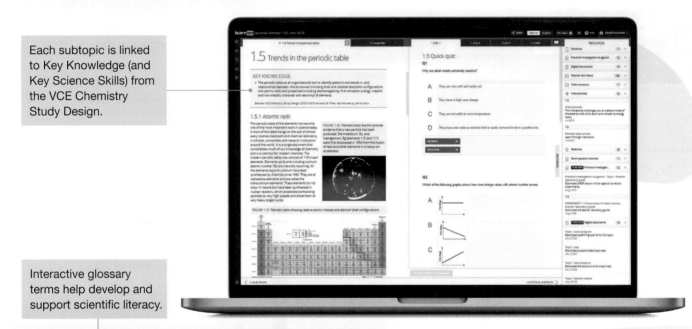

Interactive glossary terms help develop and support scientific literacy.

onResources link to targeted digital resources including video eLessons and weblinks.

Tables and images break down content, allowing students to understand complex concepts.

Pink highlight boxes summarise key information and provide tips for VCE Chemistry success.

Sample problems break down the process of answering questions using a think/write format and a supporting teacher-led video.

Practical investigations are highlighted throughout topics, and are supported by teacher-led videos and downloadable student and teacher version eLogbooks.

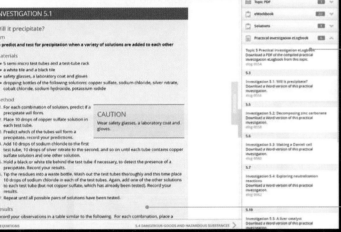

- Online and offline question sets contain practice questions and past VCAA exam questions with exemplary responses and marking guides.
- Every question has immediate, corrective feedback to help students to overcome misconceptions as they occur and to study independently — in class and at home.

Topic reviews

A summary flowchart shows the interrelationship between the main ideas of the topic. This includes links to both Key Knowledge and Key Science Skills.

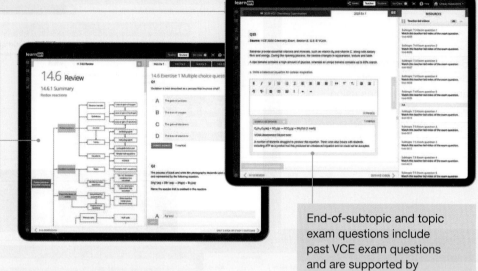

End-of-subtopic and topic exam questions include past VCE exam questions and are supported by teacher-led videos.

Area of Study reviews

Areas of study reviews include practice examinations and practice SACs with worked solutions and sample responses. Teachers have access to customisable quarantined SACs with sample responses and marking rubrics.

Practical investigations eLogbook

Enhanced practical investigation support includes practical investigation videos and an eLogbook with fully customisable practical investigations — including teacher advice and risk assessments.

A wealth of teacher resources

Enhanced teacher support resources, including:

- work programs and curriculum grids
- teaching advice
- additional activities
- teacher laboratory eLogbook, complete with solution and risk assessments
- quarantined topic tests (with solutions)
- quarantined SACs (with worked solutions and marking rubrics).

Customise and assign

A testmaker enables you to create custom tests from the complete bank of thousands of questions (including past VCAA exam questions).

Reports and results

Data analytics and instant reports provide data-driven insights into progress and performance within each lesson and across the entire course.

Show students (and their parents or carers) their own assessment data in fine detail. You can filter their results to identify areas of strength and weakness.

Acknowledgements

The authors and publisher would like to thank the following copyright holders, organisations and individuals for their assistance and for permission to reproduce copyright material in this book.

Selected extracts from the VCE Chemistry Study Design (2023–2027) are copyright Victorian Curriculum and Assessment Authority (VCAA), reproduced by permission. VCE® is a registered trademark of the VCAA. The VCAA does not endorse this product and makes no warranties regarding the correctness and accuracy of its content. To the extent permitted by law, the VCAA excludes all liability for any loss or damage suffered or incurred as a result of accessing, using or relying on the content. Current VCE Study Designs and related content can be accessed directly at www.vcaa.vic.edu.au. Teachers are advised to check the VCAA Bulletin for updates.

Images

• © weerapong/Adobe Stock Photos: **1** • © Kevin Booth/Adobe Stock Photos: **26** • © pressmaster/Adobe Stock Photos: **352** • © elcovalana/Adobe Stock Photos: **18** • © anaumenko/Adobe Stock Photos: **40** • © lainen/Adobe Stock Photos: **252** • © Danijela/Adobe Stock Photos: **105** • © bit24/Adobe Stock Photos: **42** • © Studio_East/Adobe Stock Photos: **316** • © pedro/Adobe Stock Photos: **314** • © Nandalal/Adobe Stock Photos: **290** • © JPC-PROD/Adobe Stock Photos: **443** • © Countrypixel/Adobe Stock Photos: **19** • © Andrew Lambert Photography/Science Photo Library: **113, 448** • © Pannonia/Getty Images: **427** • © moukahal/Stockimo/Alamy Stock Photo: **180** • © Brian Jackson/Alamy Stock Photo: **70** • © dpa picture alliance/Alamy Stock Photo: **14** • © LLC/Alamy Stock Photo: **221** • © Mara Radeva/Alamy Stock Photo: **4** • © Martin Bond/Alamy Stock Photo: **428** • © Pictorial Press Ltd/Alamy Stock Photo: **112** • © picturesbyrob/Alamy Stock Photo: **33** • © Science Photo Library/Alamy Stock Photo: **33, 527** • © janeff/Getty Images: **135** • © photosbyjim/Getty Images: **15** • © Richard Megna/Fundamental Photograhs: **59** • © Sara Lynn Paige/Getty Images: **41** • © Marc McCormack/Newspix: **379** • © Eye Of Science/Science Photo Library: **461** • © Anatomical Travelogue/Science Photo Library: **443** • © Don Farrall/Getty Images: **8** • © Turtle Rock Scientific/Science Source: **216, 484** • © Africa Studio/Shutterstock: **435** • © Ajamal/Shutterstock: **490** • © Aleksandar Grozdanovski/Shutterstock: **367** • © Arif biswas/Shutterstock: **74** • © bancha_photo/Shutterstock: **132** • © BradleyStearn/Shutterstock: **376** • © Chayathorn Lertpanyaroj/Shutterstock: **11** • © Chepko Danil Vitalevich/Shutterstock: **9** • © Coprid/Shutterstock: **296** • © D7INAMI7S/Shutterstock: **32** • © Danijela Maksimovic/Shutterstock: **614** • © Daxiao Productions/Shutterstock: **39** • © Decha Thapanya/Shutterstock: **509** • © Don Bendickson/Shutterstock: **273** • © Elisanth/Shutterstock: **397** • © Elizaveta Galitckaia/Shutterstock: **214** • © Fouad A. Saad/Shutterstock: **26, 499, 543** • © ggw1962/Shutterstock: **581** • © Giovanni Benintende/Shutterstock: **546** • © grebeshkovmaxim/Shutterstock, callumrc/Shutterstock: **356** • © guteksk7/Shutterstock: **309** • © Imagentle/Shutterstock: **366** • © Janaka Dharmasena/Shutterstock: **106** • © Jubal Harshaw/Shutterstock: **43** • © Kate Capture/Shutterstock: **19** • © Kateryna Kon/Shutterstock, StudioMolekuul/Shutterstock: **644** • © Kostiantyn Ablazov/Shutterstock: **213** • © Lightspring/Shutterstock: **39, 442** • © Lilyana Vynogradova/Shutterstock: **283** • © Lubo Ivanko/Shutterstock: **298** • © Luigi Bertello/Shutterstock: **100** • © magnetix/Shutterstock: **116, 540** • © Marchu Studio/Shutterstock: **355, 456** • © Marcin Sylwia Ciesielski/Shutterstock: **398** • © Mike Laptev/Shutterstock: **378** • © motorsports Photographer/Shutterstock: **63** • © Nicku/Shutterstock: **148** • © P_galasso2289/Shutterstock: **419** • © Paramonov Alexander/Shutterstock: **243** • © Pedro Bernardo/Shutterstock: **101** • © Raimundo79/Shutterstock, molekuul_be/Shutterstock: **635** • © Rattiya Thongdumhyu/Shutterstock: **76** • © riopatuca/Shutterstock: **462** • © saoirse2013/Shutterstock: **252** • © Sarah-Jane Walsh/Shutterstock: **416** • © Shulevskyy Volodymyr/Shutterstock: **426** • © Silver Spiral Arts/Shutterstock: **414** • © Sinesp/Shutterstock: **376** • © Sofiaworld/Shutterstock: **394** • © Solis Images/Shutterstock: **274** • © spline_x/Shutterstock: **440** • © ssuaphotos/Shutterstock: **56** • © stable/Shutterstock: **300** • © SUKJAI PHOTO/Shutterstock: **498** • © Tatjana Baibakova/Shutterstock: **439** • © Tewan Banditrukkanka/Shutterstock: **134** • © Triff/Shutterstock: **551** • © Tristan3D/Shutterstock: **104** • © Valentina Proskurina/Shutterstock: **639** • © wavebreakmedia/Shutterstock: **44** • © Wooden Owl/Shutterstock: **102** • © ARTYN F. CHILLMAID/Science Photo Library: **186**

Microbial electrolysis cell, 2010. Public Domain.: **142** • © NIST, 1-Propanol- NIST Chemistry WebBook, 2014. U.S. Secretary of Commerce.: **591** • © NIST, Acetic acid- NIST Chemistry WebBook, 2014. U.S. Secretary of Commerce.: **519** • © Mettler Toledo AG: **499** • © OSweetNature/Shutterstock: **642** • © supachai/Adobe Stock Photos: **178** • © VectorMine/Adobe Stock Photos: **353** • © By Chris Evans, Dr. Roger Peters, Dr. Mike Thompson, Chris Gadsby, Ken Partridge, Roy Mylan, Yehoshua Sivan, Tom Nation, Dr. David Follows, Vikash Hemnath Seeboo - D:\My Webs\index.htm, CC0 https://commons.wikimedia.org/w/index.php?curid= 22874521: **510**

Text

• © Data based on Arthur et al. Evaluating the Potential of Renewable Energy Sources in a Full-Scale Upflow Anaerobic Sludge Blanket Reactor Treating Municipal Wastewater in Ghana. Sustainability. 2023; 15(4):3743. https://doi.org/10.3390/su15043743: **17** • © Data based on World Nuclear Association https://world-nuclear. org/information-library/: **17** • © Liew F., et al. (2016) Gas Fermentation—A Flexible Platform for Commercial Scale Production of Low-Carbon-Fuels and Chemicals from Waste and Renewable Feedstocks. Front. Microbiol. 7:694. Licensed under CC BY 4.0.: **144**

Every effort has been made to trace the ownership of copyright material. Information that will enable the publisher to rectify any error or omission in subsequent reprints will be welcome. In such cases, please contact the Permissions Section of John Wiley & Sons Australia, Ltd.

How can design and innovation help to optimise chemical processes?

Source: VCE Chemistry Study Design (2024–2027) extracts © VCAA; reproduced by permission.

1 Carbon-based fuels

KEY KNOWLEDGE

In this topic you will investigate:

Carbon-based fuels

- the definition of a fuel, including the distinction between fossil fuels (coal, natural gas, petrol) and biofuels (biogas, bioethanol, biodiesel) with reference to their renewability (ability of a resource to be replaced by natural processes within a relatively short period of time)
- fuel sources for the body measured in kJ g^{-1}: carbohydrates, proteins and lipids (fats and oils)
- photosynthesis as the process that converts light energy into chemical energy and as a source of glucose and oxygen for respiration in living things:
  $6CO_2(g) + 6H_2O(l) \rightarrow C_6H_{12}O_6(aq) + 6O_2(g)$
- oxidation of glucose as the primary carbohydrate energy source, including the balanced equation for cellular respiration: $C_6H_{12}O_6(aq) + 6O_2(g) \rightarrow 6CO_2(g) + 6H_2O(l)$
- production of bioethanol by the fermentation of glucose and subsequent distillation to produce a more sustainable transport fuel: $C_6H_{12}O_6(aq) \rightarrow 2C_2H_5OH(l) + 2CO_2(g)$
- comparison of exothermic and endothermic reactions, with reference to bond making and bond breaking, including enthalpy changes (ΔH) measured in kJ, molar enthalpy changes measured in kJ mol^{-1} and enthalpy changes for mixtures measured in kJ g^{-1}, and their representations in energy profile diagrams
- determination of limiting reactants or reagents in chemical reactions
- combustion (complete and incomplete) reactions of fuels as exothermic reactions: the writing of balanced thermochemical equations, including states, for the complete and incomplete combustion of organic molecules using experimental data and data tables.

Source: VCE Chemistry Study Design (2024–2027) extracts © VCAA; reproduced by permission.

PRACTICAL WORK AND INVESTIGATIONS

Practical work is a central component of VCE Chemistry. Experiments and investigations, supported by a **practical investigation eLogbook** and **teacher-led videos**, are included in this topic to provide opportunities to undertake investigations and communicate findings.

EXAM PREPARATION

Access past VCAA questions and exam-style questions and their video solutions in every lesson, to ensure you are ready.

1.1 Overview

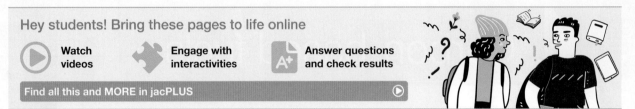

Hey students! Bring these pages to life online

▶ **Watch videos**

✦ **Engage with interactivities**

A⁺ **Answer questions and check results**

Find all this and MORE in jacPLUS ▶

1.1.1 Introduction

Due to the growing human population and invention of new technologies, the energy needs of human societies are immense and continue to increase. By 2050, the world's population is predicted to be over 9 billion. This provides an ever-increasing challenge to both feed people and meet fuel needs. Scientists are continually looking for ways to meet the increased demand for energy in a sustainable and environmentally responsible way. Currently, however, we still rely mainly on fossil fuels to meet our heating, electrical and transport needs.

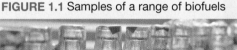

FIGURE 1.1 Samples of a range of biofuels

As we look to find new sources of energy and technologies, we must also look to improve the efficiency of the ones we currently use — that is, harnessing a greater percentage of the energy stored in the bonds of fuels to do the work we need. Engineers continually work on producing engines that combust fuel more efficiently, as well as developing technology that allows for the use of biofuel blends. Bioethanol is widely used across the globe, and biodiesel blends used in heavily populated countries like the United States — which accounts for over 20 per cent of biodiesel consumption — have increased tenfold over the last 20 years.

As we search for better ways to grow crops and extract vegetable oil for fuel, we also seek to develop more sustainable farming practices and energy usage. Biogas from agriculture is now used so that the greenhouse gas methane is combusted for energy use, rather than being released directly into the atmosphere. Anaerobic bacterial action also produces biogas from sugar cane, rice hulls (husks) and other organic waste from farming and raw food production. This topic will provide you with an overview of fuel types, food molecules and the means of obtaining energy to sustain our lives and lifestyles.

LEARNING SEQUENCE

on Resources

📋 **Solutions**   Solutions — Topic 1 (sol-0828)

🔬 **Practical investigation eLogbook**   Practical investigation eLogbook — Topic 1 (elog-1700)

📄 **Digital documents**   Key science skills — VCE Chemistry Units 1–4 (doc-37066)
                                          Key terms glossary — Topic 1 (doc-37281)
                                          Key ideas summary — Topic 1 (doc-37282)

📄 **Exam question booklet**   Exam question booklet — Topic 1 (eqb-0112)

1.2 What are fuels?

A **fuel** is a substance that can undergo a reaction to release energy. In this topic, we are exploring organic fuels that undergo chemical reactions with oxygen (O_2) gas and release energy as the waste products are formed. The energy released can then do work and sustain life.

Apart from food to live and function, our primary requirements for energy are for heating, transport and generating electricity (a **secondary fuel**).

Currently, approximately 93 per cent of Australia's energy needs are met using fossil fuels. However, there has been a decline in the total energy requirements being met by oil and coal, and growth in the consumption of energy from **renewable** means — largely from solar and wind for electricity generation.

Australia's energy requirements have increased sixfold in the last 50 years, and recent indications are that governments will implement strategies and policies to reduce the use of and reliance on fossil fuels for our energy needs.

fuel a substance that burns in air or oxygen to release useful energy

secondary fuel a fuel that is produced from another energy source

renewable (with reference to energy sources) energy sources that can be produced faster than they are used

FIGURE 1.2 Australian energy consumption by fuel type

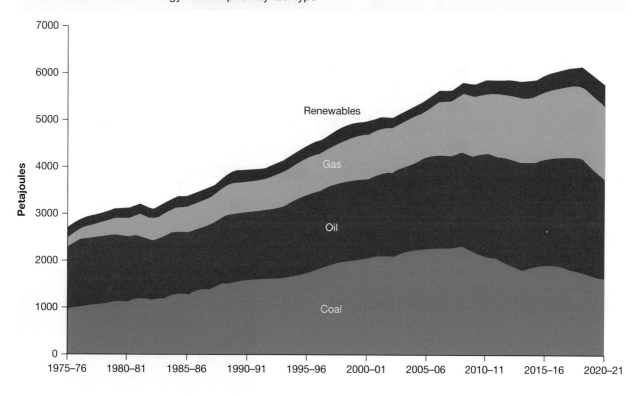

1.2.1 Fossil fuels

Fossil fuels are named as such because they are carbon-based fuels that are made over millions of years. They form because of geological processes on organic matter like algae and plants.

Fossil fuels provide approximately 80 per cent of the primary energy needs of industrialised nations. Worldwide, China consumes the largest amount of energy from fossil fuels. Although Australia consumes far less (87 per cent less) energy from fossil fuels compared to China, we sit second behind the United States for the most energy consumed per person (per capita) from fossil fuels. We also produce the most coal per person in the world.

The most common fossil fuels used are coal, natural gas, petroleum and liquefied petroleum gas (LPG).

Coal

Coal is made up of carbon, oxygen, water and traces of other elements. Over millions of years, pressure, temperature, bacterial action and moisture changed organic matter into our coal reserves. Brown coal is estimated to have formed over a period of 23–60 million years, while black coal formed over a period of 145–299 million years.

> **fossil fuels** fuels formed from once-living organisms
>
> **coal** the world's most plentiful fossil fuel; it is formed from the combined effects of pressure, temperature, moisture and bacterial decay on vegetable matter over several hundred million years

FIGURE 1.3 The steps in coal formation

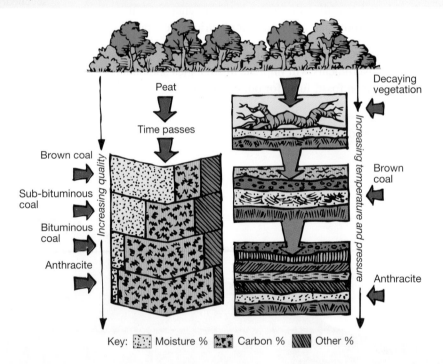

Because most of Australia's coal seams are near the surface, the majority (approximately 80 per cent) of coal is obtained by open-cut mining. This involves the top layer of soil being removed, and then explosives are used to blast the coal into pieces. This makes coal mining relatively cheap and easy in Australia.

FIGURE 1.4 An open-cut coal mine

CASE STUDY: Coal mining and use in Australia

Australia's coal-mining history dates back to 1799, in Newcastle, NSW. Australia has significant coal reserves and is the largest exporter of coal in the world, and the fourth largest producer, behind China, India and the United States. Around 90 per cent of black coal produced in Australia is exported overseas. Approximately 64 per cent of electricity generation in our National Electricity Market (NEM), which supplies 80 per cent of our country, is from coal. Australia now has only 22 operational coal-fired power stations, following the closure of the Hazelwood (Vic, in 2017) and Liddell (NSW, in 2023) power stations, and this number will continue to reduce throughout the next two decades.

FIGURE 1.5 Australian NEM electricity generation, as of March 2023

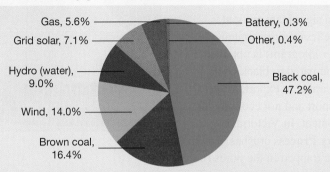

- Gas, 5.6%
- Grid solar, 7.1%
- Hydro (water), 9.0%
- Wind, 14.0%
- Brown coal, 16.4%
- Battery, 0.3%
- Other, 0.4%
- Black coal, 47.2%

Source: Based on ACCC data, *Generation capacity and output by fuel source – NEM*, accessed on 29 May 2023, https://www.aer.gov.au/wholesale-markets/wholesale-statistics/generation-capacity-and-output-by-fuel-source-nem.

When coal is burned, its stored chemical potential energy is converted into heat energy. This heat energy is used to convert water into steam, so heat energy is converted into kinetic energy. The steam flows past a turbine, so the kinetic energy of the steam is converted into mechanical energy in the spinning turbine. The turbine is connected to a generator, which converts mechanical energy into electrical energy. Electrical energy may then be used to power a wide range of appliances in the home and in industry.

FIGURE 1.6 Energy conversion in power stations

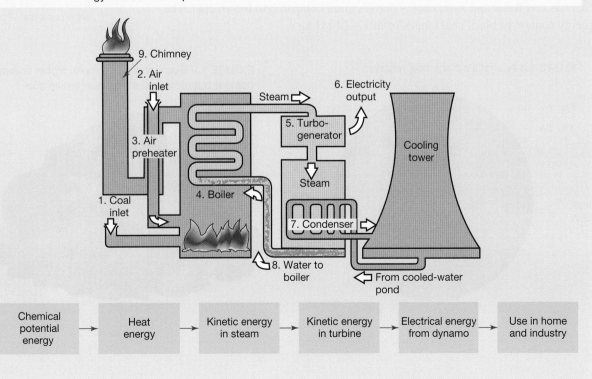

| Chemical potential energy | → | Heat energy | → | Kinetic energy in steam | → | Kinetic energy in turbine | → | Electrical energy from dynamo | → | Use in home and industry |

Brown coal

The Latrobe Valley in Victoria contains an estimated 25 per cent of the world's known reserves of brown coal. By global standards and compared to other sources of coal, Victoria's brown coal is low in impurities like sulfur and nitrogen, which means less oxides of sulfur, nitrogen and other pollutants are emitted into the atmosphere. However, brown coal is still a relatively 'heavy polluter', and pollution emissions are an environmental concern.

Victoria's brown coal has a significant amount of water in it, and therefore less energy content (6–12 **megajoules** (MJ) per kilogram) is obtained from undried (wet) coal compared to black coal. Brown coal is also less desirable to export, as wet coal can be unsafe to transport and not economical due to its high moisture content. In Victoria, a technique called the **Coldry Process** crushes brown coal to release the moisture trapped in the pores (holes) of the lumps. This makes energy production more efficient and raises the energy content per kilogram of brown coal.

Due to a combination of increased demand for renewable energy, environmental concerns and aging power plants, Victoria's fleet of brown-coal-fired power stations is predicted to be entirely closed in the next decade.

Black coal

Black coal is formed the same way as brown coal, except it is subjected to high temperatures and pressure for longer — around 6 to 10 times longer. As such, black coal has less water and more energy per kilogram compared to brown coal. The energy content of black coal ranges from 17–24 MJ kg^{-1}.

FIGURE 1.7 Almost half of the energy stored in coal is lost as heat from steam emissions by the cooling towers at Loyang B Power Station, Latrobe Valley, Victoria

megajoule a unit of energy; one megajoule (MJ) is equal to 1×10^6 joules (J)

Coldry Process a patented process that changes the naturally porous form of brown coal to produce a dry, dense pellet, via a process called 'brown coal densification'

FIGURE 1.8 Australia's black coal reserves

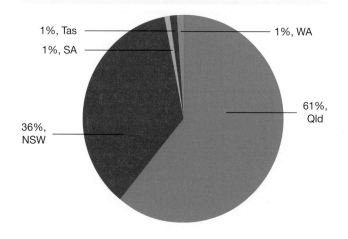

1%, Tas
1%, SA
1%, WA
36%, NSW
61%, Qld

FIGURE 1.9 Black coal has a much higher energy content but more pollutants than brown coal.

Natural gas

Natural gas is formed with oil in muds that are low in oxygen and rich in organic matter (typically ancient marine organisms). Natural gas is the lightest of the hydrocarbons produced, and is primarily composed of methane (CH_4). It is an important source of alkanes of low molecular mass. Victoria has large reserves of natural gas in the Gippsland basin. Typically, natural gas is composed of about 80 per cent methane, 10 per cent ethane, 4 per cent propane and 2 per cent butane. Nitrogen and hydrocarbons of higher molecular mass make up the remaining 4 per cent. Natural gas also contains a small amount of helium and is one of its major sources.

Natural gas is less dense than air, which means that it disperses in air. However, it is explosive in certain concentrations, so a safety measure incorporated by gas companies is to add an odour to natural gas so that leaks may be readily detected. Natural gas itself is odourless.

FIGURE 1.10 Methane gas is used in homes because it readily undergoes complete **combustion**.

Methane is the major constituent of natural gas and it burns with a hot, clean flame. Coal miners have long been aware of the dangers of methane gas. Released from coal seams during underground mining operations, methane gas has been responsible for many explosions and subsequent tragedies. Methane gas, besides being found in association with petroleum deposits, is also a by-product of coal formation. It is often **adsorbed** onto the surface of coal deposits deep underground.

Coal seam gas (CSG), also called coalbed methane, is extracted by drilling deep wells into underground coal deposits. Such wells are typically 100 to 1500 metres deep and are below the level of **aquifers** used for bore water supplies in inland Australia. The coal seams, which are nearly always filled with water, are further injected with water or chemicals to increase the pressure and crack the rocks. The accompanying decrease in pressure in the coal seam below allows the methane to desorb from the coal. It is then brought to the surface through the drilled well, along with more of the underground water. This process is called **fracking**.

natural gas a source of alkanes (mainly methane) of low molecular mass

combustion the rapid reaction of a compound with oxygen

adsorption the adhesion of atoms, ions or molecules from a gas, liquid or dissolved solid to a surface

aquifer an underground rock layer that contains water; this groundwater can be extracted using a well

fracking the process of pumping a large amount of fluid, mainly water, under high pressure into a drilled hole, in order to break rock so that it will release gas or oil

FIGURE 1.11 Coal seam gas is produced from coal deposits that lie deep underground.

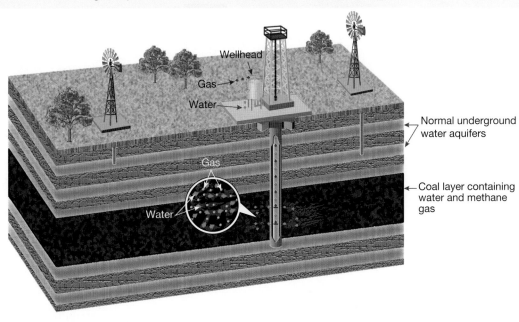

Australia has large deposits of coal seam gas, which are now being extracted from the Bowen and Surat Basins in eastern Queensland and northern New South Wales. The methane produced is relatively free from impurities, often containing only small amounts of ethane, nitrogen and carbon dioxide, and so requires minimal processing. It is used in the same way as natural gas and also contributes to a growing export industry for liquefied natural gas.

Some people consider fracking (hydraulic fracturing) to be an environmentally harmful method of extracting gas. This is because it requires high-pressured water to create fractures, or cracks, underground to release gas. There are concerns around the pollution of ground water and some evidence of earth tremors associated with fracking.

Petrochemical fuels

Petrochemical fuels are made from the refining of crude oil. Crude oil is also referred to as **petroleum**. Almost all of the contents of a barrel of crude oil are refined to make fuel, mostly for transport. These fuels include petrol (gasoline), diesel (petrodiesel), **kerosene**, **liquefied petroleum gas (LPG)** and aviation fuel.

> **petroleum** a viscous, oily liquid composed of crude oil and natural gas that was formed by geological processes acting on marine organisms over millions of years; it is a mixture of hydrocarbons used to manufacture other fuels and many other chemicals
>
> **kerosene** a mixture of hydrocarbons with molecules containing between 10 and 15 carbon atoms
>
> **liquefied petroleum gas (LPG)** a hydrocarbon fuel that consists mainly of propane and butane
>
> **fractional distillation** the process of separating component fuels based on their different boiling points

EXTENSION: Fractional distillation of crude oil

Petroleum is refined by **fractional distillation**, which separates out the component fuels based on their different boiling points. This process is performed in tall towers that are cooler at the top than at the bottom. The crude oil is heated and then introduced to the base of the tower. At this point, many of its components vaporise and these vapours rise up the tower, being cooled as they do so. When the vapours reach a point at which the tower's temperature equals their boiling temperature, condensation occurs. Specially designed trays containing bubble caps are placed inside the tower at strategic intervals. These are designed to allow the vapours to continue rising but stop condensed fractions from dripping back down to lower levels in the tower. The condensed fractions may then be removed from these trays to undergo further processing. Figure 1.12 shows a simplified outline of this process.

Fuels obtained from petroleum include petrol, liquefied petroleum gas (LPG), diesel fuel, heating oil and kerosene. Petroleum is also the raw material for a number of useful materials, including plastics, paints, synthetic fibres, medicines and pesticides.

FIGURE 1.12 A schematic of fractional distillation of crude oil showing levels of the fractionating column

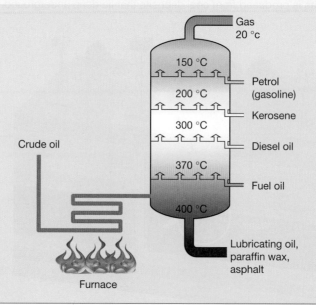

Petrol

Petrol, or gasoline, is a mixture of small hydrocarbons ranging from four to twelve carbon atoms per molecule, with five carbon atoms being the most common. Typically, petrol is a mixture of **alkanes**, **alkenes** and **cyclic hydrocarbons**. The energy content of petrol is around 44–46 MJ kg^{-1}. The majority of cars on the roads have engines designed to use petrol.

EXTENSION: Octane number

When motorists purchase fuel at a petrol station, they usually have a choice of standard or premium unleaded fuel. These fuels differ in their chemical composition and therefore have differing stabilities. The stability of a fuel is compared to an isomer of octane (C_8H_{18}), 2,2,4-trimethylpentane, which is given a Research Octane Number (RON) of 100. Fuels with high RONs are more resistant to uncontrolled combustion, called *knocking*. The benefit of high-octane fuels is that they can be used in engines designed for greater fuel efficiency and power. However, the difference between using unleaded fuel with a RON of 91 compared to premium unleaded fuel with a RON of 98 may be negligible if the fuel injection system doesn't have the programming and function to cater for it.

FIGURE 1.13 The 95 label on this petrol tank indicates the engine is designed to make use of high-octane fuel.

Diesel

Diesel has long been used as a transport fuel, particularly in large vehicles like trucks. As technology develops, diesel is becoming more popular for use in cars. This is because diesel, despite having a similar energy content per gram to petrol, has a higher density and combusts more efficiently than petrol. This means more energy per litre of diesel fuel is available. Diesel is a mixture of organic hydrocarbons ranging from 12–24 carbon atoms per molecule.

1.2.2 Biofuels

Biofuels are fuels made from waste plant and animal matter. They have been growing more popular in recent years due to the rise in oil prices and because of the impact that fossil fuel combustion has on **global warming**. The three most common **biofuels** are biogas, biodiesel and bioethanol.

Biogas

Biogas is a combustible fuel and may contain up to 65 per cent methane. It is produced when animal waste or other organic material rots in the absence of oxygen, such as when rubbish has been buried underground, or in digestive processes of mammals that involve the breakdown of food by bacteria in the gut. The most common material used for biogas production is livestock manure. The manure is fed into an airtight digester where it is allowed to ferment. The biogas produced is then collected and stored in a tank (see figure 1.14).

Biogas is commonly used to power furnaces, heaters and engines, and to generate electricity. Compressed biogas can also be used to fuel vehicles, and the residue from a biogas digester can be used as a fertiliser.

alkanes the family of hydrocarbons containing only single carbon–carbon bonds

alkenes the family of hydrocarbons that contain at least one carbon–carbon double bond

cyclic hydrocarbons also known as ring structures, because the carbon chain is a closed structure without open ends

global warming a gradual increase in the overall temperature of Earth's atmosphere

biofuel a renewable, carbon-based energy source formed in a short period of time from living matter

biogas fuel produced from the fermentation of organic matter

FIGURE 1.14 Biogas is a useful energy source.

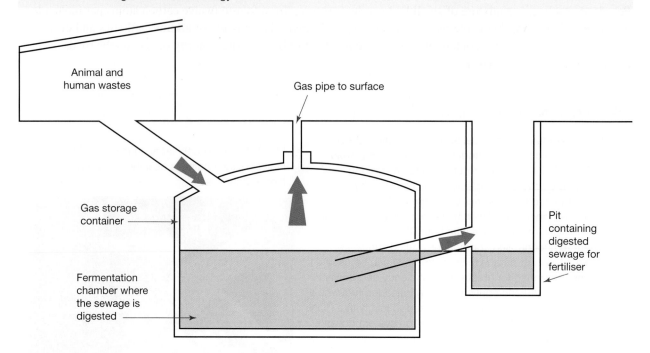

Methane

Methane from biogas (biomethane) is also referred to as renewable natural gas. Around 90 per cent of biomethane is produced by a process called upgrading. This involves the removal of other gases present in biogas — mainly carbon dioxide and some hydrogen. Efforts are being made to store the carbon dioxide component of biogas.

Another, less common way of producing biomethane is to break down solid biomass at a high temperature in an oxygen-deficient environment. This produces some methane, but mainly carbon monoxide (CO), along with hydrogen (H_2) and some CO_2. A catalyst is then used to react CO/CO_2 and H_2 together to produce methane.

feedstocks raw materials used to supply or fuel a machine or industrial process

---

CASE STUDY: Hydrogen

Almost all of the hydrogen gas made today results from a process called steam reforming. This involves reacting natural gas at a high temperature with pressurised steam to make carbon monoxide (CO) and hydrogen (H_2), with a little CO_2. Then the CO is further reacted with steam to produce more hydrogen.

$$CH_4(g) + H_2O(g) \rightarrow CO(g) + H_2(g)$$
$$CO(g) + H_2O(g) \rightarrow CO_2(g) + H_2(g)$$

There is an increase in research and the desire to produce hydrogen by steam reforming methane using renewable means and **feedstocks**. These include:
- collecting methane from biogas; for example, from landfill, crop residues, animal manure and domestic waste instead of fossil fuel deposits
- using renewable energy to generate the heat needed for the reaction
- carbon sequestration, in which CO_2 produced from the reaction is captured and stored underground, and therefore not released into the atmosphere.

Hydrogen as a fuel provides up to 142 kJ g^{-1}.

Biodiesel

Biodiesel is a diesel alternative that can be made from plant oils and animal fats. Oils and fats are naturally occurring esters formed between long-chain carboxylic acids (known as **fatty acids**) and **glycerol**. Common fatty acids are summarised in table 1.1.

TABLE 1.1 Formulas of some common fatty acids

Name	Formula
Palmitic	$C_{15}H_{31}COOH$
Palmitoleic	$C_{15}H_{29}COOH$
Stearic	$C_{17}H_{35}COOH$
Oleic	$C_{17}H_{33}COOH$
Linoleic	$C_{17}H_{31}COOH$
Linolenic	$C_{17}H_{29}COOH$

Biodiesel is produced by reacting oils or fats (also called triglycerides) with an alcohol. Although a number of small alcohols can be used, the most common is methanol. Heat and a catalyst of either concentrated sodium hydroxide or potassium hydroxide are used in this process. Biodiesel can be made on a small scale, using homemade equipment or with specially purchased kits, or on a much larger scale for commercial distribution. The chemical reaction involved converts one type of ester into another and is called **transesterification**.

biodiesel a fuel produced from vegetable oil or animal fats and combined with an alcohol, usually methanol

fatty acids long-chain carboxylic acids, usually containing an even number of 12–20 carbon atoms

glycerol an alcohol; it is a non-toxic, colourless, clear, odourless and viscous liquid that is sweet-tasting and has the semi-structural formula $CH_2OHCH(OH)CH_2OH$

transesterification the conversion of one ester (triglyceride) into another ester (biodiesel)

FIGURE 1.15 A typical transesterification reaction

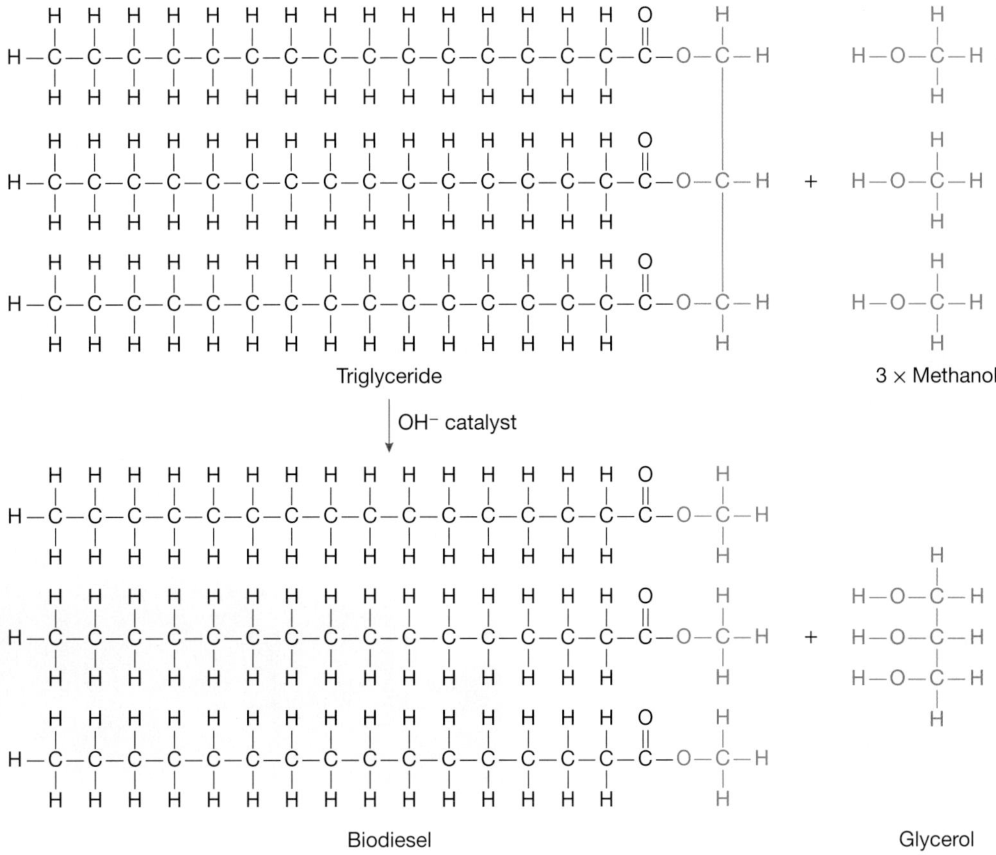

Bioethanol

Bioethanol is primarily used as a substitute for petrol in vehicles. It is obtained by fermenting sugar from sources such as waste wheat starch and molasses, a by-product of sugar production. Up to approximately 10 per cent anhydrous ethanol (E10) can be used as an additive to petrol without requiring engine modification.

There are both advantages and disadvantages to using ethanol-blended fuel. One advantage is that it reduces some pollutant emissions and contributes fewer greenhouse gases, as the production involves the utilisation of a waste product. Environmentally, the presence of oxygen in the ethanol assists the complete combustion of the petrol, and emissions of carbon monoxide and aromatic hydrocarbons are reduced. However, the cost of processing ethanol compared with petrol needs improvement, and ethanol yields less energy per gram compared to petrol. Ethanol can also contribute to the degradation of some plastic and rubber parts in vehicles. Additionally, in some countries a dilemma may arise regarding the use of land for food or fuel crops.

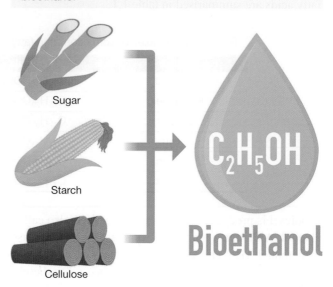

FIGURE 1.16 Raw materials used in the production of bioethanol

Sugar

Starch

Cellulose

C_2H_5OH

Bioethanol

While burning ethanol releases carbon dioxide into the atmosphere, the use of bioethanol is considered to have a lesser impact since carbon dioxide was absorbed during photosynthesis while the plant sources were grown. However, the processes used to produce the bioethanol for fuel also release some carbon dioxide, so the fuel is not entirely carbon neutral.

bioethanol ethanol produced from plants, such as sugarcane, and used as an alternative to petrol

yeast a single-celled fungus

1.2.3 The production of bioethanol

Fermentation

Fermentation is a biochemical process. This process occurs when **yeast** digests fermentable sugars (monosaccharides) to make energy. The fermentation of glucose, a monosaccharide, is represented by the following equation:

FIGURE 1.17 Straw is used as a raw material for this bioethanol plant.

$$C_6H_{12}O_6(aq) \rightarrow 2C_2H_5OH(aq) + 2CO_2(g)$$

The percentage of ethanol (alcohol) made during fermentation varies as a result of how much glucose and other fermentable carbohydrates can be extracted from the raw material. Other factors such as pH and the type of yeast itself also contribute. Producers aim for an initial concentration of between 12 and 18 per cent ethanol. If it is higher than this, the ethanol will be toxic to the yeast.

Distillation

You will explore distillation in greater detail in Unit 4. However, it's worth covering the basic principle in this topic to have an understanding of how almost all bioethanol is made.

Distillation for the purpose of making bioethanol involves the separation of ethanol from water after fermentation. The ethanol in the water mixture is heated to just above ethanol's boiling point (78.3 °C). This vaporises the ethanol, while leaving the majority of the water behind in its liquid state. The ethanol vapour rises up a tall tower where it is cooled and condensed back into a liquid, separated from the original mixture. This method, while expensive due to the heat energy required, produces up to 94 per cent (v/v) ethanol. Further water removal by dehydration methods achieves 99.8 per cent (v/v) ethanol.

FIGURE 1.18 Bioethanol is produced on a large scale using distillation and dehydration at plants such as this.

CASE STUDY: Uses of enzymes in industry

The production of bioethanol is a good case study for the increasing role of enzymes in industry. It has long been known that ethanol makes a good fuel for internal combustion engines, but its use in this context has been limited due to the cheaper availability of petrol produced from petroleum. However, due to the finite nature of petroleum and the contribution of carbon dioxide from petrol combustion to the greenhouse effect, there is now renewed interest in using ethanol as a fuel.

Currently, most ethanol is made by the following reaction:

$$C_2H_4 \ + \ H_2O \ \rightarrow C_2H_5OH$$

Ethene + Water → Ethanol

This reaction occurs in the presence of a phosphoric acid catalyst at 300 °C and a pressure of 60–70 atmospheres.

There are many problems with this production. The raw material (ethene) is produced from petroleum. Additionally, the necessary temperature and pressure conditions require a considerable energy input and plant development cost. Finally, combustion of ethanol made this way will continue to add to the greenhouse effect.

The production of ethanol using enzymes and fermentation represents a more sustainable, lower energy pathway. Ethanol produced this way is referred to as bioethanol. It also helps reduce the addition of carbon dioxide to the atmosphere, as the carbon dioxide released through combustion is removed when the next 'crop' of plants is grown. Figure 1.19 shows the current extent of this production.

To produce bioethanol, a carbohydrate source is required. A number of methods are then used to convert this into the sugars required for fermentation. Some of these methods involve adding chemicals and heating, but the use of enzymes at milder temperatures is continually being implemented and researched. Yeast is then used to ferment these sugars to ethanol. This last stage is well known, although research continues to develop more efficient strains of yeast.

The production methods for bioethanol are often classified into 'generations':
- First-generation biofuels use specially grown crops such as corn, soybean or sugarcane. Enzymes are being increasingly used to then convert the carbohydrate content of these into simple sugars for subsequent fermentation. Although the use of enzymes does produce a low-energy, green-chemistry pathway, the main debate surrounding this method centres on land use and the 'food versus fuel' debate.

- Second-generation biofuels use materials such as wood, agricultural residues and organic waste. Currently there is much research into developing suitable enzymes for this process.
- Third-generation biofuels use seaweed and microalgae as their carbohydrate source. These are also currently the subject of much research.
- Fourth-generation biofuels use genetically modified organisms to turn solar energy directly into biofuels in a way analogous to how plants turn solar energy into glucose and starch.

FIGURE 1.19 Using enzymes in the production of bioethanol

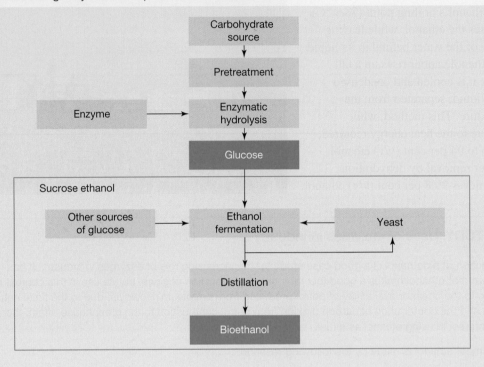

1.2.4 Costs and benefits of using renewable and non-renewable fuels

When comparing the costs and benefits of using fuels, a number of factors are considered. These include:
- energy efficiency and output
- economic costs and benefits
- environmental considerations
- sustainability.

Economic costs and benefits are not easily defined and are a source of political debate. However, there is a clear shift towards increasing investment in renewable fuel sources to sustain global energy requirements.

Energy output

Energy output and efficiency are important considerations when choosing a fuel to produce energy for work.

Producing energy from the combustion of organic fuel is a relatively inefficient process, and there is variation in the quantity of energy released per mass for different fuels. The energy content of each fuel type varies due to the type and percentage of combustible material contained within. The actual heat energy released is lower than the total energy content, as the water released during combustion retains some of the heat from the reaction; and since fuels such as bioethanol and biodiesel have oxygen present in their structure, they are already partially oxidised.

As we are comparing combustion fuels, the amount of stored chemical (potential) energy in the fuel is measured against the amount of useful energy produced.

TABLE 1.2 Comparison of chemical energy content of organic fuels

Fuel type	Fuel	Energy content (kJ g^{-1})
Fossil fuel	Coal (brown)	10–18
	Coal (black)	17–25
	Natural gas (mainly CH$_4$)	42–55
	Petrol	44–46
	Petrodiesel	42–46
Biofuel	Biogas*	25–53
	Bioethanol	30
	Biodiesel	40

*Biogas varies in its methane content. Higher values are obtained when CO$_2$ is removed via upgrading.

Source: Data based on World Nuclear Association (https://world-nuclear.org/information-library.aspx) and Arthur *et al*. Evaluating the Potential of Renewable Energy Sources in a Full-Scale Upflow Anaerobic Sludge Blanket Reactor Treating Municipal Wastewater in Ghana. *Sustainability*. 2023; 15(4):3743. https://doi.org/10.3390/su15043743.

Since the majority of commercial fuels are a mixture of different molecules — that is, they are not pure substances — using kilojoules per gram (kJ g^{-1}) is an appropriate unit rather than kilojoules per mole.

The **efficiency** of energy conversion is a concept that follows from the Second Law of Thermodynamics. It takes the amount of usable energy obtained into account and is defined as a percentage.

Energy efficiency (%):

$$\% \, \text{efficiency} = \frac{\text{energy obtained in desired form}}{\text{energy available before conversion}} \times \frac{100}{1}$$

Energy transformations are not 100 per cent efficient. This is because heat is also produced when energy conversions take place.

When electricity generation is considered, the calculations must take into account the efficiency of the generation process. For example, to produce 1 MJ of electrical energy, assuming the generation process is 30 per cent efficient, enough fuel would need to be burned to produce 3.33 MJ of heat energy.

 **Resources**

> ▶ **Video eLesson** Coal-fired power station (eles-3237)

Renewability versus sustainability

A **sustainable energy** future means providing for the needs of today's society without compromising the ability of future generations to meet their own needs. A large factor in determining if an energy source is sustainable is whether it is renewable or **non-renewable**. When categorising an energy source as either renewable or non-renewable, we must compare the rate of production versus consumption.

Fossil fuels are non-renewable fuels, as they are consumed at a much greater rate than they can be produced. This is because oil, coal and gas deposits have formed over millions of years. At the current rates of consumption, known reserves of oil and gas will begin to run out around 50 years from now, and coal in just over 100 years.

> **efficiency** (of energy conversion) the ratio between useful energy output and energy input
>
> **sustainable energy** energy that meets present needs without compromising the ability of future generations to meet their own needs
>
> **non-renewable** (with reference to energy sources) energy sources that are consumed faster than they are being formed

Fuels produced from biomass, like bioethanol, biodiesel and biogas, are classified as renewable. This is because the biomass can be grown at a rate equal to or greater than the fuel consumption.

Environmental considerations

Pollutants have long been of environmental concern and associated with the combustion of fuels. Pollutants such as SO_2, **NO_x**, CO and **particulates** are released into the atmosphere from burning coal and petrofuels.

Another environmental consequence of energy production is net carbon emissions from the release of carbon dioxide into the atmosphere. Fossil fuels have significant carbon emissions as the carbon that was stored under the ground in the form of oil, natural gas and coal is released into the atmosphere once combusted. Coal emits the most carbon dioxide for the amount of energy it produces.

FIGURE 1.20 Vehicles powered by petroleum-based fuels contribute to increased SO_2, NO_x, CO and particulates in the atmosphere.

FIGURE 1.21 CO_2 emissions (in g MJ^{-1}) of different fossil fuels for energy production

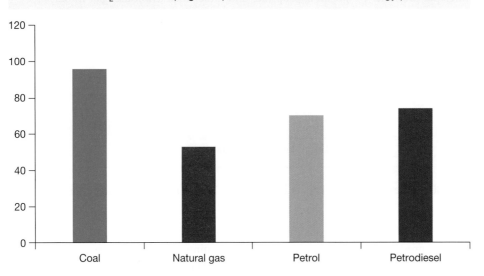

Fuels produced from biomass have close to zero net carbon emissions. This is because the carbon dioxide released into the atmosphere from combustion is offset by the carbon dioxide taken out of the atmosphere by plants and photosynthetic algae. As well as being almost **carbon neutral**, fuel produced from biomass does not release significant amounts of oxides of sulfur, nitrogen, heavy metals or particulates when combusted compared to fossil fuels — coal in particular.

NO_x a term used for oxides of nitrogen, such as NO_2 and NO, that contribute to air pollution

particulates solid and liquid particles small enough to be suspended in the atmosphere

carbon neutral no net release of carbon dioxide into the atmosphere

greenhouse effect a natural process that warms Earth's surface; when the Sun's energy reaches Earth's atmosphere, some of it is reflected back to space, and the rest is absorbed and re-radiated by greenhouse gases

greenhouse gases gases that contribute to the greenhouse effect by absorbing infrared radiation

enhanced greenhouse effect the effect of increasing concentrations of greenhouse gases in the atmosphere as the result of human activity

climate change changes in various measures of climate over a long period of time

CASE STUDY: Greenhouse gases and global warming

The **greenhouse effect** helps to keep Earth at the appropriate temperature to support life. It begins when radiation from the Sun strikes Earth and warms its surface, which then radiates heat energy back into space. Gases in the atmosphere known as **greenhouse gases** — including carbon dioxide (CO_2), methane (CH_4), nitrous oxide (N_2O) and ozone (O_3) — absorb some of this heat radiation, so the air warms up. The air may also radiate this energy back into space or down to Earth (see figure 1.23). Unfortunately, human activities have led to an increase in the amount of greenhouse gases in the atmosphere and so more heat is absorbed, which continues to adversely affect weather and climate. This results in an **enhanced greenhouse effect**, causing global warming and **climate change** (see figure 1.24).

Global warming specifically refers to Earth's rising temperature, due mainly to the increasing concentrations of greenhouse gases in the atmosphere. Climate change is a broader and more accurate term that encompasses the side effects of global warming and refers to changes in various measures of climate over a long period of time.

Greenhouse gases absorb more energy than other gases and contribute to global warming in the atmosphere. Carbon dioxide is the major greenhouse gas emitted by human activities and is generated during transportation, industrial processes, land-use change and energy production.

FIGURE 1.22 Cattle and other livestock release significant amounts of methane into the atmosphere as a result of their digestive processes.

FIGURE 1.23 The greenhouse effect allows some heat to be trapped in the atmosphere, maintaining a constant temperature.

Greenhouse effect

FIGURE 1.24 Excess production of greenhouse gases means the atmosphere retains more heat energy, increasing the average temperature of Earth.

Enhanced greenhouse effect

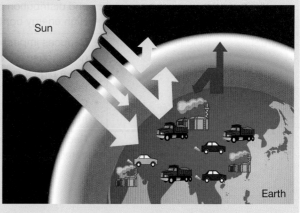

Just as extracting coal, oil and gas is harmful for the environment, biofuel production has some environmental concerns and considerations as well.

In Australia, the main feedstocks for biodiesel production are oil seeds such as canola, used cooking oil and tallow. Elsewhere in the world, soybeans and sunflower seeds, as well as palm oil, are the main sources being used to meet this demand. However, as demand increases, new economical sources will need to be found. There are also some environmental concerns, including the debate about land use: Should crops be grown for food production or fuel production? In South-East Asia, massive deforestation is occurring to make way for palm oil plantations for fuel, and this is endangering the habitats of many species — the best known of which is the Sumatran orangutan.

FIGURE 1.25 Deforestation for palm oil plantations is endangering the habitat of the Sumatran orangutan.

Not only is land required for biodiesel and bioethanol production, but a lot of water is also used to irrigate crops. It takes over 1300 litres of water to produce 1 litre of bioethanol from sugar cane, and around 850 litres of water per litre of ethanol fuel produced from corn.

Advantages and disadvantages of fuel types

While renewable fuel sources are the preferred, sustainable way of meeting global energy needs, a summary of the advantages and disadvantages of renewable and non-renewable organic fuels is shown in table 1.3.

TABLE 1.3 Advantages and disadvantages of organic fuel types

Fuel type	Fuel	Advantages	Disadvantages
Fossil fuel	Coal	• Cheap • Large reserves in Australia	• Heavy pollutants — SO_2, NO_x, particulates • Inefficient/highest CO_2 emissions produced per MJ energy
	Natural gas	• High energy content • High energy efficiency • Less pollutants compared to coal	• Moderate CO_2 emissions • CSG involves fracking • Leaks can cause explosions, therefore storage tanks and distribution networks are constantly monitored
	Petrol and diesel	• Infrastructure for fuel production and distribution is established • High energy content	• Medium to high CO_2 emissions • CO emissions in populated areas
Biofuel	Biogas	• Reduces impact on the greenhouse effect as unburnt CH_4 has a bigger impact than CO_2 released once combusted • Can be made from organic waste from farms and homes (green bin)	• Lower energy content and inefficient if not upgraded/needs to be upgraded to increase efficiency (percentage of methane)
	Biodiesel	• Reduced pollutant emissions compared to petrodiesel	• Can be problematic in lower temperatures • Production requires land, which can result in deforestation or land being used to make fuel instead of food
	Bioethanol	• Can be used in petrol blends such as E10 (10% ethanol in petrol) for existing car engines without modification • Higher octane rating than petrol so provides more power • Cheap and relatively easy to produce compared to other biofuels	• Lower energy content per mass • Requires land to grow crops

tlvd-9665

SAMPLE PROBLEM 1 Comparing fuels in terms of their renewability

Explain why fuels using a high percentage of bioethanol are more sustainable than fuels with low percentages of bioethanol.

THINK	WRITE
1. Explain what makes a fuel sustainable.	A sustainable fuel is one that can be produced indefinitely due to its raw material being able to be sourced on a continuous basis.
2. Outline the aspects of bioethanol production that make it sustainable.	Bioethanol is produced from the fermentation of plant materials and then distillation to separate the ethanol from the mixture — mainly water. Because of this, bioethanol is renewable and can be produced at a rate to meet demand.
3. Justify why fuels with a high percentage of bioethanol are more sustainable.	Fuels with a high percentage of bioethanol are more sustainable as they contain a lower percentage of petrol. Petrol is made from a mixture of hydrocarbons obtained from crude oil. Crude oil is a fossil fuel. Fossil fuels are finite, non-renewable sources of energy.

PRACTICE PROBLEM 1

Outline the advantages and disadvantages of producing and using bioethanol.

CASE STUDY: Comparison of petrodiesel and biodiesel

After petrol, diesel is the most widely used transport fuel in the world. Diesel engines — although heavier and initially more expensive — are more efficient than their petrol counterparts, have better fuel economy and tend to last longer. They produce less power than petrol engines of the same size but more torque, which makes diesel-powered vehicles slower to accelerate but ideal for hauling heavier loads. Biodiesel can easily be substituted — either straight or blended with petrodiesel — as a fuel for diesel engines and requires little or no modification to the engine. A comparison of the two fuels is given in table 1.4.

TABLE 1.4 Comparison of petrodiesel and biodiesel

Property	Petrodiesel	Biodiesel
Source	Petroleum	• Used cooking oil, tallow, oil seed crops such as canola and palm oil • Oil from algae is possible. • Methanol production requires fossil fuels but production of methanol from glycerol (a by-product) is currently under investigation.
Chemical structure	Alkanes, both straight-chain and branched (typically containing 12–24 carbon atoms per molecule)	• Esters from long-chain fatty acids (typically 15–20 carbon atoms per molecule) and methanol • Other simple alcohols

(continued)

TABLE 1.4 Comparison of petrodiesel and biodiesel *(continued)*

Property	Petrodiesel	Biodiesel
Combustion products	• Carbon dioxide • Water • Carbon monoxide • Particulate carbon (soot) • Sulfur dioxide • Nitrogen oxides	• Same as petrodiesel but generally lower in quantity • May be increased emission of nitrogen oxides
Viscosity	**Hygroscopic**, but not generally an issue as seasonal blending allows for changes in outside temperature	Hygroscopic and low outside temperatures; can lead to increased viscosity due to fuel gelling
Environmental impact	• Non-renewable • Non-biodegradable • Spills in transportation of both crude oil and refined products • Combustion emissions in transportation chain	• Renewable • Biodegradable • Issues with growing crops for food versus fuel • Deforestation issues, especially in south-east Asia

hygroscopic refers to when a substance has a tendency to absorb water vapour from the atmosphere

1.2 Activities

learnon

1.2 Quick quiz on	1.2 Exercise	1.2 Exam questions

1.2 Exercise

1. a. What is a fuel?
   b. What is the difference between a fossil fuel and a biofuel?
2. Provide two reasons for why burning natural gas to make electricity is better than using coal to make electricity.
3. Why do you think brown coal is used on such an extensive scale to generate electricity in Victoria, even though it has a relatively low energy content?
4. Fuels, and energy sources in general, may be classified as either renewable or non-renewable.
   a. Define the terms *renewable* and *non-renewable* as they apply to this context.
   b. Are all biofuels renewable? Explain.
5. Fossil fuels and biofuels can undergo complete combustion to release carbon dioxide and water. Explain why the complete combustion of fossil fuels contributes to the enhanced greenhouse effect, whereas the complete combustion of biofuels does not.
6. You have been invited to debate the statement 'The world should stop using fossil fuels and replace them with biofuels'.
   a. Outline three environmental or societal issues you would argue if you were in favour of this statement.
   b. Outline three environmental or societal issues you would argue if you were against this statement.

1.2 Exam questions

▶ Question 1 (1 mark)

Source: VCE 2021 Chemistry Exam, Section B, Q.1.a; © VCAA

Digesters use bacteria to convert organic waste into biogas, which contains mainly methane, CH_4. Biogas can be used as a source of energy.

Both biogas and coal seam gas contain CH_4 as their main component.

Why is biogas considered a renewable energy source but coal seam gas is not?

▶ Question 2 (2 marks)

Source: VCE 2020 Chemistry Exam, Section B, Q.6.d; © VCAA

Methane gas, CH_4, can be captured from the breakdown of waste in landfills. CH_4 is also a primary component of natural gas. CH_4 can be used to produce energy through combustion.

Compare the environmental impact of CH_4 obtained from landfill to the environmental impact of CH_4 obtained from natural gas.

▶ Question 3 (1 mark)

Source: VCE 2020 Chemistry Exam, Section A, Q.11; © VCAA

MC Which one of the following statements is correct?
A. Crude oil can be classified as a biofuel because it originally comes from plants.
B. Methane, CH_4, can be classified as a fossil fuel because it has major environmental impacts.
C. Ethanol, CH_3CH_2OH, can be classified as a fossil fuel because it can be produced from crude oil.
D. Hydrogen, H_2, can be classified as a biofuel because, when it combusts, it does not produce carbon dioxide, CO_2.

▶ Question 4 (4 marks)

Source: VCE 2019 Chemistry Exam, Section B, Q.10.a; © VCAA

Climate change has been identified as a threat to the environment. Fossil fuels are recognised as a significant contributor to the rise in carbon dioxide levels in the atmosphere. The replacement of fossil fuels as an energy source represents a challenge and has been the focus of research for a number of years. However, there are different opinions/views about the suitability of using a biofuel, such as biodiesel, as a replacement for fossil fuels. Some extracts representing different viewpoints are shown in the box below.

---

[1]'Biofuels are fuels that are produced from biological sources such as trees, plants or microorganisms. They are carbon neutral, because they do not result in fossil carbon being released into the atmosphere.'

[2]'All good solutions are needed in the energy transition required to achieve Europe's climate goals — and sustainable biofuels are critical to transport decarbonisation.'

[3]'Many scientists view biofuels as inherently carbon neutral: they assume the carbon dioxide (CO_2) plants absorb from the air as they grow completely offsets, or "neutralises", the CO_2 emitted when fuels made from plants burn.'

[4]'... our analysis affirms that, as a cure for climate change, biofuels are "worse than the disease."'

[5]'... although some forms of bioenergy can play a helpful role, dedicating land specifically for generating bioenergy is unwise.'

Sources: [1]CarbonNeutralEarth, <www.carbonneutralearth.com/biofuels.php>; [2]Sejersgård Fanø, quoted in Erin Voegele, 'EU reaches deal on REDII, sets new goals for renewables', *Biodiesel Magazine*, 15 June 2018, <www.biodieselmagazine.com>; [3 & 4]John DeCicco, 'Biofuels turn out to be a climate mistake – here's why', The Conversation, 5 October 2016, <http://theconversation.com/au>; [5]Andrew Steer and Craig Hanson, 'Biofuels are not a green alternative to fossil fuels', *The Guardian*, 30 January 2015, <www.theguardian.com/au>

Using the chemistry that you studied this year and the information above, discuss the carbon neutrality and the sustainability of using biodiesel as a fuel for transport.

▶

▶ **Question 5 (1 mark)**

Source: *VCE 2017 Chemistry Exam, Section A, Q.13*; © VCAA

MC Four identical vehicle models, 1, 2, 3 and 4, were tested for fuel efficiency using LPG, petrol (unleaded, 91 octane), E10 (petrol with 10% ethanol added) and petrodiesel. Carbon dioxide, CO_2, emissions per litre of fuel burnt were also determined. The following table summarises the results.

Vehicle model	Fuel	Fuel consumption (L/100 km)	CO_2 produced (g CO_2/L of fuel)
1	LPG	19.7	1665
2	petrol	14.5	2392
3	E10	14.2	2304
4	petrodiesel	9.2	2640

Using the information in the table above, which one of the following statements about petrodiesel is correct?
A. It has the highest energy content.
B. It has the poorest fuel efficiency.
C. It is a renewable energy source.
D. It has the lowest CO_2 emissions when burnt.

More exam questions are available in your learnON title.

1.3 Thermochemical reactions

KEY KNOWLEDGE

- Comparison of exothermic and endothermic reactions, with reference to bond making and bond breaking, including enthalpy changes (ΔH) measured in kJ, molar enthalpy changes measured in kJ mol^{-1} and enthalpy changes for mixtures measured in kJ g^{-1}, and their representations in energy profile diagrams
- Determination of limiting reactants or reagents in chemical reactions
- Combustion (complete and incomplete) reactions of fuels as exothermic reactions: the writing of balanced thermochemical equations, including states, for the complete and incomplete combustion of organic molecules using experimental data and data tables

Source: VCE Chemistry Study Design (2024–2027) extracts © VCAA; reproduced by permission.

The energy changes that accompany chemical reactions are vital to us. To survive, we depend on the energy content of the food we eat. Our bodies can convert the energy of the chemical bonds in food into other kinds of energy. The quality of lifestyle we lead depends on harnessing energy from different chemical sources, including coal, oil, natural gas and renewable fuels.

The study of the energy changes that accompany chemical reactions is called **thermochemistry** or **chemical energetics**. In general, all chemical reactions involve energy changes.

thermochemistry the branch of chemistry concerned with the quantities of heat evolved or absorbed during chemical reactions

chemical energetics a branch of science that deals with the properties of energy and the way it is transformed in chemical reactions

BACKGROUND KNOWLEDGE: Types of energy

Energy may take a number of different forms. These include:
- mechanical energy
- thermal (heat) energy
- chemical energy
- light energy
- sound energy
- electrical energy
- gravitational energy
- nuclear energy.

All of these forms of energy may be classified as either **potential energy** (energy that is stored, ready to do work) or **kinetic energy** (energy associated with movement, in doing work).

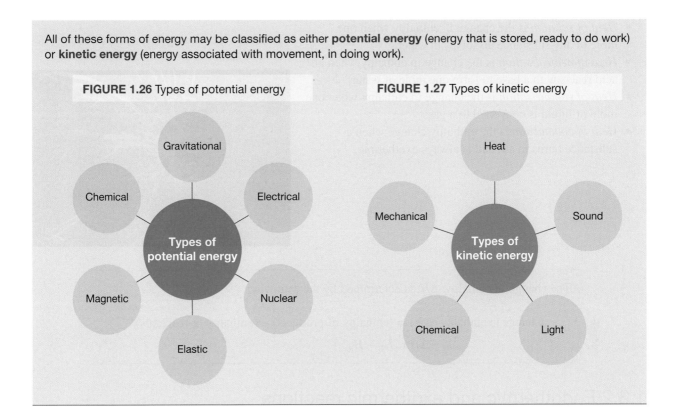

FIGURE 1.26 Types of potential energy

FIGURE 1.27 Types of kinetic energy

1.3.1 Bond making and bond breaking

When a chemical change or reaction takes place, at least one or more substances are consumed and at least one or more substances are produced. This requires the bonds in the reactant(s) to be broken and new bonds in the product(s) to be formed.

In general, all chemical reactions involve energy changes. The chemical energy stored in a substance has the potential to be converted to heat or electricity (for more on chemical energy converted to electricity, see topic 3).

A certain amount of chemical energy is stored within every atom, molecule or ion. This energy is the sum of the potential energy and kinetic energy of the substance and results from:
- the attractions and repulsions present between protons and electrons within the atom
- the attractions and, to some degree, repulsions present between atoms within the molecule
- the motion of the electrons
- the movement of the atoms.

The total energy stored in a substance is called the **enthalpy** of the substance and is given the symbol H. Enthalpy can also be referred to as the heat content of a substance. Unfortunately, we cannot directly measure the heat content of a substance, but we can measure the **change in enthalpy** when the substance undergoes a chemical reaction. In virtually all chemical reactions, the energy of the reactants and products differ, so such reactions usually involve some change in enthalpy, which is indicated by a temperature rise or fall. The change in enthalpy during a reaction is denoted by ΔH and is usually known as the **heat of reaction**, but there are some reactions for which specific names have been given.

potential energy energy that is stored, ready to do work

kinetic energy energy associated with movement, in doing work

enthalpy a thermodynamic quantity equivalent to the total heat content of a system

change in enthalpy the amount of energy released or absorbed in a chemical reaction

heat of reaction the heat evolved or absorbed during a chemical reaction taking place under conditions of constant temperature and of either constant volume or, more often, constant pressure

- *Heat of solution* is the change in enthalpy when one mole of any substance dissolves in water.
- *Heat of neutralisation* is the change in enthalpy when an acid reacts with a base to form one mole of water.
- *Heat of vapourisation* is the change in enthalpy when one mole of liquid is converted to a gas.
- *Heat of combustion* is the enthalpy change when a substance burns in air, and is always exothermic.

FIGURE 1.28 A change in enthalpy occurs when a sparkler burns.

The change in enthalpy, ΔH, is determined by the following:

Change in enthalpy (ΔH) = (enthalpy of products) − (enthalpy of reactants)

$$\Delta H = H_P - H_R$$

1.3.2 Endothermic and exothermic reactions

Chemical reactions accompanied by heat energy changes can be divided into two groups: **exothermic** reactions and **endothermic** reactions.

Enthalpy change, ΔH, is equal to heat energy produced or absorbed. When bonds are broken in reactants, energy is consumed and is therefore endothermic. When bonds are formed in the products, energy is given out and is therefore exothermic. If more energy is given out when the product bonds are formed than is absorbed when the reactant bonds are broken, the reaction is exothermic. This would result in a *negative value*.

exothermic describes a chemical reaction in which energy is released to the surroundings

endothermic describes a chemical reaction in which energy is absorbed from the surroundings

FIGURE 1.29 In endothermic reactions, the surroundings lose energy and get cooler. In exothermic reactions, the surroundings gain energy and get warmer.

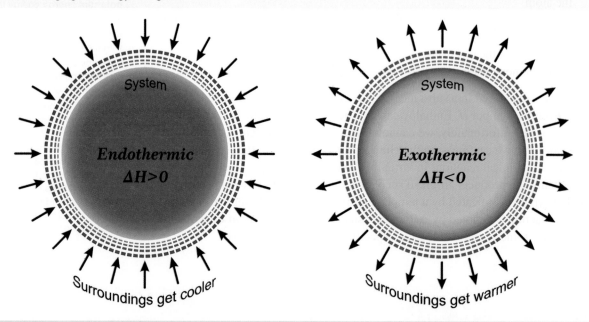

When measuring the enthalpy change of an endothermic or exothermic reaction, we assume these measurements are taken at **standard laboratory conditions (SLC)** and assign the value in **kilojoules** (kJ). ΔH values are often stated in terms of enthalpy change in kilojoules per mole (kJ mol^{-1}).

Exothermic reactions

The exothermic nature of the combustion of fuels was discussed in subtopic 1.2. The total energy stored in the bonds of the fuel and oxygen is greater than the energy stored in the bonds of the carbon dioxide and water, which results in the release of energy in the form of heat.

The heat of reaction (combustion) of methane is 55.6 kJ per gram. This means 55.6 kJ of heat is produced when 1 g of methane is reacted in an excess of oxygen.

However, when stating the value for the enthalpy of combustion (ΔH_c) of methane, a negative value of –55.6 kJ is given to indicate the exothermic nature. The molar equivalent for the enthalpy of combustion of methane is –890 kJ mol^{-1}.

Endothermic reactions

The production of hydrogen gas examined in subtopic 1.2 using steam reforming is an endothermic reaction. It requires energy (+253 kJ mol^{-1}) to form hydrogen and carbon dioxide at SLC from the reaction between methane and water. We can write this as a **thermochemical equation**, stating the positive enthalpy change.

$$CH_4(g) + 2H_2O(l) \rightarrow CO_2(g) + 4H_2(g) \quad \Delta H = +253 \text{ kJ mol}^{-1}$$

Thermochemical equations

Thermochemical equations show the amount of heat produced or absorbed by a reaction. As with other chemical equations, charge and mass must balance, but thermochemical equations must also include the enthalpy change.

When writing a thermochemical equation, the following points should be remembered:
- A positive or negative sign must be included with the ΔH value to indicate whether the reaction is endothermic or exothermic. If an enthalpy change is given as ΔH = 345 kJ mol^{-1}, the lack of sign does not mean that it is an endothermic reaction.
- Enthalpy is measured in kJ mol^{-1}. This means the coefficients in the equation represent the amount of moles of each reacting substance that the ΔH value refers to.

The hydrogen gas produced from the reaction between methane and water can also undergo a combustion reaction.

The following equation can be read as: when 2 moles of hydrogen react with 1 mole of oxygen, 2 moles of water form and 572 kJ of energy is released.

$$2H_2(g) + O_2(g) \rightarrow 2H_2O(l) \quad \Delta H = -572 \text{ kJ mol}^{-1}$$

You will notice that the key point is that the enthalpy change in a reaction is proportional to the amount of substance that reacts. If these two quantities are measured in an experiment, it is possible to write the accompanying thermochemical equation.

standard laboratory conditions (SLC) 100 kPa and 25 °C

kilojoule a unit of energy; one kilojoule (kJ) is equal to 1×10^3 joules (J)

thermochemical equations balanced stoichiometric chemical equations that include the enthalpy change

When assigning ΔH values it is important to take note of the number of moles of fuel that are combusted. If the number of moles in the equation is changed, the ΔH value will also change.

For example:

$$H_2(g) + \frac{1}{2}O_2(g) \rightarrow H_2O(g) \quad \Delta H = -282 \, kJ \, mol^{-1} \text{ (from 2 g of } H_2)$$

If twice as much hydrogen was to react, then twice the energy would be released.

$$2H_2(g) + O_2(g) \rightarrow 2H_2O(g) \quad \Delta H = -564 \, kJ \, mol^{-1} \text{ (from 4 g of } H_2)$$

Regardless of the number of moles reacting, the unit for ΔH is $kJ \, mol^{-1}$.

TIP: Remember to use the correct state symbol when referring to alcohols; they are often incorrectly assumed to be aqueous (aq) instead of pure liquid (l).

TIP: When balancing equations with alcohols, do not forget to count the oxygen in the alcohol. This is a common mistake that leads to an unbalanced equation.

1.3.3 Determining limiting reactants

A limiting reactant can be thought of as the reactant that is completely consumed in a reaction, which causes the reaction to stop. This reactant (or reagent) limits the quantity of products formed according to the mole ratios in the balanced equation.

Thermochemical equations are written for specific quantities of reactants in fixed ratios. Combustion reactions, for example, are usually written assuming an excess of oxygen. That means the fuel is the limiting reactant, and the products formed are carbon dioxide and water.

If oxygen is the limiting reactant, then the amount of energy released and the extent of **oxidation** of the fuel taking place is reduced. This means a different thermochemical equation is used, with products including solid carbon or carbon monoxide, and a lower ΔH value.

Calculations involving limiting and excess reactants are covered in topic 2. These involve determining the limiting reactant by comparing the number of moles of each reactant available to the stoichiometric ratio from the balanced equation. This allows you to identify which reactant is present in excess and which one is limiting. Oxygen can be identified as the limiting reactant in combustion reactions by recognising incomplete combustion products.

oxidation loss of electrons; an increase in the oxidation number

Determining limiting reactants

Limiting reactants for reactions involving the combustion of fuels can be determined both by using stoichiometric calculations and by identifying incomplete combustion through the formation of carbon monoxide (CO) and even soot (C) as products in combustion reactions.

For example, if 1 mol of methane is combusted with 4 mol of oxygen according to the following equation:

$$CH_4(g) + 2O_2(g) \rightarrow 2CO_2(g) + 2H_2O(g) \quad \Delta H = -889 \text{ kJ mol}^{-1}$$

then methane can be identified as the limiting reactant, since the ratio of methane to oxygen in the balanced equation is 1 : 2. Thus, the amount of oxygen required to react completely with 1 mol of methane would only be 2 mol. Methane is also confirmed as the limiting reactant by the reaction products, carbon dioxide and water, which indicate complete combustion.

If there are quantities provided for two reactants it is necessary to use the moles of the *limiting* reactant in conjunction with the ΔH value to determine total energy released or absorbed in the reaction.

1.3.4 Energy profile diagrams

As already mentioned, all chemical reactions involve energy. We have also noted that an amount of energy is required to break the reactant bonds before a reaction can proceed.

These energy changes can be summarised using **energy profile diagrams**, as shown in figure 1.30. From these diagrams we can note a number of points.
- There is always a peak between the reactants and the products. This peak represents the energy required to break the reactant bonds. The difference between the enthalpy of the reactants and the value of this peak is the **activation energy (E_a)**.
- For an exothermic reaction, the activation energy is less than the energy released when new bonds form. Consequently, there is a net release of energy (usually as heat released to the surroundings).
- In an endothermic reaction, the activation energy is greater than the energy released when new bonds form. Consequently, there is a net input of energy (in most cases, heat is absorbed from the surroundings).
- In exothermic and endothermic reactions, the activation energy represents a requirement for the progress of the chemical reaction. This must be added before a reaction proceeds. This energy is provided from the kinetic energy of the particle collisions.

Figure 1.30 shows the activation energy that has to be overcome in both energy profiles.

> **energy profile diagram** a graph or diagram that shows the energy changes involved in a reaction from the reactants through the intermediate stages to the products
>
> **activation energy (E_a)** the minimum energy required by reactants in order to react

FIGURE 1.30 A chemical reaction can be recognised as either exothermic or endothermic by its ΔH value. If the ΔH value is negative, the reaction is exothermic. If the ΔH value is positive, the reaction is endothermic.

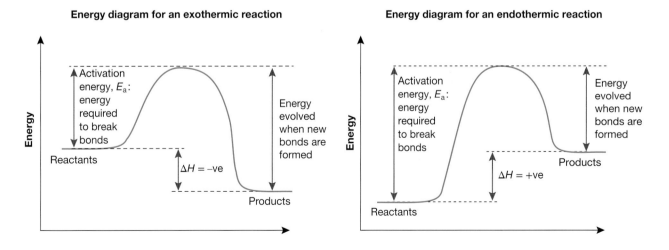

SAMPLE PROBLEM 2 Using a graph to determine activation energy and reaction type

The following diagram shows the energy profile for a particular reaction. Some values for enthalpy have been inserted on the vertical axis.

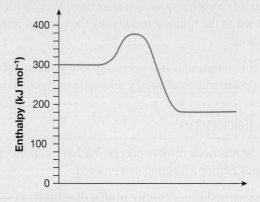

a. Is this reaction exothermic or endothermic?
b. What is the value of the activation energy?

THINK	WRITE
a. Recall that an exothermic or endothermic reaction is indicated by comparing enthalpies of the reactants and the products. Here, the products are lower in enthalpy than the reactants, so it is exothermic.	a. Exothermic
b. Recall that the activation energy is the difference in enthalpy between the reactants and the highest point of the energy profile diagram.	b. $E_a = 380 - 300$ $= 80 \, \text{kJ mol}^{-1}$
TIP: Remember that the value of the activation energy is always positive, because all reactions need energy to start them.	

PRACTICE PROBLEM 2

The following reaction profile refers to a particular reaction and has some enthalpy values indicated as shown.

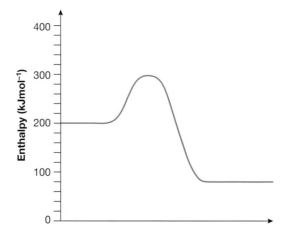

a. Is this reaction exothermic or endothermic?
b. What is the activation energy for this reaction?

---

elog-1930

tlvd-9713

EXPERIMENT 1.1

Investigating heat changes in reactions

Aim

To investigate and draw energy diagrams for some exothermic and endothermic reactions

---

EXTENSION: Multi-step reactions

Energy profiles are also used to show multi-step reactions. For example, a reaction between pollutant gases has a net, overall reaction between NO_2 and CO. It occurs via an intermediate endothermic reaction and then a further exothermic reaction as shown in the following equations for steps 1 and 2:

$$\text{Step 1:} \quad 2NO_2(g) \rightarrow NO_3(g) + NO(g) \qquad \Delta H = +95 \text{ kJ mol}^{-1}$$

$$\text{Step 2:} \quad NO_3(g) + CO(g) \rightarrow NO_2(g) + CO_2(g) \qquad \Delta H = -322 \text{ kJ mol}^{-1}$$

Adding these two equations together and cancelling NO_3 and NO_2 results in the net overall equation and the overall change in enthalpy:

$$2NO_2(g) + \cancel{NO_3}(g) + CO(g) \rightarrow \cancel{NO_3}(g) + NO(g) + \cancel{NO_2}(g) + CO_2(g) \quad \Delta H = \Delta H(1) + \Delta H(2)$$

$$= 95 - 322 \text{ kJ mol}^{-1}$$

$$NO_2(g) + CO(g) \rightarrow NO(g) + CO_2(g) \quad \Delta H = -227 \text{ kJ mol}^{-1} \text{ Overall}$$

This overall reaction can be represented on the same energy profile.

FIGURE 1.31 The change in enthalpy (ΔH overall) for the reaction is equal to the sum of the two changes in enthalpy for each step.

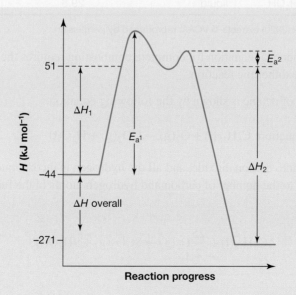

1.3.2 Combustion reactions

The combustion of all of the fuels discussed in this topic provide energy to do work via exothermic reactions. Combustion reactions are redox reactions in which fuels are oxidised and, in the process, create water plus amounts of CO_2, CO or C. If fuels have other chemicals present, like sulfur, they too will undergo oxidation reactions, as will atmospheric nitrogen (N_2).

Complete combustion

Complete combustion occurs when there is an excess of oxygen (O_2) gas. In excess oxygen, the fuel creates $CO_2(g)$ and $H_2O(l)$ and releases energy equivalent to the published ΔH value — assuming 100 per cent efficiency and conditions at SLC.

FIGURE 1.32 Complete combustion

Thermochemical equations for combustion at SLC

$$Fuel(l) \text{ or } (g) + O_2(g) \rightarrow CO_2(g) + H_2O(l) \quad \Delta H = negative$$

Table 1.5 lists the heats of combustion at SLC for common fuels.

TABLE 1.5 Heats of combustion of common fuels at SLC

Fuel	Formula	State	Heat of combustion (kJ g^{-1})	Molar heat of combustion (kJ mol^{-1})
hydrogen	H_2	gas	141	282
methane	CH_4	gas	55.6	890
ethane	C_2H_6	gas	51.9	1560
propane	C_3H_8	gas	50.5	2220
butane	C_4H_{10}	gas	49.7	2880
octane	C_8H_{18}	liquid	47.9	5460
ethyne (acetylene)	C_2H_2	gas	49.9	1300
methanol	CH_3OH	liquid	22.7	726
ethanol	C_2H_5OH	liquid	29.6	1360

Source: VCE Chemistry Data Book (2020) extracts © VCAA; reproduced by permission.

Writing a balanced, thermochemical equation for complete combustion requires the ΔH value in kJ mol^{-1} and a negative sign to show it's an exothermic reaction.

For example, the combustion of octane is shown by the following equation:

$$\text{Skeleton equation: } C_8H_{18}(l) + O_2(g) \rightarrow CO_2(g) + H_2O(l)$$

As all the carbon is oxidised into carbon dioxide, and all the hydrogen is incorporated into water, balance the CO_2 and H_2O first according to the number of carbon and hydrogen atoms in the fuel. Then balance the amount of oxygen reacting.

$$C_8H_{18}(l) + \frac{25}{2}O_2(g) \rightarrow 8CO_2(g) + 9H_2O(l)$$

or

$$2C_8H_{18}(l) + 25O_2(g) \rightarrow 16CO_2(g) + 18H_2O(l)$$

Next, assign the correct ΔH value according to the coefficient (moles) of octane to complete the thermochemical equation, remembering to use correct units and the negative sign.

$$C_8H_{18}(l) + \frac{25}{2}O_2(g) \rightarrow 8CO_2(g) + 9H_2O(l) \quad \Delta H = -5460 \text{ kJ mol}^{-1}$$

or

$$2C_8H_{18}(l) + 25O_2(g) \rightarrow 16CO_2(g) + 18H_2O(l) \quad \Delta H = -10\,920 \text{ kJ mol}^{-1}$$

dioxins highly toxic compounds formed from industrial processes and incomplete combustion of organics

Incomplete combustion

Incomplete combustion occurs when the supply of oxygen is limited, and the mixing of the reactants with oxygen is insufficient. Other factors — some of which are discussed in Unit 3, Outcome 2 — can impact the reaction with oxygen too.

Incomplete combustion results in many pollutants, such as hydrocarbons, **dioxins** and NO_x. However, we will focus on two: carbon monoxide (CO) and soot (C).

Many households use natural gas for heating and cooking. It is highly recommended that people have their heaters and gas-fuelled appliances regularly serviced, to ensure the highly toxic carbon monoxide and potentially cancer-causing soot are not being produced as a result of incomplete combustion.

Flame colour when burning methane can be an indicator of whether complete (figure 1.32) and incomplete combustion (figure 1.33) are occurring simultaneously. You can see this with a Bunsen burner when the hole on the collar is closed to restrict airflow and produce a safety flame. A blue flame colour for the combustion of the small hydrocarbons, such as methane, propane and butane, is a sign of complete combustion. Yellow, orange and red flame colours indicate incomplete combustion.

In the case of soot, black deposits can be visibly seen. The soot seen in the tractor pull competition in figure 1.34 occurs due to the incomplete combustion of diesel fuel. The ratio mixture of fuel to oxygen is too high. This is done deliberately because more power is produced, but efficiency is lower.

Let's look at the following two equations, representing the complete combustion of a hydrocarbon found in diesel fuel compared to incomplete combustion. Take note of the number of moles of oxygen reacting in each equation.

Complete combustion:

$$C_{13}H_{28}(l) + 20O_2(g) \rightarrow 13CO_2(g) + 14H_2O(l)$$

Incomplete combustion:

$$C_{13}H_{28}(l) + 13.5O_2(g) \rightarrow 13CO(g) + 14H_2O(l)$$

$$C_{13}H_{28}(l) + 7O_2(g) \rightarrow 13C(s) + 14H_2O(l)$$

FIGURE 1.33 Incomplete combustion

FIGURE 1.34 A tractor producing a cloud of thick, black, diesel smoke

Incomplete combustion is inefficient and releases less energy per kilogram of fuel used. Less oxidation means less energy released.

tlvd-9667

SAMPLE PROBLEM 3 Writing the thermochemical equation for complete combustion

Write the thermochemical equation showing the complete combustion of ethane gas.

THINK	WRITE
1. Find the formula and molar heat of combustion of ethane using the VCE Chemistry Data Book.	C_2H_6 1560 kJ mol^{-1}
2. Write out the formula and balance the equation.	$2C_2H_6(g) + 7O_2(g) \rightarrow 4CO_2(g) + 6H_2O(g)$
3. Add the value, taking care with multiples of molar heat of combustion and the corresponding units. *Alternatively:* The equation can be written for the combustion of one mole of ethane gas.	$\Delta H = 2 \times -1560$ $\quad = -3120 \text{ kJ mol}^{-1}$ $C_2H_6(g) + \frac{7}{2}CO_2 \rightarrow 2CO_2 + 3H_2O$ $\Delta H = -1560 \text{ kJ mol}^{-1}$

PRACTICE PROBLEM 3

Write the balanced equation showing the incomplete combustion of ethane gas.

 Resources

 Interactivity Combustion equations (int-1370)

1.3 Activities

learnon

Students, these questions are even better in jacPLUS

 Receive immediate feedback and access sample responses

 Access additional questions

 Track your results and progress

Find all this and MORE in jacPLUS

1.3 Quick quiz	1.3 Exercise	1.3 Exam questions

1.3 Exercise

1. State whether the following are exothermic or endothermic processes.
   a. Water changing from a liquid to a gaseous state
   b. A reaction in which the total enthalpy of the products is greater than that of the reactants
   c. Burning kerosene in a blow torch
   d. Burning fuel in a jet aircraft engine
   e. A chemical reaction that has a negative ΔH value
   f. A reaction in which the reactants are at a lower level on an energy profile diagram than the products

2. The use of hydrogen as a renewable and environmentally friendly fuel is currently the subject of much research. The main product of hydrogen combustion is water. The production of liquid water from the reaction between gaseous hydrogen and gaseous oxygen can be represented by the following thermochemical equation:

$$O_2(g) + 2H_2(g) \rightarrow 2H_2O(l) \quad \Delta H = -564 \text{ kJ mol}^{-1}$$

Calculate how much energy, in kJ, would be released or absorbed by the following reactions.

a. $2O_2(g) + 4H_2(g) \rightarrow 4H_2O(l)$

b. $H_2O(l) \rightarrow \dfrac{1}{2}O_2(g) + H_2(g)$

3. Use the heats of combustion in table 1.5 to write balanced thermochemical equations for the combustion of each of the following fuels in excess oxygen at SLC.

a. $CH_4(g)$ b. $C_3H_8(g)$ c. $CH_3OH(l)$ d. $C_2H_5OH(l)$

4. Write balanced equations for the incomplete combustion of the fuels in question 3 where the products are:

i. $CO(g)$ and $H_2O(l)$ ii. $C(s)$ and $H_2O(l)$.

5. Consider the energy profile diagram shown.
   a. Is this reaction exothermic or endothermic?
   b. For this reaction, what is the value of:
      i. ΔH
      ii. the minimum energy required to break the reactant bonds
      iii. the activation energy
      iv. the energy released when the new bonds form?

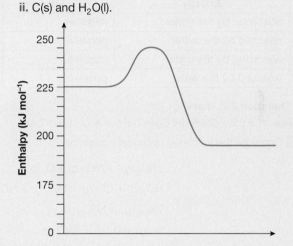

6. Consider the following energy profile diagram.

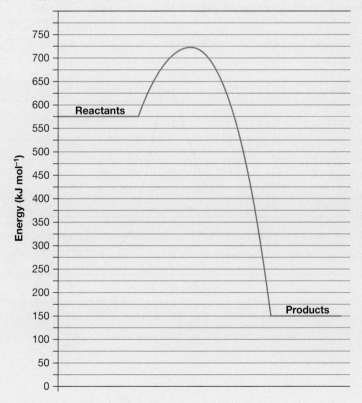

a. Give the change in enthalpy, ΔH, for the reaction.
b. Give the activation energy, E_a(reverse), for the reverse reaction in kJ mol^{-1}.

7. Sketch a labelled energy profile diagram for the following reaction

$$2NO_2(g) \rightarrow N_2(g) + 2O_2(g) \quad \Delta H = -80 \, kJ \, mol^{-1}$$

given that the enthalpy of the reactants is 110 kJ mol^{-1} and the activation energy is 70 kJ mol^{-1}.

1.3 Exam questions

▶ Question 1 (1 mark)
Source: VCE 2022 Chemistry Exam, Section A, Q.2; © VCAA

MC A fuel undergoes combustion to heat water.

Which of the following descriptions of the energy and enthalpy of combustion, ΔH, of the reaction is correct?

	Energy	ΔH
A.	absorbed by the water	negative
B.	released by the water	negative
C.	absorbed by the water	positive
D.	released by the water	positive

▶ Question 2 (1 mark)
Source: VCE 2021 Chemistry Exam, Section A, Q.18; © VCAA

MC Consider the following chemical equations.

$$2NO_2(g) \rightarrow 2NO(g) + O_2(g) \qquad \Delta H = +14 \, kJ \, mol^{-1}$$
$$NO_2(g) + CO(g) \rightarrow CO_2(g) + NO(g) \qquad \Delta H = -226 \, kJ \, mol^{-1}$$
$$2NO_2(g) \rightleftharpoons N_2O_4(g) \qquad \Delta H = -57 \, kJ \, mol^{-1}$$
$$N_2(g) + O_2(g) \rightleftharpoons 2NO(g) \qquad \Delta H = +181 \, kJ \, mol^{-1}$$

Which one of the following graphs is consistent with the chemical equations above?

A.

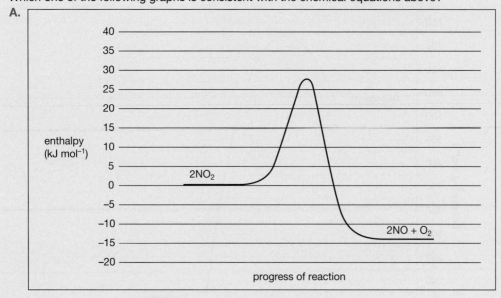

B.

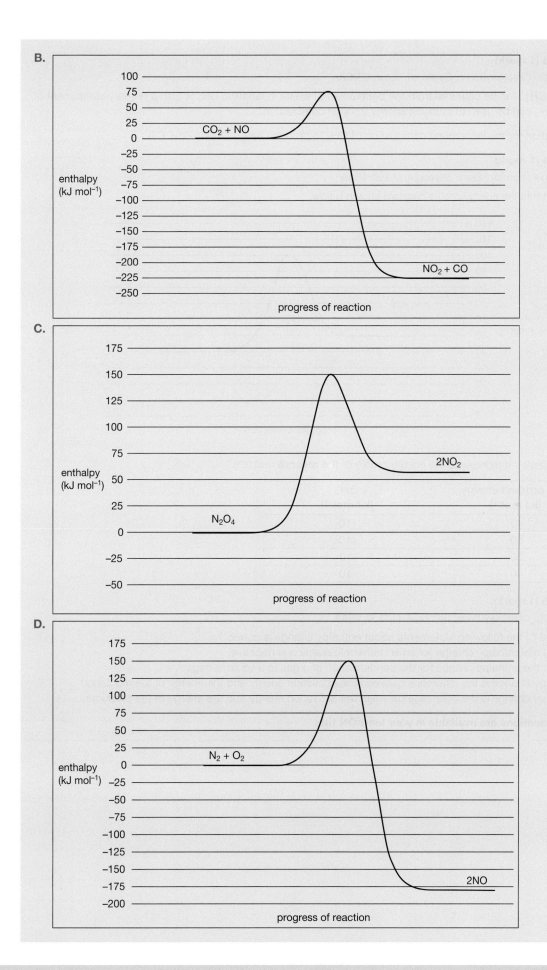

C.

D.

Question 3 (1 mark)

Source: VCE 2020 Chemistry Exam, Section B, Q.6.a; © VCAA

Methane gas, CH_4, can be captured from the breakdown of waste in landfills. CH_4 is also a primary component of natural gas. CH_4 can be used to produce energy through combustion.

Write the equation for the incomplete combustion of CH_4 to produce carbon monoxide, CO.

Question 4 (1 mark)

Source: VCE 2019 Chemistry Exam, Section A, Q.9; © VCAA

MC A reaction has the energy profile diagram shown below.

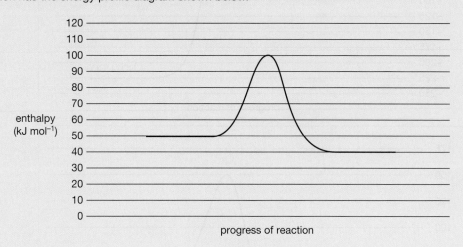

Which of the following represents the energy profile of the **reverse** reaction?

	Final product energy (kJ mol⁻¹)	ΔH (kJ mol⁻¹)
A.	40	+10
B.	50	+10
C.	50	−10
D.	40	−10

Question 5 (1 mark)

Source: VCE 2019 Chemistry Exam, Section A, Q.23; © VCAA

MC Which one of the following statements about enthalpy change is correct?
A. The sign of the enthalpy change for an endothermic reaction is negative.
B. The sign of the enthalpy change for the condensation of a gas to a liquid is negative.
C. The enthalpy change is the difference between the activation energy and the energy of the reactants.
D. The enthalpy change is the difference between the activation energy and the energy of the products.

More exam questions are available in your learnON title.

1.4 Fuel sources for plants and animals

The food we eat supplies the energy we need to power all of the billions of chemical reactions happening in our bodies every second. Where does this energy come from, how do we get it and how can we measure the energy in food?

The energy in food can be traced back to the Sun, where the process of photosynthesis converts the Sun's energy into simple **carbohydrates** like glucose. Through condensation reactions these glucose molecules are turned into starch, a carbohydrate and cellulose. Carbohydrates, **fats** and **proteins** in the food we eat provide the energy our bodies need. Carbohydrates are broken down to provide glucose, which releases energy in the process of respiration. The amount of energy a person needs depends on how active the individual is and whether the person is still growing. This energy is used for digestion, maintaining the heartbeat, breathing, brain function, nervous system movement, heat generation and maintaining constant body conditions. But how much energy is supplied by each food type and how readily available is it?

FIGURE 1.35 When exercising, muscles convert glucose into energy using oxygen in the blood in the process of respiration.

carbohydrates the general name for a large group of organic compounds occurring in food and living tissues; includes sugars, starch and cellulose

fat a triglyceride formed from glycerol and three fatty acids

proteins large molecules composed of one or more long chains of amino acids

metabolism the chemical processes that occur within a living organism to maintain life

1.4.1 Food molecules

Food molecules participate in chemical reactions that we collectively refer to as **metabolism**.

The energy that we obtain from food depends on the amount of carbohydrates, proteins, fats and oils in the meal, and how much of each nutrient is eaten. Different nutrients have different energy values. The available energy content when used in the body is measured in kilojoules per gram (kJ g^{-1}).

The energy values of food are:
* carbohydrate: 16 kJ g^{-1}
* protein: 17 kJ g^{-1}
* fat: 37 kJ g^{-1}.

Not all of the energy in the food ingested is available to the body because it is not completely digested and absorbed, and some is converted to heat. The energy values listed have been adjusted to reflect this. Not all food molecules are a source of energy, such as water, minerals, vitamins and fibre, but they are all necessary for good health. While carbohydrates, fats and proteins *are* a source of energy, some produce energy more readily than others.

FIGURE 1.36 What type of foods provide the most energy? What else must be considered when choosing food?

Carbohydrates are easily broken down to glucose, which is converted into energy using oxygen in the blood. Excess carbohydrates are stored as glycogen. Once all carbohydrates in the body, including glycogen, are consumed, the body starts to break down fat. Fat is not as efficient at providing energy as carbohydrates. Protein is not stored and is usually only used as an energy source in situations of starvation. The body can break down excess **amino acids** to produce glucose or fat if required.

FIGURE 1.37 Metabolism of food in the body

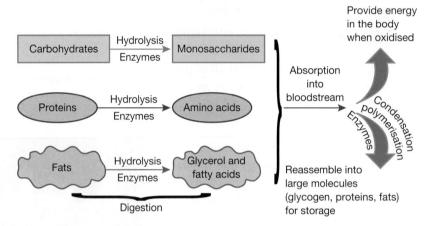

Specific reactions involved in the digestion and synthesis of key food molecules are studied further in topic 8.

Fats and oils

Collectively, fats and oils are classified as **lipids**. At room temperature, a fat is in a solid state, whereas oils exist in a liquid state. Lipids contain the elements carbon, hydrogen and oxygen. In this respect they are similar to carbohydrates, but fats and oils have a smaller percentage of oxygen and are not polymers. They are found in fish, dairy products, fruit and vegetable oils, fried foods, seeds and nuts. Fats and oils are also called **triglycerides** because they are formed by a condensation reaction between glycerol and three fatty acids.

FIGURE 1.38 Sources of some healthy unsaturated fats — omega-3 fatty acids

Fats provide about 80 per cent of the body's energy storage in adipose (fatty) tissue. They are used when food is scarce, when you haven't eaten for a while or when you are ill and don't feel like eating. While the brain is dependent on glucose, the liver, muscle and fat cells derive their energy from fat.

Fats are broken down in a complex series of steps that break the bonds in the long fatty-acid carbon chains and separate the oxygen atoms in the oxygen molecules; the atoms then recombine to produce carbon dioxide and water. The overall reaction is an exothermic oxidation process in which the fats react with oxygen to produce carbon dioxide and water. This reaction is similar to a combustion reaction. An example would be the oxidation of linoleic acid (LA), a polyunsaturated omega-6 fatty acid.

$$C_{17}H_{31}COOH + 25O_2 \rightarrow 18CO_2 + 16H_2O \quad \Delta H = -8382 \, \text{kJ mol}^{-1}$$

amino acids molecules that contain an amino and a carboxyl group

lipids substances such as fats, oils and waxes that are insoluble in water

triglycerides fats and oils formed by a condensation reaction between glycerol and three fatty acids

TIP: Always check that the oxygen atoms are equal on both sides of the equation. There are two oxygen atoms in the fatty acid. If you require half an oxygen molecule on balancing an oxidation reaction, just double all coefficients.

Proteins

Proteins are broken down into amino acids by hydrolysis. Enzymes and hydrochloric acid in the stomach assist in the digestion process. Or bodies use amino acids to make new proteins, and to synthesise fats and many other important molecules.

The body does not store an excess of amino acids. Excess amino acids are converted to glucose, which can be used for energy, and waste excreted as **urea**. Protein-rich foods include meat, eggs, nuts and **legumes**.

FIGURE 1.39 Legumes, such as the chickpeas used to make hummus, are a good vegetarian source of protein.

Carbohydrates

Carbohydrates are classified into three groups according to their molecular structure: **monosaccharides**, **disaccharides** and **polysaccharides**.

Monosaccharides (sometimes called simple sugars) are the basic building blocks of all carbohydrates. The most important monosaccharides are those containing five carbon atoms (pentoses) and those containing six carbon atoms (hexoses). They are not broken down during digestion.

Plant foods provide us with carbohydrates. Plant foods eaten in raw form contain a mixture of polysaccharides and smaller carbohydrate molecules like **fructose**, found in in foods such as fruit and honey. The digestible polysaccharides in plant foods are forms of **starch**.

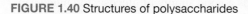

Glucose provides energy to animals. It is stored in the liver as glycogen if not immediately required and is reconverted to glucose by hydrolysis.

urea a molecule synthesised in the liver to remove ammonia from the body

legumes plants that produce pods with a seed inside

monosaccharide the simplest form of carbohydrate, consisting of one sugar molecule

disaccharide two sugar molecules (monosaccharides) bonded together

polysaccharide more than ten monosaccharides bonded together

fructose a pentose monosaccharide

starch a condensation polymer of glucose

FIGURE 1.40 Structures of polysaccharides

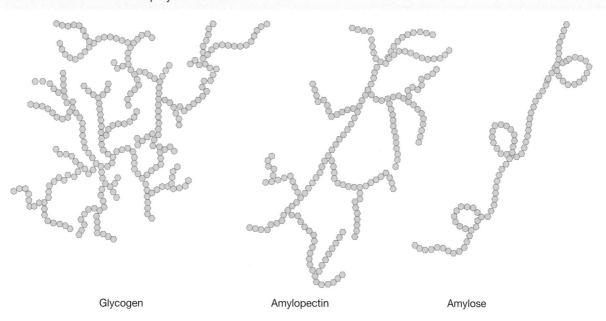

Glycogen   Amylopectin   Amylose

Indigestible carbohydrates like **cellulose** that form part of insoluble **dietary fibre** are important for our health; however, we cannot break them down in the body to use for energy.

1.4.2 Glucose — the primary energy source

The glucose formed during photosynthesis can be used by the plant to form complex carbohydrates through polymerisation reactions. When an animal eats a plant, it can use the plant's carbohydrates as an energy source. Carbohydrates provide the greatest proportion of energy in the diets of most humans. This energy is required for muscle movement and the functioning of the central nervous system. Carbohydrates are also essential parts of other important molecules such as DNA. In addition, they have a number of beneficial effects on the taste and texture of foods.

FIGURE 1.41 Sources of carbohydrates

The brain is the most energy-demanding organ, consuming about half the glucose in the body. The huge task of controlling all of the body's functions, including thinking, memory and learning, rely on glucose levels and how the brain utilises this monosaccharide. Although the brain needs a good supply of glucose, too much can cause cognitive problems and other health issues, including diabetes. Glucose is obtained through the catalysed hydrolysis of glycogen and starch. The glucose passes from the digestive system into the blood, and then to the cells of the liver and other tissues.

Photosynthesis

Photosynthesis is a process of life-sustaining redox reactions that are able to capture and transform light energy, carbon dioxide and water into glucose and oxygen.

The overall net endothermic equation for photosynthesis is:

$$6CO_2(g) + 6H_2O(l) \xrightarrow[\text{Chlorophyll}]{\text{Light}} C_6H_{12}O_6(aq) + 6O_2(g) \quad \Delta H \cong +2.8 \times 10^3 \text{ kJ mol}^{-1}$$

The difference in the energy of the products compared to the reactants can be used to calculate ΔH. However, there are other products besides glucose formed in photosynthesis, so the enthalpy change is approximate. The additional energy for photosynthesis comes from sunlight.

The efficiency of photosynthesis varies and is dependent upon a number of factors. Some of these include the type of plant or organism, the amount and type of **chlorophyll**, and other photosynthetic pigments, minerals and nutrients available.

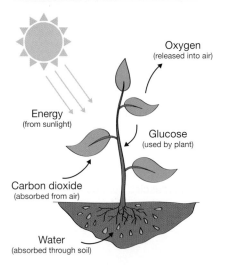

FIGURE 1.42 Photosynthesis produces glucose and oxygen from carbon dioxide and water.

Oxygen
(released into air)

Energy
(from sunlight)

Glucose
(used by plant)

Carbon dioxide
(absorbed from air)

Water
(absorbed through soil)

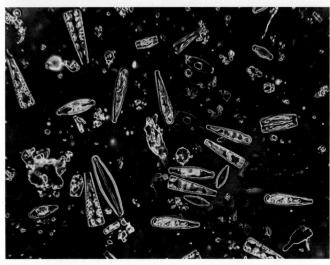

FIGURE 1.43 Phytoplankton such as these diatoms produce significant amounts of oxygen for the planet via photosynthesis.

The maximum efficiency of photosynthesis, in terms of the amount of sunlight converted into chemical energy, is approximately 26 per cent. However, the actual amount of solar energy that ends up stored in the plant (as dried biomass) is far less (1–2 per cent). There are a number of reasons for this, with a significant one being that only 34 per cent of the sunlight is absorbed by plants. This is because not all wavelengths of sunlight are absorbed and some of the light is reflected. The plants also carry out processes that use energy, including cellular respiration.

Cellular respiration

The process by which energy is obtained from glucose is a remarkable and complicated series of biochemical steps, and is called **cellular respiration**. This occurs in the cells of all living organisms. The cells release energy from the chemical bonds of food molecules, providing energy for the essential processes of life. In chemical terms it is an exothermic, redox reaction and is similar to a combustion reaction, where glucose reacts with oxygen to form carbon dioxide and water. It is also an aerobic process because oxygen is required. The energy released in this process is 2860 kJ mol^{-1}.

cellular respiration the process that occurs in cells to oxidise glucose in the presence of oxygen to carbon dioxide, water and energy

FIGURE 1.44 Cellular respiration

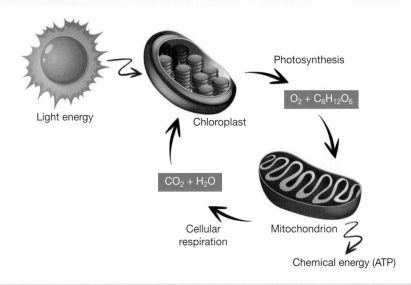

Light energy

Chloroplast

Photosynthesis

$O_2 + C_6H_{12}O_6$

$CO_2 + H_2O$

Cellular respiration

Mitochondrion

Chemical energy (ATP)

The overall thermochemical equation for cellular respiration is:

$$C_6H_{12}O_6(aq) + 6O_2(g) \rightarrow 6CO_2(g) + 6H_2O(l) \quad \Delta H = -2860 \text{ kJ mol}^{-1}$$

This equation is the reverse of the equation for photosynthesis, although the chemical pathway involved is very different.

Writing thermochemical equations

Remember to include the ΔH value and sign when asked to write a thermochemical equation.

Anaerobic respiration takes place when there is no oxygen present, and results in less energy being obtained. This occurs in tissues where there is a high demand for fast energy, such as in working muscles, but there is a shortage of oxygen to satisfy the energy needed by just using aerobic respiration. The product of anaerobic respiration is **lactic acid**, which must be oxidised to carbon dioxide and water at a later stage so that it doesn't build up. Lactic acid can cause muscle soreness because the cells cannot process waste products fast enough. The oxygen must be replaced and that is why you breathe deeply after exercise.

The chemical equation for anaerobic respiration is:

$$C_6H_{12}O_6(aq) \rightarrow 2CH_3CH(OH)COOH(aq)$$

Anaerobic respiration also occurs in plant cells and some microorganisms. This process is called fermentation. Anaerobic respiration in yeast is used during brewing and bread-making, where sugars are broken down into ethanol and carbon dioxide.

$$C_6H_{12}O_6(aq) \rightarrow 2C_2H_5OH(aq) + 2CO_2(g)$$

Ethanol is the alcohol used in beer and wine production, and is also added to petrol to be used as a fuel. In bread-making, bubbles of carbon dioxide gas form in the dough and cause the bread to rise.

FIGURE 1.45 The burning, cramping pain felt in muscles during intense exercise is caused by a build-up of lactic acid.

Aerobic respiration occurs in the presence of oxygen, and anaerobic respiration occurs in the absence of oxygen.

anaerobic respiration the breakdown of glucose in the absence of oxygen

lactic acid an organic acid, $C_3H_6O_3$, present in muscle tissue as a by-product of anaerobic respiration

ethanol an alcohol with two carbons produced from fermentation of glucose by yeast

tlvd-9668

SAMPLE PROBLEM 4 Explaining differences in fuel sources for the body

Explain why fats and oils (lipids) have more energy content per gram than carbohydrates such as glucose.

THINK	WRITE
1. How does the number of oxygen atoms compare in carbohydrates and lipids?	Carbohydrates have far more oxygen atoms in their structures than lipids do.
2. How does this affect the amount of oxygen that can react with a lipid compared to a carbohydrate?	This means less oxygen will react with a carbohydrate than with a lipid.
3. Relate the quantity of oxygen reacting to the energy released during digestion.	The greater number of C–H bonds in lipids compared to carbohydrates means that more energy is released as more electrons are donated to oxygen when reacting. The oxidation of carbon to make C–O bonds from C–H bonds releases energy, and the reduction of carbon to C–H bonds from C–O bonds requires energy.

PRACTICE PROBLEM 4

Explain why producing glucose during photosynthesis requires energy, whereas using glucose for aerobic respiration releases energy.

1.4 Activities

learnon

Students, these questions are even better in jacPLUS

- Receive immediate feedback and access sample responses
- Access additional questions
- Track your results and progress

Find all this and MORE in jacPLUS

1.4 Quick quiz on	1.4 Exercise	1.4 Exam questions

1.4 Exercise

1. Why do molecules of starch, protein and fat need to be digested?
2. Draw simplified flow charts to show the metabolism of each of the following food molecules. Where possible, include the type of reaction and state the main use of the final product.
   a. Carbohydrates
   b. Fats and oils
   c. Proteins
3. What is a hydrolysis reaction? Give an example.
4. What type of reaction is occurring as glucose is broken down to carbon dioxide?
5. An athlete is running on a treadmill. After running for some time, the athlete's legs start to cramp. What is the possible cause of this discomfort? Provide an equation to support your answer.
6. a. Write the balanced equation showing anaerobic respiration of glucose to ethanol.
   b. What is the name of this process?
   c. How is anaerobic respiration useful in bread-making?

1.4 Exam questions

Question 1 (1 mark)
Source: VCE 2020 Chemistry Exam, Section A, Q.1; © VCAA

MC Glycogen breaks down into
A. glycerol.
B. amino acids.
C. triglycerides.
D. monosaccharides.

Question 2 (1 mark)
Source: VCE 2022 Chemistry NHT Exam, Section A, Q.1; © VCAA

MC Which of the following correctly identifies the product of respiration and the small molecular product of metabolism?

	Product of respiration	Small molecular product of metabolism
A.	oxygen	glucose
B.	carbon dioxide	glucose
C.	oxygen	glycogen
D.	carbon dioxide	glycogen

Question 3 (1 mark)
Source: VCE 2016 Chemistry Exam, Section B, Q.3.a; © VCAA

The diagram below represents a certain biomolecule.

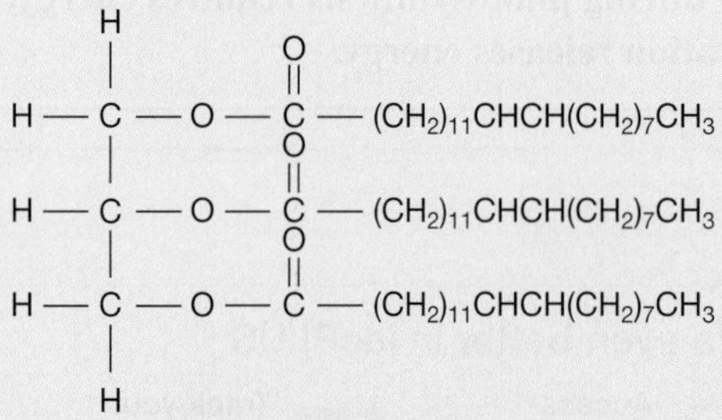

$$
\begin{array}{l}
H-C-O-C-(CH_2)_{11}CHCH(CH_2)_7CH_3 \\
H-C-O-C-(CH_2)_{11}CHCH(CH_2)_7CH_3 \\
H-C-O-C-(CH_2)_{11}CHCH(CH_2)_7CH_3
\end{array}
$$

Name the class of organic biomolecules to which the biomolecule above belongs.

Question 4 (6 marks)
Source: VCE 2012 Chemistry Exam 1, Section B, Q.1.a.i,ii,b; © VCAA

a. The cellulose that is present in plant matter cannot be directly fermented to produce bioethanol. The cellulose polymer must first be broken down into its constituent monomers.

A section of cellulose polymer is shown below.

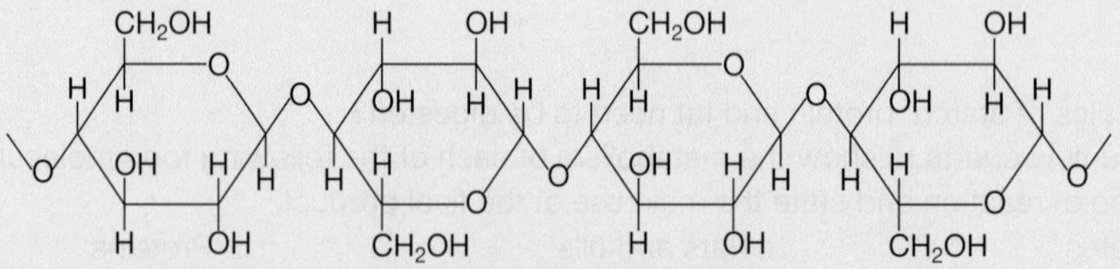

i. What is the name of the monomer from which cellulose is formed? **(1 mark)**
ii. Complete the following chemical equation to show the formation of ethanol by fermentation of the cellulose monomer. **(1 mark)**

$$C_6H_{12}O_6(aq) \longrightarrow \underline{\hspace{2cm}} + \underline{\hspace{2cm}}$$

b. Triglycerides are an important source of energy in the body. During digestion, triglycerides are broken down in the small intestine by the enzyme lipase. An incomplete chemical equation that shows the hydrolysis of a triglyceride is shown below.

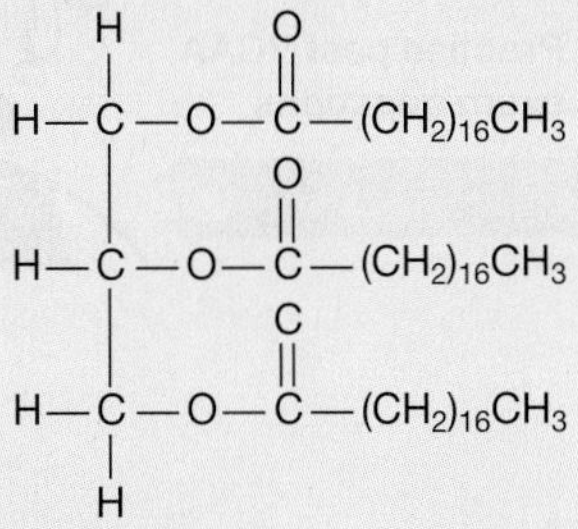

$$
\begin{array}{c}
\text{H} \quad \text{O} \\
| \quad || \\
\text{H}-\text{C}-\text{O}-\text{C}-(\text{CH}_2)_{16}\text{CH}_3 \\
| \quad \text{O} \\
| \quad || \\
\text{H}-\text{C}-\text{O}-\text{C}-(\text{CH}_2)_{16}\text{CH}_3 \quad + \quad 3\text{H}_2\text{O} \longrightarrow \quad \underline{\quad} \ \text{CH}_3(\text{CH}_2)_{16}\text{COOH} \ + \ \underline{\quad} \ \text{C}_3\text{H}_8\text{O}_3 \\
| \quad \text{C} \\
| \quad || \\
\text{H}-\text{C}-\text{O}-\text{C}-(\text{CH}_2)_{16}\text{CH}_3 \\
| \\
\text{H}
\end{array}
$$

product A   product B

i. In the spaces provided above, balance the equation by adding appropriate coefficients for product A and product B. **(1 mark)**
ii. Name the fatty acid that is produced by the hydrolysis of this triglyceride. **(1 mark)**
iii. The fatty acid produced in the above reaction is completely oxidised to produce carbon dioxide and water.

Write a balanced equation for the oxidation reaction. **(2 marks)**

▶ Question 5 (1 mark)

Source: VCE 2010 Chemistry Exam 1, Section A, Q.17; © VCAA

MC The following are **incomplete** and **unbalanced** equations representing three types of chemical reactions that involve glucose. In reactions 1 and 3, product A is the same compound. In reactions 2 and 3, product B is the same compound.

reaction 1 $C_6H_{12}O_6(aq) \rightarrow C_2H_5OH(aq) + \text{product A}$

reaction 2 $C_6H_{12}O_6(aq) \rightarrow C_{12}H_{22}O_{11}(aq) + \text{product B}$

reaction 3 $C_6H_{12}O_6(aq) \rightarrow \text{product A} + \text{product B}$

Which one of the following correctly names reaction 3 and identifies product A and product B?

	Reaction 3	Product A	Product B
A.	fermentation	water	carbon dioxide
B.	fermentation	carbon dioxide	water
C.	combustion	water	carbon dioxide
D.	combustion	carbon dioxide	water

More exam questions are available in your learnON title.

1.5 Review

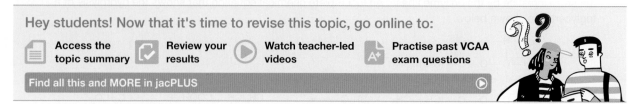

1.5.1 Topic summary

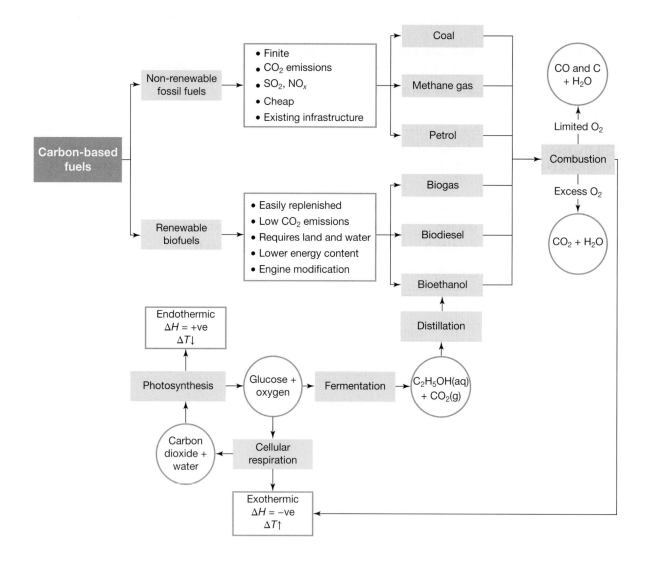

1.5.2 Key ideas summary

1.5.3 Key terms glossary online only

1.5 Activities

learnon

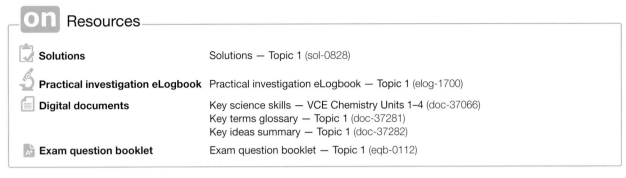

Students, these questions are even better in jacPLUS

Receive immediate feedback and access sample responses

Access additional questions

Track your results and progress

Find all this and MORE in jacPLUS

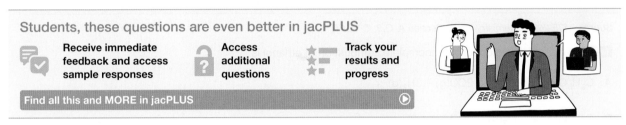

1.5 Review questions

1. **MC** Coal and ethanol are both produced from plants.

   Which of the following statements about the classification of these two fuels is correct?

   A. Coal and ethanol are both fossil fuels.
   B. Coal is a fossil fuel but ethanol is a biofuel.
   C. Coal and ethanol are both biofuels.
   D. Coal is a biofuel but ethanol is a fossil fuel.

2. **MC** When biofuels are burned, the carbon dioxide produced

   A. puts carbon atoms back into the atmosphere that were only recently removed.
   B. puts carbon atoms back into the atmosphere that were removed millions of years ago.
   C. puts oxygen atoms back into the atmosphere that were removed millions of years ago.
   D. puts carbon atoms back into the atmosphere at a slower rate than when an equivalent amount of fossil fuel is burned.

3. a. What is a fossil fuel?
   b. Give at least three examples of fossil fuels.

4. a. What is a biofuel?
   b. Give three examples of biofuels.

5. Ethanol burns in oxygen to produce water and either carbon dioxide or carbon monoxide. The particular oxide produced depends on whether the oxygen supply is plentiful or limited. The molar heats of combustion for these two reactions are $-1360 \text{ kJ mol}^{-1}$ and $-1192 \text{ kJ mol}^{-1}$ respectively.

   a. Write the thermochemical equation for the combustion of ethanol to produce carbon dioxide.
   b. Write the thermochemical equation for the combustion of ethanol to produce carbon monoxide.
   c. Use the equations from parts **a** and **b** to explain how the amount of oxygen consumed influences the oxide produced.

6. Data tables give the heat output from the complete combustion of ethane and ethene as 51.9 kJ g^{-1} and 50.3 kJ g^{-1} respectively. Write thermochemical equations for the complete combustion of these fuels, showing ΔH values in units of kJ mol^{-1}.

7. Draw and label an energy profile for each of the following. Include the formulas of the reactants and products, E_a and ΔH in your diagrams.

   a. The combustion of methane
   b. Photosynthesis

1.5 Exam questions

Section A — Multiple choice questions

All correct answers are worth 1 mark each; an incorrect answer is worth 0.

Question 1

Source: VCE 2022 Chemistry Exam, Section A, Q.3; © VCAA

MC The correct equation for the incomplete combustion of ethanol is

A. $C_2H_5OH(l) + \frac{1}{2}O_2(g) \rightarrow 2CO(g) + 3H_2(g)$

B. $C_2H_5OH(l) + \frac{3}{2}O_2(g) \rightarrow 2CO_2(g) + 3H_2(g)$

C. $C_2H_5OH(l) + 2O_2(g) \rightarrow 2CO(g) + 3H_2O(l)$

D. $C_2H_5OH(l) + 3O_2(g) \rightarrow 2CO_2(g) + 3H_2O(l)$

Question 2

Source: VCE 2018 Chemistry Exam, Section A, Q.3; © VCAA

MC Which one of the following statements about fuels is correct?

A. Petroleum gas is a form of renewable energy.
B. Electricity can only be generated by burning coal.
C. Carbon dioxide is not produced when biogas is burnt.
D. Biodiesel can be derived from both plant and animal material.

Question 3

Source: VCE 2017 Chemistry Exam, Section A, Q.5; © VCAA

MC Which one of the following is a biofuel?

A. ethanol produced from crude oil
B. ethanol produced from cellulose
C. propane produced from natural gas
D. electricity produced by hydropower

Question 4

Source: VCE 2015 Chemistry Exam, Section A, Q.5; © VCAA

MC Which one of the following statements best defines a renewable energy resource?

A. an energy resource that will not be consumed within our lifetime
B. an energy resource that does not produce greenhouse gases when consumed
C. an energy resource derived from plants that are grown for the production of liquid biofuels
D. an energy resource that can be replaced by natural processes within a relatively short time

Question 5

Source: VCE 2022 Chemistry Exam, Section A, Q.15; © VCAA

MC The molar heat of combustion of glucose, $C_6H_{12}O_6$, in the cellular respiration equation is 2805 kJ mol^{-1} at standard laboratory conditions (SLC).

Which one of the following statements about cellular respiration is correct?

A. Cellular respiration is an endothermic reaction.
B. The products of cellular respiration are carbon and carbon dioxide.
C. Cellular respiration is a redox reaction because $C_6H_{12}O_6$ accepts electrons from oxygen.
D. When one mole of oxygen is consumed in the reaction, 467.5 kJ of energy is released.

Question 6

Source: VCE 2016 Chemistry Exam, Section A, Q.17; © VCAA

MC The combustion of hexane takes place according to the equation

$$C_6H_{14}(g) + \frac{19}{2}O_2(g) \rightarrow 6CO_2(g) + 7H_2O(g) \quad \Delta H = -4158 \text{ kJ mol}^{-1}$$

Consider the following reaction.

$$12CO_2(g) + 14H_2O(g) \rightarrow 2C_6H_{14}(g) + 19O_2(g)$$

The value of ΔH, in kJ mol^{-1}, for this reaction is

A. +8316
B. +4158
C. −2079
D. −3568

Question 7

Source: VCE 2017 Chemistry Exam, Section A, Q.7; © VCAA

MC What is the total energy released, in kilojoules, when 100 g of butane and 200 g of octane undergo combustion in the presence of excess oxygen?

A. 9760
B. 14 600
C. 17 300
D. 19 500

Question 8

Source: VCE 2014 Chemistry Exam, Section A, Q.24; © VCAA

MC Methane gas may be obtained from a number of different sources. It is a major component of natural gas. Methane trapped in coal is called coal seam gas and can be extracted by a process known as fracking. Methane is also produced by the microbial decomposition of plant and animal materials. In addition, large reserves of methane were trapped in ice as methane hydrate in the ocean depths long ago.

Methane is a renewable energy source when it is obtained from

A. natural gas.
B. coal seam gas.
C. methane hydrate.
D. microbial decomposition.

▶ Question 9

Source: VCE 2006 Chemistry Exam 1, Section A, Q.13; © VCAA

MC Carbon monoxide can be oxidised to carbon dioxide.

$$2CO(g) + O_2(g) \rightarrow 2CO_2(g)$$

3 mol of CO and 2 mol of O_2 are mixed.

When the reaction is complete there will be

A. 4 mol of CO_2 produced.

B. 2 mol of CO_2 produced.

C. 1 mol of CO unreacted.

D. 0.5 mol of O_2 unreacted.

▶ Question 10

MC The thermochemical equation for the combustion of methane is:

$$CH_4(g) + 2O_2(g) \rightarrow CO_2(g) + 2H_2O(g) \quad \Delta H = -889 \, \text{kJ mol}^{-1}$$

What would be the energy released, in kJ, for the following equation?

$$2CH_4(g) + 4O_2(g) \rightarrow 2CO_2(g) + 4H_2O(l)$$

A. 889

B. 933

C. 1778

D. 1866

Section B — Short answer questions

▶ Question 11 (1 mark)

Source: VCE 2017 Chemistry Exam, Section B, Q.2.b.ii; © VCAA

A vehicle that is powered by a diesel engine is able to use either petrodiesel or biodiesel as a fuel.

Petrodiesel and biodiesel are not pure substances, but are a mixture of molecules. In general, petrodiesel consists of molecules that are shorter in length, on average, than those found in biodiesel. Biodiesel contains molecules that include functional groups.

The table below lists some of the properties of the two fuels.

Fuel	Major component	Energy content (MJ/kg)	CO₂ emission (kg CO₂/kg of fuel)
petrodiesel	$C_{12}H_{26}$	43	3.17
biodiesel	$C_{19}H_{32}O_2$	38	2.52

Assume that combustion occurs in an unlimited supply of oxygen for the following calculation.

Using the data from the table, calculate the mass of carbon dioxide, CO_2, that would be produced from 3.91 kg of biodiesel.

Source: VCE 2014 Chemistry Exam, Section B, Q.3.c.ii; © VCAA

Biodiesel may be produced by reacting canola oil with methanol in the presence of a strong base. Since canola oil contains a mixture of triglycerides, the reaction produces glycerol and a mixture of biodiesel molecules. A typical biodiesel molecule derived from canola oil has the chemical formula $C_{15}H_{30}O_2$.

The heat content of canola oil can be determined by placing it in a spirit burner in place of ethanol. A typical result is 17 kJ g^{-1}.

Suggest why the heat content of fuels such as canola oil and biodiesel are measured in kJ g^{-1} and not kJ mol^{-1}.

▶ **Question 13 (1 mark)**

Source: VCE 2008 Chemistry Exam 1, Section B, Q.7.c.ii; © VCAA

In many countries, ethanol is present in petrol as a renewable fuel additive to reduce dependence on fossil fuels.

Ethanol can be produced by fermentation of glucose.

Explain why ethanol produced by fermentation is referred to as a 'biochemical fuel'.

▶ **Question 14 (2 marks)**

Source: VCE 2007 Chemistry Exam 2, Section B, Q.4.a.ii; © VCAA

A structure for the disaccharide maltose $(C_{12}H_{22}O_{11})$ is given below.

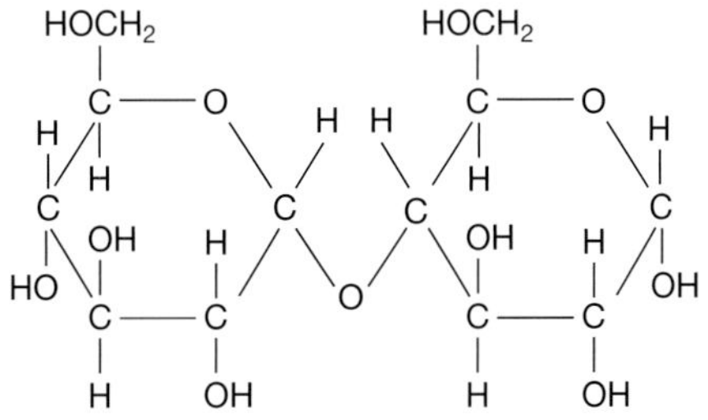

Write a balanced equation for the combustion of one mole of maltose $(C_{12}H_{22}O_{11})$ in the presence of excess oxygen.

Source: *VCE 2012 Chemistry Exam 2, Section B, Q.2.b.i;* © VCAA

The reaction between 2-bromo-2-methylpropane and hydroxide ions occurs in two steps.

step 1 $(CH_3)_3CBr(aq) \rightarrow (CH_3)_3C^+(aq) + Br^-(aq)$

step 2 $(CH_3)_3C^+(aq) + OH^-(aq) \rightarrow (CH_3)_3COH(aq)$

The energy profile diagrams for step 1 and step 2 are shown below. Both are drawn to the same scale.

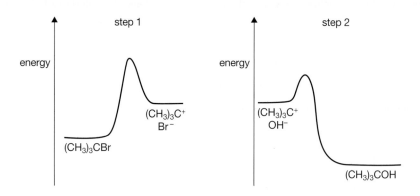

Which step involves an endothermic reaction? Provide a reason for your answer.

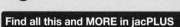 

2 Measuring changes in chemical reactions

KEY KNOWLEDGE

In this topic you will investigate:

Measuring changes in chemical reactions

- calculations related to the application of stoichiometry to reactions involving the combustion of fuels, including mass-mass, mass-volume and volume-volume stoichiometry, to determine heat energy released, reactant and product amounts and net volume or mass of major greenhouse gases (CO_2, CH_4 and H_2O), limited to standard laboratory conditions (SLC) at 25 °C and 100 kPa
- the use of specific heat capacity of water to approximate the quantity of heat energy released during the combustion of a known mass of fuel and food
- the principles of solution calorimetry, including determination of calibration factor and consideration of the effects of heat loss; analysis of temperature-time graphs obtained from solution calorimetry
- energy from fuels and food:
  - calculation of energy transformation efficiency during combustion as a percentage of chemical energy converted to useful energy
  - comparison and calculations of energy values of foods containing carbohydrates, proteins and fats and oils.

Source: VCE Chemistry Study Design (2024–2027) extracts © VCAA; reproduced by permission.

PRACTICAL WORK AND INVESTIGATIONS

Practical work is a central component of VCE Chemistry. Experiments and investigations, supported by a **practical investigation eLogbook** and **teacher-led videos**, are included in this topic to provide opportunities to undertake investigations and communicate findings.

EXAM PREPARATION

▶ Access past VCAA questions and exam-style questions and their video solutions in every lesson, to ensure you are ready.

2.1 Overview

2.1.1 Introduction

Measuring, monitoring and predicting the outcomes of chemical reactions are very important. While the affordability and availability of instrumental analysis have both improved in recent times, simple stoichiometric calculations still have a place in society and industry. Balanced chemical equations, and quantities of reactants, can be used to predict the quantities of chemical products, and the energy released or absorbed when the reactions reach completion or equilibrium. The percentage composition of fuel and food mixtures allows us to determine the amount of energy per gram that is released when they are used.

Measuring carbon dioxide (CO_2) levels from fuel consumption — and predicting future levels based on the current and expected growth of the number of vehicles using fossil fuels — allows scientists to

FIGURE 2.1 Petrol-fuelled cars contribute to atmospheric carbon dioxide and usually only have an energy efficiency of around 25 per cent.

estimate atmospheric CO_2 levels in the near future. Analysing reaction efficiency allows chemical engineers to develop better fuels that provide more energy per gram and release fewer emissions. Reducing the number of energy transformations when turning chemical potential energy into other forms helps reduce the amount of fuel we use.

LEARNING SEQUENCE

on Resources

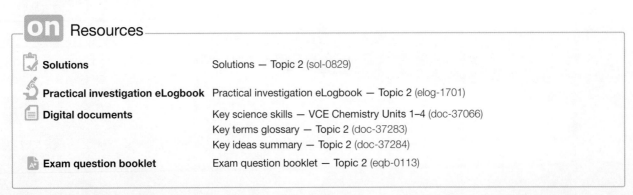

2.2 Fuel calculations

This subtopic will focus on **stoichiometric** calculations related to the **combustion** of fuels and foods for the purposes of determining energy content, gas emissions, reactant consumption and energy efficiency.

stoichiometry calculating amounts of reactants and products using a balanced chemical equation

combustion the rapid reaction of a compound with oxygen

accuracy refers to how close an experimental measurement is to a known value

precision refers to how close multiple measurements of the same investigation are to each other; a measure of repeatability or reproducibility

$$n = \frac{m}{M} \qquad m = n \times M \qquad M = \frac{m}{n}$$

where:

n = number of moles (mol)

m = weighed mass (g)

M = molar mass $\left(\dfrac{\text{g}}{\text{mol}} \right)$

2.2.1 Significant figures

Accuracy refers to how closely a measured value agrees with the correct value. **Precision** refers to how closely individual measurements agree with one another. Measurements are frequently repeated to improve accuracy and precision. Average values obtained from several measurements are usually more reliable than individual measurements. Significant figures indicate how precisely measurements have been made.

The rules applied to the use of significant figures are shown in table 2.1.

TABLE 2.1 Rules for the use of significant figures

Rule	Example
1. Non-zero numbers are always significant.	123 and –123 both have three significant figures.
2. Zeros that come after a non-zero number are significant.	100.4 and 12.01 both have four significant figures.
3. Zeros that come before the first non-zero number are not significant.	0.0010 and 2.1×10^{-3} (0.0021) both have two significant figures.
4. Exact numbers have unlimited significant figures. These are often quantities that can be counted, rather than measured.	A Year 12 Chemistry class has 25 students exactly. In a balanced equation, 1 mole of C_3H_8 produces exactly 3 moles of CO_2. A percentage is a number out of 100 exactly, so the 100 does not limit the significant figures.
5. When multiplying values together, the final calculated answer is expressed to the lowest number of significant figures (excluding exact numbers) used in the calculation.	$21.68 \times 0.15 = 3.3$ (rounded up from 3.252)
6. When adding and subtracting measurements with the same units, your answer is limited by the position of the first doubtful digit.	$50.22 - 10.1 = 40.1$ (instead of 40.12) and $100.6 + 6.234 = 106.8$ (instead of 106.834)

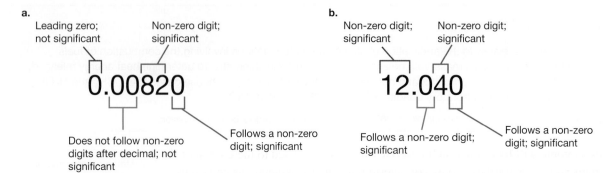

a.

Leading zero;
not significant

Non-zero digit;
significant

0.00820

Does not follow non-zero
digits after decimal; not
significant

Follows a non-zero
digit; significant

b.

Non-zero digit;
significant

Non-zero digit;
significant

12.040

Follows a non-zero digit;
significant

Follows a non-zero
digit; significant

tlvd-9640

SAMPLE PROBLEM 1 Significant figures in fuel calculations

Write the following calculations expressed to the correct number of significant figures and correct units.

a. Calculate the temperature change (ΔT) when water is heated from 18.5 °C to 26.7 °C.

b. Calculate the percentage composition by mass (%(m/m)) of 5050 mg of sulfur in 1.0 kg of fuel.

THINK	WRITE
a. 1. Determine whether the numbers are in the same unit.	**a.** The numbers are in the same unit.
2. Determine the place value of the last known digit.	The place value of the last known digit is one decimal place.
3. Complete the calculation, giving your answer to two significant figures with one decimal place.	$26.7 - 18.5 = 8.2 \,°C$
b. 1. Determine whether the numbers are in the same unit.	**b.** The numbers are not in the same unit.
2. 1 mg is exactly 10^6 times smaller than 1 kg. Divide 5050 mg by 10^6.	$\dfrac{5050}{10^6} = 0.005\,050$
3. Divide the mass of sulfur (in kg) by the mass of fuel (in kg) and express as a percentage to two significant figures, as the 1.0 kg measurement of fuel is the lowest number of significant figures used in your calculations.	$\%(m/m) = \dfrac{0.005\,050}{1.0} \times 100$ $= 0.5050$ $= 0.51\%$ (to two sig. figs.)

PRACTICE PROBLEM 1

Write the following calculations expressed to the correct number of significant figures and correct units.

a. Calculate the temperature change (ΔT) when water is cooled from 60.8 °C to 8.73 °C.

b. Calculate the percentage composition by volume (%(v/v)) of 45.87 mL of ethanol in 0.500 L of petrol.

2.2.2 Mass–mass calculations

Measuring the mass of gases produced from chemical reactions was introduced in Unit 2.

TIP: When performing calculations in Chemistry it is advisable to use the figures obtained from using your calculator throughout the question. Intermediate results should retain at least one significant figure more than given in the question. The final result should be given to the number of significant figures required.

BACKGROUND KNOWLEDGE: Standard laboratory conditions (SLC)

Amedeo Avogadro was an Italian scientist who put forward the hypothesis that 'equal volumes of all gases measured at the same temperature and pressure contain the same number of particles'. This means that, if two gases have the same temperature, pressure and volume, they must contain the same number of moles.

It has been found that 1 mole of any gas at **standard laboratory conditions (SLC)** occupies a volume of 24.8 L. This volume is called the **molar gas volume** and means that 1 mole of any gas occupies 24.8 L at 25 °C and 100.0 kPa.

standard laboratory conditions (SLC) 100 kPa and 25 °C

molar gas volume the volume occupied by a mole of a substance at a given temperature and pressure; at SLC, 1 mole of gas occupies 24.8 L

FIGURE 2.3 Under the same conditions of temperature and pressure, the volume of a gas depends only on the number of molecules it contains, and not on what the particles are.

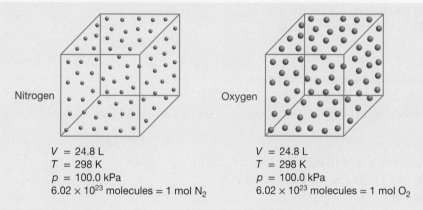

Nitrogen

$V = 24.8$ L
$T = 298$ K
$p = 100.0$ kPa
6.02×10^{23} molecules = 1 mol N_2

Oxygen

$V = 24.8$ L
$T = 298$ K
$p = 100.0$ kPa
6.02×10^{23} molecules = 1 mol O_2

The molar volume of a gas varies with temperature and pressure but, at any given temperature and pressure, it is the same for all gases. There is a direct relationship between the number of moles of a gas (n), its molar volume (V_m) and its actual volume (V), where V is measured in litres.

At a given temperature and pressure, the relationship between the number of moles of a gas (n) and its molar volume (V_m):

$$n = \frac{V}{V_m}$$

FIGURE 2.4 A mole of hydrogen gas would occupy the same volume as a mole of oxygen molecules, but because hydrogen weighs less than oxygen, it floats upwards in the air.

At SLC, $V_m = 24.8$ L mol^{-1}.

The number of moles of gas at SLC:

$$n_{SLC} = \frac{V}{24.8}$$

SAMPLE PROBLEM 2 Mass–mass stoichiometry in combustion of fuels

Calculate the mass of $CO_2(g)$ produced at SLC from the complete combustion of 5.0 g of propane gas in excess oxygen.

THINK	WRITE
1. Write a balanced equation	$C_3H_8(g) + 5O_2(g) \rightarrow 3CO_2(g) + 4H_2O(l)$
2. Calculate the molar masses (M) of the limiting reactant and product.	$M(\text{propane}) = (3 \times 12.0) + (8 \times 1.0)$ $\qquad = 44.0 \text{ g mol}^{-1}$ $M(CO_2) = 12.0 + 2 \times 16.0$ $\qquad = 44.0 \text{ g mol}^{-1}$
3. Determine the amount, in mol, of the limiting reactant (and leave the number on your calculator).	$n(C_3H_8) = \dfrac{5.0 \text{ g}}{44.0 \text{ g mol}^{-1}}$ $\qquad = 0.1136 \text{ mol}$
4. Use the mole ratios in the balanced equation to determine the amount, in mol, of $CO_2(g)$ released and leave the number on your calculator.	$\dfrac{n(CO_2)}{n(C_3H_8)} = \dfrac{3}{1}$ $\therefore n(CO_2) = \dfrac{3}{1} \times 0.1136$ $\qquad = 0.3409 \text{ mol}$
5. Use the formula $m = n \times M$ to calculate the mass in grams of $CO_2(g)$ and leave the number on your calculator.	$m(CO_2) = n \times M$ $\qquad = 0.3409 \text{ mol} \times 44.0 \text{ g mol}^{-1}$
6. Determine the least number of significant figures in your measurements from the question and the molar masses of the reactant and product, and write your answer to the correct number of significant figures with the correct unit.	$m(CO_2) = 15 \text{ g}$ (to two sig. figs.) *Note:* Because the molar masses of propane and carbon dioxide are the same, the mass of carbon dioxide is three time the mass of propane.

PRACTICE PROBLEM 2

Calculate the mass of $CO_2(g)$ produced at SLC from the complete combustion of 10.0 g of liquid octane, $C_8H_{18}(l)$, in excess oxygen.

2.2.3 Mass–volume calculations

The mass of CO_2 calculated in the previous examples will occupy a certain volume at a particular pressure and temperature. When the temperature and pressure change, so does the volume the amount of gas occupies. The concepts behind gas behaviour and the use of the **ideal gas equation** were covered in Unit 2.

ideal gas equation $PV = nRT$, where pressure is measured in kilopascals, volume is measured in litres and temperature is measured in kelvin

The ideal gas equation

$$PV = nRT$$

where:

P = **pressure** measured in kilopascals (kPa)

V = volume of gas measured in litres (L)

n = number of moles of gas

R = **molar gas constant** (8.31)

T = temperature in **kelvin** (K).

pressure the force per unit area that one region of a gas, liquid or solid exerts on another

molar gas constant (R) the constant of the universal gas equation; R = 8.31 J mol^{-1} K^{-1} when pressure is measured in kPa, volume is measured in L, temperature is measured in K and the quantity of the gas is measured in moles (n)

kelvin the SI base unit of thermodynamic temperature, equal in magnitude to the degree Celsius

tlvd-9642

SAMPLE PROBLEM 3 Mass–volume stoichiometry in combustion of fuels

What mass of completely combusted methane would produce 2.48×10^4 L of carbon dioxide at SLC?

THINK	WRITE
1. Determine if the volume is measured in standard units.	The volume is measured in standard units.
2. Use $n = \dfrac{V}{V_m}$ to calculate the number of moles of CO_2.	$n(CO_2)_{SLC} = \dfrac{V}{24.8}$ $= \dfrac{2.48 \times 10^4 \text{ L}}{2.48 \times 10^1 \text{ L mol}^{-1}}$ $= 1.00 \times 10^3$ mol
3. In complete combustion, all of the carbon from the hydrocarbon (in this case methane, CH_4) ends up oxidised to CO_2, so the number of moles of C = $n(CO_2)$ = $n(CH_4)$.	$n(CH_4) = n(CO_2)$
4. Use $m = n \times M$ to convert moles of CO_2 into mass in grams.	$m(CH_4) = n \times M$ $= 1.00 \times 10^3 \text{ mol} \times 16.0 \text{ g mol}^{-1}$ $= 1.60 \times 10^4$ g (16.0 kg)

PRACTICE PROBLEM 3

What mass of completely combusted ethanol would produce 10 L of CO_2 at SLC?

2.2.4 Volume–volume calculations

Because gases occupy the same volume at the same temperature and pressure, a volume-to-volume ratio is effectively a mole-to-mole ratio. For example, if a pure gas sample occupies a volume of 12.4 L at SLC — that is, half the molar volume (V_m) for any gas — there must be $\dfrac{12.4}{24.8} = 0.500$ moles of gas.

Consider the reaction between nitrogen and hydrogen gas:

$$N_2(g) + 3H_2(g) \rightarrow 2NH_3(g)$$

This equation says that when 1 mole of N_2 reacts with 3 moles of H_2, it will produce 2 moles of ammonia, NH_3.

According to Avogadro's hypothesis, if the gases are at the same pressure and temperature, their *molar* ratios are equal to their *volume* ratios. Therefore, we use volumes instead of moles and can say that 10 mL of N_2 reacts with 30 mL of H_2 to form 20 mL of ammonia.

$$N_2(g) + 3H_2(g) \rightarrow 2NH_3(g)$$

1 mol	3 mol → 2 mol
1 vol	3 vol → 2 vol
10 mL	30 mL → 20 mL

$$\frac{V(\text{unknown})}{V(\text{known})} = \frac{\text{coefficient of unknown}}{\text{coefficient of known}}$$

tlvd-9643

SAMPLE PROBLEM 4 Volume–volume stoichiometry with gases in combustion of fuels

If 100 m³ of ethene is burned according to the equation

$$C_2H_4(g) + 3O_2(g) \rightarrow 2CO_2(g) + 2H_2O(g)$$

calculate the volume of:
a. carbon dioxide produced
b. oxygen consumed.
(Assume all gas volumes are measured at the same temperature and pressure.)

THINK

Because all gas volumes are measured at the same temperature and pressure, the equation may be interpreted in terms of volume ratios.

WRITE

a. $V(CO_2) = 2V(C_2H_4)$
   $\therefore V(CO_2) = 2 \times 100$
   $= 200 \text{ m}^3$

b. $V(O_2) = 3V(C_2H_4)$
   $\therefore V(O_2) = 3 \times 100$
   $= 300 \text{ m}^3$

PRACTICE PROBLEM 4

Methane gas burns in air according to the following equation:

$$CH_4(g) + 2O_2(g) \rightarrow CO_2(g) + 2H_2O(g)$$

If 25 mL of methane is burned, find the volumes of the following reactants and products measured at the same temperature and pressure:
a. Oxygen
b. Carbon dioxide
c. Water.

Liquid fuel density

Fuels in a liquid state occupy a much smaller condensed volume than gases at atmospheric pressure. Due to differing intermolecular forces between molecules in liquid fuels, the same measured volume of two different fuels will provide two different amounts and masses of fuel. Density is a ratio of the mass per volume, and therefore has standard units such as grams per cubic centimetre ($g\ cm^{-3}$) and kilograms per cubic metre ($kg\ m^{-3}$).

Because we typically use millilitres and litres for measuring liquids, we often modify our density units to match.

Density units

$$\text{Density} = \frac{\text{mass}}{\text{volume}}$$

$$1\ cm^3 = 1\ mL$$

$$\therefore 1\ g\ cm^{-3} = 1\ g\ mL^{-1}$$

Changing temperature will also affect the density of a fuel and therefore the volume it occupies. For example, the average density of ethanol at 5 °C is 0.802 g mL^{-1}. At 25 °C the average drops to 0.790 g mL^{-1}.

FIGURE 2.5 The density of fuel used in Formula 1 racing varies between 700 and 800 g L^{-1}, so fuel loads are measured in kilograms before races.

---

tlvd-9644

SAMPLE PROBLEM 5 Volume–volume stoichiometry with gases and liquids in combustion of fuels

Calculate the volume of $CO_2(g)$ produced at SLC from the combustion of 1.0 L of ethanol. The density of ethanol at SLC is 0.790 g mL^{-1}.

THINK	WRITE
1. Change the volume of ethanol into mL to match the unit in the given density.	$1.0\ L = 1.0 \times 10^3\ mL$ $\text{Density }(d) = \dfrac{\text{mass }(m)}{\text{volume }(V)}$
2. Calculate the mass of ethanol in 1.0 L from the density.	$m = d \times V$ $= 0.790\ \cancel{g\ mL^{-1}} \times 1.0 \times 10^3\ \cancel{mL}$ $= 790\ g$
3. Convert the mass into an amount (moles) by using $n = \dfrac{m}{M}$.	$n(C_2H_5OH) = \dfrac{m}{M}$ $= \dfrac{790\ \cancel{g}}{46.0\ \cancel{g\ mol^{-1}}}\ mL$ $= 17.174\ mol$
4. Ethanol has two C atoms in its formula, which are oxidised into CO_2. Therefore, $n(CO_2) = 2 \times n(C_2H_5OH)$. Calculate $n(CO_2)$.	$n(CO_2) = 2 \times 17.065$ $= 34.35\ mol$
5. Use $V = n \times 24.8$ to calculate the volume at SLC.	$V(CO_2) = 34.35\ \cancel{mol} \times 24.8\ L\ \cancel{mol^{-1}}$ $= 851.8\ L$
6. Round your answer to two significant figures, as the 1.0 L measurement of ethanol is the lowest number of significant figures used in your calculations.	$V(CO_2) = 8.5 \times 10^2\ L$ (to two sigs. figs.)

PRACTICE PROBLEM 5

Calculate the volume of completely combusted ethanol, in litres, that produces 1.50×10^4 L of $CO_2(g)$ at SLC. The density of ethanol at SLC is 0.790 g mL^{-1}.

on Resources

🧩 **Interactivity** Mole relationships (int-1214)

2.2 Activities

learn on

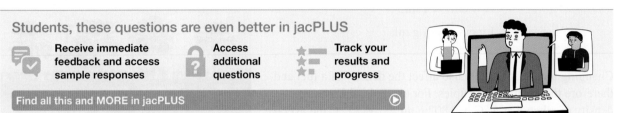

Students, these questions are even better in jacPLUS

Receive immediate feedback and access sample responses

Access additional questions

Track your results and progress

Find all this and MORE in jacPLUS ▶

| 2.2 Quick quiz **on** | 2.2 Exercise | 2.2 Exam questions |

2.2 Exercise

1. Calculate the numbers of moles of the following gases at SLC.
   a. 15 L of oxygen, O_2 b. 25 L of chlorine, Cl_2
2. Calculate the volumes of the following gases at SLC.
   a. 1.3 mol hydrogen, H_2 b. 3.6 g of methane, CH_4 c. 0.35 g of argon, Ar
3. Calculate the masses of the following gas samples at SLC.
   a. 16.5 L of neon, Ne   b. 1050 mL of sulfur dioxide, SO_2
4. What is the mass (in kg) of 850 L of carbon monoxide gas measured at SLC?
5. A 0.953 L volume of a monoatomic gas measured at SLC has a mass of 3.20 g. What is the molar mass of the gas? What is the gas?
6. a. Calculate the net change in mass of greenhouse gas (as carbon dioxide) produced by the combustion of 128 g of methane.
   b. Express your answer to part a as a percentage increase or decrease.
   c. Comment on how the units used (mass or volume) may influence the conclusions drawn from calculations such as in this question.
7. Calculate the volume of carbon dioxide produced when 50.0 mL of ethanol combusts in excess oxygen at SLC. (Density of ethanol = 0.789 g mL^{-1})
8. At high temperatures, such as those in a car engine during operation, atmospheric nitrogen burns to produce the pollutant nitrogen dioxide, according to the equation:

$$N_2(g) + 2O_2(g) \rightarrow 2NO_2(g)$$

   a. If 20 mL of nitrogen is oxidised, calculate the volume of oxygen needed to produce the pollutant. Assume that temperature and pressure remain constant
   b. What is the initial volume of reactants in this combustion reaction?
   c. What is the final volume of products in the reaction?
   d. Is there an overall increase or decrease in the volume of gases on completion of the reaction?

2.2 Exam questions

Question 1 (1 mark)
Source: VCE 2021 Chemistry Exam, Section A, Q.21; © VCAA

MC Butane, C_4H_{10}, undergoes complete combustion according to the following equation.

$$2C_4H_{10}(g) + 13O_2(g) \rightarrow 8CO_2(g) + 10H_2O(g)$$

67.0 g of C_4H_{10} released 3330 kJ of energy during complete combustion at standard laboratory conditions (SLC).

The mass of carbon dioxide, CO_2, produced was

A. 0.105 g   **B.** 3.18 g   **C.** 50 g   **D.** 204 g

Question 2 (2 marks)
Source: VCE 2020 Chemistry Exam, Section B, Q.6.b; © VCAA

Methane gas, CH_4, can be captured from the breakdown of waste in landfills. CH_4 is also a primary component of natural gas. CH_4 can be used to produce energy through combustion.

If 20.0 g of CH_4 is kept in a 5.0 L sealed container at 25 °C, what would be the pressure in the container?

Question 3 (1 mark)
Source: VCE 2018 Chemistry Exam, Section A, Q.10; © VCAA

MC Bioethanol, C_2H_5OH, is produced by the fermentation of glucose, $C_6H_{12}O_6$, according to the following equation.

$$C_6H_{12}O_6(aq) \rightarrow 2C_2H_5OH(aq) + 2CO_2(g)$$

The mass of C_2H_5OH obtained when 5.68 g of carbon dioxide, CO_2, is produced is

A. 0.168 g   **B.** 0.337 g   **C.** 2.97 g   **D.** 5.94 g

Question 4 (2 marks)
Source: VCE 2017 Chemistry Exam, Section B, Q.1.a; © VCAA

Industrially, ethanol, C_2H_5OH, is made by either of two methods.

One method uses ethene, C_2H_4, which is derived from crude oil.

The other method uses a sugar, such as sucrose, $C_{12}H_{22}O_{11}$, and yeast, in aqueous solution.

The production of C_2H_5OH from $C_{12}H_{22}O_{11}$ and yeast proceeds according to the equation

$$C_{12}H_{22}O_{11}(aq) + H_2O(l) \rightarrow 4C_2H_5OH(aq) + 4CO_2(g)$$

Determine the mass, in grams, of pure C_2H_5OH that would be produced from 1.250 kg of $C_{12}H_{22}O_{11}$ dissolved in water.

$M(C_{12}H_{22}O_{11}) = 342 \text{ g mol}^{-1}$

Question 5 (4 marks)
Source: Adapted from VCE 2016 Chemistry Exam, Section B, Q.3.e; © VCAA

Ethanol can be produced by the fermentation of sugars in plant material.
a. Write a balanced chemical equation for the fermentation of glucose. **(1 mark)**
b. The ethanol produced can be separated from the reaction mixture by distillation.

   What would be the minimum mass of pure glucose needed to produce 1.00 L of pure ethanol from fermentation?

   $d(C_2H_5OH) = 0.785 \text{ g mL}^{-1}$ **(3 marks)**

More exam questions are available in your learnON title.

2.3 Energy from food and fuels

KEY KNOWLEDGE

- Energy from fuels and food:
  - calculation of energy transformation efficiency during combustion as a percentage of chemical energy converted to useful energy
  - comparison and calculations of energy values of foods containing carbohydrates, proteins and fats and oils

Source: VCE Chemistry Study Design (2024–2027) extracts © VCAA; reproduced by permission.

In this subtopic we will be looking at the calculations associated with energy output from quantities of fuels and food. Topic 1 provided an overview of the energy content of **fuel** and food; however, energy output depends on a number of factors, including the energy density and **efficiency** of the energy transformations.

To begin with, let's look at some examples in which we calculate the energy released as heat from the combustion of specific quantities of fuel. For these calculations, heat of combustion data is required. The VCE Chemistry Data Book lists common fuels and heat of combustion data.

The published heat of combustion of ethanol at SLC is 29.6 kJ g^{-1} or 1360 kJ mol^{-1}. We can use either number to calculate the heat released by a mass or volume of ethanol.

1.00 L of ethanol (density 0.790 g L^{-1}) contains 790 g at SLC. Therefore, 1.00 L of ethanol will release 2.34×10^4 kJ.

$$29.6 \text{ kJ g}^{-1} \times 790 \text{ g} = 2.34 \times 10^4 \text{ kJ (23.4 MJ)}$$

Sample problem 6 shows the calculation of the same value using the **enthalpy** of combustion.

> **fuel** a substance that burns in air or oxygen to release useful energy
>
> **efficiency** (of energy conversion) the ratio between useful energy output and energy input
>
> **enthalpy** a thermodynamic quantity equivalent to the total heat content of a system

---

tlvd-9669

SAMPLE PROBLEM 6 Writing equations and calculating energy production in combustion

The molar heat of combustion of ethanol is tabulated as –1360 kJ mol^{-1}.
a. Write the thermochemical equation for the combustion of ethanol.
b. If the density of ethanol is 0.790 g mL^{-1}, calculate the energy evolved in MJ when 1.00 L of ethanol is burned.

THINK	WRITE
a. 1. The formula and state symbol for ethanol can be found in the VCE Chemistry Data Book. Both ethanol and methanol complete combustion equations should be learned or correctly generated.	a. $CH_3CH_2OH(l) + 3O_2(g) \rightarrow 2CO_2(g) + 3H_2O(g)$
2. The given heat value refers to 1 mole. Therefore, the value of –1360 kJ is equivalent to the 1 mole of ethanol shown in the balanced equation.	$CH_3CH_2OH(l) + 3O_2(g) \rightarrow 2CO_2(g) + 3H_2O(g)$ $\Delta H = -1360 \text{ kJ mol}^{-1}$

b. 1. Identify the given and unknown quantities.

b. Density of ethanol = 0.790 g mL^{-1}
Volume = 1.00 L
Mass of ethanol = ?

2. Recall the equation for density:

$$\text{Density} = \frac{\text{mass}}{\text{volume}}$$

Compare the units given to those required.

1.00 L ethanol = 1.00×1000
= 1000 mL ethanol

$$\text{Density} = \frac{\text{mass}}{\text{volume}}$$

$$\therefore m(\text{ethanol}) = d(\text{ethanol}) \times V(\text{ethanol})$$
$$= 0.790 \times 1000$$
$$= 790 \text{ g}$$

3. Calculate the number of moles using the mass determined in **b.2.** and M of ethanol by applying the formula $n = \dfrac{m}{M}$.

TIP: Remember to not round your answer to the number of correct significant figures until the end. Retain the number in your calculator.

$$n(\text{ethanol}) = \frac{m}{M}$$
$$= \frac{790}{46.0}$$
$$= 17.17 \text{ mol}$$

4. From the equation, 1 mole of ethanol evolves 1360 kJ. By direct proportion, 17.17 mol produces x kJ.

$$x = 17.17 \times \frac{1360}{1}$$
$$= 23\,356 \text{ kJ}$$

5. Convert kJ to MJ and give the answer to three significant figures.

$$\frac{23\,356}{1000} = 23.4 \text{ MJ}$$

Note: Alternatively, the heat of combustion provided in the VCE Chemistry Data Book can be used. Since this value is 29.6 kJ g^{-1} for ethanol, the energy evolved by 790 g is $29.6 \times 790 = 23\,384$ kJ. When rounded to the correct number of significant figures, the final result obtained is the same: 23.4 MJ.

PRACTICE PROBLEM 6

In an experiment, it was found that the combustion of 0.240 g of methanol in excess oxygen yielded 5.42 kJ.
a. Use this information to calculate the ΔH for this reaction
b. Write the thermochemical equation for this reaction.
c. If the density of methanol is 0.792 g mL^{-1}, calculate the energy evolved in MJ when 10.00 L of methanol is burned.

---

tlvd-9670

SAMPLE PROBLEM 7 Calculating the mass of a fuel to produce a set amount of heat energy

The combustion of ethene can be represented by the following thermochemical equation:

$$C_2H_4(g) + 3O_2(g) \rightarrow 2CO_2(g) + 2H_2O(l) \quad \Delta H = -1409 \, kJ \, mol^{-1}$$

Calculate the mass of ethene required to produce 500 kJ of heat energy.

THINK	WRITE
1. From the equation, 1 mole of C_2H_4 evolves 1409 kJ. By ratio, x moles are required to evolve 500 kJ.	$x = 500 \times \dfrac{1}{1409}$ $= 0.3548$
2. Use the formula $m = n \times M$, where the molar mass of ethene = 28.0 g mol^{-1}.	$m(C_2H_4) = n \times M$ $= 0.3548 \times 28.0$
3. Give the answer to three significant figures.	$m(C_2H_4) = 9.94 \, g$

PRACTICE PROBLEM 7

Use the information from sample problem 7 to calculate the volume of $CO_2(g)$ emitted to produce 1.00 MJ of heat from the complete combustion of ethene under SLC conditions.

2.3.1 Energy from incomplete combustion

Incomplete combustion, which is typically seen in combustion engines to varying degrees, results in less energy release. The reasons for this were discussed in section 1.3.2. A comparison of the oxidation of carbon to carbon dioxide and carbon to carbon monoxide highlights the different in energy released.

$$C(s) + O_2(g) \rightarrow CO_2(g) \quad \Delta H = -394 \, kJ \, mol^{-1}$$

$$C(s) + \frac{1}{2}O_2(g) \rightarrow CO(g) \quad \Delta H = -111 \, kJ \, mol^{-1}$$

So when some of the carbon in fuel doesn't fully oxidise, less heat energy is released. Methane, for example, would have a ΔH of $-608 \, kJ \, mol^{-1}$ if it only produced CO(g) and $H_2O(l)$ at SLC. This amount would be even less if solid carbon formed.

$$CH_4(g) + \frac{1}{2}O_2(g) \rightarrow CO(g) + 2H_2O(l) \quad \Delta H = -608 \, kJ \, mol^{-1}$$

$$CH_4(g) + O_2(g) \rightarrow C(s) + 2H_2O(l) \quad \Delta H = -497 \, kJ \, mol^{-1}$$

2.3.2 Energy efficiency and transformations during combustion

Energy efficiency was briefly discussed in section 1.2.4. To recap, put simply, a process is more efficient if the work done requires the use of less energy.

$$\% \text{ efficiency} = \frac{\text{amount of energy in useful form}}{\text{amount of potential energy in chemical form}} \times 100$$

We saw that the efficiency of converting the chemical potential energy stored in coal into electrical energy is less than 40 per cent. Combustion efficiency in terms of the heat released is affected by fuel type and amount, air pressure and the amount of oxygen. The more transformations that take place to turn the chemical energy into useful energy, the less efficient the overall process. Figure 2.6 shows common energy transformations involved in making useful energy.

FIGURE 2.6 The efficiency of changing one energy form into another varies.

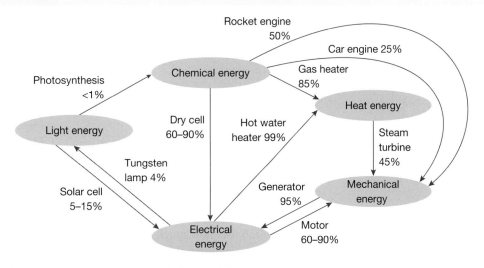

SAMPLE PROBLEM 8 Calculating the energy required given the transformation efficiency

Calculate the amount of chemical energy, in MJ, required to produce 200 MJ of useful, mechanical (kinetic) energy to drive a petrol-fuelled car with an efficiency of 25 per cent.

THINK

1. The process is only 25 per cent efficient. This means the ratio of useful energy : chemical energy is 25 : 100.

2. Write the equivalent ratios as an equation.

   $$\frac{\% \text{ efficiency}}{100\%} = \frac{\text{useful energy}}{\text{chemical energy}}$$

3. Rearrange the equation to calculate the quantity of chemical energy:

   $$\text{Chemical energy} = \frac{100\%}{\% \text{ efficiency}} \times \text{useful energy}$$

4. Calculate and round to the least number of significant figures.

WRITE

$$\frac{25}{100} = \frac{200}{x}$$

$$\frac{x}{200} = \frac{100}{25}$$

$$\therefore x = \frac{100 \times 200}{25}$$

$$x = 800 \text{ MJ}$$

$$= 8.0 \times 10^2 \text{ MJ}$$

PRACTICE PROBLEM 8

Calculate the amount of chemical energy required to make 500 MJ of mechanical energy if the net energy transformations are 30 per cent.

2.3.3 Comparing energy values of foods

Digestion is complex and involves many chemical processes that require energy — some more than others. Not all of the energy stated on nutrition labels is available to us. In some cases, particularly with uncooked food, it takes more energy to digest a quantity of food than the total energy stored in the food. However, we are able to calculate the energy contained in food samples by simply totalling the mass of nutrients in the sample and multiplying them by the stated number of kilojoules per gram.

carbohydrates the general name for a large group of organic compounds occurring in food and living tissues; includes sugars, starch and cellulose

proteins large molecules composed of one or more long chains of amino acids

fat a triglyceride formed from glycerol and three fatty acids

calorimetry a method used to determine the changes in energy of a system by measuring heat exchanges with the surroundings

serving size the recommended amount of food on a nutrition label for one serving

> The energy values of food are:
> - **carbohydrate**: 16 kJ g^{-1}
> - **protein**: 17 kJ g^{-1}
> - **fat**: 37 kJ g^{-1}.

The amount of energy produced by a food can be practically measured by a process called **calorimetry**. This measures the amount of heat released or absorbed in a chemical reaction, change of state or formation of a solution.

Nutrition labels

All packaged foods must feature a nutrition label, which lists how much of each nutrient is present in the food. The ingredients on nutrition labels are listed in descending order according to mass. The overall value of food energy stated on packaging is obtained by multiplying the energy values by the mass of protein, fat and carbohydrate, and then adding all these results together.

For example, the nutrients listed for 100 g of tomato sauce are:

Protein:	1.6 g
Fat:	0.20 g
Carbohydrate:	30.2 g

Therefore, the energy content per 100 g of the sauce is:

Protein: $1.6 \text{ g} \times 17 = 27.2 \text{ kJ}$
Fat: $0.20 \text{ g} \times 37 = 7.4 \text{ kJ}$
Carbohydrate: $30.2 \text{ g} \times 16 = 483.2 \text{ kJ}$
Total energy in 100 g $= 517.8 \text{ kJ}$

FIGURE 2.7 Learning how to read and understand food labels can help you make healthier choices.

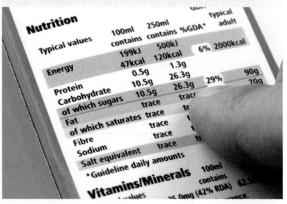

A **serving size** is how much of a food the manufacturer recommends consuming in a serving, but it is important to remember that these are often not based on dietary recommendations and may not be the amount that you consume. It is better to compare the quantity per 100 g when making food choices. Official standard serves vary depending on the type of food, but in general you should aim for:
- less than 10 g of total fat, which includes all of the different types of fats (it is healthier to choose less saturated fat where possible)
- less than 10 g of sugar. The total carbohydrate figure includes starches and sugars, but be aware that sugars may be listed under other names, often ending in -ose. Foods with no added sugar could contain a large amount of natural sugar, and low-fat foods can also contain large quantities of sugar.
- less than 400 mg of salt
- 3–6 g of fibre in breads and cereals.

2.3 Activities

2.3 Quick quiz	2.3 Exercise	2.3 Exam questions

2.3 Exercise

1. The molar heat of combustion for ethanol is -1364 kJ mol^{-1}. Calculate the mass of carbon dioxide emitted when ethanol is used to produce 1.00 MJ of heat.
2. **a.** Methanol is also a fuel. Its molar heat of combustion is -725 kJ mol^{-1}. What mass of carbon dioxide would be produced using methanol to generate 1.00 MJ of heat?
   b. What volume of carbon dioxide would be produced at SLC?
   c. Comment on your answers to question **1** (mass of carbon dioxide emitted in the combustion of ethanol) and part **a** of this question (mass of carbon dioxide emitted in the combustion of methanol) in relation to the masses of CO_2 produced.
3. A coal-fired power station using brown coal as its fuel operates at 37.0% overall energy efficiency. The brown coal has a heat value of 16.0 kJ g^{-1} and a carbon content of 29.0%. Assuming that all the carbon present forms carbon dioxide, calculate the carbon dioxide produced per MJ of electrical energy produced in units of:
   a. g MJ^{-1} **b.** L MJ^{-1} (at SLC).
4. The combustion of 3.15 g of methanol was found to yield 71.5 kJ of heat. Calculate the ΔH value for this reaction and write the thermochemical equation.
5. Calculate the energy released when 18.5 g of carbon undergoes combustion in a plentiful supply of air according to the equation:

$$C(s) + O_2(g) \rightarrow CO_2(g) \quad \Delta H = -394 \text{ kJ mol}^{-1}$$

6. Butane and octane are two hydrocarbons commonly used as fuels. The thermochemical equations for these two fuels are:

$$2C_4H_{10}(g) + 13O_2(g) \rightarrow 8CO_2(g) + 10H_2O(l) \quad \Delta H = -5760 \text{ kJ mol}^{-1}$$
$$2C_8H_{18}(g) + 25O_2(g) \rightarrow 16CO_2(g) + 18H_2O(l) \quad \Delta H = -10\,920 \text{ kJ mol}^{-1}$$

   a. Calculate the heat evolved by the combustion of 100 g of butane.
   b. Use your answer to part **a** to calculate the mass of octane required to produce the same amount of energy.
   c. Explain why it is critical to show symbols of state in thermochemical equations.
7. In 100 g of tomato sauce the energy was determined to be 548.0 kJ. The suggested average serve is 15 g. Calculate the amount of energy supplied in this serving.
8. Michael eats a hamburger for lunch. The amount of energy supplied by the hamburger is about 3020 kJ. Going for a run uses about 48 kJ per minute. How long would Michael need to run to use up the energy supplied by the hamburger?

2.3 Exam questions

▶ Question 1 (1 mark)
Source: VCE 2021 Chemistry Exam, Section A, Q.22; © VCAA

MC 1 L of octane has a mass of 703 g at SLC. The efficiency of the reaction when octane undergoes combustion in the petrol engine of a car is 25.0%.

What volume of octane stored in a petrol tank at SLC is required to produce 528 MJ of usable energy in a combustion engine?

A. 3.92 L   **B.** 11.8 L   **C.** 15.7 L   **D.** 62.7 L

▶ **Question 2 (3 marks)**

Source: VCE 2020 Chemistry Exam, Section B, Q.5.c; © VCAA

The table below shows the amount of each nutrient in 100 g of banana.

Nutrient	Per 100 g
protein	1.1 g
carbohydrates	22.8 g
fat	0.3 g
dietary fibre	2.6 g

An athlete uses 300 kJ of energy for a five-minute run. A typical ripe banana has an average mass of 116 g after it is peeled.

How many typical ripe bananas, correct to two decimal places, would the athlete need to consume to replace the energy used during the run? Show your working.

▶ **Question 3 (1 mark)**

Source: VCE 2021 Chemistry Exam, Section B, Q.5.a; © VCAA

The nutritional information for one medium serving (124 g) of sweet potato is provided in the table below.

Nutrient	Per 124 g
protein	2.0 g
fat	3.0 g
carbohydrates	18.7 g
vitamin C	3.0 mg
vitamin D	less than 0.2 mg

Calculate the energy contained in one medium serving of sweet potato.

▶ **Question 4 (5 marks)**

Source: VCE 2021 Chemistry Exam, Section B, Q.1.b; © VCAA

Digesters use bacteria to convert organic waste into biogas, which contains mainly methane, CH_4. Biogas can be used as a source of energy.

A digester processed 1 kg of organic waste to produce 496.0 L of biogas at standard laboratory conditions (SLC). The biogas contained 60.0% CH_4.

a. Write the thermochemical equation for the complete combustion of CH_4 at SLC. **(2 marks)**
b. Calculate the amount of energy that could be produced by CH_4 from 1 kg of organic waste. **(3 marks)**

▶ **Question 5 (1 mark)**

Source: VCE 2020 Chemistry Exam, Section A, Q.22; © VCAA

MC The combustion of which fuel provides the most energy per 100 g?
A. pentane (M = 72 g mol^{-1}), which releases 49 097 MJ tonne^{-1}
B. nitromethane (M = 61 g mol^{-1}), which releases 11.63 kJ g^{-1}
C. butanol (M = 74 g mol^{-1}), which releases 2670 kJ mol^{-1}
D. ethyne (M = 26 g mol^{-1}), which releases 1300 kJ mol^{-1}

More exam questions are available in your learnON title.

2.4 Calorimetry

KEY KNOWLEDGE

- The use of specific heat capacity of water to approximate the quantity of heat energy released during the combustion of a known mass of fuel and food
- The principles of solution calorimetry, including determination of calibration factor and consideration of the effects of heat loss; analysis of temperature-time graphs obtained from solution calorimetry

Source: VCE Chemistry Study Design (2024–2027) extracts © VCAA; reproduced by permission.

2.4.1 Practical calorimetry

A simple method of measuring the heat content in food is to ignite a weighed sample of food and use it as a fuel to heat a particular volume of water, and measure the water's increase in temperature. The specific heat capacity of water can then be used to determine the heat energy provided by the food sample. The **specific heat capacity** (*c*) is the energy needed to raise the temperature of 1 g of a substance by 1 °C. The energy being transferred to the water can be calculated using the heat energy released in the combustion of fuel equation.

Specific heat capacity of water

The specific heat capacity of water is 4.18 J g^{-1} °C^{-1}. This means it takes 4.18 J of heat to change the temperature of 1 g of water by 1 °C.

For example, if a 100 g water sample was heated to increase its temperature by 10.0 °C, the energy required would be the product of the mass of water, the specific heat capacity and the change in temperature. That is, $100 \times 4.18 \times 10.0 = 4.18 \times 10^3$ J. We can express this relationship in an equation.

Energy = mass of water × specific heat capacity of water × temperature increase

$$q = mc\Delta T$$

where:

q is the energy measured in joules (J)

m is the mass of water (not the food) in grams (g)

c is the specific heat capacity of water, 4.18 J g^{-1} °C^{-1}

ΔT is the change in temperature (°C).

The spirit burner

A spirit burner (figure 2.8) is a simple apparatus used to measure the heat content in a mass of fuel.

Both the mass of water and the mass of fuel need to be weighed and recorded before the spirit burner is lit. Once lit, the fuel is allowed to burn for a period of time so that a measurable change in both temperature and fuel mass are obtained (being careful not to boil the water).

This technique (figure 2.9) is certainly useful in comparing fuels for energy content; however, accurate measurements are not achievable due to the heat lost to the surrounds being substantial. That is, not all of the energy released during combustion ends up heating the water, leading to an underestimation of the energy content in the sample.

specific heat capacity (c) the energy needed to change the temperature of 1 g of a substance by 1 °C

FIGURE 2.8 A spirit burner

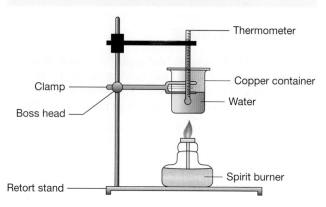

FIGURE 2.9 Calorimetry experimental set-up

The same type of set-up can be used to compare the energy content of food (figure 2.10). The food is burnt underneath the container instead of placing it above a spirit burner.

FIGURE 2.10 Measuring the energy in food

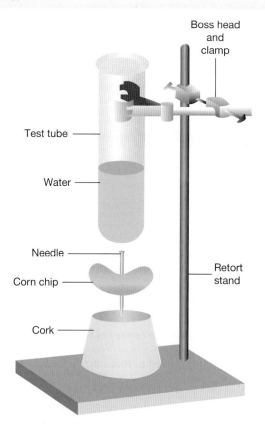

To compare the energy obtained from different foods per gram:

$$\text{Energy released from food per gram (J)} = \frac{\text{mass of water (g)} \times 4.18 \times \text{temperature rise (°C)}}{\text{mass of food sample (g)}}$$

tlvd-9646

A 1.71 g sample of a food is burned, heating 50.0 g of water. The temperature increases from 20.0 °C to 59.0 °C. Calculate the energy transferred to the water, and then estimate the energy present per gram of food.

THINK	WRITE
1. Calculate the energy that is going into the water using the equation $q = mc\Delta T$, and substituting the specific heat capacity of water ($c = 4.18$ J g^{-1} C^{-1}) and mass of the water. **TIP:** Take care to use the mass of the water to calculate the energy going into the water, not the mass of the food.	$q = mc\Delta T$ $= 50.0 \times 4.18 \times (59.0 - 20.0)$ $= 8151$ J
2. Convert to kilojoules.	8151 J $= 8.15 \times 10^3$ J $= 8.15$ kJ
3. To calculate the energy per gram of food, divide the energy by the mass of food.	Energy per gram of food $= \dfrac{8.15}{1.71}$ $= 4.77$ kJ g^{-1}

PRACTICE PROBLEM 9

A 1.08 g sample of almonds is completely burned, heating 150.0 g of water. The temperature increases from 20.0 °C to 33.0 °C. Calculate the energy transferred to the water, and then estimate the energy present per gram of almonds in kJ g^{-1}.

2.4.2 Solution calorimetry

Solution calorimetry involves the chemical reaction taking place inside a **calorimeter**. The advantage of this is that any **change in enthalpy** occurs directly in the solution, usually water. This provides better accuracy in terms of ΔT measurements. The types of reactions used in solution calorimetry are limited. For example, the combustion reactions used to burn fuel and food obviously can't take place in water! The chemical changes also need to happen spontaneously at or below the temperature of the water.

The common chemical changes that take place in solution calorimeters are **dissolution** reactions. For example, the dissolution of ammonium nitrate in water is an **endothermic** process, and its enthalpy of reaction can be calculated by dissolving a sample in a solution calorimeter and measuring the decrease in water temperature.

A simple solution calorimeter (see figures 2.11 and 2.12) is sometimes called a coffee-cup calorimeter. This is because an insulating material like polystyrene foam is used to reduce heat loss or absorption from the surrounds.

Calorimeters such as these may or may not be **calibrated**. They work on the assumption that the mass of water requires 4.18 J of heat to raise the temperature of each gram by 1 °C — the same way calculations are made when using a spirit burner.

solution calorimetry the process of using a calorimeter to measure heat changes in a solution; for example, heat of dissolution and neutralisation reactions

calorimeter an apparatus used to measure heat changes during a chemical reaction or change of state

change in enthalpy the amount of energy released or absorbed in a chemical reaction

dissolution the process of solutes dissolving in solvents to form a solution

endothermic describes a chemical reaction in which energy is absorbed from the surroundings

calibrate adjusting an instrument using standards of known measurements to ensure the instrument's accuracy

FIGURE 2.11 A simple solution calorimeter

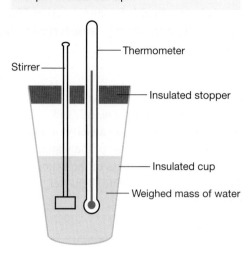

FIGURE 2.12 Solution calorimetry experimental set-up

Stirrer —

— Thermometer

— Insulated stopper

— Insulated cup

— Weighed mass of water

Determining the calibration factor

Calibrating a solution calorimeter involves measuring the amount of energy supplied and the corresponding temperature change. The amount of energy required to change the contents by one degree is called the **calibration factor** (CF) and has units J °C^{-1}. Calibration allows greater accuracy than using the heat capacity of water alone, as it takes into account heat absorbed by the entire inside of the calorimeter.

calibration factor the amount of energy required to change the contents of a calorimeter by one degree, with units J °C^{-1}

electrical calibration calibration of a calorimeter by supplying a known quantity of electricity

Electrical calibration

Electrical calibration is achieved by calibrating the calorimeter using an electrically heated coil to supply a measured quantity of electrical energy, which is converted to heat energy. The heat energy is transferred to a known mass of a substance, usually water, and then the temperature rise is measured.

FIGURE 2.13 A typical calorimeter that uses electrical calibration

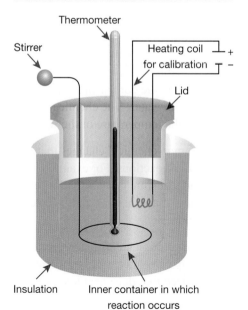

Thermometer

Stirrer

Heating coil for calibration

Lid

Insulation

Inner container in which reaction occurs

The energy released to the calorimeter is given by:

$$E = VIt$$

where:

E = energy released (joules)

V = potential difference (volts)

I = current (amps)

t = time (s).

If a current of I amps flows for t seconds at a potential difference of V volts, the calibration factor may be calculated as follows:

$$CF = \frac{\text{energy released during calibration}}{\text{temperature rise}} = \frac{VIt}{\Delta T_c}$$

The calibration factor (CF) is measured in joules per degree, J °C^{-1}.

The temperature rise (ΔT_c) during calibration is:

$$\Delta T_c = \text{final temperature } (T_f) - \text{initial temperature } (T_i)$$

 Resources

 Interactivity Calibrating a calorimeter (int-1253)

SAMPLE PROBLEM 10 Calculating the calibration factor for a calorimeter using electrical calibration

tlvd-9671

A solution calorimeter is filled with 100 mL of water and its temperature recorded as 19.50 °C. A current of 2.52 A at a potential difference of 5.68 V is passed through the water for 2.00 minutes. The final temperature is measured at 24.25 °C. Determine the calibration factor of the calorimeter in J °C^{-1}.

THINK	WRITE
1. Calculate the electrical energy that is going into the water using the equation $E = VIt$. Remember to convert time from minutes to seconds.	$E = VIt$ $= 5.68 \times 2.52 \times 120$ $= 1.72 \times 10^3$ J
2. Calculate the calibration factor (CF) by dividing the energy released by the change in temperature using the equation $\Delta T_c = T_f - T_i$. **TIP:** The calibration factor equation can be found in the VCE Chemistry Data Book. Give your answer to the smallest number of significant figures in the question.	$CF = \dfrac{\text{energy}}{\Delta T_c}$ $= \dfrac{1.72 \times 10^3}{24.25 - 19.50}$ $= 362$ J °C^{-1}

PRACTICE PROBLEM 10

Tim calibrates a calorimeter for an experiment he is about to perform. He uses 100 mL of water in an electrical calibration. The current supplied is 1.80 A and the voltage is 5.60 V over a period of 150 seconds. The temperature rises from 20.00 °C to 24.56 °C. Determine the calibration factor of the calorimeter in J °C^{-1}.

Remember 'Rules in Joules'. Both the rules $q = mc\Delta T$ and $E = VIt$ give answers in joules, so care must be taken with units if answers are required in kilojoules.

Chemical calibration

Chemical calibration uses known ΔH values for thermochemical reactions to calibrate the calorimeter. A known (measured) amount, in moles, or mass of reactant(s) is placed inside the calorimeter and ΔT is measured.

The dissolution of solids can be used, such as dissolving a mass of $NH_4NO_3(s)$. Heat of reaction, such as the **exothermic** heat of neutralisation between a strong acid and base, can also be used.

> **chemical calibration** calibration of a calorimeter using a combustion reaction with a known ΔH
>
> **exothermic** describes a chemical reaction in which energy is released to the surroundings

---

TIP: Heat of neutralisation reactions will require q to be calculated from the number of moles of the limiting reactant.

Neutralisation occurs when an acid donates one or more hydrogen ions (H^+) to a base, such as OH^- ions.

For example, $NaOH(aq)$ and $HCl(aq)$ react in a 1 : 1 stoichiometric ratio:

$$NaOH(aq) + HCl(aq) \rightarrow NaCl(aq) + H_2O(l)$$

The net reaction is:

$$OH^-(aq) + H^+(aq) \rightarrow H_2O(l)$$

---

tlvd-9647

SAMPLE PROBLEM 11 Calculating the calibration factor for a calorimeter using chemical calibration

A solution calorimeter was chemically calibrated from the dissolution of 5.00 g of ammonium nitrate in 100 g of water held at 18.0 °C. The lowest temperature reached after the dissolution was 14.5 °C. The thermochemical equation is:

$$NH_4(s) \xrightarrow{H_2O} NH_4{}^+(aq) + NO_3{}^-(aq) \quad \Delta H = +25.7 \text{ kJ mol}^{-1}$$

Calculate the calibration factor in $J\ °C^{-1}$.

THINK	WRITE
1. Calculate the number of moles of ammonium nitrate in 5.00 g using the equation $n = \dfrac{m}{M}$.	$n(NH_4NO_3) = \dfrac{m}{M}$ $= \dfrac{5.00 \text{ g}}{80.0 \text{ g mol}^{-1}}$ $= 0.0625 \text{ mol}$
2. Calculate the number of kJ absorbed from the dissolution.	$q = n \times \Delta H$ $= 0.0625 \text{ mol} \times 25.7 \text{ kJ mol}^{-1}$ $= 1.606 \text{ kJ}$
3. Calculate the calibration factor (CF) by dividing the amount of energy absorbed by the measured temperature change (ΔT).	$CF = \dfrac{q}{\Delta T}$ $= \dfrac{1.606 \times 10^3 \text{J}}{3.5 \ °C}$ $= 458 \text{ J} \ °C^{-1}$
4. Write your answer to the correct number of significant figures.	$CF = 4.6 \times 10^2 \text{ J} \ °C^{-1}$

PRACTICE PROBLEM 11

The heat of neutralisation of the exothermic reaction between HCl(aq) and NaOH(aq) releases 55.9 kJ mol⁻¹. A 50.00 mL solution of 0.500 M HCl was reacted with a 50.00 mL solution of 0.250 M NaOH in a calorimeter. The maximum temperature change measured by a digital thermometer was 1.50 °C. Calculate the calibration factor of the calorimeter.

Using a calorimeter

The steps for using a calorimeter are as follows:
1. Calibrate the calorimeter (this step can also be done at the completion of the reaction).
2. Measure the masses or volumes of the chemicals that are required for the reaction, ensuring that the volume of water or solutions used is the same as the volume used for calibration.
3. Measure the temperature of the water or solutions.
4. Add the solid or solutions to the calorimeter.
5. Record the highest or lowest temperature reached.
6. Perform the calculations.
7. If ΔH is required, remember to use the appropriate sign.

The heat of solution (when 1 mole of any substance dissolves in water) and the heat of neutralisation (when an acid reacts with a base) for a reaction may be determined using a solution calorimeter.

In an *exothermic reaction*, the heat produced by the reaction is *released into* the solution, which increases its temperature.

In an *endothermic reaction*, the heat required is *absorbed from* the thermal energy of the solution, which decreases its temperature.

In a solution calorimeter, experiments are usually carried out in aqueous solution. The change in temperature caused by the reaction (ΔT_r) is measured and then multiplied by the calibration factor to determine the heat change for the reaction.

In solution calorimetry, the energy change is calculated using the following equation:

$$\text{Energy change} = \text{CF} \times \Delta T_r$$

The ΔH is the energy change per mole. To find the ΔH, divide the energy change by the number of mole, n.

$$\Delta H = \frac{\text{energy change}}{n}$$

TIP: Make sure that you label the temperature change in the calibration as ΔT_c and the temperature change for the reaction as ΔT_r so that you are not confused about which temperature to use.

TIP: The equation $\Delta H = \dfrac{\text{energy change}}{n}$ can also be adapted from the VCE Chemistry Data Book, utilising the equation for the enthalpy of combustion, $\Delta H = \dfrac{q}{n}$, because q is the variable for energy and n is the number of mole.

2.4.3 Calorimeter calculations

Once the calorimeter is calibrated, it can be used to determine heats of reaction. It is important to use the same mass in the calorimeter as was used to determine the calibration factor. If you are determining the heat of neutralisation, use dilute solutions so that the density can be assumed to be the same as that of water.

tlvd-9672

SAMPLE PROBLEM 12 Calculating the energy change per mole and writing thermochemical equations

A pure sample of sulfuric acid with a mass of 0.231 g was combined with 100 mL of pure water in a solution calorimeter. The temperature increased from 19.90 °C to 20.42 °C. The calorimeter was previously calibrated with 100 mL of water and found to have a calibration factor of 463 J °C^{-1}. Calculate the ΔH and write the thermochemical equation of the reaction.

THINK

1. The calibration factor is provided, so it does not need to be calculated.

2. Calculate the energy change by multiplying the calibration factor (CF) by the temperature change of the reaction (ΔT_r).

3. Find the number of mole (n) of H_2SO_4 by using the equation $n = \dfrac{m}{M}$.

4. ΔH is the energy change divided by the number of mole using the equation
$\Delta H = \dfrac{\text{energy change}}{n}$.

5. Write the thermochemical equation. Remember that it is an exothermic reaction, so the sign of ΔH is negative.

WRITE

CF = 463 J °C^{-1}

Energy change = CF $\times \Delta T_r$
$= 463 \times (20.42° - 19.90°)$
$= 240.8$ J

$n(H_2SO_4) = \dfrac{0.231}{98.1}$
$= 2.35 \times 10^{-3}$ mol

$\Delta H = \dfrac{241 \text{ J}}{2.35 \times 10^{-3}}$
$= 1.02 \times 10^5$ J mol^{-1}
$= 102$ kJ mol^{-1}

$H_2SO_4(l) \rightarrow H_2SO_4(aq)$ $\Delta H = -102$ kJ mol^{-1}

PRACTICE PROBLEM 12

A sample of potassium nitrate with a mass of 5.378 g was combined with 100 mL of pure water in a solution calorimeter. The temperature decreased from 20.34 °C to 20.10 °C. The calorimeter was previously calibrated with 100 mL of water and found to have a calibration factor of 620 J °C^{-1}. Calculate the ΔH of the reaction.

A solution calorimeter containing 100 mL of water was calibrated by passing a **4.00 A** current through the instrument for **35.0 s** at a potential difference of **3.00 V**. The temperature rose by **0.700 °C**. When **6.60 g** of calcium chloride hexahydrate, $CaCl_2 \cdot 6H_2O$, was added to the calorimeter and dissolved by rapid stirring, the temperature dropped by **0.895 °C**.

Determine ΔH for the reaction:

$$CaCl_2 \cdot 6H_2O(s) \xrightarrow{H_2O} Ca^{2+}(aq) + 2Cl^-(aq)$$

THINK

1. Calculate the calibration factor (CF) for the calorimeter from the relationship using the equation $CF = \dfrac{VIt}{\Delta T_c}$.

2. Calculate the energy change during the reaction by applying the following equation: energy change = $CF \times \Delta T_r$.

3. To calculate the energy per mole, first find the number of moles using the equation $n = \dfrac{m}{M}$, where $M(CaCl_2 \cdot 6H_2O) = 219.1\,g\,mol^{-1}$.

4. The energy per mole can be calculated using $\Delta H = \dfrac{\text{energy change}}{n}$.

5. Because energy was absorbed during the reaction, the reaction is endothermic and the sign of the ΔH value is positive. Write the thermochemical equation, converting the ΔH value to kilojoules.

WRITE

$$CF = \frac{VIt}{\Delta T_c}$$

$$= \frac{3.00 \times 4.00 \times 35.0}{0.700}$$

$$= 600\,J°\,C^{-1}$$

Energy change = $CF \times \Delta T_r$
$$= 600 \times 0.895$$
$$= 537\,J$$

$$n(CaCl_2 \cdot 6H_2O) = \frac{m}{M}$$

$$= \frac{6.60}{219.1}$$

$$= 0.0301\,mol$$

$$\Delta H = \frac{\text{energy change}}{n}$$

$$= \frac{537}{0.0301}$$

$$= 17819J$$

$$= 17.8kJ$$

$$CaCl_2 \cdot 6H_2O(s) \xrightarrow{H_2O} Ca^{2+}(aq) + 2Cl^-(aq)$$

$$\Delta H = +17.8\,kJ\,mol^{-1}$$

PRACTICE PROBLEM 13

A calorimeter was calibrated electrically. The potential difference through the heating coil was **5.23 V**, producing a current of **1.83 A** for **2.00 minutes**. During this time the temperature rose from **19.40 °C** to **22.85 °C**. **5.10 g** of sodium hydroxide was then added and the temperature rose to **37.33 °C**.

Determine ΔH for the reaction:

$$NaOH(s) \xrightarrow{H_2O} Na^+(aq) + OH^-(aq)$$

elog-1932

tlvd-9714

Solution calorimetry

Aim

To determine the calibration factor of a solution calorimeter, and the heat of reaction for zinc metal and copper ions

---

EXTENSION: Bomb calorimetry

A bomb calorimeter can be used to measure enthalpy of combustion of other chemicals, such as fuels, as well as food samples. Like solution calorimeter reactions, the reaction takes place inside the calorimeter. The difference is that in a bomb calorimeter, the combustion of the sample occurs inside a metal chamber (bomb) built to withstand the pressurised excess oxygen gas supplied, as well as the products of combustion.

The sample is ignited electrically and the heat released passes through the conductive metal of the bomb directly into the contents (water) inside the calorimeter.

The bomb calorimeter can be calibrated chemically and electrically in the same way solution calorimeters are calibrated.

FIGURE 2.14 Features of a bomb calorimeter

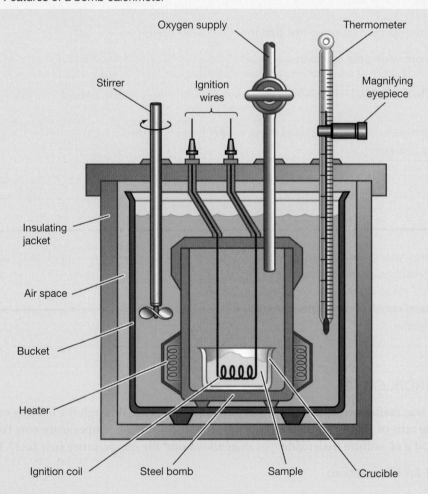

tlvd-9674

SAMPLE PROBLEM 14 Calculating the energy of food using a bomb calorimeter

A bomb calorimeter was calibrated by passing a current of 3.55 A at a potential difference of 6.40 V through a heating coil for 123.7 s. The temperature of the calorimeter rose from 21.82 °C to 26.13 °C. After the calorimeter had cooled, a dried biscuit weighing 2.34 g was then burned in the calorimeter in the presence of excess oxygen. The temperature of the calorimeter rose from 22.75 °C to 24.98 °C.

Calculate the energy content of the biscuit in J g^{-1}.

THINK	WRITE
1. To find the calibration factor, first calculate the electrical energy input using the equation $E = VIt$.	$E = VIt$ $= 6.40 \times 3.55 \times 123.7$ $= 2810$ J
2. Find the temperature change for calibration using the equation $\Delta T_c = T_f - T_i$.	$\Delta T_c = 26.13 - 21.82$ $= 4.31°C$
3. Calculate the calibration factor (CF) using the equation $CF = \dfrac{VIt}{\Delta T_c}$.	$CF = \dfrac{VIt}{\Delta T_c}$ $= \dfrac{6.40 \times 3.55 \times 123.7}{4.31}$ $= 652$ J °C^{-1}
4. The change in temperature due to combustion of the biscuit is found using the equation $\Delta T_r = T_f - T_i$.	$\Delta T_r = 24.98 - 22.75$ $= 2.23$ °C
5. The energy released by the biscuit is determined by the following equation: energy change = CF $\times \Delta T_r$.	Energy from biscuit = $CF \times \Delta T_r$ $= 652 \times 2.23$ $= 1454$ J
6. Determine the energy content per gram of the biscuit by dividing the total energy of the biscuit by the biscuit's mass.	Energy per g = $\dfrac{energy}{mass\ of\ food\ (g)}$ $= \dfrac{1454}{2.34}$ $= 621$ J g^{-1}

PRACTICE PROBLEM 14

A bomb calorimeter was calibrated by passing 1.35 A through the electric heater for 60.0 s at a potential difference of 6.44 V. The temperature of the water in the calorimeter rose from 22.85 °C to 23.30 °C. A 2.25 g piece of margarita pizza was completely burned in the calorimeter in excess oxygen. The temperature of the calorimeter rose from 22.30 °C to 39.68 °C.

Calculate the energy content of the pizza in kJ g^{-1}.

EXPERIMENT 2.2 online only

Calculating energy in food

Aim

To measure the energy content of various foods by calorimetry, and to compare the experimental value with published energy values

Heat loss and temperature–time graphs

Heat loss in a calorimeter results in inaccurate ΔT values, and therefore inaccurate ΔH values.

In solution calorimetry we assume that there is a negligible heat loss and that all of the heat from the reaction is used to heat the water. However, a small amount of heat may be lost through poor insulation, a poorly fitting lid, through holes for the thermometer and stirrer, or it may be absorbed by parts of the calorimeter. In a perfectly insulated calorimeter, the final temperature remains constant after the current is turned off (as demonstrated by the blue line in figure 2.15). However, in a poorly insulated calorimeter, the temperature rise would be less than that in a well-insulated calorimeter. While the current is flowing in a poorly insulated calorimeter, there would be heat loss throughout that time, and once the current is turned off the temperature would fall rather than remain stable. This is demonstrated by the green line in figure 2.15. The theoretical ΔH for the reaction can be calculated by extrapolating the graph to when the reaction commenced. The temperature change (ΔT_c) is the measurement from when the current was turned off in the calibration to the extrapolated line. This compensates for the error caused by loss of heat from the water to the surroundings during the time when the current was turned on.

FIGURE 2.15 Determining ΔT_c for a poorly insulated calorimeter

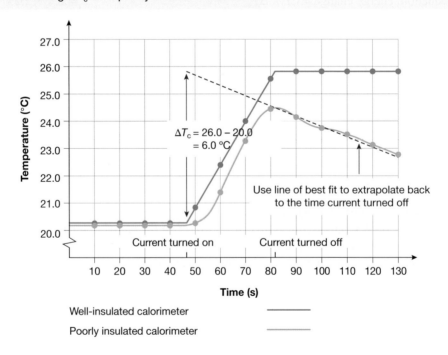

Temperature–time graphs can be used to obtain a more accurate temperature change (ΔT_c) for a poorly insulated calorimeter.

2.4 Activities

Students, these questions are even better in jacPLUS

 Receive immediate feedback and access sample responses

 Access additional questions

 Track your results and progress

Find all this and MORE in jacPLUS

| **2.4 Quick quiz** on | **2.4 Exercise** | **2.4 Exam questions** |

2.4 Exercise

1. Calculate the energy, in kJ, required to heat 1.00 L of water from room temperature (25.0 °C) to boiling point (100 °C), given the density of water is 0.997 g mL^{-1}.

2. a. Consider the apparatus shown in figure 2.10, which was used to measure the energy in a corn chip. Imagine that the test tube was replaced by a can containing 250 g of water. A corn chip of mass 2.55 g was burnt, leaving a mass of ash of 2.25 g and a temperature increase from 21.32 °C to 27.23 °C. Estimate the energy of the corn chip in kJ g^{-1}.
   b. A large amount of heat escapes using this set-up of apparatus. Suggest a modification to improve the accuracy of this experimental design.
   c. Explain why the answer is given in kJ g^{-1} and not kJ mol^{-1}.

3. a. What is calorimetry?
   b. Why do calorimeters need to be calibrated?
   c. Explain the difference there would be to the calibration factor, if any, if kelvin (K) was used instead of °C.
   d. Explain the difference there would be to the calibration factor, if any, if 50.0 mL of water was used for calibration instead of 100.0 mL.

4. Determine the temperature change in a solution calorimeter when 5.00 g of CaCl$_2 \cdot$6H$_2$O dissolves in 200 mL of water according to the following equation:

$$CaCl_2 \cdot 6H_2O(s) \xrightarrow{H_2O(l)} Ca^{2+}(aq) + 2Cl^-(aq) \quad \Delta H = +17.8 \text{ kJ mol}^{-1}$$

The calibration factor for the calorimeter under these conditions was found to be 825 J °C^{-1}.

5. A student used a simple calorimeter to determine the heat of neutralisation of a strong acid with a strong base according to the following reaction:

$$H^+(aq) + OH^-(aq) \rightarrow H_2O(l)$$

After calibrating the calorimeter, the student calculated the calibration factor to be 343 J °C^{-1}. The student then carefully added 50.0 mL of 0.100 M HCl to 50.0 mL of 0.110 M NaOH in the calorimeter. The temperature in the calorimeter rose from 20.4 °C to a maximum of 21.2 °C.
   a. Explain how the student determined the calibration factor.
   b. Calculate the heat of neutralisation for the reaction.
   c. Suggest why a higher concentration of base than acid was used.
   d. How would the heat of neutralisation have been affected if the student had added 50.0 mL of 0.100 M NaOH to 50.0 mL of 0.110 M HCl in the calorimeter?
   e. Identify the sources of error in the experiment and suggest how these could be minimised.

TOPIC 2 Measuring changes in chemical reactions 85

6. An experiment to compare the energy output of candle wax, ethanol and butane was performed by setting up the apparatus shown in the figure.

The ethanol was poured into a crucible. A small wax candle stuck onto a watch glass and a gas lighter were each used as a 'burner' after being lit. Each burner was weighed before and after it was used to heat 200 g of water. The results are shown in the following table. (Assume the formula for candle wax is $C_{20}H_{42}$.)

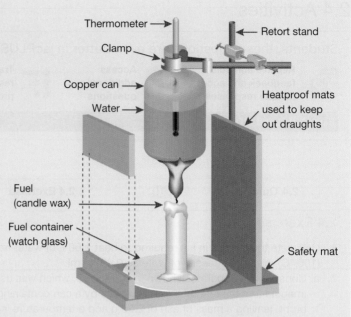

a. Complete the table.
b. Taking the specific heat capacity of water as 4.18 J g^{-1} °C^{-1}, calculate the energy produced from 1.00 g of each substance.
c. Calculate the heat of combustion (enthalpy per mole of substance used) for each substance.
d. The molar heat of combustion of ethanol has been found to be −1360 kJ mol^{-1}.
   i. Is the combustion exothermic or endothermic?
   ii. What was the percentage accuracy of the experiment shown in the figure when ethanol was used as a fuel?
   iii. List the sources of error in the experiment and describe how some of these errors can be minimised.

TABLE Results of combustion of different fuels

Property	Ethanol	Candle wax	Butane
Mass of 'burner' before heating (g)	23.77	32.72	43.94
Mass of 'burner' after heating (g)	22.54	32.50	43.71
Mass of fuel used (g)			
Mass of water (g)	200	200	200
Initial temperature of water (°C)	20.0	20.0	20.0
Highest temperature of water (°C)	35.0	30.0	29.0
Temperature rise (°C)			
Molar mass (g mol^{-1})			

2.4 Exam questions

▶ Question 1 (1 mark)
Source: VCE 2021 Chemistry Exam, Section A, Q.19; © VCAA

MC A food chemist conducted an experiment in a bomb calorimeter to determine the energy content, in joules per gram, of a muesli bar. A 3.95 g sample of the muesli bar was combusted in the calorimeter and the temperature of the water rose by 16.7 °C. The calibration factor of the calorimeter was previously determined to be 4780 J °C^{-1}.

The energy content of the muesli bar is

A. 3.51×10^5 J g^{-1} B. 2.02×10^4 J g^{-1} C. 1.13×10^3 J g^{-1} D. 7.25×10 J g^{-1}

⊙ Question 2 (3 marks)

Source: VCE 2021 Chemistry Exam, Section B, Q.1.c; © VCAA

Biogas was combusted to release 1.63×10^3 kJ of energy. This energy was used to heat 100 kg of water in a tank. The initial temperature of the water was 25.0 °C.

a. What is the maximum temperature that the water in the tank could reach? **(2 marks)**

b. State why this temperature may not be reached. **(1 mark)**

⊙ Question 3 (1 mark)

Source: VCE 2020 Chemistry Exam, Section A, Q.18; © VCAA

MC An experiment was carried out to determine the enthalpy of combustion of propan-1-ol. Combustion of 557 mg of propan-1-ol increased the temperature of 150 g of water from 22.1 °C to 40.6 °C.

The enthalpy of combustion is closest to

A. -2742 kJ mol^{-1}

B. -1208 kJ mol^{-1}

C. -1250 kJ mol^{-1}

D. -1540 kJ mol^{-1}

⊙ Question 4 (3 marks)

Source: VCE 2020 Chemistry Exam, Section B, Q.6.c; © VCAA

Methane gas, CH_4, can be captured from the breakdown of waste in landfills. CH_4 is also a primary component of natural gas. CH_4 can be used to produce energy through combustion.

A Bunsen burner is used to heat a beaker containing 350.0 g of water. Complete combustion of 0.485 g of CH_4 raises the temperature of the water from 20 °C to 32.3 °C.

Calculate the percentage of the Bunsen burner's energy that is lost to the environment.

⊙ Question 5 (1 mark)

Source: VCE 2020 Chemistry Exam, Section A, Q.9; © VCAA

MC A solution calorimeter containing 350 mL of water was set up. The calorimeter was calibrated electrically and the graph of the results is shown below.

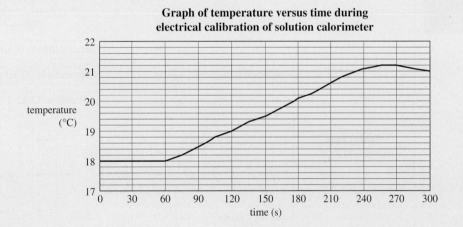

Graph of temperature versus time during electrical calibration of solution calorimeter

The calorimeter was calibrated using a current of 2.7 A, starting at 60 s. The current was applied for 180 s and the applied voltage was 5.4 V.

What is the calibration factor for this calorimeter?

A. 125 J °C^{-1}

B. 820 J °C^{-1}

C. 847 J °C^{-1}

D. 875 J °C^{-1}

More exam questions are available in your learnON title.

2.5 Review

2.5.1 Topic summary

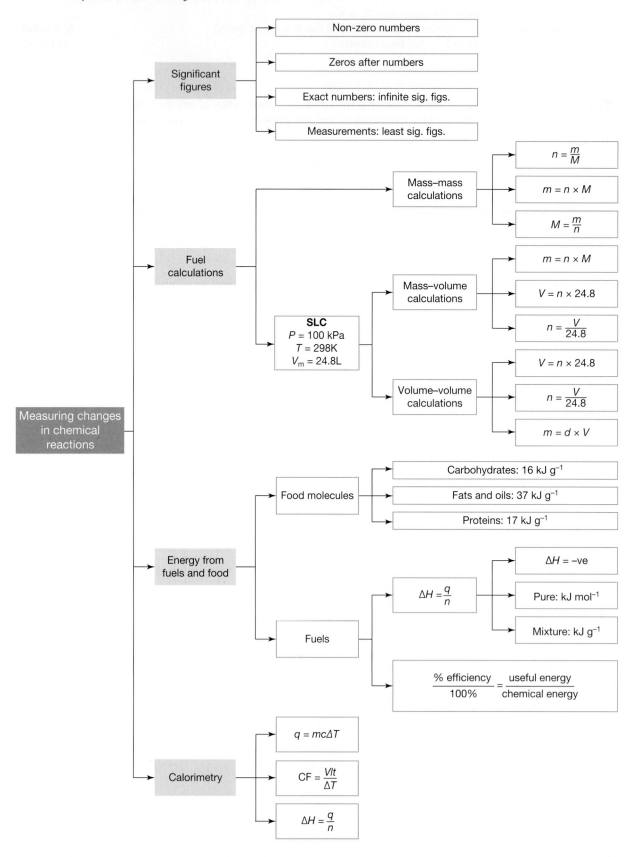

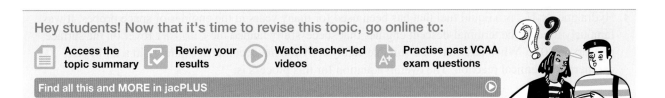
2.5.2 Key ideas summary

online only

2.5.3 Key terms glossary

online only

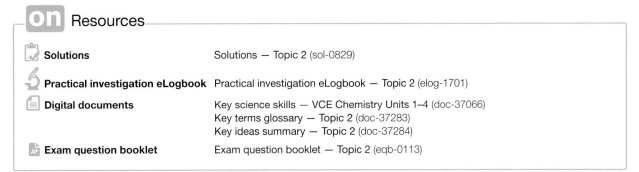

on Resources

Solutions — Solutions — Topic 2 (sol-0829)

Practical investigation eLogbook — Practical investigation eLogbook — Topic 2 (elog-1701)

Digital documents — Key science skills — VCE Chemistry Units 1–4 (doc-37066)
Key terms glossary — Topic 2 (doc-37283)
Key ideas summary — Topic 2 (doc-37284)

Exam question booklet — Exam question booklet — Topic 2 (eqb-0113)

2.5 Activities

learn on

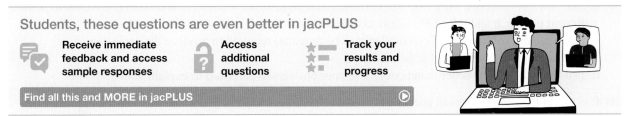
2.5 Review questions

1. Calculate the number of moles of the following gases at SLC.

   a. 1.5 L oxygen, O_2
   b. 2.56 L of hydrogen, H_2
   c. 250 mL of nitrogen, N_2

2. Calculate the volume of the following gases at SLC.

   a. 1.53 mol of hydrogen, H_2
   b. 13.6 g of methane, CH_4
   c. 2.5×10^{30} molecules of nitrogen, N_2

3. Butan-1-ol (density = 0.81 g mL^{-1}) burns according to the following equation:

$$CH_3CH_2CH_2CH_2OH(l) + 6O_2(g) \rightarrow 4CO_2(g) + 5H_2O(l)$$

When 10.0 mL of butan-1-ol is burned, calculate:

a. the mass of water produced
b. the volume of carbon dioxide produced at SLC
c. the volume of butan-1-ol needed if it is used to produce 100 mL of carbon dioxide at SLC.

4. Hydrazine, N_2H_4, is a liquid fuel that has been used for many years in the engines of space probes. It was famously used in the terminal-descent engines that successfully landed the *Curiosity* rover on the surface of Mars in 2012. When passed over a suitable catalyst, hydrazine decomposes quickly in a multi-step exothermic chemical reaction. The overall equation for this process is:

$$N_2H_4(l) \rightarrow N_2(g) + 2H_2(g) \quad \Delta H = -50.3\,\text{kJ mol}^{-1}$$

Calculate the energy released per kilogram of hydrazine in the equation.

5. Petrol and LPG are two fuels commonly used in Australia. It is claimed that LPG is better for the environment because it releases less carbon dioxide. It is also attractive to motorists because, even though more litres are used, it is cheaper than petrol. As an approximation, petrol may be assumed to be octane, whereas LPG is a mixture of propane and butane. Some relevant data is shown in the following table.

Fuel	Molar enthalpy of combustion (kJ mol^{-1})	Density (g mL^{-1})
Propane	−2217	0.51 (as LPG)
Butane	−2874	0.51 (as LPG)
Octane	−5464	10.70 (as petrol)

a. Calculate the mass of LPG (assuming it to be propane) required to produce 1.00 MJ of heat energy.
b. Calculate the mass of carbon dioxide produced from part a and express your answer as g MJ^{-1}.
c. Calculate the mass of petrol (assuming it to be octane) required to produce 1.00 MJ of heat energy.
d. Calculate the mass of carbon dioxide produced from c and express your answer as g MJ^{-1}.
e. State the net reduction (in g MJ^{-1}) of carbon dioxide emissions when LPG is used in preference to petrol.
f. Repeat parts a, b and e for LPG if it is assumed to be butane.
g. Calculate the volume of LPG (assuming it to be propane) required to produce the same energy as 1.00 L of petrol (assuming it to be octane).
h. Is it true that LPG is better than petrol? Use your answers to parts a–g to explain your response.

6. It is useful to know how much energy can be obtained from different fuels in order to determine which would be the best fuel for a particular purpose. The apparatus in the following figure can be constructed in the laboratory to measure the heat given out when a fuel such as ethanol is burned.

The heat produced when the fuel burns is absorbed by the water in the metal can. The temperature can be measured so, given that the specific heat of water is 4.18 J and the density of water is 1.00 g mL^{-1}, the heat of combustion may be determined.

The results of one experiment are shown in the following table.

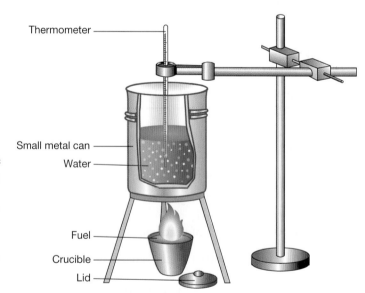

Volume of water in metal can	200 mL
Thus, mass of water in can	200 g
Rise in temperature of water	11.0 °C
Mass of ethanol burned	0.500 g

a. From the results in the table, calculate the heat produced when 1 g of ethanol is burned.

b. Calculate the heat of combustion for ethanol using the results in the table.

c. An accurate value for the heat produced when 1 mole of ethanol burns is 1360 kJ mol^{-1}. Calculate the percentage accuracy of this experiment.

d. Outline the sources of error in the experiment, and then suggest how the design of the experiment could be improved so that more accurate heats of combustion for different fuels may be determined.

7. Explain why the overall energy efficiency of a coal-fired power station that generates electricity is said to be only 30% efficient.

8. Kerosene is a hydrocarbon fuel that may be used in lamps, jet engines and camp stoves. It has a heat of combustion of 44 100 kJ kg^{-1}.

a. Explain why the heat evolved from the combustion of kerosene is measured in kJ kg^{-1} rather than kJ mol^{-1}.

b. A cup of billy tea contains 250 g of water. How many cups of tea can be made if 12.5 mL of kerosene is used to heat the water? Assume that the temperature of the water increases from 20.0 °C to 100.0 °C, the specific heat capacity of water is 4.18 J g^{-1} °C^{-1} and the heat of combustion of kerosene is 37 000 kJ L^{-1}.

9. 'Mates' savoury biscuits contain 14.7 g carbohydrate in 11 biscuits (25 g), which is described as a standard serve. What is the energy available from 100 g of biscuits?

10. A solution calorimeter containing 100 mL of water was electrically calibrated by recording the stabilised temperature for 90 seconds and then turning on the current, recording the temperature every 30 seconds for 210 seconds. The current was then turned off while still continuing to record the temperature. The following temperature–time graph was obtained.

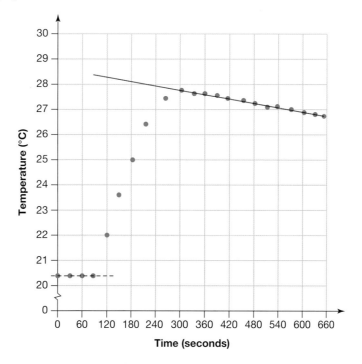

The potential difference applied was 6.30 V and the current recorded was 3.55 A. Calculate the calibration factor for this calorimeter.

2.5 Exam questions

Section A — Multiple choice questions

All correct answers are worth 1 mark each; an incorrect answer is worth 0.

Question 1

Source: *Adapted from VCE 2014 Chemistry Exam, Section A, Q.8;* © VCAA

MC When hydrochloric acid is added to aluminium sulfide, the highly toxic gas hydrogen sulfide is evolved. The equation for this reaction is

$$Al_2S_3(s) + 6HCl(aq) \rightarrow 2AlCl_3(aq) + 3H_2S(g)$$

If excess hydrochloric acid is added to 0.200 mol of aluminium sulfide, then the volume of hydrogen sulfide produced at standard laboratory conditions (SLC) will be

A. 1.65 L **B.** 4.96 L **C.** 7.44 L **D.** 14.9 L

Question 2

Source: *VCE 2013 Chemistry Exam, Section A, Q.5;* © VCAA

MC Two identical flasks, A and B, contain, respectively, 5.0 g of N_2 gas and 14.4 g of an unknown gas.

The gases in both flasks are at standard laboratory conditions (SLC).

The gas in flask B is

A. H_2 **B.** SO_2 **C.** HBr **D.** C_4H_{10}

Question 3

Source: *VCE 2018 Chemistry Exam, Section A, Q.25;* © VCAA

MC The molar heat of combustion of pentan-1-ol, $C_5H_{11}OH$, is 3329 kJ mol^{-1}.

$M(C_5H_{11}OH)$ = 88.0 g mol^{-1}

The mass of $C_5H_{11}OH$, in tonnes, required to produce 10 800 MJ of energy is closest to

A. 0.0286 **B.** 0.286 **C.** 2.86 **D.** 286

Question 4

Source: *VCE 2018 Chemistry Exam, Section A, Q.14;* © VCAA

MC An equation for the complete combustion of methanol is

$$2CH_3OH(l) + 3O_2(g) \rightarrow 2CO_2(g) + 4H_2O(g)$$

ΔH for this equation would be

A. +726 kJ mol^{-1} **B.** −726 kJ mol^{-1} **C.** +1452 kJ mol^{-1} **D.** −1452 kJ mol^{-1}

Question 5

Source: *VCE 2017 Chemistry Exam, Section A, Q.7;* © VCAA

MC What is the total energy released, in kilojoules, when 100 g of butane and 200 g of octane undergo combustion in the presence of excess oxygen?

A. 9760 **B.** 14 600 **C.** 17 300 **D.** 19 500

Question 6

Source: VCE 2016 Chemistry Exam, Section A, Q.24; © VCAA

MC Methanol is a liquid fuel that is often used in racing cars. The thermochemical equation for its complete combustion is

$$2CH_3OH(l) + 3O_2(g) \rightarrow 2CO_2(g) + 4H_2O(l) \quad \Delta H = -1450 \text{ kJ mol}^{-1}$$

Octane is a principal constituent of petrol, which is used in many motor vehicles. The thermochemical equation for the complete combustion of octane is

$$2C_8H_{18}(l) + 25O_2(g) \rightarrow 16CO_2(g) + 18H_2O(l) \quad \Delta H = -10\,900 \text{ kJ mol}^{-1}$$

The molar mass of methanol is 32 g mol^{-1} and the molar mass of octane is 114 g mol^{-1}.

Which one of the following statements is the **most** correct?

A. Burning just 1.0 g of octane releases almost 96 kJ of heat energy.
B. Burning just 1.0 g of methanol releases almost 23 kJ of heat energy.
C. Octane releases almost eight times more energy per kilogram than methanol.
D. The heat energy released by methanol will not be affected if the oxygen supply is limited.

Question 7

Source: VCE 2018 Chemistry Exam, Section A, Q.22; © VCAA

MC Four fuels undergo complete combustion in excess oxygen, O_2, and the energy released is used to heat 1000 g of water.

Assuming there is no energy lost to the environment, which one of these fuels will increase the temperature of the water from 25.0 °C to 85.0 °C?

A. 0.889 g of hydrogen, H_2
B. 3.95 g of propane, C_3H_8
C. 0.282 mol of methane, CH_4
D. 0.301 mol of methanol, CH_3OH

Question 8

Source: VCE 2017 Chemistry Exam, Section A, Q.24; © VCAA

MC A sample of olive oil with a wick in a jar is ignited and used to heat a beaker containing 500.0 g of water, H_2O.

The relevant data for the experiment is included in the table below.

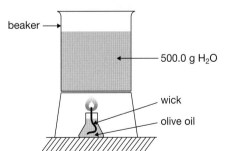

beaker → 500.0 g H₂O — wick — olive oil

Data

Initial temperature (H_2O)	21.0 °C
ΔH (olive oil)	41.0 kJ g^{-1}
Total energy lost to the environment	28.0 kJ

After complete combustion of 2.97 g of olive oil, the **final** temperature of the water, in degrees Celsius, would be

A. 44.9   **B.** 58.0   **C.** 65.9   **D.** 79.3

Question 9

Source: VCE 2019 Chemistry Exam, Section A, Q.26; © VCAA

MC The calibration factor of a bomb calorimeter was determined by connecting the calorimeter to a power supply. The calibration was done using 100 mL of water, 6.5 V and a current of 3.6 A for 4.0 minutes. The temperature of the water increased by 0.48 °C during the calibration.

4.20 g of sucrose underwent complete combustion in the bomb calorimeter. The temperature of the 100 mL of water increased from 19.6 °C to 25.8 °C.

$M(C_{12}H_{22}O_{11}) = 342 \text{ g mol}^{-1}$

The experimental heat of combustion of pure sucrose, in joules per gram, is

A. 5.9×10^6 **B.** 7.3×10^4 **C.** 1.7×10^4 **D.** 1.2×10^4

Question 10

MC The following graph of temperature versus time is obtained for a dissolving reaction taking place in a solution calorimeter.

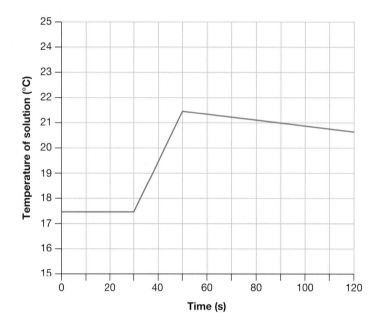

Which of the following statements are consistent with the information in the graph?

 I The initial temperature before adding the solute was 17.5 °C.
 II There was a temperature change of 21.6 °C.
 III The calorimeter was not very well insulated.

A. Statements I and II only
B. Statements I and III only
C. Statements II and III only
D. All statements are correct.

⊳ **Question 11 (9 marks)**

Source: VCE 2018 Chemistry Exam, Section B, Q.9; © VCAA

A Chemistry class conducted a practical investigation to determine the calibration factor of a calorimeter using two different methods: electrical and chemical. Each student compared the results from the two different methods and presented the investigation as a scientific poster.

The materials, set-up and methods used by the students are shown below.

Materials

calorimeter	ammeter
DC power supply	voltmeter
5 × wire leads	3 g of potassium nitrate (KNO_3)
thermometer	electronic balance
stopwatch	measuring cylinder

Calorimeter set-up

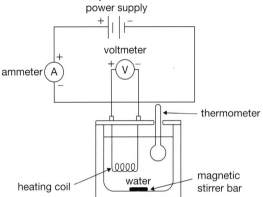

Methods

Electrical method for collecting calibration data

1. Add 100 mL of water to the calorimeter. Stir the water and record its temperature every 30 seconds for several minutes.
2. Apply a voltage of 6 V for three minutes. Stir throughout and record the temperature every 30 seconds.
3. Record the voltage and the current while the water is heating.
4. Once the power is turned off, continue to stir the water and record the temperature every 30 seconds for a further three minutes.

Chemical method for collecting calibration data

1. Measure 3.0 g of KNO_3 accurately.
2. After completing the electrical calibration, add the KNO_3 to the calorimeter.
3. Stir and record the temperature every 30 seconds.

Student A wrote the following aim.

> **Student A**
>
> **Aim**
> To compare the calibration factors obtained from two different methods
> The calibration factors were found by recording the temperature change of a solution resulting from the addition of a measured electrical input and from potassium nitrate dissolving in water.

a. The dependent variable in this investigation is the calibration factor.

   Identify the independent variable from Student A's aim. **(1 mark)**

b. Identify **one** systematic error that applies **only** to the electrical method of calibration. **(1 mark)**

c. Identify **one** limitation of the chemical method of calibration. Explain how it could affect the reliability of the results. **(2 marks)**

d. Examine the graphs below prepared by Student A and Student B for the temperature change during electrical calibration.

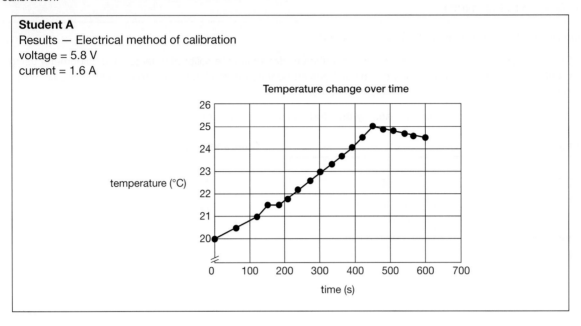

Student A
Results — Electrical method of calibration
voltage = 5.8 V
current = 1.6 A

Temperature change over time

temperature (°C)

time (s)

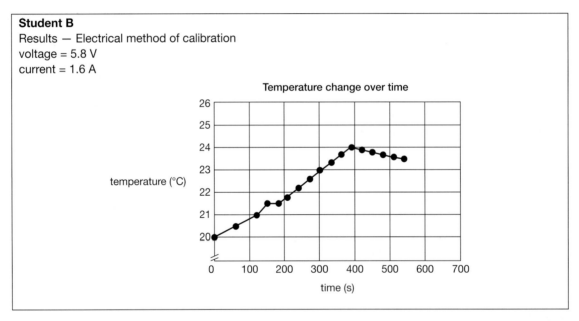

Student B
Results — Electrical method of calibration
voltage = 5.8 V
current = 1.6 A

Temperature change over time

temperature (°C)

time (s)

Identify **one** difference in the results between the students' graphs and suggest what variation in the students' experiments might account for this difference. **(2 marks)**

e. Student B's data for the chemical method of calibration is shown in the graph below.

> **Student B**
> Results — Chemical method of calibration
> Below is the chemical equation and enthalpy used to calculate the calibration factor for the chemical method.
>
> $$KNO_3(s) \xrightarrow{H_2O} K^+(aq) + NO_3^-(aq) \qquad \Delta H = 35 \, kJ \, mol^{-1}$$
>
>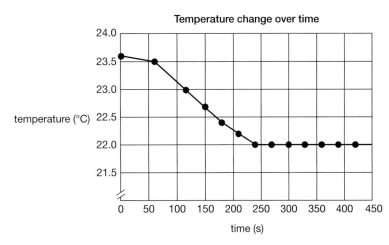

Use this data to calculate the calibration factor, in $J \, °C^{-1}$, for the chemical method of calibration. **(3 marks)**

Question 12 (4 marks)

A brand of processed meat is advertised as being 85% fat free. One hundred grams of this meat provides 1.25×10^3 kJ of energy.

a. Calculate the mass of fat in 100 g of the meat. **(1 mark)**
b. How many kilojoules are provided by this fat? **(1 mark)**
c. Assuming that protein is the only other nutrient present, what mass of protein is present in 100 g of this meat? **(1 mark)**
d. What mass is remaining and what substance might account for the remaining mass of the meat? **(1 mark)**

Question 13 (4 marks)

Source: *Adapted from VCE 2010 Chemistry Exam 2, Section B, Q.1.b.iii; ©VCAA*

Methane can be obtained from natural gas deposits **or** as a biochemical fuel from biomass.

Hydrogen and methane can be burned to produce heat energy.

Calculate the volume of hydrogen gas in L, at SLC, that produces the same amount of energy as 2.0 L of methane gas at SLC.

Question 14 (4 marks)

Palmitic acid is a very common fatty acid that is used to produce soaps and cosmetics. It has a heat of combustion of 1.00×10^4 kJ mol^{-1}.

a. Write a thermochemical equation showing the oxidation of palmitic acid.

$M(CH_3(CH_2)_{14}COOH) = 256.4$ g mol^{-1} **(2 marks)**

b. Calculate the energy released by the combustion of 20.0 g of palmitic acid. **(2 marks)**

Source: VCE 2019 Chemistry Exam, Section B, Q.6.a,f; © VCAA

There are many varieties of bread available to consumers in Australia. The nutritional values for one type of wholemeal bread are given in the table below.

	Per 100 g
Energy	1000 kJ
Protein	9.1 g
Fats and oils	2.5 g
Carbohydrates	41.5 g
Sugars	3.0 g
Fibre	6.4 g

a. Calculate the energy, in kilojoules, provided by the protein and fats and oils in 100 g of this wholemeal bread. **(1 mark)**

b. The wholemeal bread undergoes complete combustion in a bomb calorimeter containing 200 g of water. Assume that all of the energy in the combustion is transferred to the water.

  i. Calculate the mass of bread needed to raise the temperature of the water by 6 °C. **(2 marks)**

  ii. The combustion of the bread was investigated using a different method. The bread was ignited under a beaker containing 200 g of water, which was set on a tripod. The equipment used is shown below.

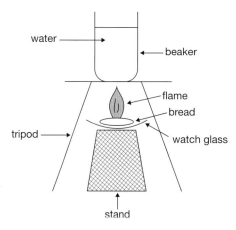

If 1.2 g of bread was needed to raise the temperature of the water by 6 °C using this different method, calculate the efficiency of the energy transfer in this combustion. **(1 mark)**

3 Primary galvanic cells and fuel cells as sources of energy

KEY KNOWLEDGE

In this topic you will investigate:

Primary galvanic cells and fuel cells as sources of energy

- redox reactions as simultaneous oxidation and reduction processes, and the use of oxidation numbers to identify the reducing agent, oxidising agent and conjugate redox pairs
- the writing of balanced half-equations (including states) for oxidation and reduction reactions, and the overall redox cell reaction in both acidic and basic conditions
- the common design features and general operating principles of non-rechargeable (primary) galvanic cells converting chemical energy into electrical energy, including electrode polarities and the role of the electrodes (inert and reactive) and electrolyte solutions (details of specific cells not required)
- the use and limitations of the electrochemical series in designing galvanic cells and as a tool for predicting the products of redox reactions, for deducing overall equations from redox half-equations and for determining maximum cell voltage under standard conditions
- the common design features and general operating principles of fuel cells, including the use of porous electrodes for gaseous reactants to increase cell efficiency (details of specific cells not required)
- the application of Faraday's Laws and stoichiometry to determine the quantity of galvanic or fuel cell reactant and product, and the current or time required to either use a particular quantity of reactant or produce a particular quantity of product
- contemporary responses to challenges and the role of innovation in the design of fuel cells to meet society's energy needs, with reference to green chemistry principles: design for energy efficiency, and use of renewable feedstocks.

Source: VCE Chemistry Study Design (2024–2027) extracts © VCAA; reproduced by permission.

PRACTICAL WORK AND INVESTIGATIONS

Practical work is a central component of VCE Chemistry. Experiments and investigations, supported by a **practical investigation eLogbook** and **teacher-led videos**, are included in this topic to provide opportunities to undertake investigations and communicate findings.

EXAM PREPARATION

▶ Access past VCAA questions and exam-style questions and their video solutions in every lesson, to ensure you are ready.

3.1 Overview

3.1.1 Introduction

Discovering the ability to transform chemical energy into electrical energy changed the world. We all but take for granted AA alkaline batteries that can be bought for less than one dollar to power a host of devices. Science and technology is still looking at new ways to use electrochemical reactions in society. How can we transform chemical energy directly into electricity in more efficient, sustainable and environmentally responsible ways in the future? Meeting our growing electricity and transportation fuel demands does have many challenges, such as making more cost-effective, ubiquitous fuel cells. Creating and storing hydrogen for durable use in fuel cells is another challenge that needs to be overcome, as we examine ways to reduce carbon dioxide emissions from energy generation. Hydrogen as a fuel is still a relatively inefficient one in transport fuel cells, despite the different ways of storing and creating hydrogen that are being researched.

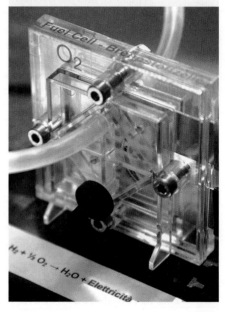

FIGURE 3.1 Laboratory experimentation for hydrogen production with a fuel cell

LEARNING SEQUENCE

  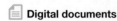 Resources

Solutions	Solutions — Topic 3 (sol-0830)
Practical investigation eLogbook	Practical investigation eLogbook — Topic 3 (elog-1702)
Digital documents	Key science skills — VCE Chemistry Units 1–4 (doc-37066)
	Key terms glossary — Topic 3 (doc-37285)
	Key ideas summary — Topic 3 (doc-37286)
Exam question booklet	Exam question booklet — Topic 3 (eqb-0114)

3.2 Redox reactions

3.2.1 What is a redox reaction?

The production of electricity from chemical reactions began in 1780 when Luigi Galvani (1739–1798), an Italian anatomist, conducted a series of experiments investigating the responses obtained from the hind legs of frogs when static electricity was applied to them. He found that the frogs' legs could be made to twitch by connecting the nerve and muscle tissues to different metals such as copper and iron. The dead frog was literally galvanised into action. Galvani thought that the muscles of the frog must contain electricity and advocated the idea of 'animal electricity'.

FIGURE 3.2 During the eighteenth century, many people believed that the nerves and muscles of animals contained a fluid that acted like an electric current. How do Galvani's results support this idea?

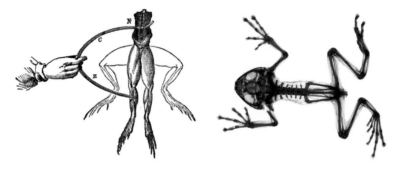

Redox reactions involve the transfer of electrons to and from substances. The term 'redox' is derived from two separate words: **reduction** and **oxidation**. Originally, these were defined in terms of loss or gain of oxygen. Now, a substance is said to be reduced when it accepts (or gains) electrons, and is said to be oxidised when it donates (or loses) electrons. In a redox reaction, reduction and oxidation always occur simultaneously. Note that oxidation and reduction are *processes*.

- Reduction is the *process* in which electrons are added to a substance.
- Oxidation is the *process* in which electrons are removed from a substance.
- This can be remembered using the acronym **OIL RIG**: Oxidation Is Loss Reduction Is Gain

Oxidising agents (also known as oxidants) and **reducing agents** (also known as reductants) are *substances* in a redox reaction. Because oxidation and reduction always occur together, an oxidising agent can be thought of as a substance that allows (or causes) another substance to undergo oxidation. It does this by accepting the electrons that are produced. In the same way, a reducing agent is a substance that permits another substance to undergo reduction, by supplying the electrons that are required. As a result, oxidising agents undergo the *process* of reduction, while reducing agents undergo the *process* of oxidation.

redox reactions reactions that involve the transfer of one or more electrons between chemical species

reduction a gain of electrons; a decrease in the oxidation number

oxidation a loss of electrons; an increase in the oxidation number

oxidising agents electron acceptors

reducing agents electron donors

- Oxidising agents are substances that cause or permit another substance to be oxidised.
- Reducing agents are substances that cause or permit another substance to be reduced.

All of these definitions can be clarified by remembering that oxidation and reduction are *processes*, whereas oxidising agents and reducing agents are *substances* involved in these processes.

Redox reactions may be represented by balanced **half-equations** and by overall equations. For example, the burning of magnesium may be represented by the overall equation:

$$2Mg(s) + O_2(g) \rightarrow 2MgO(s)$$

This equation may be deconstructed into two half-equations that illustrate the transfer of electrons:

$$Mg(s) \rightarrow Mg^{2+}(s) + 2e^- \text{ (oxidation)}$$

$$O_2(g) + 4e^- \rightarrow 2O^{2-}(s) \text{ (reduction)}$$

FIGURE 3.3 Burning magnesium powder gives out a great deal of light. It is commonly used in flash bulbs and fireworks. Here, magnesium is oxidised while oxygen is reduced.

In this reaction, magnesium is acting as a reducing agent because it is losing electrons. Magnesium is a group 2 metal, and losing two electrons allows it to attain a full outer shell configuration. Oxygen is acting as an oxidising agent because it is gaining electrons. As a member of group 16, the gain of two electrons allows it to attain a full outer shell configuration. If the oxidation half-equation is multiplied by two and the two half-equations are then added, the electrons cancel out and the *overall* balanced (in terms of both charge and species) equation is produced.

 Resources

 Video eLesson Redox-electron transfer (eles-2495)

3.2.2 Identifying redox reactions

The deconstruction of an equation into oxidation and reduction half-equations proves that a reaction is a redox reaction. While this is a relatively simple process for some reactions, there are many redox reactions that are more complex. For example, the reaction between the acidified dichromate (VI) ion and hydrogen disulfide:

$$Cr_2O_7^{2-}(aq) + 3H_2S(aq) + 8H^+(aq) \rightarrow 2Cr^{3+}(aq) + 3S(s) + 7H_2O(l)$$

is also a redox reaction, but it is much harder to produce the half-equations for this reaction. To assist in situations such as this, chemists use **oxidation numbers**. These are numbers that aid in the identification of redox reactions.

When using oxidation numbers, remember that:
- oxidation is an *increase* in the oxidation number of an atom (i.e. the number becomes more positive, such as −3 to −1 or −1 to +1)
- reduction is a *decrease* in the oxidation number of an atom (i.e. the number becomes more negative, such as +3 to +1 or +1 to −1).

If the oxidation number of an element changes between the reactant species and the product species, then that element has undergone either oxidation or reduction. Given that oxidation cannot happen without reduction, it is easy to determine if the reaction can be classified as redox.

half-equation an equation that gives one half of a redox reaction, showing the movement of electrons in either an oxidation or a reduction reaction

oxidation numbers numbers used to find an oxidising agent and a reducing agent by a change in perceived valency

Alternatively, we can determine what is oxidised and what is reduced. Oxidisation occurs if a substance is gaining oxygen, and reduction occurs if a substance is losing oxygen. Oxidation can also be determined if a substance is losing hydrogen and reduction can be determined if a substance is gaining hydrogen.

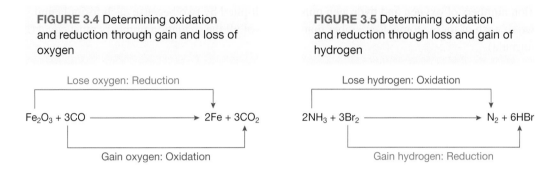

FIGURE 3.4 Determining oxidation and reduction through gain and loss of oxygen

Lose oxygen: Reduction

$Fe_2O_3 + 3CO \longrightarrow 2Fe + 3CO_2$

Gain oxygen: Oxidation

FIGURE 3.5 Determining oxidation and reduction through loss and gain of hydrogen

Lose hydrogen: Oxidation

$2NH_3 + 3Br_2 \longrightarrow N_2 + 6HBr$

Gain hydrogen: Reduction

3.2.3 Oxidation numbers

The following rules can be used to determine oxidation numbers. Remember that oxidation numbers are theoretical numbers and should not be confused with ionic charges.

TABLE 3.1 Oxidation number rules

Oxidation number rule	Oxidation number example
1. The oxidation number of an atom in its elemental form is 0.	Oxidation number: Copper metal, $Cu(s) = 0$ Nitrogen gas, $N_2(g) = 0$
2. The oxidation number of a simple ion is the charge on the ion.	$Al^{3+} = +3$ $S^{2-} = -2$
3. The oxidation number of hydrogen in non-metal compounds is +1. The oxidation number of hydrogen in metal hydrides is –1.	In HCl, H_2O and $NH_4{}^+$, H = +1. In NaH or CaH_2, H = –1.
4. The oxidation number of oxygen in a compound is usually –2, except in: • peroxide compounds, where the oxidation number is –1 • compounds with oxygen bonded to fluorine, where the oxidation number is +2.	Magnesium oxide, O = –2 Peroxide compounds: H_2O_2 and BaO_2, O = –1 Oxygen difluoride: OF_2, O = +2
5. Fluorine always has an oxidation number of –1 because it is the most electronegative element.	F = –1
6. In a neutral compound the sum of all the oxidation numbers must equal 0.	In $MgCl_2$, oxidation numbers are added as follows: $+2 + (2 \times -1) = 0$.
7. In a polyatomic ion, the sum of the oxidation numbers must equal the charge on the ion.	In $NO_3{}^-$ oxidation numbers are added as follows: $+5 + (3 \times -2) = -1$.
8. In covalent compounds that do not involve oxygen or hydrogen, the more electronegative element has the negative oxidation number. This is equal to the charge that it would have if it was a negative ion.	In ICl_3, the chlorine is the more electronegative atom. It is therefore assigned an oxidation number of –1, because this is the charge on a chloride ion. (*Note:* This is just the way the oxidation number is worked out. This molecule is a covalent, neutral molecule; it *does not* contain chloride ions.) Using rule 6, we can now calculate that the oxidation number of the iodine in ICl_3 is +3.

Using oxidation numbers

The rules for determining oxidation numbers can be used to assign an oxidation number to each atom in a compound. Calcium hydroxide, $Ca(OH)_2$, has been used in figure 3.6 to demonstrate this.

The oxidation numbers of oxygen and hydrogen must be multiplied by two because each formula unit contains two of each of these atoms. Remember that the sum of all the oxidation numbers must equal 0 (for a balanced formula).

$$Ca + (2 \times O) + (2 \times H) = +2 + (2 \times -2) + (2 \times +1)$$
$$= 2 - 4 + 2$$
$$= 0$$

FIGURE 3.6 How to determine the oxidation numbers in calcium hydroxide

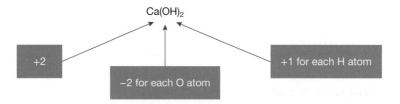

SAMPLE PROBLEM 1 Determining oxidation numbers

The main compound in limestone statues or common chalk is calcium carbonate. What is the oxidation number of carbon in the carbonate ion, CO_3^{2-}?

THINK	WRITE
1. Assign as many oxidation numbers as possible, and then find the oxidation number of the unknown atom. Oxygen's oxidation number is only ever -1 if it is hydrogen peroxide (H_2O_2) or $+2$ if it is bonded to fluorine.	The oxidation number of oxygen is -2.
2. Obtain the oxidation number for carbon by recognising the sum of the oxidation numbers for O and C are equal to the charge on the ion (-2).	(Oxidation number for C) $+ 3$ $\times$ (oxidation number for O) $= -2$ (Oxidation number for C) $+ 3 \times (-2) = -2$ (Oxidation number for C) $- 6 = -2$ $\therefore$ Oxidation number for C $= +4$

PRACTICE PROBLEM 1

The photographs obtained by the *Voyager 1* space probe showed that Io, one of the moons of Jupiter, has active volcanoes and a surface composed of sulfur and sulfur dioxide. Assign oxidation numbers to each atom in the molecule sulfur dioxide, SO_2.

FIGURE 3.7 Io has a thin atmosphere of sulfur dioxide, and sulfur compounds in liquid and solid states cover its surface.

FIGURE 3.8 Different oxidation states of chromium compounds

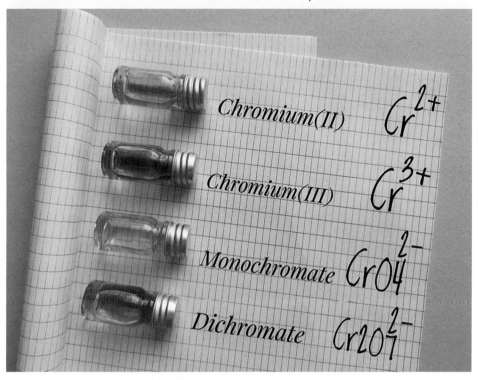

tlvd-9675

SAMPLE PROBLEM 2 Determining if a reaction is a redox reaction

Determine whether the following reaction is a redox reaction:

$$H_2(g) + I_2(g) \rightarrow 2HI(g)$$

THINK	WRITE
1. Assign oxidation numbers to each element. The oxidation number of an atom in its elemental form is 0. The oxidation number of H in non-metal compounds is +1.	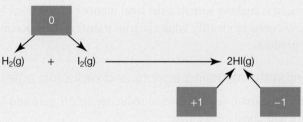
2. Determine whether a change in oxidation number has taken place.	The oxidation number of hydrogen has changed from 0 to +1, so the hydrogen has been oxidised (because its oxidation number has increased). The oxidation number of iodine has changed from 0 to –1, so the iodine has been reduced (because its oxidation number has decreased). Therefore, this is a redox reaction.

Although tungsten, W, is a rare element, it has been used extensively in the past in light globes. Tungsten is still used to make filaments for specialist incandescent globes because it has the highest melting point (3410 °C) and boiling point (5900 °C) of any metal.

FIGURE 3.9 Tungsten metal filaments used in specialist light bulbs. If hot tungsten is exposed to air, it oxidises to form tungsten oxide. To prevent this, inert argon gas is used to fill the inside of light globes.

The metal is obtained from tungsten(VI) oxide by heating it with hydrogen, according to the following equation:

$$WO_3(s) + 3H_2(g) \rightarrow W(s) + 3H_2O(g)$$

Using oxidation numbers, determine whether this equation represents a redox reaction and, if so, identify the oxidising agent and reducing agent.

3.2.4 Writing redox equations

Like all equations, redox equations must be balanced. The key to ensuring that overall redox equations are balanced is making sure that the total number of electrons lost is equal to the number of electrons gained. To do this, we need to identify what electron transfers have occurred from the changes to the formulae of the reactants and products.

Conjugate oxidising agents and reducing agents

Every time oxidising agents and reducing agents gain and lose electrons, they form a conjugate pair. A conjugate redox pair consists of either:
- an electron donor and its corresponding electron acceptor
- an electron acceptor and its corresponding electron donor.

For example, when the half-equation $Cu(s) \rightarrow Cu^{2+}(aq) + 2e^-$ occurs, the $Cu(s)$ donates two electrons and is oxidised, which means it is acting as a reducing agent. Now that a $Cu^{2+}(aq)$ ion has been formed, it can do the opposite of the $Cu(s)$ and accept two electrons back. This reduces it and allows it to behave as a conjugate oxidising agent. Therefore, species that are reduced (act as an electron acceptor, which is an oxidising agent) form conjugate reducing agents, and species that are oxidised (act as an electron donor, which is a reducing agent) form conjugate oxidising agents.

SAMPLE PROBLEM 3 Identifying the oxidising and reducing agents in a redox reaction

In the following redox reaction, identify the oxidising agent, the reducing agent and their conjugates.

$$Fe(s) + CuSO_4(aq) \rightarrow FeSO_4(aq) + Cu(s)$$

THINK

1. Recall the definitions of oxidation and reduction. Recall that all elements have an oxidation number of 0.
   TIP: Use the following acronyms:
   Oxidation **I**s **L**oss (OIL) of electrons.
   Reduction **I**s **G**ain (RIG) of electrons.

2. Oxidising agents are reduced and reducing agents are oxidised.
   TIP: When a question asks for the reducing or oxidising agent, you must always specify whether the agent is in the ion or elemental form. In this example the answer should be the copper ions, or Cu^{2+}(aq). Note that Cu(s) is actually the conjugate oxidising agent.

3. Species that are reduced (oxidising agents) form conjugate reducing agents, and species that are oxidised (reducing agents) form conjugate oxidising agents.

WRITE

Cu^{2+}(aq) forms Cu(s). Cu^{2+} has an initial oxidation number of +2 and has become less positive to have a final oxidation number for Cu(s) of 0. This means it has been reduced.
Fe(s) has an initial oxidation number of 0 and has become more positive, forming Fe^{2+}(aq) (because the charge on the sulfate group is –2). This means it has been oxidised.

Cu^{2+}(aq) acts as an oxidising agent.
Fe(s) acts as a reducing agent.

Cu(s) is the conjugate reducing agent.
Fe^{2+} is the conjugate oxidising agent.

In sample problem 3, Fe(s) is oxidised and Cu^{2+} is reduced. The relationship between oxidising and reducing agents and their conjugates is shown in table 3.2.

TABLE 3.2 The relationship between oxidising and reducing agents and their conjugates

Fe(s)	+	CuSO₄(aq)	→	FeSO₄(aq)	+	Cu(s)
Fe(s) gets oxidised. Acts as a reducing agent Forms a conjugate oxidising agent		Cu^{2+}(aq) gets reduced. Acts as an oxidising agent Forms a conjugate reducing agent		Fe^{2+} can act as a conjugate oxidising agent.		Cu(s) can act as a conjugate reducing agent.

PRACTICE PROBLEM 3

In the following redox reaction, identify the oxidising agent, the reducing agent and their conjugates.

$$2Na(s) + 2H_2O(l) \rightarrow 2NaOH(aq) + H_2(g)$$

Reactions in acidic and basic environments

Redox reactions in an acidic environment were introduced in Unit 2. The steps for writing redox equations in both acidic and basic solutions are outlined below.

Acidic solutions

Let's examine the reaction between the dichromate ion ($Cr_2O_7{}^{2-}$) and hydrogen sulphide (H_2S) in an acidic solution. The overall reaction is represented by the following equation:

$$Cr_2O_7{}^{2-}(aq) + 3H_2S(aq) + 8H^+(aq) \rightarrow 2Cr^{3+}(aq) + 3S(s) + 7H_2O(l)$$

A review of the oxidation numbers in the equation shows us that chromium in dichromate has an oxidation number of +6 on the reactant side of the equation. On the product side, the oxidation number has decreased to +3 as Cr^{3+}. Therefore, $Cr_2O_7{}^{2-}$ is the oxidising agent in this reaction.

The oxidation number of S, in H_2S, is –2. It increases to 0 as elemental sulfur (S) on the product side of the equation. H_2S is therefore the reducing agent.

Table 3.3 outlines the steps used for balancing this equation. This is sometimes referred to as the KOHES method of balancing equations.

The KOHES method of balancing redox half-equations

K –	Key elements
O –	Oxygen by adding water
H –	Hydrogen by adding H⁺
E –	balance Electrons
S –	assign States

TABLE 3.3 Balancing half-equations and redox reactions

Rule	Example
1. Identify the conjugate pairs that are involved in the reaction. Oxidation numbers may be useful in doing this. Write these pairs down with the reactant on the left and the product on the right.	Reduction: $Cr_2O_7{}^{2-} \rightarrow Cr^{3+}$ Oxidation: $H_2S \rightarrow S$
2. Balance **K**ey elements (undergoing reduction or oxidation).	Reduction: $Cr_2O_7{}^{2-} \rightarrow 2Cr^{3+}$ Oxidation: $H_2S \rightarrow S$
3. Balance **O**xygen atoms, where needed, by adding water molecules.	Reduction: $Cr_2O_7{}^{2-} \rightarrow 2Cr^{3+} + 7H_2O$ Oxidation: $H_2S \rightarrow S$
4. Balance **H**ydrogen atoms, where needed, by adding H⁺ ions.	Reduction: $Cr_2O_7{}^{2-} + 14H^+ \rightarrow 2Cr^{3+} + 7H_2O$ Oxidation: $H_2S \rightarrow S + 2H^+$
5. Balance the overall charge by adding **E**lectrons.	Reduction: $Cr_2O_7{}^{2-} + 14H^+ + 6e^- \rightarrow 2Cr^{3+} + 7H_2O$ Oxidation: $H_2S \rightarrow S + 2H^+ + 2e^-$
Once this process is done for each conjugate pair, the following steps then produce the overall equation.	At this stage, the two half-equations have been written, reduction and oxidation can be confirmed, and states would now be added. However, if the overall equation is required, the following two steps are used.
6. Multiply each half-equation from step 5 by factors that produce the same number of electrons in each half-equation.	Reduction: $Cr_2O_7{}^{2-} + 14H^+ + 6e^- \rightarrow 2Cr^{3+} + 7H_2O$ (not necessary to adjust) Oxidation: $3H_2S \rightarrow 3S + 6H^+ + 6e^-$ (multiplied by 3 so that there are 6e⁻ on both sides)

7. Add the two half-equations together, cancelling the electrons. There may be other substances that also partially cancel out at this stage.	$Cr_2O_7{}^{2-} + 14H^+ + 6e^- + 3H_2S$ $\rightarrow 2Cr^{3+} + 7H_2O + 3S + 6H^+ + 6e^-$ After cancelling the electrons and hydrogen ions, this becomes: $Cr_2O_7{}^{2-} + 3H_2S + 8H^+ \rightarrow 2Cr^{3+} + 3S + 7H_2O$
8. Identify the **States** for all species.	Adding symbols of state now gives the final equation: $Cr_2O_7{}^{2-}(aq) + 3H_2S(aq) + 8H^+(aq) \rightarrow 2Cr^{3+}(aq) + 3S(s) + 7H_2O(l)$

TIP: Ensure that you balance the charge on both sides of the overall equation. The charge on each side of the equation should be the same. They do not cancel each other out or have to equal zero.

An alternative method of remembering this process for balancing half-equations is the phrase 'I Want Half-Equations', representing the steps:
1. balance key **I**ons
2. add **W**ater to balance oxygen
3. add **H**$^+$ to balance hydrogen
4. add **E**lectrons to balance charge.

Basic (alkaline) solutions

In basic solutions we need to add an extra step to the KOHES method. This happens after step 5; that is, after you add electrons to balance the charge.

This additional step is to add the same number of OH$^-$ as you did H$^+$ to *both sides* of the half-equation. The OH$^-$ you add to the H$^+$ becomes H$_2$O(l), so that there are no H$^+$ ions left in the half-equations.

For example, let's have a look at the conjugate redox pair Ag(s) $\rightarrow$ Ag$_2$O(s).

After the addition of electrons to balance the charge in the usual fashion, we end up with:

$$2Ag(s) + H_2O(l) \rightarrow Ag_2O(s) + 2H^+(aq) + 2e^-$$

Now, we add OH$^-$ in the same quantities as H$^+$ to *both sides* of the half-equation:

$$2Ag(s) + H_2O(l) + 2OH^-(aq) \rightarrow Ag_2O(s) + 2H^+(aq) + 2OH^-(aq) + 2e^-$$

Add the H$^+$ and OH$^-$ together to make H$_2$O:

$$2Ag(s) + H_2O(l) + 2OH^-(aq) \rightarrow Ag_2O(s) + 2H_2O(l) + 2e^-$$

Now, cancel the water molecules:

$$2Ag(s) + \cancel{H_2O(l)} + 2OH^-(aq) \rightarrow Ag_2O(s) + \overset{1}{\cancel{2}}H_2O(l) + 2e^-$$

This leaves the final half-equation:

$$\mathbf{2Ag(s) + 2OH^-(aq) \rightarrow Ag_2O(s) + H_2O(l) + 2e^-}$$

The steps for balancing a redox half-equation in a basic environment can be recalled using the mnemonics '**KOHES (OH)**' or '**I Want Half-Equations (OH)**'.

 Resources

▶ **Video eLesson** Balancing redox reactions (eles-2489)

3.2 Activities

learnon

Students, these questions are even better in jacPLUS

 Receive immediate feedback and access sample responses

 Access additional questions

 Track your results and progress

Find all this and MORE in jacPLUS ▶

| 3.2 Quick quiz on | 3.2 Exercise | 3.2 Exam questions |

3.2 Exercise

1. Assign oxidation numbers to the atoms in the following substances.
   a. HBr
   b. Na_2O
   c. CH_4
   d. $NaClO_3$
   e. Al_2O_3
   f. H_3PO_4
2. Assign oxidation numbers to the atoms in the following ions.
   a. NH_2^-
   b. MnO_4^-
   c. HS^-
   d. VO^{2+}
   e. IO_3^-
   f. PO_4^{3-}
3. Identify if the following equations are redox equations. If the reaction is a redox reaction, identify the substances that have been oxidised and reduced.
   a. $2Fe(s) + 3Cl_2(g) \rightarrow 2FeCl_3(s)$
   b. $NH_3(g) + HCl(g) \rightarrow NH_4Cl(s)$
   c. $2NO(g) + O_2(g) \rightarrow 2NO_2(g)$
   d. $NaOH(aq) + HCl(aq) \rightarrow NaCl(aq) + H_2O(l)$
   e. $K_2O(s) + H_2O(l) \rightarrow 2KOH(aq)$
   f. $P_4O_{10}(s) + 6H_2O(l) \rightarrow 4H_3PO_4(aq)$
   g. $2CO(g) + O_2(g) \rightarrow 2CO_2(g)$
   h. $C_2H_4(g) + H_2(g) \rightarrow C_2H_6(g)$
4. Write half-equations for the following conjugate redox pair:

   $$Mg(s) \rightarrow MgO(s)$$

   a. in an acidic (H^+) solution
   b. in an alkaline (OH^-) solution.
5. The overall equation for a redox reaction occurring in an acidic environment is:

   $$CH_3OH(aq) + O_2(g) \rightarrow CO_2(g) + 2H_2O(l)$$

   a. Write the formula of the oxidising agent.
   b. Write a balanced half-equation for the oxidation reaction.

3.2 Exam questions

Question 1 (1 mark)

Source: VCE 2016 Chemistry Exam, Section A, Q.3; © VCAA

MC Hydrogen peroxide solutions are commercially available and have a range of uses. The active ingredient, hydrogen peroxide, H_2O_2, undergoes decomposition in the presence of a suitable catalyst according to the reaction

$$2H_2O_2(l) \rightarrow 2H_2O(l) + O_2(g)$$

In this reaction, oxygen
A. only undergoes oxidation.
B. only undergoes reduction.
C. undergoes both oxidation and reduction.
D. undergoes neither oxidation nor reduction.

Question 2 (1 mark)

Source: VCE 2015 Chemistry Exam, Section A, Q.6; © VCAA

MC In which one of the following compounds is sulfur in its lowest oxidation state?

A. SO_3

B. HSO_4^-

C. SO_2

D. Al_2S_3

Question 3 (1 mark)

Source: VCE 2015 Chemistry Exam, Section A, Q.24; © VCAA

MC The reaction between hydrogen peroxide and ammonium ions is represented by the following equation.

$$3H_2O_2(aq) + 2NH_4^+(aq) \rightarrow N_2(g) + 2H^+(aq) + 6H_2O(l)$$

Which one of the following is the correct half-equation for the reduction reaction?

A. $H_2O_2(aq) + 2H^+(aq) + 2e^- \rightarrow 2H_2O(l)$

B. $2NH_4^+(aq) \rightarrow N_2(g) + 8H^+(aq) + 6e^-$

C. $2NH_4^+(aq) + 2e^- \rightarrow N_2(g) + 4H_2(g)$

D. $H_2O_2(aq) + 2H_2O(l) \rightarrow 2O_2(g) + 6H^+(aq) + 6e^-$

Question 4 (1 mark)

Source: VCE 2014 Chemistry Exam, Section A, Q.10; © VCAA

MC Which one of the reactions of hydrochloric acid below is a redox reaction?

A. $2HCl(aq) + Fe(s) \rightarrow H_2(g) + FeCl_2(aq)$

B. $2HCl(aq) + Na_2S(s) \rightarrow H_2S(g) + 2NaCl(aq)$

C. $2HCl(aq) + MgO(s) \rightarrow MgCl_2(aq) + H_2O(l)$

D. $2HCl(aq) + K_2CO_3(s) \rightarrow CO_2(g) + 2KCl(aq) + H_2O(l)$

Question 5 (1 mark)

Source: VCE 2010 Chemistry Exam 2, Section A, Q.16; © VCAA

MC Which of the following represents a balanced reduction half-reaction?

A. $VO_2^+ + H^+ + 2e^- \rightarrow VO^{2+} + H_2O$

B. $VO_2^+ + H_2 \rightarrow VO^{2+} + H_2O + e^-$

C. $VO_2^+ + 2H^+ + e^- \rightarrow VO^{2+} + H_2O$

D. $VO_2^+ + 4H^+ + 3e^- \rightarrow VO^{2+} + 2H_2O$

More exam questions are available in your learnON title.

3.3 Galvanic cells and the electrochemical series

When a redox reaction occurs in a beaker, only heat is generated. However, redox reactions can be made more useful if the two reactants are separated.

Galvanic cells — named after Luigi Galvani — use **spontaneous reactions** between oxidising agents and reducing agents, but separate the two half-reactions into different locations, so that electrons are forced to travel through an external wire between them. They are designed to harness the electrons transferred from redox reactions to do electrical work. Batteries, including that in your mobile phone, are applications of this concept.

spontaneous reactions reactions that proceed on their own, without the need for any external supply of energy

---

EXTENSION: Free energy

You may wonder why some chemical reactions or physical changes are spontaneous while others are not. Predicting whether a chemical reaction or physical change will be spontaneous depends on whether there is a loss in free energy as a reaction or change proceeds. This is different from just a change in enthalpy (ΔH).

A change in free energy, ΔG, is a combination of enthalpy change, temperature and a quantity called entropy (S). A change in entropy (ΔS) is a change in particle order, or more accurately, disorder.

To explain entropy in simple terms, the particles in a solid ionic crystal, with a regular arrangement of ions, have low entropy. They are in a solid, ordered state. When the ionic crystal melts, the ions are free to move, the solid to liquid transition increases the state of disorder, and the entropy increases. In a gas, the molecules move completely independently of one another in a disordered manner, and the entropy of a gas is therefore always high.

Both ΔH and ΔS can increase or decrease as a result of spontaneous chemical or physical changes. Think of ice melting as an example. This requires energy, so ΔH is positive, and the particles increase their disorder because they are free to move, so ΔS is positive too. But ice doesn't spontaneously melt at all temperatures (and pressures).

FIGURE 3.10 Josiah Gibbs

An American scientist named Josiah Gibbs, who excelled in physics, chemistry, mathematics and engineering, defined free energy according to the following equation:

$$\Delta G = \Delta H - T\Delta S$$

For a chemical or physical change to be spontaneous at a particular temperature (in kelvin) and pressure, the change in free energy must be negative ($\Delta G < 0$).

Looking at the equation, if ΔH is negative and ΔS is positive, then $\Delta G < 0$ at all temperatures. This means exothermic changes that result in increased disorder will always be spontaneous.

If ΔH is positive and ΔS is negative, then $\Delta G > 0$. Therefore, changes that require energy and result in less disorder are not spontaneous at any temperature.

Converting chemical energy to thermal energy

When a zinc strip is placed in copper(II) sulfate solution (Cu^{2+} ions), the zinc is oxidised and electrons flow from the zinc metal to the copper ions. This is a spontaneous reaction and requires no energy; in fact, it releases energy (figure 3.11a).

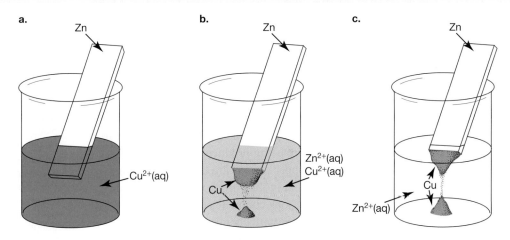

As the zinc dissolves, copper ions are reduced to copper metal and the original blue colour of the solution begins to fade (figure 3.11b). If the zinc strip remains in the solution for an extended period of time, the solution in the beaker becomes colourless. All the copper ions in the solution are reduced to form copper metal, and the zinc goes into the solution as zinc ions (figure 3.11c).

FIGURE 3.11 A zinc strip in copper(II) sulfate solution creates a spontaneous redox reaction.

All the chemical energy of the reaction is released as thermal energy (heat), and the transfer of electrons from zinc to copper ions occurs on the surface of the zinc metal. This can be seen in figures 3.12 and 3.13.

FIGURE 3.12 Chemical energy is released as thermal energy.

FIGURE 3.13 As Zn reacts with Cu^{2+} ions, it goes into solution as Zn^{2+} ions. Cu(s) is deposited on the surface of the zinc.

3.3.1 Galvanic cell design

Converting chemical energy into electrical energy

If the site of oxidation is physically separated from the site of reduction (for example, if each of the solutions is in a separate beaker), and a connecting wire is placed between them, the electrons are forced to travel through this wire to complete the redox reaction. Such movement of electrons constitutes an electric current. This arrangement converts chemical energy directly into electrical energy.

To do this for the $Zn(s)/Cu^{2+}(aq)$ reaction that we have been discussing, a strip of zinc metal is placed in a beaker containing zinc sulfate solution (see figure 3.14a). This is connected by a wire to a strip of copper placed in a beaker containing a copper sulfate solution. The wire provides a pathway for the electrons to pass from the zinc atoms to the copper cations, but the reaction cannot occur because the circuit is not complete. To complete the circuit and allow the reaction to proceed, a **salt bridge** is needed to connect the two **electrolytes**.

The salt bridge can be a simple filter paper or a U-tube with cotton wool in it. It is soaked in a salt solution, such as potassium nitrate solution, $KNO_3(aq)$, and is used to connect the two beakers (figure 3.14b). Potassium nitrate solution provides anions, NO_3^-, and cations, K^+, and the movement of these carries the current and completes the circuit. The directions that the ions move helps to maintain the electric neutrality of the beakers by supplying ions of the opposite charge. The anions flow to the anode in the first beaker, where zinc cations are being produced. The cations flow to the cathode in the second beaker, where the copper cations are being reduced and becoming copper atoms. Electrons carry the current in the wire from zinc to copper, and ions carry the current in solution. The flow of ions completes the circuit. It is important that the ions in the salt bridge do not react with chemicals in the beakers. The movement of ions in the solutions is called the **internal circuit**.

Each beaker in figure 3.14 is a **half-cell**. The metal strips are called **electrodes** and, combined with the wire, they are referred to as the **external circuit**. Electrons moving through the external circuit as a result of the redox reactions occurring in the two half-cells can be made to do useful work, such as lighting a light bulb — that is, the system can convert chemical energy into light energy.

salt bridge a component that provides a supply of mobile ions that carry the charge through the solution of a galvanic cell during a reaction

electrolytes liquids that can conduct electricity

internal circuit a circuit within a solution; anions flow to the anode and cations flow to the cathode

half-cell one half of a galvanic cell containing an electrode immersed in an electrolyte that may be the oxidising agent or the reducing agent, depending on the oxidising strength of the other cell to which it is connected

electrode a solid used to conduct electricity in a galvanic half-cell

external circuit a circuit composed of all the connected components within an electrolytic or a galvanic cell to achieve desired conditions

FIGURE 3.14 a. Two strips of different metals and solutions of each of their ions. **b.** With the addition of a wire and a salt bridge, a simple electrochemical cell — a device that converts chemical energy into electrical energy — is constructed.

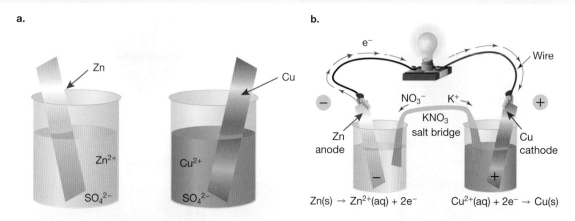

Solutions that can conduct a current are known as electrolytes. The electrode at which oxidation occurs is called the **anode**, and it has a negative charge in a galvanic cell; the electrode at which reduction occurs is called the **cathode**, and it has a positive charge in a galvanic cell. All of these components together are known as a galvanic cell or an **electrochemical cell**.

The example shown in figure 3.14 of an electrochemical cell containing the half-cells $Zn(s)/Zn^{2+}(aq)$ and $Cu(s)/Cu^{2+}(aq)$ is known as the **Daniell cell**.

> **TIP:** When constructing answers, remember:
> - the Internal circuit always involves Ions
> - the External circuit always involves Electrons.
> - RedCat: Reduction always occurs at the Cathode
> - AnOx: Oxidation always occurs at the Anode

Types of half-cells

Each half-cell in a laboratory galvanic cell contains a conjugate oxidising agent–reducing agent pair. Oxidation occurs in one of the half-cells and reduction occurs in the other. Half-cells are constructed by dipping an electrode into an electrolyte. The electrode may or may not take part in the reaction.

It is convenient to group half-cells into three types based on design. The three types are:
- the metal ion–metal half-cell
- the solution half-cell
- the gas–non-metal ion half-cell.

Metal ion–metal half-cells (figure 3.15) consist of a metal rod in a solution of its ions, usually from the sulfate salt. The sulfate ion is unreactive. Ions that are more reactive, such as bromide ions or nitrate ions, may set up a competing reaction.

Solution half-cells (figure 3.16) use an inert (unreactive) electrode in the reacting solution. The reacting solution may contain an oxidising agent — for example, $MnO_4^-(aq)$ — or a reducing agent — for example, $Fe^{2+}(aq)$.

Although gases are reactive, they are usually more difficult to manage in the laboratory. As a result, gaseous half-cells are not very common. In a gaseous half-cell, the gas bubbles over an inert electrode that is connected to the external wire (figure 3.17). Its conjugate redox non-metal ion is in solution.

anode the electrode at which oxidation occurs; in a galvanic cell it is the negative electrode, since it is the source of negative electrons for the circuit; if the reducing agent is a metal, it is used as the electrode material

cathode the electrode at which reduction occurs; in a galvanic cell it is the positive electrode, because the negative electrons are drawn towards it and then consumed by the oxidising agent, which is present in the electrolyte

electrochemical cell a cell that generates electrical energy from chemical reactions

Daniell cell one of the first electrochemical cells to produce a reliable source of electricity; it uses the redox reactions between zinc metal and copper ions to produce electricity

FIGURE 3.15 Metal ion–metal half-cell

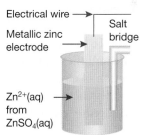

FIGURE 3.16 Solution half-cell

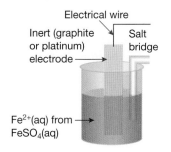

FIGURE 3.17 Gas–non-metal ion half-cell

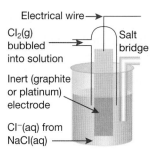

A simple galvanic cell consists of:
- two half-cells, containing two electrodes (anode and cathode) and two electrolytes
- a conducting wire
- a salt bridge, containing another electrolyte.

An electrolyte is a solution containing ions that can conduct electricity.

An electrode is a conductor through which electrons enter or leave a galvanic cell.

The anode is the electrode where oxidation occurs.

The cathode is the electrode where reduction occurs.

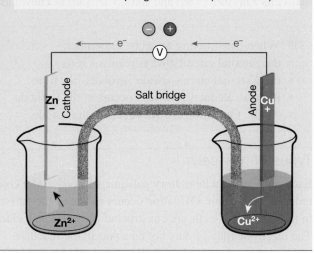

FIGURE 3.18 A simple galvanic cell (Daniell cell)

tlvd-9677

SAMPLE PROBLEM 4 Making predictions in galvanic cells

A galvanic cell was set up in the following way. A strip of clean magnesium was dipped into a beaker containing a solution of $MgSO_4$ and, in a separate beaker, an iron nail was dipped into a solution of $FeSO_4$. The iron nail and magnesium strip were connected with a wire, and the circuit was completed with a salt bridge consisting of filter paper dipped into a solution of KNO_3. The magnesium electrode was known to have a negative charge. Predict the following.
a. The substance that is oxidised and the one that is reduced
b. The anode and the cathode
c. The direction of electron flow
d. The half-cell reactions
e. The overall redox reaction

THINK

a. Oxidation occurs at the anode, and electrons always flow from the site of oxidation to the site of reduction.

b. Oxidation always occurs at the anode.

c. Electrons flow from the anode to the cathode.

d. Write out the two half-equations that represent oxidation and reduction.

e. Adding these half-equations and cancelling the electrons results in the overall redox reaction.

WRITE

a. The magnesium electrode is the anode and site of oxidation, since it is known to be the negative electrode. Electrons are produced at the magnesium electrode and consumed at the iron electrode. Therefore, magnesium is being oxidised and iron(II) ions are being reduced.

b. Magnesium is the anode and iron is the cathode.

c. Electrons flow from the magnesium electrode through the wire to the iron electrode.

d. $Mg(s) \rightarrow Mg^{2+}(aq) + 2e^-$ (oxidation)
$Fe^{2+}(aq) + 2e^- \rightarrow Fe(s)$ (reduction)

e. $Mg(s) + Fe^{2+}(aq) \rightarrow Mg^{2+}(aq) + Fe(s)$

PRACTICE PROBLEM 4

When zinc metal is dipped into a solution of silver nitrate, it forms a coating of silver, as shown in the figure.

Use this reaction to draw a diagram of a galvanic cell, using KNO_3 in the salt bridge and zinc sulfate as one of the electrolytes. Complete the following steps to construct your diagram.

a. Draw the two half-cells.
b. Write half-equations for the oxidation and reduction reactions.
c. Write the overall cell reaction.
d. Label the flow of:
   - electrons in the wire
   - anions in the salt bridge
   - cations in the salt bridge.
e. Label the anode and the cathode.

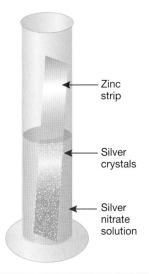

← Zinc strip

← Silver crystals

← Silver nitrate solution

TIP: In the internal circuit, anions always travel towards the anode and cations travel towards the cathode.

---

elog-1934

tlvd-9715

EXPERIMENT 3.1   online only

Investigating the Daniell cell

Aim

To set up and observe the operation of a Daniell cell

3.3.2 The electrochemical series

You may recall studying the reactivity of metals in Unit 2. The more reactive a metal is, the more likely it is to lose electrons and become a positive ion.

Zinc is more reactive than copper. This means is it much more inclined to lose two electrons (figure 3.19), to become a zinc ion, than copper is. In other words, zinc is better at undergoing oxidation than copper is.

When we look at the electrochemical series, you might notice all of the reactions are written in the same format from left to right, and the use of reversible arrows between the reactant(s) and product(s).

$$Cu^{2+}(aq) + 2e^- \rightleftharpoons Cu(s)$$

$$Zn^{2+}(aq) + 2e^- \rightleftharpoons Zn(s)$$

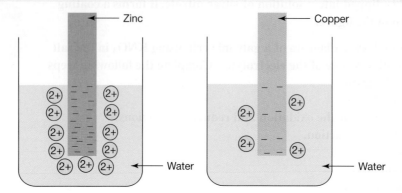

FIGURE 3.19 The build-up of negative charge on zinc is greater than copper due to it being a stronger oxidising agent and more inclined to lose electrons.

The reversible arrows indicate that it is possible for the positive ions to take back electrons and become metal again. You will learn more about reversible reactions in Outcome 2. For the purpose of this topic, we are looking at how well different chemicals give and take electrons relative to each other.

Standard electrode potentials (E^0)

The flow of electrons created by different half-cell combinations varies, and can be measured with a **voltmeter**. This is called the **cell potential difference**, cell potential or cell voltage.

We can't measure the **electrical potential** of an individual electrode/electrolyte half-cell directly, but we can compare the potential differences of different half-cell combinations and determine a **standard electrode potential** for individual half-cells.

Conditions for standard electrode potentials (E^0)

E^0 is measured at SLC using 1 M electrolyte concentrations.

---

voltmeter a device used for measuring the potential difference between two points in a circuit

cell potential difference the difference between the reduction potentials of two half-cells

electrical potential the ability of a galvanic cell to produce an electric current

standard electrode potential the voltage or potential difference due to the difference in charge on the electrode and electrolyte compared to the hydrogen half-cell

reduction potential a measure of the tendency of an oxidising agent to accept electrons and so undergo reduction

Think of it like this. Your Year 9 Physical Education teacher may have lined you up from tallest to shortest for the purpose of picking teams for a game. The teacher didn't measure the heights of each student individually, but they used the difference in height relative to each other to put them in the correct order.

The electrochemical series shows the oxidising species on the left and the reducing species on the right, and it ranks, or lines up, the **reduction potential** of half-cell reactions from highest to lowest. In other words, the chemical reactant best at taking electrons (the strongest oxidising agent) is at the top of the table. This may be the opposite to the reactivity series of metals you were shown in Unit 2. The most reactive metals, like lithium, sodium and potassium, are at the bottom of the electrochemical series, because their ions do not act as oxidising agents in spontaneous reactions.

Let's take a look at a small section of the electrochemical series (figure 3.20), omitting the E^0 values but retaining the relative position of each reaction.

For a spontaneous reaction to occur, the oxidising agent on the left-hand side of the arrows must be higher in the series than the reducing agent on the right-hand side.

In figure 3.20, Ag^+ is the highest ranked (strongest) oxidising agent. This means the silver ion is better at being reduced than Cu^{2+}, H^+ and Zn^{2+}. Since solid zinc, $Zn(s)$, is the strongest reducing agent shown, it is more inclined to lose electrons than H_2, Cu or Ag.

FIGURE 3.20 Modified electrochemical series

$$Ag^+(aq) + e^- \rightleftharpoons Ag$$

$$Cu^{2+}(aq) + 2e^- \rightleftharpoons Cu(s)$$

$$2H^+(aq) + 2e^- \rightleftharpoons H_2(g)$$

$$Zn^{2+}(aq) + 2e^- \rightleftharpoons Zn(s)$$

Therefore, the largest cell potential difference occurs between the half-cells containing silver ions and zinc metal. A galvanic cell constructed from Zn/Zn^{2+} and Ag^+/Ag half-cells will produce a higher cell voltage than other combinations listed in figure 3.20.

The **standard cell potential difference** (E^0_{cell}) is the measured cell potential difference, under standard conditions, when the concentration of each species in solution is 1 M, the pressure of a gas (where applicable) is 100 kPa and the temperature is 25 °C (298 K).

Standard hydrogen electrode

To obtain a comparative measure of the reduction potentials of different half-cells, the **standard hydrogen half-cell** is used as a standard reference electrode. This allows the determination of a redox half-cell's ability to accept electrons. It consists of hydrogen gas bubbling around an inert platinum electrode in a solution of hydrogen ions (see figure 3.21). The standard hydrogen half-cell is arbitrarily assigned a standard reduction potential of 0.00 V at 25 °C. The reaction that occurs at the electrode surface is:

$$2H^+(aq) + 2e^- \rightleftharpoons H_2(g) \quad E^0 = 0.00\,V$$

FIGURE 3.21 A diagram of a hydrogen half-cell

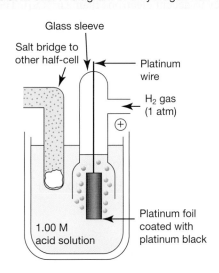

- Glass sleeve
- Salt bridge to other half-cell
- Platinum wire
- H_2 gas (1 atm)
- $\oplus$
- Platinum foil coated with platinum black
- 1.00 M acid solution

FIGURE 3.22 A hydrogen half-cell

The standard hydrogen electrode is used with other half-cells so that the reduction potentials of those cells can be measured. If a species accepts electrons more easily than hydrogen ions do, its electrode potential is *positive*. If it accepts electrons less easily than hydrogen ions do, its electrode potential is *negative*.

TIP: If a half-equation has electrons on the reactant (left-hand) side, it is a reduction half-equation.

When a standard hydrogen half-cell is connected to a standard $Cu^{2+}(aq)/Cu(s)$ half-cell, the voltmeter measures a potential difference of 0.34 volts and the copper electrode is positive (see figure 3.23). This means electrons flow towards the $Cu^{2+}(aq)/Cu(s)$ half-cell, so $Cu^{2+}(aq)$ has a greater tendency to accept electrons and is a stronger oxidising agent than $H^+(aq)$. Therefore, the measured E^0 value for the half-cell reaction

$$Cu^{2+}(aq) + 2e^- \rightleftharpoons Cu(s)$$

is positive in sign and equal to +0.34 volts.

standard cell potential difference the measured cell potential difference, under standard conditions, when the concentration of each species in solution is 1 M, the pressure of a gas (where applicable) is 100 kPa and the temperature is 25 °C (298 K)

standard hydrogen half-cell a standard reference electrode; it is assigned 0.00 volts

FIGURE 3.23 Measuring the standard half-cell potential of a Cu^{2+}/Cu half-cell

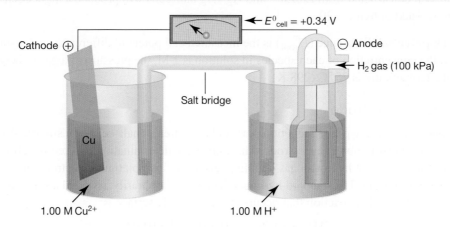

When a standard hydrogen half-cell is connected to a standard $Zn^{2+}(aq)/Zn(s)$ half-cell, the voltmeter measures a potential difference of 0.76 volts and the Zn electrode is negative (see figure 3.24). This means electrons flow to the $H^+(aq)/H_2(g)$ half-cell and $H^+(aq)$ has a greater tendency to accept electrons than $Zn^{2+}(aq)$. Therefore, the measured E^0 value for the half-cell reaction

$$Zn^{2+}(aq) + 2e^- \rightleftharpoons Zn(s)$$

is negative in sign and equal to –0.76 volts.

FIGURE 3.24 Measuring the standard half-cell potential of a Zn^{2+}/Zn half-cell

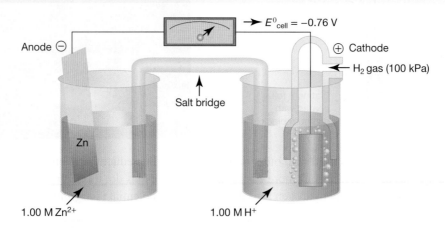

Half-cell potentials are often listed in a table such as table 3.4. These tables may be referred to as tables of standard electrode potentials or standard reduction potentials, and may also be called an **electrochemical series**. The half-cell potentials are usually arranged from the largest E^0 value to the smallest.

> **electrochemical series** a series of chemical half-equations arranged in order of their standard electrode potentials

Calculating standard cell potential difference

$$E^0_{cell} = E^0_{oxidising\ agent} - E^0_{reducing\ agent}$$

Half-cell E^0 values are measured against the standard hydrogen half-cell, which is arbitrarily assigned 0.00 volts.

TIP: An electrochemical series table can be found in the VCE Chemistry Data Book.

TABLE 3.4 The electrochemical series

	Half-reaction		E^0 (volts)
Strongest oxidising agent ↑	$F_2(g) + 2e^- \rightleftharpoons 2F^-(aq)$	Weakest reducing agent	+2.87
	$H_2O_2(aq) + 2H^+(aq) + 2e^- \rightleftharpoons 2H_2O(l)$		+1.77
	$MnO_4^-(aq) + 8H^+(aq) + 5e^- \rightleftharpoons Mn^{2+}(aq) + 4H_2O(l)$		+1.52
	$PbO_2(s) + 4H^+(aq) + 2e^- \rightleftharpoons Pb^{2+}(aq) + 2H_2O(l)$		+1.46
	$Cl_2(g) + 2e^- \rightleftharpoons 2Cl^-(aq)$		+1.36
	$Cr_2O_7^{2-}(aq) + 14H^+(aq) + 6e^- \rightleftharpoons 2Cr^{3+}(aq) + 7H_2O(l)$		+1.33
	$O_2(g) + 4H^+(aq) + 4e^- \rightleftharpoons 2H_2O(l)$		+1.23
	$Br_2(l) + 2e^- \rightleftharpoons 2Br^-(aq)$		+1.09
	$NO_3^-(aq) + 4H^+(aq) + 3e^- \rightleftharpoons NO(g) + 2H_2O(l)$		+0.95
	$NO_3^-(aq) + 2H^+(aq) + e^- \rightleftharpoons NO_2(g) + H_2O(l)$		+0.81
	$Ag^+(aq) + e^- \rightleftharpoons Ag(s)$		+0.80
	$Fe^{3+}(aq) + e^- \rightleftharpoons Fe^{2+}(aq)$		+0.77
	$O_2(g) + 2H^+(aq) + 2e^- \rightleftharpoons H_2O_2(l)$		+0.68
	$I_2(s) + 2e^- \rightleftharpoons 2I^-(aq)$		+0.54
	$O_2(g) + 2H_2O(l) + 4e^- \rightleftharpoons 4OH^-(aq)$		+0.40
Increasing oxidising strength	$Cu^{2+}(aq) + 2e^- \rightleftharpoons Cu(s)$	Increasing reducing strength	+0.34
	$SO_4^{2-}(aq) + 4H^+(aq) + 2e^- \rightleftharpoons SO_2(g) + 2H_2O(l)$		+0.20
	$Sn^{4+}(aq) + 2e^- \rightleftharpoons Sn^{2+}(aq)$		+0.15
	$S(s) + 2H^+(aq) + 2e^- \rightleftharpoons H_2S(g)$		+0.14
	$2H^+(aq) + 2e^- \rightleftharpoons H_2(g)$ (defined)		0.00
	$Pb^{2+}(aq) + 2e^- \rightleftharpoons Pb(s)$		−0.13
	$Sn^{2+}(aq) + 2e^- \rightleftharpoons Sn(s)$		−0.14
	$Ni^{2+}(aq) + 2e^- \rightleftharpoons Ni(s)$		−0.25
	$PbSO_4(s) + 2e^- \rightleftharpoons Pb(s) + SO_4^{2-}(aq)$		−0.36
	$Fe^{2+}(aq) + 2e^- \rightleftharpoons Fe(s)$		−0.44
	$Zn^{2+}(aq) + 2e^- \rightleftharpoons Zn(s)$		−0.76
	$2H_2O(l) + 2e^- \rightleftharpoons H_2(g) + 2OH^-(aq)$		−0.83
	$Al^{3+}(aq) + 3e^- \rightleftharpoons Al(s)$		−1.66
	$Mg^{2+}(aq) + 2e^- \rightleftharpoons Mg(s)$		−2.37
	$Na^+(aq) + e^- \rightleftharpoons Na(s)$		−2.71
	$Ca^{2+}(aq) + 2e^- \rightleftharpoons Ca(s)$	Strongest reducing agent	−2.87
Weakest oxidising agent	$K^+(aq) + e^- \rightleftharpoons K(s)$		−2.93
	$Li^+(aq) + e^- \rightleftharpoons Li(s)$		−3.04

Note: Standard electrode reduction potentials at a temperature of 25 °C, a pressure of 100 kPa and a concentration of 1 M for all aqueous species.

on Resources

Video eLessons	Galvanic cells 3 (eles-2596) Galvanic cells and measuring cell potential (eles-0436)
Interactivity	Electrochemical series (int-1256)

Use of standard half-cell reduction potentials

In galvanic cells, at least two oxidising agents and two reducing agents are present. The following procedure can be useful in predicting which spontaneous reaction occurs in a galvanic cell. The Daniell cell is used as an example.

1. Write the half-equations occurring in the galvanic cell in descending order of E^0. For example:

$$Cu^{2+}(aq) + 2e^- \rightleftharpoons Cu(s) \qquad E^0 = +0.34 \text{ V}$$
$$Zn^{2+}(aq) + 2e^- \rightleftharpoons Zn(s) \qquad E^0 = -0.76 \text{ V}$$

2. Circle all the species present in the galvanic cell that could participate.

$$\boxed{Cu^{2+}(aq)} + 2e^- \rightleftharpoons Cu(s) \qquad E^0 = +0.34 \text{ V}$$
$$Zn^{2+}(aq) + 2e^- \rightleftharpoons \boxed{Zn(s)} \qquad E^0 = -0.76 \text{ V}$$

3. Select the oxidising agent with the highest E^0. This is reduced at the cathode, which accepts electrons more easily than an oxidising agent with a lower E^0.

$$Cu^{2+}(aq) + 2e^- \rightarrow Cu(s) \quad E^0 = +0.34 \text{ V}$$

4. Select the reducing agent with the lowest E^0. This is oxidised at the anode, which donates electrons more easily than a reducing agent with a higher E^0 . Write this equation as an oxidation equation — that is, reverse it.

$$Zn(s) \rightarrow Zn^{2+}(aq) + 2e^-$$

5. If the number of electrons in each half-equation is different, multiply each half-equation by an appropriate factor so that the number of electrons is the same (this ensures they cancel out in the next step).
6. Write the full equation by adding the two reactions together. If there are electrons or other species (such as hydrogen or water) that appear on both sides of the equation, cancel the same number of each from both sides.

$$Cu^{2+}(aq) + Zn(s) \rightarrow Cu(s) + Zn^{2+}(aq)$$

7. Determine the cell potential difference by using the following formula:

$$\begin{aligned}\text{Cell potential difference} &= E^0_{\text{oxidising agent}} - E^0_{\text{reducing agent}} \\ &= +0.34 - (-0.76) \\ &= +1.10 \text{ V}\end{aligned}$$

In summary, when the electrode reactions are written in descending order of E^0 values, the strongest oxidising agent (highest E^0) on the left-hand side of the equation reacts with the strongest reducing agent (lowest E^0) on the right.

Predicting the products of redox reactions

In the E^0 table, the strongest oxidising agent (upper left) reacts with the strongest reducing agent (lower right).

TIP: The two half-equations of a spontaneous reaction will always make a clockwise circle in the electrochemical series in the VCE Chemistry Data Book. The largest clockwise circle from the available reactants is the one that will occur and will have the greatest cell potential.

Consider the following half-equations, in which all present species have been circled:

$$Fe^{3+}(aq) + e^- \rightleftharpoons Fe^{2+}(aq) \qquad E^0 = +0.77$$
$$Cu^{2+}(aq) + 2e^- \rightleftharpoons Cu(s) \qquad E^0 = +0.34$$
$$2H^+(aq) + 2e^- \rightleftharpoons H_2(g) \qquad E^0 = 0.00$$
$$Pb^{2+}(aq) + 2e^- \rightleftharpoons Pb(s) \qquad E^0 = -0.13$$
$$Sn^{2+}(aq) + 2e^- \rightleftharpoons Sn(s) \qquad E^0 = -0.14$$

The two possible half-equations that form the largest clockwise circle will occur. In this case, $Fe^{3+}(aq)$ will be reduced to $Fe^{2+}(aq)$, Pb(s) will be oxidised to $Pb^{2+}(aq)$, and the cell potential will be +0.90 V.

$$Fe^{3+}(aq) + e^- \rightarrow Fe^{2+}(aq) \qquad E^0 = +0.77$$
$$Cu^{2+}(aq) + 2e^- \rightleftharpoons Cu(s) \qquad E^0 = +0.34$$
$$2H^+(aq) + 2e^- \rightleftharpoons H_2(g) \qquad E^0 = 0.00$$
$$Pb^{2+}(aq) + 2e^- \leftarrow Pb(s) \qquad E^0 = -0.13$$
$$Sn^{2+}(aq) + 2e^- \rightleftharpoons Sn(s) \qquad E^0 = -0.14$$

Calculating E^0

Determine the potential difference (E^0) of a cell by finding the difference between the E^0 of the reducing agent and the E^0 of the oxidising agent. Subtract the less positive number from the more positive number and always show a + in front of the value calculated.

FIGURE 3.25 Whatever the cell, reduction always occurs at the cathode (RedCat) and oxidation always occurs at the anode (AnOx).

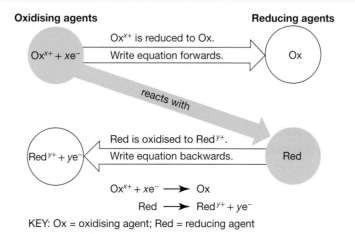

KEY: Ox = oxidising agent; Red = reducing agent

tlvd-9678

SAMPLE PROBLEM 5 Determining the best oxidising and best reducing agents

The standard half-cell potentials of some metal–metal ion half-cells are as follows:

Half-cell	E^0 (volts)
$Ag^+(aq)/Ag(s)$	+0.80
$Co^{2+}(aq)/Co(s)$	−0.28
$Ba^{2+}(aq)/Ba(s)$	−2.90

Determine which species is the best oxidising agent and which is the best reducing agent.

THINK

1. Cations are formed when atoms donate electrons. Therefore, cations can readily accept electrons back, reducing them and acting as oxidising agents.

2. Atoms are formed when cations accept electrons. Therefore, atoms can readily donate electrons back, oxidising them and acting as reducing agents.

3. In a conventional table of standard half-cell reduction potentials, the strongest oxidising agent has the most positive E^0 value, while the strongest reducing agent has the most negative E^0 value.
   TIP: You must clearly identify whether the atom or the ion is the strongest when describing the reaction occurring.

WRITE

$Ag^+(aq)$, $Co^{2+}(aq)$ and $Ba^{2+}(aq)$ are all oxidising agents.

$Ag(s)$, $Co(s)$ and $Ba(s)$ are all reducing agents.

$Ag^+(aq)$ is the strongest oxidising agent, while $Ba(s)$ is the strongest reducing agent.

PRACTICE PROBLEM 5

Consider the following conjugate redox pairs and their E^0 values.

Conjugate redox pair	E^0 (volts)
$Cl_2(g)/Cl^-(aq)$	+1.36
$I_2(s)/I^-(aq)$	+0.54
$Al^{3+}(aq)/Al(s)$	−1.66
$MnO_4^-(aq)/Mn^{2+}(aq)$	+1.52
$Pb^{2+}(aq)/Pb(s)$	−0.13

a. Which species is:
   i. the strongest oxidising agent
   ii. the strongest reducing agent
   iii. the weakest oxidising agent
   iv. the weakest reducing agent?
b. Write fully balanced half-equations for each conjugate redox pair.

SAMPLE PROBLEM 6 Predicting if a redox reaction is spontaneous

Given the two half-equations

$$Br_2(l) + 2e^- \rightleftharpoons 2Br^-(aq) \quad E^0 = +1.09 \text{ V}$$
$$Mg^{2+}(aq) + 2e^- \rightleftharpoons Mg(s) \quad E^0 = -2.37 \text{ V}$$

predict the likely spontaneous redox reaction.

THINK	WRITE
1. For a redox reaction to occur, a reducing agent must react with an oxidising agent.	$Br_2(l)$ does not react with $Mg^{2+}(aq)$ (because they can act only as oxidising agents) and $Br^-(aq)$ does not react with $Mg(s)$ (because they can act only as reducing agents).
2. For a spontaneous redox reaction to occur, the E^0 of the oxidising agent must be more positive than the E^0 of the reducing agent. **TIP:** In galvanic cells the most positive half-equation gets reduced (forward reaction in the table).	$Br_2(l)$ reacts spontaneously with Mg.

PRACTICE PROBLEM 6

Given the two half-equations

$$Fe^{3+}(aq) + e^- \rightleftharpoons Fe^{2+}(aq) \quad E^0 = +0.77 \text{ V}$$
$$Ni^{2+}(aq) + 2e^- \rightleftharpoons Ni(s) \quad E^0 = -0.25 \text{ V}$$

identify the anode and cathode, write the overall equation and calculate the standard cell potential that would be produced in a galvanic cell made from these half-cells.

TIP: When writing balanced chemical equations from redox half-equations, make sure to cancel out common species.

 Resources

 Video eLesson Predicting products of redox reactions (eles-3238)

EXPERIMENT 3.2

`online only`

Predicting redox reactions

Aim

To predict whether redox reactions will occur when a range of oxidising agents and reducing agents are mixed, and then to test these predictions experimentally

Galvanic cells and redox predictions

Aim

To set up galvanic cells, measure the cell voltages and predict the relative oxidising–reducing strength of four redox pairs

Limitations of the electrochemical series

A table such as the electrochemical series in table 3.4 provides a great deal of information about a redox reaction. This includes:

- determination of the relative strengths of oxidising agents and reducing agents
- prediction of whether a redox reaction will occur, whether by direct contact or in a suitably designed galvanic cell
- prediction of the overall reaction occurring in a cell and the potential difference of that cell.

However, the table does not tell us the rate of a reaction or if intermediates form. E^0 values and their order are temperature dependent, and the E^0 table predicts reactions only at standard conditions of 25 °C, 100 kPa and 1 M concentration for solutions.

It may be predicted that a redox reaction is possible between two reactants, but no reaction may be observed if the reaction proceeds very slowly. Redox predictions may be checked by experiment. Non-standard conditions may change redox reaction E values. If the redox reaction conditions deviate significantly from those at which standard electrode potentials are measured, the relative order of redox conjugate pairs in the table of standard electrode potentials may be altered. This could mean that previously favourable reactions become unfavourable under the new conditions.

EXTENSION: Cell potential in non-standard conditions

The potential of half-cells and overall redox reactions can be predicted in non-standard conditions by using an equation derived from the free-energy equation $\Delta G = -nFE$, where n is the number of moles of electrons, F is Faraday's constant (96 500 C mol^{-1}) and E is the potential difference.

The following derived equation is known as the Nernst equation, named after the German scientist Walther Nernst:

$$E_{cell} = E^0{}_{cell} - \frac{RT}{nF} \ln Q$$

where:

R = the gas constant (8.31 J K^{-1} mol^{-1})

T = temperature in kelvin

$E^0{}_{cell}$ is the cell potential at SLC with standard 1 M concentrations.

This can be simplified further at a constant temperature of 298 K to:

$$E_{cell} = E^0{}_{cell} - \frac{0.0592 \text{ V}}{n} \log Q$$

Q is the reaction quotient of the electrolyte product concentration and reactant concentrations raised to the power of the coefficients in the balanced equation. You will learn more about the reaction quotient, Q, in Unit 3, Area of Study 2.

$$Q = \frac{[\text{initial product}]^{\text{coefficent}}}{[\text{initial reactant}]^{\text{coefficent}}}$$

So, using these equations, the effects of changing electrolyte concentration and temperature away from the standard conditions can be predicted.

For example, what would the cell potential of a Daniell cell be if the initial Cu^{2+} ion concentration was 5.0 M instead of 1 M, and the initial Zn^{2+} concentration was 0.20 M instead of 1 M?

$$Zn(s) + Cu^{2+}(aq) \rightleftharpoons Zn^{2+}(aq) + Cu(s) \quad E^0 = 1.10 \text{ V}$$

From the electrochemical series, the number of moles of electrons transferred in the balanced equation is two. The standard cell potential is found by $E^0_{cell} = +0.34 - 0.76 = 1.10$ V, and the coefficients in front of the reactant and product electrolytes are both 1.

Substituting the values into the Nernst equation results gives:

$$\begin{aligned} E_{cell} &= 1.10 - \frac{0.0592}{2} \log \frac{[0.20]}{[5.0]} \\ &= 1.10 - 0.0296 \times -1.40 \\ &= 1.10 + 0.040 \\ &= 1.14 \text{ V} \end{aligned}$$

As the reaction proceeds, more Zn^{2+} is made and Cu^{2+} is used up. Q gets larger and E_{cell} gets smaller.

3.3 Activities

learn on

Students, these questions are even better in jacPLUS

 Receive immediate feedback and access sample responses

 Access additional questions

 Track your results and progress

Find all this and MORE in jacPLUS

3.3 Quick quiz on	3.3 Exercise	3.3 Exam questions

3.3 Exercise

1. Why is a salt bridge or porous barrier used to connect two half-cells in a galvanic cell?
2. A student was doing an experiment in the school laboratory. She placed a fresh piece of zinc metal into a beaker of silver nitrate solution and left it to stand for a short period of time. She then noted the following observations:
   - The temperature of the solution increased.
   - The zinc metal became coated with silver.

   a. Write the ionic equation for the reaction occurring in the beaker.
   b. Draw a galvanic cell that allows the energy released during the reaction to be readily used. On your diagram, identify the anode, the cathode and the polarity of these electrodes.
   c. Write half-equations for the reactions occurring at each electrode.
   d. Explain the significance of the increase in temperature of the solution.

3. A galvanic cell was set up by combining half-cells containing zinc and magnesium electrodes dipped into the appropriate sulfate solutions. A conducting wire and a salt bridge completed the circuit. After three hours, the two electrodes were removed and weighed. The mass of the Zn electrode had increased, while the mass of the Mg electrode had decreased. Draw this galvanic cell, clearly indicating the following.
   a. The anode and the cathode
   b. The ions present in the half-cells
   c. The electrolyte in the salt bridge
   d. Anion and cation flow within the salt bridge
   e. The direction of the flow of electrons
   f. The anode reaction and the cathode reaction
   g. The oxidation reaction and the reduction reaction
   h. The overall cell reaction
   i. The oxidising agent and the reducing agent
4. Write the likely spontaneous redox reactions that would occur given the following half-equations.
   a. $Cl_2(g) + 2e^- \rightleftharpoons 2Cl^-(aq)$ $E^0 = +1.36\,V$
      $Ni^{2+}(aq) + 2e^- \rightleftharpoons Ni(s)$ $E^0 = -0.23\,V$

   b. $Al^{3+}(aq) + 3e^- \rightleftharpoons Al(s)$ $E^0 = -1.67\,V$
      $Mg^{2+}(aq) + 2e^- \rightleftharpoons Mg(s)$ $E^0 = -2.34\,V$

   c. $MnO_4^-(aq) + 8H^+(aq) + 5e^- \rightleftharpoons Mn^{2+}(aq) + 4H_2O(l)$ $E^0 = +1.52\,V$
      $ClO_4^- + 2H^+(aq) + 2e^- \rightleftharpoons ClO_3^- + H_2O(l)$ $E^0 = +1.19\,V$

   d. $Fe^{2+}(aq) + 2e^- \rightleftharpoons Fe(s)$ $E^0 = -0.44\,V$
      $MnO_4^-(aq) + 8H^+(aq) + 5e^- \rightleftharpoons Mn^{2+}(aq) + 4H_2O(l)$ $E^0 = +1.52\,V$
5. Suggest two reasons why predicted spontaneous redox reactions may not be observed.

3.3 Exam questions

Question 1 (1 mark)

Source: VCE 2021 Chemistry Exam, Section A, Q.26; © VCAA

MC Different metal ion (aq)/metal (s) half-cells are combined with an In^{3+}(aq)/In(s) half-cell to create a galvanic cell at SLC, as shown in the diagram below. The equation for the In^{3+}(aq)/In(s) half-cell is

$$In^{3+}(aq) + 3e^- \rightleftharpoons In(s) \qquad E^0 = -0.34\,V$$

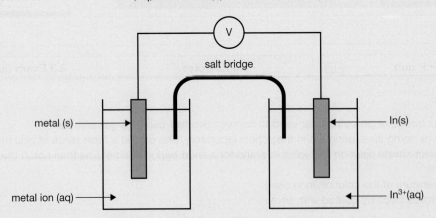

Which of the following shows the half-cells in decreasing order of voltage produced when combined with the In^{3+}(aq)/In(s) half-cell and In(s) is the negative electrode?

A. Mn^{2+}(aq)/Mn(s), Al^{3+}(aq)/Al(s), Mg^{2+}(aq)/Mg(s)

B. Mg^{2+}(aq)/Mg(s), Al^{3+}(aq)/Al(s), Mn^{2+}(aq)/Mn(s)

C. Cu^{2+}(aq)/Cu(s), Pb^{2+}(aq)/Pb(s), Ni^{2+}(aq)/Ni(s)

D. Ni^{2+}(aq)/Ni(s), Pb^{2+}(aq)/Pb(s), Cu^{2+}(aq)/Cu(s)

Question 2 (1 mark)

Source: VCE 2022 Chemistry Exam, Section A, Q.18; © VCAA

MC A student wants to investigate a galvanic cell consisting of Sn^{4+}/Sn^{2+} and Ag^+/Ag half-cells.

Which one of the following combinations of electrodes and solutions will produce an operational galvanic cell?

	Sn^{4+}/Sn^{2+} half-cell		Ag^+/Ag half-cell	
	Electrode	**Solution(s)**	**Electrode**	**Solution(s)**
A.	Sn	1 M $Sn(NO_3)_2$	graphite	1 M $AgNO_3$
B.	Sn	1 M $Sn(NO_3)_4$, 1 M $Sn(NO_3)_2$	graphite	1 M $AgNO_3$
C.	graphite	1 M $Sn(NO_3)_4$, 1 M $Sn(NO_3)_2$	Ag	1 M $AgNO_3$
D.	graphite	1 M $Sn(NO_3)_4$	Ag	1 M $AgNO_3$

Question 3 (1 mark)

Source: VCE 2020 Chemistry Exam, Section A, Q.3; © VCAA

MC A diagram of an electrochemical cell is shown below.

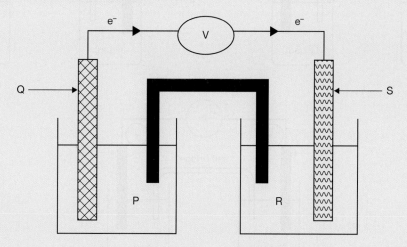

Which of the following gives the correct combination of the electrode in the oxidation half-cell and the electrolyte in the reduction half-cell?

	Electrode (oxidation half-cell)	**Electrolyte (reduction half-cell)**
A.	S	P
B.	S	R
C.	Q	R
D.	Q	P

Question 4 (1 mark)

Source: VCE 2020 Chemistry Exam, Section A, Q.30; © VCAA

MC Consider the following half-equation.

$$ClO_2(g) + e^- \rightleftharpoons ClO_2^-(aq)$$

It is also known that:
- $ClO_2(g)$ will oxidise $HI(aq)$, but not $HCl(aq)$
- $Fe^{3+}(aq)$ will oxidise $HI(aq)$, but not $NaClO_2(aq)$.

Based on this information, $Fe^{2+}(aq)$ can be oxidised by
- **A.** $Cl_2(g)$ and $I_2(aq)$.
- **B.** $Cl_2(g)$, but not $ClO_2(g)$.
- **C.** $ClO_2(g)$ and $Cl_2(g)$, but not $I_2(aq)$.
- **D.** $Cl_2(g)$, $ClO_2(g)$ and $I_2(aq)$.

Source: VCE 2018 Chemistry Exam, Section A, Q.29; © VCAA

MC The following diagrams represent combinations of four galvanic half-cells (G/G^{2+}, J/J^{2+}, Q/Q^{2+} and R/R^{2+}) that were investigated under standard conditions.

Each half-cell consisted of a metal electrode placed in a 1.0 M nitrate solution of the respective metal ion.

The diagrams show the polarity of the electrodes in each half-cell, as determined using an ammeter.

The results were then used to determine the order of the E^0 values of the half-reactions.

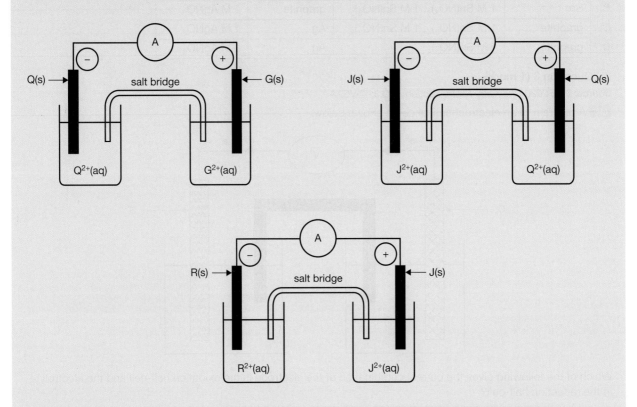

Which of the following indicates the order of the half-cell reactions, from the lowest E^0 value to the highest?

A. J/J^{2+}, R/R^{2+}, G/G^{2+}, Q/Q^{2+}

B. Q/Q^{2+}, G/G^{2+}, R/R^{2+}, J/J^{2+}

C. R/R^{2+}, J/J^{2+}, Q/Q^{2+}, G/G^{2+}

D. G/G^{2+}, Q/Q^{2+}, J/J^{2+}, R/R^{2+}

More exam questions are available in your learnON title.

3.4 Energy from primary cells and fuel cells

Primary galvanic cells and fuel cells are applications of redox reactions that allow us to supply electrical energy. Primary cells store the reactants and the products of redox reactions within. They are single-use, and are discarded once the voltage becomes too low to generate electricity. Fuel cells work so long as reactants are continually supplied from outside the cell, and the products are removed.

3.4.1 Primary cells

The electric potential difference created by the chemical reactions at the anode and cathode in cells and batteries drives the movement of electrons. The electrons build up at the anode and are drawn to the cathode, but they cannot do so by travelling through the electrolytic material inside the battery itself. Instead, the electrons flow easily through a conducting wire connecting the anode to the cathode, allowing them to reach the cathode and balance the charges within the cell or battery.

Eventually, the chemical processes creating the surplus of electrons in the anode come to a stop as the reactants are used up, and the battery dies.

There are different types of **primary cells**. The most common are the dry cell, the alkaline zinc/manganese dioxide cell and **lithium cells**.

The dry cell

An electrochemical cell in which the electrolyte is a paste, rather than a liquid, is known as a **dry cell** or **Leclanché cell**. The most commonly used dry cells are C, D or AA batteries, which have a voltage of 1.5 V. Dry cells are commonly used in torches, toys and transistor radios because they are cheap, small, reliable and easy to use. The oxidising agents and reducing agents used in such cells should:

- be far enough apart in the electrochemical series to produce a useful voltage from the cell
- not react with water in the electrolyte too quickly, or they will discharge early (therefore, highly reactive metals such as sodium, potassium and calcium are not found in such batteries)
- be inexpensive.

FIGURE 3.26 Arrangement of cells in circuits: **a.** Two 1.5 V cells connected in series make a 3.0 V battery and **b.** two 1.5 V cells connected in parallel allow a higher current at 1.5 V.

a.

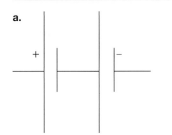

b.

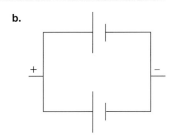

primary cell an electrolytic cell in which the cell reaction is not reversible

lithium cells cells that use lithium anodes and can produce a high voltage

dry cell an electrochemical cell in which the electrolyte is a paste, rather than a liquid; also called a Leclanché cell

Leclanché cell *see* **dry cell**

A dry cell consists of a zinc container filled with an electrolyte paste. This paste contains manganese(IV) oxide, MnO_2, zinc chloride, $ZnCl_2$, ammonium chloride, NH_4Cl, and water. A carbon rod is embedded in the paste and forms the cathode. The zinc container is the anode. The thick paste prevents the contents of the cell from mixing, so a salt bridge is not needed. Intermittent use or slight warming of the cell prevents the build-up of these products around the electrodes, increasing the life of the cell. Once the materials around the electrodes have been used up, the cell stops operating.

The electrode half-equations are as follows:

Anode (oxidation):

$$Zn(s) \rightarrow Zn^{2+}(aq) + 2e^-$$

Cathode (reduction):

$$2MnO_2(s) + 2NH_4^+(aq) + 2e^- \rightarrow Mn_2O_3(s) + 2NH_3(aq) + H_2O(l)$$

The overall cell reaction can be written as:

$$2MnO_2(s) + 2NH_4^+(aq) + Zn(s) \rightarrow Mn_2O_3(s) + 2NH_3(aq) + Zn^{2+}(aq) + H_2O(l)$$

FIGURE 3.27 a. Dry cells in a torch and b. a simplified cross-section of a dry cell

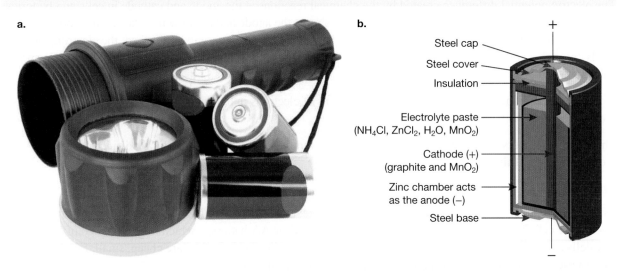

a.

b.

Steel cap
Steel cover
Insulation

Electrolyte paste
(NH_4Cl, $ZnCl_2$, H_2O, MnO_2)

Cathode (+)
(graphite and MnO_2)

Zinc chamber acts
as the anode (−)

Steel base

EXPERIMENT 3.4 on line only

Looking at a dry cell

Aim

To examine the contents of a dry cell and to investigate the redox reactions occurring in the cell

The alkaline zinc/manganese dioxide cell

Alkaline cells were developed as a consequence of the greater demand for a higher-capacity portable energy source than the dry cell could provide. An alkaline cell is designed to give a greater current output than the standard dry cell and the voltage output falls off more slowly. Alkaline batteries also have longer shelf lives than dry cells. Less electrolyte needs to be used in an alkaline cell than in a dry cell, which means that more electrode reactants can be packed into the cell. Alkaline cells are commonly used in a variety of handheld devices including remote controls, high-drain toys and head torches.

The alkaline cell is a primary cell that is more expensive than a dry cell, but lasts much longer.

The alkaline cell contains similar components to the dry cell, but has a powdered zinc anode in an electrolyte paste of potassium hydroxide. The cathode is a compressed mixture of manganese dioxide and graphite. A separator, consisting of a porous fibre soaked in electrolyte, prevents mixing of the anode and cathode components. The cell is contained within a steel shell.

FIGURE 3.28 a. Alkaline zinc/manganese dioxide cells and **b.** a simplified cross-section of an alkaline zinc/manganese dioxide cell

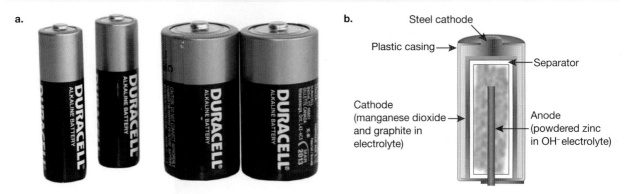

The electrode half-equations are as follows:

Anode (oxidation):

$$Zn(s) + 2OH^-(aq) \rightarrow Zn(OH)_2(s) + 2e^-$$

Cathode (reduction):

$$2MnO_2(s) + 2H_2O(l) + 2e^- \rightarrow 2MnO(OH)(s) + 2OH^-(aq)$$

The overall cell reaction is:

$$2MnO_2(s) + 2H_2O(l) + Zn(s) \rightarrow 2MnO(OH)(s) + Zn(OH)_2(s)$$

The alkaline cell has a voltage of 1.55 V, but this drops slowly with use. Although it may last up to five times longer than a dry cell, it is more difficult to make and more expensive. Both the dry cell and the alkaline cell are bulky, making them unsuitable for smaller devices such as watches and calculators.

Lithium batteries

Lithium cells are cells based on lithium anodes. Lithium is a very reactive metal, and also very light, so these batteries can produce a high cell voltage. They require a more robust construction and are far more expensive than common batteries, but have a shelf life of ten years. Owing to their relatively long shelf life, lithium cells are mainly used as power sources for electronic memory, but may also be used in electronic switchboards, navigation systems, industrial clocks and even poker machines. In many applications lithium cells outlast the probable useful lifetime of the equipment they power.

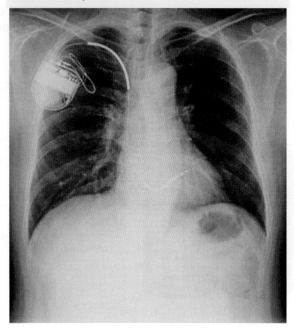

FIGURE 3.29 A pacemaker can be powered by a lithium battery. A magnetic switch operates the device, and the lithium battery usually lasts for three to five years.

Primary lithium cells include lithium manganese dioxide and lithium thionyl chloride cells. The most common is the lithium–manganese cell. This has a lithium anode, a manganese dioxide cathode and a non-aqueous electrolyte, such as propylene carbonate. The half-equations are written as follows:

Anode (oxidation):

$$Li(s) \rightarrow Li^+(l) + e^-$$

Cathode (reduction):

$$MnO_2(s) + Li^+(l) + e^- \rightarrow LiMnO_2(s)$$

3.4.2 Fuel cells

A **fuel cell** is a type of galvanic cell that converts chemical energy from a fuel into usable DC electricity and heat through redox reactions. It does not rely upon combustion as an intermediate step. By combining fuels such as hydrogen and oxygen in the presence of an electrolyte, the products of a fuel cell are electricity, heat and water. The process was first demonstrated in 1839, but fuel cell technology grew significantly in the 1960s, as part of the US space program.

Fuel cell technology has advanced with the search for energy alternatives that have greater operating efficiencies and lower costs. Power generation from fuel cells averages between 40 and 60 per cent efficiency compared with 30 to 35 per cent efficiency from fossil fuel combustion.

fuel cell an electrochemical cell that produces electrical energy directly from a fuel

Fuel cells are used, or are being investigated for use, in the following situations:

- As a portable power source for charging small appliances, such as batteries in laptops or smartphones
- For larger scale, stationary applications, including back-up power in hospitals and industry
- For transport applications such as forklifts, boats and buses. The silent operation of fuel cells is advantageous for submarines, and considerable research is being undertaken to improve the efficiency and reduce costs of fuel-cell cars.

Fuel cells differ from primary cells in a number of ways:
- Fuel and oxygen are supplied externally.
- Unreacted fuel and products are removed from the cell.
- Fuel cells don't go flat — electricity is generated for as long as reactants are supplied.

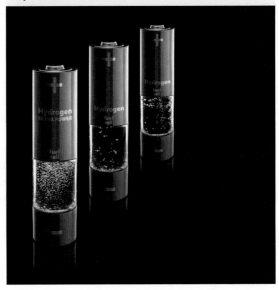

FIGURE 3.30 Will today's batteries be replaced by refillable fuel cells in the near future?

General operating principles of fuel cells

Fuel passes over the anode (and oxygen over the cathode) where it is split into ions and electrons. The electrons go through an external circuit while the ions move through the electrolyte towards the oppositely charged electrode. At this electrode, ions combine to create by-products. Depending on the input fuel and electrolyte, different chemical reactions occur.

FIGURE 3.31 A generic cross-section of a fuel cell

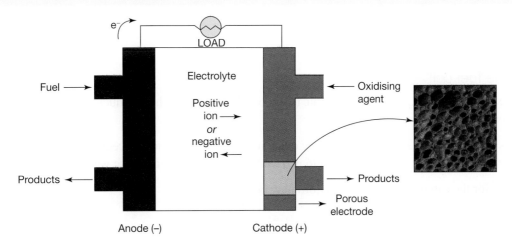

A number of factors and conditions affect the efficiency of fuel cells. These include electrodes, operating temperature, pressure and flow rate of reactant gases, as well as humidity.

Electrode design

Electrodes are designed to maximise surface area for the oxidation and reduction reactions to take place. This is achieved by making the electrodes porous (gaps between particles) and provides greater efficiency in transforming the chemical energy into electrical energy. Engineers have to research the ideal size and number of pores, and their placement, in electrode materials to maximise efficiency, heat distribution and the specific reactions at each electrode.

Sometimes, catalysts are embedded in electrodes to help the reactions take place at lower temperatures. For example, platinum, nickel, palladium and other materials are employed depending on the reactions occurring, the electrolyte employed, and the operating temperature and pH. These variables affect the voltage of the cell.

Porous electrodes help transfer gaseous reactants on to any catalysts employed.

Operating temperature

Operating temperature is another important parameter that affects fuel cell performance. Generally speaking, a higher operating temperature results in greater efficiency. For each fuel cell design, there is an optimal temperature, and the operating temperature must be specifically chosen for each fuel cell system.

Electrolytes

Although the basic operations of all fuel cells are the same, special varieties have been developed to take advantage of different electrolytes and serve different application needs. The type of electrolyte used can also have an impact on the voltage output. The types of fuel cells are usually named after the electrolyte that transports the ions.

Common fuel cell electrolytes are:
- polymer membrane/proton exchange (H^+)
- alkaline (OH^-)
- phosphoric acid ($H^+/H_2PO_4^-$)
- solid oxide (O^{2-})
- molten carbonate (CO_3^{2-}).

Ions that travel through electrolytes in fuel cells are produced at one electrode and consumed at the other. Generally, an electrolyte must meet the following requirements:
- High ionic conductivity
- Present an adequate barrier to the reactants
- Chemically and mechanically stable
- Low electronic conductivity
- Ease of manufacturability/availability
- Preferably low-cost.

Solid oxide fuel cell

The solid oxide fuel cell (SOFC) uses a ceramic (solid oxide) electrolyte that conducts O^{2-} ions at high temperatures.

Oxygen gas is reduced at the cathode to produce oxide ions that travel through a ceramic material, like zirconia, to the anode, where they react with hydrogen ions from the oxidised fuel and make water. For example:

The equation for the cathode reaction (reduction) may be written as:

$$O_2(g) + 4e^- \rightarrow 2O^{2-}_{\text{(in ceramic)}}$$

The equation for the anode reaction (oxidation) may be written as:

$$H_2(g) + O^{2-}_{\text{(in ceramic)}} \rightarrow H_2O(g) + 2e^-$$

The equation for the overall reaction may be written as:

$$2H_2(g) + O_2(g) \rightarrow 2H_2O(g)$$

SOFCs are highly efficient; however, the high operating temperatures required are a drawback.

Proton exchange membrane fuel cell

The **proton exchange membrane fuel cell (PEMFC)** offers high power density and operates at relatively low temperatures. It is used in cars, forklifts and buses, as well as some large-scale systems. Suitable fuels for the PEMFC include hydrogen gas, methanol and reformed fuels.

A typical PEMFC uses a polymer membrane as its electrolyte, which eliminates the corrosion and safety concerns associated with liquid electrolyte fuel cells. Although it is an excellent conductor of hydrogen ions, the membrane is an electrical insulator. The electrolyte is sandwiched between the anode and cathode, forming a unit less than one millimetre thick. Its low operating temperature provides instant start-up and requires no thermal shielding to protect personnel.

Hydrogen from the fuel gas stream is consumed at the anode, producing electrons that flow to the cathode via the electric load and hydrogen ions that enter the electrolyte. At the cathode, oxygen combines with electrons from the anode and hydrogen ions from the electrolyte to produce water. The PEMFC operates at about 80 °C, so the water does not dissolve in the electrolyte. Instead, it is collected from the cathode as it is carried out of the fuel cell by excess oxidising agent flow.

FIGURE 3.32 A cross-section of a proton exchange membrane fuel cell

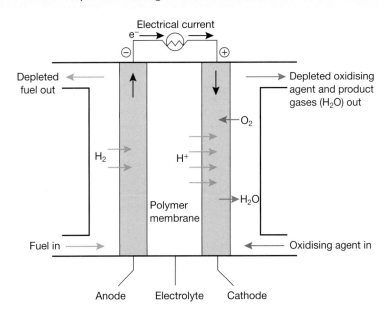

The equation for the anode reaction (oxidation) may be written as:

$$H_2(g) \rightarrow 2H^+(aq) + 2e^-$$

The equation for the cathode reaction (reduction) may be written as:

$$O_2(g) + 4H^+(aq) + 4e^- \rightarrow 2H_2O(l)$$

The equation for the overall reaction may be written as:

$$2H_2(g) + O_2(g) \rightarrow 2H_2O(l)$$

In situations in which hydrogen gas is not readily available or it is more convenient to use a different fuel source, a **fuel reformer** can be incorporated into the system. The fuel reformer can convert a variety of hydrocarbon-based fuels, such as methanol, ethanol or natural gas, into a hydrogen-rich gas stream that can be used by the PEMFC.

proton exchange membrane fuel cell (PEMFC) a fuel cell being developed for transport applications, as well as for both stationary and portable fuel cell applications

fuel reformer a device or system that converts a fuel source — typically hydrocarbons or alcohols — into a hydrogen-rich gas mixture

Direct methanol fuel cell

The **direct methanol fuel cell (DMFC)** is relatively new technology and is powered by liquid methanol, which has a higher energy density than hydrogen and is easier to transport. The anode catalyst withdraws hydrogen from the methanol. DMFCs are suitable for mobile phones, portable music devices and laptops, due to the small size of their cells and low operating temperature, and because there is no requirement for a fuel reformer, which allows devices to operate for longer periods of time.

> **direct methanol fuel cell (DMFC)** a new technology that is powered by liquid methanol

The equation for the anode reaction (oxidation) may be written as:

$$2CH_3OH(aq) + 2H_2O(l) \rightarrow 2CO_2(g) + 12H^+(aq) + 12e^-$$

The equation for the cathode reaction (reduction) may be written as:

$$O_2(g) + 4H^+(aq) + 4e^- \rightarrow 2H_2O(l)$$

The equation for the overall reaction may be written as:

$$2CH_3OH(aq) + 3O_2(g) \rightarrow 2CO_2(g) + 4H_2O(l)$$

FIGURE 3.33 A cross-section of a direct methanol fuel cell

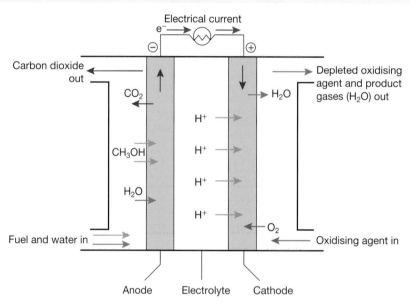

Resources

- **Video eLesson** Direct methanol fuel cell (eles-3239)
- **Weblink** Fuel cells

---

elog-1942

tlvd-9717

EXPERIMENT 3.5
online only

Investigating the hydrogen–oxygen fuel cell

Aim

To investigate the chemistry of the hydrogen–oxygen fuel cell

Alkaline fuel cell

The **alkaline fuel cell (AFC)**, also known as the hydrogen–oxygen fuel cell, uses potassium hydroxide as the electrolyte. The alkaline environment of an AFC allows the use of non-precious metal catalysts such as iron, cobalt, silver and graphene, which significantly reduces the cost of the fuel cell system.

FIGURE 3.34 A cross-section of an alkaline fuel cell

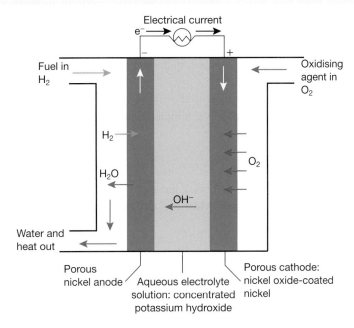

The equation for the anode reaction (oxidation) may be written as:

$$H_2(g) + 2OH^-(aq) \rightarrow 2H_2O(l) + 2e^-$$

The equation for the cathode reaction (reduction) may be written as:

$$O_2(g) + 2H_2O(l) + 4e^- \rightarrow 4OH^-(aq)$$

The equation for the overall reaction may be written as:

$$2H_2(g) + O_2(g) \rightarrow 2H_2O(l)$$

3.4.3 Fuel cells and future energy

Green energy principles

The challenges for chemists and engineers around energy production and consumption include improving efficiency and the use of renewable feedstocks. Any new innovation requires energy to be included as part of the viability of the design for commercial use.

Once upon a time, the engineering of products and synthetic chemicals gave no consideration to waste materials or wasted energy, nor the ability to recycle materials, or use waste heat and materials for other products and processes.

With respect to electricity, a tremendous amount of useful energy is lost during conversions and transmission. The amount varies depending on the distance the electricity needs to travel along power lines, but typically, only around 10 per cent of the energy from the power station makes it to the customer. Ways of reducing energy loss for electricity generation and transmission are very important for reducing emissions and waste.

alkaline fuel cell (AFC) a fuel cell that converts oxygen (from the air) and hydrogen (from a supply) into electrical energy and heat

Fuel cells reduce energy waste in a number of ways. These include:
- their use in stationary power systems, so companies can operate off the power grid and reduce transmission loss
- requiring fewer energy transformations due to chemical energy being transformed directly into electrical energy, and no moving parts like turbines and combustion engines
- using catalytic electrodes to ensure a higher percentage of useful energy
- using the heat generated from fuel cells for heating spaces and in cooling applications, or even to drive turbines to produce more electricity.

Energy production and consumption now also affords consideration to:
- the reduction of hazards and pollutants generated, including preventing waste rather than dealing with its disposal
- the degradation of chemicals so they don't last in the environment
- improved **atom economy** for reaction pathways, which includes the use of catalysts to convert more reactants into products, thereby reducing waste.

Design for energy efficiency

In the traditional bipolar stack design, the fuel cell stack has many cells stacked together so that the cathode of one cell is connected to the anode of the next cell. Most fuel cell stacks — regardless of the type of fuel cell, its size or the fuel used — are of this configuration.

FIGURE 3.35 The design of a PEMFC allows the depleted gases and excess oxidising agent gas to flow through the cell stack.

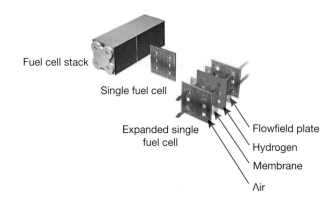

Fuel cell stack

Single fuel cell

Expanded single fuel cell

Flowfield plate

Hydrogen

Membrane

Air

Efficient fuel cell design should observe the following:
- The fuel and oxidising agent should be evenly distributed through each cell, and across their surface area.
- The temperature must be constant throughout the cell stack.
- Power loss to resistance is minimised.
- The stack must be properly sealed to ensure no gas leakage.

Electrolyte-free fuel cell

The electrolyte (layer)-free fuel cell (EFFC) is a new energy device that replaces current solid oxide fuel cells (SOFCs). It has a higher efficiency and can be used to electrolyse water to produce hydrogen fuel.

The preparation technology of an EFFC is very simple. Only one component is required: a mixture of electrode (anode and cathode) and electrolyte. This differs from the traditional fuel cell constructed from the three-layer (anode–electrolyte–cathode) structure. This new device with a single component/layer can effectively convert fuel to electricity.

atom economy a measurement of the efficiency of a reaction that considers the amount of waste produced, by calculating the percentage of the molar mass of the desired product compared to the molar mass of all reactants

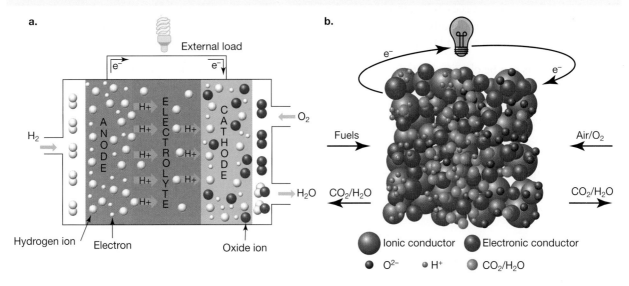

FIGURE 3.36 a. Current three-layer fuel cell technology and **b.** electrolyte (layer)-free fuel cell technology

The advantages of EFFC technology over current SOFCs include:
- increased conductivity of oxide (O^{2-}) ions
- a much lower operating temperature
- lower voltage losses
- greater efficiency
- being cheaper to produce.

Renewable feedstocks

The major fuel type used in commercial fuel cells is hydrogen, making these cells more environmentally friendly than the combustion of organic fuels; though the molten carbonate and solid oxide cells use methanol or a hydrocarbon, such as natural gas, because the high operating temperatures allow the reforming of hydrogen to occur within the fuel cell structure.

Direct methanol fuel cells use the anode catalyst to remove the hydrogen from the liquid methanol without a fuel reformer. This process is a more efficient way of producing energy than burning fuel to produce steam to drive a turbine, which means that less fuel is wasted and that lower amounts of greenhouse gases are generated.

Green hydrogen

Green hydrogen is being researched extensively for use in fuel cell technology as a means of generating electricity with net zero carbon emissions. Some biological applications of generating hydrogen can potentially have negative carbon emissions.

Several technologies exist for producing hydrogen gas from water. All involve the splitting of water into hydrogen and oxygen gas:

$$2H_2O(l) \rightarrow H_2(g) + O_2(g)$$

These technologies include **thermochemical splitting, photodecomposition** and **electrolysis**.

thermochemical splitting refers to when very high temperatures are used to decompose molecules by breaking chemical bonds

photodecomposition the use of light (photons) to break down molecules

electrolysis the process in which a non-spontaneous chemical reaction occurs by passing an electric current through a substance in solution or molten state

Hydrogen from electrolysis

Electrolysis of water is a successful method for producing green hydrogen. Water is decomposed into pure oxygen and hydrogen. Additionally, the electrolysis process can utilise the DC power from sustainable energy resources such as solar, wind and biomass.

Water electrolysis can be classified into four types based on the electrolyte, operating conditions and ionic agents (OH^-, H^+, O^{2-}) used; however, the operating principles are the same in all cases. The four types of electrolysis are:

- alkaline water electrolysis
- solid oxide electrolysis
- microbial electrolysis
- polymer electrolyte membrane water electrolysis.

FIGURE 3.37 Microbial electrolysis uses organic materials and electricity to make hydrogen.

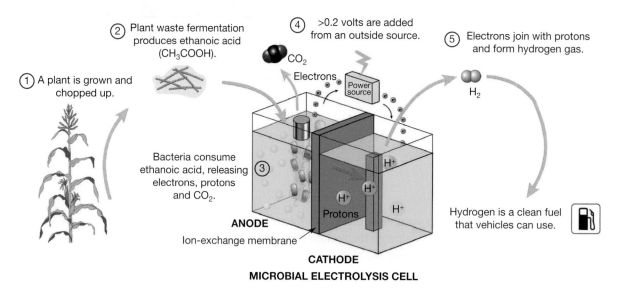

Hydrogen from alcohol

Alcohols such as ethanol can be used to produce hydrogen gas for use in fuel cells in a similar fashion to the steam reforming of fossil fuels.

$$C_2H_5OH + H_2O \rightarrow 2CO + 4H_2$$

$$CO + H_2O \rightarrow CO_2 + H_2$$

Alcohols such as methanol can also be catalytically converted directly in fuel cells, avoiding the need for hydrogen storage.

Traditionally, first-generation feedstocks such as sugar cane, cereal grains and potatoes have been used for alcohol production. In more recent times, concerns around using food sources for fuel have increased, given the world's growing population. First-generation feedstocks are processed to a large extent to increase the percentage of fermentable material; however, despite this, a large percentage of the plant material (up to 40 per cent) does not get fermented. This has prompted new research into next-generation feedstocks.

The use of algae as a bioethanol feedstock is gaining traction. This is because algae can readily and rapidly absorb carbon dioxide, and does not need as much land as terrestrial plants. Additionally, algae farms can be built in areas where the land is largely unsuitable for food production. This mitigates the issue of using crop-growing land for fuel production.

Hydrogen from biomass

Biomass can be used for hydrogen production. Enzymes convert glucose according to the following equations:

$$C_6H_{12}O_6 + 2H_2O \rightarrow 2CH_3COOH + 2CO_2 + 4H_2$$

$$C_6H_{12}O_6 + 2H_2O \rightarrow 2CH_3COOH + 2HCOOH + 2H_2$$

As with alcohol production, a range of first to third-generation feedstocks can be used. These include crop residues and algae.

3.4.4 Challenges in the production and use of renewable feedstocks

Hydrogen

One of the challenges for the use of hydrogen is storage. Hydrogen has a high energy content by weight but not by volume. This means that storing hydrogen can be difficult, as it is the lightest element and must undergo considerable compression to be contained in a suitably sized tank, which must withstand the extreme pressures required. Hydrogen can be stored as a liquid, but this requires keeping its temperature at −252.8 °C in very well-insulated containers.

Hydrogen can also be stored by combining it with certain metal or complex hydrides that can absorb it; from there, it can be released by heating it or adding water. Carbon nanomaterials or glass microspheres, as well as other chemical methods, are also being investigated as a means of storage. One such example is the use of methanoic acid (HCOOH) as a form of hydrogen storage by reacting hydrogen with carbon dioxide. The hydrogen can then be reformed.

Producing hydrogen from electrolysis of water is expensive. However, as the use of renewable energy sources such as wind and solar increases, and the technology improves, the cost of producing hydrogen from electrolysis is predicted to fall.

FIGURE 3.38 Methods of large-scale hydrogen storage

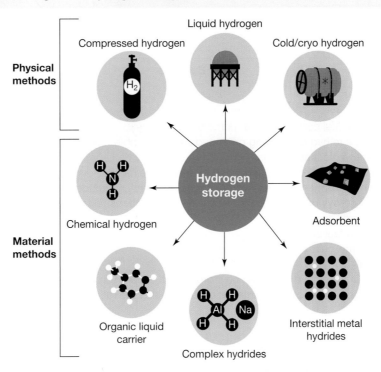

Alcohol production

As mentioned previously, a challenge with producing alcohol from plants is maximising the conversion of the organic material into alcohol. Gas fermentation has one key advantage over sugar and **cellulosic fermentation** to produce small alcohols — the ability to use all of the biomass and available carbon in it.

Biomass can be converted to a mixture of carbon monoxide (CO), carbon dioxide (CO_2), hydrogen (H_2) and nitrogen (N_2), also called synthesis gas or syngas.

Gas-fermenting bacteria, called acetogens, can be used to reduce **carboxylic acids** into their respective alcohols. These alcohols include ethanol, propanol (C_3H_7OH) and butanol (C_4H_9OH).

One major obstacle immediately present in gas fermentation is the low solubility of the gases in the liquid media. CO, H_2 and CO_2 are soluble to approximately 28 mg L^{-1}, 1.6 mg L^{-1} and 1.7 g L^{-1} (at 293 K, 100 kPa), respectively, compared to 900 g L^{-1} for glucose used for traditional fermentations.

Gas fermentation is gaining traction in the commercial world but its use as a means to industrially produce alcohol from biomass is still in the development stage.

Using algae

Using algae currently has its drawbacks. Like many of the first-generation plant feedstocks, larger species of algae that anchor to the sea floor require treatment to make the carbohydrates available and suitable for fermentation. Microscopic algae, although richer in fermentable carbohydrates, are harder to harvest from water sources than larger species.

> **cellulosic fermentation** the use of enzymes to obtain glucose from cellulose to make alcohol
>
> **carboxylic acids** the homologous series containing the —COOH functional group

3.4 Activities

learnon

Students, these questions are even better in jacPLUS

- Receive immediate feedback and access sample responses
- Access additional questions
- Track your results and progress

Find all this and MORE in jacPLUS

| 3.4 Quick quiz **on** | 3.4 Exercise | 3.4 Exam questions |

3.4 Exercise

1. List three ways in which fuel cells differ from primary galvanic cells.
2. 'Green' hydrogen refers to hydrogen being produced as a renewable feedstock for use in fuel cells.
   a. Explain what a renewable feedstock is.
   b. List three ways in which hydrogen is generated by renewable means.
   c. What are the main challenges of using hydrogen as a renewable feedstock?
3. For the following fuel cells, write half-equations to show the reactions taking place at each of the electrodes and then write the overall equation.
   a. An alkaline hydrogen fuel cell
   b. An acidic methane fuel cell
   c. An ethanol fuel cell
4. State the three ways in which hydrogen can be stored in a fuel cell vehicle.
5. Discuss two advantages and two disadvantages of using fuels cells rather than fossil fuels for energy.

3.4 Exam questions

▶ **Question 1 (9 marks)**
Source: VCE 2021 Chemistry Exam, Section B, Q.4; © VCAA

a. What is a fuel cell? **(2 marks)**

The diagram below shows part of an ethanol fuel cell, which produces carbon dioxide and uses an acidic electrolyte.

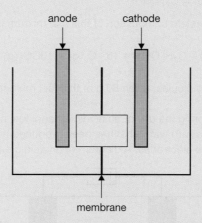

b. **i.** Name the species that crosses the membrane to enable fuel cell operation. **(1 mark)**

    ii. In the box provided on the diagram above, indicate the direction of flow of the species named in part **b.i.** **(1 mark)**

c. Write the equation for the reaction that occurs at the anode of an ethanol fuel cell, which produces carbon dioxide and uses an acidic electrolyte. **(1 mark)**

d. If an ethanol fuel cell was operating at 25 °C and at 100% efficiency, how much electrical energy could be produced from 1.0 g of ethanol? **(1 mark)**

e. Identify **two** aspects of electrode design that can improve the efficiency of a fuel cell. **(2 marks)**

f. State how the environmental impact of using an ethanol fuel cell operating at 100% efficiency can be minimised. **(1 mark)**

▶ **Question 2 (1 mark)**
Source: VCE 2019 Chemistry Exam, Section A, Q.8; © VCAA

MC Consider the following statements about galvanic cells and fuel cells.

Statement number	Statement
1	The overall reaction is exothermic.
2	Electrons are consumed at the negative electrode.
3	Both the reducing agent and the oxidising agent are stored in each half-cell.
4	The electrodes are in contact with the reactants and the electrolyte.
5	The production of electricity requires the electrodes to be replaced regularly.

Which one of the following sets of statements is correct for **both** galvanic cells and fuel cells?

A. statement numbers 2 and 3

B. statement numbers 1 and 4

C. statement numbers 2, 4 and 5

D. statement numbers 1, 3 and 5

Source: VCE 2019 Chemistry Exam, Section B, Q.4; © VCAA

Internal combustion engines are used in large numbers of motor vehicles. Historically, internal combustion engines have used fuels obtained from crude oil as a source of power. As concerns for the environment have grown, efforts have been made to obtain fuel for combustion engines from other sources.

a. One way of reducing the environmental effects of fossil fuels is to blend them with biofuels. A common method is to blend petrol with ethanol in varying ratios. A fuel can be obtained by blending 1 mole of octane, C_8H_{18}, and 1 mole of ethanol, C_2H_5OH.

The chemical equation for the complete combustion of this fuel mixture is:

$$C_8H_{18}(l) + C_2H_5OH(l) + 15\frac{1}{2}O_2(g) \rightarrow 10CO_2(g) + 12H_2O(g)$$

Calculate the energy released, in kilojoules, when 80 g of this fuel mixture undergoes complete combustion. Show your working. **(3 marks)**

b. Some car manufacturers are exploring the use of an acidic ethanol fuel cell to power vehicles. In this fuel cell, the ethanol at one electrode reacts with water that has been produced at the other electrode. A membrane is used to transport ions between the electrodes. A diagram of an acidic ethanol fuel cell is shown below.

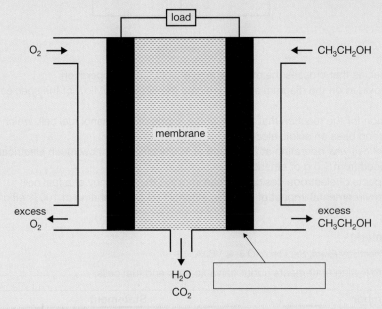

i. Identify the electrode as either the cathode or the anode as shown by the box provided in the diagram above. **(1 mark)**

ii. Write the half-equation for the reaction occurring at the anode. **(1 mark)**

iii. The combustion of ethanol and the combustion of octane release about the same amount of energy per mole of carbon dioxide produced.

Identify **two** advantages of powering a vehicle using an ethanol fuel cell instead of an internal combustion engine powered by octane. **(2 marks)**

Question 4 (1 mark)

Source: VCE 2020 Chemistry Exam, Section A, Q.13; © VCAA

MC Hydrogen, H_2, fuel cells and H_2-powered combustion engines can both be used to power cars. Three statements about H_2 fuel cells and H_2-powered combustion engines are given below:

I Neither H_2 fuel cells nor H_2-powered combustion engines produce greenhouse gases.

II Less H_2 is required per kilometre travelled when using an H_2-powered combustion engine than when using H_2 fuel cells.

III More heat per kilogram of H_2 is generated in an H_2-powered combustion engine than in H_2 fuel cells.

Which of the statements above are correct?

A. II only

B. I and II only

C. III only

D. I and III only

Question 5 (1 mark)

Source: VCE 2017 Chemistry Exam, Section A, Q.27; © VCAA

MC An increasingly popular battery for storing energy from solar panels is the vanadium redox battery. The battery takes advantage of the four oxidation states of vanadium that are stable in aqueous acidic solutions in the absence of oxygen.

A schematic diagram of a vanadium redox battery is shown below.

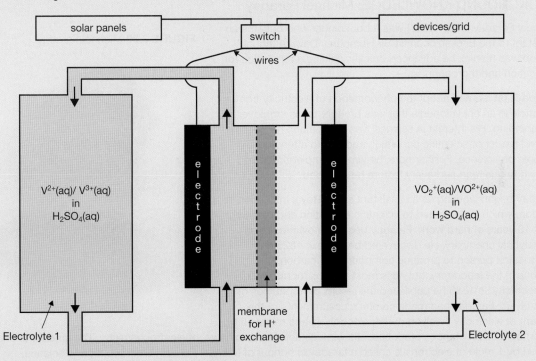

The two relevant half-equations for the battery are as follows.

$$VO_2^+(aq) + 2H^+(aq) + e^- \rightarrow VO^{2+}(aq) + H_2O(l) \qquad E^0 = +1.00\,V$$
$$V^{3+}(aq) + e^- \rightarrow V^{2+}(aq) \qquad E^0 = -0.26\,V$$

The overall reaction that occurs when the battery is discharging is

A. $VO_2^+(aq) + 2H^+(aq) + V^{2+}(aq) \rightarrow VO^{2+}(aq) + V^{3+}(aq) + H_2O(l)$

B. $VO^{2+}(aq) + H_2O(l) + V^{3+}(aq) \rightarrow VO_2^+(aq) + V^{2+}(aq) + 2H^+(aq)$

C. $VO^{2+}(aq) + V^{2+}(aq) + 2H^+(aq) \rightarrow 2V^{3+}(aq) + H_2O(l)$

D. $VO_2^+(aq) + V^{3+}(aq) \rightarrow 2VO^{2+}(aq)$

More exam questions are available in your learnON title.

3.5 Calculations involved in producing electricity from galvanic cells and fuel cells

---

KEY KNOWLEDGE

- The application of Faraday's Laws and stoichiometry to determine the quantity of galvanic or fuel cell reactant and product, and the current or time required to either use a particular quantity of reactant or produce a particular quantity of product

Source: VCE Chemistry Study Design (2024–2027) extracts © VCAA; reproduced by permission.

Half-equations from the reactions at the anode and cathode show the stoichiometric relationship between the amount, in moles, of electrons transferred and the number of moles of oxidising agent and reducing agent consumed.

The relationship between the quantities of reactants and products and the amount of electrical charge can be shown using Faraday's Laws.

BACKGROUND KNOWLEDGE: Michael Faraday

Michael Faraday (1791–1867) was a bookbinder who became an assistant to the English chemist Sir Humphrey Davy. Although the job was menial, he advanced quickly, gaining a reputation for dedication and thoroughness.

Faraday first learned about the phenomenon of electricity from an article in an encyclopedia that was brought to his employer for rebinding. His interest in science was kindled and he became an avid reader of scientific papers, in addition to attending lectures on science. Furthermore, he wrote complete notes on every book he read and every lecture he attended.

Faraday began working as an assistant at the Royal Institution in London, which is dedicated to scientific education and research. After 10 years of hard work, Faraday began his own research in analytical chemistry. He discovered benzene in 1825 and was the first person to produce compounds of carbon and chlorine in the laboratory, but he is most famous for his work on electricity. In 1833 he published the results of his studies of electrolysis. Faraday had made careful measurements of the amount of electricity involved during electrolysis and related it to the amount of substances produced. His work established two 'laws' of electrochemistry. The amount of charge carried by 1 mole of electrons is called a faraday in honour of Michael Faraday's contribution to science.

FIGURE 3.39 Michael Faraday

Michael Faraday.

Faraday's Laws were originally concerned with the quantitative relationship between electricity and extent of chemical reactions during **electrolysis**, in which an external source of electrical energy is applied to drive a non-spontaneous redox reaction. Applications of electrolysis and **electrolytic cells** will be explored further in topic 6. However, Faraday's Laws can also be used to link amount of charge and extent of reaction for spontaneous reactions, such as those in primary cells or fuel cells, where electricity is generated by the cell rather than supplied to the cell from an external source. Faraday's Laws will be explained here, and applied to galvanic and fuel cells to determine quantities relevant to the discharge of these cells, such as the amount of reactant or product, current or time.

> **electrolysis** the process in which a non-spontaneous chemical reaction occurs by passing an electric current through a substance in solution or molten state
>
> **electrolytic cells** an electric cell in which a non-spontaneous redox reaction is made to occur by the application of an external potential difference across the electrodes; also known as an electrolysis cell

3.5.1 Faraday's First Law

Faraday's First Law states that the amount of a substance deposited or liberated — that is, the amount of chemical change — is directly proportional to the quantity of electric charge passed through the cell.

Faraday's First Law

The amount of any substance deposited, evolved or dissolved at an electrode is directly proportional to the quantity of electric charge passed through the cell.

The quantity of electric charge transferred in an electric current depends on the magnitude of the current used and the time for which it flows.

The electric charge can be calculated using the formula:

$$Q = It$$

where:

Q is the electrical charge in coulombs (C)

I is the current in amperes (A)

t is the time in seconds (s).

In an experimental circuit, there is no meter that measures the charge in coulombs. However, an ammeter could be used to measure the rate at which charge flows in a circuit. A current of 1 ampere (1 A) indicates that 1 coulomb (6.24×10^{18} electrons) of charge flows every second. For example, if a current of 3.00 amperes flows for 10.0 minutes, the quantity of electricity is $(3.00 \times 10.0 \times 60) = 1.80 \times 10^3$ C. The charge flowing through a galvanic cell can be increased by increasing the duration (time) of the reaction. In an electrolytic cell, this can also be increased by increasing the magnitude of the current that is supplied.

In an experiment to investigate the relationship between the quantity of electricity and the mass of electrolytic products, copper(II) sulfate was electrolysed using copper electrodes. The copper cathode was weighed before the electrolysis. After 10.0 minutes of electrolysis with a current of 3.00 amperes, the experiment was stopped and the cathode reweighed. The mass of copper deposited was calculated, the cathode was replaced, and the experiment continued for another 10.0 minutes. The mass of copper deposited in 20.0 minutes was then found. This procedure was repeated several times and the results obtained are shown in table 3.5.

Faraday's First Law states that the amount of current passed through an electrode is directly proportional to the amount of material released from it

TABLE 3.5 Experimental results showing quantity of electricity in electrolysis and mass of copper deposited

Current (A)	Time (s)	Quantity of electricity (C)	Mass of copper (g)
3.00	600	1800	0.59
3.00	1200	3600	1.19
3.00	1800	5400	1.78
3.00	2400	7200	2.38

The graph of these results (figure 3.40) yields a straight line passing through the origin. This shows that the mass of the product is directly proportional to the quantity of electricity. The mass of copper deposited on the cathode during electrolysis is directly proportional to the quantity of electricity used.

The charge on one electron is 1.602×10^{-19} C.

Therefore, one mole of electrons has a charge of $(6.023 \times 10^{23} \times 1.602 \times 10^{-19}) = 9.649 \times 10^4$ C. This quantity of charge carried by a mole of electrons is referred to as the faraday (F), or **Faraday constant**, and is usually given the value of 96 500 C mol^{-1}.

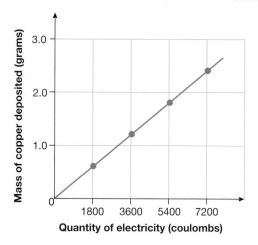

FIGURE 3.40 Graph of copper deposited versus quantity of electricity in electrolysis

TIP: The Faraday constant, 96 500 C mol^{-1}, can be found in the VCE Chemistry Data Book.

Faraday constant represents the amount of electric charge carried by 1 mole of electrons

tlvd-9853

SAMPLE PROBLEM 7 Using Faraday's First Law to calculate the amount of product evolved in a galvanic cell

A galvanic cell consists of a zinc electrode and a copper electrode. During the cell's operation, zinc metal is oxidised at the zinc electrode, while copper(II) ions are reduced at the copper electrode. Calculate the mass of copper deposited on the copper electrode when a current of 2.5 A flows through the cell for 30 minutes.

THINK	WRITE
1. Write the half-equation for the deposition of copper.	$Cu^{2+}(aq) + 2e^- \rightarrow Cu(s)$
2. Calculate the total charge transferred.	$Q = I \times t$ $= 2.5 \times 30 \times 60$ $= 4500$ C
3. Use Faraday's First Law to calculate the number of moles of electrons transferred.	$n = \dfrac{Q}{F}$ $\therefore n(e^-) = \dfrac{4500}{96\,500}$ $= 0.0466$ mol
4. Calculate the number of moles of copper deposited. Since the stoichiometric ratio of copper to electrons is 1 : 2, the number of moles of copper deposited is half the number of moles of electrons transferred.	$n(Cu) = \dfrac{n}{2}$ $= \dfrac{0.0466}{2}$ $= 0.0233$ mol
5. Calculate the mass of copper deposited.	$m = M \times n$ $\therefore m(Cu) = 63.5 \times 0.0233$ $= 1.480$ g

PRACTICE PROBLEM 7

An alkaline battery is composed of zinc anode, a manganese dioxide cathode and an electrolyte paste of potassium hydroxide. The overall reaction occurring to produce electrical energy is as follows:

$$2MnO_2(s) + 2H_2O(l) + Zn(s) \rightarrow 2MnO(OH)(s) + Zn(OH)_2(s)$$

Determine the mass of zinc that will react in this primary cell if it discharges with a current of 2.2 A for two hours.

3.5.2 Faraday's Second Law

Faraday's Second Law describes the stoichiometric relationship between the moles of substance produced and the moles of electrons required.

> **Faraday's Second Law** states that when the same quantity of electricity is passed through several electrolytes, the mass of the substances deposited are proportional to the stoichiometric coefficients in the balanced half-equations of the respective electrochemical reactions

Faraday's Second Law

To produce 1 mole of a substance, 1, 2, 3 or another whole number of moles of electrons (faradays) must be consumed, according to the relevant half-cell equation.

In the half-cell equation:

$$\underset{\text{1 mole of silver ions}}{Ag^+(aq)} \quad + \quad \underset{\text{1 mole of electrons}}{e^-} \quad \rightarrow \quad \underset{\text{1 mole of silver atoms}}{Ag(s)}$$

1 mole of electrons is needed to discharge 1 mole of $Ag^+(aq)$ ions. This liberates 1 mole of silver atoms. Thus, 1 faraday, or 96 500 coulombs, is needed to discharge 1 mole of silver atoms.

The number of faradays needed to liberate 1 mole of an element is found from the equation for the electrode reaction. Some examples are as follows:

Sodium: $Na^+ + e^- \rightarrow Na$	1 faraday must be passed to liberate 1 mole of sodium atoms (23.0 g).
Copper: $Cu^{2+} + 2e^- \rightarrow Cu$	2 faradays must be passed to liberate 1 mole of copper atoms (63.5 g).
Magnesium: $Mg^{2+} + 2e^- \rightarrow Mg$	2 faradays must be passed to liberate 1 mole of magnesium atoms (24.3 g).
Aluminium: $Al^{3+} + 3e^- \rightarrow Al$	3 faradays must be passed to liberate 1 mole of aluminium atoms (27.0 g).
Chlorine: $2Cl^- \rightarrow Cl_2 + 2e^-$	2 faradays must be passed to liberate 1 mole of chlorine molecules (71.0 g).

The number of moles of electrons, $n(e^-)$ (which carry the same number of faradays), corresponding to a given charge (in coulombs) can be determined by the equation:

$$n(e^-) = \frac{Q}{F}$$

where:

n is the number of moles of electrons

Q is the electrical charge in coulombs (C)

F is the Faraday constant, 96 500 C mol^{-1}.

tlvd-9679

SAMPLE PROBLEM 8 Comparing electrode reaction equations to determine the mole of a substance produced

When 7720 C is passed through a copper(II) sulfate solution, 0.040 mol of copper is produced. If the same amount of charge is passed through a solution containing Ag^+ ions, how many moles of Ag will be produced?

THINK	WRITE
1. Write the equation for the electrode reaction for both Cu and Ag deposition.	$Cu^{2+} + 2e^- \rightarrow Cu(s)$ $Ag^+(aq) + e^- \rightarrow Ag(s)$
2. The reaction in step 1 shows that $2F$ of charge is required to deposit 1 mol of Cu. The same $2F$ will deposit 2 mol of Ag because it only has a single charge on its ion. Hence, there is a $2:1$ ratio when equal amounts of charge are used. Give your answer to two significant figures.	$n(Ag) = 2 \times n(Cu)$ $\quad\quad\quad = 2 \times 0.040$ $\quad\quad\quad = 0.080\,mol$

PRACTICE PROBLEM 8

When 7720 C is passed through a copper(II) sulfate solution, 0.040 mol of copper is produced. If the same amount of charge is passed through a solution containing W^{3+} ions, how many moles of W will be produced?

3.5.3 Applying Faraday's First and Second Laws

Combining the two equations from Faraday's First and Second Laws results in:

$$nF = It$$

$$\therefore n(e^-) = \frac{It}{F}$$

We can use this equation to predict the amount of a fuel or reducing agent required to produce a certain amount of charge; or to calculate the current and amount of time it takes to consume and produce reactants and products, respectively, while a cell is generating electricity.

SAMPLE PROBLEM 9 Using Faraday's Second Law to calculate the amount of reactant consumed in a battery

An alkaline AA battery produces a current of 0.70 A. The equation for the reaction occurring at the anode as the cell discharges is as follows:

$$Zn(s) + 2OH^-(aq) \rightarrow ZnO(s) + H_2O(l) + 2e^-$$

Calculate the mass of zinc consumed if the battery operates for 30 minutes.

THINK	WRITE
1. Calculate the amount of charge required to deliver the 0.70 A current for 30 minutes. Remember to convert the time into seconds. Calculate the number of moles of electrons that are delivered using the equation $n(e^-) = \dfrac{It}{F}$.	$n(e^-) = \dfrac{It}{F}$ $= \dfrac{0.70 \times 30 \times 60}{96\,500}$ $= 0.013\,057 \text{ mol}$
2. Use the half-equation to calculate the number of moles of zinc consumed according to the ratio of Zn : e^- being 1 : 2.	$\dfrac{n(Zn)}{n(e^-)} = \dfrac{1}{2}$ $\therefore n(Zn) = \dfrac{1}{2} \times 0.013\,057$ $= 0.006\,528 \text{ mol}$
3. Use the equation $m = n \times M$ to calculate the mass of zinc in grams.	$m(Zn) = n \times M$ $= 0.006\,528 \times 65.4$
4. Express the answer to the least number of significant figures used in the measured values; in this case, two.	$m(Zn) = 0.4270 \text{ g}$ $= 0.43 \text{ g (to two significant figures)}$

PRACTICE PROBLEM 9

Calculate the mass of zinc consumed when an alkaline AA battery operates at 680 mA for 1.50 hours.

---

tlvd-9652

SAMPLE PROBLEM 10 Combining Faraday's Laws to calculate the amount of reducing agent required to produce a set amount of charge

Calculate the mass of oxygen gas consumed in a solid oxide fuel cell stack delivering a 5.5 A current for 6.5 hours. The reaction occurring at the cathode is as follows:

$$O_2(g) + 4e^- \rightarrow 2O^{2-} \text{ (in ceramic)}$$

THINK	WRITE
1. Calculate the amount of charge required to deliver the 5.5 A current for 6.5 hours. Remember to convert the time into seconds. Calculate the number of moles of electrons that are delivered using the equation $n(e^-) = \dfrac{It}{F}$.	$\begin{aligned} n(e^-) &= \dfrac{It}{F} \\ &= \dfrac{5.5 \times 6.5 \times 60 \times 60}{96\,500} \\ &= 1.3337 \text{ mol} \end{aligned}$
2. Use the half-equation to calculate the number of moles of oxygen gas consumed according to the ratio of $O_2 : e^-$ being 1 : 4.	$\begin{aligned} \dfrac{n(O_2)}{n(e^-)} &= \dfrac{1}{4} \\ \therefore n(O_2) &= \dfrac{1}{4} \times 1.3337 \\ &= 0.3334 \text{ mol} \end{aligned}$
3. Use the equation $m = n \times M$ to calculate the mass of O_2 in grams.	$\begin{aligned} m(O_2) &= n \times M \\ &= 0.3334 \times 32.0 \\ &= 10.67 \text{ g} \end{aligned}$
4. Express the answer to the least number of significant figures used in the measured values; in this case, two.	$= 11 \text{ g (to two significant figures)}$

PRACTICE PROBLEM 10

Calculate the mass of hydrogen gas consumed in a solid oxide fuel cell stack delivering a 4.50 A current for 95.0 minutes. The reaction occurring at the anode is as follows:

$$H_2 + O^{2-} \rightarrow H_2O + 2e^-$$

---

tlvd-9653

SAMPLE PROBLEM 11 Combining Faraday's Laws to calculate the time taken to consume reactants in a cell

A portable PEMFC has a hydrogen fuel supply of 1.50 g. Calculate the maximum time, in hours, that the cell can operate while delivering a 2.1 A current.

THINK	WRITE
1. Write the equation for the oxidation of hydrogen gas.	$H_2(g) \rightarrow 2H^+(aq) + 2e^-$
2. Calculate the number of moles of hydrogen gas using the equation $n = \dfrac{m}{M}$.	$n = \dfrac{m}{M}$ $= \dfrac{1.50}{2.0}$ $= 0.75 \text{ mol}$
3. Calculate the number of moles of electrons from the ratio in the half-equation for the oxidation of hydrogen gas.	$\dfrac{n(e^-)}{n(H_2)} = \dfrac{2}{1}$ $\therefore n(e^-) = \dfrac{2}{1} \times 0.75$ $= 1.5 \text{ mol}$
4. Rearrange the combined formulae of Faraday's Laws to $t = \dfrac{nF}{I}$ and substitute the values in.	$t = \dfrac{nF}{I}$ $= \dfrac{1.5 \times 96\,500}{2.1}$ $= 68\,929 \text{ s}$
5. Divide your answer by 3600 (60 × 60) to convert seconds into hours and express your answer to two significant figures.	$t = \dfrac{68\,929}{60 \times 60}$ $= 19 \text{ hours}$

PRACTICE PROBLEM 11

A portable PEMFC has a limiting oxygen gas supply of 2.50 g. Calculate the maximum time, in hours, that the cell can operate while delivering a 2.00 A current.

3.5 Activities

| **3.5 Quick quiz** | **3.5 Exercise** | **3.5 Exam questions** |

3.5 Exercise

1. When 2200 C is passed through a copper(II) sulfate solution, 0.011 mol of copper is produced. If the same amount of charge is passed through a solution containing Cr^{3+} ions, how many moles of Cr will be produced?
2. A given quantity of electricity is passed through two aqueous cells connected in series. The first contains silver chloride and the second contains calcium chloride. What mass of calcium is deposited in one cell if 2.00 g of silver is deposited in the other cell?
3. How long will it take, in hours, for a portable alkaline hydrogen–oxygen fuel cell to consume 2.00 g of hydrogen stored as a liquid when delivering a 2.0 A current?
4. A solid oxide fuel cell operates continuously for a 24.0-hour period. During this time, the cell delivers a 15.00 A current. The hydrogen gas used in the fuel cell is reformed from within the cell itself from methanol (CH_3OH). The equation for the reformation of hydrogen from methanol is as follows:

$$CH_3OH(l) + H_2O(l) \rightarrow CO_2(g) + 3H_2(g)$$

   a. Write a balanced half-equation for the oxidation of the hydrogen gas at the anode in the solid oxide fuel cell.
   b. Write a balanced half-equation for the reduction of oxygen gas at the cathode in the solid oxide fuel cell.
   c. Calculate the charge, in coulombs, delivered by the fuel cell.
   d. Calculate the theoretical mass of methanol required to operate the cell for the 24.0-hour period.
   e. The initial tank of methanol weighed 1.25 kg. After the 24.0-hour period, the tank had a mass of 1.04 kg. Suggest two possible reasons as to why the actual mass used was different to the theoretical mass used.
5. A particular lithium cell has an operating voltage of 3.5 V. The overall reaction taking place when discharging is as follows:

$$LiC_6 + CoO_2 \rightarrow C_6 + LiCoO_2$$

The electrodes are kept apart by a specifically designed separator that improves the safety of the cell. As the cell discharges, the anode reaction taking place is as follows:

$$LiC_6 \rightarrow Li^+ + C_6 + e^-$$

   a. Write a balanced half-equation for the cathode reaction as the cell discharges.
   b. Write the formula of the cell electrolyte.
   c. Calculate the increase in mass of solid $LiCoO_2$ if the cell delivers a 0.60 A current for 2.5 hours operating at 65% efficiency. Assume energy is lost as heat.

3.5 Exam questions

Question 1 (1 mark)
Source: Adapted from VCE 2021 Chemistry Exam, Section A, Q.9; © VCAA

MC A galvanic cell produced a charge of 4.00 C in 5.00 minutes.

This represents a production of
A. 4.15×10^{-5} mol of electrons.
B. 2.07×10^{-4} mol of electrons.
C. 1.93×10^{4} mol of electrons.
D. 2.41×10^{4} mol of electrons.

Question 2 (1 mark)

Source: Adapted from VCE 2021 Chemistry Exam, Section A, Q.20; © VCAA

MC 1 F is equivalent to the charge on 1 mol of electrons.

The mass of nickel, Ni, that can be deposited onto a platinum, Pt, electrode with 320 F of charge is
A. 9.73×10^{-2} g
B. 1.95×10^{-1} g
C. 9.39×10^{3} g
D. 1.88×10^{4} g

Question 3 (1 mark)

Source: Adapted from VCE 2018 Chemistry NHT Exam, Section A, Q.24; © VCAA

MC A galvanic cell with two platinum electrodes operates at 5.0 A for 600 s. After the circuit is disconnected, 0.54 g of metal is found to have been deposited on the cathode.

Which solution was used at the cathode?
A. 1 M $AgNO_3$
B. 1 M $Ni(NO_3)_2$
C. 1 M $Pb(NO_3)_2$
D. 1 M $Cr(NO_3)_3$

Question 4 (3 marks)

Source: Adapted from VCE 2015 Chemistry Exam, Section A, Q.26; © VCAA

The switch in the galvanic cell below may be closed to allow a current to flow through the circuit.

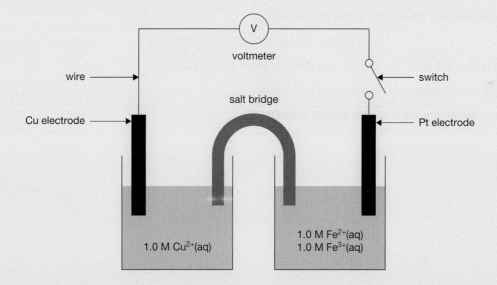

For how long would the switch need to be closed to deposit or liberate 4.0 g of copper if the current is 1.8 A?

Source: VCE 2014 Chemistry Exam, Section B, Q.10.c; © VCAA

The following diagram shows a cross-section of a small zinc–air button cell, a button cell that is used in hearing aids.

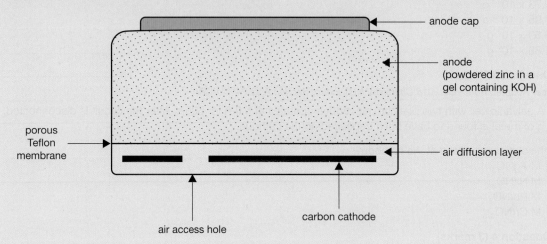

The zinc acts as the anode. It is in the form of a powder dispersed in a gel (a jelly-like substance) that also contains potassium hydroxide. The cathode consists of a carbon disc. Oxygen enters the cell via a porous Teflon membrane. This membrane also prevents any chemicals from leaking out.

The following reaction takes place as the cell discharges.

$$2Zn(s) + O_2(g) + 2H_2O(l) \rightarrow 2Zn(OH)_2(s)$$

A zinc–air button cell is run for 10 hours at a steady current of 2.36 mA.

What mass of zinc metal reacts to form zinc hydroxide?

More exam questions are available in your learnON title.

3.6 Review

3.6.1 Topic summary

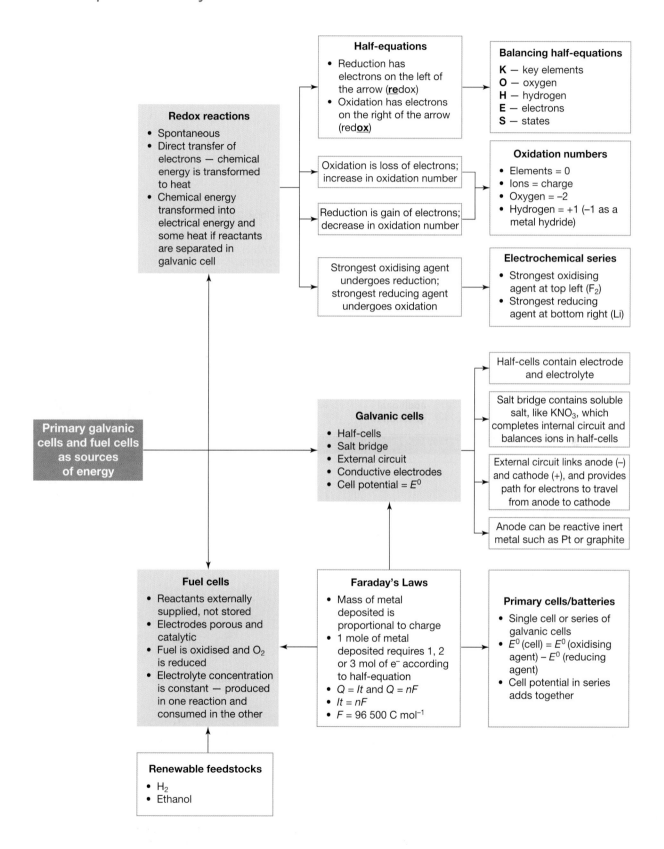

Half-equations
- Reduction has electrons on the left of the arrow (**re**dox)
- Oxidation has electrons on the right of the arrow (red**ox**)

Balancing half-equations
- **K** — key elements
- **O** — oxygen
- **H** — hydrogen
- **E** — electrons
- **S** — states

Redox reactions
- Spontaneous
- Direct transfer of electrons — chemical energy is transformed to heat
- Chemical energy transformed into electrical energy and some heat if reactants are separated in galvanic cell

Oxidation is loss of electrons; increase in oxidation number

Reduction is gain of electrons; decrease in oxidation number

Oxidation numbers
- Elements = 0
- Ions = charge
- Oxygen = –2
- Hydrogen = +1 (–1 as a metal hydride)

Strongest oxidising agent undergoes reduction; strongest reducing agent undergoes oxidation

Electrochemical series
- Strongest oxidising agent at top left (F_2)
- Strongest reducing agent at bottom right (Li)

Primary galvanic cells and fuel cells as sources of energy

Half-cells contain electrode and electrolyte

Salt bridge contains soluble salt, like KNO_3, which completes internal circuit and balances ions in half-cells

Galvanic cells
- Half-cells
- Salt bridge
- External circuit
- Conductive electrodes
- Cell potential = E^0

External circuit links anode (–) and cathode (+), and provides path for electrons to travel from anode to cathode

Anode can be reactive inert metal such as Pt or graphite

Fuel cells
- Reactants externally supplied, not stored
- Electrodes porous and catalytic
- Fuel is oxidised and O_2 is reduced
- Electrolyte concentration is constant — produced in one reaction and consumed in the other

Faraday's Laws
- Mass of metal deposited is proportional to charge
- 1 mole of metal deposited requires 1, 2 or 3 mol of e^- according to half-equation
- $Q = It$ and $Q = nF$
- $It = nF$
- $F = 96\,500$ C mol^{-1}

Primary cells/batteries
- Single cell or series of galvanic cells
- E^0 (cell) = E^0 (oxidising agent) – E^0 (reducing agent)
- Cell potential in series adds together

Renewable feedstocks
- H_2
- Ethanol

Hey students! Now that it's time to revise this topic, go online to:

Access the topic summary | Review your results | Watch teacher-led videos | Practise past VCAA exam questions

Find all this and MORE in jacPLUS

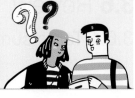

3.6.2 Key ideas summary

online only

3.6.3 Key terms glossary

online only

on Resources

Solutions	Solutions — Topic 3 (sol-0830)
Practical investigation eLogbook	Practical investigation eLogbook — Topic 3 (elog-1702)
Digital documents	Key science skills — VCE Chemistry Units 1–4 (doc-37066) Key terms glossary — Topic 3 (doc-37285) Key ideas summary — Topic 3 (doc-37286)
Exam question booklet	Exam question booklet — Topic 3 (eqb-0114)

3.6 Activities

learnon

Students, these questions are even better in jacPLUS

Receive immediate feedback and access sample responses | Access additional questions | Track your results and progress

Find all this and MORE in jacPLUS

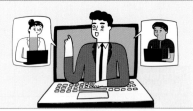

3.6 Review questions

1. In each of the following reactions, use oxidation numbers to determine which species has been reduced and which has been oxidised.
   a. $Zn(s) + 2HCl(aq) \rightarrow ZnCl_2(aq) + H_2(g)$
   b. $2NO(g) + O_2(g) \rightarrow 2NO_2(g)$
   c. $Mg(s) + H_2SO_4(aq) \rightarrow MgSO_4(aq) + H_2(g)$
   d. $2Al(s) + 3Cl_2(g) \rightarrow 2AlCl_3(s)$
2. Balance the following equations by first writing and balancing the two half-equations.
   a. $Br^-(aq) + SO_4^{2-}(aq) \rightarrow SO_2(g) + Br_2(l)$
   b. $I_2(s) + H_2S(g) \rightarrow I^-(aq) + S(s)$
   c. $Cu(s) + HNO_3(aq) \rightarrow Cu^{2+}(aq) + NO(g)$
   d. $Cu(s) + HNO_3(aq) \rightarrow Cu^{2+}(aq) + NO_2(g)$

3. Consider the reaction occurring in the diagram shown, and complete the following.

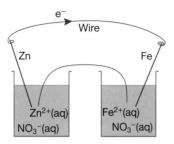

a. State the anode half-reaction.
b. State the cathode half-reaction.
c. Find the overall cell reaction.

4. By referring to a table of standard electrode potentials, state whether you would expect:

a. bromine gas to form if chlorine gas was bubbled into a solution of bromide ions
b. chlorine gas to form if liquid bromine was mixed into a solution of chloride ions
c. iron to be oxidised by acidified hydrogen peroxide solution
d. iron(II) ions to be reduced when reacted with hydrogen peroxide solution.

5. Draw a galvanic cell that uses the reaction between solid aluminium metal and an aqueous solution of blue copper sulfate. Potassium nitrate can be used in the salt bridge.

a. Clearly label the following:
   • The anode, the cathode and the appropriate electrolytes
   • Equations for the reactions at the anode and cathode, marked as either oxidation or reduction
   • The overall cell reaction
   • The direction of electron flow
   • The direction of flow of anions and cations in the salt bridge.
b. What happens to the colour of the copper sulfate solution? Explain.
c. What would happen if the salt bridge was removed? Explain.

6. a. What is a fuel cell?
   b. List the advantages of a fuel cell over a primary galvanic cell.

7. Fuel cells have been developed to run on methane. Assuming that the electrolyte is acidic:

a. write the half-equation for the oxidation reaction
b. write the half-equation for the reduction reaction.
c. Draw a diagram of this cell and label the following:
   • The methane and oxygen inlets
   • The anode and cathode and their polarities
   • The direction of electron flow
   • The ion flow in the electrolyte.

8. Chromium chloride is electrolysed using chromium electrodes. A current of 0.200 A flows for 1447 seconds. The increase in the mass of the cathode is 0.0520 g.

a. How many coulombs of electricity are used?
b. How many moles of electrons are transferred?
c. How many moles of chromium are deposited?
d. What is the charge on the chromium ion?

9. How many faradays are needed to produce:
   a. 1.0 mole of copper
   b. 2.5 moles of hydrogen gas from water
   c. 15 g of aluminium
   d. 5.3 g of sodium
   e. 87 mL of oxygen gas from water at SLC?

3.6 Exam questions

Section A — Multiple choice questions

All correct answers are worth 1 mark each; an incorrect answer is worth 0.

▶ Question 1

Source: VCE 2022 Chemistry Exam, Section A, Q.20; © VCAA

MC The equipment below was set up by a student.

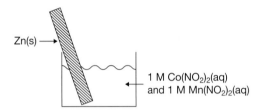

Which one of the following is correct?

A. In the beaker, the reaction between Zn(s) and Co^{2+}(aq) produces 0.48 V.
B. In the beaker, chemical energy stored in the reactants is converted to heat energy.
C. In the beaker, the concentration of ions increases because Zn(s) loses $2e^-$.
D. In the beaker, a voltage of greater than 0.42 V must be applied to Zn(s) so that it reacts with Mn^{2+}(aq).

▶ Question 2

Source: VCE 2020 Chemistry Exam, Section A, Q.26; © VCAA

MC The following reactions occur in a primary cell battery.

$$Zn + 2OH^- \rightarrow ZnO + H_2O + 2e^-$$

$$2MnO_2 + 2e^- + H_2O \rightarrow Mn_2O_3 + 2OH^-$$

Which one of the following statements about the battery is correct?

A. The reaction produces heat and Zn reacts directly with MnO_2.
B. The reaction produces heat and Zn does not react directly with MnO_2.
C. The reaction does not produce heat and Zn reacts directly with MnO_2.
D. The reaction does not produce heat and Zn does not react directly with MnO_2.

Source: VCE 2022 Chemistry Exam, Section A, Q.6; © VCAA

MC Galvanic cells and fuel cells have

A. the same energy transformations and both are reversible.
B. the same energy transformations and both produce heat.
C. different energy transformations but galvanic cells produce electricity.
D. different energy transformations but fuel cells use porous electrodes.

Source: VCE 2019 Chemistry Exam, Section A, Q.5; © VCAA

MC At the start of the day, a student set up a galvanic cell using two electrodes: nickel, Ni, and metal X.

This set-up is shown in the diagram below.

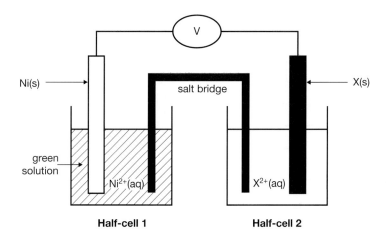

Consider the following alternative metals that could be used to replace metal X:

      1. zinc, Zn   2. lead, Pb   3. cadmium, Cd   4. copper, Cu

At the end of the day, the student checked the colour of the solution in Half-cell 1 and observed that the solution was a darker green colour.

Which of the alternative metals could cause the colour of Half-cell 1 to become a darker green?

A. metals 1 and 3
B. metals 2 and 4
C. metals 1, 2 and 3
D. metals 3 and 4

Question 5

Source: VCE 2019 Chemistry Exam, Section A, Q.18; © VCAA

MC Which one of the following galvanic cells will produce the largest cell voltage under standard laboratory conditions (SLC)?

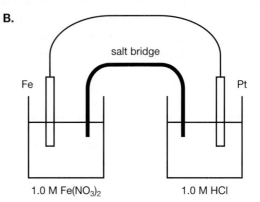

A.

salt bridge

Zn   Ni

1.0 M Zn(NO₃)₂   1.0 M Ni(NO₃)₂

B.

salt bridge

Fe   Pt

1.0 M Fe(NO₃)₂   1.0 M HCl

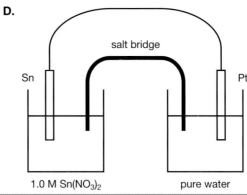

C.

salt bridge

Ag   Cu

1.0 M AgNO₃   1.0 M Cu(NO₃)₂

D.

salt bridge

Sn   Pt

1.0 M Sn(NO₃)₂   pure water

Question 6

Source: VCE 2016 Chemistry Exam, Section A, Q.25; © VCAA

MC A class of Chemistry students investigated the reaction of copper metal and iodine solution. After making predictions about the reaction, they placed a copper strip into an iodine solution and compared their predictions with their observations.

A number of groups recorded the following.

Reactants	Prediction	Observation over 10 minutes
Cu metal + I₂ solution	A reaction should occur. The expected products are Cu²⁺ and I⁻. The solution should turn from brown to blue as I₂ is consumed and Cu²⁺ is formed. The Cu metal should look corroded.	no apparent change

The predicted results were not observed. The class was asked to suggest some hypotheses to explain the unexpected result.

Which one of the following hypotheses could **not** explain the unexpected result?

A. The reaction rate might have been too slow for the time allowed.

B. An equilibrium was established and [Cu²⁺] was too low to be visible.

C. A bromine solution was accidentally used in place of the iodine solution.

D. The surface of the copper metal was greasy.

▶ Question 7

Source: Adapted from VCE 2013 Chemistry Exam, Section A, Q.28; © VCAA

MC The main reason an aqueous solution of potassium nitrate, KNO_3, is used in salt bridges is

A.	$K^+(aq)$ is a strong oxidising agent.	$NO_3^-(aq)$ is a weak reducing agent.
B.	$K^+(aq)$ is a weak reducing agent.	$NO_3^-(aq)$ is a strong oxidising agent.
C.	$K^+(aq)$ salts are soluble in water.	$NO_3^-(aq)$ salts are soluble in water.
D.	$K^+(aq)$ ions will migrate to the anode half-cell.	$NO_3^-(aq)$ ions will migrate to the cathode half-cell.

▶ Question 8

Source: VCE 2012 Chemistry Exam 2, Section A, Q.16; © VCAA

MC A galvanic cell set up under standard conditions is shown below.

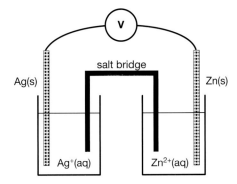

Which one of the following is correct?

As the cell discharges

	electrons would flow from the	in the salt bridge
A.	zinc electrode to the silver electrode.	anions migrate to the Ag^+/Ag half-cell.
B.	silver electrode to the zinc electrode.	cations migrate to the Zn^{2+}/Zn half-cell.
C.	silver electrode to the zinc electrode.	cations migrate to the Ag^+/Ag half-cell.
D.	zinc electrode to the silver electrode.	anions migrate to the Zn^{2+}/Zn half-cell.

▶ Question 9

Source: VCE 2018 Chemistry Exam, Section A, Q.12; © VCAA

MC The overall reaction for an acidic fuel cell is shown below.

$$2H_2 + O_2 \rightarrow 2H_2O$$

Porous electrodes are often used in acidic fuel cells because they

A. are highly reactive.

B. are cheap to produce and readily available.

C. are more efficient than solid electrodes at moving charges and reactants.

D. provide a surface for the hydrogen and oxygen to directly react together.

Source: VCE 2017 Chemistry Exam, Section A, Q.6; © VCAA

`MC` The overall equation for a particular methanol fuel cell is shown below.

$$2CH_3OH(g) + 3O_2(g) \rightarrow 2CO_2(g) + 4H_2O(l)$$

The equation for the reaction that occurs at the cathode in this fuel cell is

A. $CO_2(g) + 5H_2O(l) + 6e^- \rightarrow CH_3OH(g) + 6OH^-(aq)$
B. $CH_3OH(g) + 6OH^-(aq) \rightarrow CO_2(g) + 5H_2O(l) + 6e^-$
C. $O_2(g) + 2H_2O(l) + 4e^- \rightarrow 4OH^-(aq)$
D. $4OH^-(aq) \rightarrow O_2(g) + 2H_2O(l) + 4e^-$

Section B — Short answer questions

▶ **Question 11 (11 marks)**

Source: VCE 2019 Chemistry NHT Exam, Section B, Q.5; © VCAA

Energy can be produced in a variety of ways, including from galvanic cells, fuel cells and gas-fired power stations. Each of these methods suits particular applications.

Galvanic cells and fuel cells are methods of energy production that are based on redox reactions, similar to the reaction that would occur in Set-up A shown below. Set-up A consists of a beaker with a strip of iron, Fe, in a solution of nickel(II) nitrate, $Ni(NO_3)_2$.

Set-up A

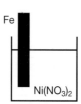

a. Identify the reducing agent for the reaction that would occur in Set-up A. **(1 mark)**

Batteries made up of primary galvanic cells, such as the one in Set-up B shown below, have traditionally been used in small electrical devices. Set-up B consists of a galvanic cell based on the redox reaction in Set-up A.

Set-up B

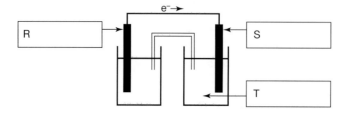

b. i. Identify an appropriate electrode material for each half-cell by writing the respective formula in boxes R and S in Set-up B. **(2 marks)**
ii. Write the formula of an appropriate solution for the half-cell in box T in Set-up B. **(1 mark)**
c. Complete the flow chart below to summarise the energy conversions that would occur in Set-up A and Set-up B. **(2 marks)**

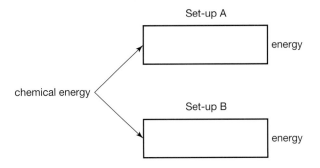

Set-up A

energy

chemical energy

Set-up B

energy

In solid oxide fuel cells (SOFC), redox reactions can be utilised to produce electrical energy for use in homes and businesses.

The diagram below represents an SOFC where the two supplied reactants are methane, CH_4, and oxygen, O_2.

The overall equation for this cell is

$$CH_4(g) + 2O_2(g) \rightarrow CO_2(g) + 2H_2O(g)$$

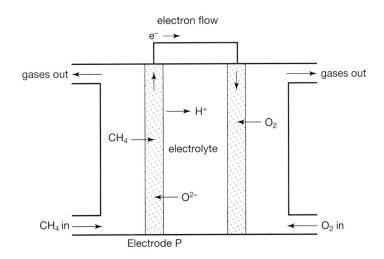

electron flow

$e^- \rightarrow$

gases out ←

gases out →

$\rightarrow H^+$

$O_2 \leftarrow$

$CH_4 \rightarrow$

electrolyte

$O^{2-} \leftarrow$

CH_4 in →

O_2 in ←

Electrode P

d. Identify Electrode P as either the anode or the cathode. **(1 mark)**
e. For the SOFC shown above, write the half-equation occurring at the cathode. States are not required. **(1 mark)**
f. In a conventional gas-fired power station, CH_4 undergoes complete combustion. The heat generated by this reaction is ultimately used to generate electrical energy.

   For a given amount of energy produced, compare the amount of greenhouse gases produced by an SOFC and a gas-fired power station based on their relative efficiencies. **(3 marks)**

Source: Adapted from VCE 2015 Chemistry Exam, Section B, Q.10.a.i,ii,b.iv; © VCAA

A car manufacturer is planning to sell hybrid cars powered by a type of hydrogen fuel cell connected to a nickel metal hydride, NiMH, battery.

A representation of the hydrogen fuel cell is given below.

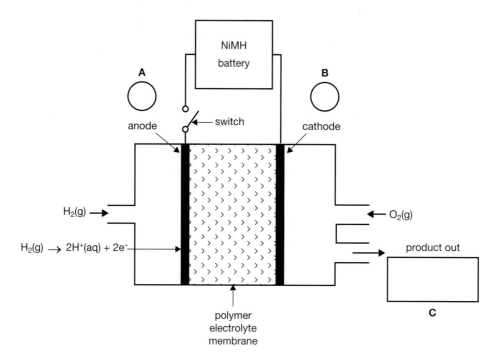

The overall cell reaction is

$$2H_2(g) + O_2(g) \rightarrow 2H_2O(g)$$

a. i. On the diagram above, indicate the polarity of the anode and the cathode in circles A and B, and identify the product of the reaction in box C. **(2 marks)**

ii. Write an equation for the reaction that occurs at the cathode when the switch is closed. **(1 mark)**

b. The storage battery to be used in the hybrid cars is comprised of a series of nickel metal hydride, NiMH, cells. MH represents a metal hydride alloy that is used as one electrode. The other electrode contains nickel oxide hydroxide, NiOOH. The electrolyte is aqueous KOH.

The simplified equation for the reaction at the anode while **recharging** is

$$Ni(OH)_2(s) + OH^-(aq) \rightarrow NiOOH(s) + H_2O(l) + e^-$$

The simplified equation for the reaction at the cathode while **recharging** is

$$M(s) + H_2O(l) + e^- \rightarrow MH(s) + OH^-(aq)$$

The battery discharged for 60 minutes, producing a current of 1.15 A.

What mass, in grams, of NiOOH would be used during this period? **(3 marks)**

Source: VCE 2021 Chemistry NHT Exam, Section B, Q.7; © VCAA

Researchers have investigated generating hydrogen, H_2, gas for hydrogen fuel cells by reacting zinc, Zn, and water, H_2O, in an electrochemical cell in series with an H_2 fuel cell. The diagram below represents an alkaline H_2 fuel cell in series with a Zn–H_2 generator cell.

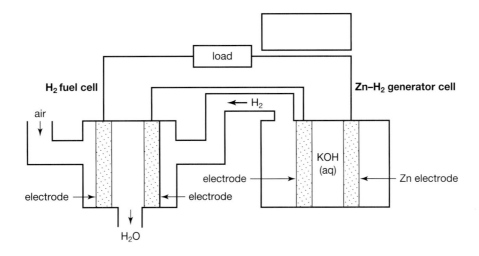

The reactions that occur at each electrode in the Zn–H_2 generator cell are given below.

$$2H_2O(l) + 2e^- \rightarrow H_2(g) + 2OH^-(aq)$$

$$Zn(s) + 2OH^-(aq) \rightarrow ZnO(s) + H_2O(l) + 2e^-$$

a. Write the overall reaction for the production of H_2 gas in the Zn–H_2 generator cell. **(1 mark)**

b. Write the half-equation that occurs at the anode of the alkaline H_2 fuel cell. **(1 mark)**

c. In the box provided in the diagram above, draw an arrow to show the direction of the flow of electrons. **(1 mark)**

d. In terms of H_2 gas flow and electron flow, explain why it is theoretically possible to connect the H_2 fuel cell in series with the Zn–H_2 generator cell. **(2 marks)**

e. Explain why the Zn–H_2 generator cell must be well-sealed to prevent contact with the atmosphere in order to produce H_2. Include any relevant equations in your answer. **(2 marks)**

f. Assuming the system is 100% efficient, describe all of the energy conversions that occur in a combined Zn–H_2 generator cell and H_2 fuel cell system. **(2 marks)**

Fuel cells have been developed to use different alkanes for fuels. An example is the propane–oxygen fuel cell. The overall reaction (assuming an acidic electrolyte) is identical to the combustion of propane in oxygen.

$$C_3H_8(g) + 5O_2(g) \rightarrow 3CO_2(g) + 4H_2O(l)$$

a. Write an equation showing the reaction occurring at the anode. **(1 mark)**

b. Write an equation showing the reaction occurring at the cathode. **(1 mark)**

c. Describe two advantages in using propane in a fuel cell instead of burning propane in a power station. **(2 marks)**

d. How do the electrodes in a fuel cell differ from those in a primary cell? **(1 mark)**

Question 15 (4 marks)

The chairperson of a leading car manufacturer says, 'Cars powered by hydrogen will be pollution-free and will therefore be carbon-neutral'.

a. Give two reasons why this statement is not true. **(2 marks)**

Hydrogen has been proposed by many car manufacturers as a future alternative to fossil fuels. One possibility is to store hydrogen as a solid hydride, which generates hydrogen gas when heated.

b. What is one advantage of using hydrogen instead of petrol as a fuel for vehicles? **(1 mark)**
c. What is one advantage of storing the hydrogen as a solid hydride rather than as a gas? **(1 mark)**

AREA OF STUDY 1 What are the current and future options for supplying energy?

OUTCOME 1

Compare fuels quantitatively with reference to combustion products and energy outputs, apply knowledge of the electrochemical series to design, construct and test primary cells and fuel cells, and evaluate the sustainability of electrochemical cells in producing energy for society.

PRACTICE EXAMINATION

STRUCTURE OF PRACTICE EXAMINATION		
Section	Number of questions	Number of marks
A	20	20
B	6	30
	Total	50

Duration: 50 minutes

Information:
- This practice examination consists of two parts. You must answer all question sections.
- Pens, pencils, highlighters, erasers, rulers and a scientific calculator are permitted.
- You may use the VCE Chemistry Data Book for this task.

 Resources

    🔗 **Weblink** VCE Chemistry Data Book

SECTION A — Multiple choice questions

All correct answers are worth 1 mark each; an incorrect answer is worth 0.

1. Which of the following statements concerning renewability is correct?
   A. Biodiesel is a renewable fuel source because carbon dioxide is taken in as the plant grows.
   B. Petrodiesel is a non-renewable fuel because it releases pollutants such as sulfur-containing compounds.
   C. Biogas is a renewable fuel source because it can be produced at the same rate as it is being used.
   D. Crude oil is a non-renewable fuel because its extraction is damaging to the environment.

2. What does the equation shown below represent?

$$6CO_2(g) + 6H_2O(l) \rightarrow C_6H_{12}O_6(aq) + 6O_2(g)$$

   A. Cellular respiration in living things
   B. Hydration of carbon dioxide to produce ethanol
   C. The production of bioethanol
   D. Reduction of carbon dioxide during photosynthesis

3. What do the arrow and the shape of the diagram represent, respectively, in the following diagram?

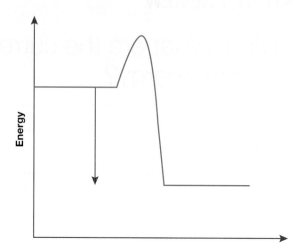

   A. Net energy released, exothermic
   B. Net energy absorbed, exothermic
   C. Net energy released, endothermic
   D. Net energy absorbed, endothermic

4. Shown below are three possible equations for the combustion of methane.

   I $CH_4(g) + 2O_2(g) \rightarrow CO_2(g) + 2H_2O(l)$

   II $CH_4(g) + 1\frac{1}{2}O_2(g) \rightarrow CO(g) + 2H_2O(l)$

   III $CH_4(g) + O_2(g) \rightarrow C(s) + 2H_2O(l)$

   20 mL of methane and 60 mL of oxygen are thoroughly mixed and then ignited in a sealed container at 20 °C.
   After combustion is complete, the container is allowed to cool back down to its original temperature.
   The gas (or gases) remaining in the container will consist of
   A. 20 mL of CO_2, 20 mL of O_2.
   B. 20 mL of CO, 30 mL of O_2.
   C. 40 mL of O_2.
   D. 20 mL of CO_2.

5. Which is the correct thermochemical equation for the complete combustion of butane?
   A. $2C_4H_{10}(g) + 13O_2(g) \rightarrow 8CO_2(g) + 10H_2O(l)$ $\Delta H = -2880$ kJ mol^{-1}
   B. $2C_4H_{10}(g) + 9O_2(g) \rightarrow 8CO(g) + 10H_2O(l)$ $\Delta H = +2880$ kJ mol^{-1}
   C. $2C_4H_{10}(g) + 13O_2(g) \rightarrow 8CO_2(g) + 10H_2O(l)$ $\Delta H = -5760$ kJ mol^{-1}
   D. $2C_4H_{10}(g) + 9O_2(g) \rightarrow 8CO(g) + 10H_2O(l)$ $\Delta H = +5760$ kJ mol^{-1}

6. A petrol engine has been found to be 35.0% efficient. If the heat content of petrol is 48.0 kJ g^{-1}, what mass of petrol is required to produce 1.00 MJ of energy using the petrol engine?
   A. 32.1 g
   B. 20.8 g
   C. 59.5 g
   D. 7.29 g

7. The amount of energy, in MJ, produced by combustion of 2.0 kg of ethane is closest to
   A. 1.0×10^2
   B. 2.1×10^2
   C. 1.0×10^3
   D. 2.1×10^3

8. Consider the following equation for incomplete combustion of carbon.

$$2C(s) + O_2(g) \rightarrow 2CO\ (g) \quad \Delta H = -221\ kJ\ mol^{-1}$$

500 mL of CO is produced at SLC. How much energy is released?
A. 4.42 kJ
B. 4.42 MJ
C. 2.21 kJ
D. 2.21 MJ

9. The following graph was obtained for a reaction that took place in a solution calorimeter.

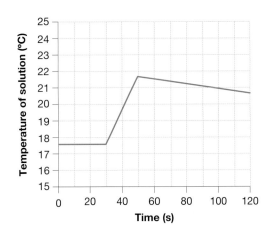

Which of the following statements are consistent with the information in the graph?
I The change in temperature was 4 °C.
II The calorimeter was not very well insulated.
III The current was turned off after 50 seconds.
A. I and II only
B. I and III only
C. II and III only
D. I, II and III

10. The equation for photosynthesis is:

$$6CO_2(g) + 6H_2O(l) \rightarrow C_6H_{12}O_6(aq) + 6O_2(g)$$

What volume of carbon dioxide, at SLC, is required to produce 200 mL of oxygen?
A. 200 mL
B. 1200 mL
C. 33.3 mL
D. 60.0 mL

11. The oxidation number of Mn, in $KMnO_4$, is
A. +1
B. +3
C. +5
D. +7

12. When metallic lead is placed in a solution of Fe^{2+} and Fe^{3+} ions, a redox reaction occurs. The correct reduction equation is
A. $Fe^{2+}(aq) \rightarrow Fe^{3+}(aq) + e^-$
B. $Fe^{3+}(aq) + e^- \rightarrow Fe^{2+}(aq)$
C. $Fe^{2+}(aq) + 2e^- \rightarrow Fe(s)$
D. $Pb^{2+}(aq) + 2e^- \rightarrow Pb(s)$

13. The following reaction is a reduction half-equation:

$$aCr_2O_7{}^{2-}(aq) + bH^+(aq) + ce^- \rightarrow dCr^{3+}(aq) + eH_2O(l)$$

The correct values for the coefficients a, b, c, d and e, respectively, are

A. 1, 14, 9, 1, 7

B. 1, 14, 9, 2, 7

C. 1, 14, 6, 2, 7

D. 2, 28, 12, 4, 14

14. In hydrogen–oxygen fuel cells, hydrogen is oxidised to produce water. A common variant of this cell operates under alkaline (basic) conditions. In such cells, what is the equation for the reaction at the cathode?

A. $O_2(g) + 2H_2O(l) + 4e^- \rightarrow 4OH^-(aq)$

B. $O_2(g) + 4H^+(aq) + 4e^- \rightarrow 2H_2O(l)$

C. $H_2(g) \rightarrow 2H^+(aq) + 2e^-$

D. $H_2(g) + 2OH^-(aq) \rightarrow 2H_2O(l) + 2e^-$

15. In a simple galvanic cell

A. electrons flow through the external circuit from anode to cathode.

B. electrons flow through the internal circuit from anode to cathode.

C. anions flow through the external circuit to the anode.

D. anions flow through the internal circuit to the cathode.

16. Nickel can displace silver ions from solution. This reaction can therefore be used to construct a galvanic cell. Using the electrochemical series, what would you predict the voltage and the material of the cathode, respectively, to be when the two half-cells are combined?

A. 0.55 V, silver

B. 1.05 V, silver

C. 0.55 V, nickel

D. 1.05 V, nickel

17. The following diagram shows a simplified fuel cell that uses the reaction between methanol and oxygen.

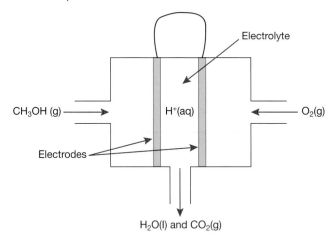

The equation for the reaction occurring at the cathode of this fuel cell is

A. $O_2(g) + 2H_2O(l) + 4e^- \rightarrow 4OH^-(aq)$

B. $O_2(g) + 2H^+(aq) + 2e^- \rightarrow H_2O_2(l)$

C. $O_2(g) + 4H^+(aq) + 4e^- \rightarrow 2H_2O(l)$

D. $CH_3OH(l) \rightarrow CO_2(g) + 4H^+(aq) + 4e^-$

18. The electrodes in fuel cells

A. are porous to allow electrons to flow through them.

B. are always made of graphite.

C. are porous so that they maximise contact with gaseous reactants.

D. act as the salt bridge.

19. In a particular type of primary cell, the following reaction occurs:

$$Zn(s) \rightarrow Zn^{2+}(aq) + 2e^-$$

This cell is used to supply 9.0 coulombs of charge. The decrease in the mass of zinc will be

A 5.1×10^{-5} g

B 1.0×10^{-4} g

C 3.0×10^{-3} g

D 6.1×10^{-3} g

20. When fuel cells of the future are compared to those in use today, which of the following statements is expected to *not* be true?

A. They will be more energy efficient.

B. They will make much more use of renewable fuels.

C. They will work with cheaper and more efficient catalysts.

D. They will make more use of renewable electricity.

SECTION B — Short answer questions

Question 21 (5 marks)

a. A sample of bioethanol undergoes complete combustion to produce 81.4 L of carbon at SLC. The density of ethanol is 0.789 g L^{-1} and $M(C_2H_5OH) = 46.0$ g mol^{-1}.

Calculate the volume of ethanol that reacted. **(4 marks)**

b. Write the equation for the production of bioethanol by fermentation. **(1 mark)**

Question 22 (6 marks)

An experiment was carried out in which ethanol was burnt to heat 150.0 mL of water at 15.0 °C to 70.0 °C. The experimental set-up is shown in the following diagram.

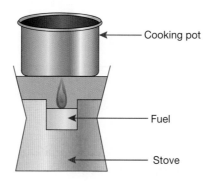

a. Calculate the mass of ethanol burnt to produce this temperature rise. **(3 marks)**

b. Is the actual value likely to be higher or lower than the calculated value? Explain your answer. **(2 marks)**

c. Comment on the reliability of the value calculated in part **a**. **(1 mark)**

Question 23 (5 marks)

Cellular respiration can be represented by the following thermochemical equation:

$$C_6H_{12}O_6(s) + 6O_2(g) \rightarrow 6CO_2(g) + 6H_2O(l) \qquad \Delta H = -2816 \text{ kJ mol}^{-1}$$

a. What is the energy value of a teaspoon (5.0 g) of glucose? **(2 marks)**

b. The table below gives information about a loaf of sliced bread.

	Quantity per serve (1 slice, 33 g)
Carbohydrate	17 g
Protein	3 g
Fat	2 g

Calculate the energy content per slice of this bread. **(2 marks)**

c. Calculate how many slices of bread a person would need to eat to meet their daily recommended energy intake of 8700 kJ. **(1 mark)**

Question 24 (8 marks)

An early type of galvanic cell developed in the early 1800s was the Daniell cell. This cell was extensively used in early forms of telegraphy.

A diagram of this cell is shown below. Note that the copper can act as both a container and an electrode in this design.

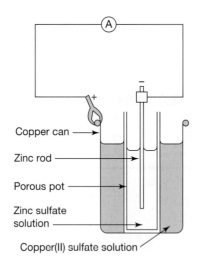

Copper can
Zinc rod
Porous pot
Zinc sulfate solution
Copper(II) sulfate solution

a. Write the equations for the reactions occurring at the cathode and the anode in this cell. Make sure you label your answers. **(2 marks)**

b. What would be the maximum predicted voltage from this cell under standard conditions? **(1 mark)**

c. If this cell provides a current of 0.80 A for 1.0 hours, calculate the change in mass of the copper can. **(4 marks)**

d. Give one disadvantage of this design. **(1 mark)**

Question 25 (3 marks)

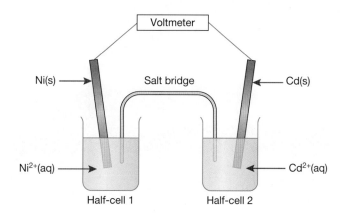

a. For the galvanic cell above, which half-cell will the cations of the salt bridge travel to? **(1 mark)**

b. What is the polarity of the nickel electrode? **(1 mark)**

c. What will happen to the mass of the Cd electrode? **(1 mark)**

Question 26 (3 marks)

The overall equation for a zinc–air button cell with a KOH electrolyte is:

$$2Zn(s) + O_2(g) + 2H_2O(l) \rightarrow 2Zn(OH)_2(s)$$

a. Write the oxidation half-equation for this cell. **(1 mark)**

b. Suggest a suitable material for the cathode. **(1 mark)**

c. Given that the cell produces a voltage of 1.4 V, determine the electrode potential for the oxidation half-equation. **(1 mark)**

UNIT 3 | AREA OF STUDY 1

PRACTICE SCHOOL-ASSESSED COURSEWORK

ASSESSMENT TASK — ANALYSIS AND EVALUATION OF A SOCIO-SCIENTIFIC ISSUE OR MEDIA COMMUNICATION

In this task you will analyse and evaluate three articles related to alternative energy sources and transport fuels.
  • Pens, pencils, highlighters, erasers, rulers and a scientific calculator are permitted.
  • You may use the VCE Chemistry Data Book to complete this task.

Total time: 60 minutes (10 minutes reading, 50 minutes writing)

Total marks: 47 marks

ALTERNATIVES FOR TRANSPORT FUELS

For this task, you will analyse four articles related to alternative energy sources and transport fuels. Answer the following questions by referring to what you have learnt in class and interpreting the four articles in the Resources panel. Remember to answer each question fully and refer to relevant data to show evidence of understanding.

▶

Articles

1. 'Will ethanol fuel a low-carbon future?'
2. 'U.S. corn-based ethanol worse for the climate than gasoline, study finds'
3. 'Research pushes auto industry closer to clean cars powered by direct ethanol fuel cells'
4. 'Hydrogen production method opens up clean energy possibilities'

Task

1. Based on the articles, why there is a need for alternative energy sources? **(2 marks)**
2. Summarise the environmental impact of the two different sources of feedstock (crop) for the production of ethanol discussed in the articles. Include the terms *renewable*, *sustainable* and *carbon neutral* in your summary. Give equations to demonstrate your arguments. **(8 marks)**
3. What is the example of CO_2 sequestration given in article 1 and why is this desirable? How viable do you think this is as a long-term solution? **(3 marks)**
4. Give two examples of the financial considerations related to the commercial production of ethanol. **(2 marks)**
5. Discuss why there is still interest in the use of commercial production of ethanol, particularly the use of bioethanol as a fuel, despite the financial issues. Include a definition of what a fuel is. **(5 marks)**
6. What is the largest hurdle facing this industry? **(1 mark)**
7. Identify two concerns raised in article 2 about corn-based ethanol. **(2 marks)**
8. Identify uses for ethanol other than blending with petrol for use in combustion engines. Why are these likely to become more important in the future? **(2 marks)**
9. Article 3 discusses the use of ethanol in a fuel cell. What is the specific advantage of ethanol compared with other fuels? **(3 marks)**
10. Identify the main innovation mentioned in article 3. **(2 marks)**
11. Explain, using reactions, the chemistry of a fuel cell that uses ethanol. **(3 marks)**
12. Discuss the differences in energy and energy conversion when ethanol is used as a transport fuel in a combustion engine versus when it is used in a fuel cell. **(2 marks)**
13. Article 4 describes the use of bioethanol as a source of hydrogen. Identify two other alternative sources of hydrogen. Describe how one of these produces hydrogen and discuss the related environmental considerations. **(4 marks)**
14. What are the advantages and disadvantages of using hydrogen in fuel cells as a transport fuel? **(4 marks)**
15. Using information from the four articles, what would you recommend that research should be focused on? Provide evidence to support your opinion. **(4 marks)**

on Resources

4 Rates of chemical reactions

KEY KNOWLEDGE

In this topic you will investigate:

Rates of chemical reactions
- factors affecting the frequency and success of reactant particle collisions and the rate of a chemical reaction in open and closed systems, including temperature, surface area, concentration, gas pressures, presence of a catalyst, activation energy and orientation
- the role of catalysts in increasing the rate of specific reactions, with reference to alternative reaction pathways of lower activation energies and represented using energy profile diagrams.

Source: VCE Chemistry Study Design (2024–2027) extracts © VCAA; reproduced by permission.

PRACTICAL WORK AND INVESTIGATIONS

Practical work is a central component of VCE Chemistry. Experiments and investigations, supported by a **practical investigation eLogbook** and **teacher-led videos**, are included in this topic to provide opportunities to undertake investigations and communicate findings.

EXAM PREPARATION

▶ Access past VCAA questions and exam-style questions and their video solutions in every lesson, to ensure you are ready.

4.1 Overview

4.1.1 Introduction

The speed, or rate, of a chemical reaction can make all the difference between it being useful or not. For example, the reactions of blasting chemicals would be useless if they did not occur at an extremely fast rate. If a potentially useful reaction is either too fast or too slow, being able to change the rate of a chemical reaction can be very useful. To achieve this, you need to understand how a reaction occurs and the different factors that affect its rate.

In this topic, you will learn about collision theory and activation energy, and use these concepts to develop a simple picture of how a reaction takes place. Factors that can affect the rate of a reaction include temperature, concentration, pressure, surface area and catalysts. These factors are applied across a range of industries in which the rates of chemical reactions are fundamental. These industries include mining, vehicle manufacture and performance, petrochemical extraction and the subsequent removal of impurities. Knowledge of rates of reaction and how these can be manipulated allows chemists to create more energy efficient, less wasteful chemical processes.

FIGURE 4.1 The city of Beirut, Lebanon, before a tragic explosion on 4 October 2020. The explosion was caused by the detonation of large quantities of ammonium nitrate stored in warehouses in the port area. Explosions release huge volumes of gas and vast amounts of energy at extremely fast rates.

LEARNING SEQUENCE

on Resources

Solutions	Solutions — Topic 4 (sol-0831)
Practical investigation eLogbook	Practical investigation eLogbook — Topic 4 (elog-1703)
Digital documents	Key science skills — VCE Chemistry Units 1–4 (doc-37066)
	Key terms glossary — Topic 4 (doc-37287)
	Key ideas summary — Topic 4 (doc-37288)
Exam question booklet	Exam question booklet — Topic 4 (eqb-0115)

4.2 Factors affecting the rate of a chemical reaction

4.2.1 How chemical reactions occur: collision theory

The Law of Conservation of Mass states that matter cannot be created or destroyed in a chemical reaction. This means that all chemical reactions involve the rearrangement of atoms that are already present. For such a rearrangement to occur, energy must be used to break existing, or 'old', bonds and will be released when 'new' bonds are allowed to form.

A chemical reaction may be pictured as particles that are moving around in constant random motion and sometimes colliding with each other.
- Some of these collisions will have enough energy to break the old bonds — these may be regarded as 'successful' collisions.
- Some orientations of the molecules during the collision process will increase the chances of the 'old' bonds being broken. This enhances the chances of a 'successful' collision.
- This is followed by the resulting pieces rearranging and forming new bonds to make the products. The greater the number of successful collisions in a given time, the faster the rate.
- Not all collisions result in a reaction; the particles may simply bounce off each other if they do not have enough energy, resulting in nothing more than changes in velocity, or they might not hit each other in the right orientation for bonds to break.

FIGURE 4.2 The chances of a reaction taking place are increased when particle collisions have the correct orientation. In the reaction ABC → AC + B, the chances of a reaction are greater if the particles collide as shown in figure **b** rather than figure **a**.

For the reaction 2ABC → 2AC + B$_2$:

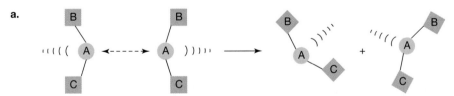

a.

Collision probably unsuccessful: colliding in this unfavourable orientation would make the bond between A and B more difficult and less likely to break than in the situation below

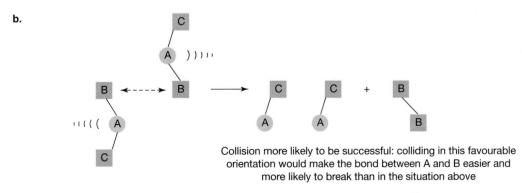

b.

Collision more likely to be successful: colliding in this favourable orientation would make the bond between A and B easier and more likely to break than in the situation above

Activation energy (E_a)

The **activation energy (E_a)** is the minimum amount of energy required by a collision in order to break the old bonds and therefore allow a reaction to begin. Collisions that do not have this minimum energy requirement will not result in a reaction. The activation energy acts as a barrier that must be overcome in order for a reaction to occur. The activation energy is shown as the difference between the reactant energy and the peak in an energy profile diagram (figure 4.3). Energy profile diagrams are discussed in detail in section 4.3.1.

activation energy (E_a) the minimum energy required by reactants in order to react

FIGURE 4.3 The E_a of a reaction represents a 'hill' that must be climbed before a reaction can occur.

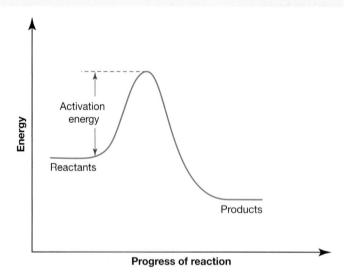

Measuring reaction rates

The progress of a reaction may be conveniently monitored by following either the decrease in the amount of a reactant or the formation of a product. Methods used to observe reaction rates include measuring the change over a period of time of:

- the volume of a gas evolved
- the mass of a solid formed
- the decrease in mass due to a gas evolved
- the intensity of colour of a solution
- the formation of a precipitate
- pH
- temperature.

It is often informative to graph such changes against time. For example, figure 4.4 shows how the concentration of hydrogen evolved varied over time in two experiments.

FIGURE 4.4 Graphs are often used to display rate data.

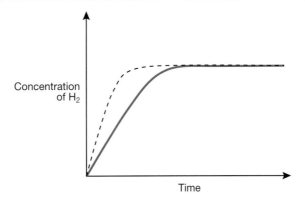

In the experiment represented by the dashed line, the rate of evolution at the beginning was faster than in the experiment represented by the solid line. This is indicated by the different gradients of the two graphs. The steeper the gradient, the faster the rate of reaction.

EXTENSION: Measuring rates

Rate measurements can often be made with simple, easily obtained apparatus. Because there are a number of factors that influence rate, such experiments must be designed carefully and take into account the variables present. These include independent, dependent and controlled variables.

To access more information on this extension concept, download the digital document from your Resources panel.

on Resources

📄 **Digital document** EXTENSION: Measuring rates (doc-37346)

4.2.2 Factors affecting the rate of a chemical reaction

There are five main factors that affect the rate of a chemical reaction: concentration, gas pressure, temperature, surface area and catalysts. Catalysts are discussed in more detail in subtopic 4.3. You should note that each of these factors will affect the frequency of successful collisions.

Concentration

Concentration refers to the amount of a substance in a given volume. Table 4.1 shows some results from an experiment involving the reaction:

$$2H_2(g) + 2NO(g) \rightarrow 2H_2O(g) + N_2(g)$$

TABLE 4.1 Rate of reaction between NO and H_2 at 800 °C

Experiment	Initial concentrations (M)		Initial rate of H_2O production (M s^{-1})
	[NO]	[H_2]	
1	6.0×10^{-3}	1.0×10^{-3}	0.64×10^{-2}
2	6.0×10^{-3}	2.0×10^{-3}	1.28×10^{-2}
3	1.0×10^{-3}	6.0×10^{-3}	1.00×10^{-3}
4	2.0×10^{-3}	6.0×10^{-3}	3.90×10^{-3}

Note: Square brackets denote concentration in mol L^{-1} (M).

In experiments 1 and 2, [NO] is the same but [H_2] is different.

In experiments 3 and 4, [H_2] is the same but [NO] is different.

Therefore, each pair of experiments allows us to analyse the effect of changing the concentration of one of the two substances. Compare experiment 1 with experiment 2, and then compare experiment 3 with experiment 4.

Increasing the concentration of either reactant causes an increase in the rate of the reaction. In terms of our model of a chemical reaction, we can explain this by the crowding together of the reacting particles as the concentration is increased. This results in an increased number of overall collisions during a given period of time. With more collisions, there will be an increase in the frequency of successful collisions, resulting in a higher rate of reaction. The relative proportion of successful collisions does not increase.

Gas pressure

For reactions involving gases, the effect of increasing pressure (which decreases the volume) is the same as increasing concentration. Both effects result in more crowding together of the particles and therefore more collisions per unit time. This ensures more successful collisions within a certain time.

Temperature

Most chemical reactions are observed to proceed more quickly as the temperature is increased. Examples from everyday life that demonstrate this are the cooking and deterioration of food, and the setting of some glues.

An understanding of collision theory explains why this is so. As temperature increases, particles on average move faster and have greater kinetic energy. When collisions occur, this increased energy means that a greater proportion of the collisions will be successful. The rate of reaction will be greater due to more existing bonds being broken in any given time period.

Another effect of increasing the temperature is that there is a greater number of collisions due to the increased movement of the particles. However, a more sophisticated analysis of the situation reveals that this has less effect on the reaction rate than the particles colliding with a higher average kinetic energy, as mentioned earlier.

FIGURE 4.5 Increasing the temperature means that particles have more kinetic energy, which results in more successful collisions.

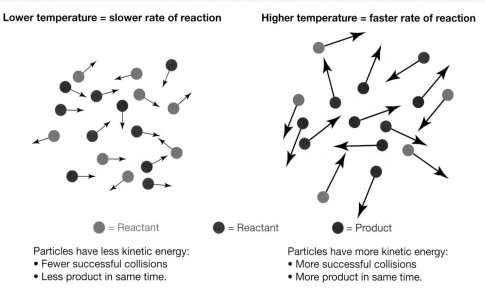

Lower temperature = slower rate of reaction   **Higher temperature = faster rate of reaction**

● = Reactant   ● = Reactant   ● = Product

Particles have less kinetic energy:
• Fewer successful collisions
• Less product in same time.

Particles have more kinetic energy:
• More successful collisions
• More product in same time.

EXTENSION: Maxwell–Boltzmann distribution curves

A **Maxwell–Boltzmann distribution curve** shows the range and distribution of particle energies within a sample. It is a statistical analysis of the energies present in the particles of a gas sample, although many of its ideas can also be applied to liquids and reactions in solution. This range of energies exists because some particles slow down as a result of the collisions they undergo, whilst others speed up.

FIGURE 4.6 Increasing the temperature of a gas sample stretches the Maxwell–Boltzmann distribution curve to the right.

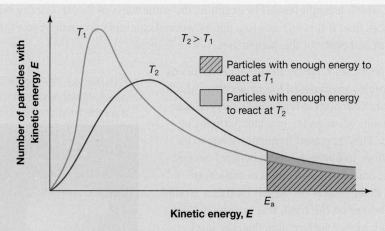

There are some points to note about a Maxwell–Boltzmann distribution curve.
• The particles in a sample have a wide range of kinetic energies. As kinetic energy is given by the formula

$$KE = \frac{1}{2}mv^2$$ (where m is mass and v is the velocity of the particles), there is also a range of velocities. This is due to the collisions that the particles are constantly undergoing.
• Only a small proportion of particles in the sample have kinetic energy that is equal to or greater than the activation energy, E_a.
• It is not symmetrical.
• The highest point represents the most probable velocity; this is not the same as the average velocity.
• The area under the graph represents the total number of particles in the sample.

Maxwell–Boltzmann distribution curve a graph that plots the number of particles with a particular energy (vertical axis) against energy (horizontal axis)

- If the temperature of a sample is increased, the graph changes in a predictable manner (see figure 4.6).
  - Increasing the temperature (From T_1 to T_2) of the gas sample stretches the graph to the right. As a result, there are more particles with higher kinetic energies. Although the area under the graph is the same (the total number of particles has not been altered), on average they all move faster and the average kinetic energy is higher.
  - Therefore, more particles have energy levels at or above activation energy (E_a) and can therefore react.

Note that, at the higher temperature, the graph is *stretched* to the right rather than *moved* to the right. The graph is always anchored to the origin, because there are always a few particles with very low or zero velocity.

elog-1726

EXPERIMENT 4.1   on line only

The effect of concentration and temperature on reaction rates

Aim

To investigate the effect of concentration changes and temperature changes on the reaction represented by the equation:

$$S_2O_3^{2-}(aq) + 2H^+(aq) \rightarrow S(s) + SO_2(g) + H_2O(l)$$

on Resources

▶ **Video eLesson**   Temperature and reaction rate (eles-1670)

Surface area

Surface area is an important factor in **heterogeneous reactions** — reactions in which the reactants are in different phases, such as a solid and a liquid.

Increasing the surface area increases the rate of reaction as more of the substance is brought into contact with other substances with which it might react. For example, the same mass of wood on a fire burns much faster if it is cut into small pieces than if it is left as a log, and powdered calcium carbonate reacts faster in acid solution than a block of calcium carbonate of the same mass.

In terms of collision theory, an increase in surface area means that more reactant particles are exposed to one another, which logically produces more collisions. More collisions produces more successful collisions between the reactant particles in a given period of time. This increased frequency leads to an increased rate of reaction. In figure 4.7, both beakers contain 25 mL of hydrochloric acid and 1 g of calcium carbonate (marble). The marble in the beaker on the left has been ground into a powder; in the beaker on the right, it is in large chunks. The powder has a much higher surface area than the large chunks, resulting in a much higher reaction rate, which is shown by the faster release of carbon dioxide bubbles.

FIGURE 4.7 The higher surface area of the powder (left) results in a higher reaction rate than the solid on the right.

The effect of increasing surface area on the rate of combustion reactions can lead to unexpected, and sometimes catastrophic, results. Solids such as coal and wheat do not normally burn very fast, but as a dust they present a huge surface area to the oxygen in air. All that is needed is a spark — for example, from a machine or from static electricity — and the resulting reaction is so fast that it causes an explosion. This effect has destroyed wheat silos and caused tragedies in underground mines.

heterogeneous reaction a reaction in which some of the substances involved are in different phases

tlvd-3053

SAMPLE PROBLEM 1 Determining conditions for fastest rates of reaction

A gas-phase reaction is carried out at four different sets of conditions of temperature and pressure as shown.
- **Set A: 25 °C and 100 kPa**
- **Set B: 50 °C and 100 kPa**
- **Set C: 50 °C and 150 kPa**
- **Set D: 25 °C and 150 kPa**

Which set of conditions will produce the fastest rate and why?

THINK	WRITE
Recall that higher temperatures and higher pressures will increase the rate of a reaction. Condition set C meets both these criteria.	Set C will produce the fastest reaction rate because it occurs at both the highest temperature and at the highest pressure. Both these factors increase the rate of a reaction.

PRACTICE PROBLEM 1

With reference to the conditions shown in sample problem 1, which set of conditions will produce the *slowest* rate of reaction?

elog-1727

EXPERIMENT 4.2

online only

Rate of hydrogen production — a problem-solving exercise

Aim

Student developed

Catalysts

Catalysts are the fifth mechanism by which the rate of a reaction may be altered. Catalysts are substances that alter the rate of a chemical reaction without being consumed.

The use of catalysts is becoming more and more important as the chemical industry moves to adopt the principles of green chemistry. Through the use of catalysts, processes can be made to occur at lower temperatures and pressures. This results in lower energy consumption with corresponding environmental and economic advantages.

Catalysts are discussed in more detail in subtopic 4.3.

4.2.3 Open and closed systems

In chemistry a *system* can be thought of as the reaction that is under study. Everything else outside this is called the surroundings or the environment. Such systems are commonly classified in one of two ways:
- A **closed system**, which allows the transfer of energy, but not matter, to or from its surroundings
- An **open system**, which allows both matter and energy transfer to or from its surroundings.

Whether a system is open or closed can influence the *apparent* rate of some reactions, including some that are familiar from everyday life. The evaporation of liquids is a good example of this.

> **closed system** a system in which energy, but not matter, can be transferred to and from its surroundings; all reactants and products are contained
>
> **open system** a system in which both energy and matter can be transferred to and from its surroundings; reactants and products are not contained

Consider an open flask containing an amount of ethanol that is sitting on a top-loading balance. Over a period of time the ethanol will evaporate until there is none left. This may be represented by the equation:

$$CH_3CH_2OH(l) \rightarrow CH_3CH_2OH(g)$$

The rate of evaporation could be obtained by measuring the decrease in mass over a suitable time interval. This is caused by some molecules of ethanol having enough kinetic energy to 'break free' from the surface and escape into the gas phase above the surface. Such molecules can then move away due to the freedom of motion that all particles in a gas phase possess and leave the flask as there is nothing to constrain them. This is an example of an open system.

If a stopper is placed in the flask, a different result occurs. This time, no loss in mass occurs. If a means of measuring gas pressure is included, an *increase* in gas pressure within the flask will be detected. In this scenario, these readings taken over time could be used to measure the rate of evaporation. This time the *apparent* rate of evaporation will decrease to zero, even though there is still liquid present. This is an example of a closed system and gives totally different results to the open system discussed above.

FIGURE 4.8 The rate of a reaction may be affected by whether it is an open system or a closed system.

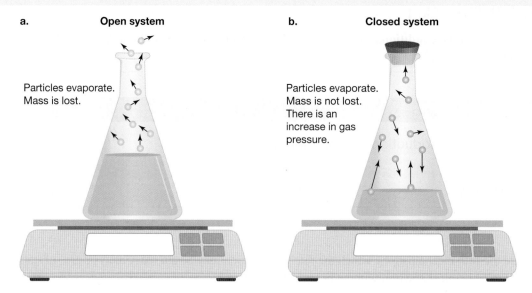

a. Open system

Particles evaporate. Mass is lost.

b. Closed system

Particles evaporate. Mass is not lost. There is an increase in gas pressure.

This difference can be explained by collision theory.

In a closed system, the gas molecules are moving in rapid, random motion and colliding with each other and the walls of the flask. They will also be colliding with the surface of the liquid. When this happens, they will be re-absorbed back into the liquid. As more and more molecules enter the gas phase through evaporation, they have a greater and greater statistical chance of colliding with the liquid surface and being re-absorbed. Eventually a stage will be reached where these two opposing processes are in balance and the rate of evaporation as measured by this apparatus will appear to be zero. Two opposing reactions (evaporation and re-absorption) are occurring, but at the same rate, resulting in no overall change. This idea is discussed further in topic 5.

Finally, it should be noted that chemists sometimes deal with a third type of system called an **isolated system**. This is a system in which neither matter *nor* energy is transferred to or from its surroundings. A well-insulated calorimeter is a good example of such a system.

isolated system a system in which neither matter nor energy is transferred to or from its surroundings

---

Open and closed systems

- An open system allows both matter and energy to be transferred between itself and its surroundings.
- A closed system allows the transfer of energy, but not matter, between itself and its surroundings.

4.2 Activities

4.2 Quick quiz on	4.2 Exercise	4.2 Exam questions

4.2 Exercise

1. State the three requirements for a reaction to occur, according to collision theory.
2. Use collision theory to explain
   a. why there is always a range of particle velocities at any temperature.
   b. why there will always be an amount of reaction, however small, at any given temperature.
3. State five methods that can be used to increase the rate of a reaction.
4. Explain why the chemicals used in fireworks are present in powdered form.
5. The reaction between two gases occurs at a measurable rate at 700 °C. If the temperature is held constant at 700 °C and the reacting mixture is compressed, predict and explain what will happen to the rate of this reaction.
6. The evolution of bubbles when a soft drink is opened is due to dissolved carbonic acid decomposing to carbon dioxide and water. The equation for this process is:

$$H_2CO_3(aq) \rightleftharpoons CO_2(g) + H_2O(l)$$

   a. This evolution is initially fast, but gradually slows with time. Why does the rate of carbon dioxide evolution decrease?
   b. How would the rate of reaction be affected if the soft drink was warm? Justify your answer.
7. An experiment was carried out to measure the rate of carbon dioxide formation when small pieces of calcium carbonate were added to hydrochloric acid solutions of different concentrations. To ensure complete reaction of the calcium carbonate, the amount of acid used was in excess each time. Use collision theory to explain why increasing the acid concentration led to an increase in the rate of gas produced.
8. In an investigation of the rate of reaction of gas produced from magnesium, Mg, and hydrochloric acid, HCl, a student has available three forms of magnesium: powder, small turnings and a strip. Also available are reagent bottles of 0.5 M HCl, 1 M HCl and 2 M HCl. The student could also use a hot water bath and a cool water bath.
   a. Which combination of reactants and conditions would produce the fastest rate of reaction?
   b. Which combination would produce the slowest rate of reaction?
9. a. Gas leaks in confined spaces can be very dangerous — a single spark can lead to an explosion. Explain, in terms of reaction rates, why this is so.
   b. This reaction is exothermic. Why is this important in producing the explosion?

4.2 Exam questions

▶ **Question 1 (1 mark)**

Source: VCE 2019 Chemistry Exam, Section A, Q.11; © VCAA

MC 5 mL of ethanol, CH_3CH_2OH, undergoes combustion in a test tube with a diameter of 1 cm. This experiment is performed in a fume cupboard. The temperature in the fume cupboard is 20 °C.

Which one of the following actions will reduce the rate of reaction?
A. Mix 2 mL of a dilute solution of sodium hydroxide, NaOH, with the ethanol.
B. Perform the experiment in a test tube with a diameter of 2 cm.
C. Increase the temperature in the fume cupboard to 25 °C.
D. Increase the volume of the ethanol to 7 mL.

⏵ **Question 2 (1 mark)**

Source: VCE 2016 Chemistry Exam, Section A, Q.27; © VCAA

MC A student set up an experiment to test the effect of different factors on the rate and extent of the reaction between a strong acid and marble chips (calcium carbonate, $CaCO_3$). In each trial, the mass of the flask and its contents was measured every 30 seconds, from the instant the reactants were mixed.

Trial 1	Trial 2
The strong acid used was hydrochloric acid, HCl. The equation for the reaction is as follows. $$2HCl(aq) + CaCO_3(s) \rightarrow CaCl_2(aq) + CO_2(g) + H_2O(l)$$ 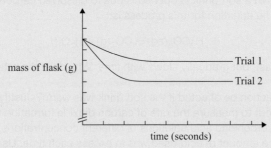	One change to the reaction conditions was made and the experiment was repeated.

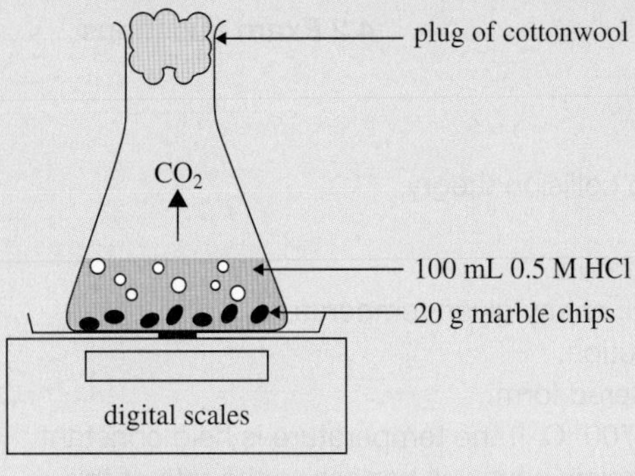

The results of the two trials were graphed on the same axes and are shown below.

In Trial 2, the student must have

A. heated the 0.5 M HCl before adding it to the flask.
B. doubled the volume of 0.5 M HCl added to the flask.
C. used 100 mL of 0.5 M H_2SO_4 instead of 100 mL of 0.5 M HCl.
D. used the same mass of marble but crushed it into a powder.

⏵ **Question 3 (4 marks)**

Source: VCE 2016 Chemistry Exam, Section B, Q.5.a; © VCAA

Bromomethane, CH_3Br, is a toxic, odourless and colourless gas. It is used by quarantine authorities to kill insect pests.

A simplified reaction for its synthesis is:

$$CH_3OH(g) + HBr(g) \rightleftharpoons CH_3Br(g) + H_2O(g) \qquad \Delta H = -37.2 \text{ kJ mol}^{-1} \text{ at 298 K}$$

The manufacturer of this chemical investigates reaction conditions that could affect the time the process takes and the percentage yield.

Predict the effect of each change given below on the rate of production of bromomethane (increase, no change or decrease). Give your reasoning.

- Increasing temperature (constant volume):

  Effect: increase   no change   decrease

- Increasing pressure (constant temperature):

  Effect: increase   no change   decrease

Question 4 (4 marks)

Source: VCE 2020 Chemistry Exam, Section B, Q.9.d; © VCAA

A student decided to investigate the effect of temperature on the rate of the following reaction.

$$2HCl(aq) + CaCO_3(s) \rightarrow CaCl_2(aq) + H_2O(l) + CO_2(g)$$

Part of the student's experimental report is provided below.

Effect of temperature on the rate of production of carbon dioxide gas

Aim

To find out how temperature affects the rate of production of carbon dioxide gas, CO_2, when a solution of hydrochloric acid, HCl, is added to chips of calcium carbonate, $CaCO_3$

Method
1. Put 0.6 g of $CaCO_3$ chips into a conical flask.
2. Put a reagent bottle containing 2 M HCl into a water bath at 5 °C.
3. When the temperature of the HCl solution has stabilised at 5 °C, use a pipette to put 10.0 mL of the HCl solution into the conical flask containing the $CaCO_3$ chips.
4. Put a balloon over the conical flask and begin timing.
5. When the top of the balloon has inflated so that it is 10 cm over the conical flask, stop timing and record the time.
6. Repeat steps 1–5 using temperatures of 15 °C, 25 °C, 35 °C and 45 °C.

Results
The following graph gives the experimental results.

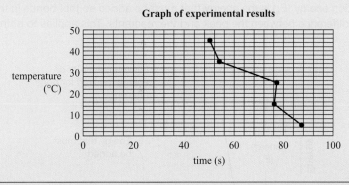

Graph of experimental results

---

a. Predict the relationship between the independent variable and the dependent variable. Explain your prediction. **(3 marks)**

b. Is the graph of the student's results consistent with your prediction? Give your reasoning. **(1 mark)**

Question 5 (1 mark)

Source: VCE 2013 Chemistry Exam, Section A, Q.14; © VCAA

$$Cu(s) + 4HNO_3(aq) \rightarrow Cu(NO_3)_2(aq) + 2NO_2(g) + 2H_2O(l)$$

MC Which one of the following will **not** increase the rate of the reaction?
A. decreasing the size of the solid copper particles
B. increasing the temperature of HNO_3 by 20 °C
C. increasing the concentration of HNO_3
D. allowing NO_2 gas to escape

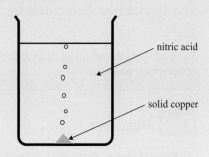

More exam questions are available in your learnON title.

4.3 Catalysts and reaction rates

4.3.1 Energy profile diagrams

You will recall that energy profile diagrams were discussed in topic 1, section 1.3.1. From a rate perspective, the most relevant feature of such diagrams is the activation energy, E_a. It does not matter if the reaction is **exothermic** or **endothermic** — the activation energy is always a requirement before a reaction can occur. It reflects the energy that must be added before existing bonds can be broken. In other words, it is the energy that must be provided before a reaction can begin. In terms of collision theory, it represents the lowest amount of energy that must be added through reactant particle collisions before bonds can be broken.

exothermic describes a chemical reaction in which energy is released to the surroundings

endothermic describes a chemical reaction in which energy is absorbed from the surroundings

These ideas are summarised in figure 4.9. Note that this figure shows only the left-hand half of an energy profile diagram because it does not matter whether a reaction is exothermic or endothermic.

FIGURE 4.9 The activation energy (E_a) is the energy that must be added so that bonds in the reactants can be broken. It is the energy difference between the peak and the reactants. This applies to both exothermic and endothermic reactions.

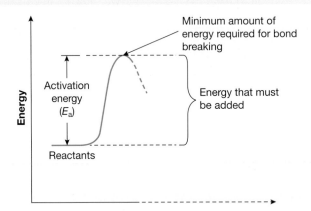

EXPERIMENT 4.3　　　　　　　　　　　　　　　　　　　　　**online only**

Investigation of two catalysts

Aim

To investigate the effect of two catalysts, iron(III) chloride and manganese dioxide, on the rate of decomposition of hydrogen peroxide

elog-1728

tlvd-9718

---

on Resources

▷ **Video eLesson**　Exothermic and endothermic reactions (eles-3240)

✦ **Interactivities**　Identifying exothermic and endothermic reactions (int-1242)
　　　　　　　　　　Constructing energy profile diagrams (int-1243)

4.3.2 How catalysts work

A **catalyst** is a substance that increases the rate of a chemical reaction without being consumed. It provides an alternative reaction pathway with a lower activation energy. This increases the proportion of collisions with energy greater than the activation energy.

- Catalysts are used to speed up a reaction.
- Adding a catalyst does not alter the value of ΔH.
- The presence of a catalyst in a chemical reaction provides an *alternative pathway* for particles to collide with sufficient energy to break bonds. This pathway has a lower activation energy, as shown in figure 4.10.

catalyst a substance that increases the rate of a reaction without a change in its own concentration

FIGURE 4.10 A catalyst acts by providing an alternative pathway with a lower activation energy for reactants to form products.

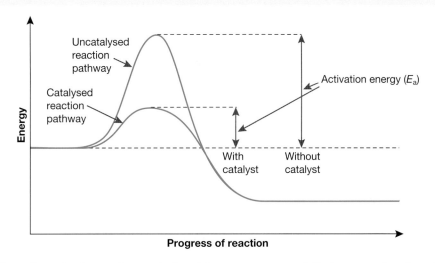

Appropriate catalysts lie at the heart of many industrial processes, especially in green chemistry industries, which attempt to reduce the use of hazardous substances. Companies spend large amounts of money on research into new and improved catalysts, and the results of such research are often among a company's most closely guarded secrets. Biological catalysts, or enzymes, are also responsible for the management of thousands of biological reactions important in maintaining life. Enzymes are discussed further in topic 11.

Catalysts

- Catalysts work by lowering the activation energy of a reaction.
- Lowering the activation energy is not the same as increasing the energy of reactant molecules.

FIGURE 4.11 Catalysts reduce the energy required for a reaction to occur.

on Resources

Video eLesson Role of catalysts (eles-3241)

tlvd-9680

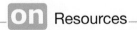

SAMPLE PROBLEM 2 Using reaction profiles to compare catalysts and reaction rates

The following diagram shows the energy profile diagram for a particular reaction. It shows the uncatalysed version as well as the reaction when two different catalysts (A and B) are used. Note that the different reaction profiles have been labelled X, Y and Z as shown.

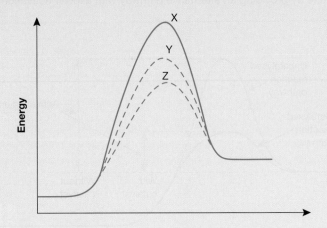

When this reaction is tested experimentally, under identical conditions with each catalyst, it is found that the reaction rate is fastest with catalyst A. Which reaction profile corresponds to catalyst A?

THINK

Catalysts increase the rate of reaction by providing a pathway that has a lower activation energy. When particles collide, less energy is required to break bonds and the frequency of successful collisions will increase.

Pathway Z has the lowest activation energy and will have the highest frequency of successful collisions, and therefore the fastest rate.

WRITE

The profile labelled Z corresponds to catalyst A.

PRACTICE PROBLEM 2

Which of the reaction profiles shown in sample problem 2 corresponds to catalyst B?

elog-1729

tlvd-9719

EXPERIMENT 4.4 online only

The effect of concentration, temperature and a catalyst on reaction rates

Aim

To investigate the effect of concentration, temperature and a catalyst on the rate of the reaction between iodide and peroxydisulfate ions

EXTENSION: Demonstrating catalysts using a Maxwell–Boltzmann curve

Catalysts provide an alternate pathway for particles to collide with sufficient energy to react, and this can be demonstrated using a Maxwell–Boltzmann curve. This means that the value of E_a on the Maxwell–Boltzmann curve is shifted to the left and that there are a greater proportion of particles under the curve to the right of this new E_a value. These are the particles that have enough energy to overcome the activation energy requirement, allowing the reaction to occur at a faster rate due to the increase in successful collisions.

FIGURE 4.12 A catalyst provides an alternative pathway with a lower activation energy, which allows more particles to overcome the new activation energy requirement.

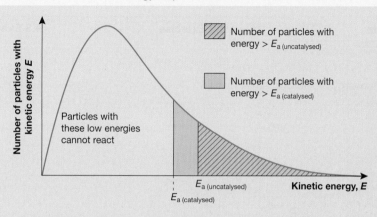

CASE STUDY: Recognising the importance and future of catalysts in green chemistry

A recent Nobel prize in chemistry was awarded for work that involved biological catalysts (enzymes).

The 2021 prize was awarded to Benjamin List and David W.C. MacMillan for 'the development of asymmetric organocatalysis'. Working independently, these two scientists have effectively discovered a third class of catalysts — organocatalysts.

FIGURE 4.13 Optical isomers are non-superimposable images of each other.

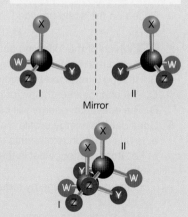

Benjamin List's discovery (or re-discovery) came from asking the question 'do amino acids have to be part of an enzyme to catalyse a chemical reaction?' It was already known that the amino acid proline was an effective catalyst, and List was able to quickly confirm this. The significant advance was the discovery that it catalyses the production of a specific optical isomer. Optical isomers (also called enantiomers) are molecules that contain a carbon atom that is bonded to four different atoms or groups of atoms, and are non-superimposable, mirror images of each other. They are often likened to your left and right hands, which cannot be placed identically on top of each other unless one is reflected in a mirror.

Such isomers are often of critical importance in the pharmaceutical industry. Often it is only one isomer that is biologically active. The other isomer is either inactive or even harmful, as evidenced by the use of thalidomide in the late 1950s and early 1960s. The ability to use an enzyme to selectively produce only one optical isomer represents a huge advance as it is currently very difficult and wasteful to do this.

David MacMillan had been working on using metals for asymmetric catalysis (that is, making optical isomers). However, the research results proved difficult to scale up to industrial application. The catalysts were simply too expensive or too difficult to use. In search of an alternative, he designed smaller molecules and tested their ability to selectively produce optical isomers, eventually finding a number that were able to do so. He subsequently coined the term 'organocatalysis' to mean the use of small molecules that were able to catalyse the production of asymmetric molecules.

Students, these questions are even better in jacPLUS

 Receive immediate feedback and access sample responses

 Access additional questions

Track your results and progress

Find all this and MORE in jacPLUS ⏵

| **4.3 Quick quiz** on | **4.3 Exercise** | **4.3 Exam questions** |

4.3 Exercise

1. **MC** Consider the energy profile diagram shown.

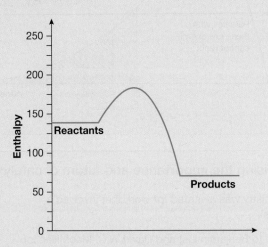

Based on the scale provided in the diagram, what is the activation energy for the forward reaction?

A. 75 B. 125 C. 150 D. 50

2. **MC** Which of the following could **not** apply to an endothermic reaction?
   A. During the reaction, the temperature changes from 23.5 °C to 18.9 °C.
   B. Heating is required to maintain the reaction.
   C. The change in enthalpy for the reaction is +236 kJ mol^{-1}.
   D. The activation energy will be lower than the enthalpy of the reactants.

3. Consider the energy profile diagram shown.
   a. Is this reaction exothermic or endothermic?
   b. For this reaction, what is the value of:
      i. ΔH
      ii. the minimum energy required to break the reactant bonds
      iii. the activation energy
      iv. the energy evolved when the new bonds form?

4. The gas silane, SiH_4, reacts spontaneously with oxygen at normal temperatures to produce silicon dioxide, SiO_2. What does this indicate about the activation energy for this reaction?

5. Why is it not possible to have a reaction with an activation energy of 200 kJ mol^{-1} and a ΔH of +350 kJ mol^{-1}?

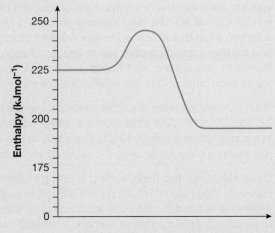

6. Many chemical reactions are reversible, meaning that they can react in a backward direction. Suggest how the activation energy of the forward reaction compares to the activation energy of the backward reaction for:
   a. an exothermic reaction.   b. an endothermic reaction.
7. Can a catalyst turn an exothermic reaction into an endothermic one? Explain.
8. Draw an energy profile for an exothermic reaction that occurs with a catalyst. On the same diagram, add an energy profile for the reaction without a catalyst.
9. Catalysts X and Y both catalyse an endothermic reaction that occurs slowly under typical laboratory conditions. It is noted that X produces a faster rate of reaction than Y.
   Summarise this information in the form of an energy profile diagram.
10. Consider a reaction that is represented by the following equation, and is catalysed by a finely divided metal powder, X.

$$AB(g) + C(g) \rightarrow CA(g) + B(g)$$

An important part of the catalytic mechanism is that the catalyst forms temporary bonds with AB on its surface.
   a. Explain why this catalyst is more effective in powdered form than in a metallic lump.
   b. Given that the catalyst temporarily holds AB on its surface in a certain way, what else is happening here to increase the rate of the reaction?
   c. Given that a catalyst lowers the activation energy, what effect do you think the temporary bonds that form between X and AB have on the bonds that need to be broken for this reaction to occur?
   d. Given that product CA does not bind to the surface of the catalyst, explain why X can keep performing its function and is not used up.

4.3 Exam questions

▶ **Question 1 (1 mark)**

Source: VCE 2018 Chemistry Exam, Section A, Q.13; © VCAA

MC The energy profile diagram below represents a particular reaction. One graph represents the uncatalysed reaction and the other graph represents the catalysed reaction.

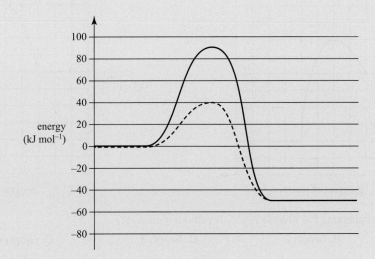

Which of the following **best** matches the energy profile diagram?

	E_a uncatalysed reaction (kJ mol⁻¹)	ΔH catalysed reaction (kJ mol⁻¹)
A.	40	−140
B.	90	−140
C.	40	−50
D.	90	−50

▶ **Question 2 (1 mark)**

Source: VCE 2022 Chemistry Exam, Section A, Q.11; © VCAA

MC The graphs shown below are energy profiles for the following reaction.

$$A + B \rightleftharpoons C \quad \Delta H < 0$$

The graphs represent the forward reaction, with and without a catalyst, and the reverse reaction, with and without a catalyst. All graphs are drawn to the same scale.

Graph 1

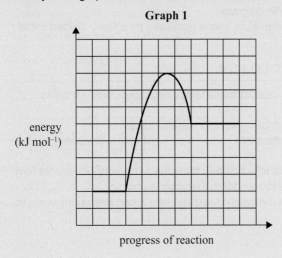

Graph 2

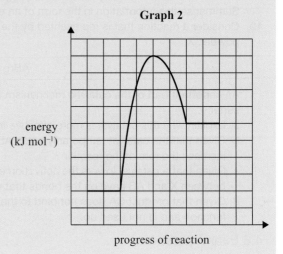

Graph 3

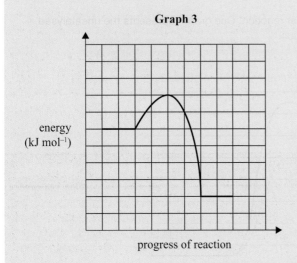

Graph 4

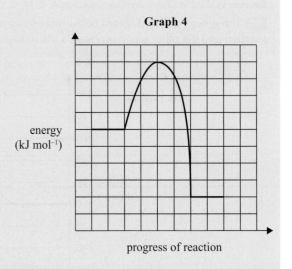

Which energy profile represents the reverse reaction without a catalyst?

A. Graph 1 **B.** Graph 2 **C.** Graph 3 **D.** Graph 4

▶ **Question 3 (1 mark)**

Source: VCE 2021 Chemistry Exam, Section A, Q.24; © VCAA

MC Which one of the following statements describes the effect that adding a catalyst will have on the energy profile diagram for an exothermic reaction?

A. The energy of the products will remain the same.

B. The shape of the energy profile diagram will remain the same.

C. The peak of the energy profile will move to the left as the reaction rate increases.

D. The activation energy will be lowered by the same proportion in the forward and reverse reactions.

▶ **Question 4 (5 marks)**

Source: VCE 2014 Chemistry Exam, Section B, Q.1; © VCAA

The decomposition of ammonia is represented by the following equation.

$$2NH_3(g) \rightleftharpoons N_2(g) + 3H_2(g) \quad \Delta H = 92.4 \, kJ \, mol^{-1}$$

a. The activation energy for the uncatalysed reaction is 335 kJ mol^{-1}.

The activation energy for the reaction when tungsten is used as a catalyst is 163 kJ mol^{-1}.

On the grid provided below, draw a labelled energy profile diagram for the uncatalysed and catalysed reactions. **(3 marks)**

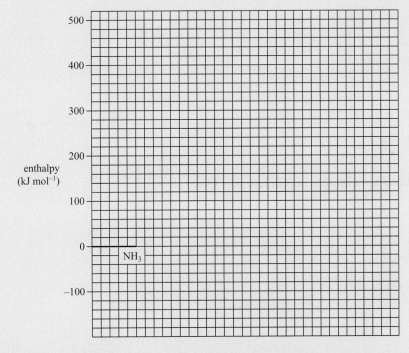

b. When osmium is used as a catalyst, the activation energy is 197 kJ mol^{-1}.

Which catalyst — osmium or tungsten — will cause ammonia to decompose at a faster rate? Justify your answer in terms of the chemical principles you have studied this year. **(2 marks)**

▶

Source: VCE 2011 Chemistry Exam 2, Section B, Q.5.b.ii; © VCAA

Nitrogen oxides are commonly found in the atmosphere in areas where there is serious atmospheric pollution.

Nitrogen monoxide, NO, is generated from the reaction between nitrogen and oxygen.

$$N_2(g) + O_2(g) \rightleftharpoons 2NO(g) \quad \Delta H = +180.8 \text{ kJ mol}^{-1}$$

NO(g) is produced in combustion engines such as a car engine. Catalysts based on platinum and palladium are used in the exhaust system to decompose NO(g) into $N_2(g)$ and $O_2(g)$.

Sketch, on the axes provided below, a fully labelled energy profile diagram for the decomposition reaction of NO. Indicate on the diagram the effect of using a catalyst in this reaction.

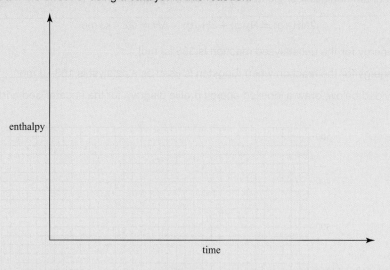

More exam questions are available in your learnON title.

4.4 Review

4.4.1 Topic summary

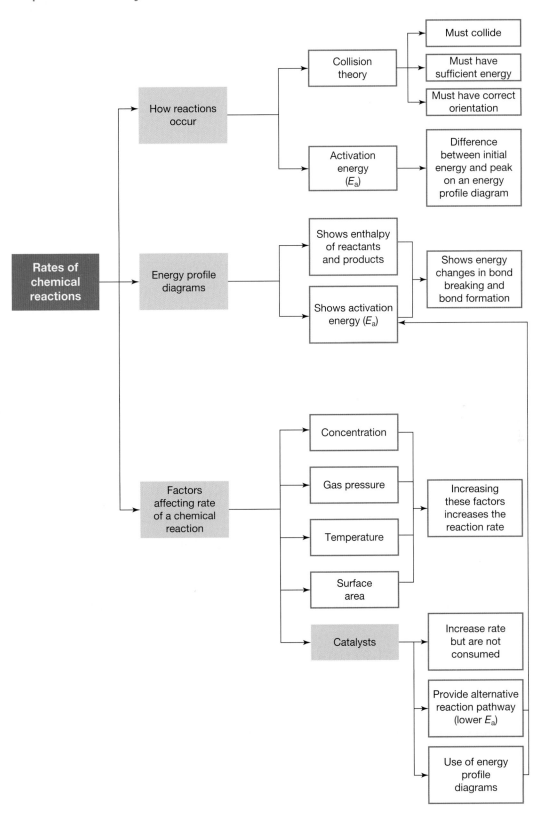

Hey students! Now that it's time to revise this topic, go online to:

Access the topic summary | Review your results | Watch teacher-led videos | Practise past VCAA exam questions

Find all this and MORE in jacPLUS

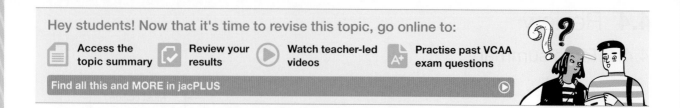

4.4.2 Key ideas summary

online only

4.4.3 Key terms glossary

online only

on Resources

Solutions	Solutions — Topic 4 (sol-0831)
Practical investigation eLogbook	Practical investigation eLogbook — Topic 4 (elog-1703)
Digital documents	Key science skills — VCE Chemistry Units 1–4 (doc-37066) Key terms glossary — Topic 4 (doc-37287) Key ideas summary — Topic 4 (doc-37288)
Exam question booklet	Exam question booklet — Topic 4 (eqb-0115)

4.4 Activities

learnon

Students, these questions are even better in jacPLUS

Receive immediate feedback and access sample responses | Access additional questions | Track your results and progress

Find all this and MORE in jacPLUS

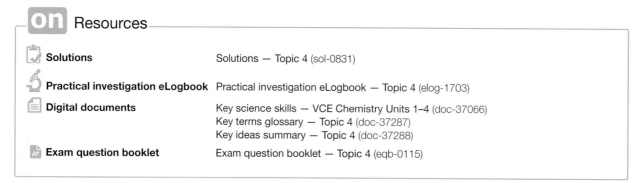

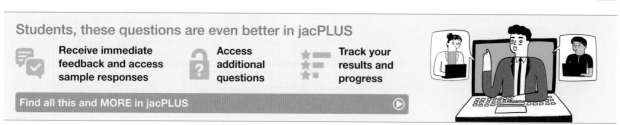

4.4 Review questions

1. Using the concepts of activation energy and reaction pathways, explain how a catalyst can speed up the rate of a chemical reaction.
2. The dehydration properties of concentrated sulfuric acid are often demonstrated using sucrose, $C_{12}H_{22}O_{11}$. Black carbon and steam, along with sulfur dioxide and considerable heat, are produced. The equation for this reaction is:

$$2C_{12}H_{22}O_{11}(s) + 2H_2SO_4(aq) + O_2(g) \rightarrow 22C(s) + 2CO_2(g) + 24H_2O(g) + 2SO_2(g)$$

   Explain why this reaction occurs much faster when caster sugar, rather than granulated sugar, is used as a source of sucrose.
3. The reaction between hydrogen gas and nitrogen(IV) oxide is represented by the equation:

$$2H_2(g) + 2NO_2(g) \rightarrow 2H_2O(g) + N_2(g)$$

   This reaction is studied in a sealed container of fixed volume.
   a. Explain why the rate of this reaction decreases with time.
   b. Explain why the progress of this reaction can be monitored by measuring the drop in pressure.

4. Comment on the rates observed for each of the following situations. For each one, use collision theory to explain the rate behaviour observed.

   a. Nail polish dries quicker on a hot day than on a cold day.
   b. A piece of steel wool burns in a Bunsen flame, but the same mass of solid steel does not.
   c. A pinch of manganese dioxide added to hydrogen peroxide continues to produce oxygen for as long as fresh hydrogen peroxide is added.
   d. The chemicals mixed by a panelbeater to make body filler harden faster on a hot day than on a cold day.
   e. The addition of vinegar to bicarbonate of soda causes an evolution of gas that is quick at first but then slows down.
   f. Photographers using infrared-sensitive film store it in a refrigerator before use.

5. The decomposition of ammonia to produce nitrogen and hydrogen according to:

$$2NH_3(g) \rightarrow N_2(g) + 3H_2(g)$$

has an activation energy of 330 kJ mol^{-1} and a ΔH value of +92 kJ mol^{-1}. If tungsten is used as a catalyst, the activation energy is 163 kJ mol^{-1}.

   a. Give a definition for the term *catalyst*.
   b. Show all of the information provided for this reaction on an energy profile diagram.
   c. The reverse reaction to this, as shown by the equation:

$$N_2(g) + 3H_2(g) \rightarrow 2NH_3(g)$$

   is a very important reaction in industry.

   i. Calculate the activation energy for the uncatalysed version of this reaction.
   ii. Calculate the activation energy when tungsten is used as a catalyst.

6. Enzymes are a very important class of biochemical molecules that are often described as biological catalysts. For example, the enzyme lipase is an important digestive enzyme that assists in the breakdown of fats according to the following generalised equation:

$$\text{Fat} + \text{water} \xrightleftharpoons{\text{Lipase}} \text{fatty acids} + \text{glycerol}$$

   a. Using the concept of activation energy, explain why the rates of reactions such as this are much greater in the presence of lipase.
   b. How does the amount of lipase present at the start of a reaction such as this compare with the amount present at its completion?

7. The rate of a chemical reaction is a very important consideration in industrial chemistry where chemicals are made on a large scale. Reactions need an acceptable rate to be economical. A number of important industrial reactions have rates that are too slow at even moderate temperatures and therefore need to be sped up. Further increasing the temperature is a common means of achieving this; however, sometimes this is an inappropriate strategy. Suggest two other methods by which an increase in reaction rate may be produced in these situations.

8. The gas phase reaction between two gases, A and B, to form C, may be represented by the following equation:

$$A(g) + B(g) \rightarrow C(g)$$

The progress of this reaction can be monitored by measuring the production of product (C).

This reaction was investigated in the presence of a fourth substance, X, and then again in the presence of substance Y. Equal amounts of A and B were used in both experiments. The concentrations of C and either X or Y are shown in the following graphs.

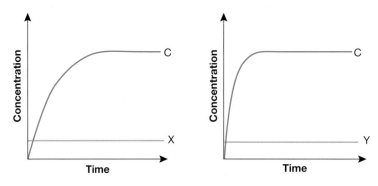

a. At which stage in each reaction is the production of C occurring at the fastest rate? Explain.
b. Explain why the concentration of C reaches the same maximum in both experiments.
c. What role can be attributed to substance Y in the second experiment? Give two pieces of evidence to support your answer.
d. Substance X may or may not be performing the same role as substance Y. Suggest an experiment that would help you decide.

4.4 Exam questions

Section A — Multiple choice questions

All correct answers are worth 1 mark each; an incorrect answer is worth 0.

Question 1

Source: *VCE 2020 Chemistry Exam, Section A, Q.27; © VCAA*

MC The heat of combustion of ethanoic acid, $C_2H_4O_2$, is –876 kJ mol^{-1} and the heat of combustion of methyl methanoate, $C_2H_4O_2$, is –973 kJ mol^{-1}. The auto-ignition temperature (the temperature at which a substance will combust in air without a source of ignition) of ethanoic acid is 485 °C and the auto-ignition temperature of methyl methanoate is 449 °C.

Which one of the following pairs is correct?

	Compound with the lower chemical energy per mole	Compound with the lower activation energy of combustion
A.	ethanoic acid	methyl methanoate
B.	ethanoic acid	ethanoic acid
C.	methyl methanoate	methyl methanoate
D.	methyl methanoate	ethanoic acid

Question 2

Source: VCE 2017 Chemistry Exam, Section A, Q.1; © VCAA

MC A catalyst

A. slows the rate of reaction.

B. ensures that a reaction is exothermic.

C. moves the chemical equilibrium of a reaction in the forward direction.

D. provides an alternative pathway for the reaction with a lower activation energy.

Question 3

Source: VCE 2015 Chemistry Exam, Section A, Q.16; © VCAA

MC Consider the following energy profile for a particular chemical reaction, where I, II and III represent enthalpy changes during the reaction.

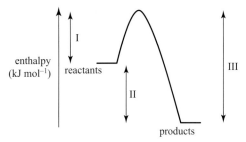

Which one of the following statements is correct?

A. The activation energy for the reverse reaction is (III–II).

B. The net energy released for the forward reaction is represented by II.

C. The energy required to break the reactant bonds is represented by II.

D. The energy released by the formation of new bonds is represented by I.

Question 4

Source: VCE 2013 Chemistry Exam, Section A, Q.15; © VCAA

$$Cu(s) + 4HNO_3(aq) \rightarrow Cu(NO_3)_2(aq) + 2NO_2(g) + 2H_2O(l)$$

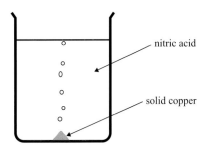

MC In the above reaction, the number of successful collisions per second is a small fraction of the total number of collisions.

The **major** reason for this is that

A. the nitric acid is ionised in solution.

B. some reactant particles have too much kinetic energy.

C. the kinetic energy of the particles is reduced when they collide with the container's walls.

D. not all reactant particles have the minimum kinetic energy required to initiate the reaction.

Question 5

Source: VCE 2015 Chemistry Exam, Section A, Q.17; © VCAA

MC The oxidation of sulfur dioxide is an exothermic reaction. The reaction is catalysed by vanadium(V) oxide.

$$2SO_2(g) + O_2(g) \rightleftharpoons 2SO_3(g)$$

Which one of the following energy profile diagrams correctly represents both the catalysed and the uncatalysed reaction?

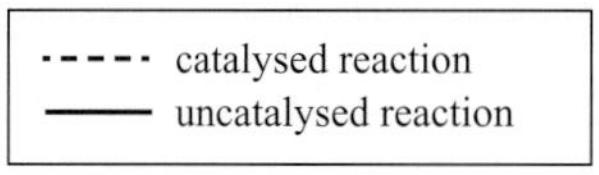

- - - - catalysed reaction
—— uncatalysed reaction

A.

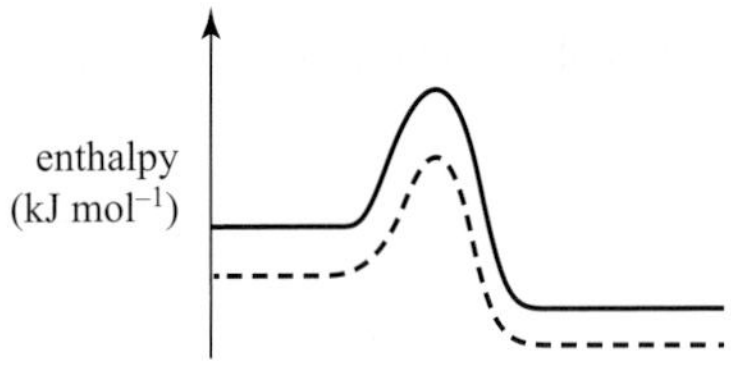

enthalpy (kJ mol^{-1})

B.

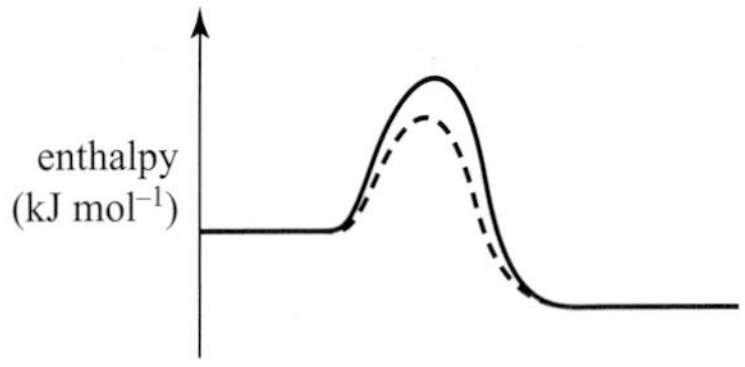

enthalpy (kJ mol^{-1})

C.

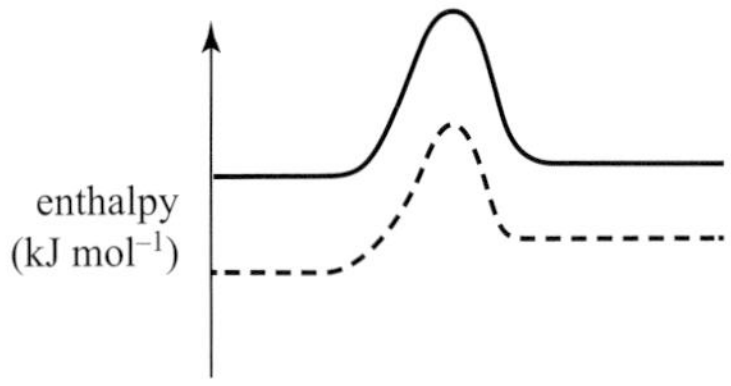

enthalpy (kJ mol^{-1})

D.

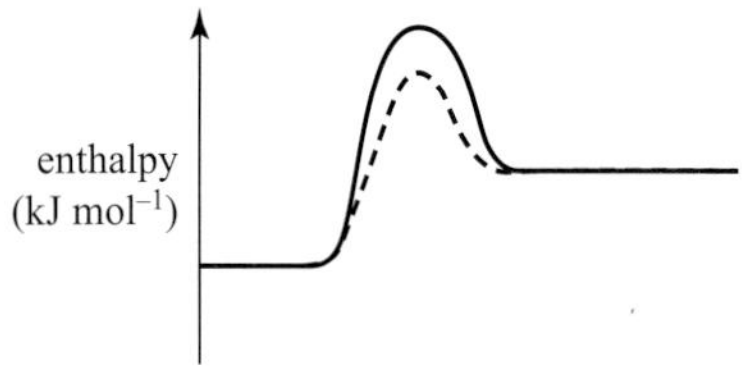

enthalpy (kJ mol^{-1})

Question 6

Source: VCE 2013 Chemistry Sample Exam for Units 3 and 4, Section A, Q.17; © VCAA

MC Consider the following statements regarding the effect of temperature on the particles in a reaction mixture.

I At a higher temperature, particles move faster and the reactant particles collide more frequently.

II At a higher temperature, more particles have energy greater than the activation energy.

Which alternative below best explains why the observed reaction rate is greater at higher temperatures?

A. I only

B. II only

C. I and II to an equal extent

D. I and II, but II to a greater extent than I

Question 7

Source: VCE Chemistry 2011 Exam 2, Section A, Q.6; © VCAA

MC In an endothermic reaction the

A. reaction system loses energy to the surroundings.

B. addition of a catalyst increases the activation energy.

C. activation energy is greater than the enthalpy of reaction.

D. energy required to break bonds in the reactants is less than the energy released when bonds are formed in the products.

Question 8

Source: VCE 2012 Chemistry Exam 2, Section A, Q.11; © VCAA

MC The following reaction is used in some industries to produce hydrogen.

$$CO(g) + H_2O(g) \rightleftharpoons CO_2(g) + H_2(g) \qquad \Delta H = -41\,kJ\,mol^{-1}$$

In trials, the reaction is carried out with and without a catalyst in the sealed container. All other conditions are unchanged. The change in hydrogen concentration with time between an uncatalysed and a catalysed reaction is represented by a graph.

Which graph is correct?

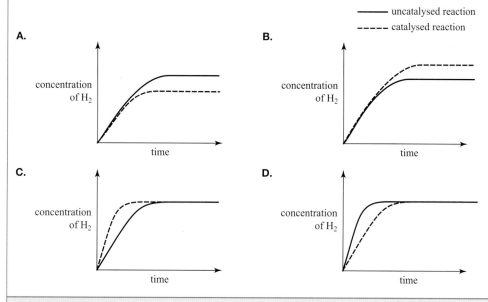

Question 9

Source: VCE 2012 Chemistry Exam 2, Section A, Q.12; © VCAA

MC Consider the following energy profile diagram for a reaction represented by the equation X + Y → Z.

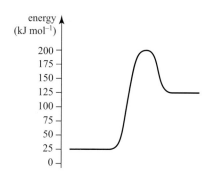

Which one of the following provides the correct values of the activation energy and enthalpy for the reaction X + Y → Z?

	Activation energy (kJ mol⁻¹)	Enthalpy (kJ mol⁻¹)
A.	+75	+100
B.	+100	+175
C.	+175	+100
D.	+200	−125

Question 10

Source: VCE 2010 Chemistry Exam 2, Section A, Q.1; © VCAA

MC The factors which influence the rate of reaction between dilute hydrochloric acid and powdered calcium carbonate were investigated.

Which one of the following changes would **not** increase the rate of the reaction?

A. stirring the mixture
B. heating the reaction mixture
C. increasing the concentration of the acid
D. replacing the powder with a lump of calcium carbonate

Section B — Short answer questions

Question 11 (1 mark)

Source: VCE 2020 Chemistry Exam, Section B, Q.1.a; © VCAA

Methanol is very useful fuel. It can be manufactured from biogas.

The main reaction in methanol production from biogas is represented by the following equation.

$$CO(g) + 2H_2(g) \rightleftharpoons CH_3OH(g) \qquad \Delta H < 0$$

This reaction requires the use of a catalyst to maximise the yield of methanol produced in optimum conditions. The energy profile diagram below represents the uncatalysed reaction.

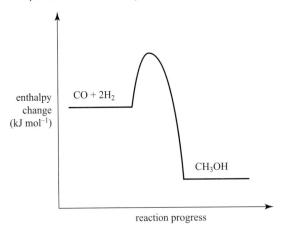

On the energy diagram above, sketch how the catalyst would alter the reaction pathway.

Source: VCE 2014 Chemistry Exam, Section B, Q.12; © VCAA

A student investigated the effect of different catalysts on the molar enthalpy of the decomposition reaction of hydrogen peroxide. The student's report is provided below.

Report — Effect of different catalysts on the enthalpy of a reaction

Background

Different catalysts, such as manganese dioxide, MnO_2, and iron(III) nitrate solution, $Fe(NO_3)_3$, will increase the rate of decomposition of hydrogen peroxide.

$$2H_2O_2(aq) \rightarrow 2H_2O(l) + O_2(g)$$

Purpose

This experiment investigated the effect of using different catalysts on the molar enthalpy of the decomposition of hydrogen peroxide.

Procedure

The temperature change was measured when MnO_2 catalyst was added to a volume of hydrogen peroxide in a beaker. The procedure was repeated using $Fe(NO_3)_3$ solution as a catalyst.

Results

	Trial 1	Trial 2
Volume H_2O_2	100 mL	200 mL
Concentration H_2O_2	2.0 M	4.0 M
Catalyst	0.5 g MnO_2	50 mL 0.1 M $Fe(NO_3)_3$
Temperature change °C	3.0	10.1

Conclusion

The change in temperature using the $Fe(NO_3)_3$ catalyst was greater than the change in temperature using the MnO_2 catalyst. This demonstrates that the molar enthalpy for the decomposition reaction depends on the catalyst used.

The student's conclusion is not valid because the experimental design is flawed.

Critically review the student's experimental design. In your response, you should:
- identify and explain **three** improvements or modifications that you would make to the experimental design
- discuss the experimental outcomes you would expect regarding the effect of different catalysts on molar heats of reaction. Justify your expectations in terms of chemical ideas you have studied this year.

Question 13 (8 marks)

In chemistry, a number of reactions are collectively referred to as 'clock reactions'. These reactions produce a sudden colour change after a period of time. One of the better known examples of such reactions is that between iodide ions and persulfate ions ($S_2O_8^{2-}$) in the presence of thiosulfate ions ($S_2O_3^{2-}$) and starch.

Two reactions are involved:

$$\text{Reaction 1: } 2I^-(aq) + S_2O_8^{2-}(aq) \rightarrow I_2(aq) + 2SO_4^{2-}(aq)$$
$$\text{Reaction 2: } I_2(aq) + 2S_2O_3^{2-}(aq) \rightarrow 2I^-(aq) + S_4O_6^{2-}(aq)$$

The iodine produced by reaction 1 is immediately removed by reaction 2. However, ($S_2O_3^{2-}$) ions are also consumed and are eventually all used up. After this time, iodine builds up and is detected by the starch present, which forms an intensely coloured dark blue complex. This occurs at iodine concentrations as low as 10^{-5} M, making starch an excellent indicator for this reaction.

If the amount of $S_2O_3^{2-}(aq)$ is kept constant, the appearance of the dark blue colour may be used to measure the rate of reaction 1.

In one such experiment using 0.2 M KI(aq) solution, 0.2 M $Na_2S_2O_8$(aq) and 0.1 M $Na_2S_2O_3$(aq) solutions, the following results were obtained.

Trial number	$V(I^-$ (aq)) solution (mL)	$V(S_2O_8^{2-}$(aq)) solution (mL)	$V(S_2O_3^{2-}$(aq)) solution (mL)	V(water) (mL)	V(starch) solution (mL)	Time for blue colour to appear (s)
1	20	20	20	40	10	220
2	20	40	20	20	10	150
3	40	20	20	20	10	142

a. Explain how these results show that increasing $(S_2O_8^{2-})$ (aq) concentration produces a faster rate in reaction 1. Compare two appropriate trials from the data provided as part of your explanation. **(2 marks)**

b. Use the results to explain how the rate of reaction is affected by the concentration of iodide ions. Select two appropriate trials to support your explanation. **(2 marks)**

c. The experiment is repeated using solutions that were stored in a refrigerator for 24 hours. These solutions were used immediately after being removed.

   How would the reaction times in the table be affected? **(1 mark)**

d. What is the purpose of the two different amounts of water used in the trials? **(2 marks)**

e. Explain why the amount of $S_2O_3^{2-}(aq)$ solution is kept constant in each trial. **(1 mark)**

Source: VCE 2012 Chemistry Exam 2, Section B, Q.1; © VCAA

Two experiments were conducted to investigate various factors that affect the rate of reaction between calcium carbonate and dilute hydrochloric acid.

$$CaCO_3(s) + 2HCl(aq) \rightleftharpoons CO_2(g) + CaCl_2(aq) + H_2O(l)$$

The two experiments are summarised in the diagrams below.

experiment 1

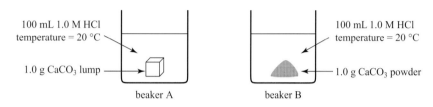

experiment 2

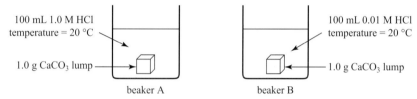

a. How could the rate of this reaction be measured in these experiments? **(1 mark)**

b. i. Identify the rate determining factor that is investigated in experiment 1. **(1 mark)**

   ii. In experiment 2, will the rate of reaction be faster in beaker A or beaker B? Explain your selection in terms of collision theory. **(2 marks)**

c. Why is the following statement **incorrect**?

   'Collision theory states that all collisions between reactant particles will result in a chemical reaction.' **(2 marks)**

An important chemical (C) is made on a large scale in industry from two immiscible liquids (A and B). Both B and C are less dense than A. The equation for the reaction involved is:

$$A(l) + B(l) \rightarrow C(s) \quad \Delta H = +ve$$

The traditional process involves pumping liquid B across the surface of a vat containing liquid A. The rate of pumping is carefully adjusted so that by the time liquid B has reached the opposite side of the vat, the reaction is complete. This process is shown in diagram 1.

A modification to this method has been proposed for making substance C. In this method, substance B is introduced into the base of the vat in the form of a fine spray, but at a rate to match the pumping rate from before. A fan blowing air across the surface assists in the collection of C as it rises to the surface. It is planned that this new process will operate at the same temperature as the original one. This process is shown in diagram 2.

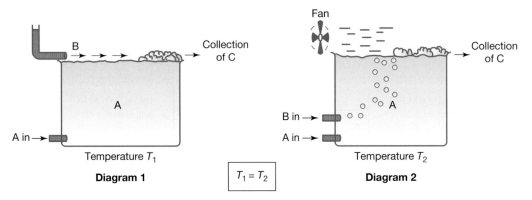

Diagram 1 — Temperature T_1

$T_1 = T_2$

Diagram 2 — Temperature T_2

a. Is the reaction shown exothermic or endothermic? **(1 mark)**

b. Which of the two methods would produce substance C the quickest? **(1 mark)**

c. Use the collision theory for reacting particles to explain your answer to part **b**. **(2 marks)**

d. In terms of substance C, explain one disadvantage of the second method and how this can be minimised. **(2 marks)**

e. As an alternative modification to the original method, it has been proposed that the temperature of the two reacting liquids (A and B) be increased, even though this will require more energy.

    Explain why this will enable liquid B to be pumped across the surface of liquid A at a faster rate. **(1 mark)**

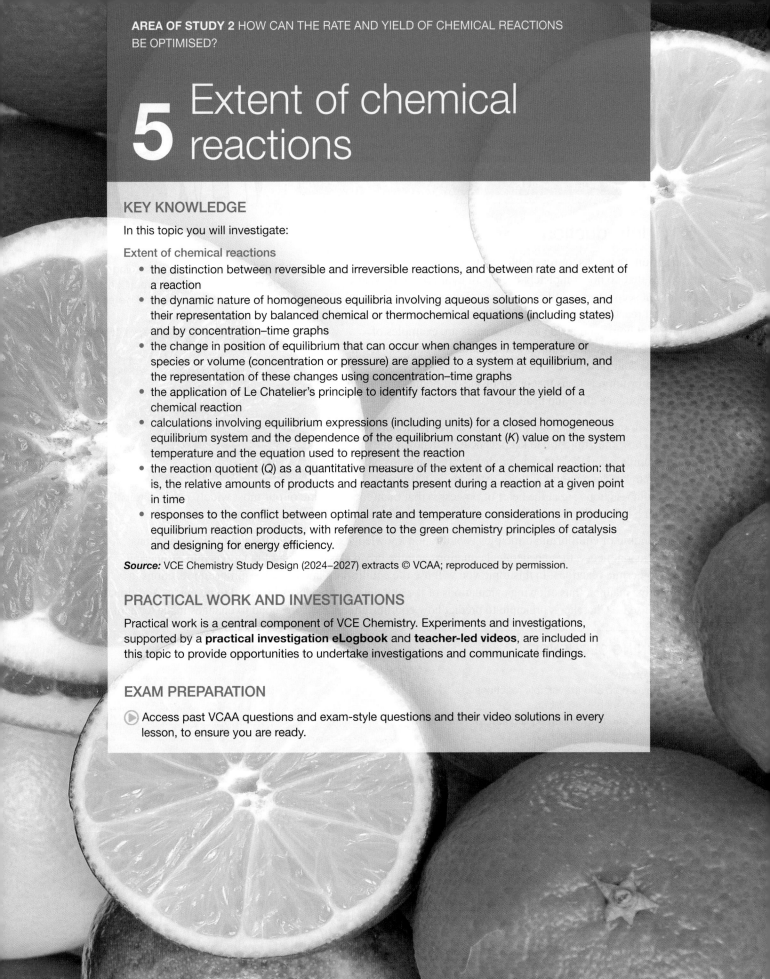

5 Extent of chemical reactions

KEY KNOWLEDGE

In this topic you will investigate:

Extent of chemical reactions

- the distinction between reversible and irreversible reactions, and between rate and extent of a reaction
- the dynamic nature of homogeneous equilibria involving aqueous solutions or gases, and their representation by balanced chemical or thermochemical equations (including states) and by concentration–time graphs
- the change in position of equilibrium that can occur when changes in temperature or species or volume (concentration or pressure) are applied to a system at equilibrium, and the representation of these changes using concentration–time graphs
- the application of Le Chatelier's principle to identify factors that favour the yield of a chemical reaction
- calculations involving equilibrium expressions (including units) for a closed homogeneous equilibrium system and the dependence of the equilibrium constant (K) value on the system temperature and the equation used to represent the reaction
- the reaction quotient (Q) as a quantitative measure of the extent of a chemical reaction: that is, the relative amounts of products and reactants present during a reaction at a given point in time
- responses to the conflict between optimal rate and temperature considerations in producing equilibrium reaction products, with reference to the green chemistry principles of catalysis and designing for energy efficiency.

Source: VCE Chemistry Study Design (2024–2027) extracts © VCAA; reproduced by permission.

PRACTICAL WORK AND INVESTIGATIONS

Practical work is a central component of VCE Chemistry. Experiments and investigations, supported by a **practical investigation eLogbook** and **teacher-led videos**, are included in this topic to provide opportunities to undertake investigations and communicate findings.

EXAM PREPARATION

▶ Access past VCAA questions and exam-style questions and their video solutions in every lesson, to ensure you are ready.

5.1 Overview

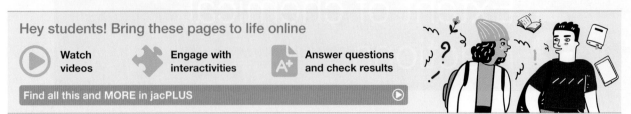

5.1.1 Introduction

Equilibrium reactions, or reversible reactions, are chemical reactions that do not completely use up all their reactants. While this might seem strange at first, such reactions are very common. Chemical reactions involving equilibria are all around us — in industry, at home and inside us. Weak acids are examples of substances that produce equilibrium reactions when dissolved in water. The sour taste of lemons is due to a weak acid — citric acid. The equilibrium reactions that occur inside our bodies play a vital role in keeping us alive and healthy.

FIGURE 5.1 Many home swimming pools rely on an equilibrium reaction between hypochlorite ions and water for critical sanitation.

Equilibrium reactions respond to changes, which is an important feature in how they function to keep us healthy. Knowledge of the equilibrium law and the ability to predict the response to change is vital to the understanding of these reactions. Additionally, equilibrium reactions are at the heart of processes that manufacture some of our most widely used chemicals. A thorough knowledge of equilibrium reactions is therefore essential to their efficient manufacture.

Understanding of equilibrium principles is built upon knowledge of reaction rates and their explanations using collision theory, thermochemical equations and the broad classification of reactions into exothermic and endothermic reactions. In this topic you will use the equilibrium law to treat equilibrium reactions in a quantitative manner, thus allowing calculations of the amounts of substances involved to be made. Additionally, you will use Le Chatelier's principle to predict how equilibrium reactions respond to changes made to them.

LEARNING SEQUENCE

Resources

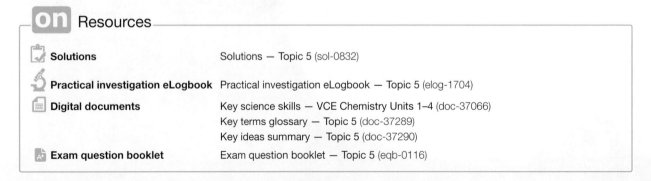

5.2 Reversible and irreversible reactions

5.2.1 Reversible reactions

In all the stoichiometric calculations you have done so far, an important assumption has been that the reaction proceeds to completion. In other words, you have assumed, subject to mole ratios and amounts present, that all reactants are converted into products. This allowed the amount of expected product to be calculated. While many reactions follow this pattern, many do not go to completion. The following two examples illustrate this point.

CASE STUDY: Reversible and irreversible reactions

Reaction 1: The decomposition of hydrogen bromide

If hydrogen bromide is placed in a suitable container and heated, it decomposes according to the following equation:

$$2HBr(g) \rightarrow H_2(g) + Br_2(g)$$

If the products are analysed after some time, it is found that the amounts of hydrogen and bromine are as predicted from a normal stoichiometric calculation. If the concentration of the hydrogen bromide is monitored against time, a graph similar to that shown in figure 5.2 is obtained.

Another way to describe this reaction is that it 'goes completely to the right'. The 'right', of course, means the product side of the chemical equation.

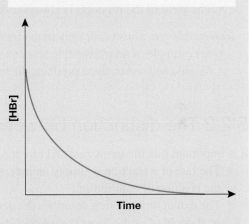

FIGURE 5.2 A typical concentration-versus-time graph for the decomposition of hydrogen bromide

[HBr]

Time

Reaction 2: The decomposition of hydrogen iodide

The equation for the decomposition of hydrogen iodide is:

$$2HI(g) \rightleftharpoons H_2(g) + I_2(g)$$

At first glance, you might expect this decomposition to be very similar to that shown for hydrogen bromide. However, when this decomposition is attempted under similar conditions to the hydrogen bromide reaction, *the yield of hydrogen and iodine is always less than the stoichiometric prediction*. This occurs no matter how long you wait. Furthermore, it appears that the concentrations of all species reach certain values and thereafter remain constant. The graph in figure 5.3 shows this effect.

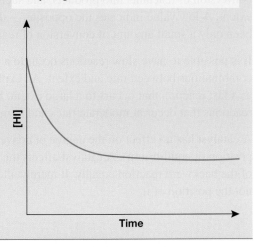

FIGURE 5.3 A typical concentration-versus-time graph for the decomposition of hydrogen iodide

[HI]

Time

Reaction 2, the decomposition of hydrogen iodide, illustrates what we call an **equilibrium reaction**, which occurs when reactions do not completely convert all the reactants into products — some reactants always remain, mixed with the products of the reaction. Since such reactions are quite common, having a method to predict their behaviour would be advantageous. The profitability of an important industrial process costing millions of dollars to research and develop could depend on such calculations.

Strictly speaking, all chemical reactions are equilibrium reactions. However, in many cases the degree of backward reaction (i.e. products re-forming reactants) is so small that it can effectively be ignored.

FIGURE 5.4 An example of a reversible reaction. The addition of ammonia to a solution containing sky-blue copper(II) ions (left beaker) forms a royal-blue product (middle beaker). Adding water to the middle beaker reforms the original sky-blue copper(II) ions (right beaker).

Reversible and irreversible reactions

Reversible reactions are equilibrium reactions in which reactants re-form into products to a significant extent.
- Their equations show a double arrow ($\rightleftharpoons$) rather than a single arrow ($\rightarrow$).
- The yield of the products is always less than the stoichiometric prediction.

Irreversible reactions occur only in the forward direction; reactants do not re-form from products.
- An example of an irreversible reaction is the combustion of fuel. When a fuel burns to produce carbon dioxide and water, these products do not react with each other to re-form the fuel.

5.2.2 The distinction between rate and extent of a reaction

It is important that the terms *rate* and *extent* of a reaction are not confused.
- The rate of a reaction is simply an indication of how fast it occurs. This shows how long it takes to establish the position of equilibrium.
- The extent of a reaction describes the degree to which reactants are converted into products. This can also be thought of as the 'position' of equilibrium or 'how far to the right' (with respect to the equation) it is situated.

This degree of conversion from reactants to products may be quantified by reference to the **equilibrium constant**, K (or K_c), for the reaction concerned (see section 5.4.1). A high K value indicates a significant conversion of reactants into products, and such a reaction would be described as having occurred to a significant extent. A low value indicates the opposite — that the reaction has occurred only to a small extent and there has been only a small amount of conversion of reactants into products.

It is possible to have slow reactions occur to a great extent as well as other combinations between rate and extent. An explosion, for example, can be described as a fast reaction that occurs to a large extent, but it is possible to have equilibrium reactions that occur at moderate rates and to moderate extents.

A catalyst has no effect on the degree of conversion of reactants into products (the position of equilibrium). A catalyst affects the rate of the forward reaction and the rate of the backward reaction equally. It merely alters the time taken to get to equilibrium, not the position of it.

equilibrium reaction a reaction in which both forward and reverse reactions are significant

equilibrium constant a value that gives an indication of the extent to which reactants are converted into products for an equilibrium reaction; it is assigned the symbol K

Rates and extents of reactions

- The rate of reaction is an indication of how fast a reaction proceeds.
- The extent of a reaction is the degree to which reactants are converted into products and describes the equilibrium position.
- Catalysts alter the rate of reactions but not their extent. They do not alter the equilibrium position.

5.2 Activities

learn on

Students, these questions are even better in jacPLUS

Receive immediate feedback and access sample responses

Access additional questions

Track your results and progress

Find all this and MORE in jacPLUS ▶

5.2 Quick quiz on	5.2 Exercise	5.2 Exam questions

5.2 Exercise

1. What is the difference between a reversible reaction and an irreversible reaction?
2. What is the difference between the rate of a reaction and the extent of a reaction?
3. Hydrogen peroxide decomposes to water and oxygen. The extent of this reaction is large. Explain why bottles of hydrogen peroxide may be kept for long periods of time but eventually need to be replaced.
4. Is it possible to achieve a yield of 100 per cent for a reversible reaction? Explain why or why not.
5. A fast reaction that occurs to a large extent will be obvious and easy to detect. Comment on each of the following scenarios, as to their obviousness and detectability.
    a. A fast reaction with a small extent
    b. An exceedingly slow reaction with a large extent
    c. A slow reaction with a small extent
6. A student decides to investigate three different reversible reactions as a prelude to her practical investigation. Each of these is set up during the same lesson and then observed again during her next lesson the next day. Her initial purpose is to attempt a classification of each reaction according to the following table.

		Extent	
		Small	**Large**
Rate	**Slow**	Type 1	Type 2
	Fast	Type 3	Type 4

Upon her return the following day, she observes the following:
- Reaction I: Large amount of product. Little detectable reactants.
- Reaction II: No apparent change. Little detectable products.
- Reaction III: A mixture of reactants and products is observed.

a. Attempt to classify each of the reactions according to the table. Note that there may be more than one classification for each reaction.
b. For those reactions with multiple classifications, suggest a possible follow-up experiment that might be able to distinguish them.

5.2 Exam questions

▶ **Question 1 (2 marks)**

Source: Adapted from VCE 2012 Chemistry Exam 2, Section B, Q.3.b; © VCAA

The following weak acids are used in the food industry.

Acid	Common use	Formula	Structure	K_a values
sorbic	preservative	$C_6H_8O_2$	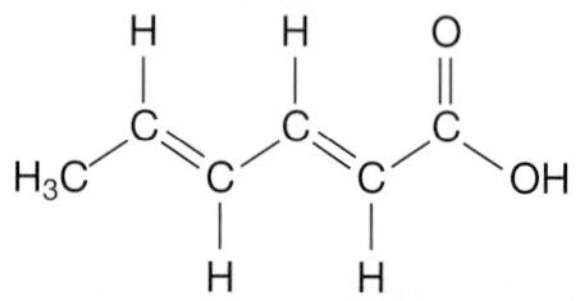	1.73×10^{-5}
malic	low-calorie fruit drinks	$C_4H_6O_5$	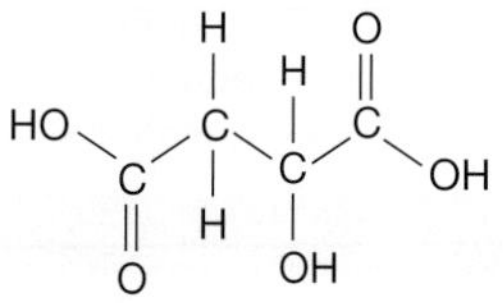	3.98×10^{-4}

K_a values are just a special type of K value. They refer to the reaction of an acid with water. K_a values may be interpreted in the same way as K values.

Which of the above acids reacts to the greatest extent with water? Explain your answer with reference to the K_a values given.

▶ **Question 2 (1 mark)**

Source: Adapted from VCE 2018 Chemistry NHT Exam, Section A, Q.3; © VCAA

MC Consider the following reaction.

$$O_2(g) + 2NO(g) \rightleftharpoons 2NO_2(g) \quad K = 9.1 \times 10^6 \ M^{-1} \text{ at } 450 \ °C$$

Which one of the following statements is **incorrect**?
A. K is very large; the reaction is reversible.
B. Adding a catalyst will increase the value of K.
C. K is very large; the reaction is effectively irreversible.
D. Adding a catalyst will not affect the value of K.

▶ **Question 3 (4 marks)**

Source: Adapted from VCE 2011 Chemistry Exam 2, Section B, Q.5.a; © VCAA

Nitrogen oxides are commonly found in the atmosphere in areas where there is serious atmospheric pollution.

Nitrogen monoxide, NO, is generated from the reaction between nitrogen and oxygen.

$$N_2(g) + O_2(g) \rightleftharpoons 2NO(g) \quad \Delta H = +180.8 \ kJ \ mol^{-1}$$

A sealed container is filled with 1.00 mole of NO(g). The temperature is maintained at 1500 °C.
a. Explain why the rate of the reaction

$$N_2(g) + O_2(g) \rightarrow 2NO(g)$$

   is initially slower than the rate of the reaction

$$2NO(g) \rightarrow N_2(g) + O_2(g) \tag{2 marks}$$

b. As this reaction proceeds, describe what happens to the rates of each of the above reactions. **(1 mark)**
c. Based on your answer to part **b**, what will eventually happen to the rates of each of these reactions? **(1 mark)**

▶ **Question 4 (1 mark)**

Explain, using collision theory, why a *closed system* is required for a reversible reaction.

Source: *Adapted from VCE 2007 Chemistry Exam 1, Section B, Q.1.a.i; © VCAA*

Carbon monoxide and hydrogen can be produced from the reaction of methane with steam according to the equation

$$CH_4(g) + H_2O(g) \rightleftharpoons CO(g) + 3H_2(g); \quad \Delta H = +206 \text{ kJ mol}^{-1}$$

Some methane and steam are placed in a closed container and allowed to react at a fixed temperature. The following graph shows the change in concentration of methane and carbon monoxide as the reaction progresses.

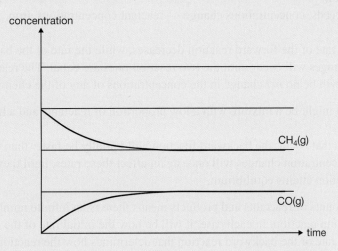

a. Use this graph to explain why this reaction can be regarded as a reversible reaction. **(1 mark)**
b. Use this graph to explain what is initially happening to the rates of the forward and reverse reactions. **(2 marks)**

More exam questions are available in your learnON title.

5.3 Homogeneous equilibria

KEY KNOWLEDGE

• The dynamic nature of homogeneous equilibria involving aqueous solutions or gases, and their representation by balanced chemical or thermochemical equations (including states) and by concentration–time graphs

Source: *VCE Chemistry Study Design (2024–2027) extracts © VCAA; reproduced by permission.*

5.3.1 Homogeneous and heterogeneous reactions

Chemical reactions may be classified in a number of ways. One simple classification is based on the physical states (or phases) of the reagents involved.

• In a **homogeneous reaction**, the reaction occurs entirely within the same physical state. The most common examples of such reactions occur either in the gaseous phase or in solution. A homogeneous equilibrium is one in which all the species in the equilibrium mixture are in the same physical state.

• A **heterogeneous reaction** is one that occurs at a boundary or interface between two physical states. An obvious example of a heterogeneous reaction is a solid reacting with either a liquid or a gas. Heterogeneous reactions can also occur between liquids that are **immiscible**, thus forming a boundary when they are mixed together.

homogeneous reaction a reaction in which all of the substances involved are in the same phase

heterogeneous reaction a reaction in which some of the substances involved are in different phases

immiscible refers to liquids that do not form a homogeneous mixture when mixed with another liquid

5.3.2 The dynamic nature of equilibrium

Although it is tempting to think that the reaction has stopped when it reaches equilibrium, further investigation reveals that this is not so. Instead, *the forward and reverse reactions are still occurring, but at the same rate*. The reagents in the reaction are thus being formed and used at the same rate, with their concentrations showing no overall change. Equilibrium is dynamic, not static.

Consider the situation when two reactants are mixed.

- Initially, due to concentration effects, the initial rate of forward reaction would be considerably greater than the backward reaction.
- As the reaction proceeds, concentrations change — reactant concentrations drop, while product concentrations rise.
- This means that the rate of the forward reaction decreases, while the rate of the backward reaction increases. These changes will occur until the two reaction rates are equal. The reaction will now be at equilibrium. There will be no *net* change in the concentrations of any of the chemicals involved.

Another starting scenario might be a mixture with a low proportion of reactants and a high proportion of products.

- Of the two reactions that occur, the backward reaction will initially be faster than the forward reaction.
- The subsequent concentration changes will once again affect these rates, until they are both occurring equally and the reaction attains equilibrium.

This sliding scale for amounts of reactants and products means there is an infinite number of possible starting conditions in an equilibrium reaction. In each case, it will be how the initial rate of the forward reaction compares with the initial rate of the backward reaction that determines how the reaction ultimately reaches its equilibrium state.

Equilibrium can also be considered in terms of successful collisions between particles. At equilibrium, the frequency of successful collisions going in one direction is balanced by the frequency of successful collisions in the opposite direction. In other words, the rates of these opposing reactions will be equal and the reaction will be at equilibrium.

The dynamic nature of equilibrium

During chemical equilibrium the forward and reverse reactions are still occurring, but at the same rate. The reaction does not stop.

 Resources

▶ **Video eLesson** Dynamic equilibrium and concentration of products and reactants (eles-3242)

Experiments using radioactive tracers verify the dynamic nature of the equilibrium state. For example, if the hydrogen iodide system described in section 5.2.1 is heated and allowed to come to equilibrium, it is possible to remove some of the iodine and replace it with the same amount of radioactive iodine — iodine containing the ^{131}I isotope. As isotopes are chemically identical, such a change would have no effect on the chemical nature of the equilibrium. If the system is examined again sometime later, the radioactive iodine is found to be distributed between the hydrogen iodide and the iodine molecules. This can be explained only if the forward and reverse reactions are still proceeding.

Figure 5.5 shows $I_2(g)$ in a stoppered flask reacting with $H_2(g)$ to form the colourless gas HI(g). The radioactive I will be found in both HI(g) and $I_2(g)$.

FIGURE 5.5 $I_2(g) + H_2(g) \rightleftharpoons HI(g)$. The forward and backward reactions are occurring at the same rate.

5.3.3 Representing chemical equilibria

Chemical equilibria can be represented using balanced chemical and thermochemical equations, and by using graphs.

Using balanced chemical and thermochemical equations

Equilibrium reactions can be represented using balanced chemical and thermochemical equations. Double arrows ($\rightleftharpoons$) are used to emphasise that the reaction is reversible. As the position of an equilibrium is affected by temperature differently for exothermic and endothermic reactions, thermochemical equations convey slightly more information about a reversible or equilibrium reaction than do equations without a ΔH value.

Because a reversible reaction involves both forward and reverse reactions, it is just as valid to write the equation the opposite way around. For example, the equations

$$2HI(g) \rightleftharpoons H_2(g) + I_2(g)$$
$$H_2(g) + I_2(g) \rightleftharpoons 2HI(g)$$

both refer to the same equilibrium reaction.

This has the potential to produce confusion when discussing forward and reverse reactions. To overcome this, the accepted procedure is to write the equation either way. It is then understood that any subsequent discussion of reactants, products, forward or reverse reactions or K refers to the equation as it has been written.

Graphical representations

Two types of graphs are frequently used to represent equilibrium situations:
- rate-versus-time graphs
- concentration-versus-time graphs.

To illustrate these, consider a situation where substance A is added to a container and allowed to come to equilibrium with products B and C at constant temperature, according to the equation:

$$A(g) \rightleftharpoons B(g) + 2C(g)$$

Rate-versus-time graphs

As discussed in topic 4, the rate of a reaction depends on concentration. As substance A is used up, its concentration drops and so does the rate of the forward reaction. Conversely, as the concentrations of substances B and C increase, so too does the rate of the backward reaction.

A general rate-versus-time graph, as shown in figure 5.6, illustrates this. There will be a net forward reaction until time t, when the two rates become equal and equilibrium is established. Thereafter, there will be no change in these rates as the net concentrations of reactants and products remain constant.

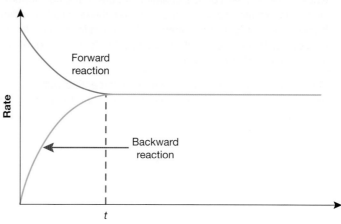

FIGURE 5.6 An example of a rate-versus-time graph

This equilibrium could just as easily be produced by mixing substances B and C and allowing a net backward reaction to produce equilibrium. In this case, the graph would show a decreasing rate for the backward reaction and an increasing rate for the forward reaction until equilibrium is once again established.

Concentration-versus-time graphs

It is also possible to represent this scenario using concentration-versus-time graphs. The forward reaction would be represented by figure 5.7.

- The final concentrations of substances A, B and C depend on their initial concentrations, the stoichiometry in the equation and the value of the equilibrium constant (i.e. the degree of conversion).
- The concentrations of the substances increase or decrease depending on whether they are produced or consumed. They also change by amounts that reflect the stoichiometry of the reaction. In this example, substance A decreases by the same amount that substance B increases, which can be seen by the 1 : 1 ratio

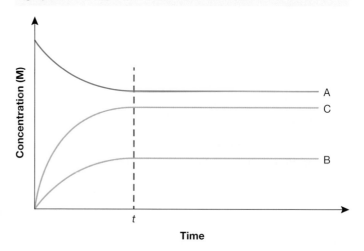

FIGURE 5.7 An example of a concentration-versus-time graph

between them in the equation. Substance C increases by twice the amount that substance B does, shown by the 2 : 1 ratio involved.

- If this reaction had a catalyst added to it, the only change to this graph would be that time, t, would be lower. In other words, equilibrium would be attained faster. However, the final concentrations of substances A, B and C would not be altered.

Concentration–time graphs are very useful when considering changes made to a reaction once it has reached equilibrium. This is discussed further in section 5.6.7.

tlvd-9681

SAMPLE PROBLEM 1 Using concentration-versus-time graphs to describe reversible reactions

The following graph shows the reversible reaction for

$$2NO_2(g) \rightleftharpoons N_2O_4(g)$$

Use this graph to answer the following questions.

a. **How does the concentration of NO_2 at t_1 compare to t_2?**

b. **Describe what is happening to the NO_2 using the information in the graph.**

c. **What is the significance of the flattening out of the NO_2 graph at the same time as the flattening out of the N_2O_4 graph?**

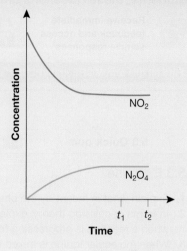

THINK	WRITE
a. This is a concentration–time graph. Horizontal sections mean that the species is at its equilibrium concentration and will not alter further unless a change is made. Here, NO_2 is at equilibrium.	a. The concentration of NO_2 at t_1 will be equal to its concentration at t_2.
b. The graph indicates that NO_2 is decreasing (but at an ever slower rate) while N_2O_4 is increasing (indicating an ever faster rate). This is due to the dynamic nature of equilibrium. The changes will occur until the two rates equalise at equilibrium.	b. NO_2 is decreasing but at an ever slower rate until the reaction reaches equilibrium. After this its value stays constant. This happens because a competing backward reaction that reforms NO_2 gradually gets faster, which reduces the rate at which NO_2 drops. Eventually the rates of the two reactions equalise and the reaction is at equilibrium. After this there will be no further concentration changes.
c. Equilibrium always occurs at a point in time when the two rates first become equal. This cannot happen at different times, otherwise concentration changes would still occur.	c. Equilibrium has been attained. This occurs when the two rates first become equal, resulting in no net change to concentrations thereafter.

PRACTICE PROBLEM 1

Answer the following questions with reference to the same reaction and graph given in sample problem 1.

a. **How does the concentration of N_2O_4 at t_1 compare to t_2?**

b. **Explain what is happening to the N_2O_4 in light of the information in the graph.**

5.3 Activities

learn on

5.3 Quick quiz on	**5.3 Exercise**	**5.3 Exam questions**

5.3 Exercise

1. Explain what is meant by the phrase 'the dynamic nature of equilibrium'.
2. In terms of collision theory, explain why one reaction always slows down while the other always speeds up when a reaction is progressing towards equilibrium.
3. When molecular iodine is mixed with iodide ions, an equilibrium is set up as triiodide ions are produced. The equation for this process is

$$I_2(aq) + I^-(aq) \rightleftharpoons I_3^-(aq)$$

   Suppose that, once this equilibrium is established, some iodine is removed and replaced by exactly the same amount of radioactive iodine.

   If this reaction is examined some time later, describe the expected observations if:
   a. equilibrium is static.
   b. equilibrium is dynamic.
4. When hydrogen iodide is placed in a sealed container and heated, it begins to decompose into its constituent elements according to the following equation:

$$2HI(g) \rightleftharpoons H_2(g) + I_2(g)$$

   Comment on the comparative rates of the forward and reverse reactions at the following stages.
   a. Just after the reaction has started
   b. As the reaction is approaching equilibrium
   c. When the reaction reaches equilibrium
   d. After the reaction has reached equilibrium
5. Refer to the concentration–time graph for the equilibrium reactions represented by the equation

$$2X(g) + Y(g) \rightleftharpoons 3Z(g)$$

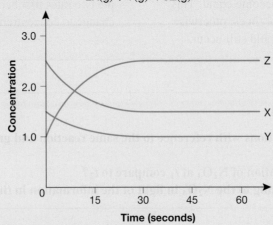

   a. Describe the appearance of this graph if it was extended to the 120-second mark.
   b. Explain the changes in the concentration of Z in comparison to the changes in the concentrations of X and Y.
   c. At time zero, describe the initial constitution of the reaction mixture and what subsequently happens in terms of reaction rates.

5.3 Exam questions

▶ Question 1 (1 mark)

Source: Adapted from VCE 2016 Chemistry Exam, Section A, Q.15; © VCAA

MC A chemist injected 0.10 mol carbon monoxide gas, CO, and 0.20 mol chlorine gas, Cl_2, into a previously evacuated and sealed 1.0 L flask.

At that instant, the following reaction began to occur.

$$CO(g) + Cl_2(g) \rightleftharpoons COCl_2(g) \quad \Delta H = -108 \text{ kJ mol}^{-1}$$

The concentrations of the three species present in the flask were monitored over time. The flask was held at a constant temperature. The following concentration–time graph was obtained.

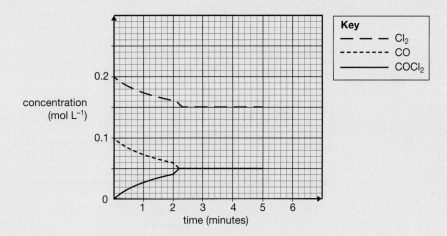

Which of the following statements is true?
- **A.** At 1 minute the rate of the forward reaction is slower than the rate of the reverse reaction.
- **B.** At 3 minutes there are no collisions between carbon monoxide and chlorine molecules.
- **C.** At 4 minutes the rate of the forward reaction is zero.
- **D.** At 5 minutes the rate of the forward reaction equals the rate of the reverse reaction.

▶ Question 2 (1 mark)

Source: VCE 2022 Chemistry Exam, Section B, Q.3.c; © VCAA

The following equation represents a gaseous reaction that takes place in a sealed container.

$$4NH_3(g) + 3O_2(g) \rightleftharpoons 2N_2(g) + 6H_2O(g) \quad \Delta H < 0$$

The concentration versus time graph for a different reaction is shown. This reaction takes place with a catalyst. Equilibrium is reached at time t_1.

The reaction is repeated without a catalyst.

On the concentration versus time graph, sketch the expected curve for the products when the reaction is performed without a catalyst.

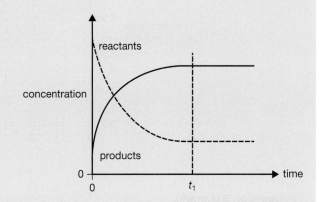

▶ Question 3 (1 mark)

Source: VCE 2020 Chemistry Exam, Section A, Q.15; © VCAA

MC For the reaction $N_2(g) + 3H_2(g) \rightleftharpoons 2NH_3(g)$
- **A.** a catalyst increases the number of collisions between the reactants.
- **B.** the rate of the forward reaction increases when the temperature increases.
- **C.** a catalyst reduces the activation energy of the forward and backward reactions by the same proportion.
- **D.** the activation energy of the forward reaction is greater than the activation energy of the reverse reaction.

Question 4 (3 marks)

Source: Adapted from VCE 2014 Chemistry Exam, Section B, Q.6.b; © VCAA

A mixture of hydrogen gas and iodine gas is injected into a vessel that is then sealed. The mixture will establish an equilibrium system as described by the following equation.

$$I_2(g) + H_2(g) \rightleftharpoons 2HI(g)$$

In an experiment, 3.00 mol of iodine and 2.00 mol of hydrogen were added to a 1.00 L reaction vessel. A graph of the decrease in the concentration of I_2 until equilibrium is effectively reached is shown in Figure 1 below.

a. On Figure 1, draw clearly labelled graphs to show how the concentrations of H_2 and HI changed over the same period of time. **(2 marks)**

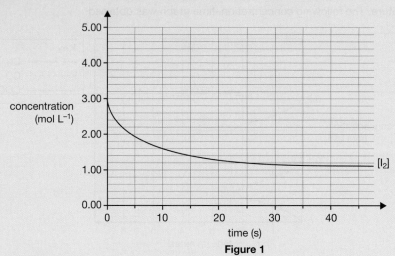

Figure 1

b. Indicate on Figure 2 how the I_2 concentration would have changed if a catalyst had been added to the vessel as well. Assume all other conditions remain the same. **(1 mark)**

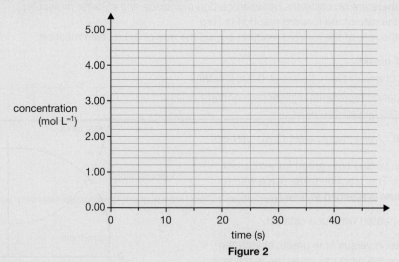

Figure 2

Question 5 (1 mark)

MC During dynamic equilibrium
A. the concentrations of the products and reactants are constantly changing.
B. the forward and reverse reactions are occurring to the same extent.
C. the product stops forming.
D. the forward and reverse reactions are occurring at the same rate.

More exam questions are available in your learnON title.

5.4 Calculations involving equilibrium systems

5.4.1 The equilibrium law and K values

Every reaction, given enough time, reaches a point at which the composition of the reaction mixture no longer changes and the system is said to be at equilibrium. If the concentrations of the substances present are measured at this stage, a large amount of seemingly unrelated data may be obtained. However, on closer analysis, a surprising result emerges.

Recall the hydrogen iodide reaction from section 5.2.1.

$$H_2(g) + I_2(g) \rightleftharpoons 2HI(g)$$

Table 5.1 shows the results of experiments where different initial amounts of the three substances involved were mixed and heated, and enough time allowed for equilibrium to be reached. The resulting equilibrium concentrations were then measured.

Note that square brackets are used to denote concentration measured in M (or mol L^{-1}).

TABLE 5.1 Data for the reaction $H_2(g) + I_2(g) \rightleftharpoons 2HI(g)$ (at 458 °C)

Equilibrium established by	Equilibrium amounts			$K = \dfrac{[HI]^2}{[H_2][I_2]}$
	$[H_2]$ (M)	$[I_2]$ (M)	$[HI]$ (M)	
Combination of hydrogen and iodine	0.002 484	0.002 514	0.016 95	46.0
	0.002 636	0.002 305	0.016 64	45.7
	0.004 173	0.001 185	0.014 94	45.1
	0.003 716	0.001 478	0.015 76	45.2
Decomposition of hydrogen iodide	0.002 594	0.002 597	0.017 63	46.3
	0.001 894	0.001 896	0.012 83	45.9
	0.001 971	0.001 981	0.013 42	46.1
	0.002 413	0.002 424	0.016 41	46.0

The right-hand column in table 5.1 shows that it is possible to write a fraction involving the equilibrium concentrations that has a constant value. A closer inspection of this **concentration fraction** reveals that it is closely related to the equation for the reaction.

- The numerator of the fraction contains the products and the denominator contains the reactants.
- The coefficients from the chemical equation become indices to their respective concentrations in this fraction.

This is the **equilibrium law**. The value of this fraction at equilibrium is called the **equilibrium constant**, which is often assigned the symbol K.

Table 5.2 shows a similar set of results for the synthesis of ammonia from nitrogen and hydrogen.

concentration fraction
essentially, the concentrations of the products divided by the concentrations of the reactants, including the coefficients of each component in the reaction

equilibrium law the relationship between the concentrations of the products and the reactants, taking into account their stoichiometric values

equilibrium constant the value of the concentration fraction at equilibrium, which gives an indication of the extent to which reactants are converted into products; it is assigned the symbol K

TABLE 5.2 Data for the reaction $N_2(g) + 3H_2(g) \rightleftharpoons 2NH_3(g)$

Run	Equilibrium amounts			$K = \dfrac{[NH_3]^2}{[N_2][H_2]^3}$
	$[N_2]$ (M)	$[H_2]$ (M)	$[NH_3]$ (M)	
1	0.0011	0.0011	2.73×10^{-7}	0.051
2	0.0025	0.0055	4.58×10^{-6}	0.050
3	0.55	0.65	0.0886	0.052
4	0.25	0.75	0.074	0.052

The equilibrium constant

For the general reaction

$$aA + bB + cC + dD + ... \rightleftharpoons zZ + yY + xX + ...$$

the equilibrium constant is

$$K = \frac{[Z]^z[Y]^y[X]^x ...}{[A]^a[B]^b[C]^c[D]^d ...}$$

The value of the equilibrium constant can be used to indicate the extent of the reaction.
- If the value is large ($K > 10^4$) we can predict that there has been a significant conversion of reactants into products by the time that equilibrium was reached.
- If the value is between $K = 10^4$ and $K = 10^{-4}$ we can predict that the extent of the reaction is moderate, with concentrations of both products and reactants present at equilibrium.
- If the value is small ($K < 10^{-4}$) we can predict that not much conversion has occurred, and the position of equilibrium favours the back reaction.

When describing an equilibrium qualitatively, the phrase 'position of equilibrium' is often used. If the value of the equilibrium constant is small, the equilibrium is said to lie to the left. If the value of the equilibrium constant is large, the equilibrium is said to lie to the right. Left and right refer to the equation as it has been written and used to evaluate K.

tlvd-3055

SAMPLE PROBLEM 2 Writing the K expression

Write the K expression for the following reaction:

$$2CH_3OH(g) + O_2(g) \rightleftharpoons 2CH_2O(g) + 2H_2O(g)$$

THINK

The K expression is a fraction related to the equation. The products form the numerator and the reactants form the denominator. The co-efficients become indices to their respective concentrations. Use square brackets to denote concentrations.

WRITE

$$K = \frac{[CH_2O]^2[H_2O]^2}{[CH_3OH]^2[O_2]}$$

PRACTICE PROBLEM 2

Write the K expression for the following reaction:

$$CH_4(g) + 2O_2(g) \rightleftharpoons CO_2(g) + 2H_2O(g)$$

The discovery of what we today call the equilibrium law is generally credited to Norwegian scientists Cato Guldberg and Peter Waage. Their findings were published in Norwegian in 1864 and in French in 1867. These papers were not widely read and so their discovery went largely unnoticed. Because of this, the law was independently discovered by the Dutch scientist Jacobus van't Hoff, who subsequently published in 1877. This prompted Guldberg and Waage to expand their ideas in a paper published in German in 1879. As a result, van't Hoff accepted that the Norwegians had made the discovery first and credit should therefore be afforded to them.

This highlights not only the international nature of science, but also the idea of ethical conduct (see key science skills) within the scientific community.

In 1901 van't Hoff was the recipient of the first ever Nobel Prize in Chemistry.

5.4.2 A closer look at equilibrium constants

The equilibrium constant is not always constant. Although this seems like a contradiction, it can change for the following reasons:

- The equilibrium constant is affected by temperature.
- The value of the equilibrium constant may depend on how the equation for the reaction is written.
- How the reaction is written also determines the units of the equilibrium constant.

To illustrate these points, consider the data in table 5.3 for the dissociation of dinitrogen tetroxide gas, N_2O_4, into nitrogen dioxide gas, NO_2, at a constant temperature.

A consistent equilibrium constant is found from each of the five experiments — as long as the same expression for K is used. However, the three cases in table 5.4 show that the same reaction can have different equilibrium constants and different units. Therefore, when discussing equilibrium constants, it is important to be clear about the equation being used to represent the reaction.

TABLE 5.3 Equilibrium data for the reaction $N_2O_4 \rightleftharpoons 2NO_2$

Experiment number	[N_2O_4] (M)	[NO_2] (M)
1	0.127	0.150
2	0.253	0.216
3	0.364	0.255
4	0.492	0.293
5	0.645	0.338

TABLE 5.4 Comparisons of K and units for dissociation of dinitrogen tetroxide gas, N_2O_4, into nitrogen dioxide gas, NO_2, at a constant temperature, using data from experiment 1

Case	K expression	K value	Units for K
CASE 1: The equation is written as $N_2O_4(g) \rightleftharpoons 2NO_2(g)$.	$K = \dfrac{[NO_2]^2}{[N_2O_4]}$	$= \dfrac{(0.150)^2}{0.127}$ $= 0.177$	$\dfrac{M^2}{M} = M$
CASE 2: The equation is written as $\frac{1}{2}N_2O_4(g) \rightleftharpoons NO_2(g)$.	$K = \dfrac{[NO_2]}{[N_2O_4]^{\frac{1}{2}}}$	$= \dfrac{0.150}{(0.127)^{\frac{1}{2}}}$ $= 0.421$	$\dfrac{M}{M^{\frac{1}{2}}} = M^{\frac{1}{2}}$
CASE 3: Using the reverse reaction $2NO_2(g) \rightleftharpoons N_2O_4(g)$ as there are two reactions involved in every equilibrium reaction	$K = \dfrac{[N_2O_4]}{[NO_2]^2}$	$= \dfrac{0.127}{(0.150)^2}$ $= 5.64$	$\dfrac{M}{M^2} = M^{-1}$

A special situation is when an equation has the same number of total moles on each side, such as in the hydrogen iodide reaction mentioned earlier:

$$2HI(g) \rightleftharpoons H_2(g) + I_2(g)$$

If we analyse the equilibrium expression for this reaction, all concentrations cancel out. The equilibrium constant (K) is therefore *without* units.

tlvd-9682

SAMPLE PROBLEM 3 Calculating the value of the equilibrium constant

At 250 °C, phosphorus(V) chloride decomposes to phosphorus(III) chloride plus chlorine, according to the following equation:

$$PCl_5(g) \rightleftharpoons PCl_3(g) + Cl_2(g)$$

In a particular investigation, a quantity of PCl₅ was heated in a 12.0 L reaction vessel to 250 °C and allowed to reach equilibrium. Subsequent analysis revealed that 0.210 mol of PCl₅, 0.320 mol of PCl₃ and 0.320 mol of Cl₂ were present. Calculate the value of the equilibrium constant.

THINK	WRITE
1. Write the equilibrium expression.	$K = \dfrac{[PCl_3][Cl_2]}{[PCl_5]}$
2. Calculate the equilibrium concentrations for each substance, recalling that $c = \dfrac{n}{V}$. **TIP:** A variation of this formula can be found in the VCE Chemistry Data Book.	$[PCl_3] = \dfrac{0.320}{12.0}$ $= 0.0267$ M $[Cl_2] = \dfrac{0.320}{12.0}$ $= 0.0267$ M $[PCl_5] = \dfrac{0.210}{12.0}$ $= 0.0175$ M
3. Substitute these values to obtain K.	$K = \dfrac{(0.0267)(0.0267)}{(0.0175)}$ $= 0.0407$
4. Work out the units.	Units: $\dfrac{M \times M}{M} = M$
5. Give the answer to three significant figures.	$K = 0.0407$ M

PRACTICE PROBLEM 3

Consider an equilibrium reaction that is represented by the following equation:

$$3A(aq) + B(aq) \rightleftharpoons 2Y(aq) + Z(aq)$$

In a particular experiment, random amounts of the above substances were mixed and allowed to come to equilibrium in 500 mL of solution. The amounts present at equilibrium were:

A: 0.351 mol   Y: 0.632 mol

B: 0.18 mol   Z: 1.21 mol

Use this information to calculate the value of the equilibrium constant.

Temperature and the equilibrium constant

Equilibrium constants are also affected by temperature. However, in many situations, such as the $N_2O_4(g)/NO_2(g)$ equilibrium discussed previously, we deal with a situation at a particular temperature. Therefore, when the equation is clearly written or understood, K is always constant. The effect of temperature on the value of K is discussed further in section 5.6.6.

In summary, when dealing with equilibrium reactions and equilibrium constants, it is important that the equation being used to represent the reaction is clearly understood. It is also assumed that, in the absence of information to the contrary, *temperature is constant.*

tlvd-9683

SAMPLE PROBLEM 4 Calculating the value of the equilibrium constant when the equation is altered

The reaction

$$2X(aq) \rightleftharpoons Y(aq) + 2Z(aq) \quad (1)$$

has an equilibrium constant of 250 M at a particular temperature.

Calculate the equilibrium constant for the reaction

$$X(aq) \rightleftharpoons \frac{1}{2}Y(aq) + Z(aq) \quad (2)$$

THINK	WRITE
1. Write and examine the K expression for each equation.	$K(1) = \dfrac{[Y][Z]^2}{[X]^2}$
2. Recognise that the desired value is the square root of the given value.	$K(2) = \dfrac{[Y]^{\frac{1}{2}}[Z]}{X} = (K(1))^{\frac{1}{2}}$
3. Evaluate, including units.	$K(2) = (250)^{\frac{1}{2}}\dfrac{M^{\frac{1}{2}}M}{M}$
	$= 15.8\ M^{\frac{1}{2}}$

PRACTICE PROBLEM 4

The reaction

$$2X(aq) \rightleftharpoons Y(aq) + 2Z(aq) \quad (1)$$

has an equilibrium constant of 250 M at a particular temperature.

Calculate the equilibrium constant for the reaction

$$Y(aq) + 2Z(aq) \rightleftharpoons 2X(aq) \quad (2)$$

5.4.3 Using stoichiometry in equilibrium law calculations

Equilibrium calculations can involve situations with more steps than the simple examples introduced in section 5.4.2. Typical examples of these more complicated types of calculations include:
- situations where initial concentrations are given and only one of the equilibrium concentrations is known
- situations where an equilibrium constant is known and you are asked to calculate something about one of the substances in the reaction. This may be a concentration, a mass or a variable relating to a gas.

The ICEBOX method uses stoichiometry in equilibrium calculations when one or more equilibrium concentration is not known. This involves setting up a table with rows labelled 'I' (initial), 'C' (change) and 'E' (equilibrium). This table can be used when working with moles (which are then converted to concentrations) or directly with concentrations as demonstrated in sample problem 5.

tlvd-9684

SAMPLE PROBLEM 5 Determining the equilibrium constant when concentration is altered

In studying the reaction represented by the equation

$$A(g) \rightleftharpoons 2Y(g) + Z(g)$$

a small amount of substance A was added to a reaction vessel, such that its initial concentration was 1.0 M. When equilibrium was subsequently attained, the concentration of product Z was measured and found to be 0.3 M.

Use this information to determine the equilibrium constant for this reaction.

THINK

1. Using stoichiometry, work out changes to concentration to determine equilibrium values. Use the ICEBOX method with row headings 'I' (initial), 'C' (change) and 'E' (equilibrium). Take care to distinguish between initial and equilibrium concentrations.

WRITE

	[A] ⇌ [2Y] + [Z]		
Initial amount	1	0	0
Change in amount	−0.3	+0.6	+0.3
Equilibrium amount	0.7	0.6	0.3

2. Write the expression and substitute values to determine K.

   TIP: Remember to check that equilibrium concentrations (not initial concentrations) are substituted into the equation.

   TIP: Don't forget units and significant figures.

$K = \dfrac{[Y]^2[Z]}{[A]}$ Units: $\dfrac{M^2 \times M}{M} = M^2$

$= \dfrac{(0.6)^2(0.3)}{(0.7)}$

$= 0.2\ M^2$

PRACTICE PROBLEM 5

For the reaction represented by

$$B(aq) \rightleftharpoons W(aq) + X(aq)$$

initial concentrations of 2.0 M and 3.0 M were recorded for substances B and X respectively. After allowing sufficient time for the reaction to reach equilibrium, the concentration of X had increased to 3.5 M.

Calculate the equilibrium constant.

SAMPLE PROBLEM 6 Calculating the number of moles of a substance present at equilibrium

The equilibrium constant for the reaction represented by the equation

$$2A(g) + 3B(g) \rightleftharpoons X(g) + 2Y(g)$$

is 300 M^{-2}. In a vessel of volume 4.00 L, equilibrium is established and the following concentrations are determined:

[A] = 0.326 M   [B] = 1.537 M   [X] = 2.541 M

Calculate the number of moles of substance Y that were present at equilibrium.

THINK	WRITE
1. Substitute all known information into the K expression.	$K = \dfrac{[X][Y]^2}{[A]^2[B]^3}$
2. Transpose and evaluate to determine [Y]. Don't forget the square root.	$300 = \dfrac{(2.541)[Y]^2}{(0.326)^2(1.537)^3}$
	$\therefore [Y]^2 = \dfrac{300 \times (0.326)^2 \times (1.537)^3}{(2.541)}$
	$= 45.6$
	$\therefore [Y] = 6.75$ M
3. Determine the number of moles of Y by using $n = c \times V$.	$n = c \times V$
	$n(Y) = 6.75 \times 4.00$
	$= 27.0$ mol

PRACTICE PROBLEM 6

The formation of ammonia gas is represented by the following equation:

$$N_2(g) + 3H_2(g) \rightleftharpoons 2NH_3(g)$$

At a particular temperature, the equilibrium constant for this reaction is 0.052 M^{-2}.

In one particular investigation, it was established that the concentrations of N_2 and H_2 at equilibrium were 0.55 M and 0.65 M respectively. Calculate the concentration of ammonia at equilibrium.

---

on Resources

▶ **Video eLesson** Calculations involving initial and equilibrium concentrations (eles-3243)

Interactivity Matching equations and equilibrium constant expressions (int-1244)

5.4 Activities

| 5.4 Quick quiz | 5.4 Exercise | 5.4 Exam questions |

5.4 Exercise

1. It has been estimated that the reaction between methane and oxygen represented by the equation

$$CH_4(g) + 2O_2(g) \rightleftharpoons CO_2(g) + 2H_2O(g)$$

has an equilibrium constant of 10^{140} at room temperature.
   a. What does this value suggest about the extent of this reaction?
   b. Do you think the use of the $\rightleftharpoons$ arrow is justified?

2. a. Write an expression for the equilibrium constant for each of the following equations.
   i. $Cl_2(g) + 3F_2(g) \rightleftharpoons 2ClF_3(g)$

   ii. $N_2(g) + O_2(g) \rightleftharpoons 2NO(g)$

   iii. $4HCl(g) + O_2(g) \rightleftharpoons 2H_2O(g) + 2Cl_2(g)$

   iv. $2COF_2(g) \rightleftharpoons CF_4(g) + CO_2(g)$

   v. $P_4(g) + 10F_2(g) \rightleftharpoons 4PF_5(g)$

   b. For each of the reactions shown in part a, state the units for K.

3. A reaction has an equilibrium expression of:

$$K = \frac{[C][D]^2}{[A]^2[B]}$$

   What is the equation for this reaction?

4. When COF_2 is held at 1000 °C, the following equilibrium occurs:

$$2COF_2(g) \rightleftharpoons CO_2(g) + CF_4(g)$$

   Calculate the value of the equilibrium constant, given the following equilibrium data:
   $[COF_2] = 0.024$ M
   $[CO_2] = [CF_4] = 0.048$ M

5. Consider the following information for the reaction

$$W(aq) + 2X(aq) \rightleftharpoons 3Y(aq) + Z(aq)$$

   - This reaction is carried out in a 250 mL beaker.
   - At equilibrium, the respective amounts of W, X and Y are 1.24 mol, 0.56 mol and 0.85 mol.
   - The molar mass of Z is 56.5 g mol^{-1}.
   - The equilibrium constant for this reaction at the temperature of the investigation is 1.89 M.

   a. Calculate the mass of Z that is present.
   b. If the equation is rewritten as follows, calculate the equilibrium constant.

$$2W(aq) + 4X(aq) \rightleftharpoons 6Y(aq) + 2Z(aq)$$

6. The equilibrium represented by the equation

$$H_2(g) + CO_2(g) \rightleftharpoons H_2O(g) + CO(g)$$

has a K value of 1.62 at 985 °C. Calculate the value of K for the following reactions.
a. $H_2O(g) + CO(g) \rightleftharpoons H_2(g) + CO_2(g)$
b. $2H_2(g) + 2CO_2(g) \rightleftharpoons 2H_2O(g) + 2CO(g)$
7. In a reaction specified by the equation

$$2A(g) \rightleftharpoons 2B(g) + C(g)$$

3.5 mol of substance A is initially introduced into a 500 mL reaction vessel and allowed to reach equilibrium. At this stage, its concentration was found to be 2.0 M. Calculate the equilibrium constant for this reaction.
8. An equilibrium reaction is represented by the following equation:

$$W(aq) + 2X(aq) \rightleftharpoons 2Y(aq) + Z(aq)$$

The magnitude of the equilibrium constant at a particular temperature is 6.3.
a. What are the units of the equilibrium constant?
b. Calculate the value of the equilibrium constant for the following reaction:

$$2Y(aq) + Z(aq) \rightleftharpoons W(aq) + 2X(aq)$$

5.4 Exam questions

▶ Question 1 (4 marks)
Source: VCE 2021 Chemistry Exam, Section B, Q.8.a; © VCAA

The reaction for the oxidation of sulphur dioxide, SO_2, is shown below.

$$2SO_2(g) + O_2(g) \rightleftharpoons 2SO_3(g) \quad \Delta H = -197 \text{ kJ mol}^{-1}$$

1.00 mol of SO_2 and 1.00 mol of oxygen, O_2, are placed into an evacuated, sealed 3.00 L container at 100 °C. After the reaction reaches equilibrium, the container contains 20.0 g of sulfur trioxide, SO_3.

Calculate the equilibrium constant, K, for this reaction at 100 °C.

▶ Question 2 (1 mark)
Source: VCE 2020 Chemistry Exam, Section A, Q.14; © VCAA

MC The magnitude of the equilibrium constant, K, at 25 °C for the following reaction is 640.

$$N_2(g) + 3H_2(g) \rightleftharpoons 2NH_3(g) \quad \Delta H = -92.3 \text{ kJ mol}^{-1}$$

For the reaction $\frac{1}{3}N_2(g) + H_2(g) \rightleftharpoons \frac{2}{3}NH_3(g)$, the magnitude of K at 25 °C is

A. 9 and $\Delta H = -30.8 \text{ kJ mol}^{-1}$
B. 213 and $\Delta H = -30.8 \text{ kJ mol}^{-1}$
C. 640 and $\Delta H = -30.8 \text{ kJ mol}^{-1}$
D. 640 and $\Delta H = -92.3 \text{ kJ mol}^{-1}$

Question 3 (1 mark)

Source: VCE 2019 Chemistry Exam, Section A, Q.28; © VCAA

MC The concentration of all of the gases in the equilibrium reactions below is 1.0 M.

Reaction 1	$CH_4(g) + 2H_2O(g) \rightleftharpoons CO_2(g) + 4H_2(g)$
Reaction 2	$N_2(g) + 3H_2(g) \rightleftharpoons 2NH_3(g)$
Reaction 3	$H_2(g) + I_2(g) \rightleftharpoons 2HI_2(g)$
Reaction 4	$2NO_2(g) \rightleftharpoons N_2O_4(g)$

In which reaction does $K = 1.0 \text{ M}^{-2}$?

A. Reaction 1   B. Reaction 2   C. Reaction 3   D. Reaction 4

Question 4 (1 mark)

Source: VCE 2019 Chemistry Exam, Section A, Q.20; © VCAA

MC The oxidation of sulfur dioxide, SO_2, to sulfur trioxide, SO_3, can be represented by the following equation.

$$2SO_2(g) + O_2(g) \rightleftharpoons 2SO_3(g) \qquad K = 1.75 \text{ M}^{-1} \text{ at } 1000\,°C$$

An equilibrium mixture has a concentration of 0.12 M SO_2 and 0.16 M oxygen gas, O_2. The temperature of the container is 1000 °C.

The equilibrium concentration of SO_3 at 1000 °C is

A. 1.5×10^{-4} M   B. 4.0×10^{-3} M   C. 1.2×10^{-2} M   D. 6.3×10^{-2} M

Question 5 (4 marks)

Source: VCE 2014 Chemistry Exam, Section B, Q.6.a; © VCAA

A mixture of hydrogen gas and iodine gas is injected into a vessel that is then sealed. The mixture will establish an equilibrium system as described by the following equation.

$$I_2(g) + H_2(g) \rightleftharpoons 2HI(g)$$

In an experiment, 3.00 mol of iodine and 2.00 mol of hydrogen were added to a 1.00 L reaction vessel. The amount of iodine present at equilibrium was 1.07 mol. A constant temperature was maintained in the reaction vessel throughout the experiment.

a. Write the expression for the equilibrium constant for this reaction. **(1 mark)**
b. Determine the equilibrium concentrations of hydrogen and hydrogen iodide, and calculate the value of the equilibrium constant. **(3 marks)**

More exam questions are available in your learnON title.

5.5 The reaction quotient (Q)

KEY KNOWLEDGE

- The reaction quotient (Q) as a quantitative measure of the extent of a chemical reaction: that is, the relative amounts of products and reactants present during a reaction at a given point in time

Source: VCE Chemistry Study Design (2024–2027) extracts © VCAA; reproduced by permission.

5.5.1 Definition of reaction quotient (Q)

From the previous section you will recall that the equilibrium law generates an expression that is a 'concentration fraction'. This is just a fraction involving concentrations that is written at a special stage during a reaction — at equilibrium. It is possible to write a similar expression at *any other stage* during a reaction. When we do this, the fraction is given the symbol Q (for reaction quotient).

The progress of a reaction towards equilibrium: making use of Q

Sometimes it is unclear when a reaction is at equilibrium. If a reaction reaches equilibrium quickly because its rate is fast, then it will soon become apparent that no further concentration changes are occurring and that chemical equilibrium has been attained. However, if concentrations are changing slowly in reactions that are slower, it may only appear that the concentrations have become constant. To overcome this problem, we use the idea of the **reaction quotient (Q)** (also known as the 'concentration fraction').

Using the idea of reaction quotient (Q), it follows that:
- if $Q = K$, the reaction is at equilibrium
- if the value of Q is *different* to the value of K, *the reaction has yet to reach equilibrium*
- if $Q > K$, a net backward reaction is occurring. The rate of the backward reaction is greater than the rate of the forward reaction as this mixture moves towards equilibrium.
- if $Q < K$, a net forward reaction is occurring. The rate of the forward reaction is greater than the rate of the backward reaction as this reaction seeks equilibrium.

reaction quotient (Q) essentially, the concentrations of the products divided by the concentrations of the reactants, including the coefficients of each component in the reaction

tlvd-3056

SAMPLE PROBLEM 7 Using the reaction quotient, Q

At 25 °C, the reaction

$$W(aq) + 2X(aq) \rightleftharpoons Y(aq)$$

has an equilibrium constant of 109 M^{-2}. In an experimental trial of this reaction, the concentrations of all species are measured and the following results obtained:

[W] = 0.13 M   [X] = 0.092 M   [Y] = 0.090 M

Is this reaction at equilibrium? If not, how does the rate of the forward reaction compare to the rate of the backward reaction?

THINK	WRITE
1. Write the formula for Q.	$Q = \dfrac{[Y]}{[W][X]^2}$
2. Calculate Q by substituting the given concentrations.	$Q = \dfrac{(0.090)}{(0.13)(0.092)^2}$ $= 82\ M^{-2}$
3. The value of K is given, $K = 109\ M^{-2}$. If $Q < K$, a net forward reaction is occurring. If $Q > K$, a net backward reaction is occurring. If $Q = K$, the reaction is at equilibrium.	As $Q < K$ there is a net forward reaction. The rate of the forward reaction is faster than the rate of the backward reaction.

PRACTICE PROBLEM 7

In another trial of the same reaction as in sample problem 7, the following concentrations were obtained:

[W] = 0.14 M   [X] = 0.085 M   [Y] = 1.0 M

Is this reaction at equilibrium? If not, how does the rate of the forward reaction compare to the rate of the backward reaction?

5.5 Activities

| 5.5 Quick quiz on | 5.5 Exercise | 5.5 Exam questions |

5.5 Exercise

1. If a reaction occurs at a slow rate, it might not be obvious if it has attained equilibrium. Describe how the use of the reaction quotient, Q, would help determine if the reaction had reached equilibrium.
2. Explain why the units of Q and K are always the same.
3. **MC** Xavier sets up an equilibrium reaction and immediately makes some measurements so that he can calculate the value of Q. He discovers that $Q > K$.

   Which of the following statements is true?
   A. Q will decrease but will always remain greater than K.
   B. Q will decrease until it equals K and then remain constant.
   C. Q will decrease and eventually become lower than K.
   D. Q will remain constant and always be greater than K.
4. The reaction represented by the equation

$$A + 2B \rightleftharpoons C + D$$

was investigated by mixing various amounts of these four substances. After a period of time, the concentration of each species was measured. This procedure was repeated a number of times to give the results shown in the following table.

Experiment	[A]	[B]	[C]	[D]
1	1.7	1.2	2.5	2.8
2	1.4	0.9	2.3	2.7
3	1.6	1.1	2.3	2.4
4	1.5	0.9	2.6	2.8
5	1.4	1.0	2.5	2.7

At the temperature of this experiment, it is known that the value of the equilibrium constant is 6.0. In which of the above experiments was the reaction mixture at equilibrium when it was analysed?
5. The reaction represented by the equation

$$CO_2(g) + H_2(g) \rightleftharpoons CO(g) + H_2O(g)$$

has an equilibrium constant, K, of 1.58 at 990 °C. In an experiment, the concentrations of these four substances were measured at a particular time. The values obtained were:
$[CO_2] = 0.00208$ M
$[H_2] = 0.00221$ M
$[CO] = 0.00270$ M
$[H_2O] = 0.00250$ M

a. Is this reaction at equilibrium?
b. Comment on the relative rates of the forward and backward reactions.

5.5 Exam questions

▶ **Question 1 (1 mark)**

Source: VCE 2015 Chemistry Exam, Section B, Q.7.a © VCAA

Consider the reaction shown in the following equation.

$$2NO(g) + Br_2(g) \rightleftharpoons 2NOBr(g) \quad \Delta H = -16.1 \text{ kJ mol}^{-1}, \quad K = 1.3 \times 10^{-2} \text{ M}^{-1} \text{ at } 1000 \text{ K}$$

Write an expression for the reaction quotient for this reaction.

▶ **Question 2 (1 mark)**

Source: VCE 2012 Chemistry Exam 2, Section A, Q.9; © VCAA

MC The following reaction is used in some industries to produce hydrogen.

$$CO(g) + H_2O(g) \rightleftharpoons CO_2(g) + H_2(g) \quad \Delta H = -41 \text{ kJ mol}^{-1}$$

Carbon monoxide, water vapour, carbon dioxide and hydrogen were pumped into a sealed container that was maintained at a constant temperature of 200 °C. After 30 seconds, the concentration of gases in the sealed container was found to be [CO] = 0.1 M, [H_2O] = 0.1 M, [H_2] = 2.0 M, [CO_2] = 2.0 M.

The equilibrium constant at 200 °C for the above reaction is $K = 210$.

Which one of the following statements about the relative rates of the forward reaction and the reverse reaction at 30 seconds is true?

A. The rate of the forward reaction is greater than the rate of the reverse reaction.
B. The rate of the forward reaction is equal to the rate of the reverse reaction.
C. The rate of the forward reaction is less than the rate of the reverse reaction.
D. There is insufficient information to allow a statement to be made about the relative rates of the forward and reverse reactions.

▶ **Question 3 (1 mark)**

Source: Adapted from VCE 2021 Chemistry Exam, Section A, Q.25; © VCAA

An equilibrium mixture of four gases is represented by the following equation.

$$A(g) + 2B(g) \rightleftharpoons C(g) + D(g) \quad \Delta H > 0$$

The graph below shows the rate of the forward and reverse reactions versus time.

A single change is made to the equilibrium mixture at time t_1 and equilibrium is re-established at time t_2.

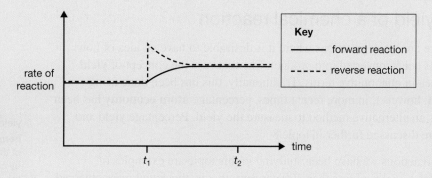

Describe how the value of Q compares to the value of K in the time period from t_1 to t_2.

▶ **Question 4 (1 mark)**
Source: Adapted from VCE 2010 Chemistry Exam 2, Section A, Q.7; © VCAA

MC The following reaction systems are at equilibrium in separate sealed containers. The volumes of the containers are halved at constant temperature.

Which reaction has the largest percentage decrease in the concentration fraction (reaction quotient) immediately after the volume change?

A. $N_2O_4(g) \rightleftharpoons 2NO_2(g)$ **B.** $H_2(g) + I_2(g) \rightleftharpoons 2HI(g)$

C. $2CO_2(g) \rightleftharpoons 2CO(g) + O_2(g)$ **D.** $CO(g) + 2H_2(g) \rightleftharpoons CH_3OH(g)$

▶ **Question 5 (1 mark)**
Source: Adapted from VCE 2009 Chemistry Exam 2, Section B, Q.3.a; © VCAA

Dimethyl ether, CH_3OCH_3, is used as an environmentally friendly propellant in spray cans. It can be synthesised from methanol according to the following equation.

$$2CH_3OH(g) \rightleftharpoons CH_3OCH_3(g) + H_2O(g) \quad \Delta H = -24 \text{ kJ mol}^{-1}$$

Write an expression for Q for this reaction.

More exam questions are available in your learnON title.

5.6 Changes to equilibrium and Le Chatelier's principle

KEY KNOWLEDGE

- The change in position of equilibrium that can occur when changes in temperature or species or volume (concentration or pressure) are applied to a system at equilibrium, and the representation of these changes using concentration–time graphs
- The application of Le Chatelier's principle to identify factors that favour the yield of a chemical reaction
- Responses to the conflict between optimal rate and temperature considerations in producing equilibrium reaction products, with reference to the green chemistry principles of catalysis and designing for energy efficiency

Source: VCE Chemistry Study Design (2024–2027) extracts © VCAA; reproduced by permission.

5.6.1 The yield of a chemical reaction

In chemistry there are many situations where it is desirable to have an idea of how much reactant has been converted into product. Chemists use the concept of **yield** to express this idea in quantitative terms. Traditionally, this has been done using **percentage yield**; however, in more recent times, percentage **atom economy** has been gaining favour as an alternative method to measure the yield. Percentage yield and atom economy are discussed further in topic 8.

The equilibrium reactions we have been studying in this topic are examples of reactions that display yields of less than 100 per cent; that is, the actual mass obtained is less than the theoretical or predicted maximum mass. However, there are often additional reasons why a given reaction does not achieve a 100 per cent yield. Some of these may be practical (to do with the method by which the chemical is made), or may involve a very slow reaction that has not been given enough time to either reach equilibrium or go to completion. This is an important consideration in equilibrium work and in the design of large-scale manufacturing techniques for chemicals produced from equilibrium reactions.

yield amount of product

percentage yield a measurement of the efficiency of a reaction, found by calculating the percentage of the actual yield compared to the theoretical yield

atom economy a measurement of the efficiency of a reaction that considers the amount of waste produced, by calculating the percentage of the molar mass of the desired product compared to the molar mass of all reactants

5.6.2 Making changes to equilibrium mixtures

An important consequence of the equilibrium law is that it is possible for every equilibrium mixture belonging to a particular reaction to be different. This is because the whole equilibrium expression is constant, while individual concentrations within this expression may vary quite considerably from one equilibrium situation to another. However, so long as the value of the whole expression is equal to the value of the equilibrium constant, a mixture will be at equilibrium.

This property can be put to use in the manufacture of some important chemicals. It is often possible to make these economically, despite the fact that the reactions from which they are formed have low equilibrium constants. By altering the concentrations of the other species involved in the equilibrium expression, it is often possible to maximise the production of the desired product, thereby increasing its yield. Many biological processes also rely on making changes to equilibrium reactions.

> **Le Chatelier's principle** states that when a change is made to an equilibrium system, the system moves to counteract the imposed change and restore the system to equilibrium

To see how changes made to an equilibrium mixture affect its components, we often make use of an important predictive tool — **Le Chatelier's principle**.

5.6.3 Introduction to Le Chatelier's principle

Le Chatelier's principle

Any change that affects the position of an equilibrium causes that equilibrium to shift, if possible, in such a way as to partially oppose the effect of that change.

Using this principle, we can make predictions about what will happen if we disturb a system that is at equilibrium.

There are three common ways in which a system at equilibrium might be disturbed:
- By adding or removing a substance that is involved in the equilibrium
- By changing the volume (at constant temperature)
- By changing the temperature.

We will now consider the effect of each of these in turn.

5.6.4 Adding or removing a substance that is involved in the reaction

Consider the reaction between carbon monoxide and water vapour to produce carbon dioxide and hydrogen.

$$CO(g) + H_2O(g) \rightleftharpoons CO_2(g) + H_2(g)$$

If extra water vapour is added to this system once it has attained equilibrium, the equilibrium will be disturbed. According to Le Chatelier's principle:
- The system will react by trying to use up some of this extra water vapour in its efforts to get back to a new equilibrium position.
- This is done by causing some of the reactants to be transformed into products. That is, more CO_2 and H_2 will be made as a result of consuming CO and H_2O.
- Although all the amounts, and hence all the concentrations, involved will now be different from their original values, the value of the equilibrium constant will remain unchanged (the same value as before the water vapour was added).

Changes like this, where a substance has been added or removed, can be identified from concentration–time graphs by a sudden spike or dip in only one of the substances involved. A concentration–time graph for this situation is shown in figure 5.8.

- Time t_1 represents the time when equilibrium was first reached. The rates of the forward and reverse reactions are equal.
- Time t_2 is when the extra water vapour was added. The forward reaction becomes faster than the reverse reaction for some time.
- Time t_3 is when equilibrium is re-established. The equilibrium may be described as having been shifted to the right (the forward reaction is favoured). The rates of the forward and reverse reactions are equal again as equilibrium is re-established.

FIGURE 5.8 A concentration–time graph of the reaction $CO(g) + H_2O(g) \rightleftharpoons CO_2(g) + H_2(g)$ with additional H_2O added

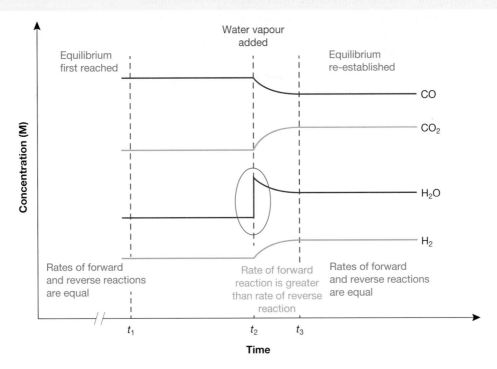

When considering these reactions, the method of addition (continuous supply) or removal (ducting gas away) may be physical, or it may be achieved by chemical means. Regardless of the method, the system always tries to oppose us in its efforts to re-attain equilibrium. However, it should be noted that equilibrium is only disturbed if products and/or reactants are added or removed. If an inert gas is added to the system, it will have no effect on the reaction.

CASE STUDY: Using Le Chatelier's principle to dissolve an insoluble salt

Le Chatelier's principle may be used to dissolve an otherwise insoluble salt. Silver chloride is only sparingly soluble in water. In other words, the equilibrium constant for the reaction

$$AgCl(s) \rightleftharpoons Ag^+(aq) + Cl^-(aq)$$

is very low ($K_s = 1.8 \times 10^{-10}$). This is an example of a heterogeneous equilibrium. In these cases, the equilibrium constant is usually denoted by K_s, a special type of K value.

However, if a chemical is added to remove the Ag^+ ions, this reaction responds by trying to replace them. To do this, more $AgCl(s)$ has to dissolve. If enough of the Ag^+ ions are removed, it is possible that all the silver chloride would dissolve. A suitable chemical for the removal of Ag^+ ions is ammonia, NH_3. Ammonia achieves this by forming a complex ion according to the following reaction:

$$Ag^+(aq) + 2NH_3(aq) \rightleftharpoons Ag(NH_3)_2^+(aq)$$

tlvd-9686

SAMPLE PROBLEM 8 Predicting the effect of altering a reactant or product on the position of equilbrium

Predict the effect of adding carbon dioxide on the position of the equilibrium for the following reaction:

$$CO(g) + H_2O(g) \rightleftharpoons CO_2(g) + H_2(g)$$

THINK

Recall that Le Chatelier's principle predicts that adding a substance involved in the equilibrium will lead to its partial removal. In this case, the reaction will respond by removing some of the added CO_2. The equilibrium will therefore shift to the left; that is, there will be a net backward reaction until the reaction is once again at equilibrium.

WRITE

Product has been added so there will be a net backward reaction until the reaction is once again at equilibrium. The new equilibrium mixture will contain more CO and H_2O than the original one.

PRACTICE PROBLEM 8

Predict the effect of removing hydrogen on the position of the equilibrium for the following reaction:

$$CO(g) + H_2O(g) \rightleftharpoons CO_2(g) + H_2(g)$$

EXTENSION: Applying Le Chatelier's principle to hide Nobel medallions

A Nobel prize is one of the highest awards that a scientist can achieve. The award consists of a pure gold medallion engraved with the winner's name, a diploma and a considerable sum of money.

As the Nazis rose to power in the 1930s, two German Nobel laureates and opponents of the regime, Max von Laue and James Franck, sent their medals to the famous physicist Niels Bohr in Copenhagen, Denmark, for safe keeping. The exporting of gold from Germany at this time was strictly forbidden and the two would have faced harsh penalties if they were discovered. When the Germans invaded Denmark in 1940, the discovery of the two medallions might have proven a death sentence for those involved. An associate of Niels Bohr, the chemist George de Hevesy, came to the rescue with a clever application of Le Chatelier's principle.

To access more information on this extension concept, download the digital document from your Resources panel.

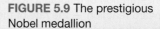

FIGURE 5.9 The prestigious Nobel medallion

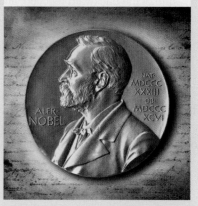

 Resources

Digital document EXTENSION: Applying Le Chatelier's principle to hide Nobel medallions (doc-37354)

Explaining Le Chatelier's principle mathematically

Consider the reaction that is represented by the following equation:

$$3A(aq) + 2B(aq) \rightleftharpoons C(aq) + 2D(aq)$$

Suppose we wish to predict the effect on substance D when more of substance A is added to the equilibrium mixture.

For this reaction we can write two expressions:

$$Q = \frac{[C][D]^2}{[A]^3[B]^2} \quad \text{and} \quad K = \frac{[C][D]^2}{[A]^3[B]^2}$$

Stage 1: Initial equilibrium	
Given the reaction is currently at equilibrium, the reaction quotient, Q, is equal to the equilibrium constant, K. The reaction quotient can be calculated at any stage during a reaction, but only when $Q = K$ has equilibrium been established.	$Q = \dfrac{[C][D]^2}{[A]^3[B]^2} = K$
Stage 2: Addition of substance A	
Adding more of substance A decreases the value of Q, since A is a reactant (and therefore in the denominator of the fraction). This means Q is now less than K (remember — K is constant!) and the reaction is no longer at equilibrium.	$\downarrow Q = \dfrac{[C][D]^2}{\uparrow[A]^3[B]^2} < K$ • [A] increases. • The value of Q decreases. • $Q < K$
Stage 3: Re-establishing equilibrium	
To re-establish equilibrium, the reaction must proceed in a way that increases the value of Q, until it is equal to K once again. For Q to increase, the concentration of products needs to increase and the concentration of reactants needs to decrease. This can be achieved by a net forward reaction — the rate of the forward reaction becomes greater than the rate of the backward reaction, and the equilibrium therefore shifts to the right.	$\uparrow Q = \dfrac{\uparrow[C]\uparrow[D]^2}{\downarrow[A]^3\downarrow[B]^2} = K$ • The reaction proceeds to the right. • [C] and [D] increase. • [A] and [B] decrease. • The value of Q increases. • $Q = K$

This is exactly what is predicted by Le Chatelier's principle. After the sudden increase in substance A, some of it is then used by the net forward reaction predicted, and more of substance D is produced. The effect of changing volume can be predicted in the same way.

5.6.5 The effect of changing volume

When considering changes in volume (at constant temperature), it is important to think of the effect on the total concentration of all the species present. Le Chatelier's principle can then be interpreted in terms of how the system changes the total number of particles present to produce an opposing trend in total concentration.

Three situations present themselves:
1. The change in volume causes an increase in the total concentration of particles. To decrease this, the system reacts in the direction that produces fewer particles. (Fewer particles means a lower overall concentration, in line with the opposition predicted by Le Chatelier's principle.)
2. The change in volume causes a decrease in the total concentration of particles. To increase this concentration, the system must react in the direction that produces more particles if it is to re-establish equilibrium.
3. Although the change in volume affects the total concentration, the system cannot change the number of particles present. This situation occurs when the total number of moles on the left-hand side of the equation equals the number of moles on the right-hand side. Mathematically, it can be shown that a volume change for such a reaction does not disturb the equilibrium.

As an example of situation 1, consider the reaction represented by the following equation:

$$A(aq) \rightleftharpoons B(aq) + C(aq)$$

It might be predicted that an increase in volume, such as by dilution, would lead to an increase in the total number of particles as the reaction attempts to re-build the total concentration. In this case, this can be achieved if the equilibrium shifts to the right (net forward reaction).

A change in volume of an equilibrium mixture is identified on concentration–time graphs as a sudden spike or dip involving all substances. This is shown in figure 5.10.

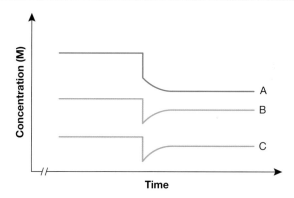

FIGURE 5.10 A concentration–time graph illustrating the effect of increasing the volume of the reaction $A(aq) \rightleftharpoons B(aq) + C(aq)$

A special note about gases

For a change in volume at constant temperature, the universal gas equation allows a concentration interpretation to be replaced by a pressure interpretation. Therefore:

Pressure is proportional to concentration in gases

$$PV = nRT$$
$$P = \left(\frac{n}{V}\right) RT$$
$$P \propto c$$

As $\dfrac{n}{V}$ is concentration (c), and $R \times T$ is constant, we can see that pressure is proportional to concentration.

For gaseous reactions, volume changes may also be interpreted in terms of partial pressures, rather than concentrations.

tlvd-9687

SAMPLE PROBLEM 9 Predicting changes in gaseous systems

a. The following gaseous system is set up and allowed to reach equilibrium.

$$2NOBr(g) \rightleftharpoons 2NO(g) + Br_2(g)$$

What would be the effect on the amount of bromine present if the volume is then increased?

b. Would a volume change affect the following reaction?

$$H_2(g) + Cl_2(g) \rightleftharpoons 2HCl(g)$$

THINK

a. A volume change requires thinking in terms of the total number of particles. Increasing volume decreases the total concentration.
The reaction will respond in the direction that results in a net gain in particles. This will partially increase the total concentration in response to the initial increase in volume. In this case, there is a net gain in the number of particles (from 2 moles to 3 moles) if the reaction moves to the right.

b. This reaction has equal numbers of particles (and hence moles) on each side of the equation (2 and 2). Hence, the equilibrium will not be affected by a change in volume.

WRITE

a. The reaction will respond by making more products (NO and Br_2). At the same time, it will use up more of the NOBr; that is, there will be a net forward reaction.

b. A volume change will not affect this reaction because there is the same number of moles on each side of the equation.

PRACTICE PROBLEM 9

a. Consider the reaction represented by the following equation:

$$I_2(aq) + I^-(aq) \rightleftharpoons I_3^-(aq)$$

What would be the effect on the amount of I_3^- caused by increasing the volume by adding more water?

b. Would a volume change affect the following reaction?

$$CaCO_3(s) \rightleftharpoons CaO(s) + CO_2(g)$$

elog-1733

tlvd-9720

EXPERIMENT 5.1   online only

Investigating changes to the position of an equilibrium

Aim

To observe the effect of a number of changes on the position of the equilibrium represented by the equation
$Fe^{3+}(aq) + SCN^-(aq) \rightleftharpoons Fe(SCN)^{2+}(aq)$

5.6.6 The effect of changing temperature

Of all the possible ways to change an equilibrium mixture, changing the temperature is the *only method that actually alters the value of the equilibrium constant, K*, because changing the temperature also changes the energy available to the system.

- At a particular temperature, the system has a certain amount of energy, which is distributed between all the species present. The equilibrium expression describes the concentrations of all species once they have settled on a way to share that energy, and the value of K is interpreted as sharing in favour of reactants or products.
- Changing the temperature changes the total amount of energy in the system. Hence, K has a different value.
- How the value of K responds to temperature change depends on whether the reaction is exothermic or endothermic. Classifying an equilibrium reaction as exothermic or endothermic relates, by convention, to the forward reaction as written in the equation. An endothermic reaction absorbs heat and has a positive ΔH value. An exothermic reaction evolves heat and has a negative ΔH value.

Varying the concentration of a species (n or V) without a change in temperature just causes a redistribution of the available energy, and a different equilibrium position is reached. However, because the available energy remains the same, the value of K is the same.

Consider an exothermic reaction, typified by the following thermochemical equation:

$$A(g) + B(g) \rightleftharpoons C(g) \quad \Delta H \text{ is negative}$$

An increase in temperature requires the application of heat.

- Le Chatelier's principle would predict that the system tries to absorb some of this added heat. Hence, the backward reaction is favoured as it is endothermic.
- A and B are produced at the expense of C, lowering the value of the equilibrium constant.

Thinking mathematically, when the value of the denominator in K increases and the value of the numerator decreases, the value of K decreases.

$$\downarrow K = \frac{\downarrow[\text{products}]}{\uparrow[\text{reactants}]}$$

Note that this equation can also be written so heat is treated as a product:

$$A(g) + B(g) \rightleftharpoons C(g) + \text{heat}$$

thus making the effect of adding heat the same as adding a product (from a Le Chatelier viewpoint). In a similar fashion, an endothermic reaction can be written with heat shown on the reactant side of the equation.

FIGURE 5.11 The general relationship between *K* and temperature for exothermic and endothermic reactions

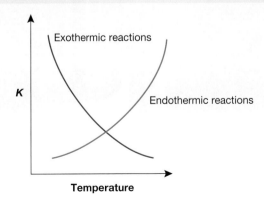

A change in temperature can be identified on concentration–time graphs by its effect on the concentrations without an obvious sudden change to any of the substances involved. For the previous reaction, the concentration–time graph might look like figure 5.12.

In an endothermic reaction under the same conditions, the forward reaction would oppose the addition of heat. Upon re-establishing equilibrium, the value of the equilibrium constant would, therefore, be higher.

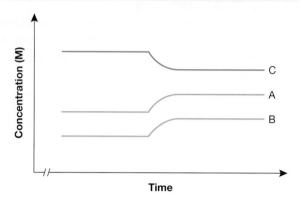

FIGURE 5.12 A concentration–time graph showing the effect of increasing temperature on the equilibrium $A(g) + B(g) \rightleftharpoons C(g)$, where ΔH is negative

The effect of temperature on K

- The value of the equilibrium constant, K, of a reaction can only be changed by changing the temperature.
- Changing the temperature changes the energy available to the system.
- If a reaction is exothermic, an increase in temperature decreases the value of the equilibrium constant.
- If a reaction is endothermic, an increase in temperature increases the value of the equilibrium constant.

FIGURE 5.13 The equilibrium reaction between brown NO_2 and colourless N_2O_4 is affected by temperature. These tubes initially contained equal amounts of gas; the tube on the left is being cooled.

on Resources

▶ **Video eLesson** Le Chatelier's principle — change in temperature (eles-3244)

tlvd-3058

SAMPLE PROBLEM 10 Determining the result of heating an equilbrium mixture on the *K* value

Suppose that, in the reaction

$$A(aq) + 2B(aq) \rightleftharpoons C(aq)$$

substance C has an easily detected red colour. It is observed that heating the equilibrium mixture causes the red colour to fade.

a. What is the resulting effect on the *K* value?

b. Is this reaction an exothermic or an endothermic reaction?

THINK

a. The fading of the red colour implies that there has been a net backward reaction. The value of *K* will, therefore, be lower.

b. *K* has dropped as a result of an increase in temperature, which means that the reaction (as written) must be an exothermic reaction.

WRITE

a. The *K* value will be lower.

b. It is an exothermic reaction.

PRACTICE PROBLEM 10

NO_2 reacts with itself to produce N_2O_4 according to the following equation:

$$2NO_2(g) \rightleftharpoons N_2O_4(g)$$

The progress of this reaction may be monitored by noting the change in the brown colour of NO_2 because N_2O_4 is colourless. It is noted that as a sample of this mixture is heated, its colour darkens. Is this reaction exothermic or endothermic?

---

elog-1734

tlvd-9721

EXPERIMENT 5.2

online only

Temperature and the equilibrium constant

Aim

To observe the effect of temperature on the value of an equilibrium constant

---

5.6.7 Identifying a change graphically

As already established, concentration–time graphs are an informative way of summarising changes made to an equilibrium system. The important features of such graphs are:

- changes in concentration that reflect the stoichiometric ratios of the equation
- the attainment of equilibrium as reflected by each concentration becoming a horizontal line and the fact that each such concentration reaches a certain constant value. This occurs at the same time for each substance.
- a sudden dip or spike in only one of the concentrations, which occurs when one substance has been added or removed. Figure 5.8 is an example of this type of graph.
- a sudden dip or spike in all the concentrations, which is reflective of a volume change. For a reaction in the aqueous phase, this would represent a dilution. For a gas phase reaction, it could represent either the compression or expansion of the sample. Figure 5.10 is an example of this type of graph.
- a change in the equilibrium values without an obvious spike or dip, which indicates that there has been a temperature change. Figure 5.12 is an example of this type of graph.
- all the reactants either increasing or decreasing together. All the products will change together in the opposite way; that is, one side of the equation will go up or down and the other side will go down or up.

EXPERIMENT 5.3 online only

Modelling an equilibrium

Aim

To use a computer spreadsheet application and reaction rate data to model a chemical reaction, and to determine the effect of temperature on the equilibrium constant for the reaction $A(aq) + B(aq) \rightleftharpoons C(aq) + D(aq)$

5.6.8 Open-versus-closed systems

All reactions can be considered either **closed systems** or **open systems** (see topic 4). All the examples used so far have been assumed to be closed systems, in which none of the substances involved are lost. In an open system this is not the case. Because of this, equilibrium reactions that occur in open systems will need to have Le Chatelier's principle applied to them. For example, the decomposition of calcium carbonate upon heating in a closed system reaches an equilibrium where the gaseous CO_2 is in equilibrium with the solid $CaCO_3$ and CaO, according to the following equation:

$$CaCO_3(s) \rightleftharpoons CaO(s) + CO_2(g)$$

However, if this reaction is carried out in an open container, the CO_2 can escape. This is effectively removing a product from the equilibrium mixture. Therefore, Le Chatelier's principle predicts that more $CaCO_3$ will decompose in an effort to replace it. As this is a continuous process, the result will be that all the $CaCO_3$ present will eventually decompose.

on Resources

Interactivity Le Chatelier's principle (int-1246)

Weblink   Le Chatelier's principle

5.6.9 Rate and temperature considerations in the production of equilibrium products

The manufacture of some important chemicals involves allowing for the fact that the temperature required to achieve a certain desired rate of production causes the value of the equilibrium constant to be too low. The two important considerations of rate and extent are therefore in conflict. Examples include ammonia (NH_3), nitric acid (HNO_3) and sulfuric acid (H_2SO_4). To make these chemicals on a large scale, the chemical engineer must have a sound knowledge of equilibrium principles and also of rate principles. In designing a plant, the engineer will ultimately be trying to achieve the following:

- Maximise the yield of the desired product by applying Le Chatelier's principle to make as much of the chemical as possible.
- Produce the desired chemical at an acceptable rate — it has to be made quickly enough to satisfy market demand if the plant is to be economical.
- Balance the previous two requirements against other variables, such as plant operating costs, to ensure that the whole process is as economical and safe as possible.
- Further balance these requirements against other variables, including sustainability, supply of raw materials and environmental concerns, as well as many others. The principles of **green chemistry** are increasingly being used in this regard.

closed system a system in which energy, but not matter, can be transferred to and from its surroundings; all reactants and products are contained

open system a system in which both energy and matter can be transferred to and from its surroundings; reactants and products are not contained

green chemistry a relatively new branch of chemistry that emphasises reducing the amounts of wastes produced, the more efficient use of energy, and the use of renewable and recyclable resources

tlvd-3059

Sulfuric acid is made in large amounts by the Contact process. This process has a number of steps but one of the most critical is the following:

$$2SO_2(g) + O_2(g) \rightleftharpoons 2SO_3(g) \quad \Delta H = -198 \text{ kJ mol}^{-1}$$

This reaction is being studied in an industrial laboratory at a certain temperature and pressure. For each of the changes listed below, state:
i. whether the yield of sulfur trioxide will increase, decrease or remain unaltered
ii. whether the rate of sulfur trioxide production will increase, decrease or remain unaltered.

a. An increase in pressure
b. A decrease in temperature at constant original pressure
c. Addition of a catalyst at the original temperature and pressure

THINK

a. i. Le Chatelier's principle predicts that increasing the pressure will force a reduction in the total number of particles present in order to partially offset this change. The reaction will proceed to the right, favouring the production of two moles of sulfur trioxide over three moles of reactants.

  ii. Increasing pressure forces all particles closer together, resulting in more collisions per unit time. Rate will therefore increase.

b. i. The reaction is exothermic. Le Chatelier's principle predicts that decreasing the temperature will increase the value of K, thus increasing the yield of sulfur trioxide.

  ii. Lowering the temperature lowers the energy of the particles, and therefore the energy of the collisions, resulting in fewer collisions that can overcome the activation energy in a given period. Additionally, the frequency of collisions is reduced, which further slows the rate.

c. i. A catalyst does not alter the position of an equilibrium. The yield of sulfur trioxide will not be affected.
  ii. Catalysts increase the rate of a reaction.

WRITE

a. i. Increase

  ii. Increase

b. i. Increase

  ii. Decrease

c. i. Unaltered

  ii. Increase

PRACTICE PROBLEM 11

Consider the following reaction:

$$CH_4(g) + H_2O(g) \rightleftharpoons CO(g) + 3H_2(g) \quad \Delta H = +206 \text{ kJ mol}^{-1}$$

For each of the changes listed below, predict:
i. whether the yield of hydrogen will increase, decrease or remain unaltered
ii. whether the rate of hydrogen production will increase, decrease or remain unaltered.

a. An increase in pressure
b. A decrease in temperature at constant original pressure
c. Addition of a catalyst at the original temperature and pressure

5.6.10 The production of ammonia — a case study on the conflict between rate and temperature

Why is ammonia so important?

Ammonia is one of the most important and widely used chemicals in the world today. In 2022, it was estimated that 150 million tonnes of it was produced. This is due to its variety of uses, the most important of which is its use to make fertilisers. Figure 5.15 shows the common uses of ammonia.

FIGURE 5.14 Ammonia is used extensively in agriculture to fertilise crops.

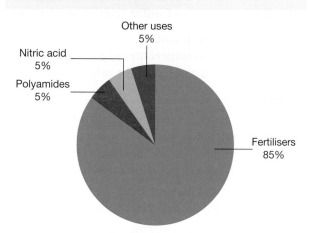

FIGURE 5.15 Uses of ammonia

Other uses 5%

Nitric acid 5%

Polyamides 5%

Fertilisers 85%

It is anticipated that the demand for ammonia will increase even further in the near- to mid-future as the world moves to reducing its dependence on fossil fuels and works to meet its carbon reduction targets. Two significant areas where demand is predicted to increase are:

- in the transport industry. The maritime sector, in particular, is investigating ways by which ships could be powered by ammonia rather than fossil fuels.
- as a means of distributing hydrogen as the world moves towards a 'hydrogen economy'. Hydrogen produced in large-scale electrolysis plants can be converted to ammonia and then transported as it can be liquefied using only mild pressure. At its destination, it can be either used directly or converted back into hydrogen.

The Haber process

The Haber process (also called the Haber–Bosch process) is the traditional method for making large-scale ammonia. It was invented in the early twentieth century and has remained essentially unchanged since.

The Haber process is a single-step process in which nitrogen and hydrogen are reacted to produce ammonia, according to the following equation:

$$N_2(g) + 3H_2(g) \rightleftharpoons 2NH_3(g) \quad \Delta H \text{ is negative}$$

This reaction is a good example of a dilemma often faced by a chemical engineer. At normal temperatures, the value of the equilibrium constant is quite high, but the rate of reaction is very slow. However, if the temperature is increased to bring about a better rate, the yield of ammonia quickly begins to suffer. This is because the reaction shown is exothermic and, as temperature is increased, the value of its equilibrium constant decreases.

FIGURE 5.16 A production plant for ammonia and nitrogen fertiliser. Note the large cooling towers.

This puts the yield and rate, two of the most important industrial factors, in conflict. The ultimate design of the plant and its operating conditions must reflect a compromise between these factors.

A closer look at these factors reveals that we can lessen the effect of this compromise in a number of ways:
- Use a suitable catalyst to help obtain the rate required. This means that the temperature needed is lower than that needed without a catalyst. The use of a lower temperature increases the yield by increasing the value of the equilibrium constant. A lower temperature reduces the rate, but this can be compensated for by using a catalyst, which increases the rate.
- Compress the gases so that the reaction is carried out at high pressure. Le Chatelier's principle predicts that, if a system is pressurised, it tries to reduce that pressure. Consequently, there should be a tendency to reduce the number of particles present. In this case, this means that the reaction proceeds to the right, as the forward reaction converts four molecules into two molecules, thus favouring the production of ammonia. This will also help increase the rate of the reaction.
- Separate the ammonia from the unreacted nitrogen and hydrogen in the exit gases from the converter. This nitrogen and hydrogen can then be recycled and may be converted into more ammonia.

In the operation of a typical 'traditional' plant, nitrogen and hydrogen are mixed in the 1 : 3 ratio required by the equation. The gases are then compressed to about 250 atm and heated to about 500 °C. (These precise conditions may vary slightly from one plant to another.) The gases then pass into the converter, which is a huge, reinforced steel cylinder containing 7 to 8 tonnes of pea-size catalyst beads. The catalyst most often used is an iron catalyst made from iron oxide, Fe_3O_4, with traces of aluminium oxide and potassium oxide.

When the gases leave this chamber they contain about 20 per cent ammonia. By cooling the mixture, the ammonia can be liquefied and separated. The unreacted nitrogen and hydrogen can then be recycled so that they pass through the converter again.

FIGURE 5.17 The Haber process

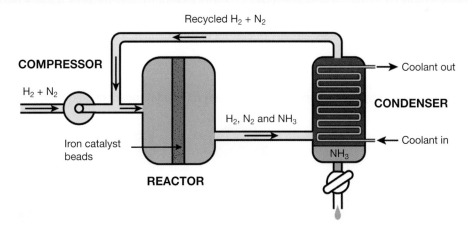

The energy costs in the operation of such a plant are one of its most important overheads. With careful planning, energy costs can be minimised, but they still represent a significant environmental and economic cost. In this case, the actual formation of the ammonia from its elements is an exothermic process. The heat generated from this reaction can be used elsewhere in the plant, rather than just being allowed to go to waste, so heat exchangers are used. For example, the incoming cold nitrogen/hydrogen mixture can be passed over pipes containing the hot gases that exit from the converter. The resultant transfer of heat helps to heat the incoming gases and cool the exit gases.

There are further compromises in operation. Better yields of ammonia might be obtained by using higher pressures, but this necessitates the use of more powerful pumping equipment and stronger reaction vessels to withstand the extra pressure. Economically, it is not worthwhile to do this, because the extra ammonia produced does not offset the extra costs involved in building such a plant.

This process, as it stands, poses a number of concerns from the viewpoint of green chemistry:
- It has low overall energy efficiency. Contributing to this are the high temperatures and pressures mentioned previously, as well as the methods used to generate the nitrogen and hydrogen feedstocks. Nitrogen is obtained from air through repeated cycles of compression, cooling and expansion until liquid air is obtained. The nitrogen is then separated by fractional distillation. The production of hydrogen is also energy intensive. The most common method used is via 'steam reforming of methane'. Not only is this process energy intensive, but it also uses methane and produces significant amounts of carbon dioxide.
- The catalysts involved still require significant temperatures for them to function efficiently. There are also energy costs involved in their manufacture, and their efficiency is seriously diminished if careful attention to purity and reaction conditions within the plant are not maintained.

It can therefore be seen that, although the atom economy (see topic 8, section 8.5.2) of the Haber process is 100 per cent, this predominant method for producing today's ammonia has a number of undesirable features when the principles of green chemistry are considered.

The future of ammonia production

Contemporary responses to the environmental issues associated with ammonia production can be divided into two categories.

The first approach is to modify and adapt existing technologies. Through improvement and combination of existing technologies, the required hydrogen can be generated by electrolysis. It has been estimated that by using 'green hydrogen', energy requirements and carbon emissions of the overall process could be halved. Such green hydrogen would be produced by the electrolysis of water, using electricity generated by solar, wind and other renewable means. Such technology is rapidly approaching the level required to produce hydrogen in the required quantities. Several smaller-scale plants are already operating in this manner. This is examined in more detail in topic 6.

FIGURE 5.18 Renewable energy can be used to reduce the environmental impact of producing ammonia

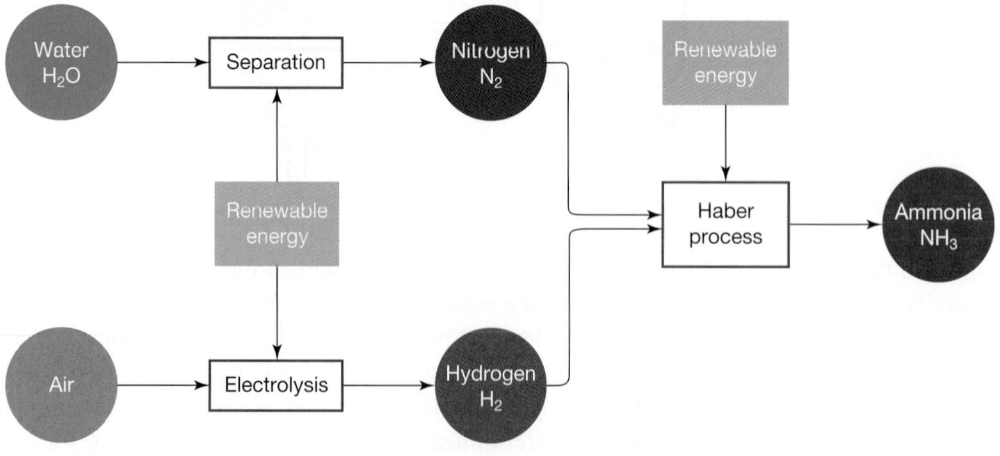

The second approach is to design completely new processes. Two promising methods are currently being researched.

Electrocatalysts and electrolysis

Inorganic electrocatalysts coupled with electrolysis methods are being developed. These would generate ammonia directly and operate at close to normal atmospheric pressures and temperatures.

Such technology could be used in large, centralised plants or be adapted to a modular design, whereby much smaller units could generate the required ammonia on site. Initial research shows that, although this process is efficient, it is slow. The development of suitable catalysts will be critical to its eventual implementation.

The use of such low temperatures and pressures would make this process much more energy efficient than the current method for producing ammonia.

FIGURE 5.19 A possible way that ammonia could be manufactured by electrolysis

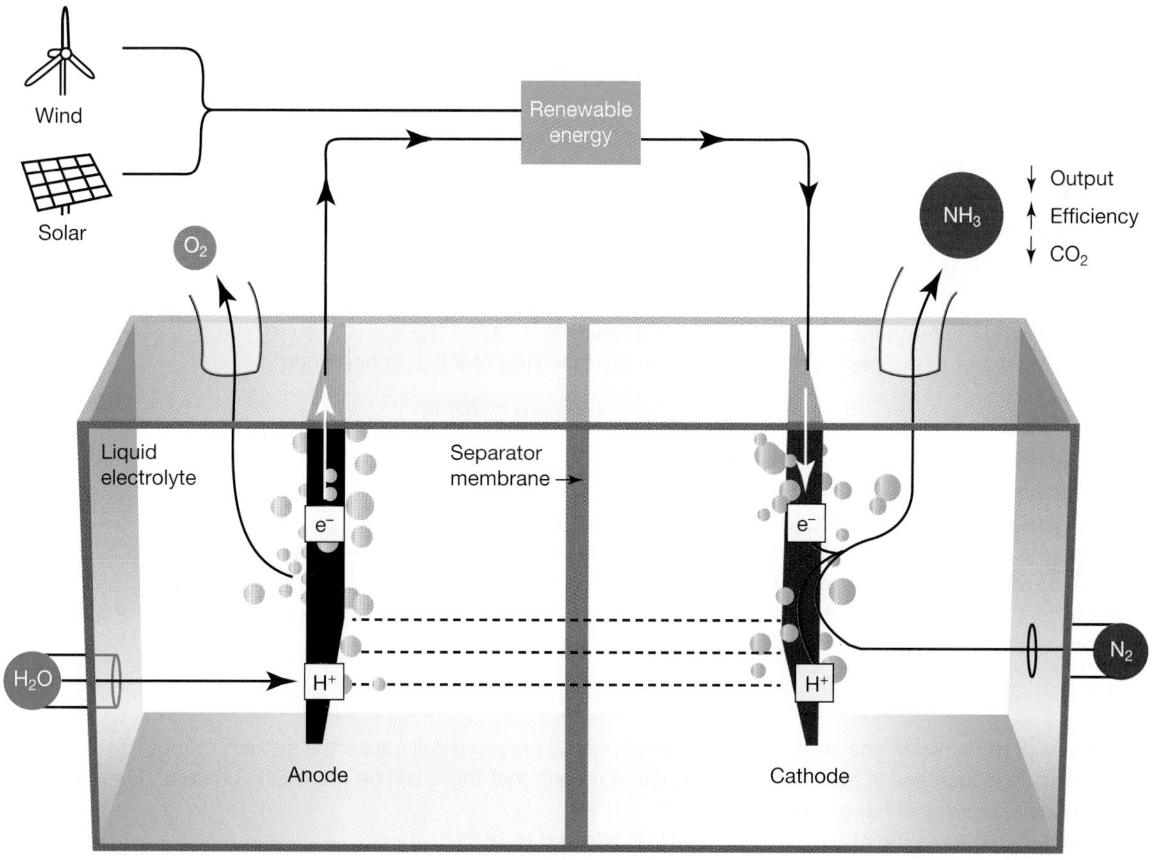

Low temperature and pressure

Biomimicry

Another approach is to copy nature. This is often described as *biomimicry*. In nature, nitrogen-fixing bacteria play a critical role in the nitrogen cycle by producing ammonia from atmospheric nitrogen. This is done by a class of enzymes called *nitrogenases*. Enzymes are biological molecules often described as biological catalysts. They have the advantages of operating at mild temperatures and normal pressures, as well as being highly selective in the reactions that they catalyse.

This approach is showing potential in the production of biocatalysts for electrolysis such as described earlier. Even more ambitious is the use of genetic engineering and directed evolution to develop enzymes that might produce ammonia directly from waste biomass. Once again, a considerable improvement in energy efficiency would be obtained if this approach proves to be successful.

All of these methods are in various stages of research, development and implementation around the world. For example, significant projects are already underway in Australia, Saudi Arabia and Japan, to name just a few.

5.6 Activities

5.6 Quick quiz [on]	5.6 Exercise	5.6 Exam questions

5.6 Exercise

1. Explain why a thermochemical equation is necessary to predict the effect of temperature on the extent of a reversible reaction.
2. Describe the effect of each of the following changes on a reaction at equilibrium.
   a. Removing a product
   b. Adding a reactant
   c. Compressing a mixture being used to make ammonia according to the following equation:

$$N_2(g) + 3H_2(g) \rightleftharpoons 2NH_3(g)$$

   d. Increasing the temperature of an exothermic reaction
   e. Adding water to double the volume in which the following reaction is occurring:

$$A(aq) + 2B(aq) \rightleftharpoons 3C(aq)$$

3. Predict the effect of the stated change on each of the systems represented by the following equations.
   a. Adding methanol: $CO(g) + 2H_2(g) \rightleftharpoons CH_3OH(g)$
   b. Increasing the pressure: $CO(g) + H_2O(g) \rightleftharpoons CO_2(g) + H_2(g)$
   c. Increasing temperature, given that the reaction is exothermic: $4HCl(g) + O_2(g) \rightleftharpoons 2H_2O(g) + 2Cl_2(g)$
   d. Removing chlorine: $PCl_5(g) \rightleftharpoons PCl_3(g) + Cl_2(g)$
4. A student was asked to explain why the use of increased pressure should favour the formation of sulfur trioxide, SO_3, according to the following equation:

$$2SO_2(g) + O_2(g) \rightleftharpoons 2SO_3(g)$$

   As part of her answer, she stated that '… the increased pressure will cause the system to move to the right. The resulting increase in the equilibrium constant means that there will be more SO_3 present'. Evaluate the chemical accuracy of her answer.
5. A reaction between the chemicals A, B and C is allowed to reach equilibrium. The equation for this reaction is:

$$A(g) + 2B(g) \rightleftharpoons C(g)$$

   At a certain time, t, a change is made. The following graph shows the concentration of each of these substances before and after this change.
   a. In which direction did this reaction react in response to the change made?
   b. Label each line in the graph with the substance that it corresponds to.

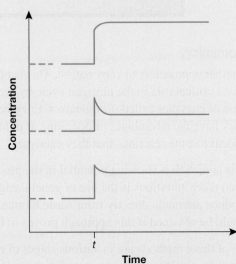

6. A reaction between the chemicals A, B and C is allowed to reach equilibrium. The equation for this reaction is:

$$A(g) + 2B(g) \rightleftharpoons C(g) \quad \Delta H > 0$$

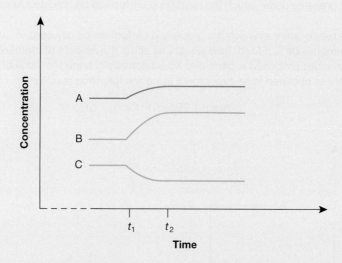

a. What change was made to the equilibrium mixture at time t_1 in the graph shown? Explain.
b. How does the rate of the forward reaction compare to the rate of the backward reaction in the time period between t_1 and t_2?
c. What will be the effect on this graph if some unreactive substance D is added after time t_2? Explain.

7. Pure hydrogen iodide is a gas that partially decomposes when heated according to the following equation:

$$2HI(g) \rightleftharpoons H_2(g) + I_2(g)$$

At 500 K the equilibrium constant for this reaction is 6.25×10^{-3}. At 600 K the equilibrium constant is 2.04×10^{-2}.

a. Is this reaction exothermic or endothermic?
b. Calculate the value of the equilibrium constant at 500 K for the following reaction:

$$H_2(g) + I_2(g) \rightleftharpoons 2HI(g)$$

8. The symbols of chemistry form an international language that can be read by anyone who understands it, regardless of the language they speak. Following is one item from a Hungarian book of chemistry questions.

> Az ammónia oxidációs folyamatát az alábbi egyenlet fejezi ki:
>
> $$4NH_3 + 5O_2 \rightleftharpoons 4NO + 6H_2O$$
>
> $$Q = -226.5 \text{ kJ/mól}$$
>
> Az egyenletekben szereplö vegyületek közül a NO képzödéshöje pozitív, az NH₃ és a víz képzödéshöje negatív. Ennek ismeretében magyarázzuk meg, miért kell a reakciót nagy reakciósebességgel lejátszatni, majd a reakciótermékeket gyorsan lehüteni?

a. Describe the reaction referred to in the question.
b. Is this reaction an exothermic or an endothermic reaction?
c. Would this reaction be best performed at a high or low temperature to obtain a high value for the equilibrium constant?
d. Would the formation of products in this reaction be favoured by high or low pressure?

9. Methanol may be prepared commercially from carbon monoxide and hydrogen using a suitable catalyst, according to the following equation:

$$CO(g) + 2H_2(g) \rightleftharpoons CH_3OH(g) \quad \Delta H = -92 \text{ kJ mol}^{-1}$$

   a. How could the pressure under which the reaction is performed be adjusted to maximise the yield of methanol?
   b. How could the temperature at which the process is performed be adjusted to make more methanol?
   c. If extra carbon monoxide is added, how would this affect the amount of methanol produced?
   d. Which of the changes proposed in parts a–c would actually change the value of the equilibrium constant?

10. The graph shown was obtained in an experiment using the following reaction:

$$A(aq) + 2B(aq) \rightleftharpoons C(aq) \quad \Delta H > 0$$

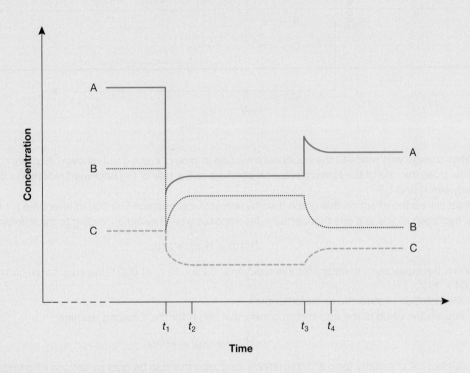

   a. What is the significance of the lines to the left of t_1?
   b. What change was made at t_1? Explain.
   c. What change was made at t_3? Explain.
   d. If a catalyst is also added at t_3, what effect will this have on the graph? Explain.
   e. Between them, a group of students use concentration values immediately before t_1, immediately before t_3 and immediately after t_4 to calculate K. How will their values compare? Explain.

5.6 Exam questions

▶ Question 1 (1 mark)

Source: VCE 2019 Chemistry Exam, Section A, Q.1; © VCAA

MC An understanding of Le Chatelier's principle is useful in the chemical industry.

The prediction that can be made using this principle is the effect of
A. catalysts on the rate of reaction.
B. catalysts on the position of equilibrium.
C. changes in temperature on the rate of reaction.
D. changes in the concentration of reactants on the position of equilibrium.

Question 2 (1 mark)

Source: VCE 2021 Chemistry Exam, Section A, Q.27; © VCAA

MC Hydrogen, H_2, and iodine, I_2, react to form hydrogen iodide, HI.

$$\frac{1}{2}H_2(g) + \frac{1}{2}I_2(g) \rightleftharpoons HI(g) \quad \Delta H = +25.9 \text{ kJ mol}^{-1}$$

The graph below shows the concentrations of H_2, I_2 and HI in a sealed container. One change was made to the equilibrium system at time t_2.

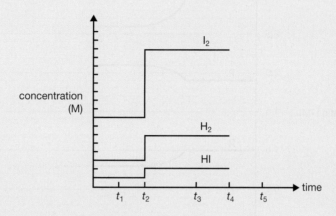

Which one of the following statements is correct?

A. A catalyst was added at time t_2.
B. The amount of HI is greater at time t_3 compared with time t_1.
C. The rate of reaction producing HI is the same at time t_1 and time t_3.
D. The rate of production of HI at time t_3 is double the rate of production of H_2 at time t_3.

Question 3 (1 mark)

Source: VCE 2020 Chemistry Exam, Section A, Q.17; © VCAA

MC The following equation represents the reaction between chlorine gas, Cl_2, and carbon monoxide gas, CO.

$$Cl_2(g) + CO(g) \rightleftharpoons COCl_2(g) \quad \Delta H = -108 \text{ kJ mol}^{-1}$$

The concentration–time graph shown represents changes to the system.

Which of the following identifies the changes to the system that took place at 1 minute and at 7 minutes?

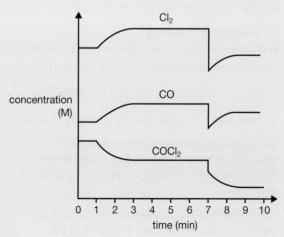

	1 minute	7 minutes
A.	increase in temperature	increase in volume
B.	decrease in temperature	decrease in volume
C.	decrease in temperature	increase in volume
D.	increase in temperature	decrease in volume

Question 4 (1 mark)

Source: VCE 2019 Chemistry Exam, Section A, Q.25; © VCAA

MC The following concentration–time graph refers to a mixture of three gases, P, Q and R, in an enclosed 5.0 L container.

At time t_1 the mixture is heated.

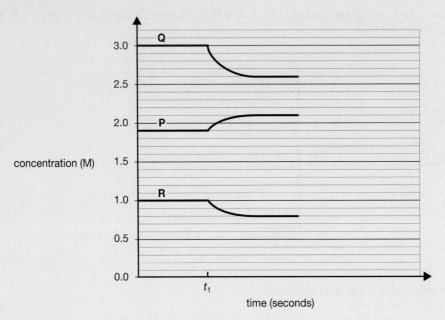

The equilibrium system that represents the graph is

A. $P(g) \rightleftharpoons 2Q(g) + R(g)$ and the forward reaction is exothermic.

B. $2Q(g) \rightleftharpoons P(g) + R(g)$ and the forward reaction is endothermic.

C. $2Q(g) + R(g) \rightleftharpoons P(g)$ and the forward reaction is exothermic.

D. $P(g) + 2Q(g) \rightleftharpoons R(g)$ and the forward reaction is endothermic.

▶ **Question 5 (3 marks)**

Source: Adapted from VCE 2015 Chemistry Exam, Section B, Q.7.c; © VCAA

Consider the reaction shown in the following equation.

$$2NO(g) + Br_2(g) \rightleftharpoons 2NOBr(g) \quad \Delta H = -16.1 \text{ kJ mol}^{-1}, \quad K = 1.3 \times 10^{-2} \text{ M}^{-1} \text{ at 1000 K}$$

A mixture of NO, NOBr and Br_2 is initially at equilibrium.

The following graph shows how the rate of formation of NOBr in the mixture changes when the volume of the reaction vessel is decreased at time t_1.

Use collision theory and factors that affect the rate of a reaction to explain the shape of the graph at the following time intervals.

a. Between t_0 and t_1 **(1 mark)**

b. At t_1 **(1 mark)**

c. Between t_1 and t_2 **(1 mark)**

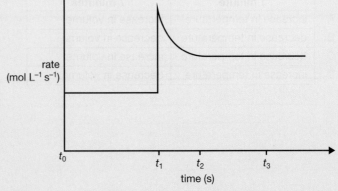

More exam questions are available in your learnON title.

5.7 Review

5.7.1 Topic summary

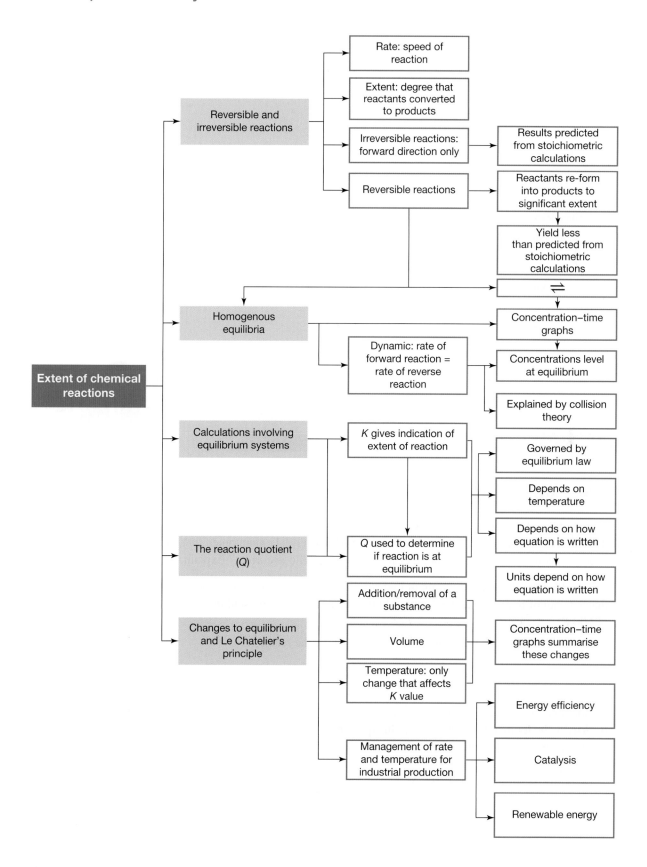

Hey students! Now that it's time to revise this topic, go online to:

Access the topic summary | Review your results | Watch teacher-led videos | Practise past VCAA exam questions

Find all this and MORE in jacPLUS

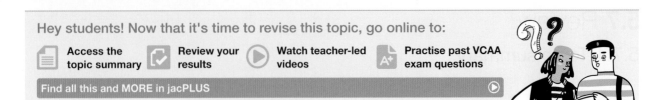

5.7.2 Key ideas summary

online only

5.7.3 Key terms glossary

online only

on Resources

Solutions	Solutions — Topic 5 (sol-0832)
Practical investigation eLogbook	Practical investigation eLogbook — Topic 5 (elog-1704)
Digital documents	Key science skills — VCE Chemistry Units 1–4 (doc-37066) Key terms glossary — Topic 5 (doc-37289) Key ideas summary — Topic 5 (doc-37290)
Exam question booklet	Exam question booklet — Topic 5 (eqb-0116)

5.7 Activities

learnon

Students, these questions are even better in jacPLUS

Receive immediate feedback and access sample responses | Access additional questions | Track your results and progress

Find all this and MORE in jacPLUS

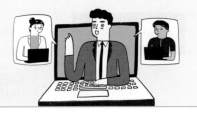

5.7 Review questions

1. Two students, Isla and Colin, were discussing the use of M as a unit. Colin stated that *K* can't have these units because M is a unit of concentration. However, Isla maintained that, although it is a unit of concentration, *K* could also sometimes have this unit. Who is correct? Explain.

2. An equilibrium mixture consisting of hydrogen, iodine and hydrogen iodide was analysed. It was found that 1.1 mol of hydrogen and 3.3 mol of hydrogen iodide were present in a 3.0 L container at a temperature of 600 K.

   a. If the value of the equilibrium constant, *K*, is 49 for the reaction

   $$H_2(g) + I_2(g) \rightleftharpoons 2HI(g)$$

   calculate the concentration of iodine.

   b. Why is the temperature specified in this question?

   c. In a further experiment, but at a different temperature, 0.250 mol of hydrogen and 0.318 mol of iodine were placed in a 1.00 L container and allowed to come to equilibrium. At equilibrium, the concentration of iodine was found to be 0.108 M. Calculate the value and units of the equilibrium constant at this new temperature.

3. In a 10 L vessel, 0.20 mol of SO_2, 0.40 mol of O_2 and 0.70 mol of SO_3 were mixed and allowed to come to equilibrium. Upon establishment of equilibrium, it was found that 0.30 mol of SO_3 remained. Calculate the equilibrium constant for this reaction, given that the equation is:

$$2SO_2(g) + O_2(g) \rightleftharpoons 2SO_3(g)$$

4. In an experiment, the equilibrium established between substances A, B and C was investigated. Certain initial concentrations of each substance were mixed and then allowed to come to equilibrium. The reaction between these three substances may be represented by the equation

$$A + yB \rightleftharpoons zC$$

where y and z are integers. The changes in concentration are shown in the following graph.

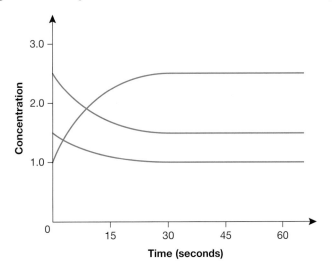

a. Identify which line belongs to which substance.
b. During the first 30 seconds of the experiment, how does the rate of the forward reaction compare with the rate of the backward reaction?
c. After 30 seconds, how does the rate of the forward reaction compare with the rate of the backward reaction?
d. From the graph, identify the values of y and z in the equation.
e. Calculate the value of the equilibrium constant for this reaction.

5. The formation of ethyl acetate, $CH_3COOCH_2CH_3$, is represented by the following equation:

$$CH_3COOH(l) + CH_3CH_2OH(l) \rightleftharpoons CH_3COOCH_2CH_3(l) + H_2O(l)$$

The value of K for this reaction is 4.

A number of experiments were conducted in which various amounts of these chemicals were mixed and allowed to react for varying periods of time. The results are shown in the following table.

Experiment	[CH_3COOH]	[CH_3CH_2OH]	[$CH_3COOCH_2CH_3$]	[H_2O]
1	1	1	1	1
2	4	0.5	2	4
3	1.5	1	3	1.5
4	0.5	1	1.5	2
5	1	1.5	2	3
6	3	2	1.5	1

a. In which experiments had equilibrium been achieved by the time of the analysis?
b. For those not at equilibrium, in which direction would the reaction proceed to establish equilibrium?

6. The following table shows the percentage formation of a product in equilibrium mixtures at different temperatures and pressures.

Pressure (×10⁵ Pa)	200 °C	300 °C	400 °C	500 °C
1	10	3.5	1.3	0.5
200	64	30	23	19
300	98	61	43	21

a. Is the reaction under study exothermic or endothermic?
b. If the equation for this reaction was written, how would the number of particles on the right-hand side compare with the number on the left?

7. Ammonia may be prepared according to the following equation:

$$N_2(g) + 3H_2(g) \rightleftharpoons 2NH_3(g) \quad \Delta H = -92 \text{ kJ mol}^{-1}$$

In a particular experiment using typical industrial equipment at a particular temperature, T, equilibrium was obtained with 4.32 mol of N_2, 2.00 mol of H_2 and 4.00 mol of NH_3 present in a 2.00 L pressure vessel.

a. Calculate the equilibrium constant at the temperature of this experiment.
b. Calculate the pressure exerted by the mixture of gases at equilibrium in terms of T.
c. If the volume of the pressure vessel is reduced (at constant temperature), what effect would this have on the amount of NH_3 present?

8. In an investigation of the decomposition of hydrogen iodide, represented by the equation

$$2HI(g) \rightleftharpoons H_2(g) + I_2(g)$$

the concentration–time graph shown was obtained.

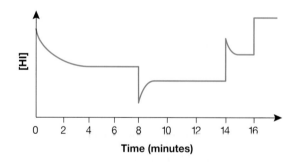

a. Describe what happened during the first two minutes of the experiment.
b. Describe what was happening from the four- to eight-minute marks.
c. Give an explanation for the cause of the dip in the graph at the eight-minute mark.
d. What do you think might have happened at the 14-minute mark?
e. What do you think might have happened at the 16-minute mark?
f. Did the change occurring at the 16-minute mark affect the equilibrium or not? Explain.

9. Upon dissolving in water, the sugar α-d-glucose undergoes conversion into an isomer called β-d-glucose. This process is called mutarotation and reaches equilibrium when 63.6 per cent of the original α-d-glucose has been converted. Calculate the value of the equilibrium constant for this process.

10. A simple form of colorimetric analysis may be used to study the equilibrium represented by the following equation:

$$Br_2(aq) + 2OH^-(aq) \rightleftharpoons OBr^-(aq) + Br^-(aq) + H_2O(l) \quad \Delta H = +15 \text{ kJ mol}^{-1}$$

As molecular bromine, Br_2, is the only coloured species in this reaction, its red-brown colour may be used to monitor various changes made to this equilibrium.

A student performed an experiment where 10 mL samples of reaction mixture were poured into identical test tubes. The following three changes were then made to different samples of the equilibrium mixture. The colours before and after the changes were compared.
- Change 1: A small amount of solid potassium bromide was added and the mixture stirred to dissolve it.
- Change 2: A small amount of solid sodium chloride was added and the mixture stirred to dissolve it.
- Change 3: The solution was warmed from room temperature to 50 °C.

For each of these three changes:

a. state whether you would expect the solution to darken, lighten or stay the same
b. use Le Chatelier's principle to explain your answer to part a.

5.7 Exam questions

Section A — Multiple choice questions

All correct answers are worth 1 mark each; an incorrect answer is worth 0.

Question 1

Source: *VCE 2021 Chemistry Exam, Section A, Q.25;* © VCAA

MC An equilibrium mixture of four gases is represented by the following equation.

$$A(g) + 2B(g) \rightleftharpoons C(g) + D(g) \quad \Delta H > 0$$

The graph below shows the rate of the forward and reverse reactions versus time.

A single change is made to the equilibrium mixture at time t_1 and equilibrium is re-established at time t_2.

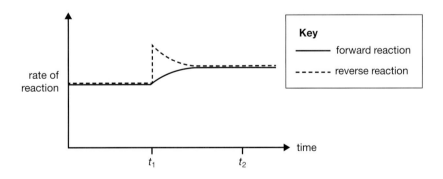

Which one of the following is consistent with the information given above?

A. Argon is added to the equilibrium mixture at time t_1.
B. At time t_1 reactants are removed from the equilibrium mixture.
C. The amount of products is higher at time t_2 compared to just before time t_1.
D. The change made at time t_1 results in an increase in the equilibrium constant at time t_2.

Question 2

Source: VCE 2018 Chemistry Exam, Section A, Q.27; © VCAA

$$Br_2(g) + I_2(g) \rightleftharpoons 2IBr(g) \quad K = 1.2 \times 10^2 \text{ at } 150\,^\circ C$$

MC Given the information above, what is K for the reaction $4IBr(g) \rightleftharpoons 2Br_2(g) + 2I_2(g)$ at 150 °C?

A. 1.6×10^{-2}

B. 4.1×10^{-3}

C. 6.9×10^{-5}

D. 8.03×10^{-5}

Question 3

Source: VCE 2018 Chemistry Exam, Section A, Q.24; © VCAA

MC The four equations below represent different equilibrium systems.

Equation 1	$2SO_2(g) + O_2(g) \rightleftharpoons 2SO_3(g)$	$\Delta H = -180\,\text{kJ mol}^{-1}$
Equation 2	$CO(g) + H_2O(g) \rightleftharpoons CO_2(g) + H_2(g)$	$\Delta H = -46\,\text{kJ mol}^{-1}$
Equation 3	$PCl_5(g) \rightleftharpoons PCl_3(g) + Cl_2(g)$	$\Delta H = 93\,\text{kJ mol}^{-1}$
Equation 4	$CH_4(g) + H_2O(g) \rightleftharpoons CO(g) + 3H_2(g)$	$\Delta H = 205\,\text{kJ mol}^{-1}$

After equilibrium was established in each system, the temperature was decreased and the pressure was increased.

In which equilibrium system would both changes result in an increase in yield?

A. Equation 1

B. Equation 2

C. Equation 3

D. Equation 4

Question 4

Source: VCE 2017 Chemistry Exam, Section A, Q.18; © VCAA

MC Ammonia, NH_3, can be produced by the reaction of hydrogen, H_2, and nitrogen, N_2. When this reaction takes place in a sealed container of fixed volume, an equilibrium system is established.

The equation for the reaction is shown below.

$$N_2(g) + 3H_2(g) \rightleftharpoons 2NH_3(g) \quad \Delta H = -92 \text{ kJ mol}^{-1}$$

If the pressure and volume remain constant when the temperature is increased, the forward reaction rate will

A. increase and the $[NH_3]$ will increase.

B. increase and the $[NH_3]$ will decrease.

C. decrease and the $[NH_3]$ will decrease.

D. decrease and the $[NH_3]$ will remain the same.

Question 5

Source: VCE 2014 Chemistry Exam, Section A, Q.1; © VCAA

MC Hydrogen is produced on an industrial scale from methane. The equation for the reaction is

$$2H_2O(g) + CH_4(g) \rightleftharpoons CO_2(g) + 4H_2(g)$$

The expression for the equilibrium constant for the reverse reaction is

A. $K = \dfrac{[H_2O]^2[CH_4]}{[H_2]^4[CO_2]}$

B. $K = \dfrac{[H_2]^4[CO_2]}{[H_2O]^2[CH_4]}$

C. $K = \dfrac{[H_2O][CH_4]}{[H_2][CO_2]}$

D. $K = \dfrac{4[H_2][CO_2]}{2[H_2O][CH_4]}$

Question 6

Source: VCE 2016 Chemistry Exam, Section A, Q.28; © VCAA

MC A team of chemists was investigating the following equilibrium reaction.

$$H_2(g) + I_2(g) \rightleftharpoons 2HI(g) \quad \Delta H \text{ is negative}$$

Hydrogen gas, H_2, and iodine gas, I_2, were injected into a sealed container and the mixture was allowed to reach equilibrium.

The effect of the following changes on the amount of HI was measured:

1. More H_2 gas was injected into the container at a constant temperature and volume.
2. The temperature of the gases was decreased at a constant volume.
3. Some argon gas, Ar, was injected into the container at a constant temperature and volume.
4. The volume of the container was decreased at a constant temperature.

Which change(s) would have resulted in the formation of a greater amount, in mol, of HI?

A. 1 and 2 only
B. 1, 2 and 4 only
C. 3 only
D. 1 and 4 only

Use the following information to answer Questions 7–9.

A solution contains an equilibrium mixture of two different cobalt(II) ions.

$$Co(H_2O)_6{}^{2+}(aq) + 4Cl^-(aq) \rightleftharpoons CoCl_4{}^{2-}(aq) + 6H_2O(l)$$

$$\text{pink} \qquad\qquad\qquad\qquad \text{blue}$$

The solution contains pink $Co(H_2O)_6{}^{2+}$ ions and blue $CoCl_4{}^{2-}$ ions, and the solution has a purple colour.

10 mL of the purple solution was poured into each of three test tubes labelled X, Y and Z.

Question 7 (1 mark)

Source: VCE 2015 Chemistry Exam, Section A, Q.19; © VCAA

MC The test tubes were placed in separate water baths, each having a different temperature. The resulting colour changes in the equilibrium mixtures were observed.

The results are shown in the following table.

Test tube	Water bath temperature	Observation
X	20 °C	solution remained purple
Y	80 °C	solution turned blue
Z	0 °C	solution turned pink

Which one of the following conclusions can be drawn from these observations?

A. Cooling significantly reduced the volume of the solution and this favoured the forward reaction.
B. Heating caused some water to evaporate and this favoured the reverse reaction.
C. Heating increased the value of the equilibrium constant for the reaction.
D. The forward reaction must be exothermic.

Question 8

Source: VCE 2015 Chemistry Exam, Section A, Q.20; © VCAA

MC Which one of the following changes would cause 10 mL of the purple cobalt(II) ion solution to turn blue?

A. the addition of a few drops of 10 M hydrochloric acid at a constant temperature
B. the addition of a few drops of 0.1 M silver nitrate at a constant temperature
C. the addition of a few drops of a catalyst at a constant temperature
D. the addition of a few drops of water at a constant temperature

Question 9

Source: VCE 2015 Chemistry Exam, Section A, Q.21; © VCAA

MC When the equilibrium system was heated, the colour changed from purple to blue.

Which one of the following concentration–time graphs best represents this change?

A.

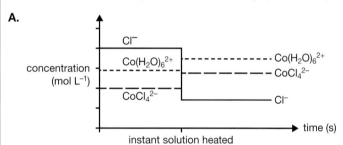

B.

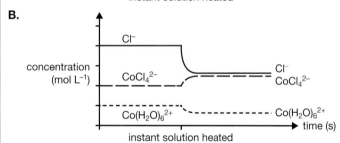

C.

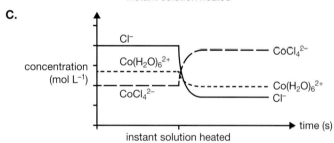

D.

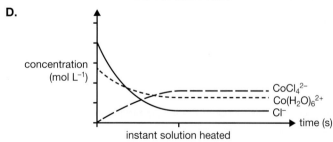

Question 10

Source: VCE 2011 Chemistry Exam 2, Section A, Q.5; © VCAA

MC It is proposed to indirectly determine the concentration of Fe^{3+} ions in a solution by using UV-visible spectroscopy to measure the concentration of red-coloured $FeSCN^{2+}$ ions generated by the equilibrium reaction

$$Fe^{3+}(aq) + SCN^-(aq) \rightleftharpoons FeSCN^{2+}(aq) \quad \Delta H = \text{negative}$$

colourless   red

This procedure would provide the **most** accurate estimate of the concentration of Fe^{3+} ions in the original solution if

A. the value of the equilibrium constant is small, an excess of SCN^- is used, and the analysis is carried out at a low temperature.

B. the value of the equilibrium constant is large, an excess of Fe^{3+} is used, and the analysis is carried out at a high temperature.

C. the value of the equilibrium constant is small, an excess of Fe^{3+} is used, and the analysis is carried out at a high temperature.

D. the value of the equilibrium constant is large, an excess of SCN^- is used, and the analysis is carried out at a low temperature.

Section B — Short answer questions

Question 11 (9 marks)

Source: VCE 2022 Chemistry Exam, Section B, Q.3.a,b; © VCAA

The following equation represents a gaseous reaction that takes place in a sealed container.

$$4NH_3(g) + 3O_2(g) \rightleftharpoons 2N_2(g) + 6H_2O(g) \quad \Delta H < 0$$

a. **i.** Write the equilibrium expression for this reaction. **(1 mark)**

    ii. State the units for this equilibrium expression. **(1 mark)**

b. The temperature of the reaction system is increased and a new equilibrium is established.

    i. How does the increase in the temperature affect the value of the equilibrium constant? Justify your answer. **(4 marks)**

    ii. Compare the rate of the forward reaction at the original equilibrium with the rate of the forward reaction at the new equilibrium after the increase in temperature. Explain the difference using collision theory. **(3 marks)**

Source: *VCE 2019 Chemistry Exam, Section B, Q.3*; © VCAA

The cobalt(II) tetrachloride ion, $CoCl_4^{2-}$, dissociates into the cobalt(II) ion, Co^{2+}, and chloride ions, Cl^-, according to the following chemical equation.

$$CoCl_4^{2-}(aq) \rightleftharpoons Co^{2+}(aq) + 4Cl^-(aq)$$

blue   pink

20 mL samples of the equilibrium mixture were heated to two temperatures, 30 °C and 80 °C. The intensity of the pink colour of the Co^{2+} product was recorded every 30 seconds by measuring the absorbance of the solution. The higher the intensity of the pink colour, the higher the absorbance.

The results of this experiment are shown in the graph below.

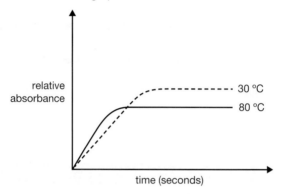

a. State whether the forward reaction is exothermic or endothermic. Justify your answer by referring to the graph. **(2 marks)**

b. When 5 mL of water was added to the equilibrium mixture, the colour of the solution immediately became a lighter pink. Describe the final colour of the solution once equilibrium is re-established. Explain your answer. **(2 marks)**

c. Five drops of silver nitrate, $AgNO_3$, solution are added to the equilibrium mixture at time t_1.

A concentration–time graph for this reaction is shown below for times between zero and t_1.

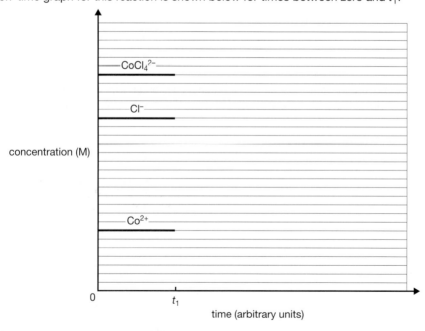

Continue the graph to show the changes that occur to the system from t_1 until equilibrium is re-established. **(3 marks)**

Question 13 (8 marks)

Source: VCE 2020 Chemistry Exam, Section B, Q.1.b; © VCAA

Methanol is a very useful fuel. It can be manufactured from biogas.

The main reaction in methanol production from biogas is represented by the following equation.

$$CO(g) + 2H_2(g) \rightleftharpoons CH_3OH(g) \quad \Delta H < 0$$

This reaction requires the use of a catalyst to maximise the yield of methanol produced in optimum conditions. The energy profile diagram below represents the uncatalysed reaction.

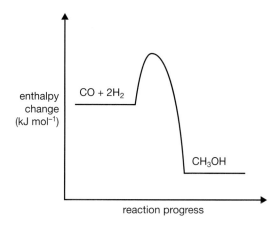

a. i. How does the reaction temperature affect the yield of methanol from biogas? In your answer, refer to Le Chatelier's principle. **(2 marks)**

   ii. How does the reaction pressure affect the yield of methanol from biogas? In your answer, refer to Le Chatelier's principle. **(2 marks)**

b. Write the expression for the equilibrium constant, K, for this reaction. **(1 mark)**

c. 0.760 mol of carbon monoxide, CO, and 0.525 mol of hydrogen, H_2, were allowed to reach equilibrium in a 500 mL container. At equilibrium the mixture contained 0.122 mol of methanol.

Calculate the equilibrium constant, K. **(3 marks)**

Question 14 (8 marks)

Source: VCE 2017 Chemistry Exam, Section B, Q.4; © VCAA

Sulfur trioxide, SO_3, is made by the reaction of sulfur dioxide, SO_2, and oxygen, O_2, in the presence of a catalyst, according to the equation below.

$$2SO_2(g) + O_2(g) \rightleftharpoons 2SO_3(g) \quad \Delta H < 0$$

In a closed system in the presence of the catalyst, the reaction quickly achieves equilibrium at 1000 K.

a. A mixture of 2.00 mol of $SO_2(g)$ and 2.00 mol of $O_2(g)$ was placed in a 4.00 L evacuated, sealed vessel and kept at 1000 K until equilibrium was reached. At equilibrium, the vessel was found to contain 1.66 mol of $SO_3(g)$.

Calculate the equilibrium constant, K, at 1000 K. **(4 marks)**

b. A manufacturer of SO_3 investigates changes to the reaction conditions used in part **a** in order to increase the percentage yield of the product in a closed system, where the volume may be changed, if required.

What changes would the manufacturer make to the temperature and volume of the system in order to increase the percentage yield of SO_3? Justify your answer. **(4 marks)**

Source: VCE 2016 Chemistry Exam, Section B, Q.5.b,c; © VCAA

Bromomethane, CH_3Br, is a toxic, odourless and colourless gas. It is used by quarantine authorities to kill insect pests.

A simplified reaction for its synthesis is

$$CH_3OH(g) + HBr(g) \rightleftharpoons CH_3Br(g) + H_2O(g) \quad \Delta H = -37.2 \text{ kJ mol}^{-1} \text{ at 298 K}$$

The manufacturer of this chemical investigates reaction conditions that could affect the time the process takes and the percentage yield.

a. Considering the system at equilibrium, predict the effect of each change given below on the percentage yield of bromomethane. Give your prediction (increase, no change or decrease) and your reasoning.

  i. Increasing pressure (constant temperature) **(2 marks)**

  ii. Continuously removing the product CH_3Br (constant volume and temperature) **(2 marks)**

b. The graph below represents the concentration of three of the species involved in the production of CH_3Br when they are at equilibrium at constant temperature. At time t_1, a small amount of HBr was suddenly added to the equilibrium mixture.

  Complete the graph after t_1, showing the changes in concentration for each of the three species. **(3 marks)**

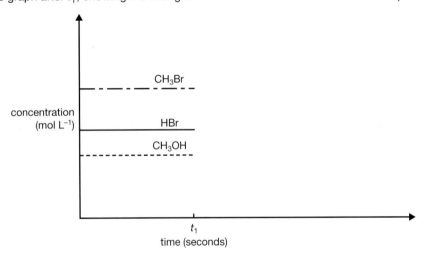

6 Production of chemicals using electrolysis

KEY KNOWLEDGE

In this topic you will investigate:

Production of chemicals using electrolysis

- the use and limitations of the electrochemical series to explain or predict the products of the electrolysis of particular chemicals, given their state (molten liquid or in aqueous solution) and the electrode materials used, including the writing of balanced equations (with states) for the reactions occurring at the anode and cathode and the overall redox reaction for the cell
- the common design features and general operating principles of commercial electrolytic cells (including, where practicable, the removal of products as they form), and the selection of suitable electrode materials, the electrolyte (including its state) and any chemical additives that result in a desired electrolysis product (details of specific cells not required)
- the common design features and general operating principles of rechargeable (secondary) cells, with reference to discharging as a galvanic cell and recharging as an electrolytic cell, including the conditions required for the cell reactions to be reversed and the electrode polarities in each mode (details of specific cells not required)
- the role of innovation in designing cells to meet society's energy needs in terms of producing 'green' hydrogen (including equations in acidic conditions) using the following methods:
  - polymer electrolyte membrane electrolysis powered by either photovoltaic (solar) or wind energy
  - artificial photosynthesis using a water oxidation and proton reduction catalyst system
- the application of Faraday's Laws and stoichiometry to determine the quantity of electrolytic reactant and product, and the current or time required to either use a particular quantity of reactant or produce a particular quantity of product.

Source: VCE Chemistry Study Design (2024–2027) extracts © VCAA; reproduced by permission.

PRACTICAL WORK AND INVESTIGATIONS

Practical work is a central component of VCE Chemistry. Experiments and investigations, supported by a **practical investigation eLogbook** and **teacher-led videos**, are included in this topic to provide opportunities to undertake investigations and communicate findings.

EXAM PREPARATION

▶ Access past VCAA questions and exam-style questions and their video solutions in every lesson, to ensure you are ready.

6.1 Overview

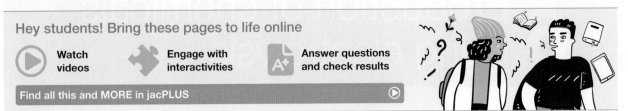
6.1.1 Introduction

Chlorine is one of the most widely used chemicals in the world. It is perhaps best known for its use in large public swimming pools. Upon addition to the pool water, chlorine reacts to form hypochlorite ions that serve as a powerful disinfectant. But chlorine has many more uses — it is used in the manufacture of organic compounds, plastics, bleach and for sterilising municipal water supplies.

Chlorine is made on a large scale by the process of electrolysis. Electrolysis is also used to make many other important chemicals and products, including sodium hydroxide, aluminium and group I and II metals. The group I metal sodium is used on a large scale as a coolant in many nuclear reactors. Electroplating is another important use of electrolysis, whereby a thin layer of metal is applied to the surface of an object (often another metal) to enhance properties such as appearance, corrosion resistance and electrical conductivity. The computer industry, for example, uses gold plating to enhance the electrical conductivity of electrical contacts on circuit boards.

FIGURE 6.1 Chlorine gas is a greenish-yellow gas that is corrosive and a toxic respiratory irritant.

This topic introduces the process of electrolysis and how the variables involved can alter the nature of the products produced. You will use your skills in stoichiometry and prior knowledge of basic redox concepts. Features of galvanic cells that you learnt about in topic 3 will be critical to your understanding of electrolytic cells. Furthermore, you will make extensive use of the electrochemical series as a tool for predicting electrode reactions and apply Faraday's Laws to make quantitative predictions concerning electrolytic cells.

on Resources

☑ **Solutions**	Solutions — Topic 6 (sol-0833)
🔬 **Practical investigation eLogbook**	Practical investigation eLogbook — Topic 6 (elog-1705)
📄 **Digital documents**	Key science skills — VCE Chemistry Units 1–4 (doc-37066)
	Key terms glossary — Topic 6 (doc-37291)
	Key ideas summary — Topic 6 (doc-37292)
🗎 **Exam question booklet**	Exam question booklet — Topic 6 (eqb-0117)

6.2 What is electrolysis?

KEY KNOWLEDGE

- This subtopic will introduce the concept of electrolysis, and the nature of the redox reactions and energy changes that it involves. It will form the basis of your understanding of the key knowledge points of the VCE Chemistry Study Design in the subtopics that follow.

6.2.1 The process of electrolysis

As we saw in the electrochemical cells covered in topic 3, in a galvanic cell, a spontaneous chemical reaction produces an electric current.

In an **electrolytic cell**, the reverse process takes place. The passage of an electric current through an aqueous or molten electrolyte causes a chemical reaction. This process is known as **electrolysis**.

In galvanic cells, chemical reactions can be used to generate a flow of electrons (an electric current). If a zinc rod is placed in copper(II) sulfate solution, a coating of copper appears on the zinc rod. This may be explained by considering the standard electrode potentials of each half-equation:

$$Cu^{2+}(aq) + 2e^- \rightleftharpoons Cu(s) \quad E^0 = +0.34 \text{ V}$$
$$Zn^{2+}(aq) + 2e^- \rightleftharpoons Zn(s) \quad E^0 = -0.76 \text{ V}$$

Because the E^0 value for the Cu^{2+}/Cu redox pair is greater than the E^0 value for the Zn^{2+}/Zn redox pair, Cu^{2+} ions react spontaneously with zinc metal. A galvanic cell constructed from these two half-cells would produce electrical energy. The overall equation for such a galvanic cell would be:

$$Cu^{2+}(aq) + Zn(s) \rightarrow Cu(s) + Zn^{2+}(aq) + \textbf{energy}$$

> **electrolytic cell** an electric cell in which a non-spontaneous redox reaction is made to occur by the application of an external potential difference across the electrodes; also known as an electrolysis cell
>
> **electrolysis** the process in which a non-spontaneous chemical reaction occurs by passing an electric current through a substance in solution or molten state

However, if a copper rod is placed in a zinc sulfate solution, no reaction occurs. This reaction is the reverse of the one that produces energy, so energy must be supplied for the reaction to occur.

$$Cu(s) + Zn^{2+}(aq) + \textbf{energy} \rightarrow Cu^{2+}(aq) + Zn(s)$$

Electrolytic cells

Electrolysis is the chemical reaction that occurs when electricity passes through a molten ionic compound or through an electrolyte solution. Electrolytes are solutes that form solutions that can conduct electricity. The apparatus in which electrolysis occurs is called an electrolytic cell.

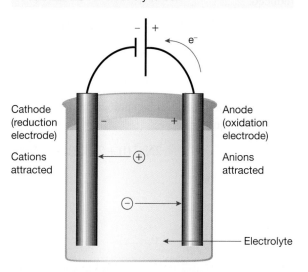

FIGURE 6.2 An electrolytic cell

Cathode (reduction electrode)

Cations attracted

Anode (oxidation electrode)

Anions attracted

Electrolyte

An electrolytic cell has three essential features:
- An electrolyte that contains free-moving ions
  - These ions can donate or accept electrons, allowing electrons to flow through the external circuit.
  - Cations are attracted to the cathode and anions are attracted to the anode.
- Two electrodes at which redox reactions occur
  - Positive ions gain electrons at the **cathode** (negative electrode). Hence, reduction occurs at the cathode.
  - Negative ions lose electrons at the **anode** (positive electrode). Hence, oxidation occurs at the anode.
- An external source of electrons, such as a battery or power pack
  - Electrons flow in one direction (this is referred to as direct current or DC) from the external power source to the negative electrode (the cathode), which is the site of reduction. Electrons are withdrawn by the power source from the positive electrode (the anode), which is the site of oxidation.
  - Cations gain electrons from the cathode.
  - Anions give up electrons to the anode.

 Resources

Video eLesson   Features of electrolytic cells (eles-3245)

Weblink   Introduction to electrolysis

6.2.2 Electrolysis of ionic compounds: molten sodium chloride

The simplest cases of electrolysis involve the electrolysis of molten ionic substances that are pure, using inert electrodes. As an example, let us consider the electrolysis of molten sodium chloride.

Solid sodium chloride does not conduct electricity — the oppositely charged sodium ions, Na^+, and chloride ions, Cl^-, are fixed in place within the ionic lattice. However, heating the solid causes the ions in the crystal to separate and they are then free to move. The molten liquid is called a melt. When an electric current is passed through molten sodium chloride, a chemical reaction can be clearly observed — a shiny bead of sodium is produced at the cathode and chlorine gas is evolved at the anode.

In the electrolysis of molten sodium chloride, the sodium ions are attracted to the negative cathode, where they are reduced.

$$\text{Cathode: } Na^+(l) + e^- \rightarrow Na(l)$$

The chloride ions are attracted to the positive anode, where they undergo oxidation.

$$\text{Anode: } 2Cl^-(l) \rightarrow Cl_2(g) + 2e^-$$

In a redox reaction the same number of electrons are consumed as are produced, so the overall equation is:

$$2Na^+(l) + 2Cl^-(l) \rightarrow 2Na(l) + Cl_2(g)$$

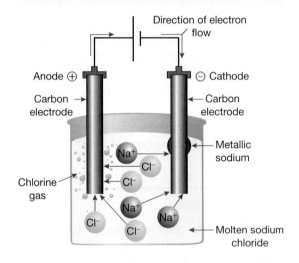

FIGURE 6.3 Electrolysis of molten sodium chloride

cathode the electrode at which reduction occurs; in an electrolytic cell it is the negative electrode, because it receives electrons from the power supply

anode the electrode at which oxidation occurs; in an electrolytic cell it is the positive electrode, because it generates electrons that then travel back to the power supply

SAMPLE PROBLEM 1 Writing equations for electrolysis in molten ionic compounds

a. **Write the equations for the reactions at each inert electrode when pure molten magnesium chloride undergoes electrolysis.**

b. **Write the overall equation for this reaction.**

THINK

a. In molten magnesium chloride there are mobile Mg^{2+} and Cl^- ions. These are the only possible reactants for electrolysis.

The Mg^{2+} ions will move to the negative electrode (the cathode), where they will accept electrons and be reduced.

The Cl^- ions will move to the positive electrode (the anode), where they will give up electrons and be oxidised.

b. Obtain the overall equation by adding the half-equations together, making sure that the electrons cancel out.

TIP: Make sure that you use the correct symbols of state. Due to the absence of water, no species are aqueous in these molten conditions.

WRITE

a. Cathode:
$$Mg^{2+}(l) + 2e^- \rightarrow Mg(l)$$
Anode:
$$2Cl^-(l) \rightarrow Cl_2(g) + 2e^-$$

b. $Mg^{2+}(l) + 2Cl^-(l) \rightarrow Mg(l) + Cl_2(g)$

PRACTICE PROBLEM 1

a. **Write the equations for the reactions at each electrode when pure molten potassium iodide undergoes electrolysis.**

b. **Write the overall equation for this reaction.**

6.2.3 Electrolysis of water

When a current is applied via two electrodes in pure water, nothing happens. This is because there are not enough ions in pure water to carry much of an electric current — no current flow means no electrolysis. But if an electrolyte such as H_2SO_4 or KNO_3 is added in *low* concentration, the resulting solution conducts electricity and electrolysis occurs. The products of the electrolysis of water, in this case, are hydrogen and oxygen.

At the cathode, water is reduced to form hydrogen:

$$\text{Cathode: } 2H_2O(l) + 2e^- \rightarrow H_2(g) + 2OH^-(aq)$$

At the anode, water is oxidised to form oxygen:

$$\text{Anode: } 2H_2O(l) \rightarrow O_2(g) + 4H^+(aq) + 4e^-$$

The region around the cathode becomes basic, owing to an increase in OH^- ions, whereas the region around the anode becomes acidic, owing to an increase in H^+ ions.

The overall cell reaction may be obtained by adding the half-equations:

$$6H_2O(l) \rightarrow 2H_2(g) + O_2(g) + 4H^+(aq) + 4OH^-(aq)$$

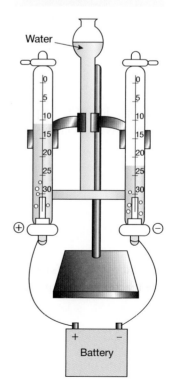

FIGURE 6.4 Electrolysis of water using a Hofmann voltameter — when an electric current is passed through water, it decomposes into oxygen and hydrogen

However, if allowed to mix freely, the hydrogen and hydroxide ions produced undergo a neutralisation reaction, reforming water:

$$4H^+(aq) + 4OH^-(aq) \rightarrow 4H_2O(l)$$

Four water molecules now appear on both sides of the equation, which cancel out to give the final overall equation:

$$2H_2O(l) \rightarrow 2H_2(g) + O_2(g)$$

6.2.4 The nature of electrolytic reactions

In the previous two sections, the electrolysis of molten sodium chloride and of water were presented as simple examples of electrolysis. The overall equations for these reactions were:

$$2Na^+(l) + 2Cl^-(l) \rightarrow 2Na(l) + Cl_2(g)$$

$$2H_2O(l) \rightarrow 2H_2(g) + O_2(g)$$

In topic 3, you learnt to use the electrochemical series to predict the occurrence and outcome of spontaneous redox reactions. You will note that the reactions above are the 'wrong way around' on the electrochemical series. They are examples of redox reactions that we would not expect to happen. *This is the nature of electrolysis reactions*. They are non-spontaneous and therefore require the input of energy to take place. This energy is supplied electrically by an external power supply.

TIP: Using the electrochemical series in the VCE Chemistry Data Book, if the two half-equations:

- make a clockwise circle, the reaction is spontaneous
- make an anticlockwise circle, the reaction is non-spontaneous.

Consider the following half-equations:

$$Cu^{2+}(aq) + 2e^- \rightleftharpoons Cu(s) \quad E^0 = +0.34$$

$$Pb^{2+}(aq) + 2e^- \rightleftharpoons Pb(s) \quad E^0 = -0.13$$

If copper ions, $Cu^{2+}(aq)$, and solid lead, Pb(s), are present, the two reactions form a clockwise circle and the reaction is spontaneous, producing 0.47 V.

$$\begin{array}{lll} \uparrow & \boxed{Cu^{2+}(aq)} + 2e^- \rightarrow Cu(s) \quad \downarrow & E^0 = +0.34 \\ & Pb^{2+}(aq) + 2e^- \leftarrow \boxed{Pb(s)} & E^0 = -0.13 \end{array}$$

If solid copper, Cu(s), and lead ions, $Pb^{2+}(aq)$, are present, the two reactions form an anticlockwise circle and the reaction is non-spontaneous, requiring greater than 0.47 V to initiate.

$$\begin{array}{lll} \downarrow & Cu^{2+}(aq) + 2e^- \leftarrow \boxed{Cu(s)} \quad \uparrow & E^0 = +0.34 \\ & \boxed{Pb^{2+}(aq)} + 2e^- \rightarrow Pb(s) & E^0 = -0.13 \end{array}$$

The difference between galvanic and electrolytic cells

- In galvanic cells, a spontaneous reaction produces electrical energy.
- In electrolytic cells, electrical energy is required to drive a non-spontaneous reaction.

6.2 Activities

6.2 Quick quiz on	6.2 Exercise	6.2 Exam questions

6.2 Exercise

1. What are the main energy transformations occurring in
   a. a galvanic cell
   b. an electrolytic cell?
2. Predict the products that are formed when molten lead(II) bromide undergoes electrolysis.
3. During the electrolysis of molten lithium iodide, what product will form around the cathode and what will form around the anode? Explain.
4. Explain how the addition of a small amount of KNO_3 allows water to conduct electricity and hence undergo electrolysis.
5. With reference to figure 6.4, and also to the equation for the decomposition of water, predict which side of the apparatus collects hydrogen gas and which collects oxygen gas.
6. For the electrolysis of water, complete the following.
   a. Write equations for the reactions that occur around each electrode.
   b. If a few drops of phenolphthalein are added to the vicinity of each electrode, describe and explain the observations that would be expected.
7. Glass is an inert substance under virtually all conditions. Explain why glass cannot be used as an electrode material.
8. a. Explain why the cathode has a negative charge in an electrolytic cell.
   b. Explain why the anode has a positive charge in an electrolytic cell.
9. Explain why the reactants in a galvanic cell must be kept separated, whereas the reactants in an electrolytic cell are usually contained within a single compartment.
10. Answer the following questions for the electrolysis of molten potassium bromide.
    a. Write the equation for the reaction occurring at the cathode.
    b. State the sign of the cathode.
    c. Write the equation for the reaction occurring at the anode.
    d. State the sign of the anode.
    e. Write the overall ionic equation for this electrolysis.

6.2 Exam questions

▶ Question 1 (1 mark)
Source: VCE 2021 Chemistry Exam, Section A, Q.7; © VCAA

MC Consider the following characteristics of electrolytic cells and galvanic cells.

Characteristic number	Electrolytic cells	Galvanic cells
1	cathode is negative	cathode is positive
2	have non-spontaneous reactions	have spontaneous reactions
3	reduction occurs at the anode	reduction occurs at the cathode
4	produce electricity	consume electricity

Which of the following combinations of characteristics of electrolytic cells and galvanic cells are correct?

A. only 1 and 2
B. only 2 and 3
C. only 3 and 4
D. only 1, 2 and 4

▶ Question 2 (2 marks)
Source: VCE 2017 Chemistry Exam, Section B, Q.8.a,b; © VCAA

Fluorine, F_2, gas is the most reactive of all non-metals. Anhydrous liquid hydrogen fluoride, HF, can be electrolysed to produce F_2 and hydrogen, H_2, gases. Potassium fluoride, KF, is added to the liquid HF to increase electrical conductivity. The equation for the reaction is

$$2HF(l) \rightarrow F_2(g) + H_2(g)$$

F_2 is used to make a range of chemicals, including sulfur hexafluoride, SF_6, an excellent electrical insulator, and xenon difluoride, XeF_2, a strong fluorinating agent.

The diagram below shows an electrolytic cell used to prepare F_2 gas.

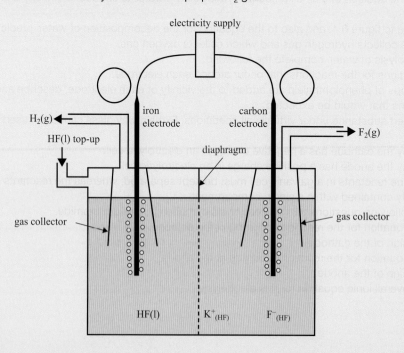

Liquid HF, like water, is an excellent solvent for ionic compounds. In the same way that water molecules in an aqueous solution form the ions $K^+(aq)$ and $F^-(aq)$, when KF is dissolved in HF, the K^+ and F^- ions form ions that are written as $K^+_{(HF)}$ and $F^-_{(HF)}$.

a. Label the polarities of each electrode in the circles provided on the diagram above. **(1 mark)**
b. Write the equation for the half-reaction occurring at the anode. **(1 mark)**

⊙ Question 3 (1 mark)

Source: VCE 2018 Chemistry Exam, Section A, Q.9; © VCAA

MC When molten sodium chloride, NaCl, is electrolysed, the product formed at the cathode is

A. sodium liquid, Na.

B. hydrogen gas, H_2.

C. chlorine gas, Cl_2.

D. oxygen gas, O_2.

⊙ Question 4 (1 mark)

Source: VCE 2013 Chemistry Exam 2, Section B, Q.7.a; © VCAA

An electrolytic process known as electrorefining is the final stage in producing highly purified copper. In a small-scale trial, a lump of impure copper is used as one electrode and a small plate of pure copper is used as the other electrode. The electrolyte is a mixture of aqueous sulfuric acid and copper sulfate.

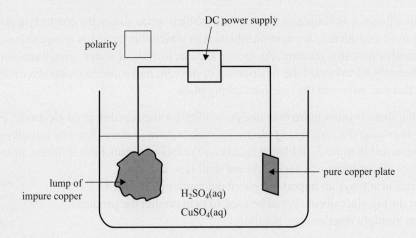

Indicate in the box labelled 'polarity' on the diagram above, the polarity of the impure copper electrode.

⊙ Question 5 (1 mark)

Source: Adapted from VCE 2012 Chemistry Exam 2, Section B, Q.9.a; © VCAA

A teacher demonstrated the process of electrolysis of a molten salt using an unknown metal salt, XBr_2.

The apparatus was set up as shown below.

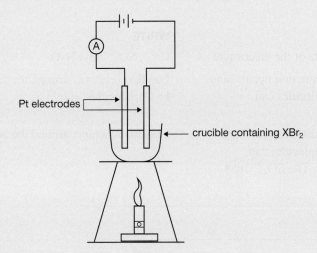

Write a balanced half-equation for the anode reaction in this electrolytic cell.

More exam questions are available in your learnON title.

6.3 Using the electrochemical series in electrolysis

6.3.1 Using the electrochemical series

Electrolytic cells are often not as simple as the examples shown so far. Often the electrolyte contains more than one substance to be considered. A common situation in which this occurs is in aqueous solutions, when the water itself is another potential reactant. As seen in the electrolysis of water, sometimes small amounts of impurities are deliberately added to aid the functioning of the cell, and sometimes the electrodes themselves may be made of metals that can influence the reactions taking place.

All of this means that there is often more than one possibility for the reaction at an electrode. For this reason, *we need to be able to predict the products of electrolysis when there is more than one possible reaction around an electrode.* As explained in topic 3, oxidising agents and reducing agents have different strengths, which can be used to produce an **electrochemical series**. As we shall now see, the electrochemical series also plays an important role in determining the redox reactions that occur during electrolysis. It can be used to help predict the products of electrolysis when multiple reactions are possible.

electrochemical series a series of chemical half-equations arranged in order of their standard electrode potentials

tlvd-3060

SAMPLE PROBLEM 2 Predicting the reactants of electrolysis

What are the possible reactants around each electrode when a dilute copper sulfate solution is electrolysed?
(Note that under conditions of dilute electrolysis in aqueous solutions, negative ions such as sulfate and nitrate will migrate towards the positive anode, but will not be further oxidised; thus, they are functionally inert.)

THINK	WRITE
1. Identify the constituents of the electrolyte.	Cu^{2+}, SO_4^{2-} and H_2O
2. Around the cathode, reduction occurs, and Cu^{2+} ions and H_2O molecules can potentially be reduced.	Possible reactants around the cathode: Cu^{2+} ions and H_2O molecules
3. Around the anode, oxidation occurs, and SO_4^{2-} ions and H_2O molecules can potentially be oxidised. The SO_4^{2-} ions can be ignored.	Possible reactants around the anode: H_2O

PRACTICE PROBLEM 2

What are the possible reactants around each electrode when a dilute potassium nitrate solution is electrolysed?

6.3.2 Predicting the products of electrolysis

In aqueous solutions, there is a mixture of at least two potential oxidising agents and two potential reducing agents. If non-inert electrodes are used, then even more possibilities may exist. Which oxidising agent and which reducing agent are the strongest?

FIGURE 6.5 Competing reactions are like wrestlers: the strongest challengers (strongest oxidising agent and strongest reducing agent) make it to the final round.

Although electrolytic products depend on a number of factors, the following procedure is useful.

Predicting the products of electrolysis

1. **List the species present**, including all metals that are used as electrodes.
2. **Write half-equations** involving these species in descending order of E^0.
3. **Circle the species present** in the electrolytic cell that could participate.
4. **Select the oxidising agent with the highest E^0** (the strongest oxidising agent). This will be reduced at the cathode, because it requires less energy for reduction than an oxidising agent with a lower E^0.
5. **Select the reducing agent with the lowest E^0** (the strongest reducing agent). This will be oxidised at the anode, because it requires less energy for oxidation than a reducing agent with a higher E^0. Don't forget to write this equation 'backwards' when you come to the next step.
6. **Write the reduction (strongest oxidising agent) and oxidation (strongest reducing agent) half-equations.** Don't forget to write the oxidation half-equation in the reverse direction as it appears in the series, and replace the equilibrium arrows with single arrows.
7. **Write the overall equation** by combining the relevant half-equations.
8. **Determine the minimum voltage** required to achieve the reaction by finding the difference between the E^0 of the reducing agent and the E^0 of the oxidising agent.

Determining the minimum voltage required

$$\text{Minimum voltage required} > \left| E^0_{\text{oxidising agent}} - E^0_{\text{reducing agent}} \right|$$

TIP: Using the electrochemical series in your VCE Chemistry Data Book, the two half-equations that will be initiated by electrolysis (if there are no species that react spontaneously) are those that make the *smallest* anticlockwise circle, as they require the least amount of energy.

Consider the electrolysis of dilute potassium iodide (as shown in figure 6.6).
1. The species present in the cell are K^+, I^- and H_2O.
2. The possible half-equations for these species in descending order of E^0 are listed using the VCE Chemistry Data Book.
3. The species present in the electrolytic cell that could participate are then circled.

$$O_2(g) + 4H^+ (aq) + 4e^- \rightleftharpoons \boxed{2H_2O(l)} \quad E^0 = +1.23 \text{ V}$$
$$I_2(s) + 2e^- \rightleftharpoons \boxed{2I^-(aq)} \quad E^0 = +0.54 \text{ V}$$
$$\boxed{2H_2O(l)} + 2e^- \rightleftharpoons H_2(g) + 2OH^-(aq) \quad E^0 = -0.83 \text{ V}$$
$$\boxed{K^+(aq)} + e^- \rightleftharpoons K(s) \quad E^0 = -2.93 \text{ V}$$

4. Recall the acronym OILRIG (oxidation is loss, reduction is gain) and that reduction occurs at the cathode, so there are two possible reactions in this cell. The oxidising agent with the highest E^0 value (the strongest oxidising agent) requires the least energy for reduction and is reduced at the cathode. So, water reacts in preference to potassium ions at the cathode.

Cathode: $2H_2O(l) + 2e^- \rightarrow H_2(g) + 2OH^-(aq) \quad E^0 = -0.83 \text{ V}$

Hydrogen gas is evolved at this electrode and the solution around the cathode becomes alkaline, owing to an increase in hydroxide ion concentration.

5. Oxidation occurs at the anode, so there are two possible reactions. The reducing agent with the lowest E^0 value (the strongest reducing agent) requires the least energy for oxidation and is oxidised at the anode. Thus, iodide ions react in preference to water molecules at the anode.

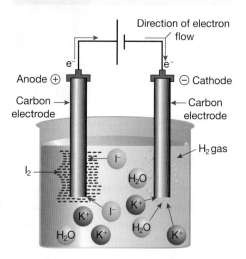

FIGURE 6.6 Electrolysis of dilute potassium iodide

Anode: $2I^-(aq) \rightarrow I_2(s) + 2e^- \quad E^0 = +0.54 \text{ V}$

The solution around the anode appears yellow-brown, owing to the formation of iodine.
6. The overall electrolytic cell reaction would be as follows:

$$2H_2O(l) + 2I^-(aq) \rightarrow H_2(g) + 2OH^-(aq) + I_2(s)$$

7. Determine the minimum voltage required to achieve this reaction. A potential difference greater than the spontaneous reverse reaction would need to be applied, so more than $+0.54 - (-0.83) = 1.37$ V should be delivered to this electrolytic cell.

on Resources

▶ **Video eLesson** Predicting products of electrolysis (eles-3246)

6.3.3 Electrolysis of aqueous solutions: more examples

As shown previously, in an aqueous solution, water must always be considered as a potential reactant at each electrode. Using the electrochemical series and the concept of *competition at each electrode*, it is possible to explain and predict the products of electrolysis reactions. Remember:

- at the cathode, reduction takes place. The strongest oxidising agent that is present will undergo reduction as it is the one that requires the least amount of energy (i.e. it is the easiest to reduce).
- at the anode, oxidation takes place. The strongest reducing agent that is present will undergo oxidation as it is the one that requires the least amount of energy (i.e. it is the easiest to oxidise).

Electrolysis of dilute sodium chloride solution with inert electrodes

In many electrolysis reactions, inert electrodes are used. These are electrodes that do not affect the reactions taking place on their surfaces, and that conduct a current but do not tend to go into solutions as ions. They are usually carbon, in the form of graphite, or platinum, which is much more expensive. As shown in figure 6.7, there is a choice of reactants at each electrode when dilute sodium chloride is electrolysed.

At the cathode:

- sodium ions are present and the cathode is in contact with water molecules from the solvent
- a gas is produced, which proves to be hydrogen
- if a few drops of phenolphthalein indicator are added to the region around it, the region turns pink. From these observations we can infer that OH^- ions are also produced.

The production of H_2 and OH^- is consistent with the water being reduced in preference to the Na^+ ions, according to the following equation:

$$\text{Cathode: } 2H_2O(l) + 2e^- \rightarrow H_2(g) + 2OH^-(aq)$$

At the anode, the choice between water and chloride ions is possible. The observed evolution of oxygen gas, together with a few drops of phenolphthalein remaining clear, support the conclusion that water is once again reacting at this electrode, but this time it is being oxidised. This is due to the relative reducing agent strengths of water and chloride ions: water is a stronger reducing agent than chloride ions.

$$\text{Anode: } 2H_2O(l) \rightarrow O_2(g) + 4H^+(aq) + 4e^-$$

FIGURE 6.7 The electrolysis of dilute sodium chloride solution using inert electrodes

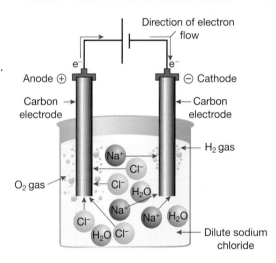

elog-1744

tlvd-9722

| **EXPERIMENT 6.1** | **online only** |

Electrolysis of aqueous solutions

Aim

To conduct electrolysis on aqueous solutions and test for the products at each electrode

Electrolysis of dilute sodium chloride solution with copper electrodes

If the previous experiment is repeated using copper electrodes instead of inert electrodes, the results change. This illustrates that the choice of electrodes has an effect on the nature of the products of an electrolysis.

This time the reactions occurring at each electrode are:

$$\text{Cathode: } 2H_2O(l) + 2e^- \rightarrow H_2(g) + 2OH^-(aq)$$

$$\text{Anode: } Cu(s) \rightarrow Cu^{2+}(aq) + 2e^-$$

It can be seen that Cu^{2+} ions are produced instead of O_2 gas at the anode. This is due to the relative reducing agent strength of copper metal being greater than water.

The products formed from the electrolysis of some more electrolytes are shown in table 6.1.

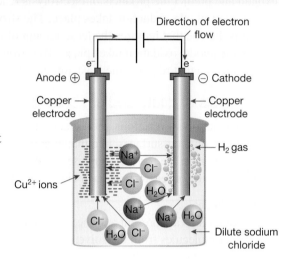

FIGURE 6.8 The electrolysis of dilute sodium chloride solution using copper electrodes

TABLE 6.1 Products of electrolysis

Electrodes	Electrolyte	Product formed at cathode	Product formed at anode
Inert (platinum or graphite)	$PbBr_2(l)$ — a melt	$Pb(s)$	$Br_2(g)$
	$NaCl(l)$ — a melt	$Na(s)$	$Cl_2(g)$
	$CuCl_2(aq)$	$Cu(s)$	$Cl_2(g), O_2(g) + 4H^+(aq)^*$
	$NaCl(aq)$	$H_2(g) + 2OH^-(aq)$	$Cl_2(g), O_2(g) + 4H^+(aq)^*$
	$KNO_3(aq)$	$H_2(g) + 2OH^-(aq)$	$O_2(g) + 4H^+(aq)$
	$CuSO_4(aq)$	$Cu(s)$	$O_2(g) + 4H^+(aq)$
	$H_2SO_4(aq)$	$H_2(g)$	$O_2(g) + 4H^+(aq)$
	$NaOH(aq)$	$H_2(g) + 2OH^-(aq)$	$O_2(g) + 2H_2O(l)$
Copper	$CuSO_4(aq)$	$Cu(s)$ deposited	$Cu(s)$ dissolves (is oxidised) to form Cu^{2+} ions

*Depending on concentration

Tips for writing oxidation reactions

- The electrochemical series is written as a series of reversible reduction reactions. Don't forget to reverse these equations when writing oxidation reactions.
- When you have decided whether an oxidation or reduction equation is required, remember to use only a single arrow in your equation.

tlvd-3061

SAMPLE PROBLEM 3 Predicting the products at each electrode and writing partial equations

A dilute solution containing tin(IV) chloride and copper sulfate is electrolysed using inert electrodes. Predict the products that will form at each electrode and write the relevant half-equations. Use these to write the overall equation and determine the minimum voltage required to achieve the reaction.

THINK	WRITE
1. Identify the constituents of the solution.	Sn^{4+}, Cl^-, Cu^{2+}, SO_4^{2-} and H_2O
2. From the VCE Chemistry Data Book, copy the half-equations for the species in descending order of standard electrode potentials. TIP: When using the electrochemical series, be careful with Sn^{2+} because it occurs in more than one half-equation. (The same applies to Fe^{2+}.)	$Cl_2(g) + 2e^- \rightleftharpoons 2Cl^-(aq)$ $E^0 = +1.36$ V $O_2(g) + 4H^+(aq) + 4e^- \rightleftharpoons 2H_2O(l)$ $E^0 = +1.23$ V $Cu^{2+}(aq) + 2e^- \rightleftharpoons Cu(s)$ $E^0 = +0.34$ V $Sn^{4+}(aq) + 2e^- \rightleftharpoons Sn^{2+}(aq)$ $E^0 = +0.15$ V $2H_2O(l) + 2e^- \rightleftharpoons H_2(g) + 2OH^-(aq)$ $E^0 = -0.83$ V
3. Circle the species present in the electrolytic cell that could participate.	$Cl_2(g) + 2e^- \rightleftharpoons \boxed{2Cl^-(aq)}$ $E^0 = +1.36$ V $O_2(g) + 4H^+(aq) + 4e^- \rightleftharpoons \boxed{2H_2O(l)}$ $E^0 = +1.23$ V $\boxed{Cu^{2+}(aq)} + 2e^- \rightleftharpoons Cu(s)$ $E^0 = +0.34$ V $\boxed{Sn^{4+}(aq)} + 2e^- \rightleftharpoons Sn^{2+}(aq)$ $E^0 = +0.15$ V $\boxed{2H_2O(l)} + 2e^- \rightleftharpoons H_2(g) + 2OH^-(aq)$ $E^0 = -0.83$ V
4. At the cathode, the strongest oxidising agent (highest E^0) undergoes reduction. Copper metal will form.	Cathode: $Cu^{2+}(aq) + 2e^- \rightarrow Cu(s)$ $E^0 = +0.34$ V
5. At the anode, the strongest reducing agent (lowest E^0) undergoes oxidation. Oxygen gas and hydrogen ions will form.	Anode: $2H_2O(l) \rightarrow O_2(g) + 4H^+(aq) + 4e^-$ $E^0 = +1.23$ V
6. Write the overall equation by combining the half-equations, ensuring that the number of electrons transferred is the same for each equation. In this case, first multiply the cathode half-equation by 2.	$2Cu^{2+}(aq) + 4e^- \rightarrow 2Cu(s)$ $2H_2O(l) \rightarrow O_2(g) + 4H^+(aq) + 4e^-$ $2Cu^{2+}(aq) + 2H_2O(l) \rightarrow 2Cu(s) + O_2(g) + 4H^+(aq)$
7. Determine the minimum voltage required by finding the difference between the E^0 of the reducing agent and the E^0 of the oxidising agent.	Minimum voltage required $> |+0.34 - 1.23|$ > 0.89 V

PRACTICE PROBLEM 3

A dilute solution containing silver nitrate and cobalt(II) chloride is electrolysed using inert electrodes. Predict the products that will form at each electrode and write the relevant half-equations. Use these to write the overall equation and determine the minimum voltage required to achieve the reaction.

6.3.4 Factors affecting electrolysis of solutions

What happens during electrolysis depends on a number of factors, including:

- the concentration of the electrolyte
- the nature of the electrolyte
- the nature of the electrodes.

In any electrolysis reaction, alteration of any of these factors can change the nature of the products. The identity of products of an electrolysis reaction under fixed conditions is found by experiment. When the products are known, the reactions occurring at the electrodes can be written.

The effect of concentration

The electrochemical series (see the VCE Chemistry Data Book) is a useful tool for predicting the products of an electrolysis reaction. However, it must be remembered that it is based on standard conditions; in particular, where the concentrations of dissolved species are 1 M. If the concentrations of reactants are different from this, the observed results might be different from those predicted.

Dilute sodium chloride

For example, in the electrolysis of dilute sodium chloride, as seen earlier, the possible reduction reactions at the cathode are as follows:

$$\text{Cathode:} \quad 2H_2O(l) + 2e^- \rightleftharpoons H_2(g) + 2OH^-(aq) \qquad E^0 = -0.83 \text{ V}$$
$$Na^+(aq) + e^- \rightleftharpoons Na(l) \qquad E^0 = -2.7 \text{ V}$$

As predicted from a table of standard redox potentials, water, rather than sodium ions, is reduced at the cathode. At the anode, chloride ions or water molecules may be oxidised:

$$\text{Anode:} \quad Cl_2(g) + 2e^- \rightleftharpoons 2Cl^-(aq) \qquad E^0 = +1.36 \text{ V}$$
$$O_2(g) + 4H^+(aq) + 4e^- \rightleftharpoons 2H_2O(l) \qquad E^0 = +1.23 \text{ V}$$

As predicted from the table of standard redox potentials, oxygen gas is evolved in preference to chlorine gas at the anode.

The overall equation is:

$$2H_2O(l) \rightarrow 2H_2(g) + O_2(g)$$

Concentrated sodium chloride

Now consider a situation with higher concentrations of chloride ions (6 M). The production of chlorine becomes more favourable. The reduction of water, rather than of sodium ions, still occurs at the cathode at higher concentrations of sodium ions. Therefore, the electrolysis of highly dilute sodium chloride produces hydrogen gas at the cathode and oxygen gas at the anode, while the electrolysis of highly concentrated sodium chloride produces hydrogen gas at the cathode and chlorine gas at the anode.

FIGURE 6.9 Chlorine gas is a greenish-yellow gas that is corrosive and a toxic respiratory irritant.

This can be understood when the E^0 values for the following reactions are examined.

$$Cl_2(g) + 2e^- \rightleftharpoons 2Cl^-(aq) \qquad E^0 = 1.36 \text{ V}$$

$$O_2(g) + 4H^+(aq) + 4e^- \rightleftharpoons 2H_2O(l) \quad E^0 = 1.23 \text{ V}$$

Values change when conditions are non-standard. In fact, they change enough to swap the order around, thus making the oxidation of chloride ions to chlorine gas the preferred reaction at the anode. Such a reversal at the cathode does not occur because the difference between H_2O and Na^+ ions is too large. Thus, when concentrated sodium chloride (>6 M) is electrolysed, the overall reaction becomes

$$2H_2O(l) + 2Cl^-(aq) \rightarrow H_2(g) + Cl_2(g) + 2OH^-(aq)$$

It should also be noted that at concentrations in between these values, a mixture of oxygen and chlorine is obtained.

EXTENSION: Electrode potentials and non-standard conditions

It is possible to predict electrode potentials under conditions different from standard conditions, using the *Nernst equation*. This takes into account variables such as concentration, the presence of solid reactants, temperature and partial pressures of any gases involved. It also involves R, the universal gas constant, and F, the Faraday constant.

Knowledge of how to use this equation is not required for VCE Chemistry.

The nature of the electrolyte

Let's compare electrolysis of dilute sodium nitrate solution to the electrolysis of dilute copper(II) nitrate solution.

- At the anode, oxygen gas is evolved in both cells:

$$2H_2O(l) \rightarrow O_2(g) + 4H^+(aq) + 4e^-$$

- At the cathode in the sodium nitrate cell, hydrogen gas is evolved, but in the copper(II) nitrate cell, two reactions are possible at the cathode:

$$Cu^{2+}(aq) + 2e^- \rightleftharpoons Cu(s) \qquad E^0 = +0.34 \text{ V}$$
$$2H_2O(l) + 2e^- \rightleftharpoons H_2(g) + 2OH^-(aq) \qquad E^0 = -0.83 \text{ V}$$

As may be predicted from a consideration of the standard electrode potentials, copper is deposited in preference to the evolution of hydrogen gas. So, the products in an electrolytic reaction depend on the nature of the electrolyte. The overall equation for the electrolysis of dilute copper(II) nitrate is:

$$2Cu^{2+}(aq) + 2H_2O(l) \rightarrow 2Cu(s) + O_2(g) + 4H^+(aq)$$

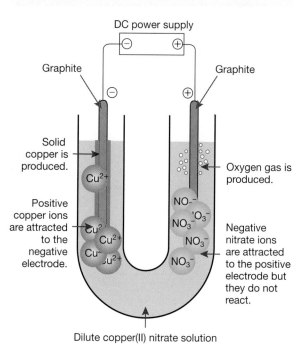

FIGURE 6.10 Electrolysis of a dilute copper(II) nitrate solution produces copper (in solid form) and oxygen gas.

DC power supply

Graphite

Graphite

Solid copper is produced.

Oxygen gas is produced.

Positive copper ions are attracted to the negative electrode.

Negative nitrate ions are attracted to the positive electrode but they do not react.

Dilute copper(II) nitrate solution

The nature of the electrodes

Inert electrodes are usually made of graphite (carbon) or platinum. However, there can sometimes be situations where the metal that the anode is made from can become a preferred reactant. Electrons will then be removed preferentially from the metal atoms in the anode rather than from the ionic species in solution or the water molecules themselves. There is no corresponding effect at the cathode because metals have no tendency to gain electrons.

As seen earlier, the electrolysis of dilute sodium chloride solution using inert electrodes results in the production of oxygen gas or chlorine gas (depending on the concentration of the solution; see table 6.1) at the anode. However, if copper electrodes are used, copper(II) ions are produced at the anode because the electrode itself acts as a stronger reducing agent than either water molecules or chloride ions. The copper anode is oxidised and dissolves to form Cu^{2+} ions.

$$Cl_2(g) + 2e^- \rightleftharpoons 2Cl^-(aq) \qquad\qquad E^0 = +1.36 \text{ V}$$
$$O_2(g) + 4H^+(aq) + 4e^- \rightleftharpoons 2H_2O(l) \qquad\qquad E^0 = +1.23 \text{ V}$$
$$Cu^{2+}(aq) + 2e^- \rightleftharpoons Cu(s) \qquad\qquad E^0 = +0.34 \text{ V}$$
$$2H_2O(l) + 2e^- \rightleftharpoons H_2(g) + 2OH^-(aq) \qquad\qquad E^0 = -0.83 \text{ V}$$

The overall equations for these reactions are:

$$\text{Inert electrodes: } 2H_2O(l) \rightarrow 2H_2(g) + O_2(g)$$
$$\text{Copper electrodes: } Cu(s) + 2H_2O(l) \rightarrow Cu^{2+}(aq) + H_2(g) + 2OH^-(aq)$$

However, these are only the *initial* reactions that occur in this experiment. Once Cu^{2+} ions are present in solution, this becomes the strongest oxidising agent and will be preferentially reduced at the cathode, effectively just shifting copper mass from the anode to the cathode. This is a vastly different result to the production of hydrogen and oxygen gas from the electrolysis of the same solution but with inert electrodes.

FIGURE 6.11 If copper electrodes are used to electrolyse sodium chloride solution, the copper will be preferentially oxidised. The copper anode will be eroded, and copper metal will be deposited at the cathode.

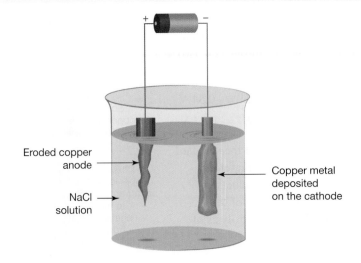

Eroded copper anode

NaCl solution

Copper metal deposited on the cathode

SAMPLE PROBLEM 4 Writing partial equations for reactions at each electrode in electrolysis

A dilute solution containing cobalt(II) nitrate is electrolysed using lead electrodes. Predict the products that will initially form at each electrode and write the relevant half-equations. Use these to write the overall equation and determine the minimum voltage required to achieve the reaction.

THINK	WRITE
1. Identify the constituents of the solution.	Co^{2+}, NO_3^- and H_2O. Note that NO_3^- is inert and that Pb from the electrode could react.
2. From the electrochemical series in the VCE Chemistry Data Book, copy the half-equations for the species in descending E^0.	$O_2(g) + 4H^+(aq) + 4e^- \rightleftharpoons 2H_2O(l)$ $E^0 = +1.23$ V $Pb^{2+}(aq) + 2e^- \rightleftharpoons Pb(s)$ $E^0 = -0.13$ V $Co^{2+}(aq) + 2e^- \rightleftharpoons Co(s)$ $E^0 = -0.28$ V $2H_2O(l) + 2e^- \rightleftharpoons H_2(g) + 2OH^-(aq)$ $E^0 = -0.83$ V
3. Circle the species present in the electrolytic cell that could participate.	$O_2(g) + 4H^+(aq) + 4e^- \rightleftharpoons \boxed{2H_2O(l)}$ $E^0 = +1.23$ V $Pb^{2+}(aq) + 2e^- \rightleftharpoons \boxed{Pb(s)}$ $E^0 = -0.13$ V $\boxed{Co^{2+}(aq)} + 2e^- \rightleftharpoons Co(s)$ $E^0 = -0.28$ V $\boxed{2H_2O(l)} + 2e^- \rightleftharpoons H_2(g) + 2OH^-(aq)$ $E^0 = -0.83$ V
4. At the cathode, the strongest oxidising agent (highest E^0) initially present in solution undergoes reduction.	Cathode: $Co^{2+}(aq) + 2e^- \rightarrow Co(s)$
5. At the anode, the strongest reducing agent (lowest E^0) undergoes oxidation. Lead ions will form.	Anode: $Pb(s) \rightarrow Pb^{2+}(aq) + 2e^-$
6. Write the overall equation for the reaction initially observed by combining these half-equations, ensuring that the number of electrons transferred is the same for each equation.	$Co^{2+}(aq) + Pb(s) \rightarrow Co(s) + Pb^{2+}(aq)$
7. Determine the minimum voltage required by finding the difference between the E^0 of the reducing agent and the E^0 of the oxidising agent.	Minimum voltage required $> |-0.28 - (-0.13)|$ > 0.15 V

PRACTICE PROBLEM 4

A solution of lead(II) nitrate is electrolysed using copper electrodes. Write the half-equations for the reaction occurring at each electrode. Use these to write the overall equation and determine the minimum voltage required to achieve the reaction.

EXPERIMENT 6.2

Factors affecting electrolysis (student design)

Aim
Student developed

6.3 Activities   **learn** on

Students, these questions are even better in jacPLUS

 Receive immediate feedback and access sample responses

 Access additional questions

 Track your results and progress

Find all this and MORE in jacPLUS ▶

| 6.3 Quick quiz on | 6.3 Exercise | 6.3 Exam questions |

6.3 Exercise

1. Give two initial observations that would be noted during the electrolysis of dilute sodium chloride with copper electrodes.
2. a. Write the equations for the initial reaction at each electrode when a 1 M solution of hydrochloric acid is electrolysed using silver electrodes.
   b. Use these equations to write the overall equation for the initial reaction.
3. Write the half-equations that occur at the cathode and the anode when a dilute solution of Na_2SO_4 is electrolysed using inert electrodes.
4. Predict the products and the minimum cell voltage required for the electrolysis of a 1 M solution of aluminium chloride.
5. Using inert electrodes, predict the products formed from the electrolysis of:
   a. molten copper(II) fluoride
   b. 1 M solution of copper(II) fluoride.
6. A solution containing lead(II), magnesium and copper(II) ions is electrolysed for a long time.
   a. What will be the first product formed at the cathode?
   b. If the electrolysis is continued until all the ions responsible for the first product are used up, what will be the next product observed at the cathode?
   c. If the electrolysis is continued further until the second product is observed to stop forming, what will be the third product formed at the cathode?
7. Magnesium can be obtained commercially from sea water. During the last stage of this process, molten magnesium chloride undergoes electrolysis in a cell that contains an iron cathode and a graphite anode.
   a. Why can iron be used for the cathode but not the anode?
   b. Draw a fully labelled diagram of an electrolytic cell that could be used to produce magnesium. Include equations.
8. A dilute solution of copper(II) sulfate is electrolysed using platinum electrodes.
   a. Write the half-equations for the reactions at each electrode.
   b. How will the concentration of Cu^{2+} ions change during this process?

   The platinum electrodes are replaced by copper electrodes.
   c. Write the half-equations for the reactions that now occur at each electrode.
   d. How will the concentration of Cu^{2+} ions change during this time?
9. Sometimes reaction products from an electrolysis reaction may be different to those predicted. How might this happen?
10. Why is it not possible to electrolyse a solution containing both tin(II) chloride and iron(III) chloride?

6.3 Exam questions

Question 1 (1 mark)

Source: VCE 2020 Chemistry Exam, Section A, Q.6; © VCAA

MC Which one of the following pairs of statements is correct for both electrolysis cells and galvanic cells?

	Electrolysis cell	Galvanic cell
A.	Both electrodes are always inert.	Both electrodes are always made of metal.
B.	Electrical energy is converted to chemical energy.	The voltage of the cell is independent of the electrolyte concentration.
C.	Chemical energy is converted to electrical energy.	The products are dependent on the half-cell components.
D.	The products are dependent on the half-cell components.	Chemical energy is converted to electrical energy.

Question 2 (1 mark)

Source: VCE 2022 Chemistry Exam, Section A, Q.13; © VCAA

MC An electrolysis cell is set up with inert platinum, Pt, electrodes.

Which one of the following will produce a gas at the cathode when undergoing electrolysis in the cell?

A. potassium iodide, $KI(aq)$
C. lead bromide, $PbBr_2(l)$
B. sodium chloride, $NaCl(l)$
D. copper sulfate, $CuSO_4(aq)$

Question 3 (3 marks)

Source: VCE 2020 Chemistry Exam, Section B, Q.4.c; © VCAA

Research scientists are developing a rechargeable lithium–carbon dioxide, $Li–CO_2$, battery. The rechargeable $Li–CO_2$ battery is made of lithium metal, carbon in the form of graphite (coated with a catalyst) and a non-aqueous electrolyte that absorbs CO_2.

A diagram of the rechargeable $Li–CO_2$ cell is shown below. One $Li–CO_2$ cell generates 4.5 V.

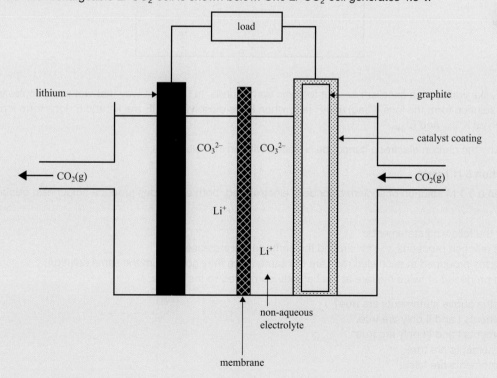

Explain why it is unsafe to use an aqueous electrolyte in the design of the $Li–CO_2$ battery. Include appropriate equations in your answer.

Question 4 (3 marks)

Source: VCE 2017 Chemistry Exam, Section B, Q.8.d; © VCAA

Fluorine, F_2, gas is the most reactive of all non-metals. Anhydrous liquid hydrogen fluoride, HF, can be electrolysed to produce F_2 and hydrogen, H_2, gases. Potassium fluoride, KF, is added to the liquid HF to increase electrical conductivity. The equation for the reaction is

$$2HF(l) \rightarrow F_2(g) + H_2(g)$$

F_2 is used to make a range of chemicals, including sulfur hexafluoride, SF_6, an excellent electrical insulator, and xenon difluoride, XeF_2, a strong fluorinating agent.

The diagram below shows an electrolytic cell used to prepare F_2 gas.

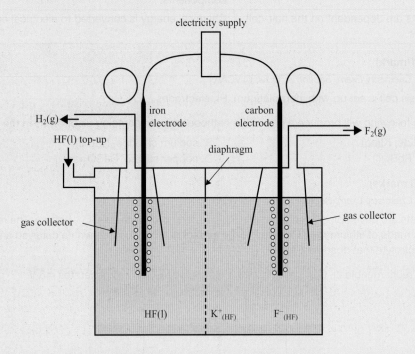

Liquid HF, like water, is an excellent solvent for ionic compounds. In the same way that water molecules in an aqueous solution form the ions $K^+(aq)$ and $F^-(aq)$, when KF is dissolved in HF, the K^+ and F^- ions form ions that are written as $K^+_{(HF)}$ and $F^-_{(HF)}$.

Explain why the carbon electrode cannot be replaced with an iron electrode.

Question 5 (1 mark)

MC When a 0.1 M solution of sodium chloride is electrolysed, both electrodes produce colourless, odourless gases.

Consider the following statements:
 I   As the reaction proceeds, the pH around the cathode will increase.
 II The gases produced at each electrode are the same since they come from the same solution.
 III Gas is produced at twice the rate at the cathode compared to the anode.

Which of the above statements are true?
A. Statements I and II only are true.
B. Statements I and III only are true.
C. All statements are true.
D. All statements are false.

More exam questions are available in your learnON title.

6.4 Commercial electrolytic cells

TIP: Some examples of commercial electrolytic cells are provided in this subtopic. Although you are not required to know the details of any specific cell, you should examine these examples and make sure that you understand the principles behind their operation. It is expected that you should be able to apply these principles in an examination, rather than just restating facts that you have learnt.

6.4.1 Electrolysis of molten sodium chloride

Sodium metal is made commercially by the electrolysis of molten sodium chloride in a specially designed electrolytic cell called a Downs cell.

Sodium is a soft and very reactive metal that is used in sodium vapour lamps, in the manufacture of esters and as a coolant in nuclear power plants.

It is necessary that the sodium chloride be molten rather than in aqueous solution as the water would otherwise be reduced in preference to the sodium ions. The design of the cell means the products of the electrolysis (sodium and chlorine gas) are continually removed so that they do not react to re-form sodium chloride. Other operational features include the use of density differences to collect the liquid sodium and of added chemicals to reduce the melting point of the sodium chloride.

The Downs cell

Electrolysis of molten sodium chloride results in the production of sodium and chlorine. The electrolytic cell in which this process is carried out commercially is called the **Downs cell**.
- Current flows through molten NaCl because it exists as separate Na^+ and Cl^- ions.
- The Downs cell allows fresh sodium chloride to be added when required.
- The electrodes are separated by an iron mesh screen that keeps the products apart so that they do not react to re-form sodium chloride.
- Chlorine gas is evolved at the anode in a specially designed compartment above the anode as a valuable by-product.

$$\text{Anode (oxidation): } 2Cl^-(l) \rightarrow Cl_2(g) + 2e^-$$

- Molten sodium is deposited at the cathode. Since molten sodium is less dense than the electrolyte, it floats to the surface where it overflows into a separate container.

$$\text{Cathode (reduction): } Na^+(l) + e^- \rightarrow Na(l)$$

Downs cell an electrolytic cell used for the commercial production of sodium and chlorine

- Overall cell reaction:

$$2NaCl(l) \rightarrow 2Na(l) + Cl_2(g)$$

FIGURE 6.12 Operation of a Downs cell

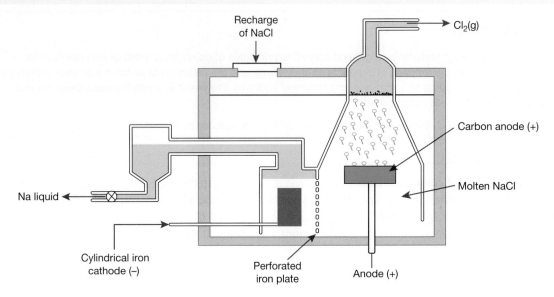

- A Downs cell operates at about 600 °C to maintain the salt in a molten state.
- Sodium chloride melts at 801 °C but calcium chloride is added to the sodium chloride to reduce the melting point; that is, calcium chloride acts as a flux. This is possible because calcium ions have a lower E^0 value than sodium ions and will therefore not interfere with the production of sodium.

While used to produce sodium on a large scale, the chlorine produced is regarded as a by-product as it can be made more economically by other means.

Magnesium metal can also be produced in a very similar process.

6.4.2 Producing aluminium

BACKGROUND KNOWLEDGE: The history of aluminium

Although aluminium is the most abundant metallic element in Earth's crust, it was difficult to extract before 1886. The most common process involved its extraction from the ore and conversion into $AlCl_3$. This was then chemically reduced using either sodium or potassium, metals that were also difficult to produce. Compared with today's methods, this process was on a small scale and very expensive. One hundred and thirty years ago, only the wealthy could afford aluminium. Napoleon III of France was famous for serving food to special guests at banquets on aluminium plates, while ordinary guests were served food on plates made from gold! The aluminium extraction breakthrough came in 1886 with the development of what we now call the Hall–Héroult cell. Paul Héroult, a French scientist, and Charles Hall, a US inventor and chemist, almost simultaneously filed patent applications for the industrial electrolytic production of aluminium, despite working completely independently of each other. It has resulted in an enormous growth in primary aluminium production, from 13 tonnes per year in 1885 to more than 73 million tonnes per year today. Aluminium is now used whenever a lightweight metal is required, from aluminium cans through to aircraft parts.

FIGURE 6.13 Today, aluminium products, such as this aluminium scooter, are common.

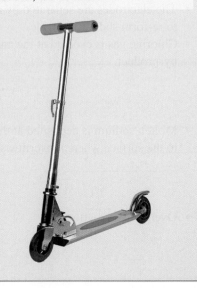

The Hall–Héroult cell

The industrial production of aluminium involves electrolysis of alumina, Al_2O_3, that is dissolved in molten cryolite, Na_3AlF_6 (figure 6.14). The electrolysis takes place in a steel vessel. The cell is lined with carbon and contains the molten cryolite and dissolved alumina mixture maintained at a temperature of about 980 °C. Carbon blocks suspended above the cell and partially immersed in the electrolyte act as anodes, which take part in the chemical reactions in the cell, while the carbon lining of the cell acts as the cathode.

FIGURE 6.14 Schematic diagram of a cross-section of a Hall–Héroult cell for the electrolytic production of aluminium

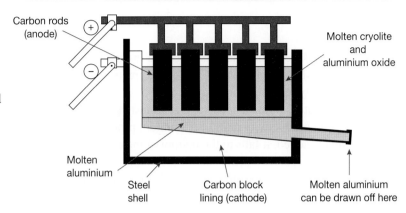

Aluminium cannot be reduced by electrolysis of an aqueous solution of a soluble aluminium salt because water, a stronger oxidising agent than aluminium ions, is preferentially reduced.

$$2H_2O(l) + 2e^- \rightarrow H_2(g) + 2OH^-(aq) \quad E^0 = -0.83 \text{ V}$$

$$Al^{3+}(aq) + 3e^- \rightarrow Al(s) \quad\quad\quad\quad E^0 = -1.67 \text{ V}$$

Cryolite acts as a solvent and an electrolyte due to its unique combination of properties. It has a melting point less than half that of alumina (960 °C compared with alumina's melting point of 2020 °C), a low vapour pressure and a density lower than molten aluminium (2.05 g cm^{-3} compared with aluminium's density of 2.30 g cm^{-3}). Cryolite can dissolve sufficient alumina to allow deposition of aluminium at about 980 °C. This reduction in temperature is also aided by the addition of small amounts of calcium fluoride (CaF_2), lithium fluoride (LiF) and aluminium fluoride (AlF_3), which also increase the efficiency of the process.

Key features of the Hall–Héroult cell include the following:
- The carbon anode and cathode are both made from petroleum coke with pitch as a binder.
- Alumina, Al_2O_3, is fed into the electrolyte at regular intervals where it dissolves, forming aluminium ions, Al^{3+}, and oxide ions, O^{2-}. The direct current applied across each cell moves the ions in opposite directions.
- At the anode, the oxide ions are oxidised to form oxygen gas. The oxygen then immediately reacts with the carbon anode to form carbon dioxide.

$$\text{Anode: } 2O^{2-}(l) \rightarrow O_2(g) + 4e^-$$

$$O_2(g) + C(s) \rightarrow CO_2(g)$$

The overall reaction at the anode can therefore be written as follows:

$$\text{Anode: } C(s) + 2O^{2-}(l) \rightarrow CO_2(g) + 4e^-$$

As the carbon anodes are gradually consumed during the process, they are lowered to maintain the optimum distance between the anode and cathode surfaces, until they are used up and replaced. The anodes are generally replaced every three weeks so that the process is continuous.
- The positively charged aluminium ions that are dissolved in the cryolite are drawn to the negatively charged cathode where they form aluminium.

$$\text{Cathode: } Al^{3+}(l) + 3e^- \rightarrow Al(l)$$

- The density difference between cryolite and the newly formed molten aluminium allows the aluminium to settle at the bottom of the cell, where it is regularly drained. After draining, the molten aluminium can be cast.
- The overall reaction is as follows:

$$2Al_2O_3(l) + 3C(s) \rightarrow 4Al(l) + 3CO_2(g)$$

Carbon dioxide is the main gas produced in this process. Other gases produced include fluorides such as CF_4 and C_2F_6, which — like CO_2 — are greenhouse gases. These are initially confined by gas hoods, then continuously removed and treated.
- The amount of alumina added to a cell must be strictly controlled. If too little alumina is added, maximum yields and productivity rates of aluminium production become economically unfavourable. If too much alumina is added, it falls to the bottom of the cell instead of dissolving (because it is denser than molten aluminium). There, it settles below the aluminium and interferes with the flow of current.
- Hall–Héroult cells operate continuously at a low voltage of about 4–5 V, but require a high current of 50 000–280 000 A. The electrical resistance to the flow of this current generates enough heat to keep the electrolyte in a liquid state.
- Although aluminium is a very light and versatile metal, it is estimated that aluminium production requires ten times the energy of steel production.

6.4.3 The industrial electrolysis of brine

Chlorine gas, hydrogen gas and sodium hydroxide are three important industrial chemicals. Industrial production occurs simultaneously by electrolysis of a concentrated aqueous sodium chloride solution (brine), using a **membrane cell**.

Although chloride ions are weaker reducing agents than water molecules, chlorine may be produced electrolytically from aqueous solutions of sodium chloride. This is done by altering the operating conditions of electrolytic cells to favour the reduction of chloride ions rather than water molecules. The main way that this is achieved is by using a concentrated solution, as discussed in subtopic 6.3.

FIGURE 6.15 In the manufacture of soap, oils and fats are reacted with sodium hydroxide solution. Sodium hydroxide has many other industrial applications and may be produced in large quantities by electrolysis.

Membrane cells

Early cells for the electrolysis of brine used either mercury or asbestos in their design. Membrane cells were developed in response to the potential health hazards involved with such cells. Industrial membrane cells can be very large.

A membrane cell is characterised by its plastic, semi-permeable membrane that separates the anode half-cell from the cathode half-cell of the electrolytic cell. This semi-permeable membrane allows the smaller sodium ions to pass through from one compartment to the other, but prevents the movement of larger species, such as water molecules and hydroxide ions. As a result, sodium ions and hydroxide ions are trapped in the cathode compartment, thus producing sodium hydroxide solution and hydrogen gas, which is evolved at the cathode. Chlorine gas is produced at the anode. The relevant equations are as follows:

$$\text{Cathode: } 2H_2O(l) + 2e^- \rightarrow 2OH^-(aq) + H_2(g)$$

$$\text{Anode: } 2Cl^-(aq) \rightarrow Cl_2(g) + 2e^-$$

membrane cell an electrolytic cell used for the electrolysis of brine

As with all electrolytic cells, the products are prevented from coming into contact with each other. The overall reaction for the production of chlorine via the membrane cell process with brine is as follows:

$$2NaCl(aq) + 2H_2O(l) \rightarrow 2NaOH(aq) + H_2(g) + Cl_2(g)$$

FIGURE 6.16 A membrane cell

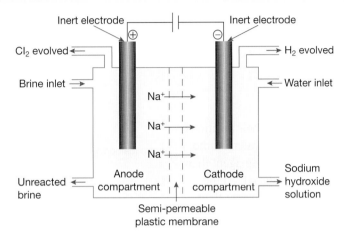

6.4.4 Electroplating

Electroplating is the process of coating an object in a metal. Inexpensive silver-plated jewellery can be produced through electroplating. 'Gold' rings that turn fingers green are actually copper rings that have been electroplated with gold.

In the electroplating process, the article to be plated is used as the cathode and the metal being plated onto the article is used as the anode. The electrolytic solution or 'bath' contains a salt of the metal being plated. A low-voltage electric current causes metal ions from the bath to gain electrons at the cathode and to deposit as a metal coating on the cathode (the object). It also causes metal atoms in the anode to lose electrons and go into the bath as ions. As the plating proceeds, the anode gradually disappears and maintains the metal ion's concentration in the bath.

Electroplating operations have traditionally used many toxic solutions, such as cyanides, and produced a lot of toxic waste. Due to the costs of this waste treatment and the environmental implications, along with increasing regulation from environmental agencies, there is now a lot of research into the use of alternative electrolyte solutions.

Objects to be plated are thoroughly cleaned of all grease and dirt using concentrated acidic or basic solutions. The cleaning solutions eventually become ineffective, owing to contamination, and must be disposed of.

A number of factors contribute to the quality of the metal coating formed in electroplating. These include:
- the carefully controlled concentration of the cations to be reduced in the plating solution. Unwanted side reactions must be avoided.
- the careful consideration of the type and concentration of electrolyte
- the solution, which must contain compounds to control the acidity and increase the conductivity
- the compounds, some of which make the metal coating brighter or smoother
- the shape of the anode, which must often be shaped like the object at the cathode to achieve an even metal coating.

electroplating the process of adding a thin metal coating by electrolysis

Silver-plating

In silver-plating, objects such as cutlery are coated at the cathode. The plating solution contains silver ions, $Ag^+(aq)$. The anode is pure silver. When current flows, silver is deposited on the metal object.

$$\text{Cathode: } Ag^+(aq) + e^- \rightarrow Ag(s)$$

At the same time, silver atoms at the anode form silver ions.

$$\text{Anode: } Ag(s) \rightarrow Ag^+(aq) + e^-$$

Both these reactions are possible due to their positions on the electrochemical series relative to water. At the cathode, silver ions are a stronger oxidising agent than water and so they are preferentially reduced. At the anode, silver metal is a stronger reducing agent than water. Therefore, the silver metal reacts in preference to water and is oxidised to silver ions.

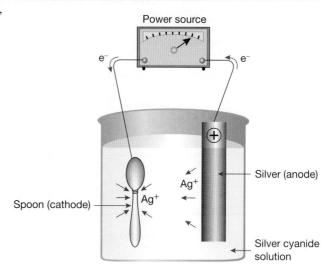

FIGURE 6.17 Silver-plating mechanism for cutlery

The plating is only a few hundredths of a centimetre thick. If the experimental conditions are right, the metal coating adheres strongly and may be polished. However, if conditions are not satisfactory, the metal becomes powder-like and drops off.

elog-1746

tlvd-9723

EXPERIMENT 6.3

online only

Electroplating

Aim

To plate a piece of copper with nickel metal

EXTENSION: Protecting metal structures from corrosion

Corrosion of metals causes millions of dollars of damage worldwide each year, and, not surprisingly, may make structures and objects unsafe. This is especially true for iron and steel objects and structures, due to the huge amount of iron that is used throughout the world.

Corrosion of metals is due to oxidation. Galvanic cells are often unwittingly set up when there are impurities or stress points in the iron, and water with dissolved electrolyte (for example, salt) and oxygen is present. In the rusting of iron, for example, the first step is the oxidation of iron:

$$\text{Anode: } Fe(s) \rightarrow Fe^{2+}(aq) + 2e^-$$

One method to prevent the corrosion of metals such as iron is to force the iron structure or object to act as a *cathode*. This is called cathodic protection.

FIGURE 6.18 Corrosion of iron occurs when metal is exposed to oxygen and water containing electrolytes.

There are two main ways that this can be achieved:
1. *By connecting the iron to a more reactive metal.* This essentially makes a galvanic cell in which the more reactive metal is the anode and the iron is the cathode. This is called sacrificial protection.
2. *By setting up an electrolytic cell.* In the case of iron, the structure to be protected is connected to the negative terminal of a DC supply, thus making it the cathode. The positive terminal is connected to some less valued metal which then corrodes and is replaced when necessary. This method is called *impressed current cathodic protection* (or ICCP).

To access more information on this extension concept, download the digital document from your Resources panel.

on Resources

📄 **Digital document** EXTENSION: Protecting metal structures from corrosion (doc-37372)

🧩 **Interactivity** Electroplating simulation (int-1258)

6.4 Activities

learn on

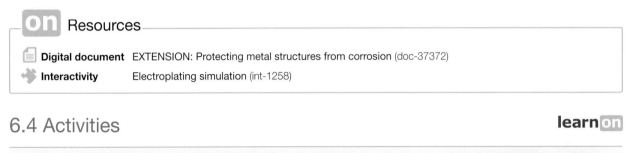

Students, these questions are even better in jacPLUS

Receive immediate feedback and access sample responses

Access additional questions

Track your results and progress

Find all this and MORE in jacPLUS

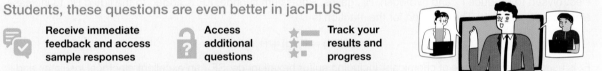

6.4 Quick quiz on	6.4 Exercise	6.4 Exam questions

6.4 Exercise

1. Explain why the addition of calcium chloride to the Downs cell does not interfere with the production of sodium.
2. List all the products produced from the commercial electrolysis of a brine solution using a membrane cell.
3. Why is it important that the membrane used in a membrane cell be impervious to OH^- ions?
4. a. What is the main advantage of the electrolytic production of aluminium from alumina, Al_2O_3, dissolved in cryolite rather than from straight molten alumina?
   b. Why can't a solution of alumina dissolved in water at normal temperatures be used instead?
5. Membrane cells operate using a concentrated solution of sodium chloride. Explain what would happen if this solution was allowed to become diluted.
6. The addition of cryolite, Na_3AlF_6, in the Hall–Héroult process introduces $Na^+(l)$ and $F^-(l)$ into the mixture. Why is there no issue with contamination?
7. In the production of aluminium in a Hall–Héroult cell, the cathode and anode are made of carbon. The anode needs to be replaced every three weeks, whereas the cathode can last up to five years. Explain this difference using an appropriate equation.
8. Using a fully labelled diagram, explain how you would plate a piece of lead with nickel by electroplating.
9. In electroplating, why is the object being electroplated made the cathode?
10. In silver-plating, what substance would you choose for:
    a. the anode
    b. the electrolyte?

6.4 Exam questions

Question 1 (1 mark)

Source: VCE 2021 Chemistry Exam, Section A, Q.21; © VCAA

MC An electrolysis cell with a 5 V power supply is shown.

Using the electrochemical series, which one of the following changes to the electrolysis cell may reduce the amount of Ni electroplated onto the Pt electrode?

A. replacing the Ni electrode with a Cu electrode
B. replacing $Ni(NO_3)_2$(l) with 1 M $Ni(NO_3)_2$(aq)
C. replacing the Pt electrode with Pb(s)
D. replacing $Ni(NO_3)_2$(l) with $NiCl_2$(l)

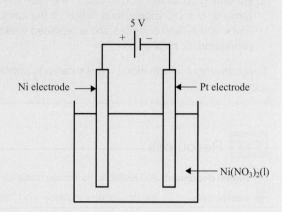

Question 2 (1 mark)

Source: VCE 2017 Chemistry Exam, Section B, Q.8.c; © VCAA

Fluorine, F_2, gas is the most reactive of all non-metals. Anhydrous liquid hydrogen fluoride, HF, can be electrolysed to produce F_2 and hydrogen, H_2, gases. Potassium fluoride, KF, is added to the liquid HF to increase electrical conductivity. The equation for the reaction is

$$2HF(l) \rightarrow F_2(g) + H_2(g)$$

F_2 is used to make a range of chemicals, including sulfur hexafluoride, SF_6, an excellent electrical insulator, and xenon difluoride, XeF_2, a strong fluorinating agent.

The diagram below shows an electrolytic cell used to prepare F_2 gas.

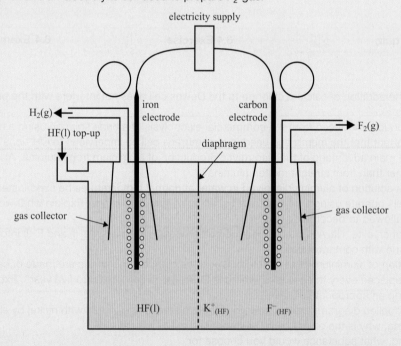

Liquid HF, like water, is an excellent solvent for ionic compounds. In the same way that water molecules in an aqueous solution form the ions K^+(aq) and F^-(aq), when KF is dissolved in HF, the K^+ and F^- ions form ions that are written as $K^+_{(HF)}$ and $F^-_{(HF)}$.

Suggest why the diaphragm, shown in the diagram above, is important for the safe operation of the electrolytic cell.

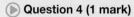 **Question 3 (1 mark)**

Source: VCE 2019 Chemistry Exam, Section A, Q.17; © VCAA

MC The tradition of bronzing baby shoes dates back for generations. Before electroplating, the shoe is painted with a conductive material. The copper, Cu, electrode and copper sulfate, $CuSO_4$, solution cell used for electroplating a shoe is shown.

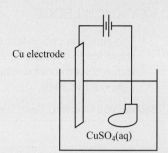

During the electroplating process
A. the copper electrode is oxidised and its mass is unchanged.
B. the shoe is coated with copper metal at the cathode.
C. the copper electrode is the oxidising agent.
D. oxygen is produced at the cathode.

Question 4 (1 mark)

In the Downs cell for the production of sodium, a molten electrolyte containing sodium chloride is used even though this requires more energy to heat the electrolyte to its molten state than simply dissolving it in water. Explain why this is necessary.

Question 5 (8 marks)

Source: VCE 2014 Chemistry Exam, Section B, Q.9; © VCAA

Magnesium is one of the most abundant elements on Earth. It is used extensively in the production of magnesium–aluminium alloys. It is produced by the electrolysis of molten magnesium chloride.

A schematic diagram of the electrolytic cell is shown below.

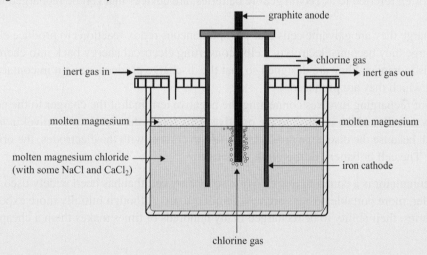

The design of this cell takes into account the following properties of both magnesium metal and magnesium chloride:
- Molten magnesium reacts vigorously with oxygen.
- At the temperature of molten magnesium chloride, magnesium is a liquid.
- Molten magnesium has a lower density than molten magnesium chloride and forms a separate layer on the surface.

a. Write a balanced half-equation for the reaction occurring at each of
   i. the cathode **(1 mark)**
   ii. the anode. **(1 mark)**
b. Explain why an inert gas is constantly blown through the cathode compartment. **(1 mark)**
c. The melting point of a compound can often be lowered by the addition of small amounts of other compounds. In an industrial process, this will save energy. In this cell, NaCl and $CaCl_2$ are used to lower the melting point of $MgCl_2$.
   Why can NaCl and $CaCl_2$ be used to lower the melting point of $MgCl_2$ but $ZnCl_2$ cannot be used? **(2 marks)**
d. What difference would it make to the half-cell reactions if the graphite anode were replaced with an iron anode? Write the half-equation for any different half-cell reaction. Justify your answer. **(3 marks)**

More exam questions are available in your learnON title.

6.5 Rechargeable batteries (secondary cells)

KEY KNOWLEDGE

- The common design features and general operating principles of rechargeable (secondary) cells, with reference to discharging as a galvanic cell and recharging as an electrolytic cell, including the conditions required for the cell reactions to be reversed and the electrode polarities in each mode (details of specific cells not required)

Source: VCE Chemistry Study Design (2024–2027) extracts © VCAA; reproduced by permission.

TIP: Some examples of secondary cells are provided in this subtopic. Although you are not required to know the details of any specific cell, you should examine these examples and make sure that you understand the principles behind their operation. It is expected that you should be able to apply these principles in an examination, rather than just restating facts that you have learnt.

6.5.1 What is a secondary cell?

A secondary cell is essentially a galvanic cell combined with an electrolytic cell.

Secondary cells, often referred to as **rechargeable** batteries, are devices that can be recharged when they become 'flat'.
- During discharge they are galvanic cells that use spontaneous redox reactions to produce electricity.
- During recharge they become electrolytic cells, converting electrical energy back into chemical energy.
- To enable this to happen, they are designed so that the discharge products remain in contact with the electrodes at which they are produced.
- The process of recharging involves connecting the negative terminal of the charger to the negative terminal of the battery or cell, and the positive to the positive. This forces the electrons to travel in the reverse direction and, because the discharge products are still in contact with the electrodes, the original reactions are reversed. The cell or battery is, therefore, recharged.

The lead–acid accumulator is a common example of a secondary cell that has been widely used for many years. Other smaller, more portable designs are now familiar to us. Although initially more expensive than non-rechargeable batteries, their ability to be recharged many hundreds of times makes them a cheaper alternative in the long term.

6.5.2 Lead–acid accumulator

Developed in the late nineteenth century, the **lead–acid accumulator** has remained the most common and durable of battery technologies. Lead–acid accumulators are secondary cells. They have a relatively long life and high current, and are cheap to produce. Largely used in transport applications, they rely on a direct current generator or alternator in the vehicle to apply enough voltage to reverse the spontaneous electrochemical reaction that provides electricity for the car.

A 12-volt lead–acid storage battery consists of six 2-volt cells connected in series. The cells do not need to be in separate compartments, although this improves performance (see figure 6.19).

secondary cell a cell that can be recharged once its production of electric current drops; often called a rechargeable battery

rechargeable describes a battery that is an energy storage device; it can be charged again after being discharged by applying DC current to its terminals

lead–acid accumulator a battery with lead electrodes using dilute sulfuric acid as the electrolyte; each cell generates about 2 volts

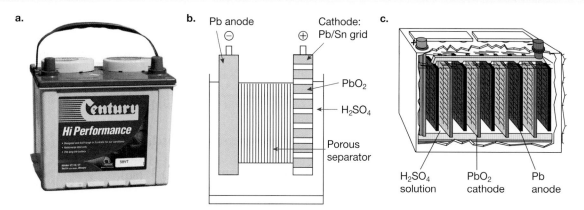

Each cell consists of two lead electrodes or grids. The grid structure provides a larger surface area for electrode reactions. The grid that forms the anode (the negative terminal) of the cell is packed with spongy lead. The grid that forms the cathode (the positive terminal) is packed with lead(IV) oxide, PbO_2. The electrodes are both immersed in approximately 4 M sulfuric acid and are separated by a porous plate.

The discharging process in a lead–acid accumulator

When a lead–acid accumulator discharges, it produces electric power to start the car. Discharge results from a spontaneous redox reaction.

The half-equations at each electrode are as follows:

$$\text{Anode (oxidation): } Pb(s) + SO_4{}^{2-}(aq) \rightarrow PbSO_4(s) + 2e^-$$
$$\text{Cathode (reduction): } PbO_2(s) + 4H^+(aq) + SO_4{}^{2-}(aq) + 2e^- \rightarrow PbSO_4(s) + 2H_2O(l)$$

At the anode, lead is oxidised to Pb^{2+} ions. These react immediately with the sulfuric acid solution to produce insoluble lead(II) sulfate, which deposits on the grid.

At the cathode, lead(IV) oxide is reduced to Pb^{2+} ions, which again react with the sulfuric acid to form a lead(II) sulfate deposit on the grid. The overall equation for the discharging reaction is as follows:

$$PbO_2(s) + 4H^+(aq) + Pb(s) + 2SO_4{}^{2-}(aq) \rightarrow 2PbSO_4(s) + 2H_2O(l)$$

Note that the pH of the cell increases during the discharge cycle.

The recharging process in a lead–acid accumulator

The products of the discharge process remain as a deposit on the electrodes. This means that the reactions at these electrodes can be reversed by passing a current through the cell in the opposite direction. The battery is then said to be **recharging**. When the battery is recharged, the electrode reactions are reversed by connecting the terminals to another electrical source of higher voltage and reversing the direction of the electric current through the circuit. Recharging occurs while a car is in motion.

recharging forcing electrons to travel in the reverse direction; because the discharge products are still in contact with the electrodes, the original reactions are reversed

While recharging, the flow of electrons is reversed and the electrode forming the negatively charged anode in the discharging process becomes the negatively charged cathode, where reduction occurs:

$$\text{Cathode (reduction): } PbSO_4(s) + 2e^- \rightarrow Pb(s) + SO_4^{2-}(aq)$$

The electrode previously forming the positively charged cathode in the discharging process now becomes the positively charged anode, where oxidation occurs, in the recharging process:

$$\text{Anode (oxidation): } PbSO_4(s) + 2H_2O(l) \rightarrow PbO_2(s) + SO_4^{2-}(aq) + 4H^+(aq) + 2e^-$$

The overall reaction for the recharging process is therefore the reverse of the discharging process:

$$2PbSO_4(s) + 2H_2O(l) \rightarrow PbO_2(s) + 4H^+(aq) + Pb(s) + 2SO_4^{2-}(aq)$$

Note that the pH of the cell decreases during the recharge cycle.

This reaction is not spontaneous, so a direct current must be applied in order for it to proceed. This is achieved by the alternator (a motor-driven electrical source of higher voltage than the battery), which has a potential difference of 14 V. The recharging process converts electrical energy into chemical energy and is an example of an electrolytic reaction.

FIGURE 6.20 Discharge/recharge effect on electrodes

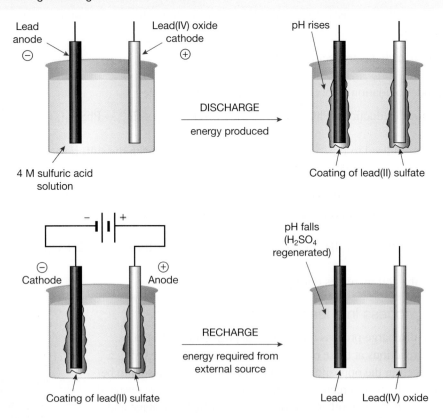

In theory, a lead storage battery can be recharged indefinitely, but in practice, it may only last for about four years. This is because small amounts of lead(II) sulfate continually fall from the electrodes and drop to the bottom of the cell. Eventually, the electrodes lose so much lead(II) sulfate that the recharging process is ineffective.

6.5.3 Nickel–metal hydride (NiMH) rechargeable cell

Although the nickel–cadmium (NiCd) cell was the first rechargeable cell to find widespread use in many common household devices, it was eventually replaced by the nickel–metal hydride (NiMH) cell. This shares a number of features with the NiCd cell but is environmentally safer due to the absence of cadmium.

An additional problem with NiCd cells was the so-called memory effect. If the cell was only partially discharged before recharging occurred, it would not receive a full charge. NiMH cells show much less of this effect. They also have nearly 50 per cent more charge per gram, can recharge faster and can run longer on each charge.

The reactions involved during discharge of an NiMH cell to produce an electric current are as follows.

Oxidation takes place at the negative electrode (anode):

$$\text{Anode: } M\text{H(s)} + \text{OH}^-\text{(aq)} \rightarrow M\text{(s)} + \text{H}_2\text{O(l)} + \text{e}^- \text{ (note the } M \text{ here refers to a metal)}$$

Reduction takes place at the positive electrode (cathode):

$$\text{Cathode: NiO(OH)(s)} + \text{H}_2\text{O(l)} + \text{e}^- \rightarrow \text{Ni(OH)}_2\text{(s)} + \text{OH}^-\text{(aq)}$$

The overall equation for the discharging reaction is as follows:

$$\text{NiO(OH)(s)} + M\text{H(s)} \rightarrow \text{Ni(OH)}_2\text{(s)} + M\text{(s)}$$

The metals used in NiMH batteries are often alloys of lanthanum and rare-earth elements. The electrolyte is potassium hydroxide and the voltage produced is 1.2 V.

Nickel–metal hydride batteries have many advantages but also some disadvantages. They suffer from self-discharge — a problem that is worse at higher temperatures — and require more complicated charging devices to prevent over-charging.

They were commonly used in laptops, electric shavers and toothbrushes, cameras, camcorders, mobile phones and medical instruments. Today, an even newer type of rechargeable battery, the lithium-ion battery, is rapidly gaining popularity.

FIGURE 6.21 A nickel–metal hydride cell for a digital camera

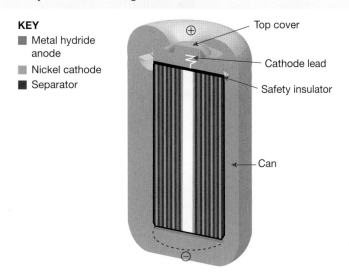

KEY
- ■ Metal hydride anode
- ■ Nickel cathode
- ■ Separator

Top cover

Cathode lead

Safety insulator

Can

6.5.4 Lithium-ion batteries

A new type of cell popular in portable devices and electric vehicles is the lithium-ion battery. These batteries have a good shelf life and a very high energy density; they supply a voltage of about 3.7 V. However, due to technical considerations, they cannot be over-discharged or over-charged. This means that they need to be equipped with a protection circuit that prevents these situations from occurring.

Features of the lithium-ion cell include the following:
- The negative electrode is graphite impregnated with lithium.
- The positive electrode is cobalt(IV) oxide that has been doped with lithium ions. Cobalt (and sometimes other transition metal oxides) are chosen due to their multiple possible oxidation states.
- Various non-aqueous electrolytes are used and a solid separator that is permeable to lithium ions is inserted between the electrodes to prevent them from coming into contact with each other.

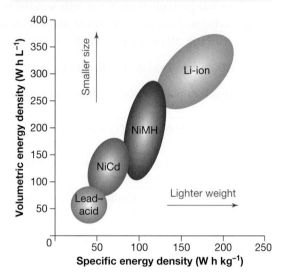

FIGURE 6.22 Lithium-ion batteries offer a much better energy density than other battery technologies.

FIGURE 6.23 A simplified diagram of a lithium-ion battery during discharge. Recharge forces the lithium ions to move in the opposite direction.

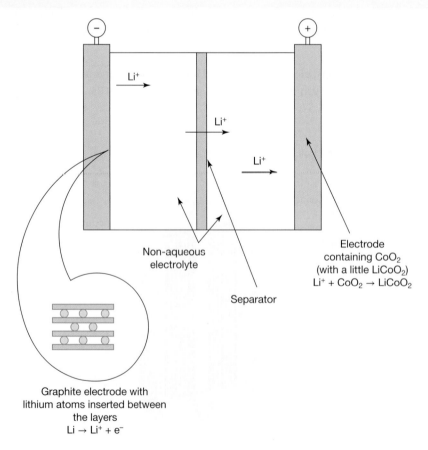

Non-aqueous electrolyte

Separator

Electrode containing CoO_2 (with a little $LiCoO_2$) $Li^+ + CoO_2 \rightarrow LiCoO_2$

Graphite electrode with lithium atoms inserted between the layers $Li \rightarrow Li^+ + e^-$

The reactions that occur in this cell are complex. The following are simplified versions of the reactions that occur during discharge.

At the anode (oxidation), lithium from the graphite reacts according to the following equation:

$$\text{Anode (oxidation): } Li \rightarrow Li^+ + e^-$$

The lithium ions produced during discharge move through the non-aqueous electrolyte to the cathode. At the cathode (reduction), lithium ions react with cobalt(IV) oxide after having arrived from the anode:

$$\text{Cathode (reduction): } CoO_2 + Li^+ + e^- \rightarrow LiCoO_2$$

Note that in this reaction, cobalt is reduced from a +4 oxidation state to a +3 oxidation state.

$$\text{Overall cell reaction: } Li + CoO_2 \rightarrow LiCoO_2$$

You will note that the product of this reaction ($LiCoO_2$) remains in contact with its electrode. It can easily be made to react in the opposite direction by reversing the direction of electron flow with a charger. The reactions that take place during charging are the reverse of those shown previously.

FIGURE 6.24 Electric vehicles are becoming more and more popular on our roads. The batteries in these vehicles utilise Li-ion technology.

6.5.5 A closer look at the discharging and recharging processes

All the examples shown so far are secondary cells, meaning they can be recharged. It is worthwhile to compare the nature of the opposing processes of discharge and recharge. Remember that what makes a battery rechargeable is that the discharge reactions can be reversed. Several conditions are required for this to happen:
- A device (charger) must be used to force the electrons to flow in the opposite direction to discharge.
- This device must operate at a slightly higher voltage than the normal operating voltage of the cell.
- The products of the discharge must be able to be reversed by the charging flow of electrons.
- The products of discharge must be 'available' for recharging. They must not be lost through unwanted side reactions or solid coatings (e.g. $PbSO_4$) physically falling off electrode surfaces.

Rechargeable batteries illustrate both the similarities and differences that exist between galvanic cells (discharging) and electrolytic cells (recharging). These are summarised in table 6.2.

TABLE 6.2 Comparison of galvanic and electrolytic cells

Feature	Galvanic cell	Electrolytic cell
Type of redox reaction	Spontaneous	Non-spontaneous
Energy produced or required	Produced	Required
Where oxidation occurs	Anode	Anode
Where reduction occurs	Cathode	Cathode
Anode polarity	Negative	Positive
Cathode polarity	Positive	Negative
How cell polarity is determined	Depends on reactions occurring within the cell	External power source

tlvd-3063

SAMPLE PROBLEM 5 Writing discharge reactions for rechargeable batteries

Flow batteries were invented in Australia in the 1950s. They are currently being investigated as a large-scale means of storing energy from renewable resources, such as wind and solar.

The diagram shows a type of rechargeable battery that utilises the different oxidation states of vanadium. It is called a flow battery because the electrode reactants are stored in tanks and pumped over the surface of their respective electrodes so that reaction can occur.

The relevant standard electrode potentials for the reactions that take place are as follows:

$$VO_2^+(aq) + 2H^+(aq) + e^- \rightleftharpoons VO_2^+(aq) + H_2O(l) \quad E^0 = +1.00 \text{ V}$$
$$V^{3+}(aq) + e^- \rightleftharpoons V^{2+}(aq) \quad\quad\quad\quad\quad\quad E^0 = -0.26 \text{ V}$$

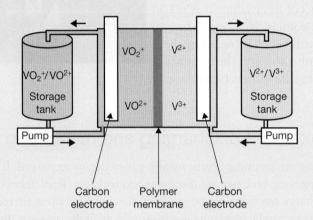

Carbon electrode   Polymer membrane   Carbon electrode

Write the equation for the reaction that occurs at the cathode during discharge.

THINK	WRITE
E^0 values indicate that VO_2^+ reacts with V^{2+} in the spontaneous discharge reaction. Recall that reduction occurs at the cathode, and that it is the VO_2^+ that is reduced. Write the equation, making sure to use a single arrow.	$VO_2^+(aq) + 2H^+(aq) + e^- \rightarrow VO^{2+}(aq) + H_2O(l)$

PRACTICE PROBLEM 5

Write the equation for the reaction that occurs at the anode during the recharging of the redox flow battery shown in sample problem 5. (Remember that during recharge the reactions are non-spontaneous.)

6.5 Activities

6.5 Quick quiz **on**	6.5 Exercise	6.5 Exam questions

6.5 Exercise

1. During the recharging process for a lead–acid accumulator, will the pH of the contents rise or fall? Explain.
2. Explain why the positive terminal of the charging device must be connected to the positive terminal of the battery, and the negative to the negative, when a rechargeable battery is to be recharged.
3. The nickel–cadmium rechargeable cell was a widely used predecessor to today's nickel–metal hydride cells and lithium-ion cells. The following figure shows some of the components of this type of cell.

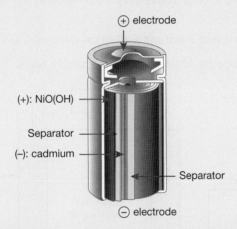

⊕ electrode

(+): NiO(OH)

Separator

(–): cadmium

Separator

⊖ electrode

The cell contains cadmium and NiO(OH) as its reactants. These are kept apart by a porous separator that has been soaked in KOH. During *discharge*, the reaction at the anode is as follows:

$$Cd(s) + 2OH^-(aq) \rightarrow Cd(OH)_2(s) + 2e^-$$

The overall cell reaction is as follows:

$$Cd(s) + 2NiO(OH)(s) + 2H_2O(l) \rightarrow Cd(OH)_2(s) + 2Ni(OH)_2(s)$$

   a. Write the equation that occurs at the cathode during recharging.
   b. During recharging, what terminal of the recharging device should the discharge anode be connected to?
   c. Write the equation for the overall cell reaction during the recharging process.
   d. Write the equation for the reaction that occurs at the anode during recharging.
4. Batteries based on vanadium chemistry are increasingly being used to store energy from solar panels. During the day, solar cells store energy in the battery as it is charged. At night, the battery functions as a galvanic cell, producing electricity to power the household. The two relevant half-equations for the functioning of this battery are as follows:

$$VO_2^+(aq) + 2H^+(aq) + e^- \rightarrow VO^{2+}(aq) + H_2O(l) \quad E^0 = +1.00 \text{ V}$$
$$V^{3+}(aq) + e^- \rightarrow V^{2+}(aq) \quad\quad\quad\quad\quad\quad E^0 = -0.26 \text{ V}$$

   a. Write the overall equation for this battery as it is discharging.
   b. Write the overall equation for this battery as it is recharging.
   c. Write the half-equations for the reactions occurring at each electrode as the cell recharges.

5. When a lead–acid accumulator is charged, and especially if the voltage is too high, some of the water in the electrolyte solution undergoes an electrolytic reaction to form hydrogen and oxygen gas.
   a. Determine the half-cell reaction at each electrode, then write the overall equation for this reaction.
   b. How would the operation of the cell be affected by this reaction?
   c. What safety precautions should be taken during the recharging of a lead–acid accumulator?

6.5 Exam questions

▶ Question 1 (1 mark)

Source: VCE 2022 Chemistry Exam, Section A, Q.14; © VCAA

MC The discharge reaction in a vanadium redox battery is represented by the following equation.

$$VO_2^+(aq) + 2H^+(aq) + V^{2+}(aq) \rightarrow V^{3+}(aq) + VO^{2+}(aq) + H_2O(l)$$

When the vanadium redox battery is recharging

A. H^+ is the reducing agent.
B. H_2O is the oxidising agent.
C. VO^{2+} is the reducing agent.
D. VO_2^+ is the oxidising agent.

▶ Question 2 (4 marks)

Source: VCE 2021 Chemistry Exam, Section B, Q.2.a,b; © VCAA

Research scientists are developing a rechargeable magnesium–sodium, Mg–Na, hybrid cell for use in portable devices.

The Mg–Na hybrid cell uses magnesium metal and sodium ion electrodes and a hybrid organic/salt electrolyte, X.

A simplified diagram of the rechargeable Mg–Na hybrid cell is shown below.

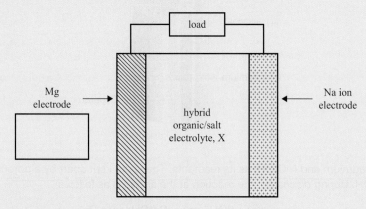

a. The equation for the overall reaction during **recharge** is

$$2NaX + Mg^{2+} \rightarrow Mg + 2Na^+ + 2X$$

   i. Identify the polarity of the Mg electrode when the cell is discharging by placing a positive (+) or a negative (–) sign in the box provided in the diagram above. **(1 mark)**
   ii. Write the half-cell equation of the reaction that occurs at the Mg electrode when the cell is **discharging**. **(1 mark)**
b. A pacemaker is a small electronic device that is implanted in the body to regulate a person's heart rate.

   If the Mg–Na hybrid cell were to be used to power pacemakers, what would be **two** potential safety hazards of having this cell in the body? **(2 marks)**

▶ Question 3 (1 mark)

Source: VCE 2020 Chemistry Exam, Section A, Q.8; © VCAA

MC Which one of the following is the most correct statement about fuel cells and secondary cells?
A. Fuel cells can be recharged like secondary cells.
B. Fuel cells produce thermal energy, whereas secondary cells do not produce thermal energy.
C. The anode in a fuel cell is positive, whereas the anode in a secondary cell is negative.
D. Fuel cells deliver a constant voltage during their operation, whereas secondary cells reduce in voltage as they discharge.

▶ **Question 4 (4 marks)**

Source: VCE 2020 Chemistry Exam, Section B, Q.4.a,b,d; © VCAA

Research scientists are developing a rechargeable lithium–carbon dioxide, $Li–CO_2$, battery. The rechargeable $Li–CO_2$ battery is made of lithium metal, carbon in the form of graphite (coated with a catalyst) and a non-aqueous electrolyte that absorbs CO_2.

A diagram of the rechargeable $Li–CO_2$ cell is shown below. One $Li–CO_2$ cell generates 4.5 V.

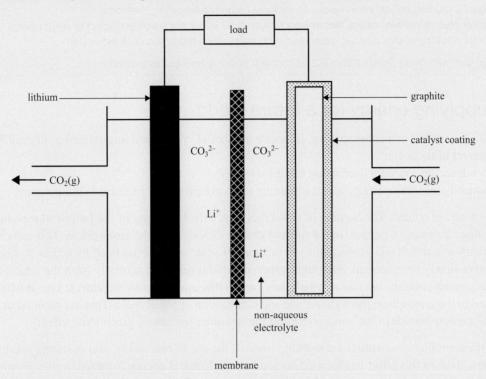

a. When the $Li–CO_2$ cell generates electricity, the two half-cell reactions are

$$4Li^+ + 3CO_2 + 4e^- \rightarrow 2Li_2CO_3 + C$$
$$Li \rightarrow Li^+ + e^-$$

   Write the equation for the overall recharge reaction. **(1 mark)**
b. During discharge, lithium carbonate, Li_2CO_3, deposits break away from the electrode. Describe how this might affect the performance of the battery. **(2 marks)**
c. Could the $Li–CO_2$ battery be used to reduce the amount of $CO_2(g)$ in the atmosphere? Give your reasoning. **(1 mark)**

▶ **Question 5 (1 mark)**

Source: VCE 2018 Chemistry Exam, Section A, Q.16; © VCAA

MC The silver oxide–zinc battery is rechargeable and utilises sodium hydroxide, NaOH, solution as the electrolyte. The battery is used as a backup in spacecraft, if the primary energy supply fails.

The overall reaction during discharge is

$$Zn + Ag_2O \rightarrow ZnO + 2Ag$$

When the silver oxide–zinc battery is being **recharged**, the reaction at the anode is

A. $2Ag + 2OH^- \rightarrow Ag_2O + H_2O + 2e^-$
B. $Ag_2O + H_2O + 2e^- \rightarrow 2Ag + 2OH^-$
C. $ZnO + H_2O + 2e^- \rightarrow Zn + 2OH^-$
D. $Zn + 2OH^- \rightarrow ZnO + H_2O + 2e^-$

More exam questions are available in your learnON title.

6.6 Contemporary responses to meeting society's energy needs

KEY KNOWLEDGE

- The role of innovation in designing cells to meet society's energy needs in terms of producing 'green' hydrogen (including equations in acidic conditions) using the following methods:
  - polymer electrolyte membrane electrolysis powered by either photovoltaic (solar) or wind energy
  - artificial photosynthesis using a water oxidation and proton reduction catalyst system

Source: VCE Chemistry Study Design (2024–2027) extracts © VCAA; reproduced by permission.

6.6.1 Supplying energy for a future world

As the world's population continues to grow, it faces a number of critical and interrelated problems. Two of the most fundamental of these are:
- the dependence on a rapidly decreasing fossil fuel supply
- the enhanced greenhouse effect, which is leading to an overall global increase in temperature.

The two are of course related. The burning of fossil fuels since the beginning of the Industrial Revolution has been adding huge amounts of carbon (in the form of carbon dioxide) into the atmosphere. This carbon has come from the remains of animals and plants that lived millions of years ago and has been locked away since then as fossil fuels have slowly been formed. Although carbon dioxide is removed naturally from the atmosphere by plants through photosynthesis, the rate at which they can do this cannot match the current rate at which this gas is being added to the atmosphere through our use and combustion of fossil fuels. This has resulted in a net gain to the level of carbon dioxide in the atmosphere and the so-called 'enhanced greenhouse effect'.

To alleviate these problems, scientists are working towards the use of renewable, non-polluting energy sources. In recent years, most of this effort has focused on generating electrical energy from renewable resources, such as the Sun (photovoltaic cells) and wind (wind turbines). A number of other location-specific energy sources, such as wave energy, tidal energy, geothermal energy and kinetic energy from running water, have also been utilised.

The two most common of these energy sources, solar and wind, suffer from one very obvious drawback. Electricity cannot be made if the Sun does not shine or the wind does not blow! The existence of a stable electricity grid is critical to many industries that use electricity for the production of goods, services and materials.

In addition, society needs some of its energy in forms other than electricity — most notably as fuels. Rapidly emerging as a possibility in this area is hydrogen. Progress in the areas of both fuel cell and electrolytic cell technology has meant that it is becoming possible to produce hydrogen in increasing amounts from renewable electricity, and to then transport it to where it is required, to be used as a non-polluting fuel. It is hoped that hydrogen might also be used to stabilise electrical grids powered by solar and wind technologies. During the day and in times of high winds, if energy demand is low, the excess energy could be used to generate hydrogen. Then, at night or in times of low wind, and if electrical demand is high, fuel cells could be used to convert this hydrogen back into electrical energy.

FIGURE 6.25 Scientists are working towards the use of renewable, non-polluting energy sources to meet society's energy needs.

This possibility is helped by the fact that there is a degree of common componentry between some fuel cells and some electrolytic cells. Innovative research is continuing into designing cells that can operate in both modes, depending on the energy needs at a particular point in time.

FIGURE 6.26 Cross-section of **a.** a fuel cell and **b.** an electrolysis cell, with similar components

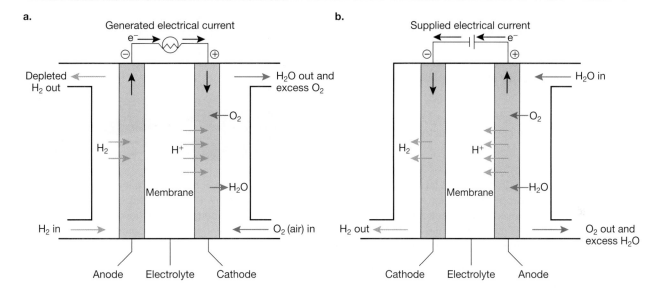

As was seen in topic 5, another advantage of hydrogen is that it can be used to make ammonia, which can also be used as a non-polluting fuel.

Green hydrogen

For hydrogen to be adopted as a clean, non-polluting future fuel, it needs to be manufactured in a clean and non-polluting fashion. This is now becoming increasingly possible through the use of electrolytic cells powered by renewable energy, as well as other technologies that produce hydrogen directly from sunlight. Hydrogen produced this way is termed **green hydrogen**.

Hydrogen, however, has not always been 'green'. Traditionally, it has been produced from the steam reforming of fossil fuels, often methane. This is a two-step process represented by the following equations:

$$CH_4(g) + H_2O(g) \rightleftharpoons CO(g) + 3H_2(g) \quad \Delta H = +206 \, kJ \, mol^{-1}$$
$$CO(g) + H_2O(g) \rightleftharpoons CO_2(g) + H_2(g) \quad \Delta H = -41 \, kJ \, mol^{-1}$$

As can be seen, this method also generates undesirable carbon dioxide as a waste product. It has been estimated that, allowing for inefficiencies in the process, every 1 kg of hydrogen produced this way generates 9.3 kg of carbon dioxide. Hydrogen produced in this manner has come to be known as *grey* hydrogen.

A modification to this process is to capture the carbon dioxide and to store it in a form that prevents its release into the atmosphere. Hydrogen produced in this fashion is known as *blue* hydrogen. Although advantageous in one way, blue hydrogen still has the problem that its generation is dependent on fossil fuels. Additionally, methane is a very potent greenhouse house, and so-called 'fugitive' leaks will always occur in obtaining and transporting it.

green hydrogen hydrogen that does not contribute to the enhanced greenhouse effect in its production

CASE STUDY: The colours of hydrogen

Hydrogen, of course, is not coloured! It is a colourless, odourless and tasteless gas that was discovered by Henry Cavendish in 1766. It is a flammable gas that is lighter (or less dense) than air, and is the simplest element. The so-called 'colours' refer to the way in which it is obtained.

FIGURE 6.27 The 'colours' of hydrogen refer to the way it is obtained.

Over the past few years, it has become commonplace to refer to these methods through the use of colour, with 'brighter' colours referring to methods that are more environmentally friendly and 'duller' colours referring to those that are not.

The following are the generally accepted colours used, but some may vary from country to country.

Green hydrogen

Green hydrogen is hydrogen that is produced with no net greenhouse gas emissions. As such, it is the ultimate goal for hydrogen production in a world based on a hydrogen economy. The best prospective technologies to achieve this are electrolysis of water coupled with renewable energy (see section 6.6.2), and with photoelectrolysis (see section 6.6.3). However, at the present point in time, only a small percentage of the world's total production of hydrogen is achieved in this manner.

The term 'green hydrogen' is also sometimes used to describe hydrogen produced from biogas (mainly methane) via steam reforming that uses renewable sources to supply the energy required.

Blue hydrogen

The steam reforming of methane referred to in the previous section is an established technology. To be classified as blue hydrogen, the carbon dioxide produced needs to be captured and prevented from entering the atmosphere. This is called *sequestering*. Steam reforming is an energy-intensive process and, to qualify as blue hydrogen (with no greenhouse gas emissions), such plants would need to be powered by renewable energy.

As mentioned earlier, the main problems associated with blue hydrogen are its dependence on methane as a fossil fuel and the unavoidable release of fugitive leaks containing methane into the atmosphere.

Grey hydrogen

Grey hydrogen is also produced by the steam reforming of natural gas, but during its production the carbon dioxide generated is released into the atmosphere. This method therefore contributes directly to the enhanced greenhouse effect. The majority of today's hydrogen is currently produced by this method.

Brown hydrogen

The gasification of coal is an established technology. Along with hydrogen, carbon monoxide and carbon dioxide are produced. Other aspects of this production method make it a very polluting process. Brown hydrogen is made using brown coal.

Black hydrogen

Black hydrogen is the same as brown hydrogen, except that black coal is used.

Turquoise hydrogen

Turquoise hydrogen is produced by the *pyrolysis* of natural gas, for which the reactors involved are powered by renewable energy. Pyrolysis refers to the use of heat to thermally decompose the methane in natural gas according to the following equation:

$$CH_4(g) \rightarrow 2H_2(g) + C(s)$$

It is important that this process is carried out in the absence of oxygen so that solid carbon, rather than gaseous carbon dioxide, is produced.

Once again, although there are no direct greenhouse emissions from this process, there are emissions associated with the mining and transport of the natural gas.

Pink/purple/red hydrogen

Pink, purple or red hydrogen refers to hydrogen that is produced via a number of methods (including electrolysis) using nuclear energy.

White hydrogen

White hydrogen refers to hydrogen that occurs naturally (miniscule) or hydrogen that is obtained as a by-product of some other process.

Transitioning to a hydrogen-based society

While the production and use of green hydrogen will be the ultimate goal, there are hurdles that must be overcome before hydrogen can be supplied in the amounts that society will require. The other significant issues to be faced are:
- developing and building the necessary infrastructure
- developing and enhancing the technology required to extract the energy from the hydrogen in the most efficient manner.

It is anticipated that some of the other colours of hydrogen will play an important role in the transition process from a fossil-fuelled society to a hydrogen-based one. For example, the necessary infrastructure could be built and fed by blue hydrogen. Then, as green hydrogen becomes cheaper and more abundant, this could be added until it eventually prevails as the main source of hydrogen.

6.6.2 Producing green hydrogen by electrolysis

The electrolysis of water using renewable electricity to produce green hydrogen offers a method for the future that is consistent with the principles of green chemistry. There are several different designs for electrolytic cells that can do this. These include alkaline electrolysis cells (AECs), polymer electrolyte membrane electrolysis cells (PEMECs) and solid oxide electrolysis cells (SOECs).

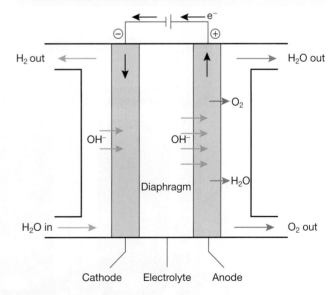

FIGURE 6.28 Cross-section of an AEC

Alkaline electrolysis cells

Alkaline electrolysis cells (AECs) are large cells used to electrolytically split water into hydrogen and oxygen. Many large electrolysers exist that produce hydrogen to meet the needs of specific users. This existing technology can produce hydrogen at the rate of 60 kg h^{-1}, but there are a number of drawbacks to their use if they are to meet the demands for green hydrogen in the future. These include:

- corrosion due to the corrosive nature of the potassium hydroxide electrolyte
- the requirement for a consistent and stable supply of electricity to function properly. This makes them less suitable to being powered by a renewable energy supply than some other designs.
- a degree of crossover of the produced gases, hydrogen and oxygen, resulting in lower efficiency and lower purity in the hydrogen produced.

Polymer electrolyte membrane electrolysis cells

The polymer electrolyte membrane electrolysis cell (PEMEC) is another design that has existed for some time but is now the subject of renewed interest. This is due to its suitability for use with renewable energy sources, such as solar- and wind-powered electricity. It also has a number of other advantages that could make it an ideal method to generate the future hydrogen requirements of a hydrogen-based society. These include:

- *adaptability of scale.* The design is flexible in that it can be used for small-scale, on-site generation, or it can be modularised and scaled up to suit larger applications. PEMECs are currently in operation in Australia, Canada and Germany. A larger plant is being built in Spain and an even larger one is planned for China.
- *the production of high-purity hydrogen.* This is important for three reasons. The first is that high-purity hydrogen is required for use in the fuel cells that are anticipated to be found in transport applications and other electrical generation applications in the future. The second is safety. It is important to prevent hydrogen mixing with the other gas produced during electrolysis: oxygen. Such mixtures could become explosive. Many cell designs suffer from this crossover effect to some degree or other, but PEMECs are particularly good at preventing it. The third reason is for production of chemicals that require hydrogen as a raw material. Foremost among these is ammonia, which has tremendous potential for use as a transport fuel, but is also important in the production of fertilisers.
- *sharing common componentry with polymer electrolyte membrane fuel cells (PEMFCs).* Advances and research into one type of cell are often transferable to the other type. An innovative approach currently under investigation is to design a cell that could operate as either a fuel cell or an electrolytic cell, depending upon the requirements at a particular point in time.
- *an operational temperature of less than 100 °C.* This means less energy is required to power it.
- *the potential for significant increases in efficiency with further research and development.* It is hoped that figures of 85 per cent might be achieved by 2030.

The main drawback at present for PEMECs is cost. The membrane is very expensive, and expensive noble metal catalysts such as gold, indium and platinum are required as electrode catalysts.

FIGURE 6.29 PEM electrolyser stacks

How PEMECs work

Figure 6.30 shows the essential features of a PEMEC. The most critical components of this design are the electrodes and the membrane.

FIGURE 6.30 The essential features of a PEMEC designed to produce hydrogen

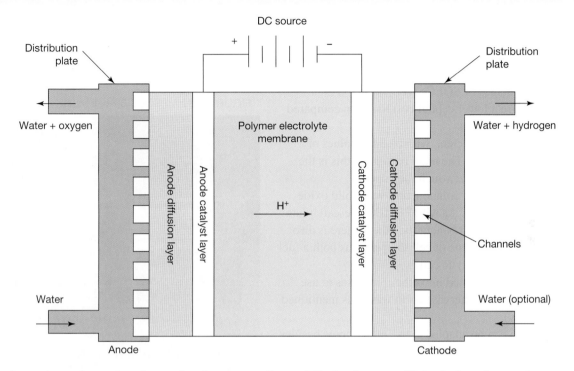

The electrodes each consist of a catalyst layer, as well as a diffusion layer to efficiently introduce and remove reactants and products from the catalyst layer. Typical catalysts include platinum for the cathode and iridium for the anode, although other metals such as ruthenium, indium and gold have been trialled. This is one of the main reasons for the expense of these cells.

The thin membrane between the cathode and the anode is also important for the efficient functioning of this cell. Its most obvious property is that it must conduct hydrogen ions (protons) between the anode and the cathode. It therefore needs to be able to withstand acidic conditions. It also must exhibit low permeability to the gases that are produced in order to prevent remixing and contamination. Currently, these membranes are also expensive, and much current research is focused on finding more cost-effective membranes, as well as more efficient and cheaper catalysts.

As can be seen from figure 6.30, water is added to the anode side of the cell, where it undergoes oxidation to produce oxygen gas. The equation for this reaction is as follows:

$$2H_2O(l) \rightarrow O_2(g) + 4H^+(aq) + 4e^-$$

The unused water is removed along with the oxygen produced.

The hydrogen ions produced from this reaction travel through the membrane until they come into contact with the cathode. They then undergo reduction to form hydrogen gas, according to the following reaction:

$$2H^+(aq) + 2e^- \rightarrow H_2(g)$$

In some designs, water is also added to this side of the cell to assist with the removal of the hydrogen gas.

In practice, cells such as these are combined into stacks to generate larger amounts of hydrogen. As mentioned, these cells produce very pure hydrogen and are suitable for being powered by renewable energy sources.

Solid oxide electrolysis cells

Solid oxide electrolysis cells (SOECs) are another type of electrolysis cell being investigated for future hydrogen production. These make use of external heat to significantly reduce the electrical energy requirements of the cells, and typically operate at temperatures of 700–800 °C. They feature a ceramic oxide electrolyte and are best suited to applications in which a constant supply of electricity is to be provided. It is anticipated that they could find application in large chemical synthesis locations in which the waste heat generated in so-called 'downstream' processes could be used to achieve the required temperatures. Research on SOECs is currently aimed at overcoming their poor lifetimes, due to mechanical issues such as electrode cracking and brittleness in the ceramic solid oxide electrolyte.

The main advantages of SOECs are:
- they require a lower capital cost to set up compared to some alternatives
- they are highly efficient, with claimed values of above 80 per cent. The main reason for this is the high temperatures at which they operate.
- they have common componentry with solid oxide fuel cells (SOFCs). As with PEMECs, research into one type of cell can benefit the other. There is also potential for this type of cell to operate as both a fuel cell and an electrolytic cell.
- they can be integrated into other processes to use heat that would otherwise go to waste (as mentioned previously)
- the design can be adapted to use other inputs besides water, and a range of other products can therefore be produced. For example, carbon dioxide could be used to make carbon monoxide, or a combination of water and carbon dioxide could be used to produce a hydrogen–carbon monoxide mixture. This mixture could then be processed into other fuels and chemicals.

FIGURE 6.31 An SOEC 60-cell stack

Operational SOECs have been demonstrated on a number of different scales and are now beginning to find their way into commercial applications.

How SOECs work

> **TIP:** Knowledge of the exact functioning of this type of cell is not prescribed in the VCE Chemistry Study Design. However, it is expected that you would be able to apply your knowledge of electrolysis and PEMECs to other designs such as this.

The general structure and functioning of this type of cell can be inferred from a knowledge of other types of electrolysis cells, such as the PEMECs discussed previously, and of electrolysis principles in general. These principles are:
- a cathode at which reduction occurs
- an anode at which oxidation occurs
- an electrolyte to conduct ions from one electrode to the other
- the use of electrical energy to produce non-spontaneous reactions.

SOECs feature a solid oxide electrolyte that conducts oxide ions. However, for this to occur efficiently, temperatures in the range of 700–800 °C are currently required. Such cells are therefore suited to situations that are able to supply the necessary thermal energy at a low cost.

These cells are very efficient and can be reversed to act as fuel cells under the right conditions. Although they currently have reliability issues, a big incentive for their further development is that they can operate using water (as steam) to produce hydrogen, or carbon dioxide to produce carbon monoxide.

Using this information, together with the general principles listed earlier, it is possible to deduce the equations for the reactions at each electrode and the polarity involved.

At the cathode, reduction must occur. This requires electrons, so it will need to be connected to the negative terminal of the power supply. Knowing that oxide ions are produced, the equation for the reaction occurring at the anode would be:

$$H_2O(g) + 2e^- \rightarrow H_2(g) + O^{2-}(aq) \quad \text{(if water is used)}$$

OR

$$CO_2(g) + 2e^- \rightarrow CO(g) + O^{2-}(aq) \quad \text{(if carbon dioxide is used)}$$

At the anode, oxidation takes place and the electrons produced from the oxide ions then travel through the external circuit to the positive terminal of the power supply. The equation is therefore:

$$2O^{2-} \rightarrow O_2(g) + 4e^-$$

Figure 6.32 shows a typical diagram of this type of cell.

FIGURE 6.32 Diagram of a typical SOEC. Note the similarities to the PEMEC.

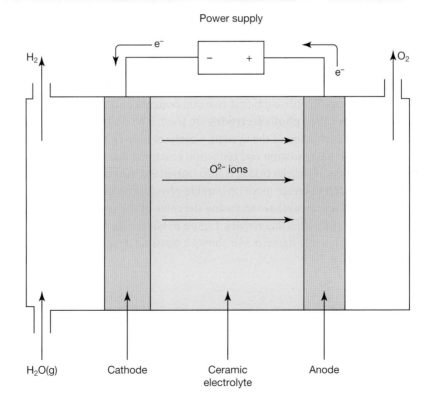

6.6.3 Producing green hydrogen by artificial photosynthesis

An innovative approach to producing green hydrogen is through **biomimicry**. To this end, natural photosynthesis has provided inspiration for scientists to copy and adapt the way this process harnesses sunlight to produce energy-rich compounds. The goal of artificial photosynthesis is to simulate the natural process that has evolved over millions of years in plants, in order to meet society's future needs for clean, renewable energy sources. The term **photosynthesis** refers to using light to synthesise simple substances into more complicated and more energy-rich substances. In the case of plants, the simple molecules in carbon dioxide and water are combined using light as the energy source to produce glucose and oxygen. The light in solar energy is transformed into stored chemical energy in glucose. Artificial photosynthesis does not aim to produce glucose, but rather other substances that can be used as fuels. Examples of such fuels include hydrogen and methanol, with the potential to produce other fuels as well.

Artificial photosynthesis research is currently based on optimising and integrating three essential components. These are:

- the light capture and electron transport system
- the water-splitting system
- the carbon dioxide reduction system.

Figure 6.33 illustrates how this might be done.

FIGURE 6.33 An artificial photosynthesis system has three important components and will be able to produce fuels such as hydrogen and methanol.

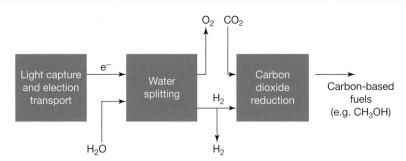

To produce hydrogen, it is necessary to utilise the first two components of this system. To achieve this, special electrodes called **photoelectrodes** are used. The design of such electrodes is critical to the functioning of the system. Catalysts play an important role in increasing the rates of both the oxidation and reduction reactions that are involved. The physical construction of the electrodes is also important. Advances in nanotechnology are being utilised to create more favourable physical properties in the electrode materials themselves, as well as increasing the rates of the reactions involved through the increased surface area that results. Figure 6.34a shows a general illustration of the process involved, while figure 6.34b shows a possible design.

biomimicry the act of copying and adapting processes that occur in nature

photosynthesis in the presence of light, carbon dioxide + water → glucose + oxygen

photoelectrodes electrodes that achieve redox reactions utilising light as an energy source

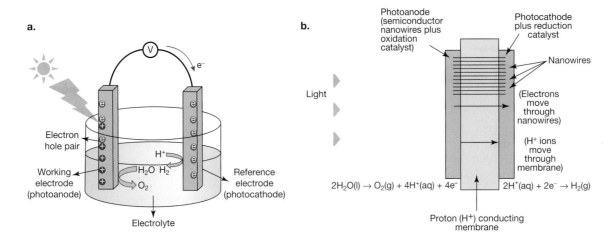

FIGURE 6.34 a. A general illustration of an artificial photosynthesis system to produce hydrogen **b.** A possible design

When the photoanode is exposed to light, free electrons are produced in the semiconductor material that it contains. These then travel to the other electrode. At the same time, there are spaces left behind in the semiconductor called 'holes', where these electrons came from. These holes are unstable and are filled by electrons that come from the oxidation of water, according to the following equation:

$$2H_2O(l) \rightarrow O_2(g) + 4H^+(aq) + 4e^-$$

The hydrogen ions produced travel through the electrolyte to the other electrode. This electrode is called either a balance electrode or a photocathode. Here, they are reduced, utilising the electrons from the photoanode according to the following equation:

$$2H^+(aq) + 2e^- \rightarrow H_2(g)$$

Because of the processes involved, this application of artificial photosynthesis is also called a *water oxidation and proton reduction catalyst system*.

Currently, much research is being conducted into producing green hydrogen via this method. A large part of this centres around the catalysts for the reactions involved. To date, the most effective catalysts are derivatives of rare and noble metals. These are expensive, and the discovery of cheaper and more effective catalysts would represent a major advance in this technology. The integration of the last component (the carbon dioxide reduction system) is also being extensively investigated. This could provide fuels that could be utilised in some current technologies with only minor adaptations, and would also have the benefit of removing carbon dioxide from the atmosphere.

6.6 Activities

learn on

6.6 Quick quiz on	6.6 Exercise	6.6 Exam questions

6.6 Exercise

1. Use of the following two technologies is currently under extensive research for the purposes of providing society's future energy and fuel needs:
   - polymer electrolyte membrane electrolysis cells (PEMECs) powered by renewable energy
   - artificial photosynthesis.
   a. State two features that these technologies have in common.
   b. Give one feature in which they are different.

2. The solid oxide electrolytic cell (SOEC) is a type of electrolytic cell that can produce hydrogen from water. It contains ceramic components and typically operates at temperatures of around 800 °C. It functions best under conditions of constant output (load).
   a. Would you expect this type of cell to be suited to large-scale or small-scale operation? Explain your answer.
   b. Under normal operation, the equation for the reaction occurring at the cathode is as follows:

   $$H_2O + 2e^- \rightarrow H_2 + O^{2-}$$

   Write the partial equation for the reaction occurring at the anode (symbols of state are not required).
   c. State the polarity of the cathode and anode in this cell.
   d. How could a cell (or series of cells) such as this be located so that the required operating temperature could be reached with minimal extra energy costs?

3. The following diagram shows the basic features of a polymer electrolyte membrane electrolysis cell (PEMEC).

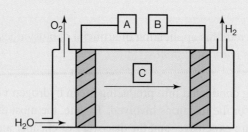

   a. The boxes A and B represent the polarity of each electrode. Which electrode is positive and which is negative?
   b. Box C represents the species that is transferred between the electrodes when the cell is functioning. State the identity of this species.
   c. Write the equation for the reaction taking place at the cathode.
   d. Write the equation for the reaction taking place at the anode.
   e. State one disadvantage that PEMECs currently have.

4. Hydrogen is being seriously explored as a fuel to replace our current dependence on fossil fuels. State three barriers that must be overcome before its use becomes widespread.

5. It is becoming commonplace to describe hydrogen in terms of colours, according to its method of production. These include grey, blue and green. From an environmental viewpoint and for the future:
   a. explain why blue hydrogen is regarded as superior to grey hydrogen
   b. explain why green hydrogen is regarded as superior to blue hydrogen.

6.6 Exam questions

▶ Question 1 (8 marks)

Imagine that you have been asked to consult for a firm that is installing either polymer electrolyte membrane electrolysis cells (PEMECs) or solid oxide electrolysis cells (SOECs) to generate hydrogen in the following locations:

I   On the west coast of Western Australia. Many locations along this coast experience high winds due to the winds blowing unimpeded across the Indian Ocean from South Africa.

II Various locations in central Australia. Such locations often have plentiful sunshine but are remote and sparsely populated.

III In cities such as Melbourne and Sydney that have large industrial areas. In these areas, processes are performed that often generate a large amount of waste heat.

IV In Tasmania. Tasmania has many mountains and valleys that, together with its high rainfall, have contributed to a well-developed hydroelectrical capacity.

For each of the above locations, state whether a PEMEC or a SOEC would be your recommendation. In each case, justify your choice.

▶ Question 2 (7 marks)

Worldwide, many research projects are investigating the concept of artificial photosynthesis, with some small-scale projects showing encouraging results. The following figure shows the essential components of this process.

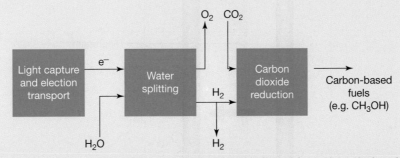

a. In what significant way is this method for generating hydrogen different from electrolysis using a renewable energy source? **(2 marks)**

b. State two major environmental benefits that this process could have if it is adopted on a large scale. **(2 marks)**

c. Electrocatalysts will play an important role in this process.
   What is the function of a catalyst? **(1 mark)**

d. Natural photosynthesis produces glucose ($C_6H_{12}O_6$) from water and carbon dioxide.
   Why do you think the term *artificial photosynthesis* is justified, even though glucose is not produced? **(1 mark)**

e. State one other product, besides hydrogen, that would be produced by this process. **(1 mark)**

▶ Question 3 (10 marks)

The following diagram shows a polymer electrolyte membrane electrolysis cell (PEMEC).

a. State what the boxes labelled A to E represent.   **(5 marks)**

b. Write the equation for the reaction taking place at the cathode when the cell operates under acidic conditions. **(1 mark)**

c. Write the equation for the reaction taking place at the anode when the cell operates under acidic conditions.   **(1 mark)**

d. Explain why the structure of each electrode must contain an electrical conductor and a catalyst.   **(2 marks)**

e. Besides its ability to conduct the species represented by E, what other important function does the structure labelled B perform?   **(1 mark)**

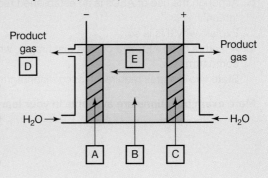

▶ **Question 4 (6 marks)**

The water oxidation and proton reduction catalyst system is sometimes referred to as 'artificial photosynthesis'.

a. Explain why the term *water oxidation* is appropriate and support your answer with a relevant equation. **(2 marks)**

b. Explain why the term *proton reduction* is appropriate and support your answer with a relevant equation. **(2 marks)**

c. Catalysts on each electrode will play an important role in the development of this technology, and scientists are currently conducting research to find better electrode catalysts for each electrode. Different catalysts will be required for each electrode.
Explain why the catalysts for each electrode will not be the same. **(2 marks)**

▶ **Question 5 (7 marks)**

The diagram of an alkaline electrolysis cell (AEC) shown illustrates an established technology from which hydrogen can be produced by electrolysis. These cells are large and use aqueous solutions of either sodium hydroxide or potassium hydroxide as an electrolyte. They work best when operating under constant conditions of production and power supply.

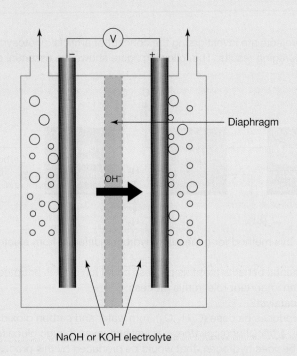

NaOH or KOH electrolyte

a. Name the two gases produced by AECs and state the electrode at which each is produced. **(2 marks)**
b. Write the relevant equations for the production of each of the gases you named in part **a.** **(2 marks)**
c. Although the use of AECs is an established technology, the gases produced cannot currently be termed 'green'.
Explain why this is so. **(1 mark)**
d. A feature of an AEC is a porous diaphragm, which is used to separate the cell into anode and cathode compartments.
State two features required in such diaphragms. **(2 marks)**

More exam questions are available in your learnON title.

6.7 Applications of Faraday's Laws

> **KEY KNOWLEDGE**
>
> - The application of Faraday's Laws and stoichiometry to determine the quantity of electrolytic reactant and product, and the current or time required to either use a particular quantity of reactant or produce a particular quantity of product
>
> *Source:* VCE Chemistry Study Design (2024–2027) extracts © VCAA; reproduced by permission.

Faraday's Laws were introduced and explained in subtopic 3.5. The application of these laws was focused on galvanic (primary) cells and fuel cells, connecting the amount of charge to the quantity of reactant or product, or to discharge time.

The application of these laws to electrolytic and rechargeable (secondary) cells is further explored and consolidated here.

6.7.1 Applying Faraday's Laws to electrolysis

Most calculations in electrolysis involve:
- finding the quantity of reactant or product involved (this can be a mass or a volume of gas)
- finding the current or time required to use or produce a certain amount of a substance
- determining the charge on an ion that is involved.

Faraday's Laws

$$\text{Faraday's First Law:} \qquad Q = It$$

$$\text{Faraday's Second Law:} \qquad n(\text{e}^-) = \frac{Q}{F}$$

where:

Q = the electrical charge (C)

I = the current (A)

t = the time (s)

$F = 96500 \text{ C mol}^{-1}$.

Together, these relationships summarise Faraday's two laws of electrolysis.

tlvd-9650

> **SAMPLE PROBLEM 6 Using Faraday's First Law to calculate the amount of product evolved**
>
> **When a current of 3.2 A is passed through a solution for 10.0 minutes, 0.010 mol of gas B is evolved. What amount of gas B will be evolved if a current of 2.0 A is used for 15 minutes?**
>
> | THINK | WRITE |
> |---|---|
> | 1. Charge used is calculated from the formula $Q = It$. The formula can be found in the VCE Chemistry Data Book. **TIP:** When using $Q = It$, remember that units must be considered. While t should be in seconds because we are using a ratio to determine the amount of gas B evolved, as long as the same units for time are used, they do not need to be converted to seconds. | $Q = It$
 $= 3.2 \times 10.0 \times 60$
 $= 1920 \text{ C (original)}$
 $Q = 2.0 \times 15 \times 60$
 $= 1800 \text{ C (new)}$ |

2. A charge of 1920 C resulted in 0.010 mol of gas B being evolved, so it is necessary to find the amount of gas B evolved when a charge of 1800 C is applied. Recall that amount evolved is proportional to charge flowing through the cell (Faraday's First Law). Give your answer to two significant figures.

Amount of gas B $= 0.010 \times \dfrac{1800}{1920}$
$= 0.0094$ mol (2 sig. figs)

PRACTICE PROBLEM 6

When a current of 3.4 A is passed through a solution for 7.0 minutes, 0.015 mol of metal X is deposited. What amount of X will be deposited if a current of 2.5 A is used for 20 minutes?

---

tlvd-9689

SAMPLE PROBLEM 7 Calculating the mass and volume of substances produced at the anode and cathode

A solution of copper(II) sulfate is electrolysed for 30.0 minutes using a current of 0.500 A. Calculate:
a. the mass of copper deposited on the cathode
b. the volume (at SLC) of oxygen gas evolved at the anode.

THINK

a. 1. To determine the mass of copper deposited, first calculate the amount of charge and convert it to faradays (same as moles of charge). Remember to convert time to seconds.

2. Determine the number of moles of electrons and the equation at the cathode.

3. Use stoichiometry involving electrons to calculate moles of copper produced. One mole of copper requires two moles of electrons.

4. Determine the mass of the copper produced using the molar mass formula.
TIP: Formulas and the value of the Faraday constant can be found in the VCE Chemistry Data Book.

WRITE

a. $Q = It$
$= 0.500 \times (30.0 \times 60)$
$= 900 \, C$

$n(e^-) = \dfrac{Q}{F}$
$= \dfrac{900}{96\,500}$
$= 9.33 \times 10^{-3}$ mol
$Cu^{2+} + 2e^- \rightarrow Cu(s)$

$n(Cu) = \dfrac{n(e^-)}{2}$
$= \dfrac{(9.33 \times 10^{-3})}{2}$
$= 4.67 \times 10^{-3}$ mol

$n = \dfrac{m}{M}$
$\therefore m(Cu) = n \times M$
$= (4.67 \times 10^{-3}) \times 63.5$
$= 0.296$ g

b. 1. Determine the equation at the anode.

b. $2H_2O(l) \rightarrow O_2(g) + 4H^+(aq) + 4e^-$

2. Using the number of electrons determined in part **a**, step **2**, use stoichiometry involving electrons to calculate moles of oxygen produced. There is one mole of oxygen to four moles of electrons.

$$n(O_2) = \frac{n(e^-)}{4}$$
$$= \frac{(9.33 \times 10^{-3})}{4}$$
$$= 2.33 \times 10^{-3} \text{ mol}$$

3. Determine the volume of oxygen produced using the molar volume formula.

$$n = \frac{V}{V_m}$$
$$V(O_2)_{SLC} = n \times V_m$$
$$= (2.33 \times 10^{-3}) \times 24.8$$
$$= 0.0578 \text{ L or } 57.8 \text{ mL}$$

PRACTICE PROBLEM 7

A solution of copper(II) sulfate is electrolysed for 17.5 minutes using a current of 0.500 A. Calculate:
a. the mass of copper deposited on the cathode
b. the volume (at SLC) of oxygen gas evolved at the anode.

tlvd-9690

SAMPLE PROBLEM 8 Calculating the time involved for an electrolysis reaction

A solution containing Ni^{2+} ions undergoes electrolysis. Calculate the time (in minutes) required to deposit 5.00 g of nickel if a current of 6.4 A is used.

THINK	WRITE
1. Calculate the number of moles of nickel deposited.	$n(Ni) = \dfrac{m}{M}$ $= \dfrac{5.00}{58.7}$ $= 0.0852 \text{ mol}$
2. Determine the equation at the cathode and the quantity of charge required.	$Ni^{2+}(aq) + 2e^- \rightarrow Ni(s)$ $n(Ni) = \dfrac{n(e^-)}{2}$ $\therefore n(e^-) = 2 \times 0.0852$ $= 0.170 \text{ mol}$ 0.170 F of charge is required. $0.170 \times 96\,500 = 1.64 \times 10^4 \text{ C}$ $\therefore 1.64 \times 10^4 \text{ C of charge is required.}$
3. Determine the time required from the charge required and current used. *Note:* Don't forget to change the answer from seconds into minutes.	$Q = It$ $\therefore t = \dfrac{Q}{I}$ $= \dfrac{1.64 \times 10^4}{6.4} \text{ seconds}$ $= 2.6 \times 10^3 \text{ seconds}$ $= \dfrac{2.6 \times 10^3}{60} = 43 \text{ minutes}$

PRACTICE PROBLEM 8

How many minutes are required to deposit 0.500 g of silver from a solution of Ag^+ ions, using a current of 2.50 A?

---

tlvd-9691

SAMPLE PROBLEM 9 Determining the charge on an ion using Faraday's Laws

When molten calcium chloride is electrolysed by a current of 0.200 A flowing for 965 seconds, 0.0401 g of calcium is formed. What is the charge on a calcium ion?

THINK	WRITE
1. To determine the charge on the calcium ion, first determine the number of moles of calcium.	$n(Ca) = \dfrac{m}{M}$ $= \dfrac{0.0401}{40.1}$ $= 0.00100 \, mol$
2. Determine the amount of electricity used and the number of moles of electrons. Compare the units given to those required. Time must be in seconds.	$Q = It$ $= 0.200 \times 965$ $= 193 \, C$ $n(e^-) = \dfrac{Q}{F}$ $= \dfrac{193}{96\,500}$ $= 0.00200 \, mol$
3. According to step 2, 0.00200 moles of electrons are needed to produce 0.00100 moles of calcium. Let the charge on ions be $x+$. Use stoichiometry to calculate x.	$Ca^{x+}(l) + xe^- \rightarrow Ca(s)$ $\dfrac{n(e^-)}{x} = n(Ca)$ $\therefore x = \dfrac{n(e^-)}{n(Ca)}$ $= \dfrac{0.00200}{0.00100}$ $= 2$ The charge on the calcium ion is 2+ (Ca^{2+}).

PRACTICE PROBLEM 9

In order to determine the charge on an aluminium ion experimentally, a molten aluminium chloride solution is electrolysed by a current of 0.300 A flowing for 965 seconds. The mass of aluminium formed is 0.0270 g. What is the charge on the aluminium ion?

6.7.2 Faraday's Laws in industry

Calculations based on Faraday's Laws are critical to industrial electrolytic processes. Due to the large scale of these processes, small variations and inefficiencies can result in the loss of many thousands of dollars. In order to determine the efficiency of a particular process, knowledge of the theoretical maximum amount is required.

tlvd-9692

SAMPLE PROBLEM 10 Calculating the theoretical mass produced in a Hall–Héroult cell

A typical Hall–Héroult cell in an aluminium plant operates at an average current of 1.70×10^4 A. Calculate the theoretical mass of aluminium produced in a Hall–Héroult cell over 24 hours.

THINK	WRITE
1. Determine the equation at the cathode.	$Al^{3+}(l) + 3e^- \rightarrow Al(l)$
2. Calculate the amount of charge used by applying the formula $Q = It$. Compare the units given to those required. Time must be in seconds. Then calculate the number of moles of electrons.	$Q = It$ $= 1.70 \times 10^4 \times (24 \times 60 \times 60)$ $= 1.47 \times 10^9\,C$ $n(e^-) = \dfrac{Q}{F}$ $= \dfrac{1.47 \times 10^9}{96\,500}$ $= 1.52 \times 10^4\,mol$
3. Using the number of moles of electrons determined in step **2**, use stoichiometry involving electrons to calculate the moles of aluminium produced. There is one mole of aluminium to three moles of electrons.	$n(Al) = \dfrac{1.52 \times 10^4}{3}$ $= 5.07 \times 10^3$ moles of Al
4. Convert moles to mass of aluminium using the molar mass formula, $n = \dfrac{m}{M}$.	$m(Al) = n(Al) \times M(Al)$ $= 5.07 \times 10^3 \times 27.0$ $= 1.37 \times 10^5\,g$ $= 1.37 \times 10^2\,kg$

PRACTICE PROBLEM 10

Calculate the mass of magnesium produced over 24 hours when a current of 10 000 A is used to electrolyse molten magnesium ions.

6.7 Activities

6.7 Quick quiz on	6.7 Exercise	6.7 Exam questions

6.7 Exercise

1. Calculate the charge involved (in coulombs) in the following situations.
   a. A current of 1.10 A flows for 12.0 seconds
   b. A current of 4.20 A flows for 3.0 minutes
   c. A current of 0.920 A flows for 2.0 hours
   d. A current of 0.215 A flows for 1.0 days

2. Calculate the number of moles of electrons involved when the following amounts of charge flow in an electrolytic cell.
   a. 515 C   b. 46 500 C   c. 2.58 F

3. A solution of silver nitrate is electrolysed for 20.0 minutes using a current of 0.600 A. Calculate:
   a. the mass of silver deposited at the cathode
   b. the volume (at SLC) of oxygen gas evolved at the anode.

4. How long will it take to deposit 1.00 g of cobalt in an electrolytic cell that uses a current of 3.50 A? The equation for the reduction is as follows:

$$Co^{2+}(aq) + 2e^- \rightarrow Co(s)$$

5. A current of 4.25 A is passed through molten Al_2O_3 for 13.5 hours using graphite electrodes.
   a. How many grams of aluminium would be produced?
   b. What volume of carbon dioxide gas would be produced, once cooled to 29.0 °C and 152 kPa?

6. When a current of 10.0 A was passed through a concentrated solution of sodium chloride using carbon electrodes, 2.80 L of chlorine (at SLC) was collected. How long (in minutes) did the electrolysis take?

7. Calculate the current required to produce 2.00 kg of magnesium metal by the electrolysis of molten magnesium chloride, $MgCl_2$, over a period of 4 days and 2 hours.

8. When a solution containing gold ions is electrolysed by a current of 0.100 A flowing for 965 seconds, 0.197 g of gold is formed. What is the charge on the gold ion?

9. A given quantity of electricity is passed through two aqueous cells connected in series. The first contains silver nitrate and the second contains calcium chloride. What mass of calcium is deposited in one cell if 2.00 g of silver is deposited in the other cell?

10. Calculate the time taken to deposit gold from a solution of gold(I) cyanide to a thickness of 0.0100 mm onto a copper disc that has a surface area of 3.14 cm^2 if a current of 0.750 A is used. (The density of gold is 19.3 g cm^{-3}.)

6.7 Exam questions

▶ Question 1 (1 mark)
Source: VCE 2021 Chemistry Exam, Section A, Q.9; © VCAA

MC An electrolysis cell consumed a charge of 4.00 C in 5.00 minutes.

This represents a consumption of
A. 4.15×10^{-5} mol of electrons.
B. 2.07×10^{-4} mol of electrons.
C. 1.93×10^4 mol of electrons.
D. 2.41×10^4 mol of electrons.

Question 2 (1 mark)

Source: VCE 2019 Chemistry Exam, Section A, Q.24; © VCAA

MC The diagram shows an electroplating cell.

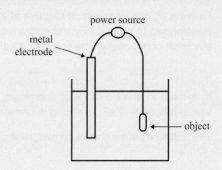

The cell contains 1 L of an electroplating solution. The electroplating cell is run for one hour at 3 A.

Which one of the following electroplating solutions will deposit the largest mass of metal onto the object?

A. 1 M $AgNO_3$
B. 1 M $Cd(NO_3)_2$
C. 1 M $Pb(NO_3)_2$
D. 1 M $Al(NO_3)_3$

Question 3 (1 mark)

Source: VCE 2018 Chemistry NHT Exam, Section A, Q.24; © VCAA

MC An electroplating cell containing two platinum electrodes and an electroplating solution is operated at 5.0 A for 600 s. After the cell is turned off, 0.54 g of metal is found to have been deposited on the cathode.

Which electroplating solution was used in this process?

A. 1 M $AgNO_3$
B. 1 M $Ni(NO_3)_2$
C. 1 M $Pb(NO_3)_2$
D. 1 M $Cr(NO_3)_3$

Question 4 (1 mark)

Source: VCE 2017 Chemistry Exam, Section A, Q.30; © VCAA

MC The diagram shows the basic set-up of an electroplating cell.

Initially the cell is set up with a lead, Pb, electrode as Electrode Z and 1.0 M lead nitrate, $Pb(NO_3)_2$, as the electroplating solution. The cell runs for a set time and current, with 1.0 g of Pb deposited onto Electrode Z.

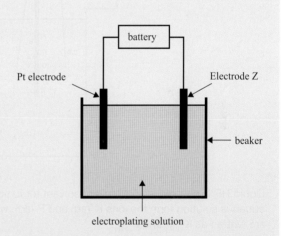

Four subsequent electroplating cells are set up, each containing a platinum, Pt, electrode, a different Electrode Z and an appropriate 1.0 M electroplating solution. These four electroplating cells are operated for the same time and at the same current as the original Pb electroplating cell.

Which combination of Electrode Z and electroplating solution would be expected to deposit **more** metal by mass onto Electrode Z than the original Pb electroplating cell?

	Electrode Z	Electroplating solution
A.	chromium, Cr	1.0 M $Cr(NO_3)_3$
B.	silver, Ag	1.0 M $AgNO_3$
C.	gold, Au	1.0 M $AuCl_3$
D.	tin, Sn	1.0 M $SnSO_4$

Source: VCE 2017 Chemistry Exam, Section B, Q.8.e; © VCAA

Fluorine, F_2, gas is the most reactive of all non-metals. Anhydrous liquid hydrogen fluoride, HF, can be electrolysed to produce F_2 and hydrogen, H_2, gases. Potassium fluoride, KF, is added to the liquid HF to increase electrical conductivity. The equation for the reaction is

$$2HF(l) \rightarrow F_2(g) + H_2(g)$$

F_2 is used to make a range of chemicals, including sulfur hexafluoride, SF_6, an excellent electrical insulator, and xenon difluoride, XeF_2, a strong fluorinating agent.

The diagram below shows an electrolytic cell used to prepare F_2 gas.

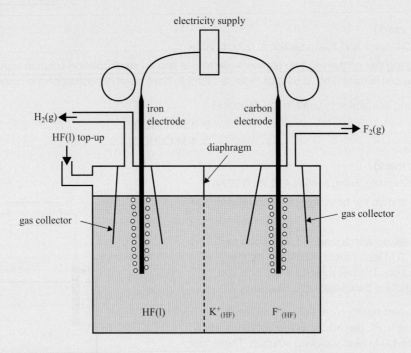

Liquid HF, like water, is an excellent solvent for ionic compounds. In the same way that water molecules in an aqueous solution form the ions $K^+(aq)$ and $F^-(aq)$, when KF is dissolved in HF, the K^+ and F^- ions form ions that are written as $K^+_{(HF)}$ and $F^-_{(HF)}$.

Calculate the volume of F_2 gas, measured at standard laboratory conditions (SLC), that would be produced when a current of 1.50 A is passed through the cell for 2.00 hours.

More exam questions are available in your learnON title.

6.8 Review

6.8.1 Topic summary

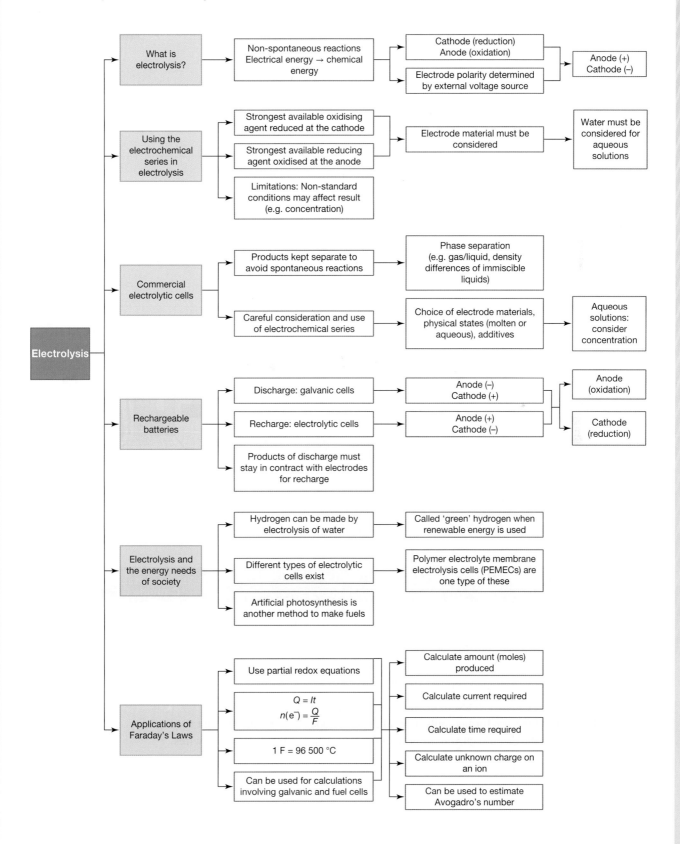

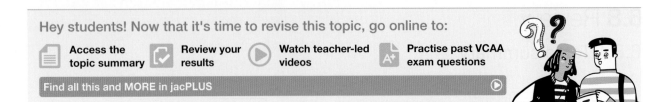

6.8.2 Key ideas summary

online only

6.8.3 Key terms glossary

online only

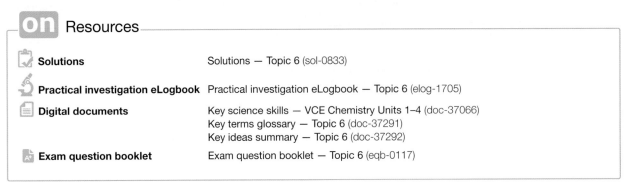

on Resources

Solutions Solutions — Topic 6 (sol-0833)

Practical investigation eLogbook Practical investigation eLogbook — Topic 6 (elog-1705)

Digital documents Key science skills — VCE Chemistry Units 1–4 (doc-37066)
Key terms glossary — Topic 6 (doc-37291)
Key ideas summary — Topic 6 (doc-37292)

Exam question booklet Exam question booklet — Topic 6 (eqb-0117)

6.8 Activities

learn on

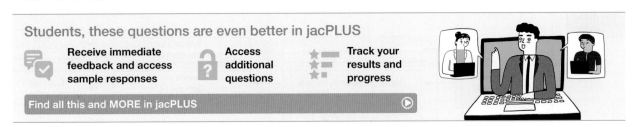

6.8 Review questions

1. Complete the following table to summarise what happens at each electrode during electrolysis of NaCl in different forms.

Electrolyte type	Reaction at		
	Electrodes	Anode (+)	Cathode (–)
Molten salt	Inert		
0.1 M aqueous salt solution	Inert		
6 M aqueous salt solution	Inert		

2. In the electrolysis of molten sodium chloride, explain:
   a. why electricity is conducted in the molten state but not in the solid state
   b. why the products are formed only around the electrodes and not throughout the liquid
   c. what causes the electric current to flow in the liquid and in the connecting wires.

3. For each of the following, predict the products at the anode and cathode, determine the minimum cell voltage required for the electrolysis (using the electrodes shown) of 1.0 M aqueous solutions, and write an overall equation.

   a. Potassium hydroxide (inert electrodes)
   b. Magnesium iodide (inert electrodes)
   c. Zinc bromide (inert electrodes)
   d. Zinc bromide (gold electrodes)
   e. Zinc bromide (silver electrodes)
   f. Sodium chloride (iron electrodes)

4. An aqueous solution of $NiBr_2$ is electrolysed using inert electrodes.

   a. Sketch the cell showing:

      i. the direction of current flow in the external circuit and through the electrolyte
      ii. the cathode and anode, and their polarities.

   b. Write half-equations for the expected reactions at each electrode, and then write the overall equation.
   c. Calculate the minimum voltage needed to electrolyse the solution under standard conditions (SLC).
   d. Explain how the products of electrolysis would differ if nickel electrodes were used.

5. Sodium is made commercially by the electrolysis of molten sodium chloride in a Downs cell. This cell contains an iron cathode and a carbon anode, and design features to collect and keep the products of electrolysis separate. A number of methods can be used to reduce the melting temperature of the sodium chloride and save on energy costs. A common method is to add an amount of calcium chloride to the melt. The following diagram shows the essential features of this cell.

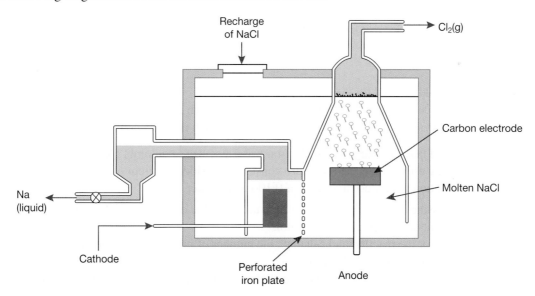

   a. Which electrode forms the positive electrode and which electrode forms the negative electrode?
   b. Write the equation for the half-reaction occurring at each electrode.
   c. Suggest why the perforated iron plate, shown in the diagram, is important for the safe operation of this cell.
   d. Explain why the carbon electrode cannot be replaced with an iron electrode.
   e. Explain why the addition of calcium chloride does not interfere with the production of sodium.
   f. Calculate the volume of Cl_2 gas, measured at SLC, that would be produced when this cell operates with a current of 2.50 A for 12.00 hours.

6. Sketch an electrolytic cell that could be used to plate copper onto a tin key ring. On your sketch, label the:
   - anode and cathode
   - direction of electron flow
   - nature of each electrode
   - electrode polarity
   - equations occurring at each electrode.

7. Chromium chloride is electrolysed using chromium electrodes. A current of 0.200 A flows for 1447 seconds. The increase in the mass of the cathode is 0.0520 g.

   a. How many coulombs of electricity are used?
   b. How many moles of electrons are transferred?
   c. How many moles of chromium are liberated?
   d. What is the charge on the chromium ion?

8. A given quantity of electricity is passed through three cells connected in series. These cells contain solutions of silver nitrate, tin(II) chloride and magnesium chloride respectively, all at 1 M concentration. All cells have inert electrodes. After a period of time it is observed that 2 g of silver has been deposited in the first cell.

   a. What mass of tin would have been deposited in the second cell?
   b. What mass of magnesium would have been deposited in the third cell?

9. The Edison cell is a 1.3 V storage battery that can be recharged, even after long periods of being left uncharged. Its electrolyte is 21% potassium hydroxide solution and the reaction on discharge is as follows:

$$Fe(s) + 2NiO(OH)(s) + 2H_2O(l) \rightarrow Fe(OH)_2(s) + 2Ni(OH)_2(s)$$

   a. Give electrode reactions occurring during:
      i. discharging
      ii. recharging.
   b. What materials would be used for the electrodes?
   c. In the discharge process, which electrode is the anode and which is the cathode?
   d. In the recharge process, which electrode is the anode and which is the cathode?

10. After Millikan showed that the charge on an electron was 1.6×10^{-19} coulomb, electrolytic reactions were used to obtain accurate estimates of the Avogadro constant. Consider a current of 0.100 A flowing through a copper(II) nitrate solution to produce a deposit of 0.100 g of copper.

    a. Find the charge passing through the cell if the time taken for the deposit was 50 minutes and 40 seconds.
    b. Calculate the amount of copper produced in mol.
    c. Write the equation for the reaction and calculate the number of moles of electrons consumed.
    d. Calculate the charge on 1 mole of electrons.
    e. Calculate the Avogadro constant, given that the charge on an electron is 1.6×10^{-19} coulombs.

6.8 Exam questions

Section A — Multiple choice questions

All correct answers are worth 1 mark each; an incorrect answer is worth 0.

Question 1

Source: VCE 2022 Chemistry Exam, Section A, Q.22; © VCAA

MC Lithium-ion batteries are used in a range of electronic devices, including mobile phones. The discharge reaction for this type of battery is

$$LiC_6(s) + CoO_2(s) \rightarrow C_6(s) + LiCoO_2(s)$$

Which of the following is correct about lithium-ion batteries?

	During discharge, reduction occurs at the	During recharge, reduction occurs at the
A.	anode	cathode
B.	cathode	anode
C.	anode	anode
D.	cathode	cathode

Question 2

Source: VCE 2016 Chemistry Exam, Section A, Q.19; © VCAA

MC An electroplating process uses a solution of chromium(III) sulfate, $Cr_2(SO_4)_3$, to deposit a thin layer of chromium on the surface of an object. A current of 5.00 A is maintained.

How long does it take, in seconds, to deposit 0.0192 mol chromium onto the surface?

A. 371   **B.** 1110   **C.** 1860   **D.** 5570

Question 3

Source: VCE 2019 Chemistry Exam, Section A, Q.7; © VCAA

MC A molten mixture of equal parts aluminium fluoride, AlF_3, and sodium chloride, NaCl, undergoes electrolysis.

Which one of the following statements about this reaction is correct?

A. Sodium metal will be produced at the cathode and fluorine gas will be produced at the anode.
B. Sodium metal will be produced at the anode and chlorine gas will be produced at the cathode.
C. Aluminium metal will be produced at the cathode and chlorine gas will be produced at the anode.
D. Aluminium metal will be produced at the anode and fluorine gas will be produced at the cathode.

Question 4

Source: VCE 2017 Chemistry Exam, Section A, Q.20; © VCAA

MC The reaction below represents the discharge cycle of a standard lead–acid rechargeable car battery.

$$Pb(s) + PbO_2(s) + 4H^+(aq) + 2SO_4{}^{2-}(aq) \rightarrow 2PbSO_4(s) + 2H_2O(l)$$

During the recharge cycle, the pH

A. increases and solid Pb is a reactant.
B. increases and solid PbO_2 is produced.
C. decreases and chemical energy is converted to electrical energy.
D. decreases and electrical energy is converted to chemical energy.

Use the following information to answer Questions 5–7.

An electrolytic cell is set up to obtain pure copper from an impure piece of copper called 'blister copper'.

The electrolyte solution contains both copper(II) sulfate and sulfuric acid. The blister copper, Electrode I, contains impurities such as zinc, cobalt, silver, gold, nickel and iron. The cell voltage is adjusted so that only copper is deposited on Electrode II. Sludge, which contains some of the solid metal impurities present in the blister copper, forms beneath Electrode I. The other impurities remain in solution as ions.

The diagram below represents the cell.

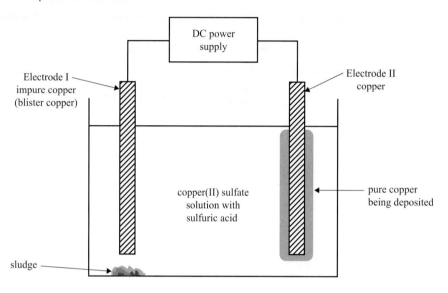

⊳ Question 5

Source: VCE 2015 Chemistry Exam, Section A, Q.28; © VCAA

MC The solid metal impurities that are found in the sludge are

A. gold, nickel and cobalt.
B. cobalt, nickel and iron.
C. nickel and iron.
D. silver and gold.

⊳ Question 6

Source: VCE 2015 Chemistry Exam, Section A, Q.29; © VCAA

MC Which of the following correctly shows both the equation for the reaction occurring at the cathode and the polarity of Electrode I?

	Cathode reaction	Polarity of Electrode I
A.	$Cu^{2+}(aq) + 2e^- \rightarrow Cu(s)$	positive
B.	$Cu(s) \rightarrow Cu^{2+}(aq) + 2e^-$	negative
C.	$Cu^{2+}(aq) + 2e^- \rightarrow Cu(s)$	negative
D.	$Cu(s) \rightarrow Cu^{2+}(aq) + 2e^-$	positive

Question 7

Source: VCE 2015 Chemistry Exam, Section A, Q.30; © VCAA

MC Which one of the following graphs best shows the change in mass of Electrode I over a period of time, starting from the moment the power supply is connected?

A.

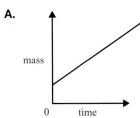

B.

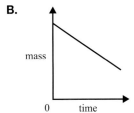

C.

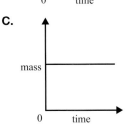

D.

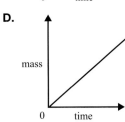

Question 8

Source: VCE 2013 Chemistry Sample Exam for Units 3 and 4, Section A, Q.30; © VCAA

MC A series of electrolysis experiments is conducted using the apparatus shown below.

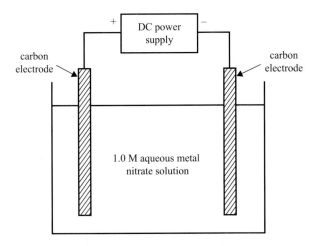

An electric charge of 0.030 faraday was passed through separate solutions of 1.0 M $Cr(NO_3)_3$, 1.0 M $Cu(NO_3)_2$ and 1.0 M $AgNO_3$. In each case the corresponding metal was deposited on the negative electrode.

The amount, in mol, of each metal deposited is

	Amount, in mol, of chromium deposited	Amount, in mol, of copper deposited	Amount, in mol, of silver deposited
A.	0.030	0.030	0.030
B.	0.010	0.015	0.030
C.	0.090	0.060	0.030
D.	0.030	0.020	0.010

Source: VCE 2011 Chemistry Exam 2, Section A, Q.19; © VCAA

MC An ornament was coated with a metal, M, by electrolysis of a solution of the metal ion, M^{x+}. During the electrolysis, a current of 1.50 amperes was applied for 180 seconds. The ornament was coated in 0.0014 mol of metal.

The value of x in M^{x+} is

A. 1   **B.** 2   **C.** 3   **D.** 4

Source: VCE 2009 Chemistry Exam 2, Section A, Q.20; © VCAA

MC Lithium metal is manufactured by electrolysis of lithium salts.

Which of the following would be the best choice for the electrolyte and the anode in a commercial cell?

	Electrolyte	Anode
A.	LiCl solution	iron rod
B.	molten LiCl	iron rod
C.	LiCl solution	carbon rod
D.	molten LiCl	carbon rod

Section B — Short answer questions

▶ **Question 11 (8 marks)**

Source: VCE 2022 Chemistry Exam, Section B, Q.2.b,c; © VCAA

A coal-fired power station is used to generate electricity. Carbon dioxide, CO_2, gas is produced as part of the process.

a. Hydrogen, H_2, can be produced using electricity generated by renewable sources. A simplified diagram of an acidic electrolyser used to produce hydrogen is shown below.

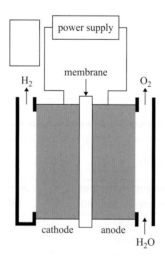

   i. Draw an arrow in the box provided on the diagram above to show the direction of flow of electrons through the wire. Justify your answer. **(2 marks)**

   ii. State **two** functions of the membrane. **(2 marks)**

b. i. Write the overall equation for the reaction that takes place in the acidic electrolyser shown in the diagram above when it is operating at 80 °C. **(1 mark)**

ii. How many moles of H_2 could be produced by the acidic electrolyser using 1625.0 A in 1.25 hours, assuming 100% efficiency? **(3 marks)**

▶ Question 12 (8 marks)

Source: VCE 2020 Chemistry Exam, Section B, Q.2; © VCAA

The electrolysis of carbon dioxide gas, CO_2, in water is one way of making ethanol, C_2H_5OH.

The diagram below shows a CO_2–H_2O electrolysis cell. The electrolyte used in the electrolysis cell is sodium bicarbonate solution, $NaHCO_3$(aq).

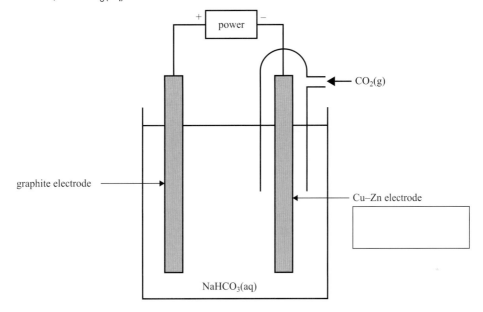

The following half-cell reactions occur in the CO_2–H_2O electrolysis cell.

$$O_2(g) + 2H_2O(l) + 4e^- \rightleftharpoons 4OH^-(aq) \qquad\qquad E^0 = +0.40 \text{ V}$$

$$2CO_2(g) + 9H_2O(l) + 12e^- \rightleftharpoons C_2H_5OH(l) + 12OH^-(aq) \qquad E^0 = -0.33 \text{ V}$$

a. Identify the Cu–Zn electrode as either the anode or the cathode in the box provided in the diagram above. **(1 mark)**

b. Determine the applied voltage required for the electrolysis cell to operate. **(1 mark)**

c. Write the balanced equation for the overall electrolysis reaction. **(1 mark)**

d. Identify the oxidising agent in the electrolysis reaction. Give your reasoning using oxidation numbers. **(2 marks)**

e. A current of 2.70 A is passed through the CO_2–H_2O electrolysis cell. The cell has an efficiency of 58%.

Calculate the time taken, in minutes, for this cell to consume 6.05×10^{-3} mol of CO_2(g). **(3 marks)**

Source: VCE 2019 Chemistry Exam, Section B, Q.7; © VCAA

The zinc–cerium battery is a commercial rechargeable battery that comprises a series of cells.

During recharging, the cells use energy from wind farms or solar cell panels.

During discharging, energy is supplied to electric grids to power local factories and homes.

The electrolytes are stored in separate storage tanks, and are pumped into and out of each cell when in use. A membrane separates the two electrodes that are immersed in 1 M methanesulfonic acid, CH_3SO_3H.

A diagram representing a zinc–cerium cell is shown below.

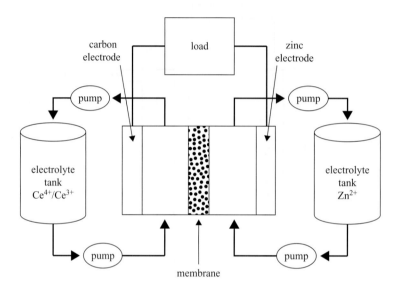

The following half-cell reactions occur in the zinc–cerium cell.

$$Zn(CH_3SO_3)_2(aq) + 2H^+(aq) + 2e^- \rightleftharpoons Zn(s) + 2CH_3SO_3H(aq) \qquad E^0 = -0.76\,V$$

$$Ce(CH_3SO_3)_4(aq) + H^+(aq) + e^- \rightleftharpoons Ce(CH_3SO_3)_3(aq) + CH_3SO_3H(aq) \qquad E^0 = 1.64\,V$$

a. Write the equation for the overall discharge reaction. **(1 mark)**
b. Identify the oxidising agent during discharging and justify your answer using oxidation numbers. **(2 marks)**
c. Determine the theoretical voltage produced by a single cell as it discharges. **(1 mark)**
d. Write the ionic equation for the reaction occurring at the positive electrode during recharging. **(1 mark)**
e. Other than transporting ions between the electrodes, describe **one** function of the membrane in the zinc–cerium cell. **(1 mark)**
f. Specify **one** factor that would limit the life of the zinc–cerium cell. **(1 mark)**
g. Experts have regarded the zinc–cerium cell as a hybrid of a fuel cell and a secondary cell.

Why would this be the case? **(1 mark)**

Source: VCE 2016 Chemistry Exam, Section B, Q.11; © VCAA

A student investigated the electroplating of a metal with nickel. The following is her report.

Electroplating a brass key with nickel

Aim

To investigate whether Faraday's Laws apply to the electroplating of a brass key with nickel

Procedure

Step 1 — The apparatus was set up as in the diagram below. The electrolyte solution was supplied. The brass key was sanded, weighed and placed in the solution, as shown below.

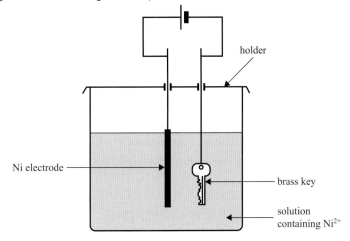

Step 2 — The current was turned on for exactly 20 minutes. The current and voltage were measured when the power was turned on.

Step 3 — The key was removed from the solution, patted dry with a paper towel and weighed. Steps 1–3 were repeated for two more keys.

Results

Three trials of the experiment were conducted, X, Y and Z.

Trial	Initial mass of brass key (g)	Final mass of brass key (g)	Mass of nickel deposit (g)	Current (A)	Voltage (V)
X	2.774	2.907	0.133	0.52	2.4
Y	3.068	3.269	0.201	0.54	2.2
Z	3.122	3.310	0.188	0.50	1.9

Predicted mass for Trial X using Faraday's Laws

$$m(Ni) = \frac{0.52 \times 20 \times 60}{96\,500} \times \frac{58.7}{2} = 0.19 \text{ g}$$

Conclusion

Faraday's Laws apply to the electroplating of a brass key with nickel.

Evaluate the student's experimental design and report. In your response:
- identify and explain **one** strength of the experimental design
- suggest **two** improvements or modifications that you would make to the experimental design and justify your suggestions
- comment on the validity of the conclusion based on the results obtained.

Question 15 (7 marks)

Source: VCE 2016 Chemistry Exam, Section B, Q.8; © VCAA

The lithium-ion battery is a secondary cell that is now widely used in portable electronic devices.

In these batteries, lithium ions, Li^+, move through a special non-aqueous electrolyte between the two electrodes. The batteries are housed in sealed containers to ensure that no moisture can enter them.

Both electrodes are made up of materials that allow the lithium ions to move into and out of their structures. The anode consists of LiC_6, where lithium is embedded in the graphite structure. Lithium cobalt oxide, $LiCoO_2$, is commonly used as the material in the cathode. The reaction at the cathode is quite complex. When the cell discharges, Li^+ ions move out of the anode and enter the cathode.

During discharge, the half-cell reaction at the anode is

$$LiC_6 \rightarrow Li^+ + e^- + C_6$$

a. During discharge, what is the polarity of the graphite electrode? **(1 mark)**

b. Write the half-equation for the reaction that occurs at the cathode of a lithium-ion battery when it is recharged. **(1 mark)**

c. In a lithium-ion battery, lithium metal must not be in contact with water.

Explain why and justify your answer with the use of appropriate equations. **(3 marks)**

d. Identify **one** design feature of the lithium-ion battery that enables it to be recharged. **(1 mark)**

e. What is **one** advantage of using a secondary cell compared to using a fuel cell? **(1 mark)**

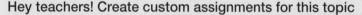

AREA OF STUDY 2 How can the rate and yield of chemical reactions be optimised?

OUTCOME 2

Analyse chemical systems to predict how the rate and extent of chemical reactions can be optimised, explain how electrolysis is involved in the production of chemicals, and evaluate the sustainability of electrolytic processes in producing useful materials for society.

PRACTICE EXAMINATION

STRUCTURE OF PRACTICE EXAMINATION		
Section	Number of questions	Number of marks
A	20	20
B	5	30
Total		50

Duration: 50 minutes

Information:

- This practice examination consists of two parts. You must answer all question sections.
- Pens, pencils, highlighters, erasers, rulers and a scientific calculator are permitted.
- You may use the VCE Chemistry Data Book for this task.

 Resources

🔗 **Weblink** VCE Chemistry Data Book

SECTION A — Multiple choice questions

All correct answers are worth 1 mark each; an incorrect answer is worth 0.

1. At a particle level, reactions occur when
   A. particles collide at a certain angle.
   B. particles collide with high energy.
   C. particles collide with sufficient energy and at the correct orientation.
   D. a catalyst is used.

2. Which of the following is *not* a suitable method for determining changes in the rate of a reaction?
   A. Measuring the volume of a gas evolved every 10 seconds for 1 minute
   B. Measuring the change in intensity of the colour of a solution using colorimetry every 30 seconds for 5 minutes
   C. Measuring the change in pH every second using a pH probe until a colour change is observed
   D. Measuring the time it takes for a gas to stop forming in a reaction

3. What are the expression for the equilibrium constant and the units, respectively, for the reaction below?

$$2HI(g) \rightarrow H_2(g) + I_2(g)$$

A. $\dfrac{[H_2][I_2]}{[HI]^2}$, M

B. $\dfrac{[H_2][I_2]}{[HI]^2}$, no units

C. $\dfrac{[HI]^2}{[H_2][I_2]}$, M

D. $\dfrac{[HI]^2}{[H_2][I_2]}$, no units

4. When a system is at equilibrium, it can be said that
   A. the forward and backward reaction are occurring at the same rate.
   B. the forward and backward reaction have stopped.
   C. the forward and backward reaction are both occurring to the same extent.
   D. the concentration of reactants and products is equal.

5.

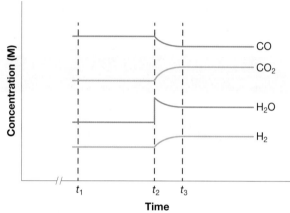

What is the correct equation for the reaction illustrated?
   A. $CO_2(g) + H_2(g) \rightleftharpoons CO(g) + H_2O(l)$
   B. $CO_2(g) + 2H_2(g) \rightleftharpoons 2H_2O(l) + CO(g)$
   C. $CO(g) + H_2O(l) \rightleftharpoons CO_2(g) + H_2(g)$
   D. $CO(g) + H_2O(g) \rightarrow CO_2(g) + H_2(g)$

6. The figure shows a concentration-versus-time graph for the reaction represented by the following equation:

$$2A(g) \rightleftharpoons B(g)$$

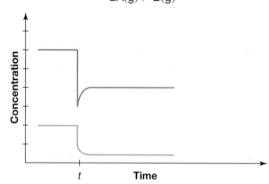

After reaching equilibrium, a change was made at time t.
What was this change?
   A. The addition of a catalyst
   B. Some removal of one of the chemicals involved
   C. A doubling of the volume of the container
   D. An increase in temperature

7. Consider the following reaction:

$$2NO(g) + Br_2(g) \rightleftharpoons 2NOBr(g)$$

If an inert gas is injected into the system when it is at equilibrium, what will happen?

   A. The rate of the reaction will increase.

   B. The position of equilibrium will shift to the right.

   C. The position of equilibrium will shift to the left.

   D. The position of equilibrium will remain the same.

8. Which of the following can change the value of the equilibrium constant, K?

   A. Changing pressure

   B. Changing temperature

   C. Changing volume

   D. Adding or removing a substance

9. The equilibrium constant, K, for the following reaction at 25 °C is 4.2 M^{-1}.

$$2SO_2(g) + O_2(g) \rightleftharpoons 2SO_3(g)$$

What is the numerical value of K for the reaction below?

$$2SO_3(g) \rightleftharpoons 2SO_2(g) + O_2(g)$$

   A. −4.2 M   **B.** 4.2 M   **C.** 2.1 M   **D.** 0.24 M

10. The reaction represented by the equation

$$A + B \rightleftharpoons 2C + D$$

has a rate that allows it to reach equilibrium over a period of hours.
After a period of time, a student measures the concentrations of the substances involved. The following results were obtained.

$$[A] = 0.15 \, M \quad [B] = 0.050 \, M \quad [C] = 0.20 \, M \quad [D] = 0.35 \, M$$

The value of the equilibrium constant for this reaction is 1.6 M.
Which of the following statements is true?

   A. The reaction is at equilibrium.

   B. The rate of the forward reaction is equal to the rate of the reverse reaction.

   C. The reaction is not at equilibrium and the rate of the forward reaction is greater than the rate of the reverse reaction.

   D. The reaction is not at equilibrium and the rate of the reverse reaction is greater than the rate of the forward reaction.

11. Currently there is much research into using the principles of green chemistry to make many important chemicals using methods that are more environmentally friendly.
This research would be least likely to involve

   A. biomimicry.

   B. the use of renewable energy.

   C. the development of low temperature catalysts.

   D. using higher pressures for more efficient reactions.

12. Which of the following is a *correct* statement regarding electrolysis?

   A. A spontaneous chemical reaction produces an electric current.

   B. Chemical energy is converted into electrical energy.

   C. Electrons flow in the internal circuit.

   D. The passage of an electric current through an electrolyte causes a chemical reaction.

13. Which of the following is an *incorrect* statement regarding electrolytic cells?

   A. Oxidation occurs at the anode.

   B. The cathode is positive.

   C. Anions travel to the anode.

   D. Electrons travel from the anode to the cathode.

14. Using inert electrodes, what will the product at the anode be when $CuSO_4$(aq) is electrolysed?

   A. Cu(s)   B. SO_2(g)   C. Cu^{2+}(aq)   D. O_2(g)

15. Using copper electrodes, what will the product at the anode be when $CuSO_4$(aq) is electrolysed?

   A. Cu(s)   B. SO_2(g)   C. Cu^{2+}(aq)   D. O_2(g)

16. During the commercial production of sodium, a small amount of calcium chloride is added to the molten sodium chloride to reduce its melting point.
   Why does this not interfere with the production of sodium?

   A. Calcium ions are more difficult to reduce than sodium ions at the anode.

   B. Calcium ions are more difficult to reduce than sodium ions at the cathode.

   C. Sodium ions are easier to oxidise at the anode than calcium ions.

   D. The two metals have different densities and do not mix once they are formed.

17. Which of the following is required for a battery to be rechargeable?

   A. The products of discharge must be solid.

   B. The products of discharge must remain in contact with the electrodes.

   C. The electrolyte must be acidic.

   D. Water must not be present in the battery.

18. Hydrogen produced by which of the following methods would be regarded as 'green' hydrogen?

   A. Steam reforming of methane with capture of the carbon dioxide produced

   B. Steam reforming of methane without capture of the carbon dioxide produced

   C. Electrolysis of water using electricity from a renewable source

   D. Electrolysis of water using electricity generated from gas turbines rather than coal

19. In a polymer electrolyte membrane electrolysis cell

   A. water moves through a thin membrane from the cathode to the anode.

   B. water moves through a thin membrane from the anode to the cathode.

   C. hydrogen is produced at the anode.

   D. protons move through a thin membrane from the anode to the cathode.

20. A solution of iron(II) sulfate is electrolysed by a current of 6.0 A for 5.0 minutes.
   What is the mass of iron deposited?

   A. 1.04 g

   B. 0.35 g

   C. 0.52 g

   D. 0.017 g

SECTION B — Short answer questions

Question 21 (3 marks)

Explain how using a catalyst affects:

a. the rate of a reaction   **(2 marks)**

b. the extent of a reaction.   **(1 mark)**

Question 22 (11 marks)

Consider the following reaction:

$$4NH_3(g) + 5O_2(g) \rightleftharpoons 4NO(g) + 6H_2O(g) \qquad \Delta H = -900 \, kJ \, mol^{-1}$$

An initial mixture in which all concentrations were 2.0 M was allowed to reach equilibrium. At equilibrium, the concentration of NO(g) was found to be 1.4 M.

a. Calculate the equilibrium constant, K, for the equation. **(4 marks)**
b. What can be said about the position of equilibrium? **(1 mark)**
c. Predict the effect of an increase in temperature, with reasoning, on:
   i. the rate of reaction **(1 mark)**
   ii. the position of equilibrium. **(2 marks)**
d. Predict the effect of a decrease in pressure, with reasoning, on:
   i. the rate of reaction **(1 mark)**
   ii. the position of equilibrium. **(2 marks)**

Question 23 (7 marks)

Using graphite electrodes, 1.0 M $MgCl_2$(aq) undergoes electrolysis.

a. Write the oxidation and reduction half-equations when electrolysis is occurring. **(2 marks)**
b. Determine the overall equation for the reaction. **(1 mark)**
c. Determine the minimum cell voltage required. **(1 mark)**
d. How would the products differ if concentrated $MgCl_2$(aq) was electrolysed? **(1 mark)**
e. Write the half-equation for the negative electrode if the magnesium chloride was molten. **(2 marks)**

Question 24 (5 marks)

Chromium chloride undergoes electrolysis using chromium electrodes. A current of 0.400 A flows for 1.00 hour. The increase in mass of the cathode is 0.259 g.

What is the charge on the chromium ion?

Question 25 (4 marks)

In a zinc–cerium secondary cell, the following reactions occur during discharge:

$$\text{Electrode 1: } Zn^{2+}(aq) + 2e^- \rightarrow Zn(s)$$
$$\text{Electrode 2: } Ce^{4+}(aq) \rightarrow Ce^{3+}(aq) + e^-$$

a. When recharging:
   i. which electrode do electrons flow towards **(1 mark)**
   ii. which electrode is negative **(1 mark)**
   iii. what will happen to the mass of electrode 1? **(1 mark)**
b. Write the overall reaction when the cell is recharging. **(1 mark)**

PRACTICE SCHOOL-ASSESSED COURSEWORK

ASSESSMENT TASK — COMPARISON AND EVALUATION OF CHEMICAL CONCEPTS, METHODOLOGIES AND METHODS, AND FINDINGS FROM AT LEAST TWO PRACTICAL ACTIVITIES

For this task, you will produce a graphic organiser based on your learnings from class practicals relating to optimising the rate and yield of a chemical product.

- Pens, pencils, highlighters, erasers, rulers and a scientific calculator are permitted.
- You may use your practical logbook and the VCE Chemistry Data Book to complete this task.

Total time: 50 minutes
Total marks: 54 marks

GRAPHIC ORGANISER

In completing this task, please consider the following:
- How reactions occur (collision theory)
- Factors that affect the rate of reactions
- Chemical equilibrium
- Equilibrium law
- Electrolysis — process
- Electrolysis — commercial and innovative examples
- Faraday's Laws

The start of the graphic organiser has been prepared for you.

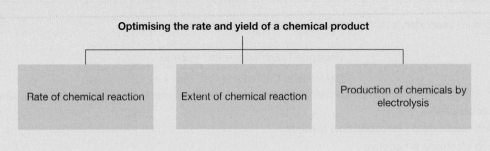

Optimising the rate and yield of a chemical product

| Rate of chemical reaction | Extent of chemical reaction | Production of chemicals by electrolysis |

on Resources

📄 **Digital document** U3AOS1 School-assessed coursework (doc-39698)

How are carbon-based compounds designed for purpose?

Source: VCE Chemistry Study Design (2024–2027) extracts © VCAA; reproduced by permission.

7 Structure, nomenclature and properties of organic compounds

KEY KNOWLEDGE

In this topic you will investigate:

Structure, nomenclature and properties of organic compounds

- characteristics of the carbon atom that contribute to the diversity of organic compounds formed, with reference to valence electron number, relative bond strength, relative stability of carbon bonds with other elements, degree of unsaturation, and the formation of structural isomers
- molecular, structural and semi-structural (condensed) formulas and skeletal structures of alkanes (including cyclohexane), alkenes, benzene, haloalkanes, primary amines, primary amides, alcohols (primary, secondary and tertiary), aldehydes, ketones, carboxylic acids and non-branched esters
- the International Union of Pure and Applied Chemistry (IUPAC) systematic naming of organic compounds up to C8, with no more than two functional groups for a molecule, limited to non-cyclic hydrocarbons, haloalkanes, primary amines, alcohols (primary, secondary and tertiary), aldehydes, ketones, carboxylic acids and non-branched esters
- trends in physical properties within homologous series (boiling point and melting point, viscosity), with reference to structure and bonding.

Source: VCE Chemistry Study Design (2024–2027) extracts © VCAA; reproduced by permission.

PRACTICAL WORK AND INVESTIGATIONS

Practical work is a central component of VCE Chemistry. Experiments and investigations, supported by a **practical investigation eLogbook** and **teacher-led videos**, are included in this topic to provide opportunities to undertake investigations and communicate findings.

EXAM PREPARATION

▶ Access past VCAA questions and exam-style questions and their video solutions in every lesson, to ensure you are ready.

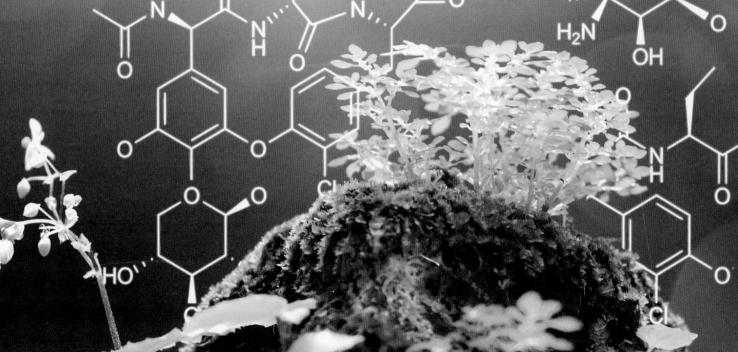

7.1 Overview

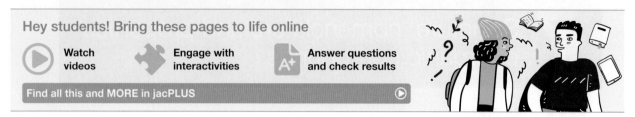
7.1.1 Introduction

Plastics, fuels and medicines, as well as simple and complex life forms, make up only a fraction of the millions of natural and synthetic organic compounds that exist. To understand organic chemistry, we need to learn about carbon and its unique physical and chemical properties. We need to understand the structure of carbon compounds and learn how to represent and name these molecules.

Organic compounds are marketed and sold to us every day, but they might not be instantly recognisable. For example, Prozac is the brand name given to a medication that treats a variety of conditions, including depression and anxiety. It has a molecular formula of $C_{17}H_{18}F_3NO$. Calling it 'Prozac' is a lot easier than using its systematic name: N-methyl-3-phenyl-3-4-(trifluoromethyl) phenoxypropan-1-amine!

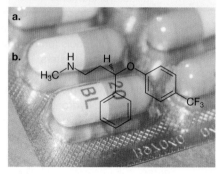

FIGURE 7.1 Prozac **a.** as capsules prescribed by doctors and **b.** as a chemical structure

However, as Prozac and other commercial brand names are used to market the same chemical, a naming system maintained by the International Union of Pure and Applied Chemistry (IUPAC) is used to avoid confusion.

In this topic, you will examine carbon and its tendency to bond with itself and other elements in many stable forms that result in compounds with diverse physical properties. These compounds are named systematically and can be drawn as full structures, semi-structures and skeletal structures.

LEARNING SEQUENCE

7.2 Characteristics of the carbon atom

KEY KNOWLEDGE

- Characteristics of the carbon atom that contribute to the diversity of organic compounds formed, with reference to valence electron number, relative bond strength, relative stability of carbon bonds with other elements and degree of unsaturation

7.2.1 Carbon: a remarkable element

The carbon atom is the one constant in the millions of organic compounds either found in natural substances or made (synthesised) in the laboratory. It not only forms the basis of all life on Earth, but also is the primary component of fossil fuels and the plastics and many compounds we use every day. This incredible variety can even be seen in samples of pure carbon, which exist in different chemical and physical forms. The different physical forms in which an element can exist are called **allotropes**.

FIGURE 7.2 Allotropes — different physical forms of carbon: **a.** diamond **b.** glassy carbon **c.** carbon nanotube **d.** graphite pencil

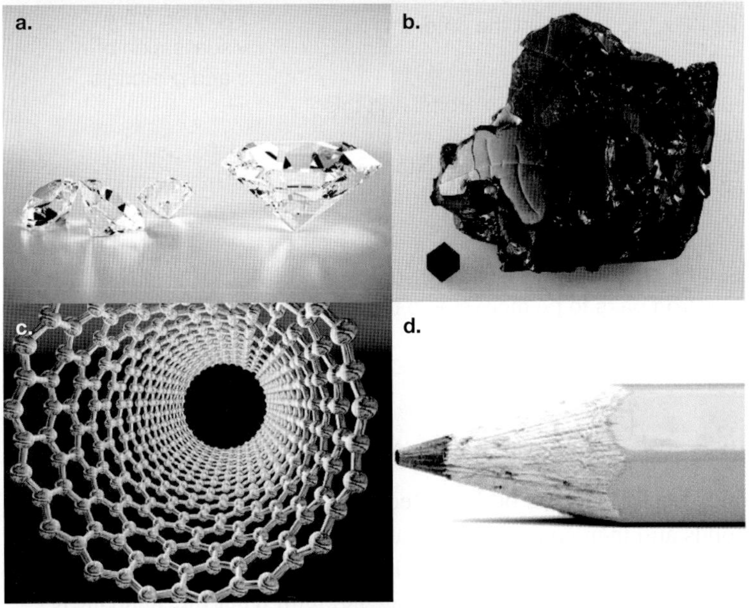

FIGURE 7.3 A carbon atom demonstrating four valence electrons

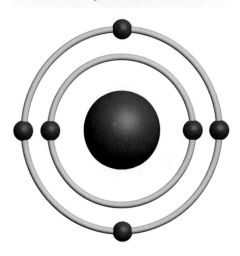

Carbon's ability to form so many compounds is due to its arrangement of electrons. Carbon is located at the top of group 14 of the periodic table of elements and in the second period. This location is determined not only by the six protons in a carbon nucleus (atomic number), but also the **electron configuration**; that is, the number of electrons in each shell.

A neutral carbon atom has six electrons. Only two of carbon's six electrons occupy the first electron shell; therefore, the remaining four valence electrons (**valence number** = 4) are found in the second, outermost shell. These four valence electrons are available for bonding.

Carbon can therefore use four covalent bonds to combine with other carbon atoms or non-metals to form small or extremely large molecules. Millions of different compounds can be formed by carbon; these are studied in organic chemistry.

allotropes the different physical forms in which an element can exist

electron configuration the number of electrons and shells they occupy (e.g. 2,4 for a carbon atom)

valence number the number of electrons occupying the orbitals in the outermost electron shell

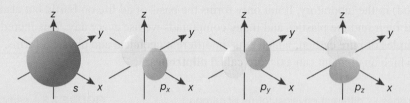
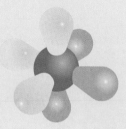
7.2.2 Stability of carbon bonds

Energy transfer is involved when chemical bonds are formed and broken. Recall from Unit 3 the energy profiles showing activation energy required to break bonds and then an amount of energy lost when new bonds are formed. Carbon forms stable, lower energy compounds when its valence shell resembles that of neon (2, 8); therefore, it bonds with other atoms to achieve this electron arrangement. How and with what this happens is varied and complex.

Bond energy

The strength of a covalent bond is measured by the energy needed to pull the atoms apart. This is called the **bond energy** and has a unit of kilojoule(s) per mole ($kJ\ mol^{-1}$).

Bond energy:
- can be defined as the amount of energy required to break the bonds of a mole of molecules into its individual atoms
- depends on the atoms involved in sharing a covalent bond and is affected by the distance between the two atoms. If atoms are too close they repel, and if they are too far away they are unable to share the electrons. The distance between the nuclei of the atoms sharing the electrons is known as the **bond length**. Shorter bonds are usually more difficult to break. For example, when a carbon atom is bonded to a halogen atom, C–F bonds are stronger than C–Cl bonds, which are stronger than C–Br bonds.

excited state refers to when electrons move to higher energy orbitals when energy is applied

bond energy the amount of energy required to break the bonds of a mole of molecules into its individual atoms

bond length the distance between two nuclei involved in covalent bonding, which depends on the size of the atoms

It can be seen in table 7.1 that carbon–carbon bonds are relatively strong, as are carbon–hydrogen bonds, so hydrocarbon molecules are quite stable.

TABLE 7.1 Comparison of bond energies (in kJ mol^{-1}). Carbon atoms form relatively strong bonds with carbon atoms as well as with other elements.

Hydrogen bonds		Carbon bonds		Nitrogen bonds		Oxygen bonds		Same elements	
H–H	436	C–H	414	N–H	391	O–H	463	C–C	346
H–C	414	C–C	346	N–C	286	O–C	358	C=C	614
H–N	391	C–N	286	N–N	158	O–N	214	C≡C	839
H–O	463	C–O	358	N–O	214	O–O	144	Cl–Cl	242
H–F	567	C–F	492	N–F	278	O–F	191	Br–Br	193
H–Cl	431	C–Cl	324	N–Cl	192	O–Cl	206	I–I	151

Strength of a covalent bond

The strength of a covalent bond is indicated by the bond energy; that is, the amount of energy needed to separate the two atoms completely.

It is possible to calculate the energy needed to break the bonds in a molecule by using the data in table 7.1. Remember that the bond energy values are stated per mole and that the compound must be in the gaseous state. So, to calculate the energy required to convert liquid water into hydrogen and oxygen atoms, energy is first required to vaporise the water, and then more energy is required to separate the atoms.

Bond angle and stability

The ability of carbon to form millions of compounds is also dependent upon the geometry (spatial arrangement) of atoms attached to it. The shape of molecules is determined by the maximum repulsion of electron pairs in a molecule and results in **covalent bonds** with greater stability. For example, when carbon forms four, single covalent bonds, the bonds separate so that the angle between the bonds is 109.5°, forming a tetrahedral molecule shape.

FIGURE 7.6 The tetrahedral geometry of CH$_4$

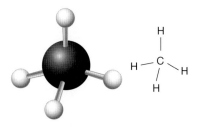

Multiple carbon-to-carbon bonds

The majority of carbon atoms bond at this approximate angle of 109.5°. However, carbon can react in such a way that C=C double covalent bonds form. In this scenario, other atoms bonded to the two carbon atoms have bonds that are spaced at an approximate 120° angle, resulting in a planar geometry.

The bonds in a C=C double bond are shorter and stronger than C–C single bonds. It takes almost twice the amount of energy to break a C=C bond than it does a C–C bond. This is not surprising given there are four electrons providing stability in a C=C bond compared to just two in a C–C bond. The bonds in the simplest C=C molecule, C$_2$H$_4$, occur in the same plane. This is why it is often referred to as a flat molecule (figure 7.7).

covalent bonds bonds that involve the sharing of electron pairs between atoms

Carbon-to-carbon triple bonds have a linear geometry and 180° between bonds (see figure 7.8). C≡C triple bonds are stronger than C=C bonds. They are the shortest of the carbon-to-carbon bonds, with a length of approximately 120 picometres (1.2×10^{-10} m), compared to 134 pm for double and 154 pm for single carbon-to-carbon bonds.

FIGURE 7.7 The planar, flat geometry of C_2H_4

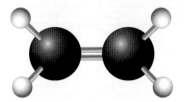

FIGURE 7.8 The planar, flat geometry of C_2H_2

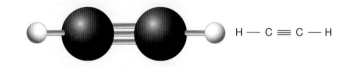

tlvd-9693

SAMPLE PROBLEM 1 Calculating the energy required to break covalent bonds in a gas

Calculate the energy, in kJ, required to break all covalent bonds in 1.6 g of methane (CH_4) gas.

THINK	WRITE
1. The standard unit for bond energy is kJ mol^{-1}, but the question has given a mass of CH_4 and not an amount in mol. Therefore, we need to convert mass into moles using $n = \dfrac{m}{M}$.	$n = \dfrac{m}{M}$ $= \dfrac{1.6\,\text{g}}{16.0\,\text{g mol}^{-1}}$ $= 0.10\,\text{mol}$
2. There are four C–H bonds in CH_4. Use table 7.1 to find the value stated for a C–H bond and multiply it by four.	Energy in C–H bonds $= 4 \times 414$ $= 1656\,\text{kJ mol}^{-1}$
3. To find the total bond energy to break all the bonds in 1.6 g of methane, multiply bond energy per mol calculated in step 2 by the number of moles in 1.6 g of methane. **TIP:** Remember to give your answer to the correct number of significant figures.	$0.10\,\text{mol} \times 1656\,\text{kJ mol}^{-1} = 166\,\text{kJ}$ $= 1.7 \times 10^2\,\text{kJ}$

PRACTICE PROBLEM 1

Calculate the energy required, in kJ, to break all covalent bonds in 64 g of methane (CH_4) gas.

7.2.3 Degree of unsaturation

If a molecule has only single carbon-to-carbon bonds it is said to be **saturated**, but if double or triple carbon-to-carbon bonds are present, the molecule is described as **unsaturated**. The degree of unsaturation can be measured by reacting a compound with iodine. The **iodine number** or value is the mass of iodine that reacts with 100 g of the compound. The higher the number, the greater the number of carbon-to-carbon double bonds. The procedure for measuring iodine number is discussed in topic 9.

saturated describes hydrocarbons containing only single carbon–carbon bonds

unsaturated describes hydrocarbons containing at least one double or triple carbon–carbon bond

iodine number the mass of iodine that reacts with 100 g of a compound

Iodine number

The iodine number refers to the mass of iodine that reacts with 100 g of a compound.

tlvd-8928

SAMPLE PROBLEM 2 Determining the degree of unsaturation of a molecule

Determine the degree of unsaturation (iodine number) for palmitoleic acid ($C_{16}H_{30}O_2$), a monounsaturated fatty acid that is found in plants.

THINK	WRITE
1. The iodine number is the number of grams of iodine that reacts with 100 g of a compound. Convert 100 g of palmitoleic acid to moles.	$$n(C_{16}H_{30}O_2) = \frac{m}{M}$$ $$= \frac{100}{(16 \times 12.0) + (30 \times 1.0) + (2 \times 16.0)}$$ $$= \frac{100}{254.0}$$ $$= 0.3937 \, mol$$
2. Palmitoleic acid has one double bond, so will react with one mole of iodine (I_2). So, $n(C_{16}H_{30}O_2) = n(I_2)$.	$$n(I_2) = n(C_{16}H_{30}O_2)$$ $$= 0.3937 \, mol$$
3. Calculate the mass of iodine that reacts.	$$m(I_2) = n \times M$$ $$= 0.3937 \times (126.9 \times 2)$$ $$= 99.9 \, g/100 \, g$$

PRACTICE PROBLEM 2

Linoleic acid ($C_{18}H_{32}O_2$) is a polyunsaturated fatty acid containing two double bonds. Calculate the iodine number for linoleic acid.

7.2 Activities

learnon

Students, these questions are even better in jacPLUS

 Receive immediate feedback and access sample responses

 Access additional questions

Track your results and progress

Find all this and MORE in jacPLUS

7.2 Quick quiz on	7.2 Exercise	7.2 Exam questions

7.2 Exercise

1. **MC** Bond energy is defined as the energy required to
   A. combine atoms to make a mole of a compound.
   B. break the bonds in a mole of a compound.
   C. burn a mole of a compound.
   D. evaporate a mole of a compound.

2. **MC** The length of a carbon-to-carbon double bond is
   A. longer than the length of a carbon-to-carbon single bond.
   B. shorter than the length of a carbon-to-carbon single bond.
   C. the same as the length of a carbon-to-carbon single bond.
   D. unable to be determined.
3. **MC** The molecule that makes up the gas ethene is a
   A. planar molecule with a single covalent bond between the carbon atoms.
   B. tetrahedral molecule with single covalent bonds between the carbon atoms.
   C. planar molecule with a double covalent bond between the carbon atoms.
   D. linear molecule with double covalent bonds between the carbon atoms.
4. **MC** The degree of unsaturation is measured by
   A. determining the number of moles of iodine that reacts with 100 g of an organic compound.
   B. determining the mass of iodine that reacts with 100 g of an organic compound.
   C. determining the number of moles of an organic compound that reacts with 100 g of iodine.
   D. determining the mass of an organic compound that reacts with 100 g of iodine.
5. Give two reasons why the element carbon is able to form a diverse range of carbon compounds.
6. What is the difference in C–H bond angles between C_2H_2 and C_2H_4 molecules?
7. If the bond energy of H–F is 567 kJ mol^{-1}, what is the overall bond energy of two moles of HF?
8. What amount of energy per mole would be released if all covalent bonds in CH_2F_2 were broken?
9. Refer to table 7.1. List the bonds and bond energies present in the following molecule from smallest to largest.

$$
\begin{array}{c}
\quad\;\; H \quad H \quad H \quad H \\
\quad\;\; | \qquad | \qquad | \qquad | \\
H - C - C = C - C - O - H \\
\quad\;\; | \qquad\qquad\qquad | \\
\quad\;\; H \qquad\qquad\quad\; H
\end{array}
$$

10. Use the data in table 7.1 to calculate whether it takes more energy to break all the bonds in ethene or to break all the bonds in ethane.

7.2 Exam questions

Question 1 (1 mark)

Source: VCE 2022 Chemistry Exam, Section A, Q.29; © VCAA

MC One mole of methane, CH_4, reacts with one mole of halogen, X_2. X can be fluorine, F, chlorine, Cl, or bromine, Br. The general equation for the reaction is given below.

$$CH_4(g) + X_2(g) \xrightarrow{\text{catalyst}} CH_3X(g) + HX(g) \quad \Delta H < 0$$

Which one of the following statements is true?
A. The strength of the bonds from weakest to strongest is C–Br < C–Cl < C–F.
B. Since hydrogen has the smallest atomic radius, the C–H bond is the weakest bond.
C. The C–Br bond is stronger than the C–H bond because of the size of the bromine atom.
D. The C–Br, C–Cl and C–F bonds are equal in strength because Br, Cl and F are halogens.

Question 2 (1 mark)

MC Which of the following statements is *incorrect*?
A. Single carbon-to-carbon bonds have a lower bond energy and longer bond length than double carbon-to-carbon bonds.
B. Double carbon-to-carbon bonds have a higher bond energy and longer bond length than triple carbon-to-carbon bonds.
C. Triple carbon-to-carbon bonds have a higher bond energy and shorter bond length than double carbon-to-carbon bonds.
D. Double carbon-to-carbon bonds have a higher bond energy than single carbon to carbon bonds, but less bond energy than triple carbon-to-carbon bonds.

▶ **Question 3 (1 mark)**

Explain why C=C bonds are stronger than C–C bonds.

▶ **Question 4 (2 marks)**

Use the following table to calculate the energy, in kJ, required to break all covalent bonds in 25.0 g of C_2H_4 gas.

Bond	Bond energy (kJ mol^{-1})
C–H	414
C–C	346
C=C	614

▶ **Question 5 (4 marks)**

Water and hydrogen peroxide are both made up of the elements hydrogen and oxygen. Bond energies (in kJ mol^{-1}) for hydrogen and oxygen bonds are provided in the following table.

Hydrogen bonds		Oxygen bonds	
H–H	436	O–H	463
H–C	414	O–C	358
H–N	391	O–N	214
H–O	463	O–O	144
H–F	567	O–F	191
H–Cl	431	O–Cl	206

a. i. Calculate the energy required to separate the atoms in one mole of liquid water if the heat of vaporisation for water is 40.8 kJ mol^{-1}. **(1 mark)**
   ii. Explain why the bond energy is so high compared to the heat of vaporisation. **(1 mark)**
b. Which of the following is easier to break: the O–H bond in water, H_2O, or the O–O bond in hydrogen peroxide, H_2O_2? **(1 mark)**
c. Explain which molecule would be less stable. **(1 mark)**

More exam questions are available in your learnON title.

7.3 Structure and systematic naming

KEY KNOWLEDGE

- Molecular, structural and semi-structural (condensed) formulas and skeletal structures of alkanes (including cyclohexane), alkenes and benzene
- The International Union of Pure and Applied Chemistry (IUPAC) systematic naming of organic compounds up to C8 with no more than two functional groups for a molecule, limited to non-cyclic hydrocarbons

Source: Adapted from VCE Chemistry Study Design (2024–2027) extracts © VCAA; reproduced by permission.

7.3.1 Representing organic compounds

We model the way atoms are bonded and arranged in a molecule in a number of ways. The simplest molecular models are Lewis (electron dot) diagrams and structural diagrams, which were covered in Unit 1. As molecules become larger and more complex, we look for easier ways to represent all of the bonded atoms.

TABLE 7.2 Different ways of representing butane molecules

Formula	Example	Description
Molecular	C_4H_{10}	The number and kinds of atoms in a molecule
Empirical	C_2H_5	The simplest whole-number ratio of atoms in a molecule
Structural	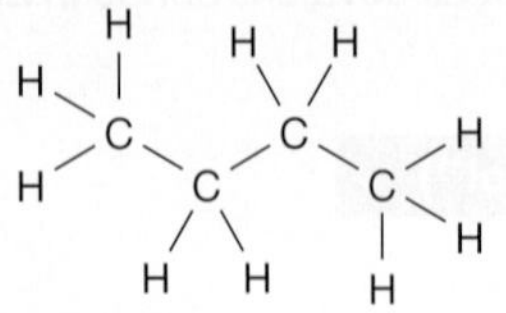 or, in 2D, 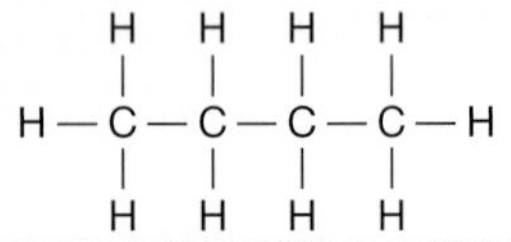	The actual arrangement of atoms in a molecule. As molecules become longer, the second example of a structural formula tends to be used. Structural formulas can be modified to write semi-structural and skeletal formulas.
Semi-structural or condensed	$CH_3CH_2CH_2CH_3$ or $CH_3(CH_2)_2CH_3$	Can be written on a single line, with each carbon atom being followed by the atoms that are joined to it. Repeated CH_2 groups can be collected in brackets with a subscript as shown.
Skeletal		Skeletal structural formulas are a further simplification of semi-structural formulas. Skeletal structures use lines and vertices to simplify a structural formula by omitting the carbon and the hydrogen atoms bonded to it. *Note:* <ul><li>It is assumed that a carbon atom (and enough hydrogen atoms to satisfy carbon's valency) is present at each vertex (and at the ends).</li><li>Double bonds and other different types of atoms are specifically shown.</li></ul>Skeletal structures preserve the bond angles in a carbon chain, and are the preferred method for representing complex organic molecules that are large and often contain ring or cyclic structures.
3D structural or shape diagram	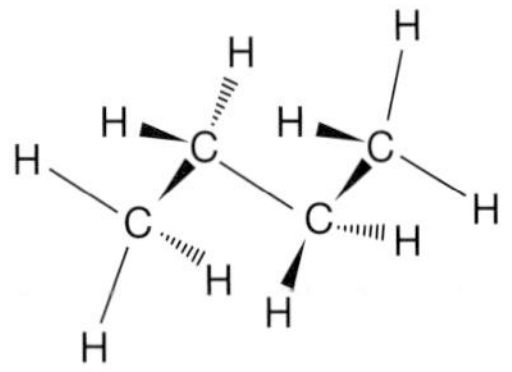	Shows the 3D arrangement of atoms (wedge–dash), with the bonds represented as follows: <ul><li>The continuous line is in the plane of the paper.</li><li>The dashed line extends to the back of the plane of the paper.</li><li>The solid wedge comes out of the plane of the paper.</li></ul>

---

tlvd-9694

SAMPLE PROBLEM 3 Drawing semi-structures and skeleton structures

For the structure shown, draw:
a. a semi-structural formula
b. a skeletal structure.

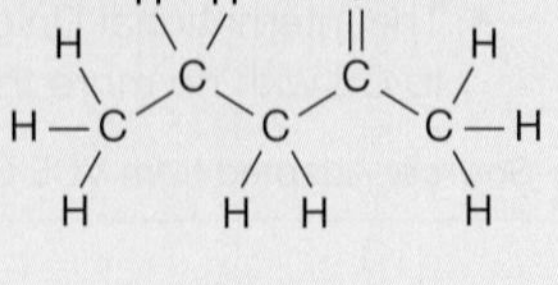

THINK

Recall that semi-structures condense the carbon chain by removing the covalent bonds while preserving the order of the atoms, and skeletal structures remove the C and their H atoms from the structural diagram.

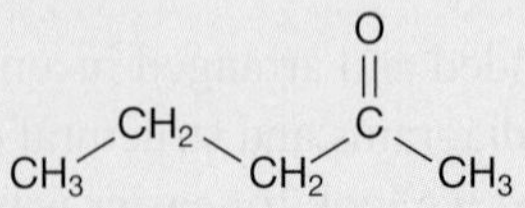

WRITE

Draw the structure, but condense all hydrogen atoms connected to each carbon.

a. For the semi-structure, remove all of the covalent bonds and write out the sequence of groups in the chain.

b. For the skeletal structure, remove all of the C and H atoms attached to the covalent bonds but retain the bonds.

a. $CH_3CH_2CH_2COCH_3$

b.

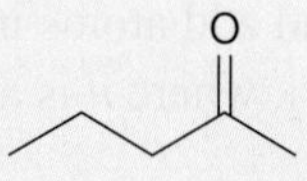

PRACTICE PROBLEM 3

For the structure shown, draw:
a. **a semi-structural formula**
b. **a skeletal structure.**

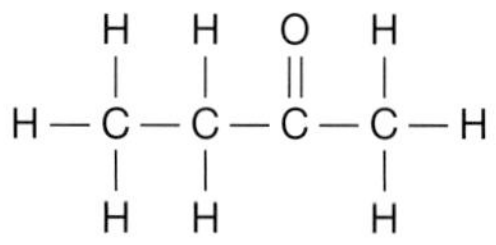

7.3.2 Hydrocarbon families

Hydrocarbons are the simplest organic compounds and are composed solely of carbon and hydrogen; they were introduced in Unit 1. They are obtained mainly from crude oil and are used as fuels or solvents, or in the production of plastics, dyes, pharmaceuticals, explosives and other industrial chemicals. The organic families studied in this topic contain various percentages of carbon and hydrogen atoms. They include the alkanes, alkenes and alkynes, which are classified as **aliphatic** compounds. If a compound contains one or more benzene rings, it is described as an **aromatic** compound.

A family of carbon compounds that are structurally related and in which members of the family can be represented by a general formula is called a **homologous series**. Successive members of a homologous series have formulas that differ by CH_2. Each is named for the number of carbon atoms in the longest chain.

aliphatic describes organic compounds in which carbon atoms form open chains

aromatic describes a compound that contains at least one benzene ring and is characterised by the presence of alternating double and single bonds within the ring

homologous series a series of organic compounds that have the same structure but in which the formula of each molecule differs from the next by a CH_2 group

FIGURE 7.9 The structural arrangement of ethane (alkane), ethene (alkene), ethyne (alkyne) and the benzene ring (aromatic). Carbon can form ring structures as well as single or multiple bonds with itself.

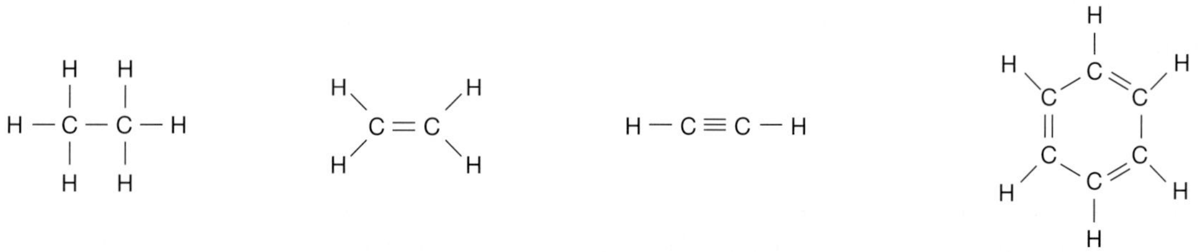

Ethane   Ethene   Ethyne   Benzene

Alkanes

Alkanes with carbon atoms in long chains are known as straight-chain hydrocarbons.

Alkanes:
- are classified as saturated hydrocarbons because only single covalent bonds exist between atoms, and there are no available multiple carbon bonds to break and add atoms into the molecule
- have the general formula C_nH_{2n+2}, where n is an integer.

FIGURE 7.10 Propane, the third member of the alkane homologous series, is used to fly hot air balloons.

> **Alkanes**
> - Alkanes are saturated hydrocarbons with only single bonds between the carbon atoms.
> - Alkanes have the general formula C_nH_{2n+2}.

The first four alkanes are gases at room temperature, and are summarised in table 7.3.

TABLE 7.3 The first four members of the alkane homologous series

Alkane	Semi-structural (condensed) formula	Source	Uses
Methane	CH_4	Natural gas or biogas	• Fuel • Synthesis of other chemicals
Ethane	CH_3CH_3	Natural gas	• Manufacture of ethene • Refrigerant in cryogenic systems
Propane	$CH_3CH_2CH_3$	Natural gas processing or petroleum refining	• Fuel (e.g. in gas cylinders for heating) • Propellant for aerosols
Butane	$CH_3CH_2CH_2CH_3$	Natural gas processing or petroleum refining	• Fuel (e.g. cigarette lighters and portable stoves) • Synthesis of other chemicals • Propellant for aerosols

The next four members of the alkane homologous series are:
- pentane, C_5H_{12}
- hexane, C_6H_{14}
- heptane, C_7H_{16}
- octane, C_8H_{18}.

Naming alkanes

The name of each alkane has two parts. The prefix (the start) of each name tells us how many carbon atoms are in the straight chain. The suffix (the end) -*ane* of each name tells us that the hydrocarbon is an alkane. The prefixes are used to name the number of carbon atoms in a chain in the majority of the homologous series studied in this topic.

alkanes the family of hydrocarbons containing only single carbon–carbon bonds

 Resources

▶ **Video eLesson** Naming alkanes (eles-2484)

Alkenes

Alkenes are a homologous series with double bonds between carbon atoms. They are formed when two hydrogen atoms are removed from alkanes.

Alkenes:
- are unsaturated, due to the presence of at least one double carbon–carbon bond
- have the general formula C_nH_{2n}.

The first two members of the alkene series are ethene, C_2H_4, and propene, C_3H_6. Their structural formulas are shown in figure 7.11.

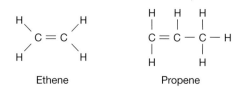

FIGURE 7.11 Structural formulas of ethene and propene

Ethene   Propene

FIGURE 7.12 These tomatoes are the same age but the red one has been ripened using ethene gas.

Alkenes

- Alkenes are unsaturated hydrocarbons with a double bond between two carbon atoms.
- Alkenes have the general formula C_nH_{2n}.

---

alkenes the family of hydrocarbons that contain at least one carbon–carbon double bond

Ethene is also commonly known as ethylene. It is produced naturally by some plants and aids in ripening fruit. It can also be produced artificially by heating petroleum in the absence of air in a process called cracking. Ethene is an important raw product for making many chemicals and plastics.

Naming alkenes

Table 7.4 shows the first seven members of the alkene series. Alkenes are named using the same general rules described for alkanes except that the suffix -*ene* is added instead of -*ane*, and the number of the carbon atom after which the double bond is positioned is indicated. The longest unbranched chain must contain the double bond, so the molecule $CH_3CH_2CH=CHCH_3$ is named pent-2-ene. The '2' indicates the position of the double bond between carbon atoms 2 and 3 (the lower number is used in the formula, and numbering starts from the carbon atom closest to the double bond), and 'pent' indicates that five carbon atoms are present in the unbranched chain. Some people prefer to name it 2-pentene. Either way, the number of carbon atoms in the chain and the position of the carbon bond are indicated. For consistency, we will use the first naming method.

TABLE 7.4 Members of the alkene homologous series

Systematic name	Formula	Semi-structural formula with double bonds
Ethene	C_2H_4	$H_2C=CH_2$
Prop-1-ene	C_3H_6	$H_2C=CHCH_3$
But-1-ene	C_4H_8	$H_2C=CHCH_2CH_3$
Pent-1-ene	C_5H_{10}	$H_2C=CH(CH_2)_2CH_3$
Hex-1-ene	C_6H_{12}	$H_2C=CH(CH_2)_3CH_3$
Hept-1-ene	C_7H_{14}	$H_2C=CH(CH_2)_4CH_3$
Oct-1-ene	C_8H_{16}	$H_2C=CH(CH_2)_5CH_3$

 Resources

▶ **Video eLesson** Homologous series of alkenes (eles-2477)

Alkynes

Alkynes:

- contain a carbon–carbon triple bond, and are therefore unsaturated
- have the general formula C_nH_{2n-2}
- include, for example, ethyne, C_2H_2 ($HC{\equiv}CH$), and propyne, C_3H_4 ($HC{\equiv}CCH_3$). Ethyne is used to produce ethane and in oxyacetylene torches for welding to join metals. It can heat objects up to 3000 °C.

FIGURE 7.13 Prop-1-yne

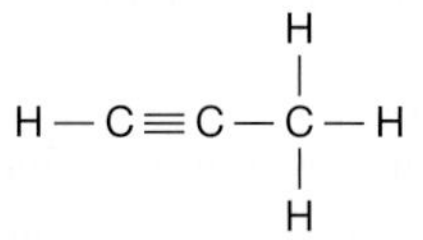

$$H-C{\equiv}C-\underset{\underset{H}{|}}{\overset{\overset{H}{|}}{C}}-H$$

Alkynes

- Alkynes are unsaturated hydrocarbons with a triple bond between two carbon atoms.
- Alkynes have the general formula C_nH_{2n-2}.

Naming alkynes

Alkynes are named using the same general rules as for alkenes except that the *-ene* is replaced with *-yne*.

Cyclic hydrocarbons

Cyclic hydrocarbons are also known as ring structures because the carbon chain is a closed structure without open ends.

Single-ringed cycloalkanes have the same molecular formula as alkenes due to all carbon atoms being covalently bonded to two others either side to form the closed ring. As well as having a different molecular formula to straight-chain alkanes, the prefix '*cyclo-*' is used to indicate the ring structure.

For example, the cyclic hydrocarbon cyclohexane has the molecular formula C_6H_{12} and is a colourless, flammable liquid that is used as a reactant in the production of nylon.

FIGURE 7.14 The polystyrene foam and the epoxy resin in surfboards make them light and strong, respectively. Both chemicals contain cyclic hydrocarbon groups.

Note that the formula for cycloalkanes does not follow the general formula of non-cyclic alkanes because there are fewer hydrogen atoms than in the non-cyclic alkanes.

FIGURE 7.15 Structural diagrams and model of cyclohexane

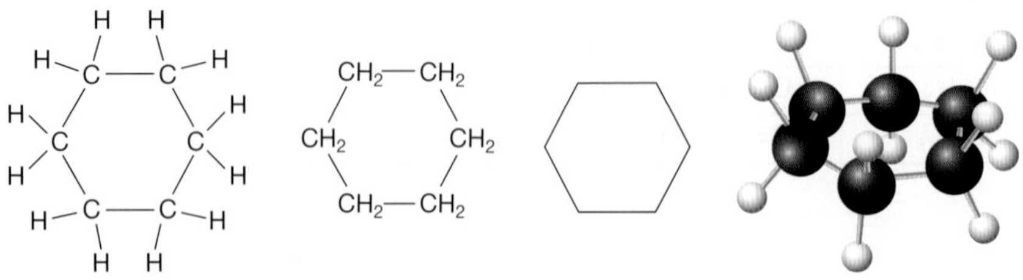

alkynes the family of hydrocarbons with a carbon–carbon triple bond

cyclic hydrocarbons also known as ring structures, because the carbon chain is a closed structure without open ends

Another important group of cyclic hydrocarbons are the **arenes**. These compounds are derived from **benzene**. The benzene molecule, C_6H_6, consists of six carbon atoms arranged in a ring with one hydrogen atom bonded to each carbon. Originally, it was thought that there were alternating single and double carbon–carbon bonds in the ring. However, the lack of reactivity, high stability and same bond lengths between the carbon atoms did not support this theory. Currently, benzene is considered to be a molecule with six electrons from the three double bonds shared by all of the carbon atoms in the ring. The attraction of the electrons to all of the carbon atoms gives the molecule stability.

Benzene is a very important compound in organic chemistry. Even though benzene itself is carcinogenic, many of the chemicals produced from it are not. In fact, many foods and pharmaceuticals, such as paracetamol, contain benzene rings. There are various ways of representing the benzene ring, as shown in figure 7.16.

FIGURE 7.16 Representations of benzene

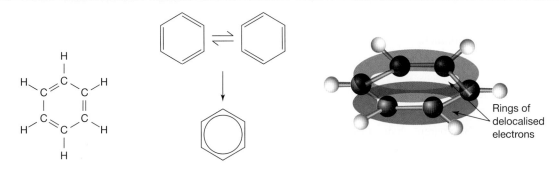

elog-1896

EXPERIMENT 7.1 online only

Constructing models of hydrocarbons

Aim

To construct models of alkanes, alkenes, alkynes and some of their isomers

Alkyl groups

Alkyl groups are hydrocarbon branches coming off the longest carbon chain of an organic molecule. Alkyl branches use the same prefixes to represent the number of carbons in the branch as those used in straight-chain molecules, but the suffix changes to '-*yl*'. To branch off the main chain, alkyl groups have one less hydrogen atom than the alkanes that share the same prefix in their name.

arenes aromatic, benzene-based hydrocarbons

benzene an aromatic hydrocarbon with the formula C_6H_6

alkyl groups hydrocarbon branches joined to the parent hydrocarbon chain (e.g. CH_3 (methyl), CH_2CH_3 (ethyl))

TABLE 7.5 The first three alkane and alkyl groups

Alkane	Semi-structural formula	Alkyl groups	Semi-structural formula
Methane	CH_4	Methyl	$-CH_3$
Ethane	CH_3CH_3	Ethyl	$-CH_2CH_3$
Propane	$CH_3CH_2CH_3$	Propyl	$-CH_2CH_2CH_3$

7.3.3 Systematic naming of organic compounds

Due to the vast number of organic compounds, a systematic method of naming is required. The International Union of Pure and Applied Chemists (IUPAC) is the organisation that prescribes the rules so that each compound has its own unique name that can be used consistently throughout the world. Each family of organic compounds follows a pattern of naming as shown in the following examples.

Alkyl groups are named in both number and position on a carbon chain. As an example, let's follow the IUPAC rules for naming the following branched alkane.

1. Count the longest carbon chain. This will determine the prefix used to name the main/parent chain.

$$CH_3 - CH_2 - CH - CH_2 - CH_2 - CH_3$$
$$|$$
$$CH_3$$

2. Identify alkyl groups branching off the main chain.

$$\boxed{CH_3 - CH_2 - CH - CH_2 - CH_2 - CH_3}$$
$$|$$
$$CH_3$$

3. Starting at the end, number the chain that gives the lowest number for an alkyl branch.

$$\boxed{\underset{1}{CH_3} - \underset{2}{CH_2} - \underset{3}{CH} - \underset{4}{CH_2} - \underset{5}{CH_2} - \underset{6}{CH_3}}$$
$$|$$
$$(CH_3) \longleftarrow \text{Methyl group}$$
$$\text{Alkyl group}$$

3-methyl

4. Write the number and the name of the alkyl group(s) attached in alphabetical order.

5. Write the name of the parent chain at the end — in this case, six carbons separated by single bonds.

3-methylhexane

When there are more than one of the same type of alkyl group branching off the parent chain, prefixes are used to indicate how many there are, and numbers are used to indicate which carbon atom they are attached to in the parent chain. For example, consider the following multi-branched alkane.

1. The longest chain is five carbons.

$$\begin{array}{ccc} CH_3 & & CH_3 \\ | & & | \\ \boxed{CH_3 - C - CH_2 - CH - CH_3} \\ | \\ CH_3 \end{array}$$

2. Three methyl (CH_3) groups are branching off the chain.

3. The chain is numbered from left to right because this gives two alkyl groups coming off C2 instead of C4 if named from the opposite end.

$$\begin{array}{ccc} (CH_3) & & (CH_3) \\ | & & | \\ \boxed{\underset{1}{CH_3} - \underset{2}{C} - \underset{3}{CH_2} - \underset{4}{CH} - \underset{5}{CH_3}} \\ | \\ (CH_3) \end{array}$$

2,2,4-trimethyl

4. List the numbers of the C atoms the CH_3 groups are branching off separated by commas, and then a hyphen before the prefix 'tri-' to indicate there are three methyl groups.

5. Add the parent chain name to the end.

2,2,4-trimethylpentane

TIP: If there are molecules with two or more branches of the same type, the branch type is named and a prefix (for example, *di-*, *tri-*) is used to indicate the number of branches. Branches are listed in alphabetical order, ignoring the prefix (i.e. ethyl is written before methyl or dimethyl).

SAMPLE PROBLEM 4 Writing the systematic (IUPAC) name of hydrocarbons

Write the systematic (IUPAC) name of the following hydrocarbon.

$$
\begin{array}{ccccc}
CH_3 & & CH_3 & & \\
| & & | & & \\
CH_2 & & CH_2 & CH_3 & \\
| & & | & | & \\
CH_3-CH-CH_2-CH-CH_2-CH-CH_3
\end{array}
$$

THINK	**WRITE**
1. Examine the molecule to find the longest carbon chain.	
2. Locate the alkyl groups.	
3. Number the chain from an open end that gives one or more alkyl groups the lowest number.	
4. Write the number(s) of the alkyl groups separated by commas and then hyphenated to the prefix to indicate the number of the same type of alkyl group.	2,6-dimethyl 4-ethyl
5. List the different alkyl groups in alphabetical order, ignoring the prefix.	4-ethyl-2,6-dimethyl
6. Add the name of the parent chain to the end of the name.	4-ethyl-2,6-dimethyloctane

PRACTICE PROBLEM 4

Write the IUPAC name of the following hydrocarbon.

$$
\begin{array}{ccc}
& & CH_3 \\
& & | \\
CH_3 & CH_3 & CH_2 \\
| & | & | \\
CH_3-CH-CH_2-CH \\
& & | \\
& & CH_2 \\
& & | \\
& & CH_3
\end{array}
$$

 Resources

Interactivity Systematic naming of alkanes, alkenes and alkynes (int-1231)

7.3 Activities

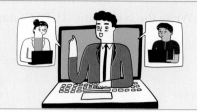

Students, these questions are even better in jacPLUS

Receive immediate feedback and access sample responses

Access additional questions

Track your results and progress

Find all this and MORE in jacPLUS

7.3 Quick quiz on | **7.3 Exercise** | **7.3 Exam questions**

7.3 Exercise

1. MC Which compound has the same molecular formula as cyclohexane?
   A. Hexane
   B. Benzene
   C. Hex-1-yne
   D. Hex-1-ene
2. MC Which of the following formulas belongs to an unsaturated hydrocarbon?
   A. $CH_3CH_2CH_2CH_3$
   B. $CH_3CH_2CHCH_2$
   C. $CH_3CH_2CH_2OH$
   D. $CH_3OCH_2CH_3$
3. MC The correct names for the semi-structural formulas $CH_3CH(CH_3)CH_2CH_3$ and $CH_3CH{=}CHCH_2CH_3$ are, respectively,
   A. methylbutane and pent-3-ene.
   B. methylbutane and pent-2-ene.
   C. methylbutene and pent-3-ane.
   D. methylbutene and 2-methylbutene.
4. What is the molecular formula of the alkane containing 18 carbon atoms?
5. For propane, show the
   a. molecular formula
   b. empirical formula
   c. semi-structural formula
   d. skeletal structure.
6. What are the systematic names of the following molecules?
   a. $CH_3CH(CH_3)C(CH_3)_2CH_2CH_3$
   b.
   c.
7. Consider the hydrocarbon shown.
   a. Name the alkyl groups present.
   b. Write the systematic name.

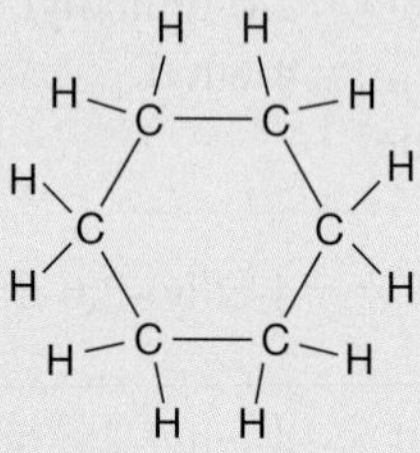

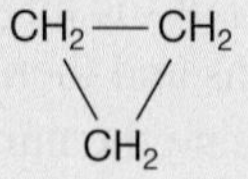

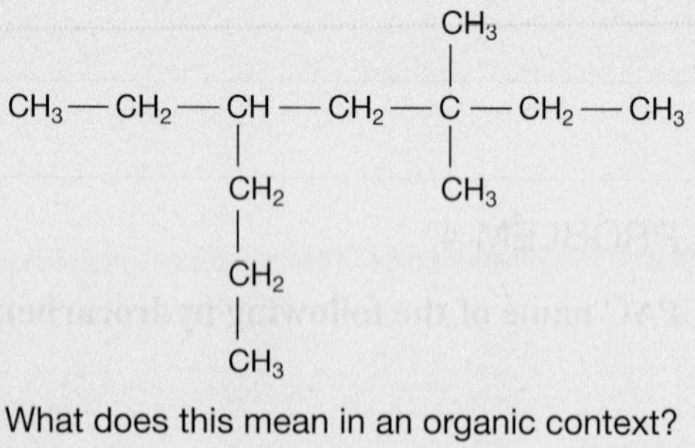

8. Ethene is described as an unsaturated compound. What does this mean in an organic context?
9. Draw the structural formula for 2-methyl hex-1-ene.
10. Write the systematic name of the compound with the structural formula shown.

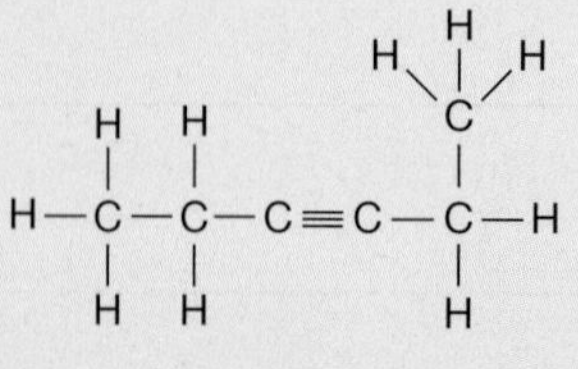

7.3 Exam questions

▶ Question 1 (1 mark)

Source: VCE 2016 Chemistry Exam, Section A, Q.2; © VCAA

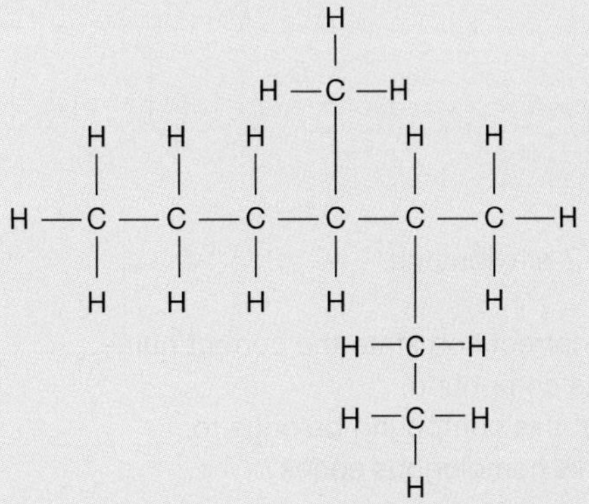

MC What is the correct systematic name for the compound shown above?

A. 4-methyl-5-ethylhexane
B. 2-ethyl-3-methylhexane
C. 4,5-dimethylheptane
D. 3,4-dimethylheptane

▶ Question 2 (1 mark)

Source: VCE 2013 Chemistry Sample Exam for Units 3 and 4, Section A, Q.1; © VCAA

MC What is the correct systematic name for the following compound?

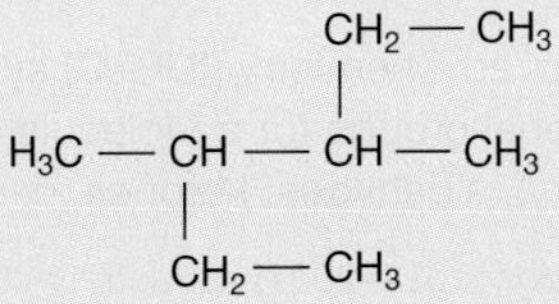

A. 2-ethyl-3-methylpentane
B. 3-methyl-4-ethylpentane
C. 3,4-dimethylhexane
D. 2,3-diethylbutane

▶ Question 3 (1 mark)

Source: VCE 2012 Chemistry Exam 1, Section A, Q.1; © VCAA

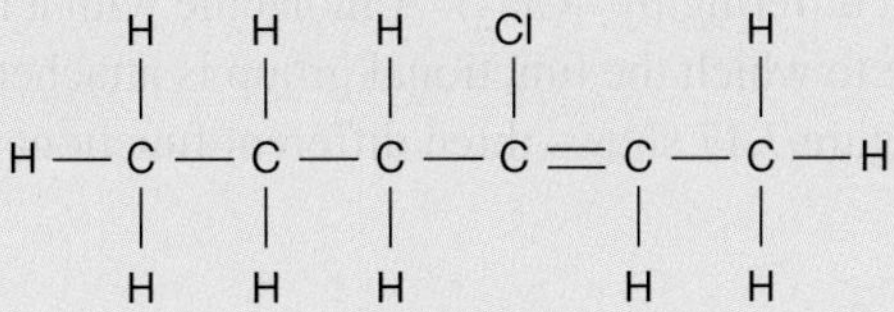

MC The correct systematic name for the compound shown above is

A. 2-chlorohex-2-ene.
B. 3-chlorohex-2-ene.
C. 3-chlorohex-3-ene.
D. 4-chlorohex-5-ene.

Question 4 (1 mark)

MC Which of the following skeletal formulas is *not* represented by the molecular formula C_5H_{12}?

A.

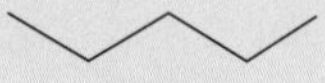

B.

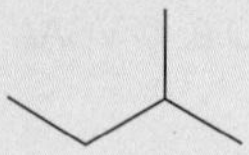

C.

D.

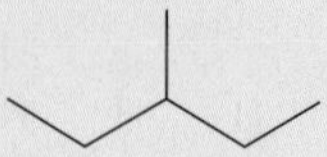

Question 5 (6 marks)

A student identified a compound as 2-ethylbutane.
a. Draw this molecule. **(1 mark)**
b. Explain why the name given is incorrect and state the correct name. **(2 marks)**
c. State the molecular formula of the compound. **(1 mark)**
d. Name the homologous series that this compound belongs to. **(1 mark)**
e. What is the general formula for this homologous series? **(1 mark)**

More exam questions are available in your learnON title.

7.4 Functional groups

KEY KNOWLEDGE

- Molecular, structural and semi-structural (condensed) formulas and skeletal structures of haloalkanes, primary amines, primary amides, alcohols (primary, secondary and tertiary), aldehydes, ketones, carboxylic acids and non-branched esters
- The International Union of Pure and Applied Chemistry (IUPAC) systematic naming of organic compounds up to C8, with no more than two functional groups for a molecule, limited to haloalkanes, primary amines, alcohols (primary, secondary and tertiary), aldehydes, ketones, carboxylic acids and non-branched esters

Source: Adapted from VCE Chemistry Study Design (2024–2027) extracts © VCAA; reproduced by permission.

7.4.1 Identifying functional groups

The **functional group** of a hydrocarbon is an atom or a group of atoms that determines the function (chemical nature) of a compound.

As with the alkanes and alkenes, compounds containing the same functional group form a homologous series (a family with similar properties and differing by $-CH_2$). A molecule with a functional group attached is usually less stable than the carbon backbone to which the functional group is attached and therefore more likely to participate in chemical reactions. Figure 7.17 shows three different functional groups attached to the basic carbon skeleton of methane.

FIGURE 7.17 Different functional groups (–OH, –Cl, –COOH) attached to the basic carbon skeleton of methane

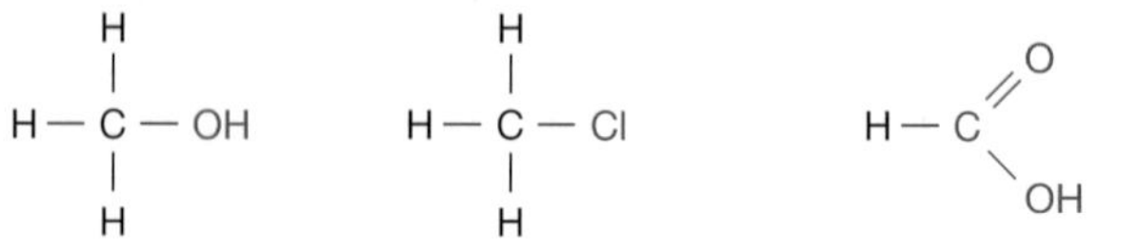

CH_3OH — Methanol

CH_3Cl — Chloromethane

$HCOOH$ — Methanoic acid

functional group an atom or group of atoms that is attached to or part of a hydrocarbon chain, and influences the physical and chemical properties of the molecule

7.4.2 Haloalkanes

Haloalkanes are a class of molecules that have one or more **halogens** attached to the carbon chain. They are used in a variety of applications, including solvents, refrigeration and medicine.

Haloalkanes are often represented as **R–X**. The **R** is used to represent the hydrocarbon chain of any length and the **X** is used to represent any of the halogens, such as fluorine (F), chlorine (Cl) and bromine (Br).

Naming haloalkanes

The naming system of haloalkanes follows the same rules as naming hydrocarbons. However, like all functional groups, halogen functional groups take priority over alkyl groups when numbering the longest carbon chain. We use the same prefixes to name compounds with more than one of the same halogens.

Another feature of the naming is the replacement of '*-ine*' with '*-o*'. Fluorine becomes fluoro, chlorine becomes chloro and bromine becomes bromo when they are part of a haloalkane.

Consider the following molecule:

$$Cl - \underset{\underset{H}{|}}{\overset{\overset{H}{|}}{C}} - \underset{\underset{H}{|}}{\overset{\overset{H}{|}}{C}} - \underset{\underset{H}{|}}{\overset{\overset{Cl}{|}}{C}} - \underset{\underset{H}{|}}{\overset{\overset{H}{|}}{C}} - H$$

It has a four-carbon chain with two chlorine atoms attached on C1 and C3. The systematic name of this compound is 1,3-dichlorobutane.

When there are different halogens on the carbon chain, they are numbered as usual but written alphabetically, like alkyl groups.

For example:

$$Cl - CH_2 - \underset{\underset{CH_3}{|}}{CH} - \underset{\underset{Br}{|}}{CH} - CH_3$$

3-bromo-1-chloro-2-methylbutane

Haloalkanes

- Haloalkanes are hydrocarbons with one or more halogens attached to the carbon chain.
- Haloalkanes have the general formula R–X, where R is a hydrocarbon chain of any length and X is a halogen.
- In naming, halogen functional groups take priority over any alkyl groups, and the *-ine* ending becomes *-o*.

7.4.3 Amines and amides

Amines are weak bases and have a variety of uses, including in the manufacturing of dyes and nylon, in pest control and in the pharmaceutical industry.

- Primary amines (**R–NH$_2$**) contain the amino functional group, –NH$_2$.
- When naming amines, the '*-e*' in the alkane name is replaced by '*-amine*'. A number is given before this suffix in compounds with three or more carbon atoms to indicate the position on the carbon chain.

> **halogens** elements in group 17 of the periodic table: F, Cl, Br, I and At
>
> **amines** organic compounds containing the amino functional group, –NH$_2$

Amides also have a wide range of uses, including being constituents of Kevlar and paracetamol, as well as proteins in the body.

- They contain the amide functional group, –CONH–.
- Primary amides can be thought of as having the functional group –CONH$_2$.
- Primary amides are named the same way as amines, but replacing the '-*amine*' suffix with '-*amide*'. No number is required for amides as this functional group is always positioned at the end of the chain, similar to carboxylic acids (see section 7.4.7).

FIGURE 7.18 a. Butan-1-amine and **b.** butanamide

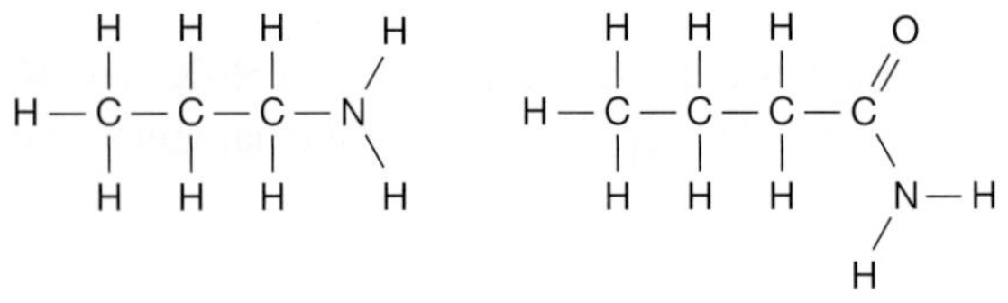

TABLE 7.6 The first eight compounds of the amine homologous series

Systematic name	Semi-structural formula
Methanamine	CH_3NH_2
Ethanamine	$CH_3CH_2NH_2$
Propan-1-amine	$CH_3(CH_2)_2NH_2$
Butan-1-amine	$CH_3(CH_2)_3NH_2$
Pentan-1-amine	$CH_3(CH_2)_4NH_2$
Hexan-1-amine	$CH_3(CH_2)_5NH_2$
Heptan-1-amine	$CH_3(CH_2)_6NH_2$
Octan-1-amine	$CH_3(CH_2)_7NH_2$

FIGURE 7.19 Wing suits are made from polymers containing amide links.

Amines and amides

- Amines are hydrocarbons with the amino functional group, –NH$_2$. They have the general formula R–NH$_2$.
- Amides are hydrocarbons with the amide functional group, –CONH–. Primary amides have the general formula R–CONH$_2$.

amides organic compounds containing the amide functional group, –CONH–

alcohols compounds in which a hydroxyl group (–OH) is the parent functional group

7.4.4 Alcohols

Organic hydroxyl compounds containing the –OH group belong to the homologous series called **alcohols**. A study of the properties of the –OH group is important to chemists because of the industrial importance of compounds containing this functional group, and because of its wide occurrence in biological molecules. Ethanol is the most common alcohol and it has many uses. It is present in beer, wine and spirits, and is used in the preparation of ethanoic acid (acetic acid) and for sterilising wounds. Methylated spirits (containing 95 per cent ethanol) is a very useful solvent and is used in the manufacture of varnishes, polishes, inks, glues and paints.

FIGURE 7.20 Ethanol is used in hand sanitiser.

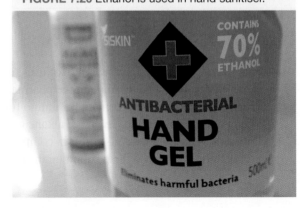

Other alcohols that are **volatile** at low temperatures are also good solvents and are used in deodorants, colognes and aftershave lotions. Glycerol, $C_3H_5(OH)_3$, is used for making fats and soaps. **Alkanols** are a sub-group of alcohols; they consist of an alkane with a hydroxyl group.

Naming alcohols

For naming purposes, alcohols have the general formula **R–O–H**. The first three members of the alcohol homologous series are shown in figure 7.21. Again, the standard prefixes are used to name the number of carbon atoms in the chain. In the first two members (methanol and ethanol), the hydroxyl group is not numbered because its position is always C1. However, for the third member, the –OH group could be on C1 or C2, and therefore needs to be stated in the name. There are two ways of naming molecules with an –OH group; using the suffix '*-ol*' or the prefix '*hydroxy-*'. The suffix is used when there are no other functional groups with a higher priority (see section 7.4.10), so the third member is named propan-1-ol rather than 1-hydroxypropane. When the hydroxyl group branches off C2 we simply change the ending of the name to *-2-ol,* as shown in figure 7.21.

FIGURE 7.21 Members of the alcohols: methanol, ethanol, propan-1-ol and butan-2-ol

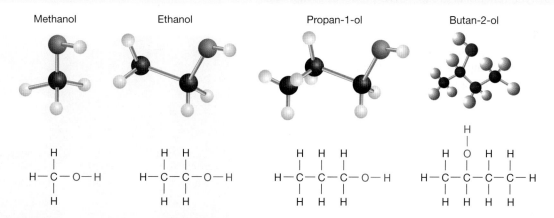

TABLE 7.7 The first eight alkanols. The –OH group is attached to the first carbon atom in each case.

Systematic name	Semi-structural formula
Methanol	CH_3OH
Ethanol	CH_3CH_2OH
Propan-1-ol	$CH_3(CH_2)_2OH$
Butan-1-ol	$CH_3(CH_2)_3OH$
Pentan-1-ol	$CH_3(CH_2)_4OH$
Hexan-1-ol	$CH_3(CH_2)_5OH$
Heptan-1-ol	$CH_3(CH_2)_6OH$
Octan-1-ol	$CH_3(CH_2)_7OH$

volatility describes how readily a liquid substance will form a vapour

alkanols alkanes with a hydroxyl group replacing a hydrogen atom

Classifying alcohols

Alcohols are classified as primary, secondary or tertiary, based on the number of carbon atoms connected to the carbon atom attached to the hydroxyl functional group.

- Primary (1°) alcohol: The C–OH is attached to one other carbon atom.
- Secondary (2°) alcohol: The C–OH is attached to two other carbon atoms.
- Tertiary (3°) alcohol: The C–OH is attached to three other carbon atoms.

FIGURE 7.22 Primary, secondary and tertiary alcohols

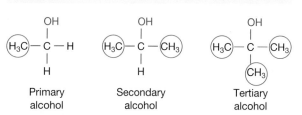

Primary alcohol   Secondary alcohol   Tertiary alcohol

Alcohols

Alcohols are hydrocarbons with the –OH functional group. They have the general formula R–OH.
- Primary (1°) alcohols have the C–OH group attached to one other C atom.
- Secondary (2°) alcohols have the C–OH group attached to two other C atoms.
- Tertiary (3°) alcohols have the C–OH group attached to three other C atoms.

 Resources

▶ **Video eLesson** Molecular representations of butan-1-ol (eles-2485)

7.4.5 Aldehydes

The aldehyde functional group –CHO produces compounds that have characteristic odours. The familiar smells of vanilla and cinnamon are caused by aldehydes. Aldehydes with low molecular weights, such as methanal (formaldehyde) and ethanal, have unpleasant odours; formaldehyde was previously used as a preservative but is now suspected to be carcinogenic. Aldehydes with higher molecular weights have sweet, pleasant smells and are used in perfumes. Other uses of aldehydes include solvents and the manufacture of plastics, dyes and pharmaceuticals.

FIGURE 7.23 $C_7H_{15}CHO$ is added to fragrances to provide hints of citrus.

FIGURE 7.24 The first aldehydes: methanal and ethanal

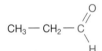

Methanal   Ethanal

Naming aldehydes

Aldehydes are generally written as **R–CHO** and are named by replacing the last -*e* on the name of the corresponding alkane with -*al*. For example, propane becomes propanal. Aldehydes have a C=O bond at the end of the carbon chain, at C1. Aldehydes are always named from C1, so they do not need a number in the name to reference where the functional group is.

Aldehydes

- Aldehydes are hydrocarbons with the –CHO functional group, with a C=O double bond at C1.
- Aldehydes have the general formula R–CHO.

7.4.6 Ketones

Ketones contain the carbonyl (C=O) functional group and are used extensively to produce pharmaceuticals, perfumes, solvents and polymers. They have important physiological properties and are found in medicinal compounds and steroid hormones, including cortisone. The most familiar ketone is propanone (acetone), which has unlimited solubility in water and is a solvent for many organic compounds. It evaporates readily because of its low boiling point, which contributes to its usefulness.

FIGURE 7.25 Propanone, butanone and pentan-3-one

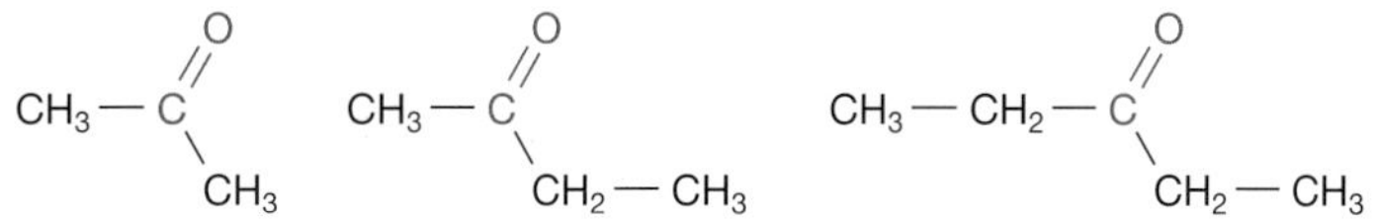

Propanone   Butanone   Pentan-3-one

Naming ketones

The '*-e*' is replaced by '*-one*' and the general formula of a ketone is **R–CO–R'**. Hence, the carbonyl functional group is never found at C1.

Ketones

- Ketones are compounds with the carbonyl functional group, C=O.
- Ketones have the general formula R–CO–R'.

7.4.7 Carboxylic acids

Carboxylic acids are the homologous series with the carboxyl functional group, –COOH. Carboxylic acids are generally weak acids and occur widely in nature. Some common examples include citric acid, found in citrus fruits such as oranges and lemons; malic acid, found in apples; and ascorbic acid (vitamin C), found in a number of foods. Other carboxylic acids, such as stearic acid, and oleic and palmitic acids, are used in the formation of animal and vegetable fats. Carboxylic acids are also used to make soaps and polyesters. The carboxylic acids with chains of four to eight carbons have a strong, unpleasant smell and are found in cheese, perspiration and rancid butter.

FIGURE 7.26 Quarantine dogs can be trained to find fruit by sniffing out distinctive carboxylic acids, such as malic acid, which is found in apples.

Naming carboxylic acids

Carboxylic acids are generally written as **R–COOH**, and the C in the carboxyl functional group is always assigned C1 for naming purposes. Because of this, the prefix '*1-*' is left off the start of the names. The '*-e*' in the name is replaced by the suffix '*-oic acid*'.

Carboxylic acids

- Carboxylic acids are hydrocarbons with the carboxyl functional group, –COOH.
- Carboxylic acids have the general formula R–COOH. The C in the carboxyl functional group is always assigned C1.

TABLE 7.8 Some carboxylic acids and their uses

Systematic name	Semi-structural formula	Non-systematic name	Occurrence and uses
Methanoic acid	HCOOH	Formic acid	• Used by ants as a defence mechanism • Used in textile processing • Used as a grain preservative
Ethanoic acid	CH_3COOH	Acetic acid	• Found in vinegar • Used in making artificial textiles
Propanoic acid	CH_3CH_2COOH	Propionic acid	• Calcium propanoate is used as an additive in bread manufacture
Butanoic acid	$CH_3CH_2CH_2COOH$	Butyric acid	• Present in human sweat • Responsible for the smell of rancid butter
Benzoic acid	—COOH	Benzoic acid	• Used as a preservative

 Resources

Interactivity Matching carboxylic acids and formula (int-1232)

7.4.8 Esters

The homologous series of esters contain the functional group –COO–. Esters are used in a variety of applications. The small, volatile esters are used in artificial flavours and smells in food and fragrances. Larger esters occur naturally as fats and oils. They can also be used in the manufacture of materials as diverse as Perspex and artificial arteries used in open heart surgery.

The ester functional group, –COO–, also referred to as an ester link, forms via a condensation reaction between hydroxyl and carboxyl functional groups. Esters have the general formula **R–COO–R'**.

Naming esters

The name of an unbranched ester is a product of the carboxylic acid and the primary alcohol that produces it. The first part of the name comes from the hydrocarbon or alkyl part of the alcohol. For example, if methanol is used to make an ester the first part of the name will be methyl, and if ethanol is used the first part of the name will be ethyl.

The second part of the ester name comes from the carboxylic acid. The '-*oic acid*' suffix is removed and replaced with '-*oate*'. For example, if methanoic acid is used to make an ester, the second part of the name will be methanoate.

Figure 7.28 shows propanoic acid and ethanol being used to make ethyl propanoate. This reaction can be summarised as:

$$CH_3CH_2COOH \quad + \quad CH_3CH_2OH \quad \rightleftharpoons \quad CH_3CH_2COOCH_2CH_3 \quad + \quad H_2O$$

Propanoic acid + ethanol ⇌ ethyl propanoate + water

FIGURE 7.28 Formation of ethyl propanoate from propanoic acid and ethanol

Propanoic acid   Ethanol   Ethyl propanoate

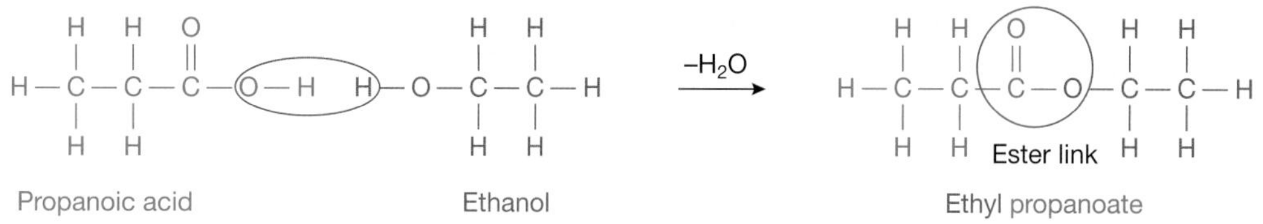

Esters

- Esters are hydrocarbons with the ester functional group, $-COO-$. They have the general formula $R-COO-R'$.
- Esters form through condensation reactions between an alcohol and a carboxylic acid. The alcohol gives the first part of the name and the carboxylic acid gives the second part of the name, with the suffix '-oate'.

7.4.9 Functional group summary

TABLE 7.9 Functional group summary

Formula	Name	Homologous series	Method of naming	Example
$-Cl$ $-Br$ $-I$	Chloro Bromo Iodo	Haloalkanes	Prefix *chloro-* Prefix *bromo-* Prefix *iodo-*	CH_3Cl Chloromethane CH_3CH_2Br Bromoethane ICH_2CH_2I 1,2-diiodoethane
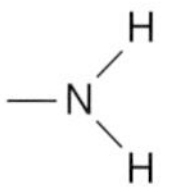 Amine (structure)	Amine	Amine	Suffix *-amine*	CH_3NH_2 Methanamine $CH_3CH_2NH_2$ Ethanamine
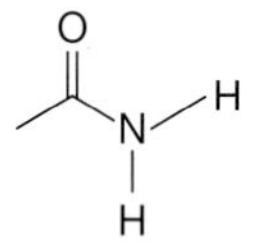 Primary amide (structure)	Primary amide	Primary amide	Suffix *-amide*	CH_3CONH_2 Ethanamide $CH_3CH_2CONH_2$ Propanamide
$-O-H$	Hydroxyl	Alcohol	Suffix *-ol*	$CH_3CH_2CH_2OH$ Propan-1-ol (1-propanol)
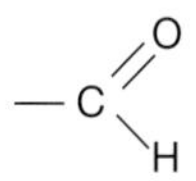 Aldehyde (structure)	Aldehyde	Aldehyde	Suffix *-al*	CH_3CHO Ethanal (acetaldehyde)
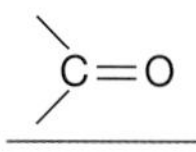 Carbonyl (structure)	Carbonyl	Ketone	Suffix *-one*	$CH_3CH_2COCH_2CH_3$ Pentan-3-one (3-pentanone)
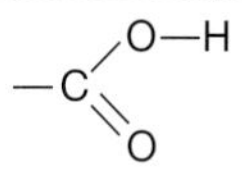 Carboxyl (structure)	Carboxyl	Carboxylic acid	Suffix *-oic acid*	CH_3COOH Ethanoic acid (acetic acid)
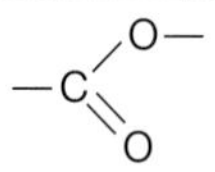 Ester (structure)	Ester	Ester	As alkyl alkanoate	$CH_3CH_2COOCH_3$ Methyl propanoate

SAMPLE PROBLEM 5 Naming the functional groups and homologous series of a molecule to determine the systematic name

For the following molecule:
a. name the functional group and the homologous series
b. write its systematic (IUPAC) name.

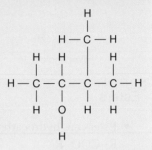

THINK

a. 1. Identify and circle the longest chain. This is the parent chain, and as it is composed of four carbons, the prefix is '*but–*'.

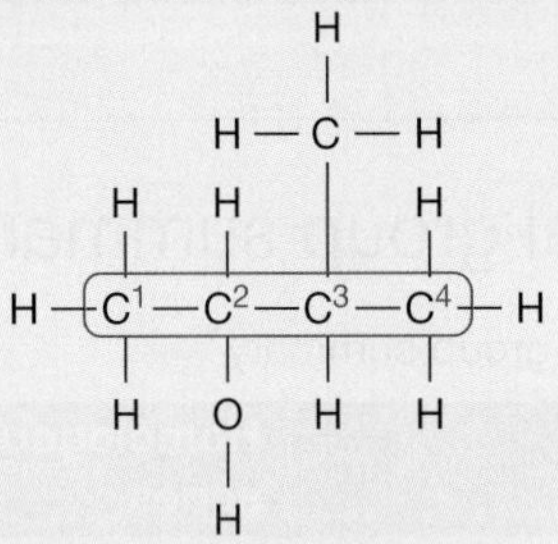

2. Find groups of atoms that are not alkyl groups. Remember, alkyl groups are hydrocarbon branches coming off the longest carbon chain of an organic molecule.
The functional group is the hydroxyl group in the alcohol homologous series. The name will end in '*-ol*'.

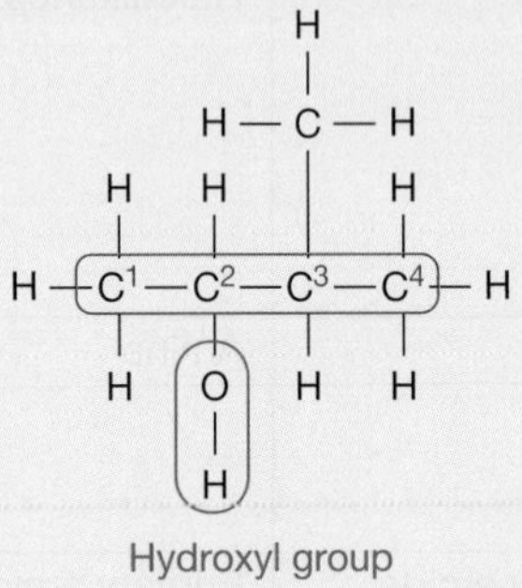

b. 1. The longest chain is composed of four carbons, making the molecule butanol.
The hydroxyl group branches from C2.

WRITE

a.

```
            H
            |
        H — C — H
   H    H   |    H
   |    |   |    |
H —(C¹ — C² — C³ — C⁴)— H
   |    |   |    |
   H    O   H    H
        |
        H
```

```
            H
            |
        H — C — H
   H    H   |    H
   |    |   |    |
H —(C¹ — C² — C³ — C⁴)— H
   |   ╭─╮  |    |
   H   │O│  H    H
       │|│
       │H│
       ╰─╯
```
Hydroxyl group

Hydroxyl functional group
Alcohol homologous series

b.

```
            H
            |
        H — C — H
   H    H   |    H
   |    |   |    |
H —(C¹ — C² — C³ — C⁴)— H
   |   ╭─╮  |    |
   H   │O│  H    H
       │|│
       │H│
       ╰─╯
```
Hydroxyl group

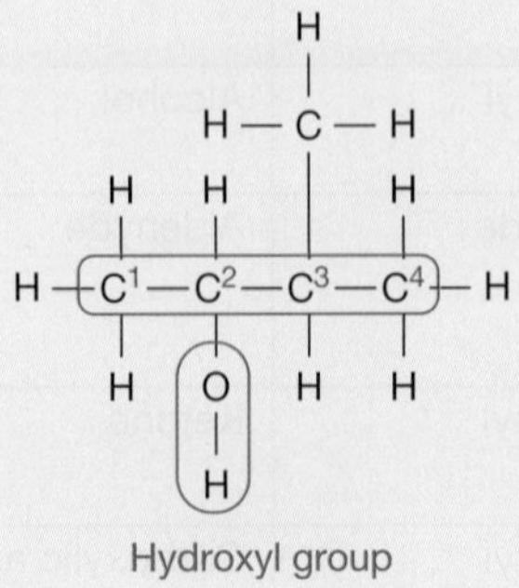

2. Identify any alkyl groups branching off the longest carbon chain.
Methyl is branching from C3.
Ensure that all parts of the molecule are accounted for.

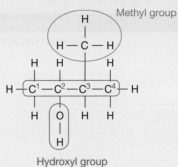

Methyl group

Hydroxyl group

3. Name the molecule.

3-methylbutan-2-ol

PRACTICE PROBLEM 5

For the following molecule:
a. **name the functional group and the homologous series**
b. **write its systematic (IUPAC) name.**

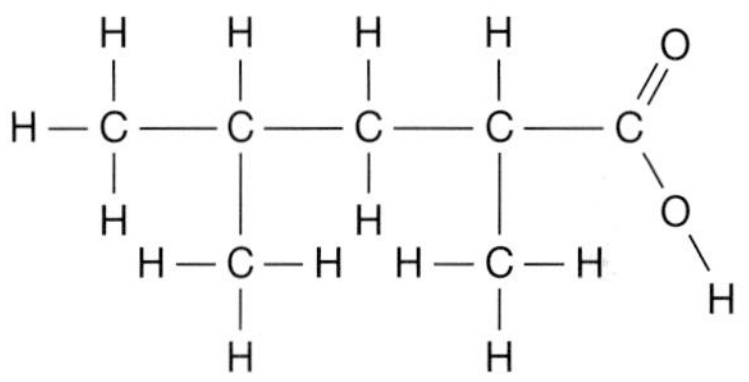

---

elog-1897

EXPERIMENT 7.2

Constructing models of organic compounds

Aim

To construct models of various organic compounds

---

on Resources

Interactivity Organic molecule structures (int-1234)

7.4.10 Naming compounds with two functional groups

When molecules have two or more functional groups, the naming becomes more complex. A lot of the molecules with many functional groups are referred to by their commercial or simplified names. Pharmaceuticals are typically branded or referred to using non-preferred IUPAC names.

IUPAC naming of compounds with two functional groups

Compounds with two or more functional groups are classified by the principal group (the main functional group) defining the series to which they belong. Table 7.10 shows the priority scale for the groups.

TABLE 7.10 Functional group priority scale

Group	Homologous series	Suffix	
Carboxyl 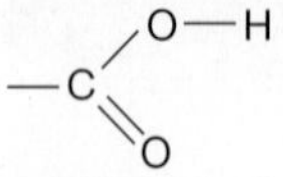 $-C \overset{O-H}{\underset{O}{\diagup}}$	Carboxylic acid	-oic acid	Highest priority name ↑
Ester 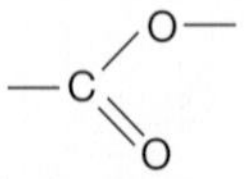 $-C \overset{O-}{\underset{O}{\diagup}}$	Ester	-oate	
Aldehyde 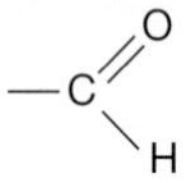 $-C \overset{O}{\underset{H}{\diagup}}$	Aldehyde	-al	
Carbonyl 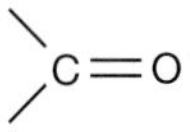 $\diagdown C = O \diagup$	Ketone	-one	
Hydroxyl –O–H	Alcohol	-ol	
Amine R–NH$_2$	Amine	-amine	
Alkyne and alkene C≡C C=C	Alkyne and alkene	-yne and -ene	
Alkane C–C	Alkane	-ane	Lowest priority name

The parent name of the compound is derived from the principal group according to the following rules:
- The numbering system used is that of the principal group.
- An alcohol is regarded as a hydroxyl side group, an amine is regarded as an amino side group and a ketone is regarded as a carbonyl side group.
- A compound containing both alcohol and aldehyde functional groups is named as an aldehyde with a hydroxyl side group.
- A compound containing an alcohol, a ketone and an acid is named as an acid with hydroxyl and carbonyl side groups.

Some examples are shown in figure 7.29.

FIGURE 7.29 a. 3-aminobutan-2-ol **b.** 3-hydroxypropanoic acid **c.** Pent-2-en-1-ol

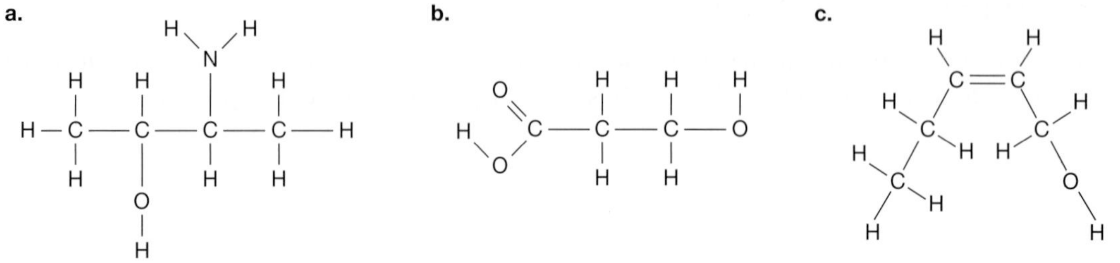

SAMPLE PROBLEM 6 Naming organic compounds with two functional groups

Name the following organic compound.

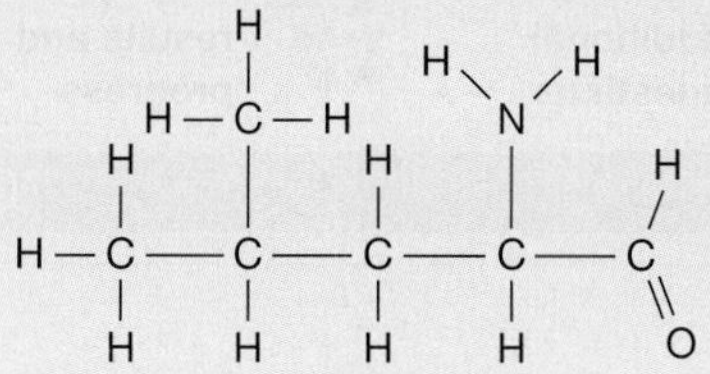

THINK

1. Circle or highlight the longest chain. The parent chain has five carbons and so is *pent–*.

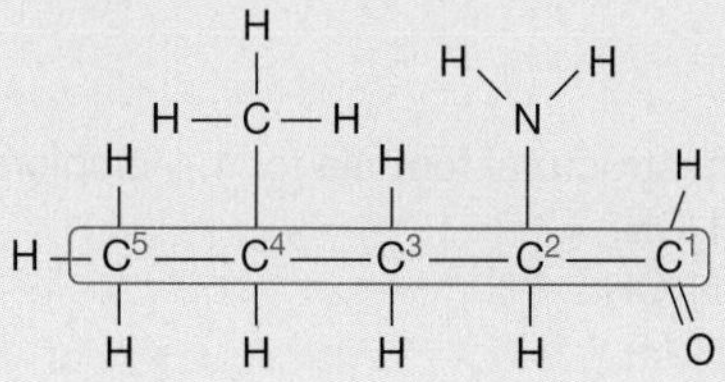

2. Identify the functional groups on the molecule and assign priority.
   The aldehyde functional group takes priority over the amino functional group. The name will end in '*-al*': pentanal.

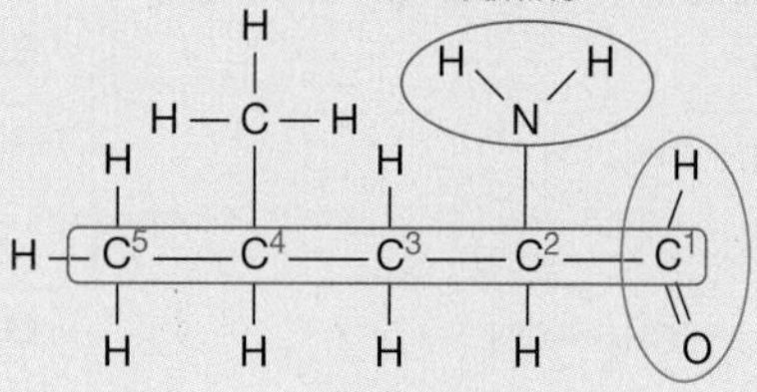

3. The longest chain including the priority functional group (aldehyde) has five carbons. Determine the branch position of the second functional group, amino.
   The amino is branching at C2.

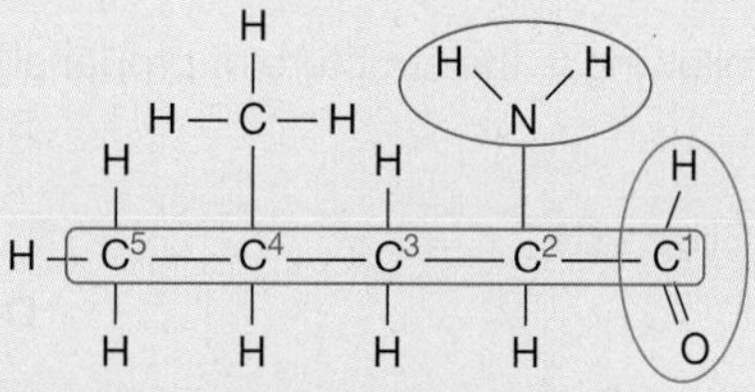

4. Identify any alkyl groups branching off the longest carbon chain.
   Methyl is branching from C4.
   Ensure that all parts of the molecule are accounted for.

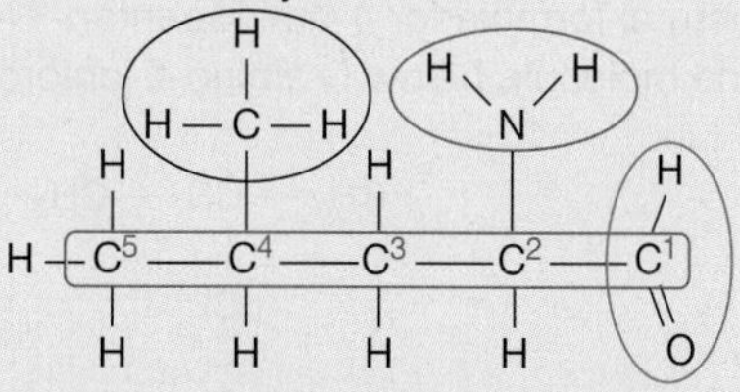

5. Write the name with substituents in alphabetical order.

WRITE

2-amino-4-methylpentanal

PRACTICE PROBLEM 6

Name the following organic compound.

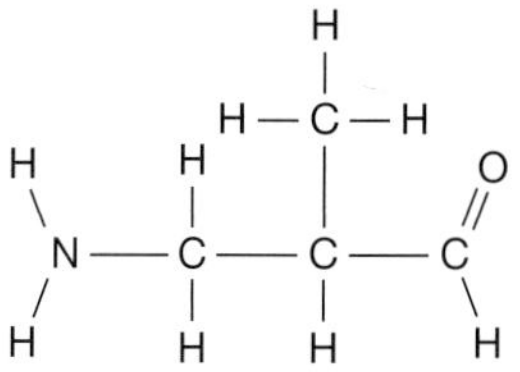

7.4 Quick quiz **on**	7.4 Exercise	7.4 Exam questions

7.4 Exercise

1. **MC** The correct semi-structural formula for 1,4-dichlorobutan-2-ol is
   A. $CH_3CHClCHClCH_2OH$.
   B. $CH_2ClCH_2CHClCH_2OH$.
   C. $CH_2ClCHClCHOHCH_3$.
   D. $CH_2ClCHOHCH_2CH_2Cl$.

2. **MC** Which of the following is a tertiary alcohol?

   A.
   $$CH_3-CH_2-\overset{\overset{\displaystyle OH}{|}}{CH}-CH_3$$

   B.
   $$CH_3-\overset{\overset{\displaystyle CH_3}{|}}{\underset{\underset{\displaystyle CH_3}{|}}{C}}-OH$$

   C. $CH_3-CH_2-CH_2-CH_2-OH$

   D.
   $$CH_3-\overset{\overset{\displaystyle }{|}}{\underset{\underset{\displaystyle CH_3}{|}}{CH}}-CH_2-OH$$

3. **MC** Which of the following is the structure of propanal?

   A.
   $$CH_3-\overset{\overset{\displaystyle O}{\|}}{C}-O-CH_2-CH_3$$

   B.
   $$H-\overset{\overset{\displaystyle O}{\|}}{C}-CH_2-CH_3$$

   C.
   $$CH_3-\overset{\overset{\displaystyle O}{\|}}{C}-H$$

   D.
   $$CH_3-\overset{\overset{\displaystyle O}{\|}}{C}-CH_2-CH_3$$

4. Write the semi-structural formula for 4-aminopentan-1-ol.

5. A student named the molecule below 5-amino-2-chlorohexane. Is this name correct? Explain your answer.

   $$CH_3-\overset{\overset{\displaystyle }{|}}{\underset{\underset{\displaystyle Cl}{|}}{CH}}-CH_2-CH_2-\overset{\overset{\displaystyle }{|}}{\underset{\underset{\displaystyle NH_2}{|}}{CH}}-CH_3$$

6. Study the following molecule.

   a. Write its systematic (IUPAC) name.
   b. Write its semi-structural formula.
   c. Write its molecular formula.
   d. Write its empirical formula.

7. Draw the structure and semi-structure of a four-carbon amide with 2,3-dihydroxy groups.

8. Using systematic nomenclature, name the compound represented by $CH_2ClCHClCH_3$.

9. a. Name the following molecules.
   b. Draw semi-structures for the molecules.
   c. Draw skeletal structures for the molecules.

i.

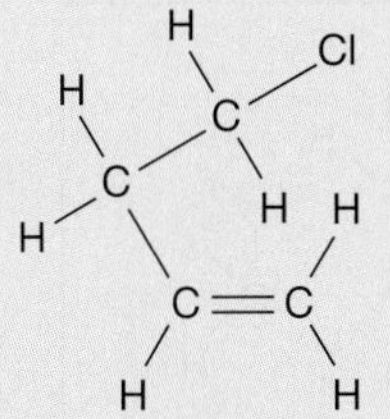

ii. CH_3—CH—CH—CH—CH_3
    with Br, Cl, CH_3 substituents

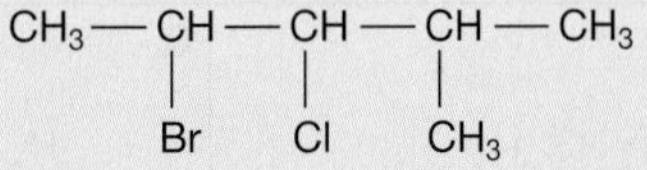

iii.

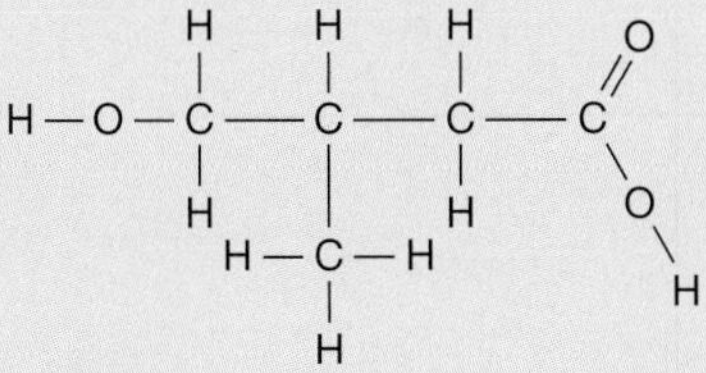

iv.

CH_3—CH—C (=O)—CH_2—CH_3
with Br substituent

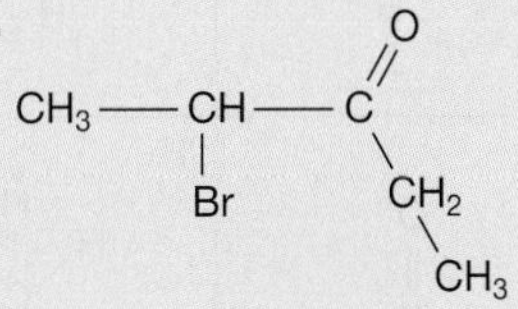

v.

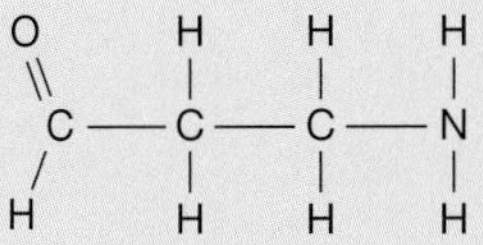

10. Draw the structural formula of the ester ethyl methanoate.

7.4 Exam questions

▶ **Question 1 (1 mark)**

Source: VCE 2014 Chemistry Exam, Section B, Q.5.a; © VCAA

A 2% solution of glycolic acid (2-hydroxyethanoic acid), $CH_2(OH)COOH$, is used in some skincare products.

Draw the structural formula of glycolic acid.

▶ **Question 2 (1 mark)**

Source: VCE 2013 Chemistry Exam, Section A, Q.9; © VCAA

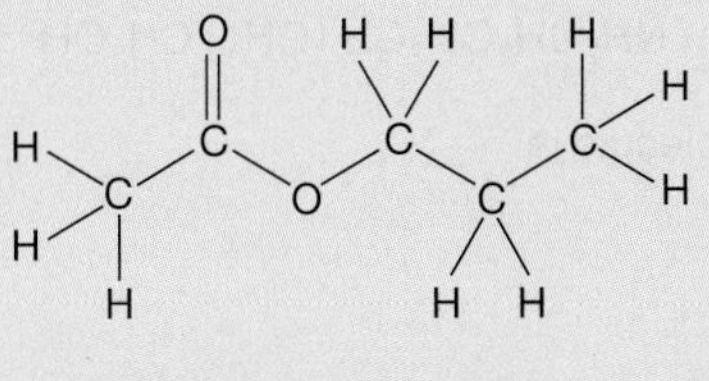

MC The systematic IUPAC name for the molecule shown above is
A. ethyl ethanoate.
B. ethyl propanoate.
C. propyl ethanoate.
D. methyl propanoate.

Question 3 (1 mark)

Source: Adapted from VCE 2021 Chemistry Exam, Section B, Q.7; © VCAA

MC Five isomers with the molecular formula $C_5H_{10}O_2$ are shown.

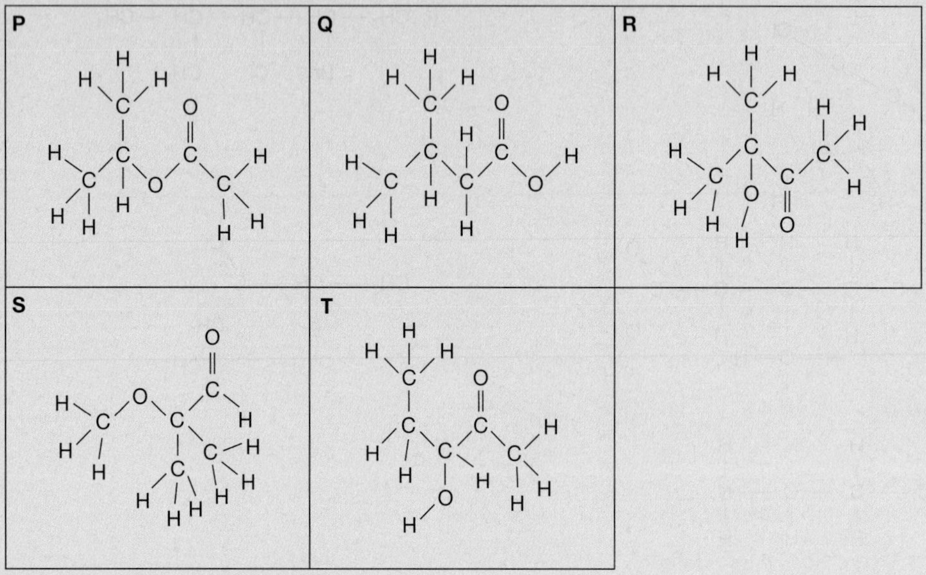

Which of the following statements is true?

A. P is an ester and its systematic name is propyl ethanoate.

B. R is a ketone and its systematic name is 3-hydroxy-3-methylbutan-2-one.

C. R, S and T are all ketones.

D. Q is a carboxylic acid and its systematic name is 2-methylbutanoic acid.

Question 4 (1 mark)

Source: VCE 2018 Chemistry NHT Exam, Section A, Q.12; © VCAA

MC The semi-structural formula for an isomer of $C_5H_{13}NO$ is

$$NH_2CH_2CH_2CH(CH_3)CH_2OH$$

The correct systematic name for this molecule is

A. 4-amino-pentan-1-ol

B. 4-amino-2-methyl-butan-1-ol

C. 4-hydroxy-3-methyl-butan-1-amine

D. 1-hydroxy-2-methyl-4-amino-butane

Question 5 (3 marks)

The skeletal structure of an organic molecule is shown.

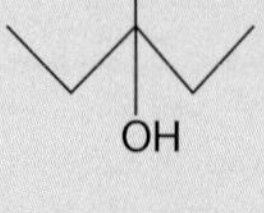

a. Draw its structural formula. (1 mark)

b. Write its molecular formula. (1 mark)

c. Write its semi-structural formula. (1 mark)

More exam questions are available in your learnON title.

7.5 Isomers

7.5.1 Introduction to isomers

Another reason for the enormous number of organic compounds is the existence of **isomers**.
- Isomers are two or more compounds with the same molecular formula but different arrangements of atoms.
- **Structural isomers** occur when atoms are arranged in different orders. In many cases, three-dimensional space diagrams (using wedges and dashes) are used to show the different positions.
- The properties of substances are affected by the type of isomerism present.
- Structural isomers are chain, positional and functional isomers.

on Resources

▶ **Video eLesson** Isomers (eles-2478)

7.5.2 Structural isomers

Structural isomers, also known as constitutional isomers, are those in which the connectivity (or arrangement) of atoms or groups of atoms are different. For the first three alkanes (methane, ethane and propane) there is only one way of arranging the atoms and that is the *straight-chain* (unbranched) arrangement. In butane, C_4H_{10}, there are two ways of arranging the carbon and hydrogen atoms and, therefore, there are two structures. One is the straight-chain structure and the other is the *branched-chain* structure.

Each of the two structures of butane satisfies the valence of carbon and hydrogen atoms, and each is a neutral and stable molecule (see figure 7.30). Their chemical and physical properties are similar but not identical. For instance, straight-chain butane has a boiling point of −1 °C, while the branched-chain molecule, 2-methylpropane, has a boiling point of −12 °C. Butane and 2-methylpropane are called structural isomers because they have the same molecular formula but different arrangements of atoms.

Three types of structural isomers are chain, positional and functional isomers. An easy way of determining structural isomers is to go through the systematic naming process. A different name means a different structure.

Chain isomers

Chain isomers are structures that are different because of the size of the parent chain and the alkyl branches, if any, attached. Butane and 2-methylproane (methylpropane) are examples of chain isomers.

The number of possible ways of combining the atoms to form chain isomers increases with the number of carbon atoms in the molecule. Pentane, C_5H_{12}, has three isomers, and heptane, C_7H_{16}, has nine, while decane, $C_{10}H_{22}$, has 75 isomers. For $C_{15}H_{32}$, there are 4347 possible isomers and for $C_{40}H_{82}$, more than 6×10^{13} isomers are possible. Isomerism is responsible for the enormous number of organic compounds that are known.

FIGURE 7.30 Structural isomer models of **a.** butane and **b.** 2-methylpropane

a.

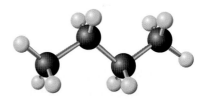

b.

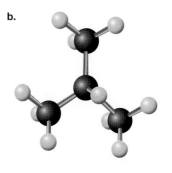

> **isomers** molecules with the same formula but a different arrangement of atoms and different properties
>
> **structural isomers** molecules that have the same molecular formula but different structural formulas
>
> **chain isomers** a type of structural isomer that involves branching

Positional isomers

Positional isomers occur when the functional group is located on different carbon atoms. Propan-1-ol and propan-2-ol are examples.

tlvd-9698

SAMPLE PROBLEM 7 Drawing and naming structural isomers

Draw and name all structural isomers of C$_4$H$_9$Cl.

THINK	WRITE
1. This is a haloalkane, so the easiest place to start is to draw the straight chain and put the chlorine atom on C1, and name it.	H H H H | | | | Cl — C — C — C — C — H | | | | H H H H 1-chlorobutane
2. Change the position of the Cl atom to make a different structure with a different name. **TIP:** There is no 3-chlorobutane or 4-chlorobutane because when they are flipped over they are actually just 1-chlorobutane and 2-chlorobutane.	H Cl H H | | | | H — C — C — C — C — H | | | | H H H H 2-chlorobutane
3. Move the –CH$_3$ group to make it a methyl group and have the Cl branch off the same carbon. Make sure that it has a different name in case you have drawn the same structure but a different way.	Cl H | H | | | H — C — C — C — H | | | H | H H — C — H | H 2-chloro-2-methylpropane
4. Make the last change possible by moving the chlorine back to C1 of the methylpropane.	H H H | | | Cl — C — C — C — H | | | H | H H — C — H | H 1-chloro-2-methylpropane

PRACTICE PROBLEM 7

Draw and name all structural isomers of C$_4$H$_9$OH with one hydroxyl functional group.

Functional isomers

If isomers have **functional isomerism** they have the same molecular formula but different functional groups in their structures. Figure 7.31 shows some examples.

positional isomers isomers in which the position of the functional group differentiates the compounds

functional isomerism refers to when isomers contain different functional groups

FIGURE 7.31 Functional isomers of **a.** C_3H_6O: propanal and propanone, and **b.** $C_3H_6O_2$: methyl ethanoate and propanoic acid

a.

Propanal

Propanone

b.

Methyl ethanoate

Propanoic acid

7.5.3 Isomer summary

Table 7.11 summarises the different types of structural isomers.

TABLE 7.11 Types of structural isomers

Structural isomers	Example
Chain isomers: different branching in carbon chain	 Butane Methylpropane
Positional isomers: different positions of the functional group, which is usually indicated by a number in the name	Propan-1-ol Propan-2-ol
Functional isomers: same atoms but different functional groups	Propanal Propanone (acetone)

Resources

Interactivity   Identifying structural isomers (int-1233)

Weblink   Structural isomers

elog-1898

tlvd-9724

EXPERIMENT 7.3   on line only

Constructing models of structural isomers

Aim

To construct models of structural isomers

7.5 Quick quiz on	7.5 Exercise	7.5 Exam questions

7.5 Exercise

1. **MC** Which of these compounds is *not* a structural isomer of the ester represented by $CH_3CH_2COOCH_3$?
   A. $CH_3COOCH_2CH_3$
   B. $HCOOCH_2CH_2CH_3$
   C. $CH_3CH_2COOCH_2CH_3$
   D. $CH_3CH_2CH_2COOH$

2. Draw the structural formula (showing all bonds) for the two carboxylic acids that have the molecular formula $C_4H_8O_2$.

3. Draw and systematically name *all* structural isomers with the molecular formula C_3H_7Cl.

4. Draw *all* isomers of C_4H_{10}.

5. Draw and give systematic names to *all* structural isomers with the molecular formula C_4H_8O that contain the carbonyl functional group.

6. For the following molecules, identify if they are isomers, and if so, state the type of isomerism present.

   a.

   b.

   c.

   d.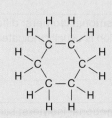

   e. $(CH_3)_2CHCH_2CH_3$ $CH_3(CH_2)_3CH_3$

7. Propanol is used as a disinfectant or antifreeze.
   a. Draw the structures of the two isomers of propanol and name them.
   b. Identify which is a primary alcohol and which is a secondary alcohol.
   c. State the type of structural isomerism present.

8. Using examples of molecules with four carbon atoms and the following types of isomerism:
   a. chain
   b. positional
      i. explain the difference between isomers of this type
      ii. draw and name the isomers used.

7.5 Exam questions

▶ **Question 1 (1 mark)**

MC C_4H_8 reacts with HCl to form C_4H_9Cl.

The number of possible isomers is

A. 2   B. 3   C. 4   D. ≥ 5

Question 2 (1 mark)

Source: *VCE 2020 Chemistry Exam, Section A, Q.7; © VCAA*

MC How many structural isomers have the molecular formula C_3H_6BrCl?

A. 4
B. 5
C. 6
D. 7

Question 3 (2 marks)

Source: *Adapted from VCE 2017 Chemistry Exam, Section B, Q.5.a.i, ii; © VCAA*

There are a number of structural isomers for the molecular formula C_3H_6O. Three of these are propanal, propanone and prop-2-en-1-ol.

a. Write the semi-structural formula for the ketone isomer propanone. **(1 mark)**
b. Draw the structural formula for the isomer prop-2-en-1-ol. **(1 mark)**

Question 4 (1 mark)

Source: *VCE 2010 Chemistry Exam 1, Section A, Q.11; © VCAA*

MC For which one of the following molecular formulas is there only one possible structure?

A. C_2HCl_3
B. $C_2H_4Cl_2$
C. $C_2H_2Cl_2$
D. C_4H_9OH

Question 5 (1 mark)

Source: *VCE 2015 Chemistry Exam, Section B, Q.5.b; © VCAA*

Draw the full structural formula of an isomer of butan-2-ol.

More exam questions are available in your learnON title.

7.6 Trends in physical properties

KEY KNOWLEDGE

- Trends in physical properties within homologous series (boiling point and melting point, viscosity), with reference to structure and bonding

Source: VCE Chemistry Study Design (2024–2027) extracts © VCAA; reproduced by permission.

The physical and chemical properties of organic compounds provide information that helps us understand and evaluate the interactions between organic chemicals. Depending on the types of atoms present in compounds, these interactions determine how organic molecules react to produce important chemicals for fuels, pharmaceuticals, manufacturing, industry and biological processes. The number of all the possible organic reactions that can occur is essentially infinite because there are so many combinations of organic compounds. However, certain general patterns involving addition, decomposition, combination, substitution or rearrangement of atoms or groups of atoms can be used to describe many common and useful reactions. It is not unusual to find that different pathways can produce the same organic substance.

7.6.1 Intermolecular forces

The properties of substances are determined by intermolecular forces. Intermolecular forces are those that act between molecules. They are influenced by the elements, bonds and shapes of molecules. These forces, along with the kinetic energies of the particles, determine properties such as density and melting and boiling points.

In molecules (where atoms are connected by *intra*molecular covalent bonds), *inter*molecular forces may be of three types: dispersion forces, dipole–dipole attractions and hydrogen bonding.

Dispersion forces

In any molecule, electrons can momentarily be distributed unevenly within the molecules, inducing a temporary dipole. Neighbouring molecules with similar temporary dipoles are attracted weakly to each other. This results in weak dispersion forces between the molecules. The strength of the dispersion forces is affected by the size and shape of molecules. In non-polar molecules, such as methane, CH_4, waxes and oils, there are no other types of intermolecular forces present, so the strength of the dispersion forces determines the overall strength of the intermolecular bonding. Dispersion forces are also called van der Waals forces.

Consider figure 7.32, which demonstrates a temporary dipole resulting in intermolecular attractions (dispersion forces).

- Molecule A has a temporary polarity due to uneven distribution of electrons.
- As the non-polar molecule B approaches A, its electrons are redistributed, because there is a tendency for them to be attracted to the end of A. This sets up an induced dipole in B.

These intermolecular attractions are called dispersion forces. Dispersion forces are weak and temporary, because electrons tend to redistribute themselves at different instances.

FIGURE 7.32 Temporary dipoles giving rise to intermolecular attractions (dispersion forces)

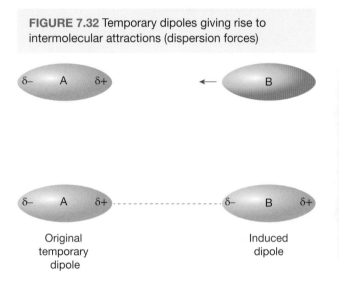

FIGURE 7.33 Candle wax consists of long hydrocarbon molecules held together by dispersion forces, which increase with molecular size.

Dipole–dipole attractions

Molecules such as HCl, HBr and CH_3Cl are polar and have permanent dipoles. The partial positive charge on one molecule is electrostatically attracted to the partial negative charge on a neighbouring molecule. Dipole–dipole attractions are stronger intermolecular forces than dispersion forces.

FIGURE 7.34 In dipole–dipole attractions, the central polar molecule is attracted to other polar molecules around it. They, in turn, are attracted by their neighbours.

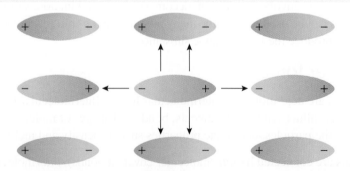

Hydrogen bonding

When hydrogen forms a bond with one of fluorine, oxygen or nitrogen (highly electronegative atoms), its electrons move slightly towards that atom. This causes the hydrogen nucleus to be exposed or unshielded. The molecule that forms is a dipole. Hydrogen bonding occurs between this dipole and another molecule that must also contain an electronegative atom, such as oxygen or nitrogen.

FIGURE 7.35 In an HF molecule, the highly electronegative F atom attracts the electrons; this leaves the H nucleus exposed, creating a dipole.

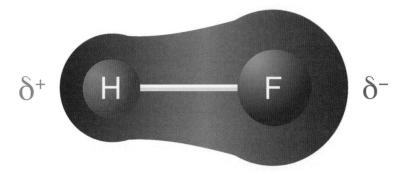

Hydrogen bonds are *stronger* intermolecular forces than both dispersion forces and dipole–dipole attractions. Hydrogen bonding occurs between water molecules and also between organic compounds such as alcohols (for example methanol, CH_3OH), carboxylic acids (for example ethanoic acid, CH_3COOH) and organic amines (for example methylamine, CH_3NH_2).

Hydrogen bonding is stronger than other dipole–dipole attractions due to the larger dipole moment that exists within these molecules, and because of the small size of the hydrogen atom involved. This allows the molecules to get closer to each other than in dipole–dipole attractions, thus increasing the force of attraction.

FIGURE 7.36 Examples of hydrogen bonding showing the lone pairs in one molecule are attracted to the unshielded hydrogens in another molecule.

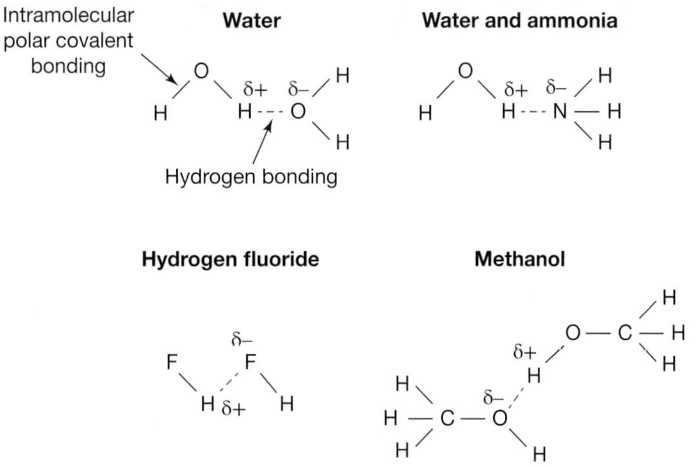

Relative strength of intermolecular forces

Hydrogen bonding is stronger than dipole–dipole attractions, which are stronger than dispersion forces.

7.6.2 Physical properties of organic compounds

Physical properties are measurable and used to describe how a substance behaves and exists without changing its chemical composition. For example, physical states such as solid (s), liquid (l) and gas (g), density and colour are observable and measurable. There are many physical properties associated with chemical substances, but boiling point, viscosity and solubility are examined in detail here.

Boiling point and melting point

Boiling and melting points depend on the strength of intermolecular forces. Remember that hydrogen bonding is stronger than dipole–dipole attractions, which are stronger than dispersion forces.

- The presence of different functional groups increases melting points because of the potential hydrogen bonding or dipole–dipole attractions.
- As the number of carbon atoms increases (i.e. the mass increases) there are more dispersion forces between the molecules, which may have a greater effect than the hydrogen bonding and dipole–dipole attractions.
- The closer the molecules can be arranged in solids, the greater the dispersion forces, so compounds with longer-chain molecules will have higher melting temperatures than branched molecules of the same mass. Similarly, more symmetrical molecules can pack more closely, which also results in a higher melting point.

An increase in intermolecular forces makes it harder to separate particles; therefore, they need higher temperatures to boil.

Boiling point and intermolecular forces

The stronger the intermolecular forces, the higher the boiling point.

TABLE 7.12 Summary of intermolecular forces in the homologous series

Type of compound	Intermolecular forces present
Alkanes, alkenes, alkynes	Dispersion only
Haloalkanes, aldehydes, ketones, esters	Dispersion, dipole–dipole
Alcohols, carboxylic acids, amines, amides	Dispersion, hydrogen bonding

---

EXTENSION: Why boiling points can change

Standard boiling points are the temperatures at which liquids can vaporise at atmospheric pressure. For a liquid to boil, it must overcome the pressure of the atmosphere. This means that when the air pressure varies, so does a substance's boiling point. Kinetic energy of the particles causes an outward pressure and, if larger than atmospheric pressure, particles vaporise. At high altitudes with less air and therefore less air pressure, liquids need less energy to boil. For example, it is difficult to hard-boil an egg on Mount Everest because water boils at around 71 °C at that altitude.

Viscosity

Viscosity is the resistance to flow of a liquid, and is affected by intermolecular forces and the shapes of the molecules present. Honey has high viscosity and water has low viscosity.

- The increased number of intermolecular forces in larger molecules, together with the possibility of branched molecules becoming tangled, results in higher viscosity.
- Viscosity decreases as temperature increases because the molecules attain enough energy to overcome the forces holding them together.

FIGURE 7.37 Honey is a viscous liquid due to strong intermolecular forces.

Viscosity and intermolecular forces

The stronger the intermolecular forces, the higher the viscosity.

Solubility

Only some organic compounds dissolve in water. For a substance to dissolve, the molecules must be able to interact with the water molecules, causing them to separate so that new interactions can be formed.

- Non-polar molecules cannot interact with water but can be attracted to non-polar solvents through dispersion forces. This is sometimes referred to as the 'like-dissolves-like rule'.
- Polar molecules are slightly soluble because some dipole–dipole attractions with water molecules can take place.

viscosity the resistance to flow of a liquid

The most likely molecules to dissolve in water are those that can form hydrogen bonds with water. However, as the size increases, and the non-polar section of the molecule increases, the solubility in water decreases.

7.6.3 Trends in homologous series

Within a homologous series, trends in physical properties are apparent. For example, in the alkanes the melting and boiling points increase with the size of the hydrocarbon. Their solubility in water is virtually non-existent due to the non-polar nature of hydrocarbons and the weak dispersion forces between molecules. The presence of functional groups containing atoms other than hydrogen affects the properties of organic compounds.

Hydrocarbons

The alkane, alkene and alkyne homologous series' have similar physical properties. Alkanes are colourless compounds that are less dense than water and have weak intermolecular attractive forces. Alkanes consist of non-polar molecules. The first four in the series are gases. As the size of the molecule increases, so does the influence of the dispersion forces; therefore, the melting and boiling points increase.

TABLE 7.13 The melting and boiling points of alkanes increase with increasing molecular size.

Alkane	Formula	Semi-structural formula	Melting point (°C)	Boiling point (°C)	State
Methane	CH_4	CH_4	−183	−164	Gas
Ethane	C_2H_6	CH_3CH_3	−182	−87	Gas
Propane	C_3H_8	$CH_3CH_2CH_3$	−190	−42	Gas
Butane	C_4H_{10}	$CH_3(CH_2)_2CH_3$	−135	−1	Gas
Pentane	C_5H_{12}	$CH_3(CH_2)_3CH_3$	−130	36	Liquid
Hexane	C_6H_{14}	$CH_3(CH_2)_4CH_3$	−94	68	Liquid
Heptane	C_7H_{16}	$CH_3(CH_2)_5CH_3$	−90	98	Liquid
Octane	C_8H_{18}	$CH_3(CH_2)_6CH_3$	−57	126	Liquid

Effect of side-chains or branching

The degree of branching affects the boiling point; as the amount of branching increases, the boiling point decreases. This is due to the inability of molecules to get close to each other. The dispersion forces operate over a small distance only, so the attraction is diminished. However, a higher degree of symmetrical branching has the opposite effect on melting point.

TABLE 7.14 Physical properties of branched butane isomers

Name	Skeletal structure	Melting point (°C)	Boiling point (°C)	Flashpoint (°C)
2-methylbutane		−159.8	27.8	−51
2,2-dimethylbutane		−99.9	49.7	−47.8
2,3-dimethylbutane		−128.8	57.9	−28.9

Haloalkanes

The existence of a halogen in an organic molecule can result in a polar molecule. Compared to the corresponding alkane, this would increase the strength of the intermolecular forces due to dipole–dipole attractions that would be present, and increased dispersion forces due to the greater mass of the molecule. When oxygen, fluorine or nitrogen is involved, hydrogen bonding will be present.

Alcohols

The hydroxyl group (–OH) in alcohols has a significant effect on properties. It can form hydrogen bonds with other alcohol or water molecules. Consequently, alcohols have a higher boiling point than corresponding alkanes, and smaller alcohols (three or fewer carbon atoms) are soluble in water. The boiling point of primary alcohols increases with increasing chain length due to the increasing number of dispersion forces, whereas the solubility decreases with increasing chain length due to the increasing length of the non-polar (hydrophobic) section of the molecules. The effect of the increased number of dispersion forces explains why volatility (tendency to vaporise) decreases with molecular size, whereas viscosity increases.

Many alcohols are highly flammable (with **flashpoints** below 37.8 °C), especially methanol (11 °C) and ethanol (17 °C). The flammability of alcohols decreases as the molecules increase in size and mass due to the increased strength of attraction between the molecules. Volatility also decreases as the size of the molecule increases.

FIGURE 7.38 This bar is made almost entirely of ice. The alcoholic drinks do not freeze even though they are served in ice vessels because alcohol has a lower freezing point than water.

flashpoints the temperature at which a particular organic compound gives off sufficient vapour to ignite in air

Two ethanol molecules   Ethanol and water molecules

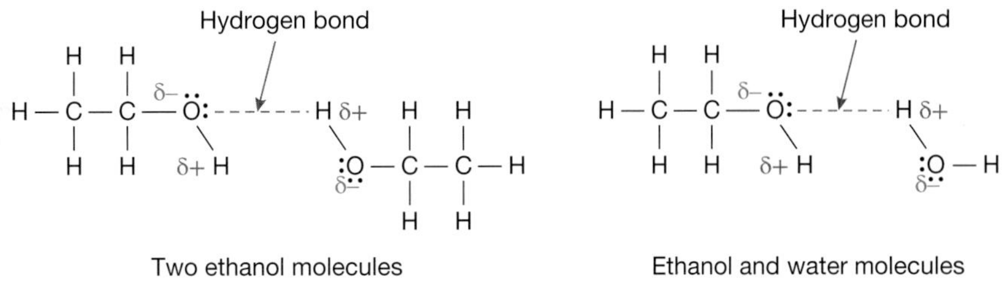

The carboxyl functional group

Like the alcohols, the first few members of the carboxylic acid homologous series are very soluble in water due to their capacity for strong hydrogen bonding with water molecules (figure 7.40).

FIGURE 7.40 Carboxylic acids form hydrogen bonds with water.

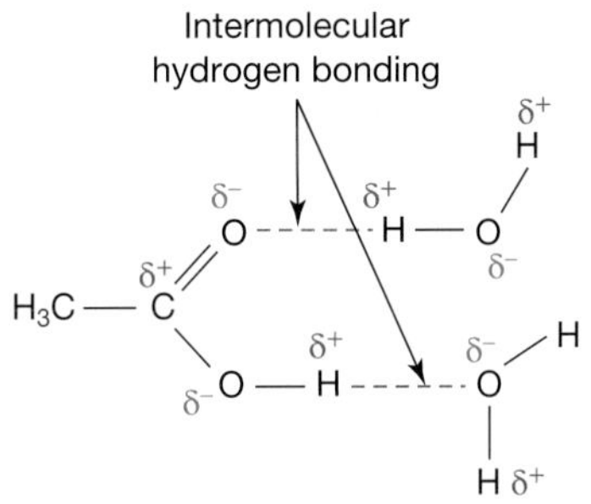

Carboxylic acids have much higher boiling points than the previously discussed homologous series' because carboxylic acid molecules can form two hydrogen bonds with each other (figure 7.41).

Carboxylic acids are weak acids that only partially ionise in water. They are still stronger acids than their corresponding alcohols because the −OH group is more polarised in the −COOH group by the presence of the highly electronegative O atom of C=O. This double-bonded O atom attracts the electrons away from the −OH group. Therefore, the H (from the hydroxyl group) is more weakly bonded to O and is more easily donated (figure 7.42).

FIGURE 7.41 A dimer (a pair of molecules) is formed by carboxylic acid molecules.

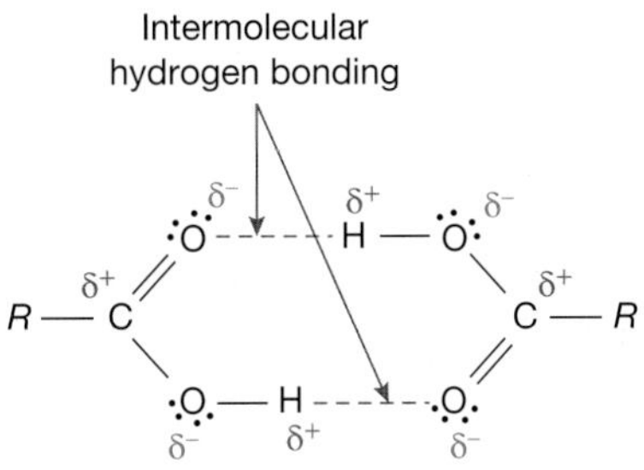

FIGURE 7.42 Polarisation of the −OH group in a carboxylic acid and an alcohol

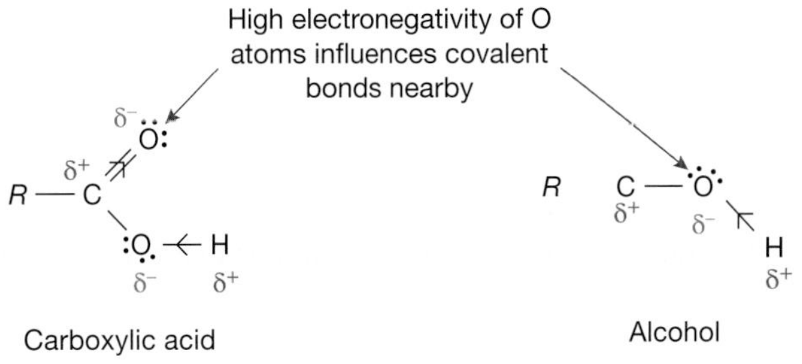

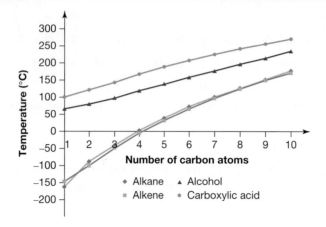

Aldehydes and ketones

Aldehydes and ketones are volatile compounds and are commonly found in perfumes and flavourings. Smaller molecules of these compounds are soluble because they can form hydrogen bonds with water, but solubility decreases with increasing length of the non-polar chain. These molecules cannot hydrogen bond with each other, but the polarity of the molecules means that the boiling point is higher than for similar-sized alkanes, but lower than for alcohols and carboxylic acids, which have hydrogen bonding between hydroxyl groups.

FIGURE 7.44 The general formulas for an aldehyde and a ketone. R represents an alkyl group.

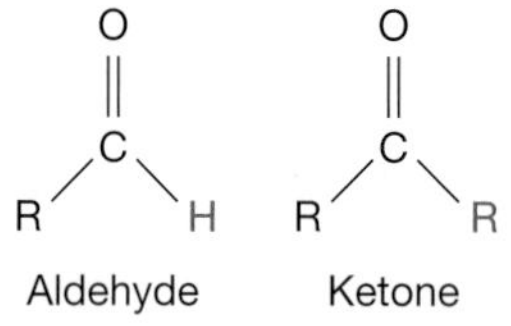

Aldehyde   Ketone

Esters

Esters are commonly found in plants and are responsible for many distinctive odours and flavours. Esters have lower boiling points than carboxylic acids because esters cannot form hydrogen bonds with each other (they do not have an O–H bond). Esters with very short carbon chains are soluble in water, whereas those with longer chains are less soluble.

FIGURE 7.45 An ester forming a hydrogen bond with water

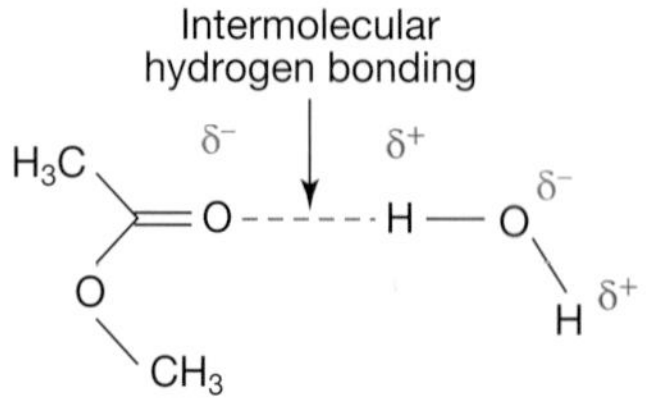

Intermolecular hydrogen bonding

FIGURE 7.46 Smells in fruits are due to a variety of chemicals, but the odours of some esters are more significant than others. One ester that contributes to the smell of pears is pentyl ethanoate.

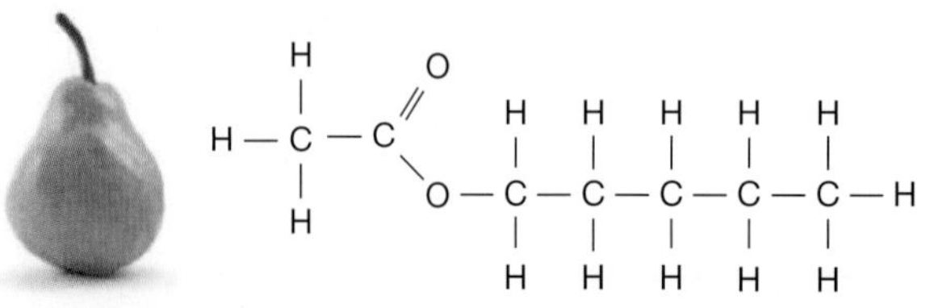

Amines and amides

Amines are stable compounds that generally have strong or unpleasant odours, similar to rotting fish. They are weak bases because they can accept a proton (figure 7.47).

FIGURE 7.47 Amines are weak bases.

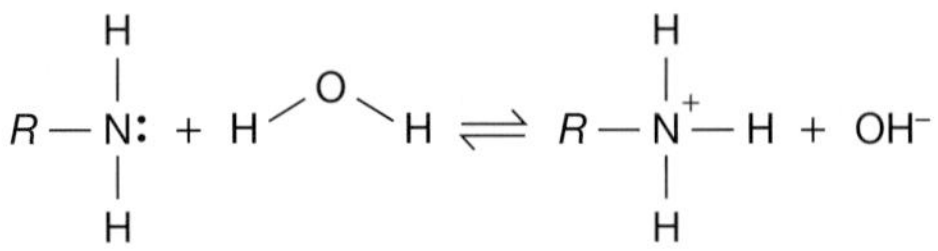

$$R-\underset{\underset{H}{|}}{\overset{\overset{H}{|}}{N}}{:}\ +\ H\overset{O}{\diagdown}H\ \rightleftharpoons\ R-\underset{\underset{H}{|}}{\overset{\overset{H}{|}}{N}}{\overset{+}{}}-H\ +\ OH^-$$

Hydrogen bonding is possible in amines (due to the presence of N–H bonds), but their boiling points are lower than the corresponding alcohols. The first two members of the homologous series are gases at room temperature, whereas the larger members are liquids. As with the other polar compounds containing hydrogen bonding, the solubility decreases with chain length.

Amides have higher melting and boiling temperatures than similar-sized organic compounds due to their capacity to form multiple hydrogen bonds between molecules. Methanamide is a liquid at room temperature, but larger amides are solids because of the increased number of dispersion forces. While smaller amides are soluble, they are less soluble than comparable amines and carboxylic acids; their solubility is similar to that of esters.

TIP: When explaining physical properties, ensure that the structure of the molecule is used to justify the type of intermolecular forces existing, and how the difference in strength of the intermolecular bonds results in the different properties.

tlvd-9699

SAMPLE PROBLEM 8 Explaining differences in boiling points, referring to intermolecular forces

Propane has a boiling point of –42 °C, whereas the boiling point of propan-1-ol is 97 °C. Explain this difference by referring to the intermolecular forces in both compounds.

THINK	WRITE
1. Draw the structures of propane and propan-1-ol.	Propane: 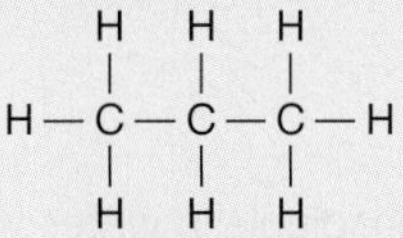 $$H-\underset{\underset{H}{\mid}}{\overset{\overset{H}{\mid}}{C}}-\underset{\underset{H}{\mid}}{\overset{\overset{H}{\mid}}{C}}-\underset{\underset{H}{\mid}}{\overset{\overset{H}{\mid}}{C}}-H$$ Propan-1-ol: 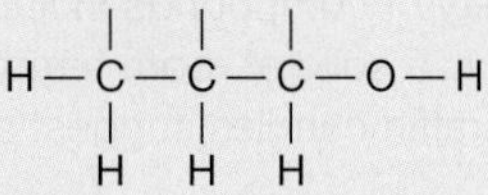 $$H-\underset{\underset{H}{\mid}}{\overset{\overset{H}{\mid}}{C}}-\underset{\underset{H}{\mid}}{\overset{\overset{H}{\mid}}{C}}-\underset{\underset{H}{\mid}}{\overset{\overset{H}{\mid}}{C}}-O-H$$
2. Consider the type of intermolecular forces that exist between molecules with the –OH functional group compared to hydrocarbons.	Propane: Intermolecular forces are dispersion forces only. These forces are weak and temporary. As a result, it only takes a small amount of energy in the form of heat to vaporise the liquid. Propan-1-ol: Intermolecular forces include hydrogen bonds due to the polar hydroxyl functional group on one end of the molecule, as well as dispersion forces.
3. Explain how the different intermolecular forces affect boiling point.	Hydrogen bonds are much stronger than dispersion forces, so significantly more heat is required to vaporise propan-1-ol than propane.

PRACTICE PROBLEM 8

By referring to intermolecular forces, explain the difference in the boiling point of 1-chloropropane (46.6 °C) and propane (–42 °C).

elog-1899

tlvd-9725

EXPERIMENT 7.4 online only

Comparing physical properties of alkanes, haloalkanes, alcohols and esters

Aim

To compare the properties of alkanes, haloalkanes, alcohols and esters, exploring melting point, boiling point, density and viscosity

7.6 Activities

learn on

Students, these questions are even better in jacPLUS

Receive immediate feedback and access sample responses

Access additional questions

Track your results and progress

Find all this and MORE in jacPLUS

| 7.6 Quick quiz   on | 7.6 Exercise | 7.6 Exam questions |

7.6 Exercise

1. Describe the intramolecular and intermolecular bonding that exists in hydrocarbons.
2. Identify the types of intermolecular forces (dispersion forces, dipole–dipole attractions or hydrogen bonding) acting between molecules of the following compounds.
   a. CH_3OH
   b. CH_3CH_3
   c. CH_3CH_2Cl
   d. CH_3NH_2
3. Explain why methane and ethane are insoluble in water, whereas methanol and ethanol are soluble.
4. Explain which has the higher boiling point: butanamide, $CH_3CH_2CH_2CONH_2$, or ethyl ethanoate, $CH_3COOCH_2CH_3$.
5. Candles can be made from a variety of compounds including soy, and bee and paraffin waxes. Paraffin contains a mixture of alkanes and is a solid at room temperature, and melts at 50–60 °C.
   a. What does the solid state of paraffin candles suggest about the size of the mix of alkanes used to make it?
   b. Why are essential oils, used to add scent to a candle, able to be mixed with paraffin wax?
6. Hexane is often used as an industrial solvent. Explain why each of the following is or is not an appropriate use of hexane.
   a. Removing salt from water
   b. Removing oil from soy beans
   c. Removing oil contaminants in water

7. The structures of dichloromethane and carbon tetrachloride are shown.

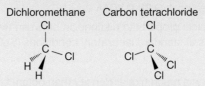

Dichloromethane   Carbon tetrachloride

Explain the following table of physical properties based on the two structures.

Physical property	Dichloromethane	Carbon tetrachloride
Boiling point	39.6 °C	76.7 °C
Solubility in water at SLC	17.5 g L^{-1}	0.81 g L^{-1}

8. Explain why ethyl ethanoate has a higher solubility in water but a lower boiling point than ethyl butanoate.
9. Alcohols are a family of organic compounds in which the –OH group is attached to a hydrocarbon chain.
   a. What term is given to this type of group?
   b. Explain why the boiling point of alcohols increases with the size of the molecules, but the solubility of alcohols decreases with increasing size.
   c. Draw and name a primary alcohol and a secondary alcohol that each contain three carbons, in order of decreasing boiling point.
10. Consider the list of compounds given. All are liquids at room temperature.
    - Octane, C_8H_{18}
    - Methylbutane, C_5H_{12}
    - Pentane, C_5H_{12}
    - Hexane, C_6H_{14}
    Which compound would be expected to:
    a. have the highest melting point
    b. have the highest viscosity
    c. be the most volatile?

7.6 Exam questions

▶ Question 1 (1 mark)

Source: VCE 2020 Chemistry Exam, Section A, Q.16; © VCAA

MC The following table provides information about three organic compounds: X, Y and Z.

Compound	Structural formula	Molar mass (g mol^{-1})	Boiling point (°C)
X		60	97
Y		60	118
Z		60	?

Which one of the following is the best estimate for the boiling point of Compound Z?
A. 31 °C
B. 101 °C
C. 114 °C
D. 156 °C

ⓘ Question 2 (1 mark)

Source: VCE 2018 Chemistry Exam, Section A, Q.21; © VCAA

MC A student wants to use a physical property to distinguish between two alcohols, octan-1-ol and propan-1-ol. Both alcohols are colourless liquids at standard laboratory conditions (SLC).

The student should use
A. density because propan-1-ol has a much higher density than octan-1-ol.
B. boiling point because octan-1-ol has a higher boiling point than propan-1-ol.
C. electrical conductivity because octan-1-ol has a higher conductivity than propan-1-ol.
D. spectroscopy because it is not possible to distinguish between the alcohols using their physical properties.

ⓘ Question 3 (1 mark)

Source: VCE 2011 Chemistry Exam 1, Section A, Q.1; © VCAA

MC Which one of the following compounds is **least** soluble in water at room temperature?
A. ethane
B. ethanol
C. ethylamine
D. ethanoic acid

ⓘ Question 4 (3 marks)

Source: VCE 2010 Chemistry Exam 1, Section B, Q.9.c; © VCAA

The boiling points of several alkanols are provided in the following table.

Alkanol	methanol	ethanol	propan-l-ol	butan-l-ol	pental-l-ol
Boiling point (°C)	64.5	78.3	97.2	117.2	138.0

Butane and propan-1-ol have similar molar masses. The boiling point of butane is −138.4 °C and that of propan-1-ol is 97.2 °C.

Explain, in terms of intermolecular forces, the difference between the boiling points of these two compounds.

ⓘ Question 5 (5 marks)

Two compounds have the molecular formula C_2H_6O.

Use the following data to draw the structures of compound A and compound B.

	Compound A	Compound B
Boiling point	−24 °C	78 °C
Solubility in water at SLC	71 g L^{-1}	Miscible

More exam questions are available in your learnON title.

7.7 Review

7.7.1 Topic summary

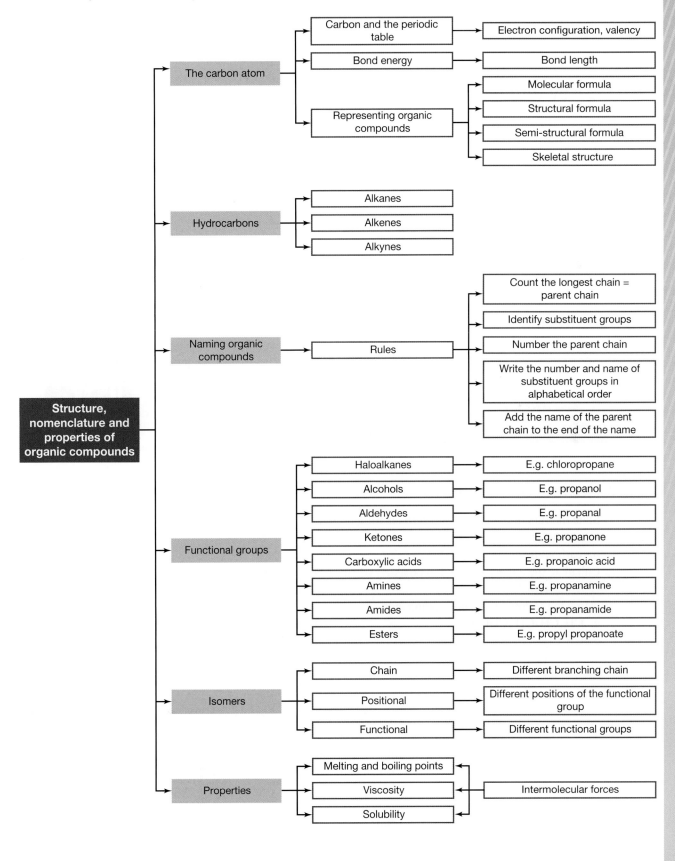

7.7.2 Key ideas summary

online only

7.7.3 Key terms glossary

online only

 Resources

Solutions	Solutions — Topic 7 (sol-0834)
Practical investigation eLogbook	Practical investigation eLogbook — Topic 7 (elog-1706)
Digital documents	Key science skills — VCE Chemistry Units 1–4 (doc-37066)
	Key terms glossary — Topic 7 (doc-37293)
	Key ideas summary — Topic 7 (doc-37294)
Exam question booklet	Exam question booklet — Topic 7 (eqb-0118)

7.7 Activities

learn on

7.7 Review questions

1. State the functional groups and suffixes for naming:
   a. alcohols
   b. carboxylic acids.

2. Calculate the energy required to break all covalent bonds in 20.0 g of CCl_4. The C–Cl bond energy is 324 kJ mol^{-1}.

3. Name the following organic compounds.
   a. $CH_3CH_2CH=CHCH_3$
   b. $CH_2=CHCl$
   c. CF_2Cl_2
   d. CH_3CH_2OH
   e.
   $$CH_3 - \overset{\overset{\displaystyle CH_3}{|}}{\underset{\underset{\displaystyle CH_3}{|}}{C}} - CH_3$$
   f. $CH_3(CH_2)_3\underset{\underset{\displaystyle OH}{|}}{C}HCH_3$

   g. $CHCl_2CHBr_2$

4. Draw the structural formulas for the following compounds.
   a. 2-methylpentan-3-ol
   b. Butan-2-ol
   c. 2,2-dimethylbutan-1-ol
   d. 2,3-dimethylpentan-2-ol
   e. Pentanoic acid
   f. 2-methylpropanoic acid
   g. Butanoic acid
   h. Pentanal
   i. 2-chloropropan-1-amine
   j. Butyl methanoate
   k. Butanone
   l. 2-pentyne

5. Write the IUPAC name for $(CH_3)_2CHCH(NH_2)CH_3$.
6. Complete the following table, identifying the intermolecular forces present in each compound.

Compound	Intermolecular forces
$CH_3CH_2CH_2OH$	
$CH_3CH_2OCH_2CH_3$	
$CH_3CH_2CH_2F$	
$CH_3CH_2N(CH_3)_2$	

7. Place the following compounds in increasing order of solubility, from most soluble to least soluble.

$CH_3CH_2CH_2OH$, CH_3CH_2OH, $CH_3CH_2CH_3$, CH_3CH_3Cl

8. Which of the two compounds, propane, C_3H_8, or propan-1-ol, C_3H_7OH, would be expected to have the higher boiling point? Explain your reasoning.
9. Which of the following compounds *only* have dispersion forces operating between their molecules?
   i. CH_4
   ii. CH_3OH
   iii. $C_6H_{12}O_6$
   iv. C_2H_2
   v. C_2H_5OH
   vi. CH_3COOH
   vii. CH_3Cl
   viii. C_6H_6
   ix. $C_2H_2Cl_3$
10. C_4H_9Cl has four isomers.
    a. Draw the structures of each of these isomers and name each of them.
    b. Draw the skeletal structures.
    c. Write out the semi-structural formulas.

7.7 Exam questions

Section A — Multiple choice questions

All correct answers are worth 1 mark each; an incorrect answer is worth 0.

Question 1

Source: *VCE 2022 Chemistry Exam, Section A, Q.16; © VCAA*

MC The correct IUPAC name for $CH_3CH_2CHClCHOHCH_3$ is

A. 3-chloropentan-4-ol **B.** 3-chloropentan-2-ol **C.** 2,3-chloro-pentanol **D.** 3,2-chloro-pentanol

Question 2

Source: *VCE 2020 Chemistry Exam, Section A, Q.4; © VCAA*

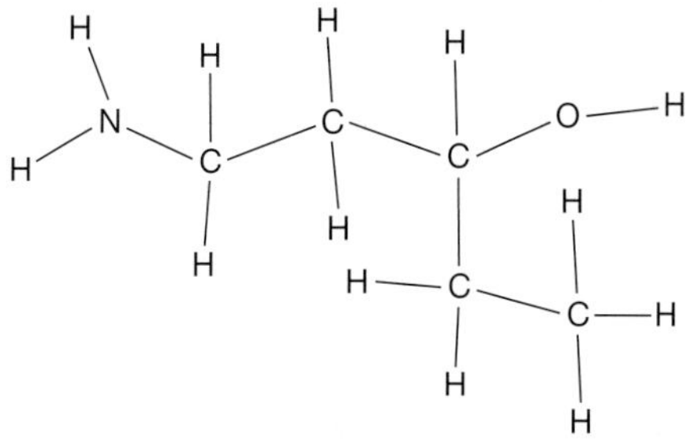

MC What is the IUPAC name of the molecule shown above?

A. 3-hydroxy-3-ethyl-propan-1-amine
B. 3-amino-1-methylpropan-1-ol
C. 3-hydroxypentan-1-amine
D. 1-aminopentan-3-ol

Question 3

Source: VCE 2019 Chemistry Exam, Section A, Q.3; © VCAA

MC A compound has the following skeletal formula.

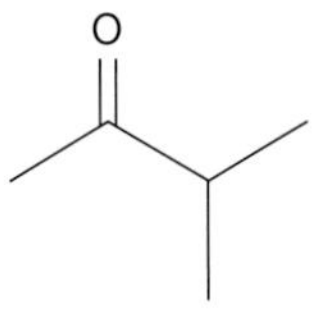

The molar mass of the compound is

A. 71 g mol^{-1}
B. 74 g mol^{-1}
C. 85 g mol^{-1}
D. 86 g mol^{-1}

Question 4

Source: VCE 2012 Chemistry Exam 1, Section A, Q.9; © VCAA

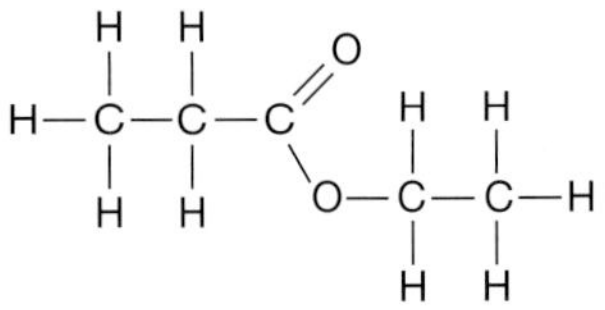

MC Which one of the following is the correct systematic name of this compound?

A. ethyl propanoate
B. ethyl ethanoate
C. propyl ethanoate
D. propyl pentanoate

Question 5

Source: VCE 2018 Chemistry NHT Exam, Section A, Q.5; © VCAA

MC Pentane, hexane, heptane and octane are non-branched alkanes.

Which one of the following statements gives a valid comparison?

A. Octane has a greater viscosity and a higher boiling point than hexane.
B. Pentane has a greater viscosity and a lower boiling point than octane.
C. Heptane has a lower viscosity and a higher boiling point than octane.
D. Heptane has a lower viscosity and a lower boiling point than pentane.

Question 6

Source: VCE 2011 Chemistry Exam 1, Section A, Q.4; © VCAA

MC The compound that is **not** an isomer of 2,2,4-trimethylpentane is

A. octane.
B. 3-ethylhexane.
C. 2,4-dimethylpentane.
D. 2,4-dimethylhexane.

Source: VCE 2011 Chemistry Exam 1, Section A, Q.3; © VCAA

MC Consider the following structures.

I
$$\begin{array}{c} CH_3 \\ | \\ H-C-CH_2-CH_2-CH-CH_3 \\ | \qquad\qquad | \\ CH_3 \qquad\qquad CH_3 \end{array}$$

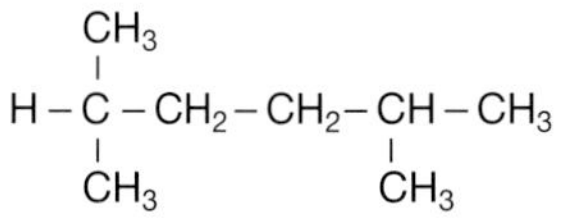

II
$$\begin{array}{c} CH_3 \\ | \\ H_3C-C-CH_2-CH-CH_3 \\ | \qquad\quad | \\ CH_3 \quad\; CH_3 \end{array}$$

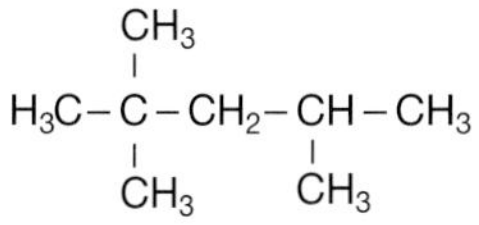

III
$$\begin{array}{c} CH_3 \qquad\; CH_3 \\ | \qquad\qquad | \\ H_3C-CH-CH_2-C-CH_3 \\ | \\ CH_3 \end{array}$$

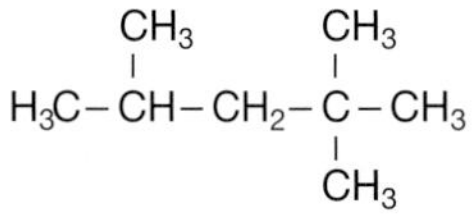

IV
$$\begin{array}{c} CH_3 \qquad\qquad CH_3 \\ | \qquad\qquad\quad | \\ H_3C-CH-CH_2-CH_2-C-H \\ | \\ CH_3 \end{array}$$

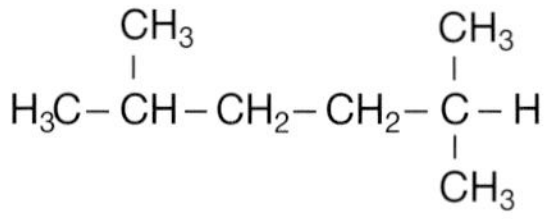

Which of the structures is that of 2,2,4-trimethylpentane?

A. I and III only

B. I and IV only

C. II and III only

D. II and IV only

▶ **Question 8**

Source: VCE 2012 Chemistry Exam 1, Section A, Q.2; © VCAA

MC The number of structural isomers of C_4H_9Cl is

A. 2

B. 3

C. 4

D. 5

▶ **Question 9**

MC The structural arrangement of 2,2-dimethylbutan-1-ol is

A.
$$\begin{array}{c} H \quad CH_3 \; CH_3 \; H \\ | \qquad | \qquad | \quad\; | \\ H-C-C-C-C-OH \\ | \qquad | \qquad | \quad\; | \\ H \quad\; H \quad\; H \quad\; H \end{array}$$

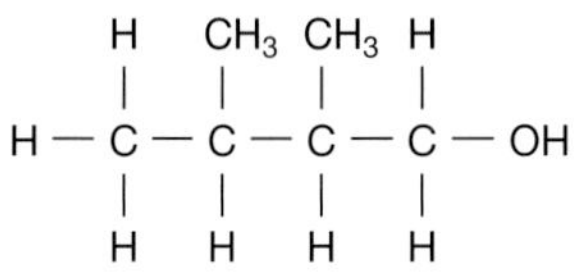

B.
$$\begin{array}{c} H \quad\; H \quad\; CH_3 \; H \\ | \qquad | \qquad | \quad\; | \\ H-C-C-C-C-OH \\ | \qquad | \qquad | \quad\; | \\ H \quad\; H \quad\; CH_3 \; H \end{array}$$

C.
$$\begin{array}{c} H \quad\; H \quad\; CH_3 \; H \\ | \qquad | \qquad | \quad\; | \\ HO-C-C-C-C-H \\ | \qquad | \qquad | \quad\; | \\ H \quad\; H \quad\; CH_3 \; H \end{array}$$

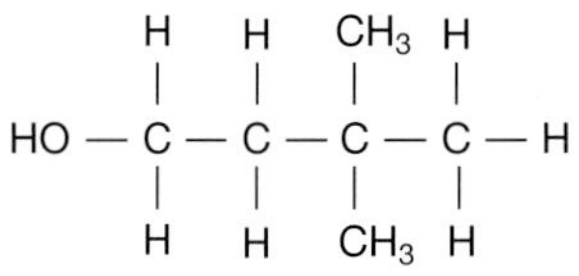

D.
$$\begin{array}{c} H \quad\; H \\ | \qquad | \\ H-C-C-OH \\ | \qquad | \\ CH_3 \; CH_3 \end{array}$$

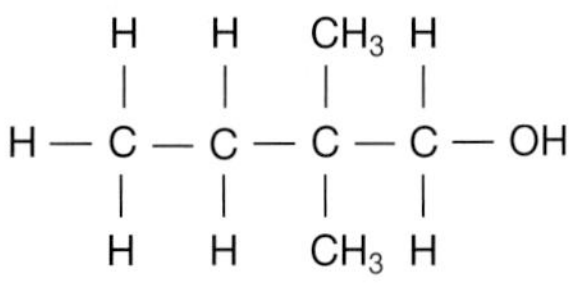

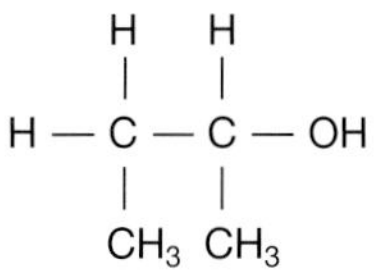

Question 10

MC What is the IUPAC name of the following molecule?

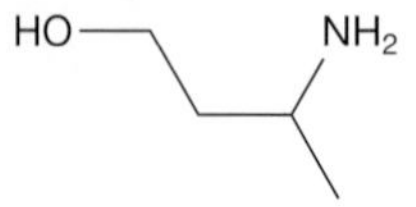

A. 1-amino-2-methylbutan-4-ol

B. 2-aminobutan-4-ol

C. 4-amino-3-methylbutan-1-ol

D. 3-aminobutan-1-ol

Section B — Short answer questions

Question 11 (3 marks)

Source: VCE 2018 Chemistry Exam, Section B, Q.1.a; © VCAA

Organic compounds are numerous and diverse due to the nature of the carbon atom. There are international conventions for the naming and representation of organic compounds.

a. Draw the structural formula of 2-methyl-propan-2-ol. **(1 mark)**

b. Give the molecular formula of but-2-yne. **(1 mark)**

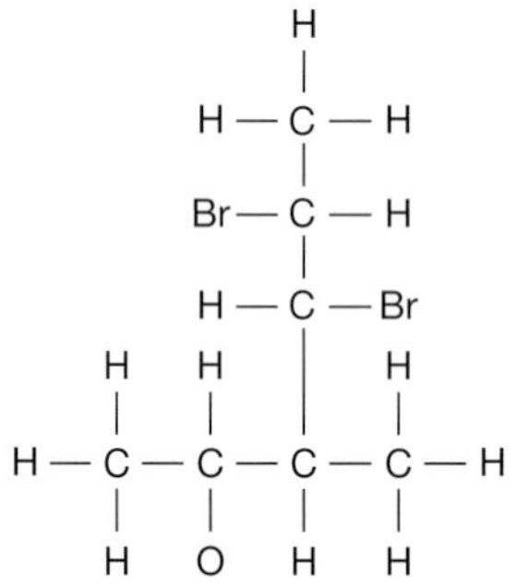

c. Give the IUPAC name of the compound that has the structural formula shown above. **(1 mark)**

Question 12 (4 marks)

Source: Adapted from VCE 2019 Chemistry Exam, Section B, Q.2.a; © VCAA

A sequence of reactions can be applied to pent-2-ene to produce a variety of desirable products.

a. Draw the skeletal formula for pent-2-ene. **(1 mark)**

Two structural isomers are possible when pent-2-ene is reacted with water at a high temperature in the presence of an acid catalyst. Each isomer is a secondary alcohol with the formula $C_5H_{11}OH$.

b. Give the IUPAC name of both isomers. **(2 marks)**

c. When one of these isomers is reacted with acidified dichromate ions, pentan-2-one is formed.
Draw the semi-structural formula of pentan-2-one. **(1 mark)**

▶ **Question 13 (4 marks)**

Source: *Adapted from VCE 2017 Chemistry Exam, Section B, Q.7a; © VCAA*

A table of carboxylic acids and their melting points is shown below.

Carboxylic acid	Melting point (°C)
heptanoic acid	−11
pentanoic acid	−34.5
2-methylhexanoic acid	−56

With reference to their structure and bonding, explain the difference in melting points between

a. heptanoic acid and pentanoic acid   (2 marks)
b. heptanoic acid and 2-methylhexanoic acid.   (2 marks)

▶ **Question 14 (6 marks)**

Cyclohexane and cyclohexene are examples of cyclic hydrocarbons.

a. Draw the structures of cyclohexane and cyclohexene   (2 marks)
b. Draw a straight-chain structure of a non-cyclic isomer of each.   (2 marks)
c. Name the homologous series to which the isomers in part **b** belong.   (2 marks)

▶ **Question 15 (5 marks)**

The diverse nature of organic compounds can be attributed to carbon and its unique chemical properties. In fact, there are millions of organic compounds found in nature and synthesised in the laboratory. To be able to identify these molecules, IUPAC has developed naming conventions and models.

Consider the alcohol 2-methylpropan-2-ol.

a. Draw the structural formula of 2-methylpropan-2-ol.   (1 mark)
b. What is the name of the functional group?   (1 mark)
c. Is this a primary, secondary or tertiary alcohol? Justify your answer.   (1 mark)
d. Write the semi-structural formula of 2-methylpropan-2-ol.   (1 mark)
e. Draw the skeletal structure of 2-methylpropan-2-ol.   (1 mark)

8 Reactions of organic compounds

KEY KNOWLEDGE

In this topic you will investigate:

Reactions of organic compounds

- organic reactions and pathways, including equations, reactants, products, reaction conditions and catalysts (specific enzymes not required):
  - synthesis of primary haloalkanes and primary alcohols by substitution
  - addition reactions of alkenes
  - the esterification between an alcohol and a carboxylic acid
  - hydrolysis of esters
  - pathways for the synthesis of primary amines and carboxylic acids
  - transesterification of plant triglycerides using alcohols to produce biodiesel
  - hydrolytic reactions of proteins, carbohydrates and fats and oils to break down large biomolecules in food to produce smaller molecules
  - condensation reactions to synthesise large biologically important molecules for storage as proteins, starch, glycogen and lipids (fats and oils)
- calculations of percentage yield and atom economy of single-step or overall reaction pathways, and the advantages for society and for industry of developing chemical processes with a high atom economy
- the sustainability of the production of chemicals, with reference to the green chemistry principles of use of renewable feedstocks, catalysis and designing safer chemicals.

Source: VCE Chemistry Study Design (2024–2027) extracts © VCAA; reproduced by permission.

PRACTICAL WORK AND INVESTIGATIONS

Practical work is a central component of VCE Chemistry. Experiments and investigations, supported by a **practical investigation eLogbook** and **teacher-led videos**, are included in this topic to provide opportunities to undertake investigations and communicate findings.

EXAM PREPARATION

▶ Access past VCAA questions and exam-style questions and their video solutions in every lesson, to ensure you are ready.

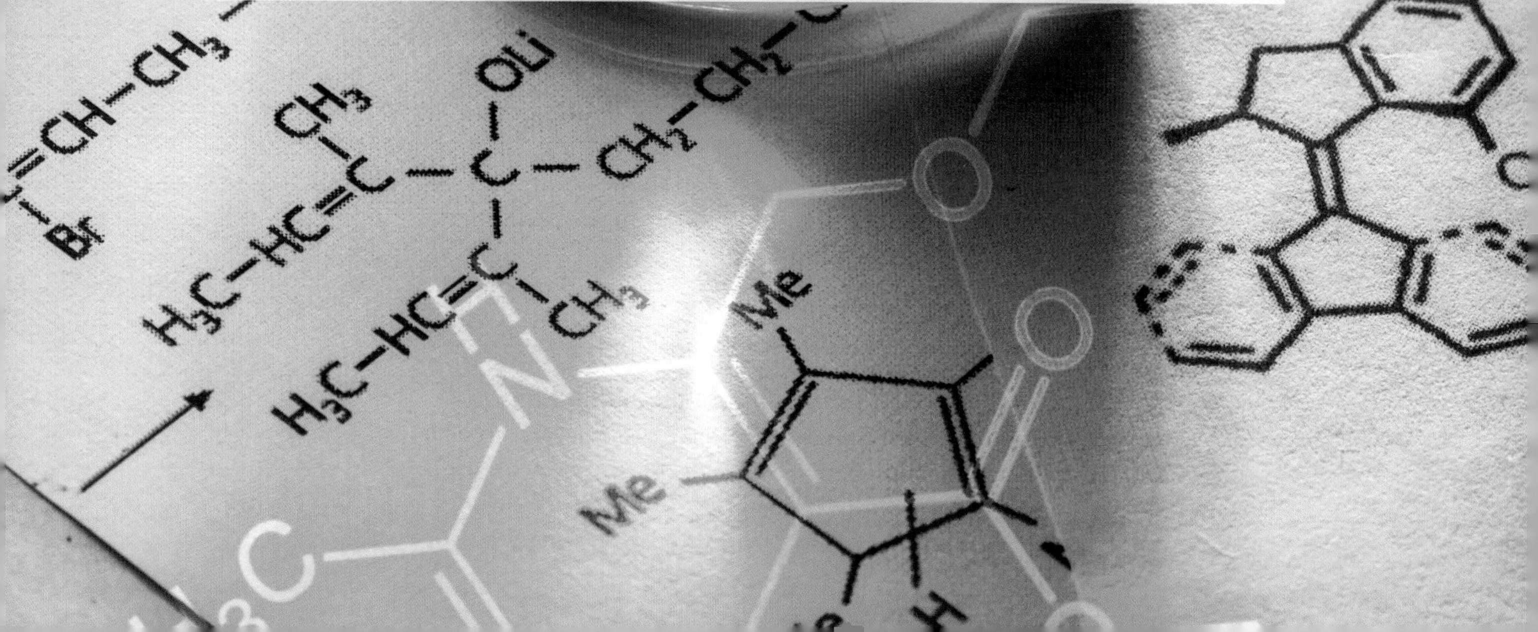

8.1 Overview

Organic reactions are involved in the production of diverse materials, including fuels and fabrics, plastics and pharmaceuticals, and foods and flavourings. Our own bodies only function effectively because of the enormous number of organic chemical reactions that are occurring every second of every day.

Chemical processes can include one step or many steps in a reaction pathway. Natural and synthetic organic reaction pathways can range from simple to quite complex. Chemists and engineers are constantly looking for ways to make new materials, improve products and reduce environmental impact, and finding the right pathway can be a challenging and time-consuming process. For example, some engineers are looking to replace plastic bottles made from fossil fuels with polymers derived from sugars in biomass. Others have reduced the many steps involved in adding nitrogen to drugs, fertilisers and pesticides to a single step, making it possible to increase production and yield.

FIGURE 8.1 The variety of jellybean flavours and colours that exists is due to esters. Esters are synthesised in a condensation reaction between an alcohol and a carboxylic acid.

This topic builds on the chemical interactions between the hydrocarbons and homologous series studied in topic 7. Reactions for many diverse processes, including substitution, addition, oxidation, hydrolysis and condensation reactions, are examined. You will learn about different reaction pathways to make a variety of organic compounds, including amines, carboxylic acids and esters. You will examine reaction yields and atom economy because chemists are increasingly looking for ways to improve the efficiency of reactions and reduce chemical waste.

LEARNING SEQUENCE

on Resources

8.2 Substitution, addition and oxidation reactions

As described in topic 7, **functional groups** influence physical properties, but they also influence chemical properties; that is, the type of chemical reactions that occur. Many of the products we use every day are the result of organic reactions. Common reaction types include substitution, addition, redox, condensation and hydrolysis.

8.2.1 Substitution reactions

Substitution reactions occur when one or more atoms on a molecule are replaced by others, as opposed to being added in like those in addition reactions. The types of atoms or groups involved in these reactions generally depend on the atoms they replace on the molecule. **Halogens** are very good at substituting, as are polar functional molecules or groups with lone pairs of electrons, such as H_2O and NH_3.

Substitution reactions of alkanes

Alkanes are relatively unreactive, but they do undergo substitution reactions with halogens. In these reactions, the halogen atoms replace one or more hydrogen atoms. This is also described as a **halogenation** reaction. For example, the successive chlorination of methane to form chloromethanes occurs as follows:

$$CH_4 \quad + \quad Cl_2 \quad \xrightarrow{\text{UV light}} \quad \underset{\text{Chloromethane}}{CH_3Cl} \quad + \quad HCl$$

$$CH_3Cl \quad + \quad Cl_2 \quad \xrightarrow{\text{UV light}} \quad \underset{\text{Dichloromethane}}{CH_2Cl_2} \quad + \quad HCl$$

$$CH_2Cl_2 \quad + \quad Cl_2 \quad \xrightarrow{\text{UV light}} \quad \underset{\substack{\text{Trichloromethane} \\ \text{(chloroform)}}}{CHCl_3} \quad + \quad HCl$$

$$CHCl_3 \quad + \quad Cl_2 \quad \xrightarrow{\text{UV light}} \quad \underset{\substack{\text{Tetrachloromethane} \\ \text{(carbon tetrachloride)}}}{CCl_4} \quad + \quad HCl$$

These reactions require energy in the form of UV light to catalyse the reactions. The formation of different chloromethanes depends on the amount of chlorine present. These chloromethanes have different boiling points because of their different structures, so mixtures can be separated using fractional distillation. Distillation will be discussed in topic 9.

The UV light breaks the **covalent bond** between the chlorine atoms to produce unstable chlorine free radicals. In general, when exposed to light:

$$R{-}H + X_2 \xrightarrow{\text{UV light}} R{-}X + HX$$

The most commonly used halogens for this reaction are Cl_2 and Br_2; F_2 is too reactive and I_2 is too unreactive.

functional group an atom or group of atoms that is attached to or part of a hydrocarbon chain, and influences the physical and chemical properties of the molecule

substitution reaction a reaction in which one or more atoms of a molecule are replaced by different atoms

halogens elements in group 17 of the periodic table: F, Cl, Br, I and At

alkanes the family of hydrocarbons containing only single carbon–carbon bonds

halogenation a reaction in which one or more halogen atoms are added

covalent bonds bonds that involve the sharing of electron pairs between atoms

These haloalkanes are **primary haloalkanes**. In primary haloalkanes, the halogen atom is attached to a carbon atom that is only attached to one other carbon atom. The general formula for a primary haloalkane is $R–CH_2X$. Methyl halides (halomethanes) are classed as primary haloalkanes.

FIGURE 8.2 Examples of primary haloalkanes

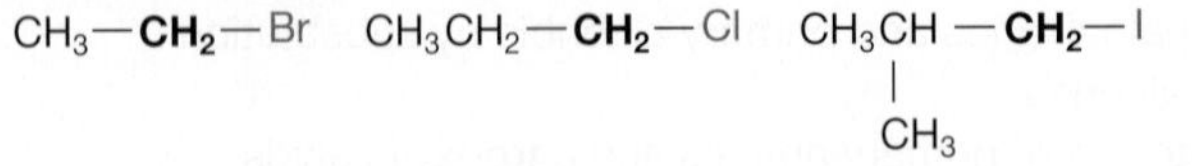

$$CH_3—\mathbf{CH_2}—Br \quad CH_3CH_2—\mathbf{CH_2}—Cl \quad CH_3CH—\mathbf{CH_2}—I$$
$$\underset{\displaystyle CH_3}{|}$$

Substitution reactions of haloalkanes

Haloalkanes are widely used in chemical processes, but most do not occur naturally and must be produced synthetically. One of the first haloalkanes used was chloroform. It was used during the American Civil War (1861–65) as an anaesthetic for amputations and treatment of soldiers. Now, haloalkanes are widely used in medicine, agriculture and the production of polymers.

Haloalkanes are particularly useful as precursors (reactants) to the preparation of further substances. **Alcohols** can be prepared from haloalkanes in substitution reactions by reacting them with solutions of either sodium or potassium hydroxide. For example, propan-2-ol can be made by reacting either 2-chloropropane or 2-bromopropane with dilute sodium hydroxide.

FIGURE 8.3 Coral secretes natural haloalkanes to deter starfish like the crown of thorns from feeding on it.

$$\underset{\displaystyle Br}{\underset{|}{CH_3CHCH_3}} + NaOH \longrightarrow \underset{\displaystyle OH}{\underset{|}{CH_3CHCH_3}} + NaBr$$

Either 1-chloropropane or 1-bromopropane could be used to make propan-1-ol.

$$CH_3CH_2CH_2Cl + NaOH \longrightarrow CH_3CH_2CH_2OH + NaCl$$

Another reaction to produce an alcohol is to add water and heat, but this is quite a slow reaction; in the following example, the other product is hydrobromic acid:

$$CH_3CH_2Br + H_2O \longrightarrow CH_3CH_2OH + HBr$$

If, however, an amine is required, then the process requires the addition of excess ammonia:

$$CH_3CH_2CH_2Br \xrightarrow{\text{Excess } NH_3} CH_3CH_2CH_2NH_2$$

or

$$CH_3CH_2CH_2Br + 2NH_3 \rightleftharpoons CH_3CH_2CH_2NH_2 + NH_4^+ + Br^-$$

The more ammonia, the more the forward reaction is favoured.

primary haloalkane a haloalkane in which the halogen atom is attached to a carbon that is only attached to one other carbon atom

alcohols compounds in which a hydroxyl group (–OH) is the parent functional group

Substitution reaction

In a substitution reaction, an atom or functional group is replaced by another atom or functional group.

tlvd-10340

SAMPLE PROBLEM 1 Writing the reaction steps and reagents to make a primary alcohol from an alkane

Write the steps and reagents to produce 1-propanol from propane.

THINK

1. Propane must be converted to chloropropane before 1-propanol can be produced in a substitution reaction.

2. A mixture of different chloropropane compounds is formed, which needs to be separated by fractional distillation to obtain 1-chloropropane.

3. React the 1-chloropropane with NaOH in another substitution reaction to produce propan-1-ol.

WRITE

Separate 1-chloropropane by fractional distillation.

PRACTICE PROBLEM 1

Write the steps and reagents to produce methanol from methane.

On Resources

✦ **Interactivity** Comparing substitution and addition reactions (int-1235)

8.2.2 Addition reactions of alkenes

Alkenes are more reactive than alkanes. They are unsaturated hydrocarbons and undergo **addition reactions** in which the C=C bond is broken and new single bonds are formed. This is because the energy required to break the double bond is less than the energy released in the formation of two single bonds. For example, hydrogenation of ethene produces ethane and releases energy:

$$H_2C{=}CH_2(g) + H_2(g) \xrightarrow{\text{Catalyst}} H_3CCH_3(g)$$

Substances that undergo addition reactions with alkenes include hydrogen (H_2), chlorine (Cl_2), bromine (Br_2), hydrochloric acid (HCl), hydrogen bromide (HBr), hydrogen iodide (HI) and water (H_2O). The addition of H_2 requires the presence of a catalyst, such as finely divided platinum (Pt), palladium (Pd) or nickel (Ni). The others react without the need for catalysts.

alkenes the family of hydrocarbons that contain at least one carbon–carbon double bond

addition reaction a reaction in which one molecule bonds covalently with another molecule without losing any other atoms

The following equations are examples of addition reactions with alkenes. Note that the reactants Br_2, HCl and H_2 in these reactions add *across* the double bond. Therefore, 1,2-dibromopropane is the only product.

$$H_2C{=}CHCH_3 \quad + \quad Br_2 \quad \longrightarrow \quad CH_2BrCHBrCH_3$$
Propene $\qquad\qquad\qquad$ Bromine $\qquad\qquad\qquad$ 1,2-dibromopropane

$$H_2C{=}CH_2 \quad + \quad HCl \quad \longrightarrow \quad CH_3CH_2Cl$$
Ethene $\qquad\qquad\qquad$ Hydrogen chloride $\qquad\qquad$ Chloroethane

$$H_2C{=}CHCH_2CH_3 \quad + \quad H_2 \quad \xrightarrow{\text{Pd}} \quad CH_3CH_2CH_2CH_3$$
But-1-ene $\qquad\qquad\qquad$ Hydrogen $\qquad\qquad\qquad$ Butane

The reaction of propene with HCl could result in two possible products — 1-chloropropane, $CH_2ClCH_2CH_3$, and 2-chloropropane, $CH_3CHClCH_3$— when the HCl reacts across the double bond.

TIP: Draw the structures of molecules when answering questions in organic chemistry to ensure there are the correct number and type of bonds. Carbon atoms must have four bonds.

The reaction of an alkene with bromine is used as a test for unsaturation. When red-brown bromine water is shaken with an unsaturated hydrocarbon, the reaction mixture becomes colourless due to the formation of the dibromo derivative (a compound formed from another structurally similar compound).

Ethene is used as a raw material in a fast method to produce the large amounts of ethanol needed for industrial use. Ethene is mixed with steam and passed over a phosphoric acid catalyst at 330 °C. The reaction of the direct catalytic hydration of ethene in the vapour phase is an addition reaction. This reaction is demonstrated using molecular models in figure 8.4.

$$H_2C{=}CH_2(g) \quad + \quad H_2O(g) \quad \xrightarrow{H_3PO_4} \quad CH_3CH_2OH(g)$$
Ethene $\qquad\qquad\qquad$ Steam $\qquad\qquad\qquad$ Ethanol

FIGURE 8.4 A model equation for the direct hydration of ethene to form ethanol

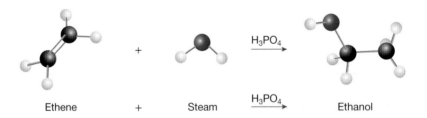

Ethene $\qquad\qquad$ + $\qquad\qquad$ Steam $\qquad \xrightarrow{H_3PO_4} \qquad$ Ethanol

---

Addition reactions

Alkenes undergo addition reactions across the double carbon–carbon bond, forming a single product.

Alkenes can also undergo addition polymerisation reactions to form polymers in the presence of a catalyst. This occurs when the carbon-to-carbon double bonds in alkene molecules are broken, and each molecule then joins with other alkene molecules on either side, forming a long chain.

on Resources

Video eLessons Addition reactions of alkenes (eles-3248)
$\qquad\qquad\qquad\qquad$ Catalysis: hydrogenation of ethene (eles-1673)
$\qquad\qquad\qquad\qquad$ Bromination of ethene (eles-1674)
$\qquad\qquad\qquad\qquad$ Indirect hydration of ethene (eles-1677)

elog-1904

Modelling substitution and addition reactions

Aim

To use molecular models to compare a substitution reaction of chlorine and ethane, an alkane; and an addition reaction of chlorine and ethene, an alkene

8.2.3 Oxidation reactions of alcohols

Alcohols are widely used in pharmaceuticals, and as fuels or solvents. They can participate in a variety of reactions, including substitution, oxidation and condensation. The oxidation of alcohols is an important reaction in chemistry. Alcohols are flammable, and can be burned to produce carbon dioxide and water. Primary and secondary alcohols can also be oxidised with oxidising agents like permanganate or dichromate ions in acidic mediums.

Primary alcohols

Primary alcohols are easily oxidised in the laboratory, first to aldehydes and then to carboxylic acids, using either acidified permanganate, acidified dichromate or another oxidising agent. If an excess of alcohol is used, the aldehyde can be distilled off as soon as it forms. Having an excess of an oxidising agent, and allowing the aldehyde to remain, results in the formation of the carboxylic acid.

FIGURE 8.5 Alcohols, like ethanol, are used in sanitisers because they destroy the outer coating of viruses. They are also effective in disrupting the cell membranes of bacteria.

primary alcohol an alcohol in which the carbon atom that carries the –OH group is attached to only one other carbon atom

FIGURE 8.6 The stages of oxidation of carbon compounds. Notice the increase in oxidation number of carbon from left to right. [O] is the symbol for an oxidising agent.

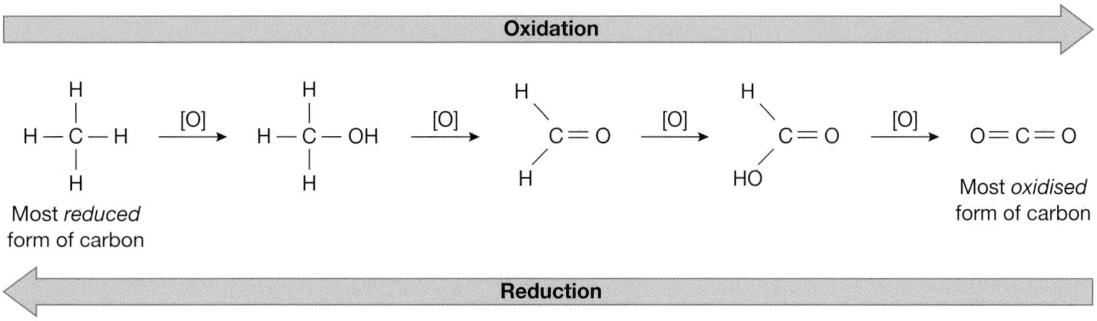

Oxidation →

$$H-\overset{\overset{\displaystyle H}{|}}{\underset{\underset{\displaystyle H}{|}}{C}}-H \xrightarrow{[O]} H-\overset{\overset{\displaystyle H}{|}}{\underset{\underset{\displaystyle H}{|}}{C}}-OH \xrightarrow{[O]} \overset{H}{\underset{H}{>}}C=O \xrightarrow{[O]} \overset{H}{\underset{HO}{>}}C=O \xrightarrow{[O]} O=C=O$$

Most *reduced*
form of carbon

Most *oxidised*
form of carbon

← Reduction

In the oxidation reactions shown in figure 8.6, there is an increase in the oxygen-to-hydrogen ratio; that is, there is more oxygen and less hydrogen. The product of these reactions is a carboxylic acid. For example, propan-1-ol can be converted into propanoic acid as shown in the following reaction:

$$CH_3CH_2CH_2OH \xrightarrow{H^+/Cr_2O_7{}^{2-}} CH_3CH_2COOH$$

The oxidation of ethanol in wine takes place when it is left exposed to air for some time. Such wine has a sour taste of ethanoic acid, commonly known as acetic acid. The oxidation reaction that takes place is:

$$CH_3CH_2OH(aq) + O_2(g) \longrightarrow CH_3COOH(aq) + H_2O(l)$$

The reaction is catalysed by the presence of the microorganism *Acetobacter aceti* in the exposed wine.

Secondary alcohols

Secondary alcohols are oxidised to ketones, which do not undergo further oxidation.

For example, propan-2-ol is oxidised to propanone.

$$CH_3CHOHCH_3 \xrightarrow{H^+/Cr_2O_7{}^{2-}} CH_3COCH_3$$

Tertiary alcohols

Tertiary alcohols cannot be oxidised, regardless of the oxidising agent. The absence of a relatively easily removed hydrogen atom on the central carbon atom is the reason for this, as it prevents a double bond from forming with the oxygen.

FIGURE 8.7 Oxidation of alcohols

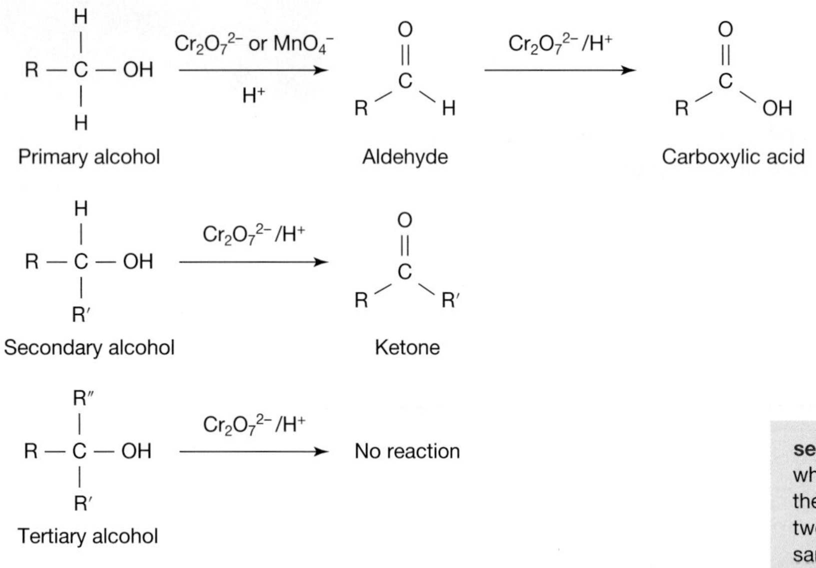

secondary alcohol an alcohol in which the carbon atom that carries the –OH group is joined directly to two alkyl groups, which may be the same or different

Oxidation of primary alcohols

Oxidation of a primary alcohol results in the production of a carboxylic acid.

8.2.4 Reaction pathways

Reaction pathways are important in synthetic chemistry. A reaction pathway shows how to synthesise a particular chemical product from raw materials using a series of steps, and the necessary catalysts and conditions.

Choosing a particular reaction pathway to manufacture chemical products depends on several factors:
- Cost and availability of raw materials
- Cost, type, availability and safety of catalysts
- Energy cost and availability
- Percentage yield
- Rate of reaction
- Yield in equilibrium reactions
- Atom economy
- By-products/wastes — possibility of use in other industries, disposal of waste
- Safety of chemicals and processes for workers and the environment
- Environmental impact of sourcing and producing chemicals
- Technology required for bulk processing.

Synthesis of primary amines

Haloalkanes are far more reactive than alkanes, so the first step in this pathway is to convert the alkane to a haloalkane using a substitution reaction with a halogen in the presence of UV light (or the alkene to a haloalkane), as discussed earlier. The haloalkane is heated with a solution of concentrated ammonia in ethanol. A mixture of **amines** is produced, which is separated by fractional distillation. Possible pathways include:

Alkane → haloalkane → amine

Alkene → haloalkane → amine

Alkene → alkane → haloalkane → amine

FIGURE 8.8 Producing an amine from an alkene

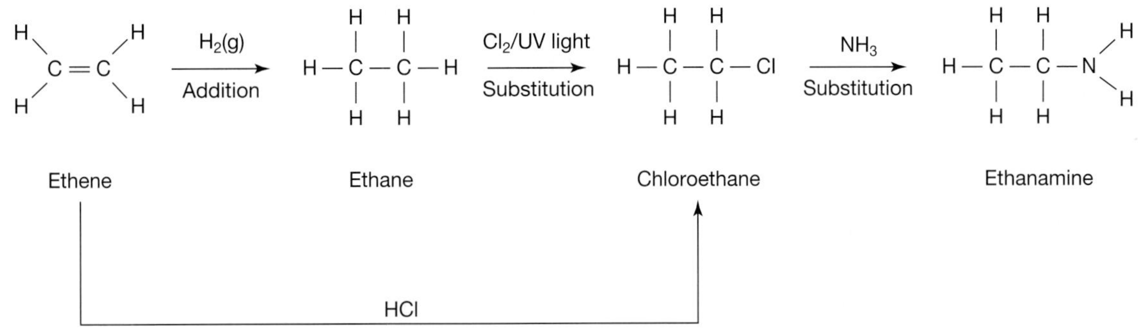

Synthesis of carboxylic acids

To arrive at the carboxylic acid and primary alcohol reactants from alkanes, reactive haloalkanes are required as precursors. The further down the group a halogen is, the faster the rate of reaction when OH^- replaces it. For example, part of a primary haloalkane, Br, is 500 times more reactive than Cl, which means it is a lot easier to substitute OH^- for Br^- than it is Cl^-. Nonetheless, Cl can still readily leave the molecule and be replaced by OH^-.

As seen earlier, substituted haloalkanes are prepared using diatomic halogens and UV light to catalyse the reaction; that is:

Alkane → haloalkane → alcohol → carboxylic acid (→ ester)

amines organic compounds containing the amino functional group, $-NH_2$

TIP: Remember the sequence of the synthesis of carboxylic acids and esters from alkanes by using the mnemonic **AH ACE!**

FIGURE 8.9 Synthesis of ethanoic acid from ethane or ethene

Ethane → Chloroethane (Cl₂/UV light, Substitution) → Ethanol (OH⁻, Substitution) → Ethanoic acid (H⁺/MnO₄⁻ or Cr₂O₇²⁻, H⁺, Oxidation)

Ethene → (H₂ Addition) → Ethane; Ethene → (HCl Addition) → Chloroethane

Reaction pathway

A reaction pathway shows the raw materials and sequence of steps required to synthesise a chemical product.

SAMPLE PROBLEM 2 Writing the reaction steps and reagents to make a primary amine from an alkane

tlvd-10341

Write the reaction steps and reagents used to produce methanamine from methane.

THINK	WRITE
1. Methane must be converted to chloromethane in a substitution reaction before methanamine can be produced.	$H-CH_3 + Cl_2 \xrightarrow{\text{UV light}} H-CH_2Cl + HCl$
2. A mixture of different chloromethane compounds is formed, which needs to be separated by fractional distillation to obtain chloromethane.	Separate chloromethane from other chloromethanes (e.g. dichloromethane, trichloromethane and tetrachloromethane) by fractional distillation.
3. React the chloromethane with excess NH₃ in another substitution reaction to produce methanamine.	$H-CH_2Cl + 2NH_3 \rightarrow H-CH_2-NH_2 + NH_4Cl$

PRACTICE PROBLEM 2

Write the reaction pathway used to produce propanoic acid from propene.

---

on Resources

Interactivity Identifying compounds in reaction pathways (int-1236)

8.2 Activities

8.2 Quick quiz on	8.2 Exercise	8.2 Exam questions

8.2 Exercise

1. Use structural formulas to show how you would make 1,1-dichloroethane from ethane.
2. Write the structural equations for the reactions of chlorine, hydrogen and hydrogen chloride with propene. Name the compounds formed.
3. Red bromine, Br_2, liquid is decolourised in an addition reaction with an alkene. With reference to ethane and ethene, explain how this reaction could demonstrate which substance is unsaturated.
4. a. Name the following alcohol.

$$CH_3$$
$$|$$
$$H_3C - CH - CH_2 - CH_2 - OH$$

   b. Give the name and structure of the product formed by the complete oxidation of this alcohol by acidified potassium dichromate.
5. Why can't secondary alcohols be used to make carboxylic acids?
6. **MC** Which of the following is a primary haloalkane?
   A. $C_6H_5CHClCH_3$
   B. $CH_3CHClCH_2CH_3$
   C. $(CH_3)_3CCH_2Cl$
   D. $(CH_3)_3CCl$
7. List the conditions and reagents required to convert ethene into:
   a. ethane
   b. ethanol
   c. chloroethane
   d. ethanamine.
8. Complete the following equations. (*Note*: The equation in part **b** does not require balancing.)
   a. $CH_3CH=CHCH_3 + HCl \rightarrow$
   b.

$$\begin{array}{cc} H & H \\ | & | \\ H - C - C - OH \\ | & | \\ H & H \end{array} \xrightarrow{K_2Cr_2O_7/H^+}$$

   c.

$$\begin{array}{c} H \\ | \\ H - C - Br \ + \ Br - Br \\ | \\ H \end{array} \xrightarrow{UV\ light}$$

   d.

$$\begin{array}{c} H \qquad\qquad H \\ \diagdown \qquad\qquad | \\ C = C - C - H \\ \diagup \qquad | \quad | \\ H \qquad H \quad H \\ \qquad | \\ \qquad H - C - H \\ \qquad | \\ \qquad H \end{array} \ + \ H - H \longrightarrow$$

9. Use semi-structural formulas to show the reaction pathway used to convert methane to methanamine.

8.2 Exam questions

▶ Question 1 (1 mark)

Source: VCE 2015 Chemistry Exam, Section A, Q.13; © VCAA

MC What is the name of the product formed when chlorine, Cl_2, reacts with but-1-ene?
A. 1,2-dichlorobutane
B. 1,4-dichlorobutane
C. 2,2-dichlorobutane
D. 2,3-dichlorobutane

▶ Question 2 (1 mark)

MC Butanoic acid can be produced by the reaction of acidified potassium dichromate with
A. $CH_3CHOHCH_2CH_3$
B. $CH_3CH_2CH_2OH$
C. $CH_3CH_2CH_2CH_2OH$
D. $CH_3CH_2CH_2CH_2Cl$

▶ Question 3 (4 marks)

Source: VCE 2016 Chemistry Exam, Section B, Q.7. a.i–iv; © VCAA

Butanoic acid is the simplest carboxylic acid that is also classified as a fatty acid. Butanoic acid may be synthesised as outlined in the following reaction flow chart.

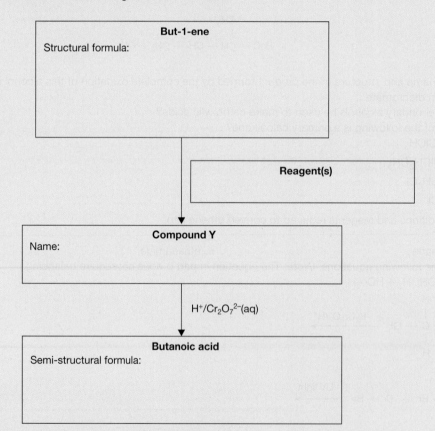

a. Draw the structural formula of but-1-ene in the box provided. **(1 mark)**
b. State the reagent(s) needed to convert but-1-ene to Compound Y in the box provided. **(1 mark)**
c. Write the systematic name of Compound Y in the box provided. **(1 mark)**
d. Write the semi-structural formula of butanoic acid in the box provided. **(1 mark)**

Source: Adapted from VCE 2017 Chemistry Exam, Section B, Q.1.b,c; © VCAA

a. i. Complete the following reaction by writing the formula for the reactant in the box provided. **(1 mark)**

$$C_2H_4 + \boxed{} \longrightarrow C_2H_5Cl$$

ii. Classify this type of reaction. **(1 mark)**

b. C_2H_5Cl can be converted into ethanamine, $CH_3CH_2NH_2$.

i. Complete the following reaction by writing the formula for the reactant in the box provided. **(1 mark)**

$$C_2H_5Cl + \boxed{} \longrightarrow CH_3CH_2NH_2$$

ii. Classify this type of reaction. **(1 mark)**

Question 5 (6 marks)

Use structural formulas to show the conditions and reagents required in the reaction pathway used to convert propane to propanoic acid.

More exam questions are available in your learnON title.

8.3 Condensation and hydrolytic reactions of esters

KEY KNOWLEDGE

- Organic reactions and pathways, including equations, reactants, products, reaction conditions and catalysts (specific enzymes not required):
  - the esterification between an alcohol and a carboxylic acid
  - hydrolysis of esters
  - transesterification of plant triglycerides using alcohols to produce biodiesel

Source: Adapted from VCE Chemistry Study Design (2024–2027) extracts © VCAA; reproduced by permission.

Condensation reactions occur when two smaller molecules combine to form a larger molecule, with the loss of a small molecule like H_2O. Condensation reactions are used in the production of esters and biodiesel, as well as the essential biomolecules: proteins, carbohydrates and lipids.

On the other hand, a **hydrolytic reaction**, also known as hydrolysis, involves the breakdown of a larger molecule by the addition of water, with an –OH going to one part and –H to the other. Esters can be hydrolysed to their component acids and alcohols, and nutrients are broken down in the process of digestion into smaller molecules that can be absorbed into the bloodstream.

condensation reaction a reaction in which molecules react and link together by covalent bonding with the elimination of a small molecule, such as water or hydrogen chloride, from the bond that is formed

hydrolytic reaction the chemical breakdown of a compound due to a reaction with water; also known as hydrolysis

Condensation reaction:
- A larger molecule is synthesised from smaller molecules.
- Water is released.
- New covalent bonds are formed.

Hydrolytic reaction:
- A larger molecule is broken down into smaller molecules.
- Water is added.
- Covalent bonds are broken.

8.3.1 Synthesis of esters by condensation reactions

Esters

An **esterification** reaction is a type of condensation reaction. When a carboxylic acid reacts with an alcohol, an ester is produced. A typical esterification reaction is the formation of ethyl ethanoate, $CH_3COOCH_2CH_3$, by heating ethanol, C_2H_5OH, and ethanoic acid, CH_3COOH, in the presence of an acid catalyst, such as concentrated sulfuric acid. During the condensation reaction, a water molecule is produced. Smaller esters are only partially insoluble in water and can be purified by mixing them with cold water. The esters form a sweet-smelling insoluble layer on top of the water, while sulfuric acid and unreacted ethanol and ethanoic acid dissolve.

FIGURE 8.10 The acid present in vinegar is ethanoic acid, CH_3COOH, commonly known as acetic acid.

FIGURE 8.11 A typical esterification reaction

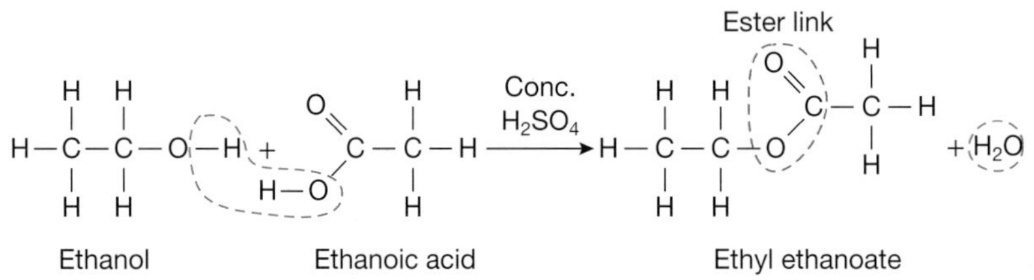

The process is called a condensation reaction because a small molecule (water) is split off as the two molecules join together. In general:

$$R'OH \ + \ RCOOH \ \rightarrow \ RCOOR' \ + \ H_2O$$

Alcohol   Carboxylic acid   Ester   Water

esterification the process of ester formation

amides organic compounds containing the amide functional group, –CONH–

TIP: When writing condensation reactions it is important to specify that the acid acting as a catalyst is concentrated and not in aqueous form.

EXTENSION: Reaction with amines

Small amines react in a similar manner to ammonia, which is a weak base. Therefore, carboxylic acids can donate a proton to a primary amine. An example is the acid–base reaction of ethanoic acid with methanamine to produce the salt methylammonium ethanoate. This salt is heated to over 100 °C, water is removed and an **amide** is formed.

FIGURE 8.12 Reaction of a carboxylic acid with an amine

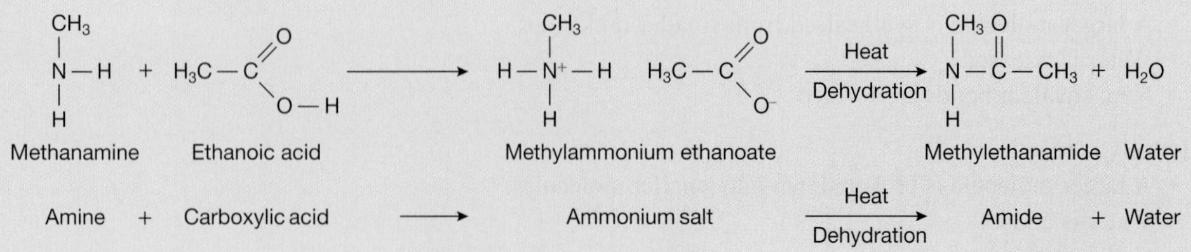

EXPERIMENT 8.2   online only

Esterification

Aim

To prepare a small amount of the ester ethyl ethanoate from ethanol and ethanoic acid

on Resources

▶ **Video eLessons**   Esterification (eles-1668)

Ester formation (eles-3249)

8.3.2 Hydrolytic reactions of esters

Esters undergo hydrolysis in aqueous acids or bases to reverse the condensation reaction. The products are carboxylic acids and alcohols. Alkaline hydrolysis is possible using hydroxide ions, resulting in a salt and alcohol being formed; this reaction is one-way, rather than reversible like the reaction with dilute acid. For example, ethyl ethanoate reacts with dilute sodium hydroxide to form sodium ethanoate and ethanol. The reaction is very slow if only water is used.

FIGURE 8.13 Hydrolysis of an ester

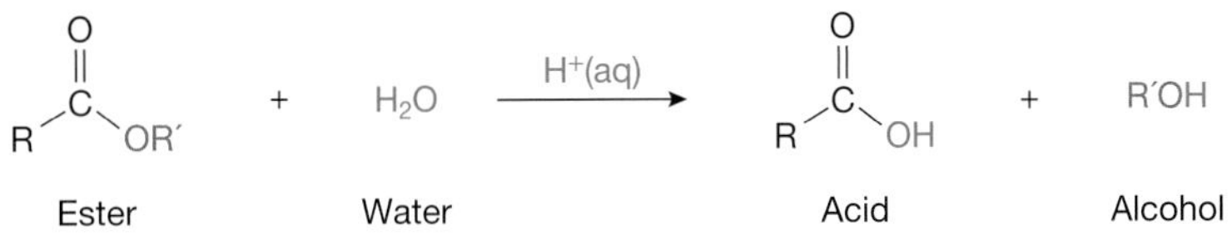

$$\underset{\text{Ester}}{R-\overset{\overset{\displaystyle O}{\|}}{C}-OR'} \quad + \quad \underset{\text{Water}}{H_2O} \quad \xrightarrow{H^+(aq)} \quad \underset{\text{Acid}}{R-\overset{\overset{\displaystyle O}{\|}}{C}-OH} \quad + \quad \underset{\text{Alcohol}}{R'OH}$$

Ester hydrolysis in alkaline solutions is also known as saponification, from the Latin word *sapo* meaning 'soap'. The ester linkages in fats are hydrolysed in basic solutions to make soap.

FIGURE 8.14 Soaps are manufactured through ester hydrolysis.

Hydrolysis of an ester

Hydrolysis of an ester using a dilute acid catalyst results in the production of an alcohol and a carboxylic acid.

on Resources

 Weblink   Organic chemistry reactions

FIGURE 8.15 Organic reaction pathways summary

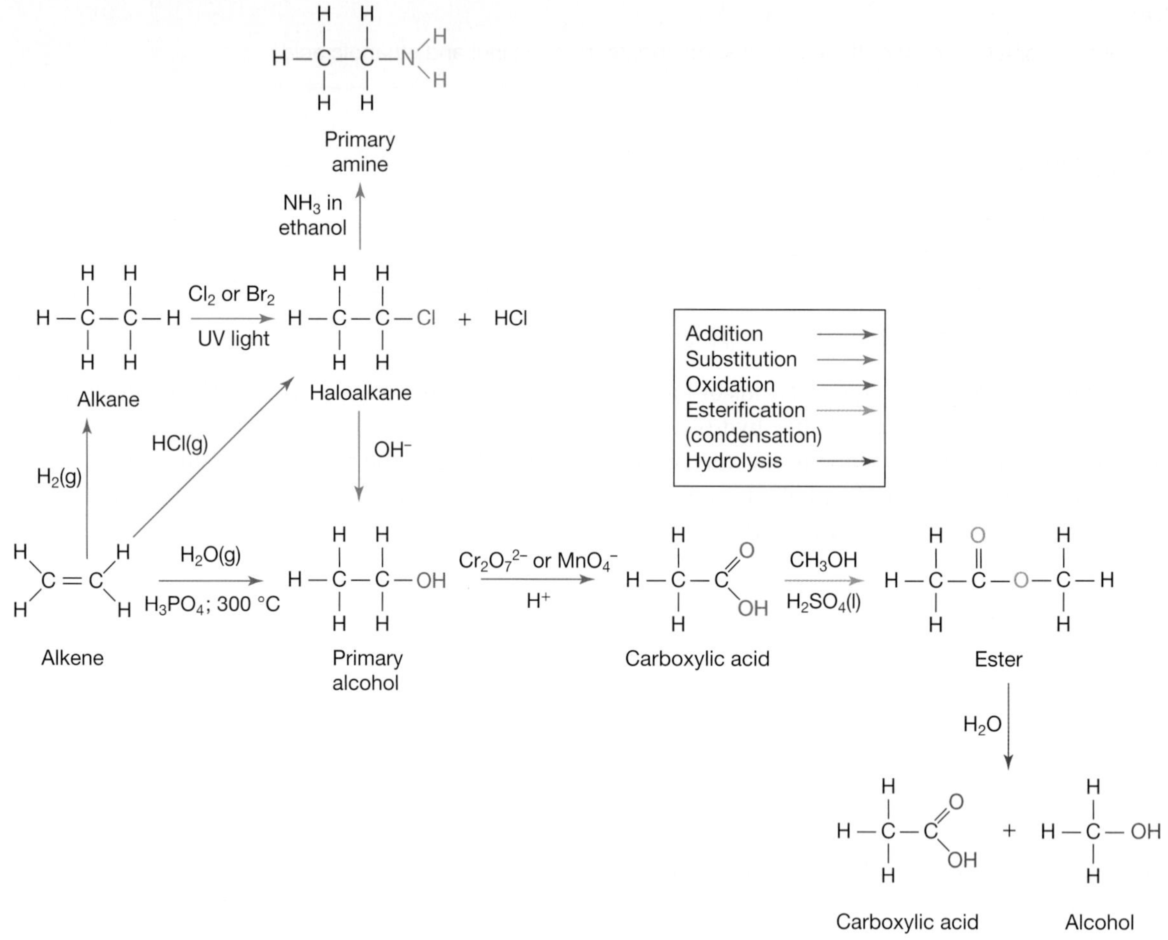

Addition	→
Substitution	→
Oxidation	→
Esterification (condensation)	→
Hydrolysis	→

Biodiesel

Biodiesel is a diesel alternative that can be made from plant oils and animal fats. Typical sources include canola, palm oil and animal tallow. It can also be made from used cooking oil, such as that used in restaurant fryers. The CSIRO has estimated that Australia could reduce its petrodiesel demand by 4 to 8 per cent if all current sources of plant oil, tallow and waste cooking oil were used.

FIGURE 8.16 A bus that runs on biodiesel made from recycled waste cooking oil that is collected from local restaurants

biodiesel a fuel produced from vegetable oil or animal fats and combined with an alcohol, usually methanol

Triglycerides

Triglycerides are naturally occurring esters formed between three long-chain carboxylic acids, known as fatty acids, and **glycerol**, an alcohol. **Fatty acids** usually contain an even number of 12–20 carbon atoms. Examples of fatty acids include stearic acid and oleic acid, which have eighteen carbon atoms in their chains.

Common fatty acids are summarised in table 8.1. The triglyceride is formed in a condensation reaction in which carboxyl groups in the fatty acids react with the three hydroxyl groups in the glycerol to form three ester links, –COO–, and water is a product of the reaction. Triglycerides are hydrophobic and less dense than water because of the large non-polar, hydrophobic sections of the molecules.

TABLE 8.1 Formulas of some fatty acids

Name	Formula
Palmitic	$C_{15}H_{31}COOH$
Palmitoleic	$C_{15}H_{29}COOH$
Stearic	$C_{17}H_{35}COOH$
Oleic	$C_{17}H_{33}COOH$
Linoleic	$C_{17}H_{31}COOH$
Linolenic	$C_{17}H_{29}COOH$

Glycerol is an alcohol; it is a non-toxic, colourless, clear, odourless and viscous liquid that is sweet-tasting and has the semi-structural formula $CH_2OHCH(OH)CH_2OH$. Figure 8.17 shows the structures of glycerol and a typical fatty acid.

FIGURE 8.17 Structures of **a.** glycerol and **b.** a typical fatty acid

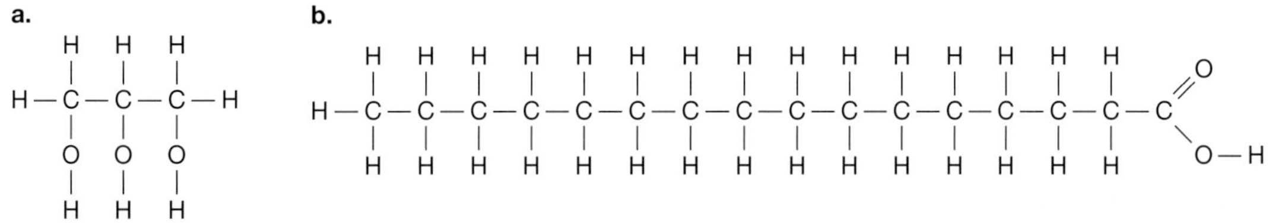

Production of biodiesel

Biodiesel is produced by reacting a triglyceride with an alcohol in a condensation reaction. Although several small alcohols can be used, the most common is methanol. If methanol is used the product is a **methyl ester**. Heat and either concentrated sodium or potassium hydroxide, which acts as a catalyst, are used in this process.

Biodiesel can be made on a small scale, using homemade equipment or with specially purchased kits, or on a much larger scale for commercial distribution. The chemical reaction involved converts one type of ester into another and is called **transesterification**. A typical transesterification is shown in figure 8.18. The other product formed, glycerol, can be sold as a by-product for use in cosmetics and foods, and as a precursor for certain explosives.

triglycerides fats and oils formed by a condensation reaction between glycerol and three fatty acids

glycerol an alcohol; it is a non-toxic, colourless, clear, odourless and viscous liquid that is sweet-tasting and has the semi-structural formula $CH_2OHCH(OH)CH_2OH$

fatty acids long-chain carboxylic acids, usually containing an even number of 12–20 carbon atoms

methyl ester the product of a condensation reaction between a triglyceride and methanol

transesterification the conversion of one ester (triglyceride) into another ester (biodiesel)

The main component of fats and oils is triglycerides, which are formed from three fatty acid groups and glycerol.

FIGURE 8.18 A typical transesterification reaction

Triglyceride

3 × Methanol

OH⁻ catalyst

Biodiesel

Glycerol

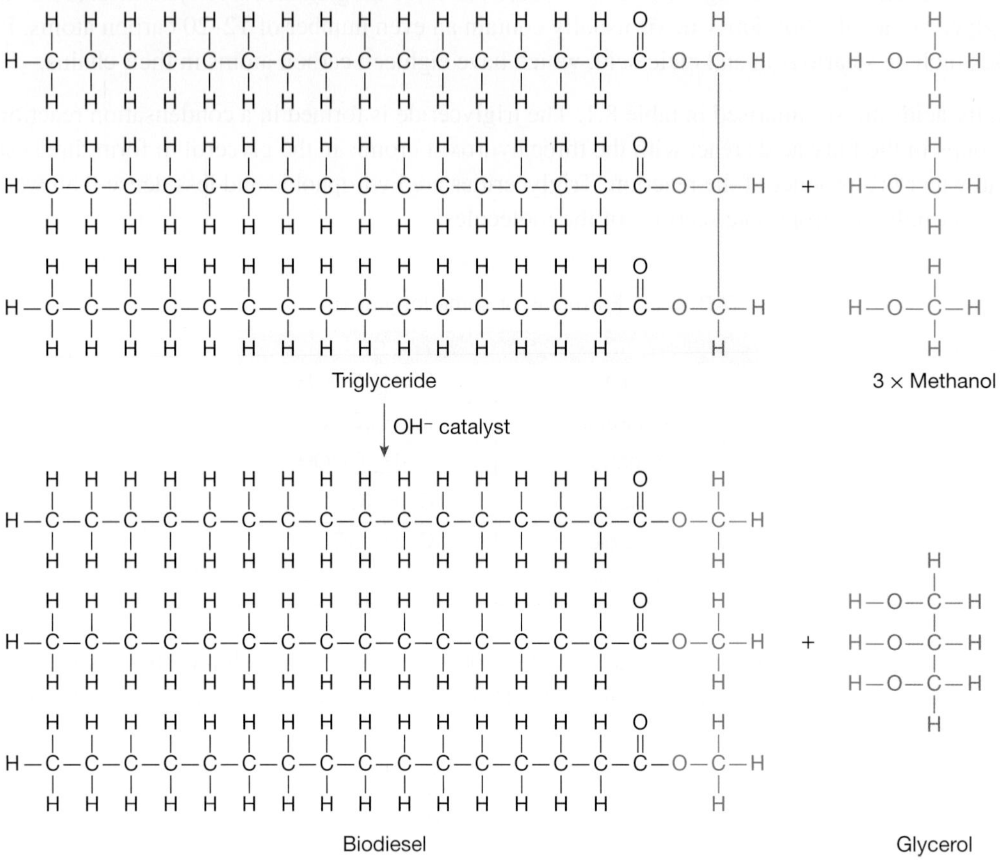

Obtaining methanol

While biodiesel is regarded as a more environmentally friendly choice than diesel, the most economical method of biodiesel production requires the use of non-renewable fossil fuels to make the methanol required for the transesterification process. In this process, steam reforming is used to produce 'synthesis gas', which then undergoes further reactions to make methanol. When natural gas (methane) is used as the feedstock, the overall equation for this process is:

$$CH_4(g) + H_2O(g) \rightarrow CH_3OH(g) + H_2(g)$$

Several methods of producing methanol economically from renewable resources are being investigated. The most exciting of these involves using the glycerol produced in the transesterification reaction as the initial feedstock for producing synthesis gas to feed into the methanol production process. Another method involves using a catalyst to facilitate the direct conversion of glycerol to methanol. A few new methods are also being developed to convert the cellulose in waste or low-quality plant material into biodiesel.

Biodiesel

Biodiesel is produced by a condensation reaction between a triglyceride and an alcohol, often methanol.

Students, these questions are even better in jacPLUS

Receive immediate feedback and access sample responses

Access additional questions

Track your results and progress

Find all this and MORE in jacPLUS ▶

| 8.3 Quick quiz **on** | 8.3 Exercise | 8.3 Exam questions |

8.3 Exercise

1. What are the similarities and differences between condensation and hydrolytic reactions of esters?
2. List the conditions and reagents required to convert ethene into propyl ethanoate.
3. Name the alcohol and carboxylic acid used to produce the following esters.
   a. Propyl ethanoate
   b. Ethyl butanoate
   c. Ethyl propanoate
4. Complete the equations, including the products and conditions, for the hydrolysis of the following esters in the laboratory.
   a.
   $$CH_3-O-\overset{\overset{\displaystyle O}{\|}}{C}-CH_3$$
   b.
   $$CH_3-CH_2-O-\overset{\overset{\displaystyle O}{\|}}{CH}$$
5. Name the following esters.
   a. This ester smells like pineapple.

   b. This ester smells like apple.

6. Describe where the raw materials required to produce biodiesel are obtained.
7. Explain the difference between the terms *esterification*, *condensation* and *transesterification*.
8. Write the semi-structural formula for biodiesel if it is formed from palmitic acid and methanol.
9. State one advantage and one disadvantage of the use of biodiesel as a fuel.

8.3 Exam questions

▶ **Question 1 (6 marks)**
Source: VCE 2017 Chemistry Exam, Section B, Q.1.b–d; © VCAA

a. i. Complete the reaction by writing the formula for the reactant in the box provided below. **(1 mark)**

$$C_2H_4(g) \; + \; \boxed{} \; \xrightarrow{\text{catalyst}} \; C_2H_5OH(g)$$

   ii. Classify this type of reaction. **(1 mark)**
b. C_2H_5OH can be converted into ethanoic acid, CH_3COOH, in the presence of Reagent X.
   Write the formula for Reagent X in the box provided below. **(1 mark)**

$$\boxed{\text{Reagent X}}$$

$$C_2H_5OH \; \xrightarrow{} \; CH_3COOH$$

c. CH_3COOH can be used in the production of esters.
   i. Write a balanced chemical equation for the reaction of CH_3COOH with propan-1-ol using semi-structural formulas for all organic compounds. **(2 marks)**
   ii. Write the IUPAC name for the ester product of the equation written in part **c. i.** **(1 mark)**

▶ Question 2 (5 marks)

Source: VCE 2015 Chemistry Exam, Section B, Q.5.a; © VCAA

A reaction pathway is designed for the synthesis of the compound that has the structural formula shown below.

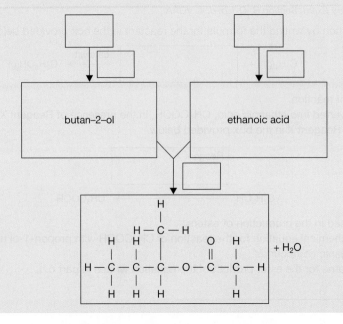

The table below gives a list of available organic reactants and reagents.

Letter	Available organic reactants and reagents
A	acidified $KMnO_4$
B	concentrated H_2SO_4
C	H_2O and H_3PO_4
D	 H H H H | | | | H—C—C=C—C—H | | H H
E	 H H \ / C=C / \ H H
F	 H H H H | | | | H—C—C—C—C—O—H | | | | H H H H
G	 H H | | H—C—C—O—H | | H H

Complete the reaction pathway design flow chart below. Write the corresponding letter for the structural formula of all organic reactants in each of the boxes provided. The corresponding letter for the formula of other necessary reagents should be shown in the boxes next to the arrows.

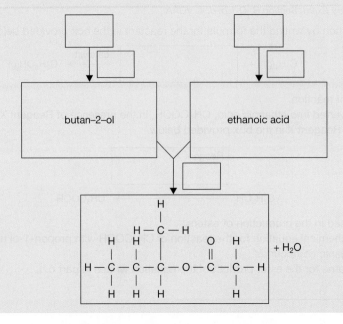

▶ **Question 3 (1 mark)**

Source: VCE 2016 Chemistry Exam, Section B, Q.3.c; © VCAA

The diagram below represents a certain biomolecule.

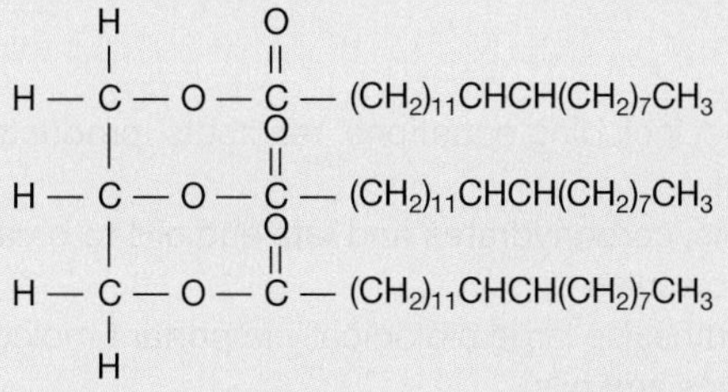

$$
\begin{array}{c}
\text{H} \quad\quad\quad \text{O} \\
| \quad\quad\quad\quad \| \\
\text{H} - \text{C} - \text{O} - \text{C} - (\text{CH}_2)_{11}\text{CHCH}(\text{CH}_2)_7\text{CH}_3 \\
| \quad\quad\quad\quad \text{O} \\
\quad\quad\quad\quad\quad \| \\
\text{H} - \text{C} - \text{O} - \text{C} - (\text{CH}_2)_{11}\text{CHCH}(\text{CH}_2)_7\text{CH}_3 \\
| \quad\quad\quad\quad \text{O} \\
\quad\quad\quad\quad\quad \| \\
\text{H} - \text{C} - \text{O} - \text{C} - (\text{CH}_2)_{11}\text{CHCH}(\text{CH}_2)_7\text{CH}_3 \\
| \\
\text{H}
\end{array}
$$

This biomolecule can be hydrolysed to form glycerol and erucic acid, a fatty acid.

Erucic acid can be extracted from plants. It can react with methanol to make methyl erucate, which can be used as the biofuel known as biodiesel.

Write the semi-structural formula of methyl erucate.

▶ **Question 4 (1 mark)**

Source: VCE 2009 Chemistry Exam 1, Section A, Q.13; © VCAA

MC Cinnamic acid is an organic substance that partly contributes to the flavour of oil of cinnamon. A structure of cinnamic acid is given below.

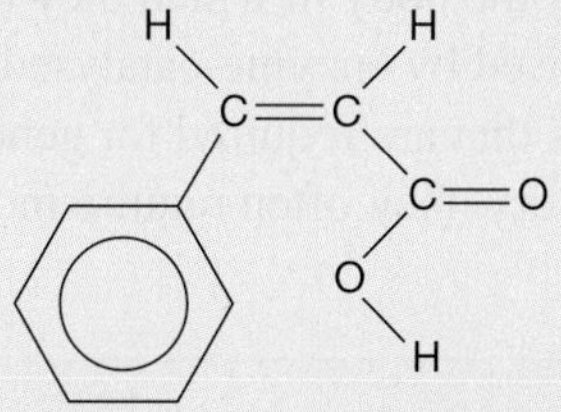

Which of the following reagents would you expect to react with cinnamic acid under the conditions given below?

	CH_2CH_2 and catalyst	$Br_2(aq)$ at room temperature	CH_3OH and H_2SO_4 catalyst
A.	Yes	Yes	Yes
B.	Yes	No	Yes
C.	No	Yes	Yes
D.	No	Yes	No

▶ **Question 5 (5 marks)**

Source: Adapted from VCE 2014 Chemistry Exam, Section B, Q.3.c.i; © VCAA

A triglyceride has the following formula.

$$
\begin{array}{l}
\text{CH}_2\text{OOCC}_{13}\text{H}_{31} \\
| \\
\text{CHOOCC}_{13}\text{H}_{27} \\
| \\
\text{CH}_2\text{OOCC}_{13}\text{H}_{29}
\end{array}
$$

a. Write the equation for the reaction of this triglyceride with methanol to form biodiesel. **(2 marks)**
b. Circle and name the functional group present in the biodiesel molecules. **(1 mark)**
c. Calculate the molar mass of the smallest biodiesel molecule. **(2 marks)**

More exam questions are available in your learnON title.

8.4 Hydrolytic and condensation reactions of biomolecules

KEY KNOWLEDGE

- Organic reactions and pathways, including equations, reactants, products, reaction conditions and catalysts (specific enzymes not required):
  - hydrolytic reactions of proteins, carbohydrates and fats and oils to break down large biomolecules in food to produce smaller molecules
  - condensation reactions to synthesise large biologically important molecules for storage as proteins, starch, glycogen and lipids (fats and oils)

Source: Adapted from VCE Chemistry Study Design (2024–2027) extracts © VCAA; reproduced by permission.

Our bodies are like a giant chemical laboratory, with thousands of chemical reactions occurring at the same time. These processes are referred to as the **metabolism** of the body and affect how nutrients are digested, and how proteins, carbohydrates, fats and oils are broken down into their component parts. When we eat food, the body breaks down the large molecules into smaller molecules, which either provide energy or are built up into different substances that are required by the body. The reactions in which complex carbohydrates, lipids and proteins are broken down into smaller molecules are hydrolytic reactions. In hydrolysis, water is a reactant, and each reaction occurs in the presence of a specific **enzyme**, which is a biological catalyst. Enzymes are proteins that catalyse specific reactions but can only function effectively at body temperature and an optimal pH, depending on the part of the body in which they function. The products of these reactions are then synthesised by enzyme-catalysed condensation reactions into complex biological molecules that are required for general well-being. If these reactions are repeated in the laboratory, they often require more extreme conditions.

metabolism the chemical processes that occur within a living organism to maintain life

enzyme a protein that acts as a biological catalyst

Enzymes

Specific enzymes are required for hydrolytic and condensation reactions in the body.

FIGURE 8.19 Metabolism of food in the body

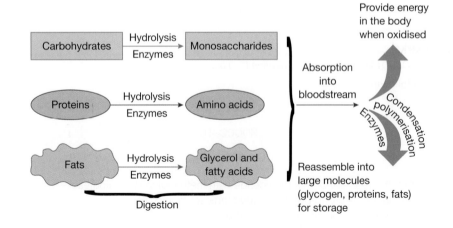

8.4.1 Hydrolytic reactions of proteins

Proteins are made up of the elements carbon, hydrogen, oxygen, nitrogen and sulfur. As well as acting as enzymes, **proteins** have many other roles in the body. Virtually all body structures and activities depend on proteins, and eating protein is not just about building muscles — it provides the raw materials for many structural and functional components in the body. Some proteins are antibodies that prevent infection, some assist with the formation of new molecules by interpreting the genetic information stored in DNA, and others act as hormones and transmit signals to coordinate biological processes between different cells, tissues and organs. Proteins provide structural components for cells; they form muscle fibres, which assist in movement; and they carry atoms and small molecules around the body. Sources of protein include meat, fish, poultry, eggs, dairy products, soy products and beans.

FIGURE 8.20 Protein-rich foods

Proteins are **polypeptides**, which are condensation polymers made up of **amino acid** monomers. Hundreds of amino acids are known, but only 20 have been found in proteins in the human body. Just as esters are broken down by hydrolysis, proteins can also be broken down by the addition of water to form amino acids. At the molecular level, when hydrolysis occurs, the functional groups on the amino acid residues (sections of a protein) form covalent bonds with the atoms from the water, resulting in breakage of the peptide link to produce amino and carboxyl groups. This is shown in figure 8.21. If this hydrolysis was carried out in a laboratory, it would be necessary to heat the protein with 6 M hydrochloric acid at over 100 °C for about 24 hours.

proteins large molecules composed of one or more long chains of amino acids

polypeptide many amino acid residues bonded together

amino acids molecules that contain an amino and a carboxyl group

Hydrolysis of proteins

In this hydrolysis reaction, water is added across the peptide link (–CONH–) with the help of an enzyme. This breaks the covalent bonds in the structure of the protein to form amino acids.

FIGURE 8.21 Enzyme-catalysed hydrolysis of a section of a protein

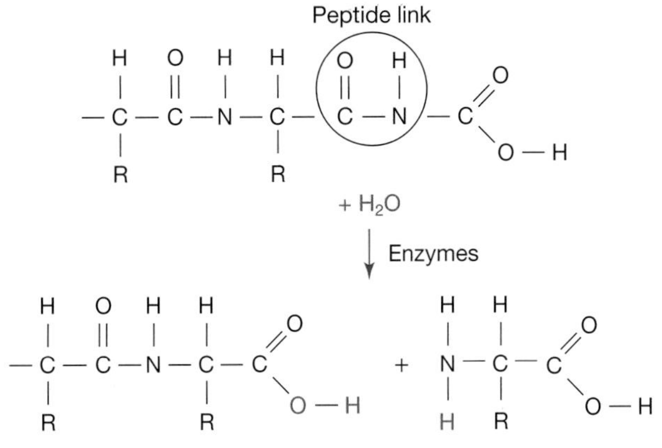

Peptide link

$$
\begin{array}{c}
\underset{R}{\overset{H}{\underset{|}{C}}} - \underset{}{\overset{O}{\overset{||}{C}}} - \underset{}{\overset{H}{\underset{|}{N}}} - \underset{R}{\overset{H}{\underset{|}{C}}} - \overbrace{\overset{O}{\overset{||}{C}} - \overset{H}{\underset{|}{N}}} - C\overset{O}{\underset{O-H}{}}
\end{array}
$$

+ H₂O

↓ Enzymes

Peptide — short chain of amino acids

Amino acid

Amino acids are compounds that contain amino ($-NH_2$) and carboxyl ($-COOH$) functional groups.

Most amino acids have four groups bonded to a central carbon atom. These are:

- a carboxyl group ($-COOH$)
- an amino group ($-NH_2$)
- an R group (the amino acid side chain)
- a hydrogen atom.

The amino acids in proteins are called alpha amino acids because the amine and carboxyl groups are bonded to the same carbon atom.

FIGURE 8.22 The general structure of an amino acid

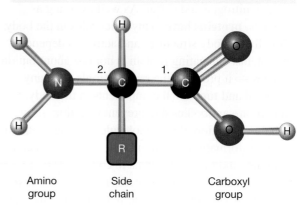

Amino group   Side chain   Carboxyl group

TIP: Do not confuse the amino functional group ($-NH_2$) with the amide group ($-CONH-$).

TABLE 8.2 The 20 amino acids that make up the proteins in the human body

Name	Symbol	Structure
Alanine	Ala	CH_3 $\|$ $H_2N-CH-COOH$
Arginine	Arg	$CH_2-CH_2-CH_2-NH-\overset{\overset{\displaystyle NH}{\|\|}}{C}-NH_2$ $\|$ $H_2N-CH-COOH$
Asparagine	Asn	$CH_2-\overset{\overset{\displaystyle O}{\|\|}}{C}-NH_2$ $\|$ $H_2N-CH-COOH$
Aspartic acid	Asp	CH_2-COOH $\|$ $H_2N-CH-COOH$
Cysteine	Cys	CH_2-SH $\|$ $H_2N-CH-COOH$
Glutamic acid	Glu	CH_2-CH_2-COOH $\|$ $H_2N-CH-COOH$
Glutamine	Gln	$CH_2-CH_2-\overset{\overset{\displaystyle O}{\|\|}}{C}-NH_2$ $\|$ $H_2N-CH-COOH$
Glycine	Gly	H_2N-CH_2-COOH

Name	Symbol	Structure
Histidine	His	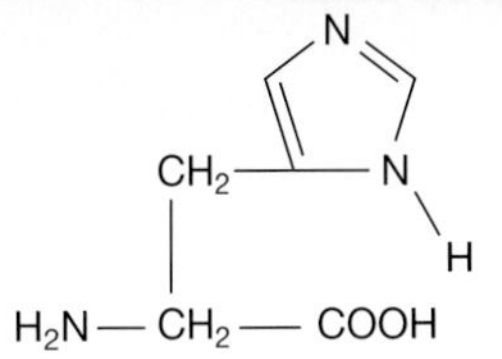
Isoleucine	Ile	$CH_3 - CH - CH_2 - CH_3$ $\quad\quad\; \vert$ $H_2N - CH - COOH$
Leucine	Leu	$CH_3 - CH - CH_3$ $\quad\quad\; \vert$ $\quad\quad CH_2$ $\quad\quad\; \vert$ $H_2N - CH - COOH$
Lysine	Lys	$CH_2 - CH_2 - CH_2 - CH_2 - NH_2$ $\; \vert$ $H_2N - CH - COOH$
Methionine	Met	$CH_2 - CH_2 - S - CH_3$ $\quad\quad\; \vert$ $H_2N - CH - COOH$
Phenylalanine	Phe	$CH_2 -$ ⬡ $\quad \vert$ $H_2N - CH - COOH$
Proline	Pro	$COOH$ $HN \diagdown$ / ⬠
Serine	Ser	$CH_2 - OH$ $\quad \vert$ $H_2N - CH_2 - COOH$
Threonine	Thr	$CH_3 - CH - OH$ $\quad\quad\; \vert$ $H_2N - CH - COOH$
Trytophan	Trp	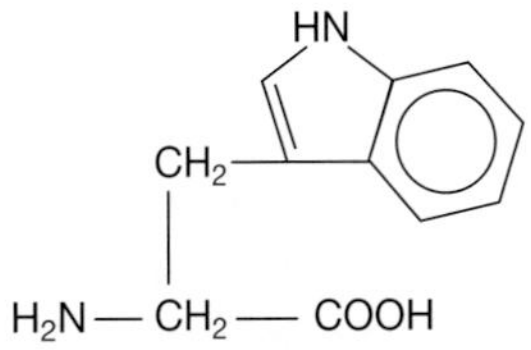
Tyrosine	Tyr	$CH_2 -$ ⬡ $- OH$ $\quad \vert$ $H_2N - CH_2 - COOH$
Valine	Val	$CH_3 - CH - CH_3$ $\quad\quad\; \vert$ $H_2N - CH - COOH$

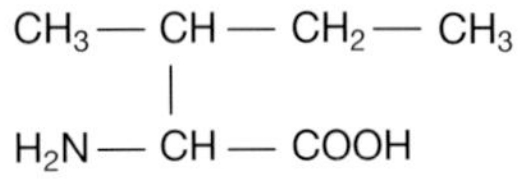

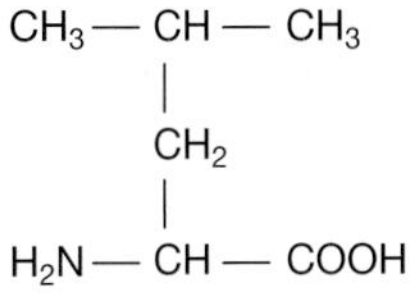

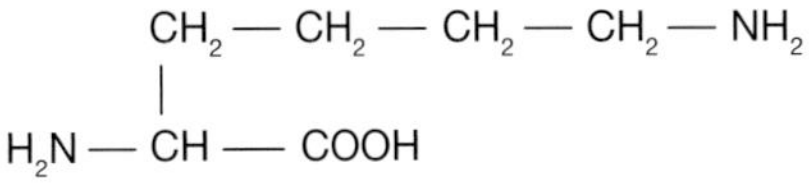

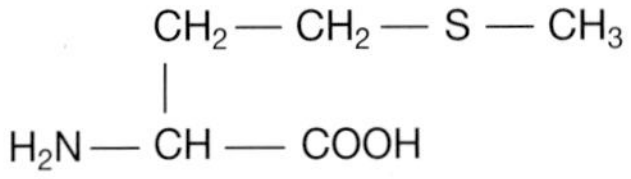

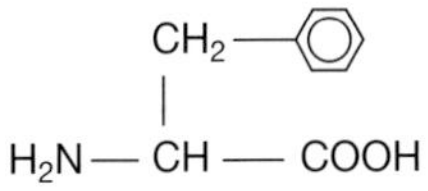

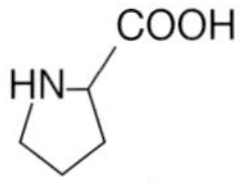

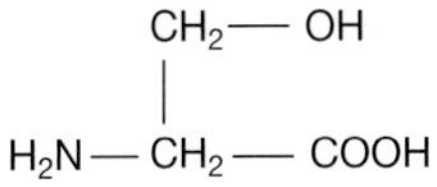

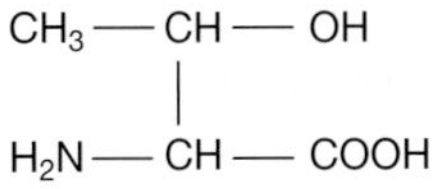

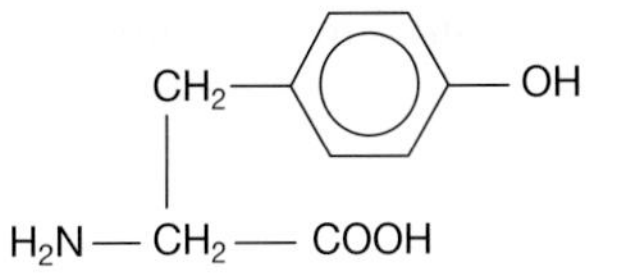

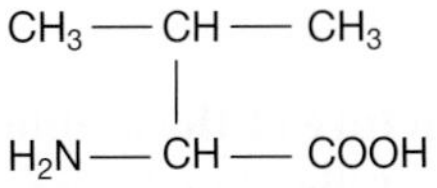

Resources

 Video eLesson Hydrolysis of proteins (eles-3259)

tlvd-9700

SAMPLE PROBLEM 3 Showing the structure of amino acids formed during hydrolysis

A dipeptide is a peptide made up of two amino acids. Show the structure of the amino acids formed from the hydrolysis reaction of the following dipeptide.

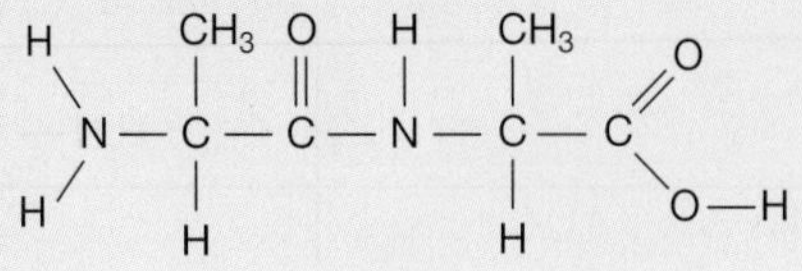

[Structure of the dipeptide]

THINK	WRITE
1. A hydrolysis reaction requires water, so add water.	[Dipeptide structure] + H_2O
2. Circle the peptide link and draw a vertical line through the middle to indicate separation of the two parts of the molecule.	[Dipeptide structure with peptide link circled and vertical line] + H_2O
3. Separate the two parts of the molecule.	[Two separated molecule parts]
4. Remember that each amino acid needs to have a carboxyl group and an amino group. The −O−H from the water is added to the C=O side, and the −H from the water is attached to the nitrogen atom to form an amino group.	[Two complete amino acids]

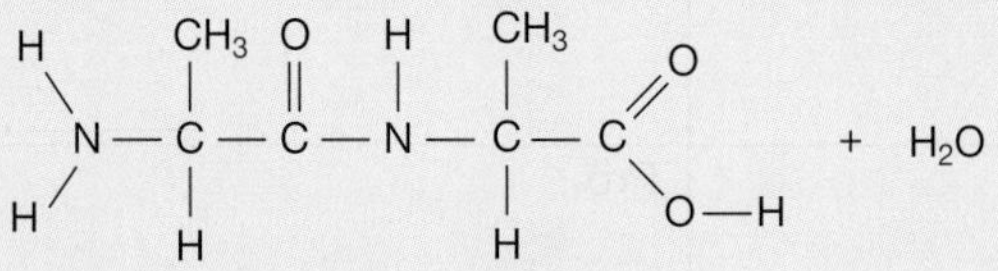

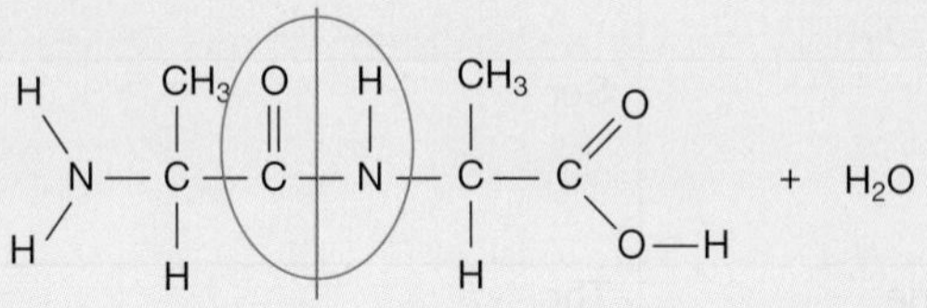

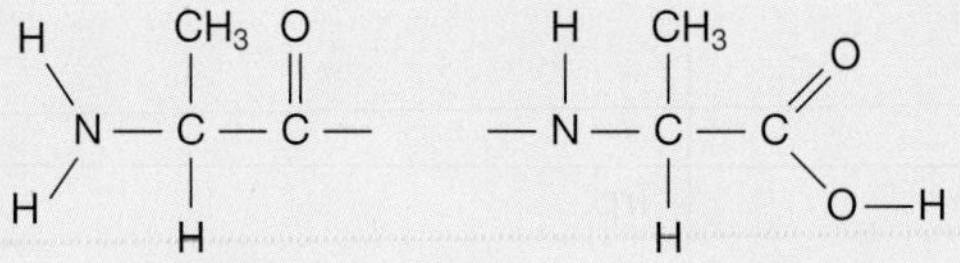

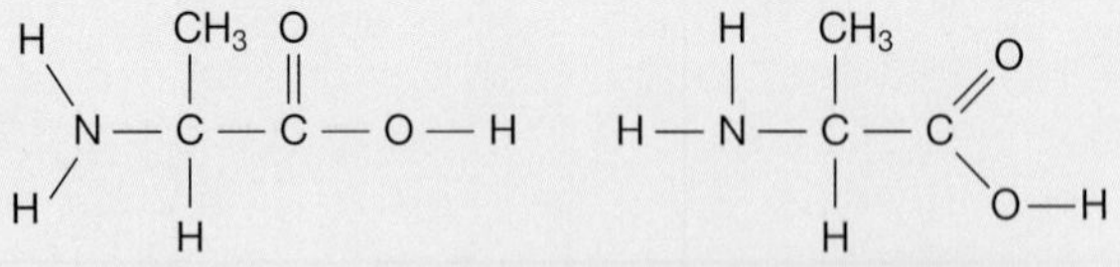

PRACTICE PROBLEM 3

Show the structure of the amino acids formed from the hydrolysis reaction of the following dipeptide.

[Structure of dipeptide with OH and SH groups]

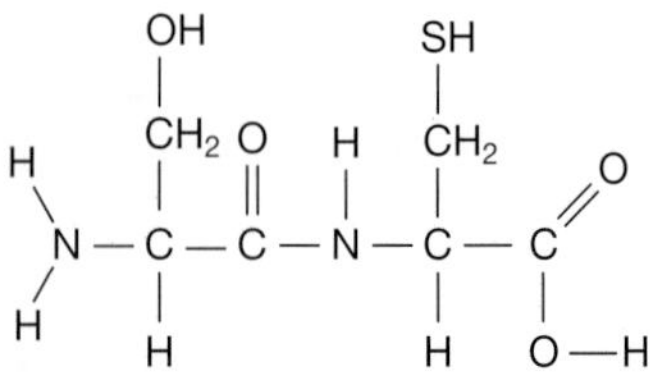

8.4.2 Hydrolytic reactions of carbohydrates

Whereas the structure of a protein is classified as a polymer, **carbohydrates** can be either small molecules or polymers. Carbohydrates are made up of the elements carbon, hydrogen and oxygen, and have a general formula that can be represented as $C_x(H_2O)_y$. Most carbohydrates in our food originate from plants, in the form of glucose, and are produced through the process of photosynthesis. This sugar, **glucose**, is a simple carbohydrate that is stored in the liver or muscles. It is readily available as an energy source because it requires less oxygen to react than either protein or fat.

$$6CO_2(g) \quad + \quad 6H_2O(l) \quad \xrightarrow[\text{Sunlight}]{\text{Chlorophyll}} \quad C_6H_{12}O_6(aq) \quad + \quad 6O_2(g) \quad \Delta H = +2820\,\text{kJ mol}^{-1}$$

Carbon dioxide $+$ Water $\longrightarrow$ Glucose $+$ Oxygen

Glucose can be used by a plant to form complex carbohydrates through polymerisation reactions. When an animal eats a plant, it can use the plant's carbohydrates as an energy source. Carbohydrates provide the greatest proportion of energy in the diets of most humans. The sugar in sports drinks and in so many other foods helps replace the fuel that your body uses during exercise.

Carbohydrates are classified into three groups according to their molecular structure: **monosaccharides**, **disaccharides** and **polysaccharides**. Monosaccharides (sometimes called simple sugars) are the basic building blocks of all carbohydrates.

carbohydrates the general name for a large group of organic compounds occurring in food and living tissues; includes sugars, starch and cellulose

glucose a simple carbohydrate stored in the liver or muscles

monosaccharide the simplest form of carbohydrate, consisting of one sugar molecule

disaccharide two sugar molecules (monosaccharides) bonded together

polysaccharide more than ten monosaccharides bonded together

FIGURE 8.23 Sources of carbohydrates

FIGURE 8.24 Structures of some common monosaccharides

Glucose   Galactose   Fructose

Disaccharides are the result of two monosaccharides reacting together. Table sugar is the disaccharide sucrose. Polysaccharides are polymers that consist of large numbers of monosaccharide monomers that have combined in condensation polymerisation reactions. They are sometimes called complex carbohydrates. Starch, glycogen and cellulose are important polysaccharides in plant and animal systems.

FIGURE 8.25 Fructose is a monosaccharide and is the main carbohydrate in fruit.

Starch

The most important complex carbohydrate digested by humans is **starch**. Starch is a white, granular polysaccharide that is the major storage form of glucose for plants. It is found in corn, wheat, seeds and the fleshy part of vegetables and fruit, and is the second most common organic compound. Starch is initially hydrolysed in the mouth by an enzyme present in the saliva called salivary amylase. Salivary amylase does not completely hydrolyse starch to glucose, but it can split the bonds between every second pair of glucose units, producing maltose, a disaccharide. Hydroxyl functional groups form from the glycosidic (ether) linkages that are broken.

> **starch** a condensation polymer of glucose

$$2(C_6H_{10}O_5)_n + nH_2O \xrightarrow{\text{Salivary amylase}} nC_{12}H_{22}O_{11}$$

Starch   Maltose

FIGURE 8.26 Enzyme-catalysed hydrolysis of starch

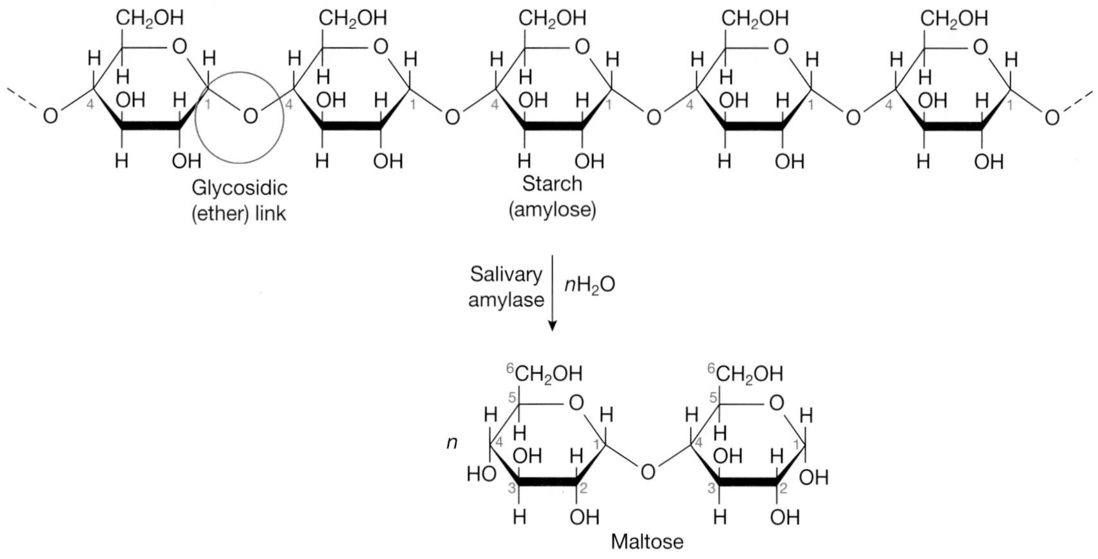

Glycosidic (ether) link

Starch (amylose)

Salivary amylase | $n\text{H}_2\text{O}$

Maltose

Carbohydrate digestion is completed in the small intestine, where the disaccharides are changed by enzymes into their constituent monosaccharides: glucose, fructose and galactose.

The formation of glucose eliminates the glycosidic linkages and creates two hydroxyl functional groups on the monosaccharides.

Glycosidic links

The glucose monomers in polysaccharides are held together by glycosidic links (–O–).

FIGURE 8.27 Enzyme-catalysed hydrolysis of maltose

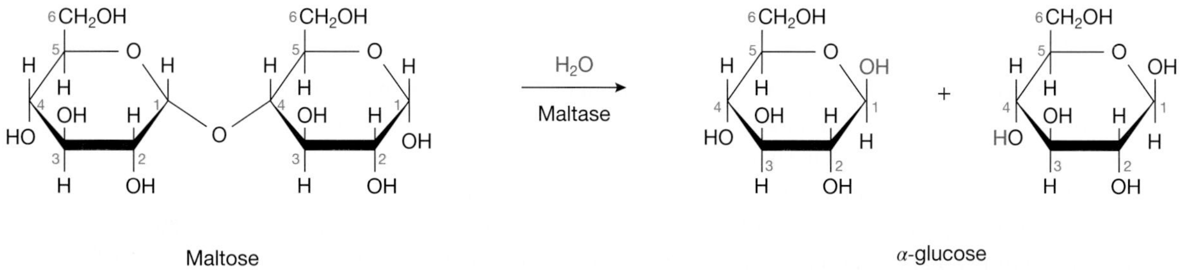

Maltose

H_2O
Maltase

α-glucose

---

elog-1906

tlvd-9727

EXPERIMENT 8.3

online only

Studying starch — hydrolysis of starch

Aim

To demonstrate the hydrolysis of starch by hydrochloric acid

Glycogen

If glucose is not immediately required by the body as an energy source, it is stored in the liver and, to a lesser extent, in the body tissues, as **glycogen**. Glycogen molecules are more highly branched and shorter than starch molecules. The liver reconverts glycogen to glucose by hydrolysis for use by the body. In this way, the liver keeps the glucose concentration of the blood relatively constant. If the body needs energy in a hurry or when the body is not getting glucose from food, glycogen is hydrolysed to provide the needed glucose.

Hydrolysis of starch and glycogen

Complete hydrolysis of starch and glycogen polymers results in glucose molecules.

CASE STUDY: Cellulose

Cellulose, a polysaccharide, is the main structural component of the cell wall in plant cells. Brussels sprouts, cabbage, celery and kale are all high in cellulose. Both starch and cellulose are condensation polymers of glucose, but their arrangement is different. As humans do not have the enzyme required to hydrolyse the type of glycosidic link found in cellulose, we are unable to utilise it as a source of energy. Nevertheless, cellulose known as fibre or roughage is still important in the human diet because it assists the passage of food through the digestive system.

FIGURE 8.28 Green vegetables provide fibre, which is an important component of our diet.

8.4.3 Hydrolytic reactions of fats and oils

Fats and oils (triglycerides) belong to a group of compounds called **lipids**, which also include a small number of other compounds such as waxes, fat-soluble vitamins, monoglycerides and diglycerides. Our bodies need lipids because, as well as functioning as an energy store, they form components in cell membranes, and are important for hormone production, insulation, protection of vital organs and transport of fat-soluble vitamins. Fats and oils contain the elements carbon, hydrogen and oxygen. In this respect they are similar to carbohydrates, but fats and oils have a smaller percentage of oxygen. Unlike proteins and complex carbohydrates, they are not polymers. Triglycerides are hydrophobic and less dense than water due to the large non-polar, hydrophobic sections of the molecules. They are found in fish, dairy products, oils, fried foods, seeds and nuts.

glycogen the storage form of glucose in animals

cellulose the most common carbohydrate and a condensation polymer of glucose; humans cannot hydrolyse cellulose, so it is not a source of energy

lipids substances such as fats, oils and waxes that are insoluble in water

Digestion of lipids mostly takes place in the alkaline conditions of the small intestine, where the lipid is mixed with bile. Bile is made by the liver and released from the gall bladder. Lipids are insoluble, so they clump together in an aqueous environment, and bile is an emulsifier, which means that it increases the surface area of the fat by breaking it into smaller droplets. Once emulsified, the fat undergoes hydrolysis, catalysed by the water-soluble lipase enzymes from the pancreas. This produces glycerol, free fatty acids and monoglycerides, which are absorbed by cells lining the small intestine and converted back into triglycerides that the body requires.

FIGURE 8.29 Fats are solid at room temperature and oils are liquid.

FIGURE 8.30 In this x-ray of a finger, the subcutaneous fat layer is visible as the yellow layer under the skin and around the bone. This fat stores triglycerides.

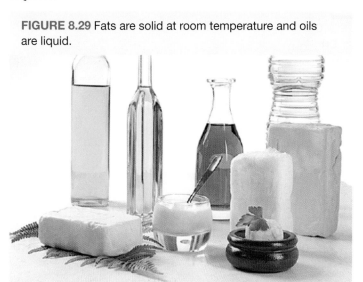

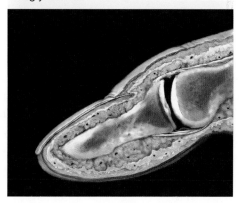

FIGURE 8.31 Triglycerides are hydrolysed to produce glycerol and three fatty acids.

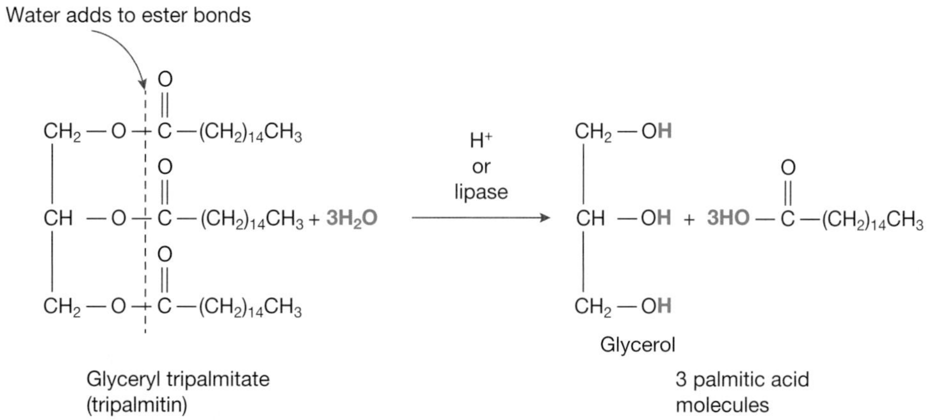

Water adds to ester bonds

$CH_2-O \vdash \overset{\overset{O}{\|}}{C}-(CH_2)_{14}CH_3$

$CH \ -O \vdash \overset{\overset{O}{\|}}{C}-(CH_2)_{14}CH_3 + 3H_2O$

$CH_2-O \vdash \overset{\overset{O}{\|}}{C}-(CH_2)_{14}CH_3$

Glyceryl tripalmitate
(tripalmitin)

$\xrightarrow{\begin{array}{c}H^+\\ or\\ lipase\end{array}}$

CH_2-OH

$CH \ -OH \ + \ 3HO-\overset{\overset{O}{\|}}{C}-(CH_2)_{14}CH_3$

CH_2-OH

Glycerol

3 palmitic acid
molecules

Hydrolysis of triglycerides

The hydrolysis of triglycerides produces three fatty acids and glycerol.

SAMPLE PROBLEM 4 Showing the hydrolysis of a polysaccharide

Show the hydrolysis of the following section of a glycogen molecule (glucose residue). R and R′ refer to the continuation of the polymer chain.

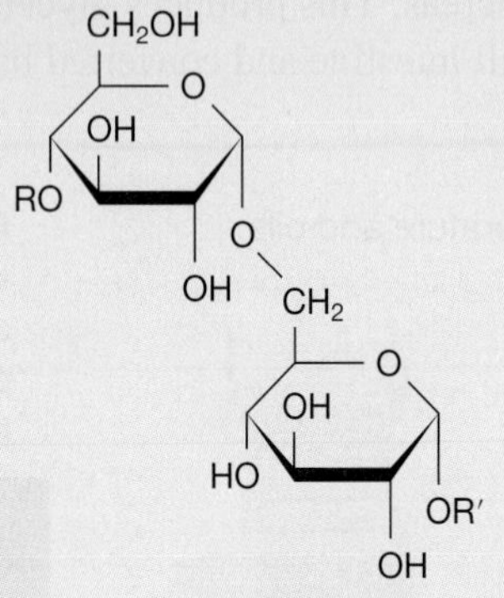

Glycogen

THINK

1. Remember that the glycosidic link (–O–) is created from a condensation reaction in which water is formed from two hydroxyl groups. Circle the glycosidic (ether) link.

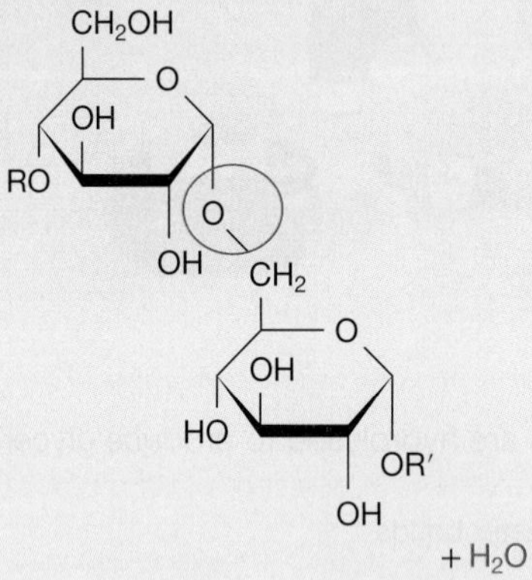

2. Separate the two molecules, copying them exactly as shown in the original diagram.

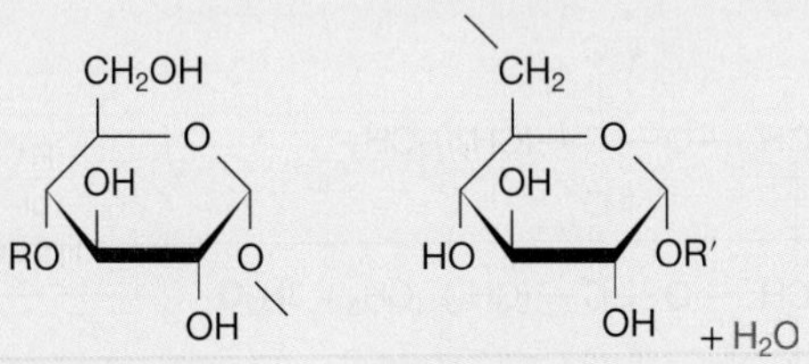

3. Add an –H to the oxygen atom and an –OH to the other bond. Check that the remainder of the groups on the molecules are the same as those originally present.

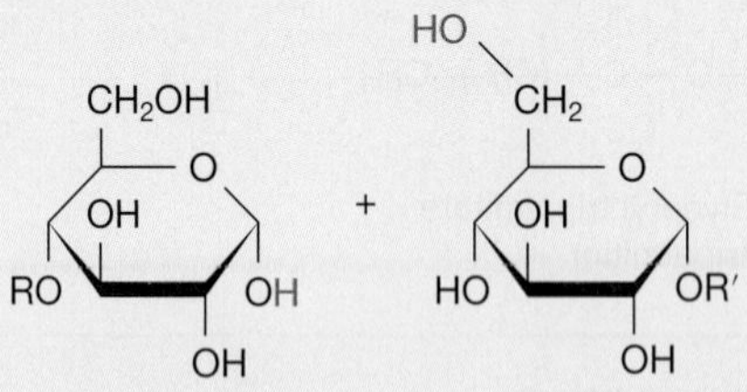

WRITE

$+ H_2O$

$+ H_2O$

PRACTICE PROBLEM 4

Draw the structural formulas of the products of the hydrolysis of the following triglyceride.

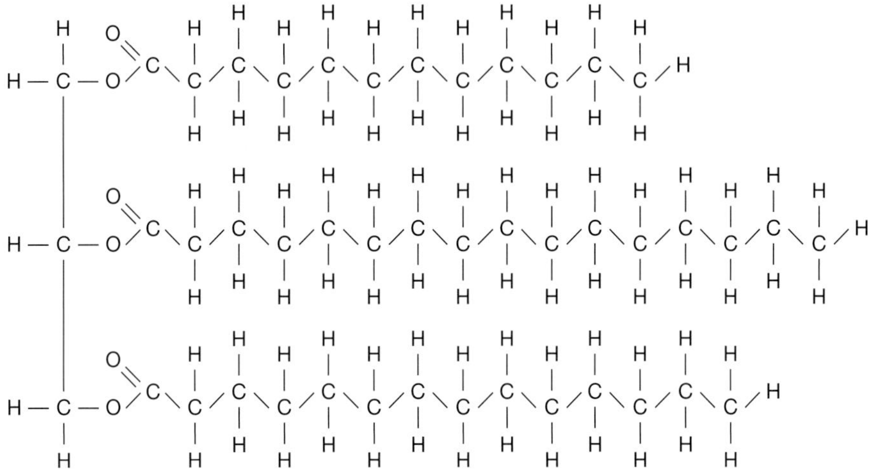

8.4.4 Synthesis of proteins

Once proteins, carbohydrates and lipids from food are broken down in the digestive system by hydrolytic reactions, the body then builds them up by condensation reactions into different proteins, carbohydrates and lipids that are required for growth, repair and energy in the body.

The amino acids produced by the hydrolysis of proteins in the food that we eat are converted into the specific proteins that the body needs in a polymerisation reaction forming **peptide links**.

Water is produced when a peptide link is formed, so this reaction between amino acids to form peptides is an example of a condensation reaction. When two or more amino acids combine to form a peptide, what remains of each amino acid in the peptide is called an amino acid **residue**.

> **peptide link** the link formed when a carboxyl group reacts with an amino group in a condensation reaction between two amino acids
>
> **residue** what remains when two or more amino acids combine to form a peptide
>
> **dipeptide** formed when two amino acids combine

FIGURE 8.32 Formation of a peptide link

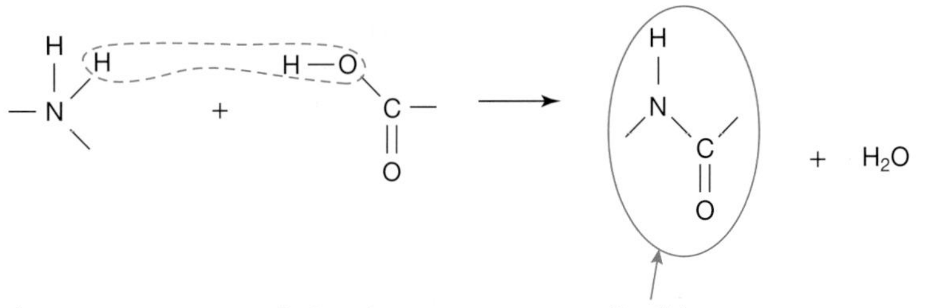

Amino group of one amino acid

Carboxyl group of another amino acid

Peptide or amide

FIGURE 8.33 Two amino acid residues

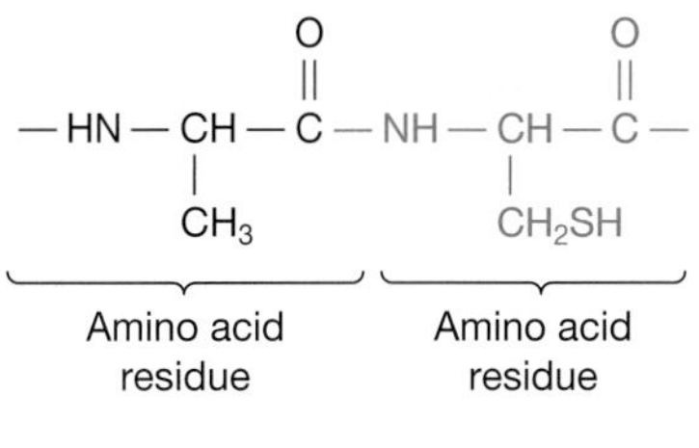

Amino acid residue

Amino acid residue

When two amino acids combine, a **dipeptide** is produced. Two different dipeptides can be made depending on the initial alignment of the amino acids. The simplest amino acids, glycine and alanine, can combine in two ways. This is important because alanine-glycine has a different structure from glycine-alanine, as shown in figure 8.33.

TIP: Remember to include the water molecule as a product when writing equations for condensation reactions.

FIGURE 8.34 Two different dipeptides produced during the condensation reaction between glycine and alanine

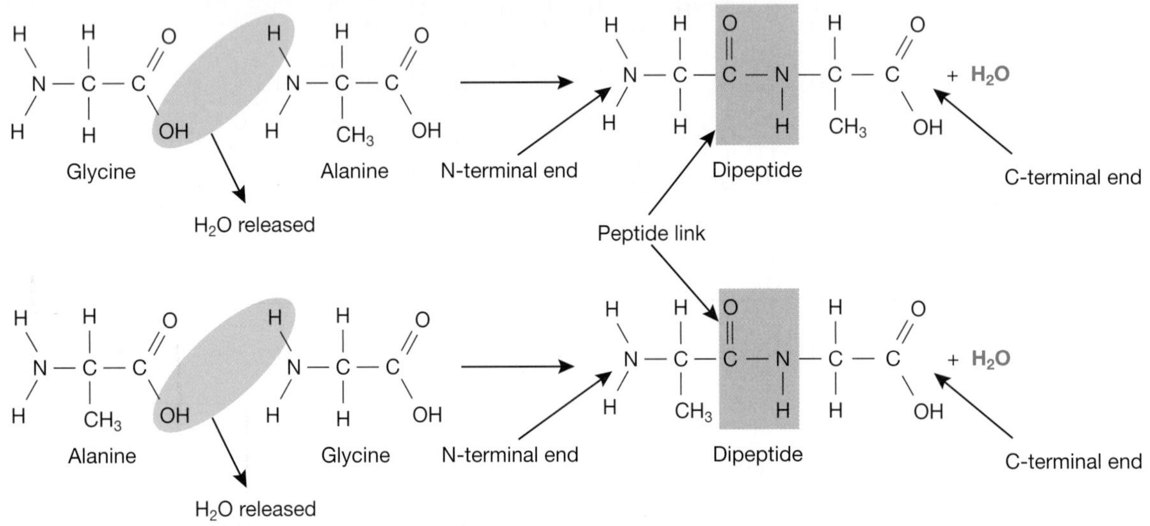

---

tlvd-9701

SAMPLE PROBLEM 5 Drawing the structure of a dipeptide formed in a condensation reaction

Show the structure of the dipeptide formed from the condensation reaction of two alanine molecules.

THINK

1. Draw the two alanine molecules, with the carboxyl functional group of one molecule next to the amino functional group of the other.

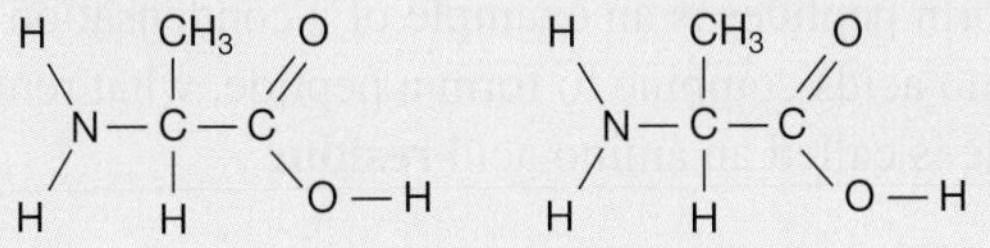

2. Circle the −O−H of the carboxyl group and the −H from the amine group. This is the water molecule that is removed in the condensation reaction.

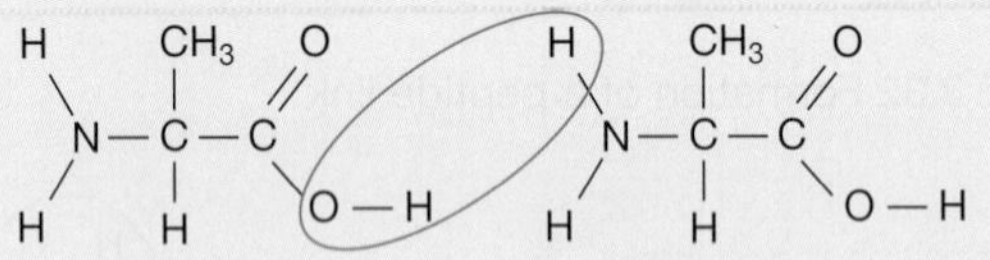

3. Join the remaining half bonds between the carbon atom and the nitrogen atom.

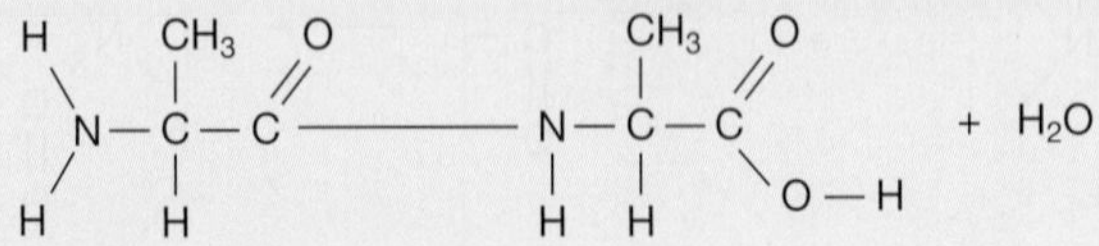

4. Draw the structure of the dipeptide with the double bond to the oxygen and the single bond from the hydrogen to the nitrogen drawn vertically. Include the water molecule produced if an equation is required.
   Check that a peptide link has been formed (−CONH−).

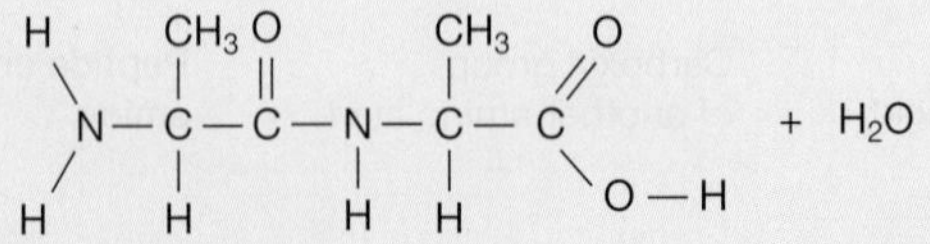

WRITE

PRACTICE PROBLEM 5

Show the structure of a dipeptide formed from the condensation reaction of a serine and valine molecule.

Depending on the number of amino acid residues per molecule, the peptides formed from the condensation reactions of amino acids are known as dipeptides, tripeptides and so on until we reach polypeptides, which are the result of condensation polymerisation reactions. Note that, when drawing segments of proteins, it is necessary to show the open bonds at each end by using dashes where the next amino acid residue would be attached. Peptides of molar mass up to about 5000 g mol^{-1} (about 50 amino acid units) are known as polypeptides; even larger peptides are called proteins. The peptide chain of covalently bonded nitrogen and carbon atoms is referred to as the **backbone**, and the R-groups are described as **side chains**.

> **backbone** a peptide chain of covalently bonded nitrogen and carbon atoms
>
> **side chain** an R group attached to an amino acid

Amino acids

- Proteins are made from amino acids.
- Amino acids combine in condensation polymerisation reactions to form proteins.

FIGURE 8.35 Formation of a polypeptide

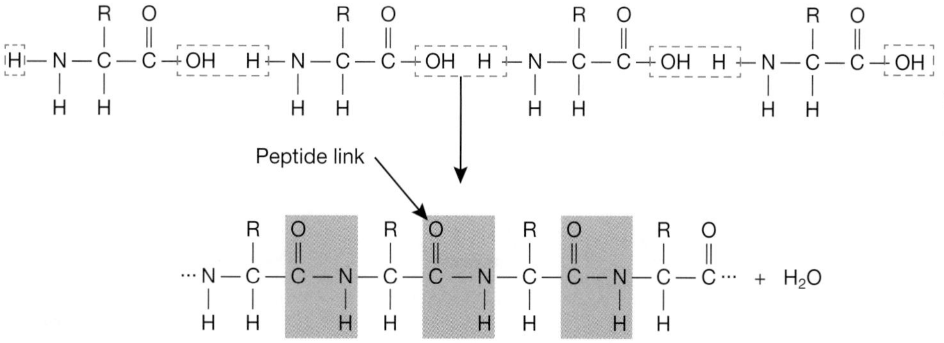

TIP: Remember to use open bonds on each end when drawing a section of a polymer.

on Resources

▶ **Video eLesson** Condensation of amino acids (eles-3258)

8.4.5 Synthesis of carbohydrates

Polysaccharides are polymers made up of monosaccharide or disaccharide monomers. Two monosaccharides can form a disaccharide in a condensation reaction; water is produced as a by-product.

Polysaccharides such as starch, glycogen and cellulose are polymers consisting of large numbers of monosaccharide monomers that have combined in condensation polymerisation reactions.

FIGURE 8.36 a. Formation of a disaccharide **b.** Two other disaccharides

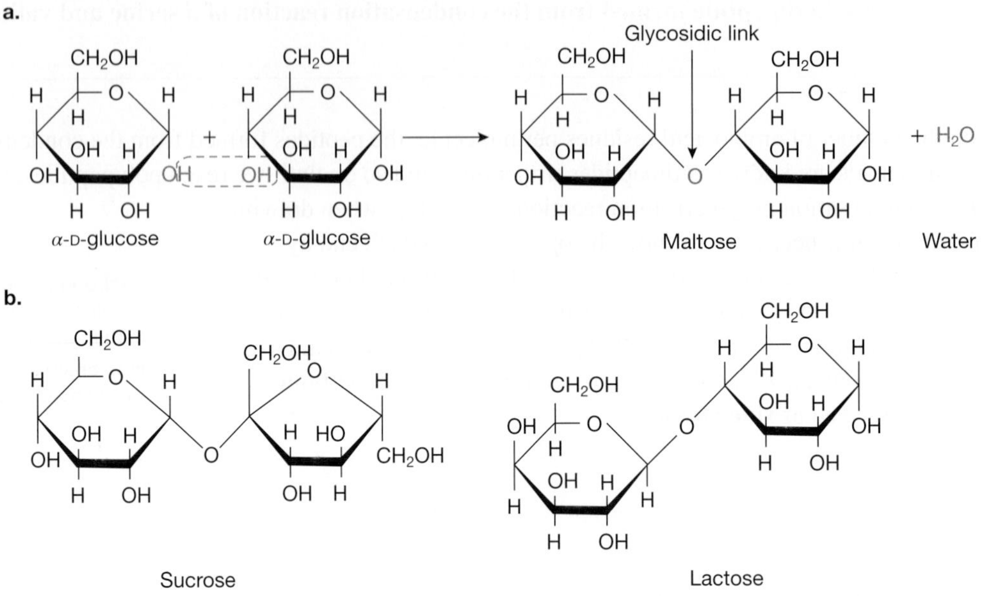

a.

CH₂OH   CH₂OH

α-D-glucose   α-D-glucose

Glycosidic link

CH₂OH   CH₂OH

Maltose   + H₂O

Water

b.

CH₂OH   CH₂OH

Sucrose

CH₂OH   CH₂OH

Lactose

Synthesis of starch

Starch is a condensation polymer of glucose, and the glucose monomers join together by the combination of the hydrogen atom on one glucose monomer with the hydroxyl functional group on another, and water is eliminated. This reaction occurs in plants using the glucose they form during photosynthesis.

FIGURE 8.37 Section of starch polymer

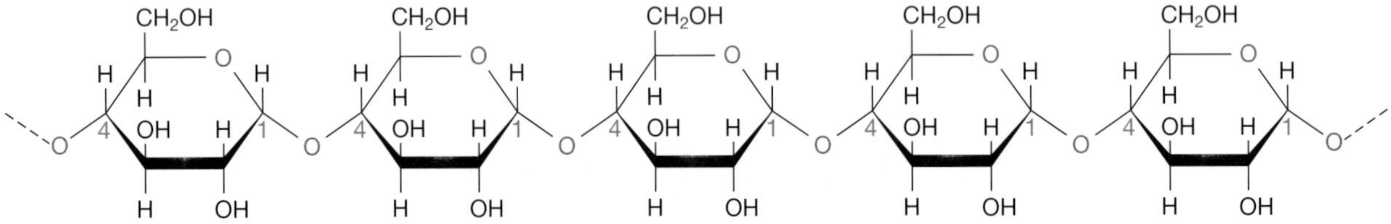

CH₂OH   CH₂OH   CH₂OH   CH₂OH   CH₂OH

FIGURE 8.38 Using iodine to test for starch. Iodine is pipetted onto the cut surface of a potato. If starch is present, the colour changes from orange to black.

Starch is the energy storage molecule in plants and is formed by the condensation polymerisation of glucose monomers.

elog-1907

tlvd-9728

EXPERIMENT 8.4

online only

Studying starch — starch in foods

Aim

To test for starch in foods

tlvd-10343

SAMPLE PROBLEM 6 Calculating the molar mass of starch formed in a condensation reaction

Calculate the molar mass of a starch molecule made from 500 glucose monomers.

THINK	WRITE
1. Calculate the mass of 500 glucose monomers.	Molecular mass of glucose: $C_6H_{12}O_6 = (6 \times 12.0) + (12 \times 1.0) + (6 \times 16.0)$ $= 180.0 \text{ g mol}^{-1}$ Mass of 500 glucose monomers $= 500 \times 180.0$ $= 90\,000 \text{ g mol}^{-1}$
2. Determine how many water molecules are lost.	$500 - 1 = 499$ water molecules lost between 500 monomers in this condensation reaction
3. Calculate the mass of 499 water molecules.	Mass of 499 water molecules $= 499 \times 18.0$ $= 8982 \text{ g mol}^{-1}$
4. Calculate the molar mass of this starch molecule.	Molar mass of starch molecule $= 90\,000 - 8982$ $= 81\,018$ $= 8.10 \times 10^4 \text{ g mol}^{-1}$

PRACTICE PROBLEM 6

Calculate the molar mass of a starch molecule made from 650 glucose monomers.

Synthesis of glycogen

Glycogen has a similar structure to starch but it is more highly branched. It is also formed by the condensation polymerisation of glucose monomers.

The polysaccharides starch and glycogen are produced from monomers of glucose by condensation polymerisation reactions.

EXPERIMENT 8.5

online only

elog-1908

Constructing models of carbohydrates

Aim

To construct models of carbohydrates

8.4.6 Synthesis of lipids

Fats and oils are formed by a condensation reaction between glycerol and three fatty acids. The fatty acids can be the same or different. There are three hydroxyl groups on a glycerol molecule. A hydrogen atom from each of these three hydroxyl groups combines with an –OH group from three fatty acids, which may be the same or different, to form three water molecules. A triglyceride is formed with three ester links.

FIGURE 8.39 Reaction of glycerol and three fatty acids to form a triglyceride

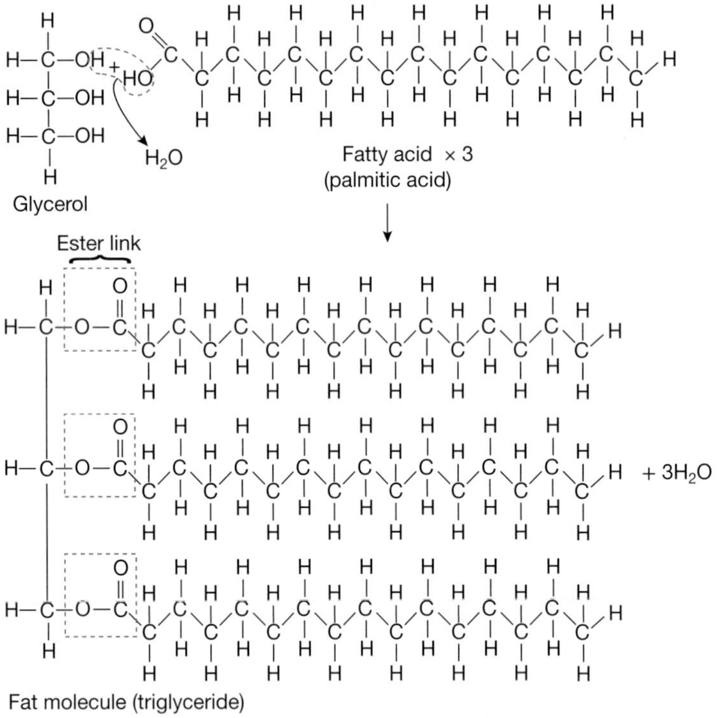

Fats and oils are formed by a condensation reaction between glycerol and three fatty acids.

Condensation of biomolecules

- Condensation of amino acids results in proteins (polymer).
- Condensation of glucose molecules results in polysaccharides (polymer).
- Condensation of three fatty acids and glycerol results in triglycerides (large molecule).

8.4 Activities

| 8.4 Quick quiz | 8.4 Exercise | 8.4 Exam questions |

8.4 Exercise

1. **MC** What would be the formula of a trisaccharide formed from the condensation of three glucose molecules?
   A. $C_{18}H_{32}O_{16}$ **B.** $C_{18}H_{34}O_{17}$ **C.** $C_{18}H_{36}O_{18}$ **D.** $C_{18}H_{38}O_{19}$

2. Identify the following as either condensation or hydrolysis reactions.
   a. Water produced
   b. Disaccharide to polysaccharide
   c. Dipeptide to amino acid
   d. Glucose to starch
   e. Maltose to glucose
   f. Water consumed
   g. Triglyceride to glycerol and fatty acids

3. What are the main products of the hydrolysis of:
   a. fat
   b. starch
   c. protein
   d. maltose
   e. oil
   f. glycogen?

4. Draw an example of the simplest amino acid, glycine, and label the two functional groups present.

5. **a.** Draw a possible dipeptide formed between alanine and aspartic acid.
   b Circle the peptide link.
   c. Name the type of reaction.
   d. Calculate the molar mass of the dipeptide formed.

6. **a.** Write an equation for the hydrolysis of the following polypeptide.

   b. Name the amino acids produced.

7. **a.** Name the functional group that is lost when a protein is hydrolysed.
   b. Name the functional groups that are produced when a protein is hydrolysed.

8. **a.** Write the molecular, semi-structural and skeletal structure for glycerol.
   b. What family of organic compounds does glycerol belong to?

9. **a.** Provide a chemical equation for the hydrolysis of the fat with the following formula.

$$
\begin{array}{l}
H_2C - O - \overset{\displaystyle O}{\overset{\displaystyle \|}{C}} - (CH_2)_{14}CH_3 \\
\quad\quad\quad\quad O \\
\quad\quad\quad\quad \| \\
HC - O - C - (CH_2)_{14}CH_3 \\
\quad\quad\quad\quad O \\
\quad\quad\quad\quad \| \\
H_2C - O - C - (CH_2)_{14}CH_3
\end{array}
$$

   b. Identify the functional groups involved in this reaction.

8.4 Exam questions

▶ Question 1 (4 marks)

Source: VCE 2020 Chemistry Exam, Section B, Q.5.b,d.i; © VCAA

Bananas provide essential vitamins and minerals, such as vitamin B_6 and vitamin C, along with dietary fibre and energy. During the ripening process, the banana changes in appearance, texture and taste.

a. The structure of a disaccharide found in a ripe banana is shown below.

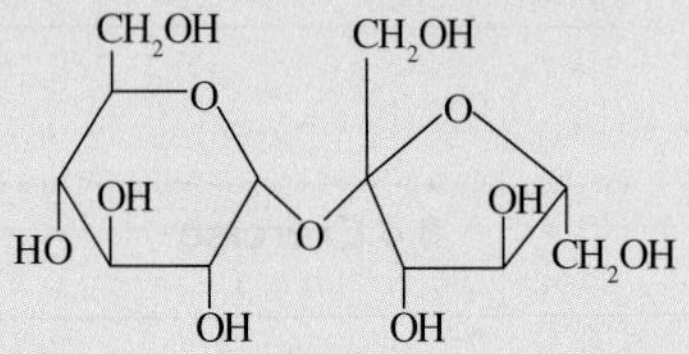

i. On the structure above, circle and name the link that joins the two sugar units that make up the disaccharide. **(1 mark)**

ii. Name the two sugar units that make up this disaccharide. **(1 mark)**

iii. Banana skins are primarily composed of cellulose.
   Suggest why cellulose cannot be used as a source of energy in the human body. **(1 mark)**

b. During the ripening process, the enzyme amylase breaks down starch molecules into disaccharides and monosaccharides.
   What name is given to this type of reaction? **(1 mark)**

▶ Question 2 (1 mark)

Source: VCE 2013 Chemistry Exam, Section B, Q.10.a.i; © VCAA

Olive oil, which has been part of the human diet for thousands of years, is derived from the fruit of the olive tree.

The main fatty acid that makes up olive oil is oleic acid, $CH_3(CH_2)_7CH=CH(CH_2)_7\ COOH$.

The triglyceride formed from three oleic acid molecules is glycerol trioleate, $C_{57}H_{104}O_6$. The molar mass of glycerol trioleate is 884 g mol^{-1}.

An incomplete semi-structural formula of glycerol trioleate is provided below.

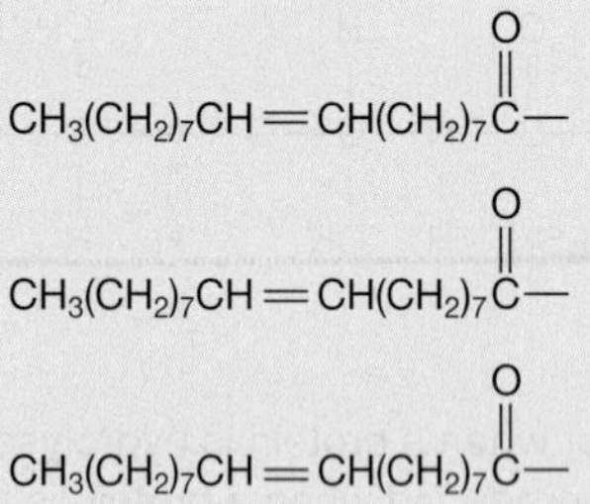

Complete the semi-structural formula of glycerol trioleate.

▶ Question 3 (6 marks)

Source: Adapted from VCE 2013 Chemistry Exam, Section B, Q.3.a–c; © VCAA

Spider webs are very strong and elastic. Spider web silk is a protein that mainly consists of glycine and alanine residues.

a. Assuming that these amino acid residues alternate in a spider web, draw a section of the spider web protein that contains at least **three** amino acid residues. **(2 marks)**

b. What is the name of the bond between each amino acid residue? **(1 mark)**

c. What type of polymerisation reaction occurs in the formation of spider web silk? **(1 mark)**

d. Proteins can be completely hydrolysed to their component amino acids by treatment with concentrated acids.
   Identify the two functional groups that are formed as a result of this hydrolysis. **(2 marks)**

▶ **Question 4 (6 marks)**

Source: Adapted from VCE 2018 Chemistry NHT Exam, Section B, Q.4; © VCAA

The structures or formulas of a number of important biomolecules are shown below.

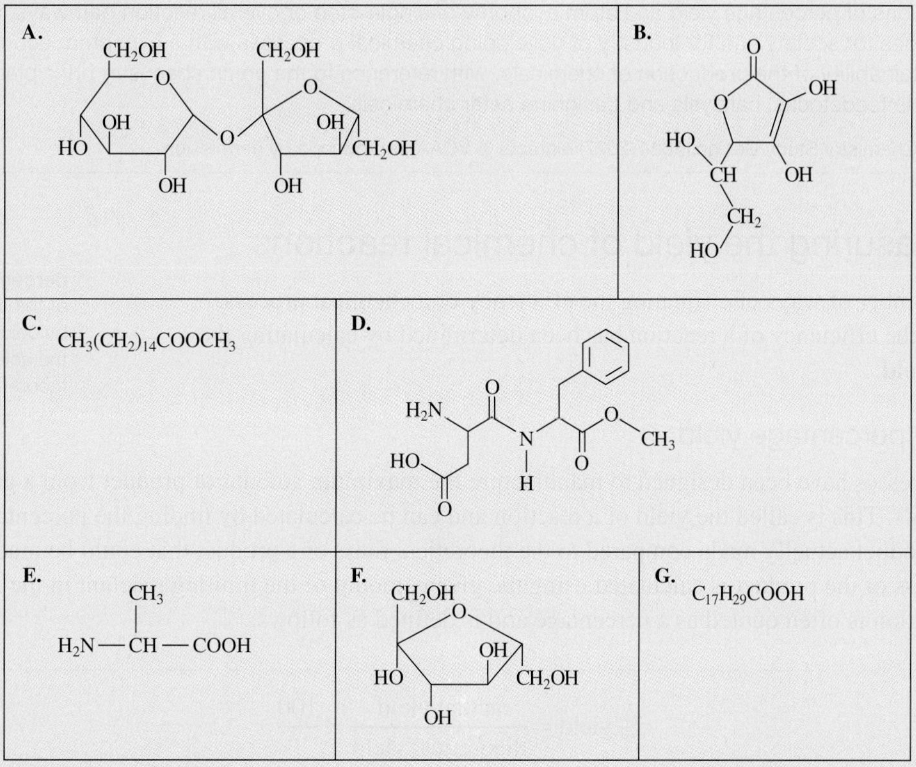

A.

CH$_2$OH   CH$_2$OH

B.

C. CH$_3$(CH$_2$)$_{14}$COOCH$_3$

D.

H$_2$N

HO

CH$_3$

E.

CH$_3$

H$_2$N — CH — COOH

F.

CH$_2$OH

HO   CH$_2$OH

OH

G. C$_{17}$H$_{29}$COOH

For each of the following characteristics of biomolecules, write the letter or letters in the space provided for the corresponding biomolecule or biomolecules shown. Each biomolecule may be used more than once or may not be used at all.

Characteristic	Biomolecule letter(s) (A.–G.)
contains a glycosidic linkage	
can produce an ester when reacted with an alcohol in the presence of a concentrated acid	
is soluble in water (give letters for **two** examples)	
contains an ester linkage (give letters for **two** examples)	
can be a key constituent of biodiesel	
has phenylalanine as a component	

▶ **Question 5 (1 mark)**

Source: VCE 2014 Chemistry Exam, Section A, Q.21; © VCAA

MC Maltotriose is a trisaccharide that is formed when three glucose molecules link together. The molar mass of glucose, C$_6$H$_{12}$O$_6$, is 180 g mol^{-1}.

The molar mass of maltotriose is

A. 472 g mol^{-1} B. 486 g mol^{-1} C. 504 g mol^{-1} D. 540 g mol^{-1}

More exam questions are available in your learnON title.

8.5 The production of chemicals and green chemistry

KEY KNOWLEDGE

- Calculations of percentage yield and atom economy of single-step or overall reaction pathways, and the advantages for society and for industry of developing chemical processes with a high atom economy
- The sustainability of the production of chemicals, with reference to the green chemistry principles of use of renewable feedstocks, catalysis and designing safer chemicals

Source: VCE Chemistry Study Design (2024–2027) extracts © VCAA; reproduced by permission.

8.5.1 Measuring the yield of chemical reactions

There are a number of ways of evaluating the efficiency of a chemical process. Traditionally, the efficiency of a reaction has been determined by calculating the **percentage yield**.

> **percentage yield** a measurement of the efficiency of a reaction, found by calculating the percentage of the actual yield compared to the theoretical yield

Calculating percentage yield

Chemical processes have been designed to manufacture the maximum amount of product from a given amount of raw materials. This is called the yield of a reaction and can be calculated by finding the percentage of the mass of the product actually made compared to the theoretical mass of a product that could be made. The theoretical mass of the product is calculated using the given amount of the limiting reactant in the reaction. The yield of a reaction is often quoted as a percentage and is defined as follows:

$$\% \text{ yield} = \frac{\text{actual yield}}{\text{theoretical yield}} \times \frac{100}{1}$$

In many cases, reactions display yields of less than 100 per cent; that is, the actual mass obtained is less than the theoretical or predicted maximum mass. However, there are often additional reasons why a given reaction does not achieve a 100 per cent yield. Some of these may be practical (losses during the method by which the chemical is made), or may involve a very slow reaction that has not been given enough time to either reach equilibrium, go to completion, or there may be side reactions occurring. This is an important consideration in the design of large-scale manufacturing techniques for chemical production that aim to achieve high yields.

---

tlvd-9702

SAMPLE PROBLEM 7 Determining the percentage yield of a reaction

2.18 g of ethanol, C_2H_5OH, is reacted with excess oxygen to produce 3.63 g of carbon dioxide according to the following equation:

$$C_2H_5OH(l) + 3O_2(g) \rightarrow 3H_2O(g) + 2CO_2(g)$$

What is the percentage yield of this reaction?

THINK

1. The percentage yield will be the mass of CO_2 actually produced (actual yield) divided by the theoretical mass of CO_2 expected to be produced (theoretical yield) according to the mole ratios in the equation and multiplied by 100.

WRITE

$$\% \text{ yield} = \frac{\text{actual yield}}{\text{theoretical yield}} \times \frac{100}{1}$$

2. Calculate the number of moles of ethanol by using the formula $n = \frac{m}{M}$.

$$n(C_2H_5OH) = \frac{m}{M}$$
$$= \frac{2.18\,g}{(2 \times 12.0 + 6 \times 1.0 + 16.0)\,g\,mol^{-1}}$$
$$= 0.0474\,mol$$

3. Determine the number of CO_2 moles expected. From the equation given, one mole of ethanol produces two moles of carbon dioxide.

$$n(CO_2) = 2 \times 0.0474$$
$$= 0.0948\,mol$$

4. Calculate the theoretical yield of CO_2 by rearranging the molar mass formula.

$$n = \frac{m}{M}$$
$$m(CO_2) = n \times M$$
$$= 0.0948\,mol \times (12.0 + 2 \times 16.0)\,g\,mol^{-1}$$
$$= 4.17\,g$$

5. Determine the percentage yield by dividing the actual yield by the theoretical yield and multiplying by 100. Express the answer to three significant figures and do not round answers at each step.

$$\%\ yield = \frac{3.63\,g}{4.17\,g} \times 100$$
$$= 87.0\%$$

PRACTICE PROBLEM 7

2.5 g of methanol is reacted with excess oxygen to produce 3.1 g of carbon dioxide according to the following equation:

$$2CH_3OH(l) + 5O_2(g) \rightarrow 2CO_2(g) + 4H_2O(g)$$

Calculate the percentage yield of carbon dioxide.

Calculating percentage yield for multi-step pathways

To calculate the percentage yield for a multi-step reaction, multiply the yields for each step. So, for example, if the yield of the first step is 30 per cent, and the yield of the second step is 50 per cent, then the overall yield is:

$$30\% \times 50\% = \frac{30}{100} \times \frac{50}{100} \times \frac{100}{1} = 15\%$$

8.5.2 A sustainable approach

Green chemistry

In the past, the chemical industry has focused on getting the best yields from the chemical synthesis of products — that is, getting the most products. A lesser focus has been on how much waste is generated, the impact of chemical processes on the environment, and the long-term viability of those processes. Although getting high yields is important, there are other issues that need to be addressed in the production of chemicals for a sustainable future.

These issues are addressed in the principles of green chemistry, and include:

- ensuring a high atom efficiency to minimise waste
- choosing catalysts to minimise waste and energy usage
- designing products that will not persist, but break down into harmless products
- developing safer, simpler and energy-efficient processes
- designing effective and less-hazardous chemicals
- reducing the amount of wastes produced
- using renewable reactants and raw materials to make the processes more sustainable.

Green chemists believe it is necessary to change from secondary prevention (the costly cleaning up of wastes after they have been generated) to primary prevention (the development of manufacturing processes that are essentially non-polluting). This process of clean production not only targets the elimination of pollution, but also aims to encourage profitable manufacturing with more efficient use of raw materials and energy.

Green chemistry will play an increasingly important role in the design of all new chemical processes, whether these be for a new chemical or for a better way of making an existing chemical. Reasons for this include economics, statutory and financial deterrents, economic incentives, energy costs, public pressure, and availability and sustainability of resources, to name just a few.

FIGURE 8.40 Green chemistry aims to design chemical processes that have minimal impact on the environment.

8.5.3 Atom economy

Calculating the percentage yield does not give an indication of how effectively the reactants have been used to generate the product with minimal waste. **Atom economy** is another method for measuring the efficiency of a reaction that takes into account the amount of waste produced.

This is a very useful concept when planning or reviewing a process. The optimal situation is one in which the yield of a reaction is maximised, and as many atoms as possible of the reactants are incorporated into the final product. It is preferable to decrease the amount of waste produced rather than have to deal with it at the end of the process.

Measuring atom economy

Atom economy is a means of quantifying how much desired product and how much waste is produced in a chemical reaction or process. It indicates the number of reactant atoms that end up in the product(s) and is usually expressed as a percentage. It is calculated by using the following formula:

$$\% \text{ atom economy} = \frac{\text{molar mass of desired product}}{\text{molar mass of all reactants}} \times \frac{100}{1}$$

From this formula, we can see that if a process has an atom economy of 100 per cent, there will be no waste products formed. From an environmental viewpoint, obviously the higher this percentage, the better; the less waste produced, the less energy and resources must be employed to deal with it. It is easy to see how this concept fits in with green chemistry. A process should not be just about making as much of a product as possible, but should also consider the wastes produced. It is better not to produce waste in the first place than to have to deal with it afterwards.

atom economy a measurement of the efficiency of a reaction that considers the amount of waste produced, by calculating the percentage of the molar mass of the desired product compared to the molar mass of all reactants

tlvd-9703

SAMPLE PROBLEM 8 Determining the percentage atom economy in a reaction

What is the percentage atom economy for the synthesis of ethanamine from chloroethane?

$$C_2H_5Cl(g) + NH_3(g) \rightarrow C_2H_5NH_2(g) + HCl(g)$$

THINK	WRITE
1. Calculate the relative molecular mass (M_r) of the product ethanamine.	$M_r(C_2H_5NH_2) = (2 \times 12) + (5 \times 1) + 14.0 + (2 \times 1)$ $= 45$
2. Calculate the M_r of the reactants and add them together.	$M_r(C_2H_5Cl) + M_r(NH_3) = (2 \times 12.0) + (5 \times 1.0) + 35.5 + 14.0 + (3 \times 1.0)$ $= 81.5$
3. Calculate the percentage atom economy by dividing the M_r of the desired product by the total M_r of the reactants and multiplying by 100%.	$\% \text{ atom economy} = \dfrac{\text{molar mass of desired product}}{\text{molar mass of all reactants}} \times \dfrac{100}{1}$ $= \dfrac{45.0}{81.5} \times 100$ $= 55\%$

PRACTICE PROBLEM 8

Calculate the percentage atom economy when methyl ethanoate is produced from the condensation of methanol and ethanoic acid.

TIP: Write ester functional group semi-structures as COO and not OCO.

Calculating the atom economy of two-step reactions

To calculate the atom economy of a two-step reaction, the intermediate product(s) from the first reaction that are used in the second are removed from the total mass of the reactants.

For example, butyl ethanoate is a clear and colourless liquid that is used as a flavouring in sweets and ice creams, and as a solvent. It can be manufactured using butanol and ethanoic acid, but there is an alternative two-stage process:

$$CH_3COOH + SOCl_2 \rightarrow CH_3COCl + SO_2 + HCl$$

$$CH_3CH_2CH_2CH_2OH + CH_3COCl \rightarrow CH_3COOCH_2CH_2CH_2CH_3 + HCl$$

CH_3COCl (acetyl chloride) is an intermediate product that is produced in the first step and consumed in the second step, so it is not included in the calculations.

The atom economy is calculated as follows:

$$\text{Molar mass of reactants} = 60.0 + 74.0 + 119.1 = 253.1 \text{ g mol}^{-1}$$

$$\text{Molar mass of desired product} = 116.0 \text{ g mol}^{-1}$$

$$\% \text{ atom economy} = \frac{\text{molar mass of desired product}}{\text{molar mass of all reactants}} \times \frac{100}{1}$$

$$= \frac{116.0}{253.1} \times \frac{100}{1}$$

$$= 45.83\%$$

The concept of high atom efficiency has many benefits for society. An example of improved atom economy in practice is the multi-step manufacture of the common painkiller ibuprofen. It shows a more sustainable and environmentally friendly production pathway for an existing chemical.

Calculations of atom economy of an overall reaction pathway

CASE STUDY: The manufacture of ibuprofen

Ibuprofen is one of the most widely used drugs in the world today, with an estimated 20 000 tonnes currently being produced each year. It is used as an analgesic (painkiller) and anti-inflammatory, and is available as an 'over-the-counter' drug worldwide. It is popular because it is cheap and has relatively few side effects.

FIGURE 8.41 Dr Stewart Adams, the discoverer of ibuprofen

Its discovery dates to the late 1950s and early 1960s. A British pharmaceutical chemist, Dr Stewart Adams (figure 8.41), working for the Boots company in England, was tasked with trying to find a drug to treat rheumatoid arthritis that had fewer side effects than those already in use. Initially, over 600 possible compounds were considered, and Adams and a team of co-workers patiently tested all of these over a period of ten years. During this time, four went to clinical trials but failed. Eventually, a compound called 2-(4-isobutylphenyl) propionic acid, later to be known as ibuprofen, was successful. A patent was filed for by the Boots company in 1961 and subsequently granted in 1962. Following further trials, the drug was finally approved for prescription use in Britain in 1969.

The original pathway for producing ibuprofen

The original pathway for the production of ibuprofen used by the Boots company was a six-step process, as shown in figure 8.42.

Although this scheme appears complicated, you will notice three things:
- The changes in each step occur on the right-hand side of each molecule.
- Many chemicals are added as further reactants in several steps. These are shown to the side of the appropriate arrow.
- The molar masses of all the reactants and the product are shown for convenience. The molar mass of $AlCl_3$ is not shown as it is a catalyst.

The molar masses shown allow us to calculate the atom economy of this process as follows:

$$\% \text{ atom economy} = \frac{\text{molar mass of desired products}}{\text{molar mass of all reactants}} \times \frac{100}{1}$$

$$= \frac{206.29}{(134.22 + 102.09 + 122.55 + 68.05 + 19.02 + 33.03 + 38.04)} \times \frac{100}{1}$$

$$= \frac{206.29}{517.00} \times \frac{100}{1}$$

$$= 40\%$$

FIGURE 8.42 The original (Boots) pathway for producing ibuprofen

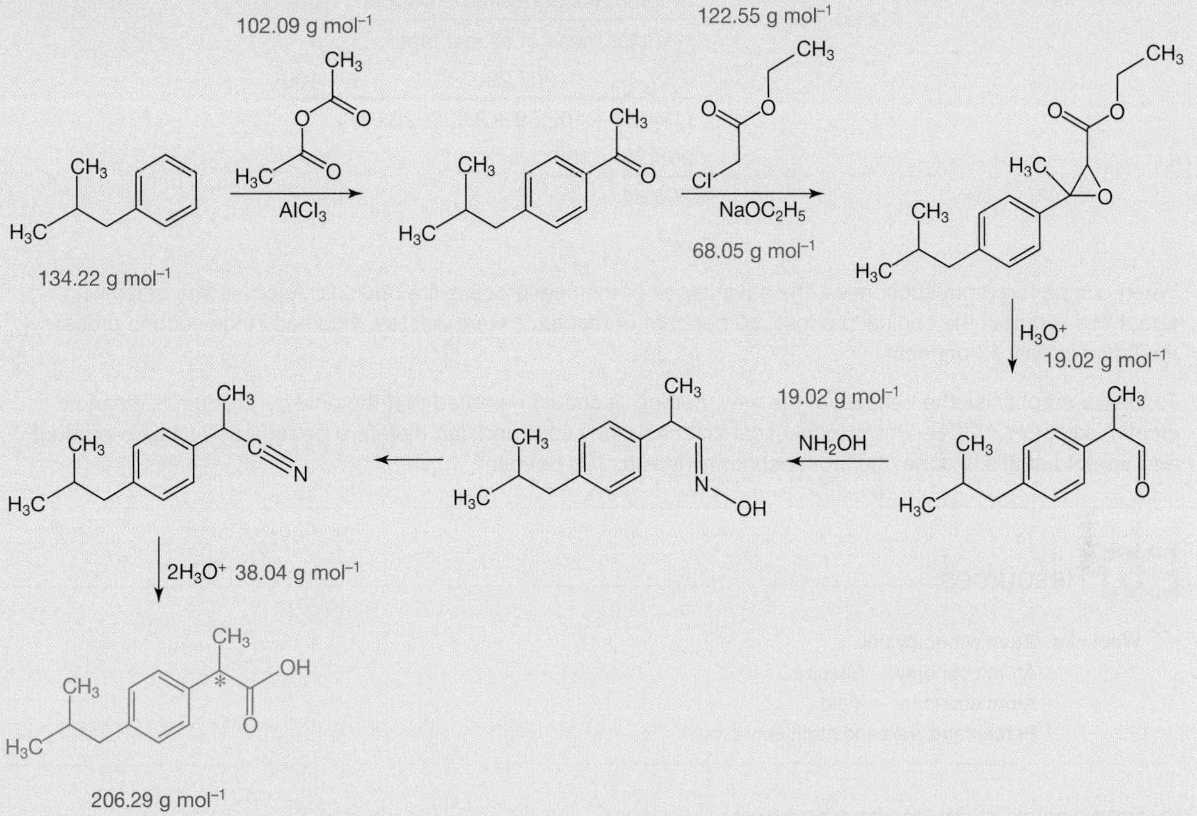

Note that this figure represents a theoretical maximum. In practice, an amount of ibuprofen may be lost due to various mechanical and sometimes chemical means, resulting in a lower figure.

The modern pathway for producing ibuprofen

In the mid-1980s, a consortium was formed between the Boots, Hoescht and Celanese companies (BHC) to research an alternative method for the manufacture of ibuprofen. This was commercialised in 1992 and is now the major pathway by which the drug is made. This pathway involves only three steps and produces much less waste than the original one. Figure 8.43 shows the new pathway.

FIGURE 8.43 The modern (BHC) pathway for producing ibuprofen

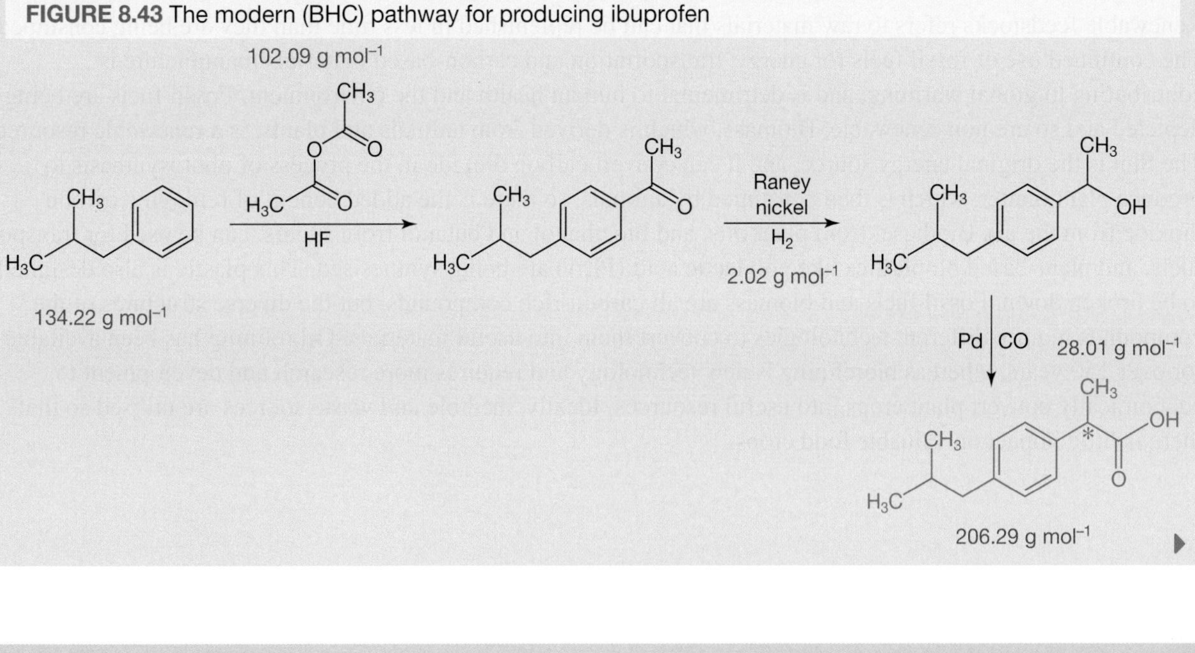

In a similar manner to that shown for the original pathway, the atom economy of this process can be calculated.

$$\% \text{ atom economy} = \frac{\text{molar mass of desired products}}{\text{molar mass of all reactants}} \times \frac{100}{1}$$

$$= \frac{206.29}{(134.22 + 102.09 + 2.02 + 28.01)} \times \frac{100}{1}$$

$$= \frac{206.29}{266.34} \times \frac{100}{1}$$

$$= 77\%$$

When comparing atom economies, the advantages of the new process are obvious. Another way of thinking about this is that in the original process, 60 per cent of reactants were wasted, whereas in the second process, the figure is only 23 per cent.

To further emphasise the benefits of the new method, it should be noted that the only by-product is ethanoic (acetic) acid, CH_3COOH. This is a chemical that has many uses and can therefore be regarded as a co-product and on-sold. If this is done, the atom economy rises to 100 per cent.

 Resources

🔗 **Weblinks** Atom economy quiz
Atom economy — Bitesize
Atom economy — Yield
Percentage yield and atom economy

Advantages of high atom economy

As mentioned earlier, developing reactions and processes that have a high atom economy is one of the principles of green chemistry. Developing processes with a high atom economy has many benefits for industry and society in general. The most obvious is the production of less waste, with immediate implications for the economics of the process and its environmental effects. A high atom economy, however, is only part of the green chemistry approach. Atom economy, for example, does not consider the cost or availability of raw materials, the toxicity of the chemicals used and produced, or the energy required.

Using renewable feedstocks

Renewable feedstocks refers to raw materials that can be replenished in less time than they are being consumed. The continued use of fossil fuels for energy, transportation and carbon-based materials manufacture is contributing to global warming, and is detrimental to human health and the environment. Fossil fuels are being depleted and so are non-renewable. Biomass, which is derived from animals and plants, is a renewable resource. The Sun is the original energy source, and it can convert carbon dioxide in the process of photosynthesis to produce plant matter, which is then consumed by animals, so there is the added benefit of removing carbon dioxide from the air. Biodiesel from plant oils, and bioethanol and butanol from sugars, can be used for transport fuels, and plant-based bioplastics like polylactic acid (PLA) are being synthesised. This plastic is also designed to be broken down. Fossil fuels and biomass are all carbon-rich compounds, but the diverse structures of the compounds require different technologies to convert them into useful materials. Oil refining has been available for over 150 years, whereas biorefining is new technology and requires more research and development to economically convert plant crops into useful resources. Ideally, inedible and waste sources are utilised so that there is little impact on valuable food crops.

Choosing catalysts

Using catalysts to increase the rate of reactions can enable processes to be carried out in milder conditions of temperature and pressure, thus conserving energy. In addition, only small amounts of catalyst are usually required, and they are not consumed in the reaction. These points are in accord with green chemistry principles, but not all catalysts are 'green'. A problem arises with the use of heavy metals as catalysts, first because they are being depleted, but also because they are frequently hazardous to human health and the environment. Chemists are investigating the use of more benign and less toxic substances — for example, the more plentiful iron or aluminium catalysts, or clays and zeolites.

FIGURE 8.44 Zeolite catalyst crystals, coloured scanning electron micrograph (SEM). Zeolites are hydrated aluminosilicate minerals and have a microporous structure. This zeolite, boron-beta-zeolite, is used in the petrochemical industry as a catalyst for the preparation of amines.

Heterogeneous catalysts — that is, those that are in a different state than the reactants and products — are preferable because they can be relatively easily separated from the products of the reaction, therefore reducing the number of steps in the process. An example of using catalysts (catalytic process) instead of adding chemical reagents (stoichiometric process) is the hydrogenation of ketones. Generally, hydrogen will not react with ketones under normal conditions. If sodium borohydride is added to the ketone followed by the addition of water (see figure 8.45), the waste products are borane and sodium hydroxide. If palladium on carbon is used as a catalyst for the same reaction, there is no waste. Another possibility is that scientists may need to look at new reactions instead of trying to find catalysts for old reactions.

FIGURE 8.45 Stoichiometric versus catalytic hydrogenation of ketones

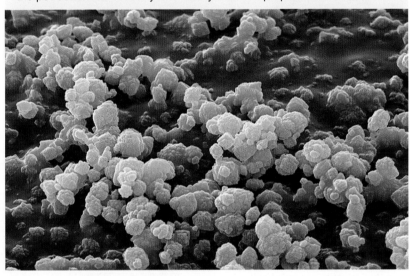

Stoichiometric

4 [ketone] $+ \ NaBH_4 \ + \ 4H_2O \longrightarrow 4$ [alcohol] $+ \ H_3BO_3 \ + \ NaOH$

81% atom economy

Catalytic

[ketone] $+ \ H_2 \xrightarrow{\text{Pd-on-C catalyst}}$ [alcohol]

100% atom economy

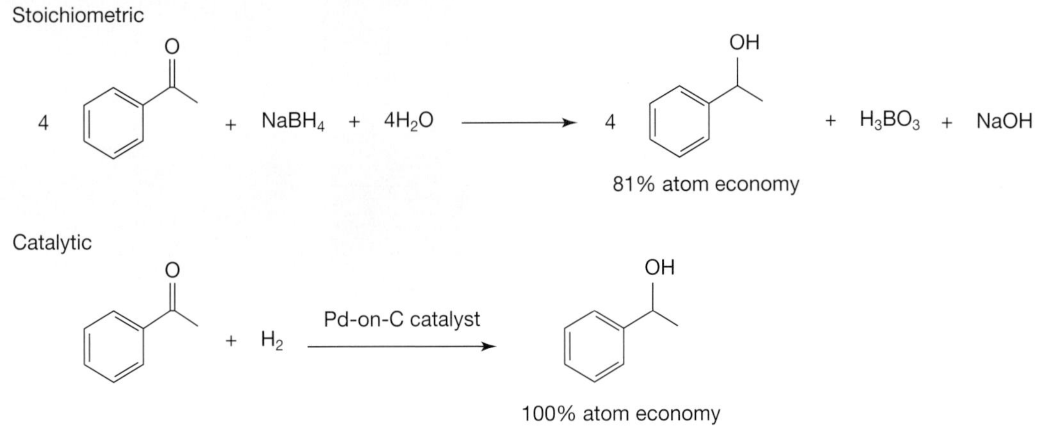

Another innovation is the use of microorganisms or biocatalysts (enzymes) to accelerate reactions. Biocatalysts require milder conditions, usually produce less waste, are less hazardous, use fewer steps and conserve energy compared with synthetic catalysts. Catalysts are highly selective and need to be designed for specific processes, and reaction pathways can be more accurately controlled, reducing undesired side reactions. Water is usually the solvent, which is not hazardous but limits the processes that biocatalysts can be used for unless other chemicals are added. Another disadvantage of using biocatalysts is that they are usually only effective for specific temperatures, pH and the presence of other chemicals. An example is bioleaching, which uses microorganisms from mining environments to separate metals from ores (refer to topic 3, *Jacaranda Chemistry 1 VCE Units 1 & 2 Third Edition*). Further investigation is being undertaken into the use of biocatalysts in chemical production.

Catalysts

Catalysts are useful to minimise energy use, but environmental issues need to be considered — for example, conservation of resources and disposal of waste materials. The use of biocatalysts can have less impact on the environment.

Designing safer chemicals

It is difficult to imagine that consideration of the toxicity of chemicals was not of high priority in past chemical production. Green chemistry rightly insists that chemicals should be designed with minimal toxicity without diminishing their effectiveness. Removing hazardous substances from chemical processes is critical for the health of workers and for preventing damage to the environment. Chemists need to improve their knowledge of toxicity so that it is considered at the beginning of molecular design. It is essential to find which part of a molecular structure is effective and which part is toxic so that chemists can design less hazardous chemicals. These new chemicals must also be economical to produce if they are to be acceptable to chemical industries. They must not be a physical hazard, explosive or flammable, persist in the environment nor accumulate in the food web. It may be necessary to find completely different raw materials to replace hazardous chemicals. An example is the development of new firefighting foam to replace fluorosurfactants that are hazardous to human health and detrimental to the environment.

Chemists need to design processes in which the chemicals used and the products formed are not hazardous to organisms or the environment.

FIGURE 8.46 A worker opening a barrel of toxic waste. The elimination of hazardous waste is one of the important principles of green chemistry and central to the idea of a circular economy.

8.5 Activities

8.5 Quick quiz on	8.5 Exercise	8.5 Exam questions

8.5 Exercise

1. Calculate the percentage yield of a reaction that produces 34.5 tonnes of product out of a theoretical maximum of 40.0 tonnes.
2. Hydrogen can be made by reacting methane with steam.

$$CH_4(g) + H_2O(g) \rightarrow 3H_2(g) + CO(g)$$

   a. Calculate the % atom economy for making hydrogen using this process.
   b. If a use for all of the carbon monoxide produced is found, what would the % atom economy now be? Explain.
3. The production of quicklime (CaO) takes place when calcium carbonate is heated to a high temperature. Carbon dioxide is also produced. If this occurs in an open system so that the carbon dioxide can escape, the reaction essentially goes to completion.

   If 100.1 g of calcium carbonate produces 50.3 g of quicklime, calculate the percentage yield. The equation for this reaction is:

$$CaCO_3(s) \rightarrow CaO(s) + CO_2(g)$$

4. The complete combustion of 82.2 g of propane produces a 73.2% yield. How many grams of CO_2 would be produced?
5. **MC** Which of the following reactions produces the lowest % atom economy of $BaCl_2$ ($M_r = 208.3$)?
   A. $BaO + 2HCl \rightarrow BaCl_2 + H_2O$
   B. $BaCO_3 + 2HCl \rightarrow BaCl_2 + CO_2 + H_2O$
   C. $Ba(OH)_2 + 2HCl \rightarrow BaCl_2 + 2H_2O$
   D. $Ba + 2HCl \rightarrow BaCl_2 + H_2$
6. Two methods of producing hydrogen are shown below. Which has the higher % atom economy?

$$\text{Method 1: } CH_4(g) + H_2O(g) \rightarrow CO(g) + 3H_2(g)$$
$$\text{Method 2: } C(s) + H_2O(g) \rightarrow CO(g) + H_2(g)$$

7. **MC** Ammonia is a colourless gas with a pungent smell. It is produced industrially by reacting nitrogen and hydrogen using a catalyst at high temperature and pressure. It is an equilibrium reaction that favours the reverse reaction.

$$N_2(g) + 3H_2(g) \rightleftharpoons 2NH_3(g)$$

   Which of the following is correct regarding this reaction?
   A. The atom economy is low and the percentage yield is high.
   B. The atom economy is low and the percentage yield is low.
   C. The atom economy is high and the percentage yield is high.
   D. The atom economy is high and the percentage yield is low.
8. You have been asked to manufacture a new product, 'Superclean', to use for washing cars. Describe the factors that will need to be considered to design:
   a. the chemicals
   b. the process.

9. Does it matter whether you use the total mass of the reactants or the total mass of the products when calculating the atom economy of a reaction?
10. Catalysts are important for green chemistry because they can speed up reactions without the need for high temperature. However, some catalysts are 'greener' than others. Describe why some catalysts are in accord with green chemistry principles and others are less so.

8.5 Exam questions

▶ Question 1 (1 mark)

Source: VCE 2018 Chemistry NHT Exam, Section A, Q.13; © VCAA

MC Which one of the following reactions has the lowest percentage atom economy for the production of ethanol, C_2H_5OH?

A. $C_2H_4(aq) + H_2O(l) \rightarrow C_2H_5OH(aq)$
B. $C_6H_{12}O_6(aq) \rightarrow 2C_2H_5OH(aq) + 2CO_2(g)$
C. $C_2H_5Cl(aq) + NaOH(aq) \rightarrow C_2H_5OH(aq) + NaCl(aq)$
D. $C_2H_5NH_2(aq) + HNO_2(aq) \rightarrow C_2H_5OH(aq) + H_2O(l) + N_2(g)$

▶ Question 2 (3 marks)

Source: Adapted from VCE 2009 Chemistry Exam 1, Section B, Q.2.ai,iii,iv; © VCAA

A sample of aspirin was prepared by reacting 2.20 g of salicylic acid with 4.20 mL of ethanoic anhydride in a conical flask. After heating for 20 minutes the reaction mixture was cooled and white crystals precipitated. The crystals were then collected, dried to constant mass and weighed.

The equation for the reaction is

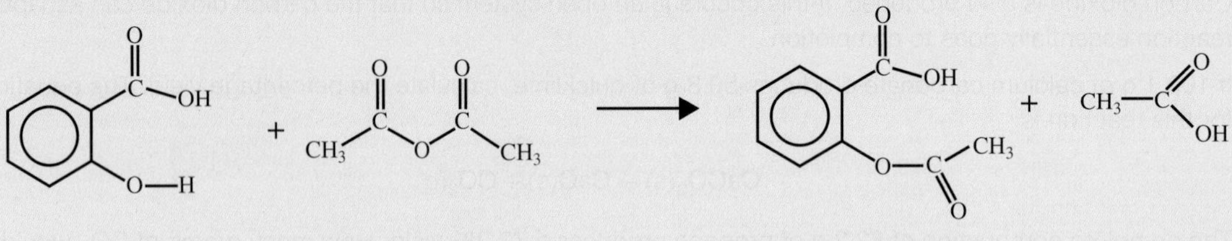

salicylic acid (s)　　　　ethanoic anhydride (l)　　　　　　　　aspirin (s)

The following results were obtained.

mass of salicylic acid	2.20 g
volume ethanoic anhydride	4.20 mL
mass product	2.25 g

Use the following data to answer the questions below.

	molar mass (g mol^{-1})
aspirin	180
ethanoic anhydride	102
salicylic acid	138

a. Calculate the initial amount, in moles, of salicylic acid used in this preparation. **(1 mark)**
b. Given that salicylic acid is the limiting reagent, what is the maximum mass of aspirin that can theoretically be produced? **(1 mark)**
c. Determine the percentage yield in this preparation. **(1 mark)**

▶ **Question 3 (1 mark)**

MC Which of the following reaction pathways used to produce ethanamine has the greatest percentage atom economy?

A. $C_2H_4 + H_2O \rightarrow C_2H_5OH$

  $C_2H_5OH + NH_3 \rightarrow C_2H_5NH_2 + H_2O$

B. $C_2H_5Cl + NH_3 \rightarrow C_2H_5NH_2 + HCl$

C. $C_2H_6 + Cl_2 \rightarrow C_2H_5Cl + HCl$

  $C_2H_5Cl + NH_3 \rightarrow C_2H_5NH_2 + HCl$

D. $C_2H_5Cl + NaOH \rightarrow C_2H_5OH + NaCl$

  $C_2H_5OH + NH_3 \rightarrow C_2H_5NH_2 + H_2O$

▶ **Question 4 (14 marks)**

There are two main ways of producing ethanol: hydration and fermentation. The process of hydration occurs by reacting ethene from crude oil with steam and a phosphoric acid catalyst at 300 °C to produce pure ethanol. Fermentation is a slow, exothermic reaction that involves the use of yeast to break down glucose to produce a low concentration of ethanol. The fermentation is then followed by distillation to separate the ethanol.

$$\text{Equation 1: } H_2C\text{=}CH_2(g) + H_2O(g) \overset{H_3PO_4}{\rightleftharpoons} CH_3CH_2OH(g)$$

$$\text{Equation 2: } C_6H_{12}O_6(aq) \xrightarrow{\text{yeast}} 2C_2H_5OH(aq) + 2CO_2(g)$$

a. Calculate the atom economy for each reaction. **(2 marks)**

b. Complete the following table. **(8 marks)**

	Hydration	Fermentation
Advantages		
Disadvantages		

c. Is there a conflict between the atom economy and the equilibrium nature of the reaction? Explain your answer. **(2 marks)**

d. Which process would you recommend for the industrial production of ethanol and why? **(2 marks)**

Question 5 (10 marks)

Prop-2-en-1-ol is a toxic liquid that was formerly used as a herbicide. In chemical processes, it can be used to manufacture glycerol, flame-resistant materials and various polymers.

Two methods of production of this compound are shown by the following equations:

$$\text{Method 1: } CH_2\text{=}CHCH_2Cl + H_2O \rightarrow CH_2\text{=}CHCH_2OH + HCl$$

$$\text{Method 2: } CH_2\text{=}CHCH_3 + CH_3COOH + 1\frac{1}{2}O_2 \rightarrow CH_2\text{=}CHCH_2OCOCH_3 + H_2O$$

$$CH_2\text{=}CHCH_2OCOCH_3 + H_2O \rightarrow CH_2\text{=}CHCH_2OH + CH_3COOH$$

a. Using green chemistry principles, discuss the advantages and disadvantages of these two methods and suggest which method would be preferable. **(4 marks)**

b. Suggest what further information would be useful in making this decision. **(6 marks)**

More exam questions are available in your learnON title.

8.6 Review

8.6.1 Topic summary

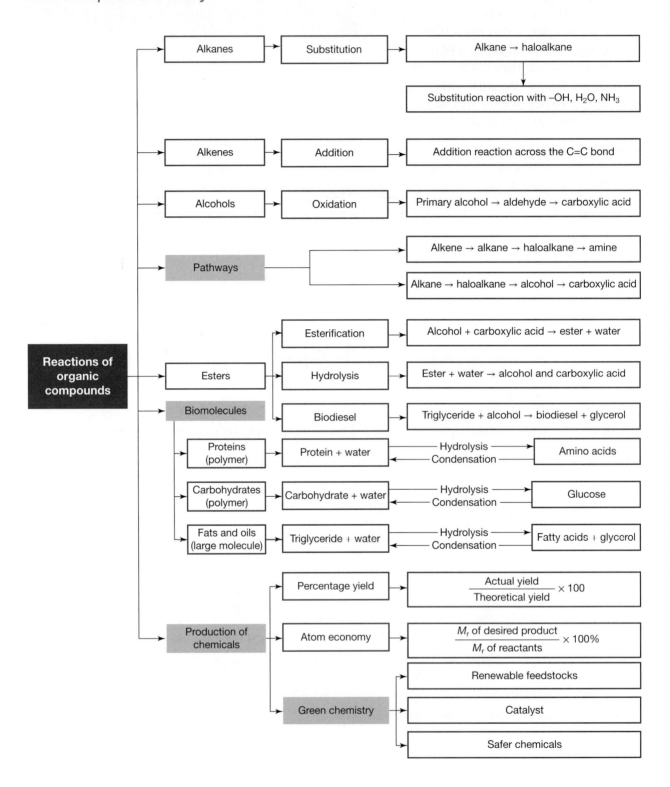

Hey students! Now that it's time to revise this topic, go online to:

Access the topic summary | Review your results | Watch teacher-led videos | Practise past VCAA exam questions

Find all this and MORE in jacPLUS

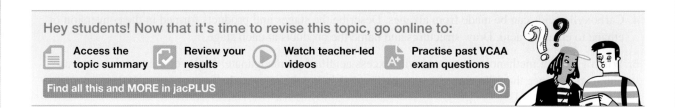

8.6.2 Key ideas summary

online only

8.6.3 Key terms glossary

online only

on Resources

Solutions	Solutions — Topic 8 (sol-0835)
Practical investigation eLogbook	Practical investigation eLogbook — Topic 8 (elog-1707)
Digital documents	Key science skills — VCE Chemistry Units 1–4 (doc-37066) Key terms glossary — Topic 8 (doc-37295) Key ideas summary — Topic 8 (doc-37296)
Exam question booklet	Exam question booklet — Topic 8 (eqb-0119)

8.6 Activities

learnon

Students, these questions are even better in jacPLUS

Receive immediate feedback and access sample responses | Access additional questions | Track your results and progress

Find all this and MORE in jacPLUS

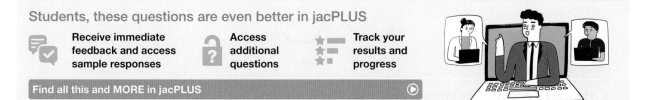

8.6 Review questions

1. Write equations for the reactions between:
   a. ethene and HI
   b. propene and H_2 (Pt catalyst)
   c. ethene and Br_2
   d. but-1-ene and Cl_2
   e. methane and excess O_2
   f. ethane and Cl_2
   g. ethene and H_2O
   h. but-2-ene and H_2.
2. Write the formulas for substances X, Y and Z shown in the following diagram.

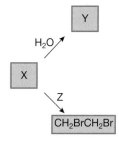

3. What type of reaction is the esterification process? What are the reactants and products?

4. Carboxylic acids can be made from alkanes. Describe the stages and products formed in the conversion of ethane to ethanoic acid. Draw structures and name the products at each stage.

5. When 11.5 g of methanol was treated with excess acidified permanganate, 13.2 g of methanoic acid was obtained. Balance the following equation by first balancing the relevant half-equations, and then calculate the percentage yield.

$$CH_3OH + MnO_4^- \rightarrow HCOOH + Mn^{2+}$$

6. 2-methylpropan-1-ol can be used to manufacture diesel and jet fuel. The first step in the process is the production of 2-methylpropene (C_4H_8).

$$\underset{\text{2-methylpropan-1-ol}}{C_4H_{10}O(l)} \quad \rightarrow \quad \underset{\text{2-methylpropene}}{C_4H_8(g)} \quad + \quad H_2O(g)$$

Calculate the % atom economy for this reaction.

7. Explain why the atom economy of the following reaction is 100%.

$$H_2C{=}CH_2 + H_2O \rightarrow CH_3CH_2OH$$

8. The disaccharide sucrose is common table sugar, obtained from sugarcane and added to a wide range of foods. Most of us eat more than the recommended level. The structure of sucrose is shown in the following figure.

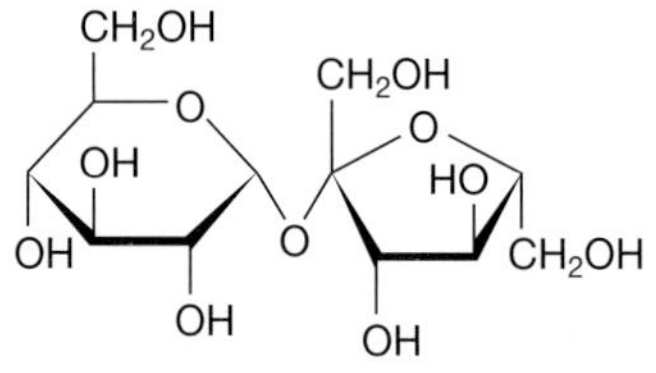

a. Draw the structure of the monosaccharide with the four-carbon ring.
b. What type of reaction is involved in the formation of sucrose?
c. Name the other molecule that is produced in this reaction.
d. Name the types of functional groups present in sucrose.
e. State the molecular formula of this disaccharide.

9. Write a formula equation showing the hydrolysis of the following peptide.

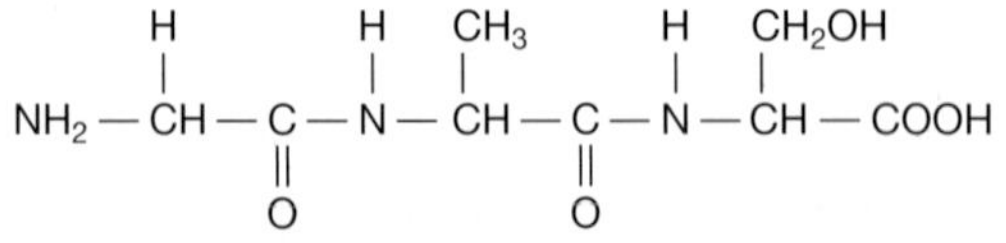

10. Draw a semi-structural equation showing the hydrolysis of the following triglyceride.

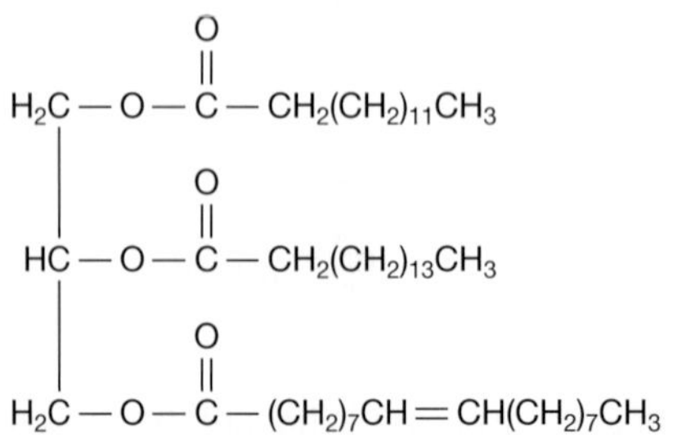

8.6 Exam questions

Section A — Multiple choice questions

All correct answers are worth 1 mark each; an incorrect answer is worth 0.

▶ Question 1

Source: VCE 2020 Chemistry Exam, Section A, Q.1; © VCAA

MC Glycogen breaks down into

A. glycerol.

B. amino acids.

C. triglycerides.

D. monosaccharides.

▶ Question 2

Source: VCE 2017 Chemistry Exam, Section A, Q.3; © VCAA

MC A hydrolytic reaction occurs when

A. a dipeptide is formed.

B. a triglyceride is formed.

C. water is a reaction product.

D. glucose is formed from maltose.

▶ Question 3

Source: VCE 2022 Chemistry Exam, Section A, Q.27; © VCAA

MC Which one of the following reactions has the highest atom economy in the production of an organic molecule?

A. complete combustion of propyne, C_3H_4

B. reaction of iodine, I_2, with propane, C_3H_8

C. reaction of bromine, Br_2, and propene, C_3H_6

D. formation of a dipeptide from alanine, $C_3H_7NO_2$

▶ Question 4

Source: VCE 2016 Chemistry Exam, Section A, Q.18; © VCAA

MC The molecule with the structural formula shown below reacts with hydrogen bromide, HBr, to form $C_5H_{11}Br$.

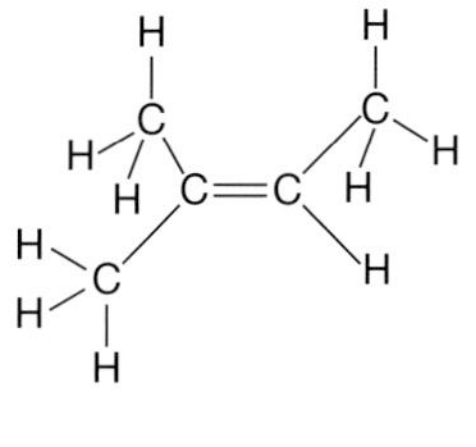

The number of different structural isomers theoretically possible to be produced by this reaction is

A. 1

B. 2

C. 3

D. 4

Question 5

Source: VCE 2016 Chemistry Exam, Section A, Q.12; © VCAA

MC A condensation reaction involving 200 glucose molecules, $C_6H_{12}O_6$, results in a polysaccharide.

The molar mass, in g mol^{-1}, of the polysaccharide is

A. 36 000
B. 35 982
C. 32 418
D. 32 400

Question 6

Source: VCE 2016 Chemistry Exam, Section A, Q.22; © VCAA

MC When ethene is mixed with chlorine in the presence of UV light, the following reaction takes place.

$$CH_2CH_2(g) + Cl_2(g) \xrightarrow{\text{UV light}} CH_2ClCH_2Cl(l)$$

Reactions of organic compounds can be classified in a number of ways. The following list shows four possible classifications:

1. addition
2. substitution
3. redox
4. condensation

Which classification(s) applies to the reaction between ethene and chlorine?

A. 1
B. 1 and 2
C. 1 and 3
D. 4

Question 7

Source: VCE 2014 Chemistry Exam, Section A, Q.19; © VCAA

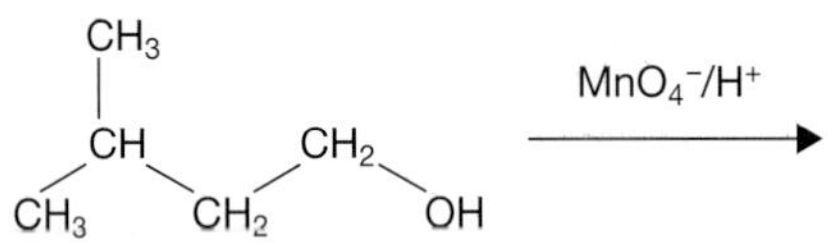

MC What is the systematic name for the product of the reaction above?

A. 2-methylpentanoic acid
B. 4-methylpentanoic acid
C. 2-methylbutanoic acid
D. 3-methylbutanoic acid

Question 8

Source: VCE 2022 Chemistry Exam, Section A, Q.10; © VCAA

MC The molar mass of glycerol, $C_3H_8O_3$, is 92.0 g mol^{-1}.

The production of 65.0 g of $C_3H_8O_3$ from tripalmitin, $C_{51}H_{98}O_6$, which is a triglyceride.

A. requires 12.7 g of water.
B. requires 38.2 g of water.
C. produces 12.7 g of water.
D. produces 38.2 g of water.

Question 9

Source: VCE 2009 Chemistry Exam 1, Section A, Q.18; © VCAA

MC A product derived from palm tree oil is used as an alternative fuel in diesel engines.

Palm oil is converted to biodiesel by the following reaction.

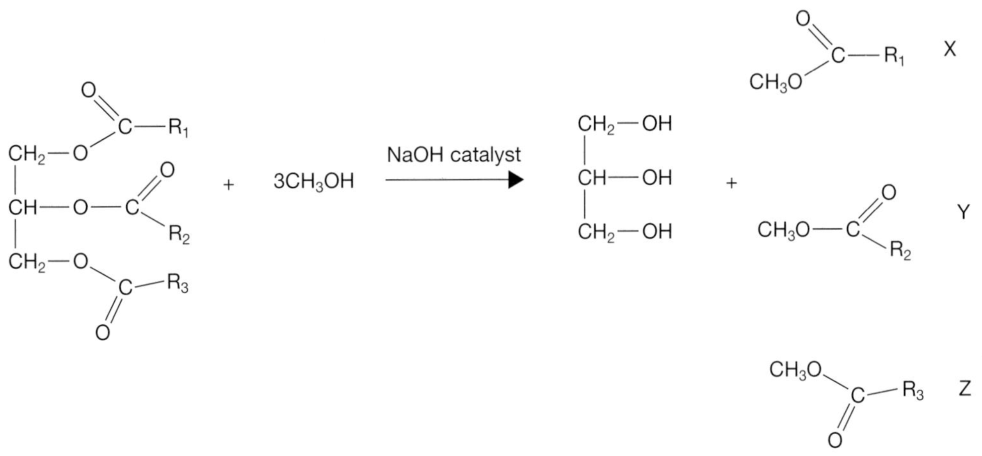

Glycerol is separated from the reaction mixture. The mixture of the compounds labelled X, Y and Z is used as palm oil biodiesel.

The term that best describes the mixture of compounds in palm oil biodiesel is

A. methyl esters.
B. carbohydrates.
C. carboxylic acids.
D. monoglycerides.

Question 10

Source: VCE 2008 Chemistry Exam 1, Section A, Q.18; © VCAA

MC Starch consists mainly of amylose, which is a polymer made from glucose, $C_6H_{12}O_6$. A particular form of amylose has a molar mass 3.62×10^5 g mol^{-1}.

A molecule of this amylose can be described as

A. an addition polymer of 2235 glucose molecules.
B. an addition polymer of 2011 glucose molecules.
C. a condensation polymer of 2235 glucose molecules.
D. a condensation polymer of 2011 glucose molecules.

▶ Question 11 (7 marks)

Source: VCE 2022 Chemistry Exam, Section B, Q.1; © VCAA

A reaction pathway to produce a primary alcohol is shown below.

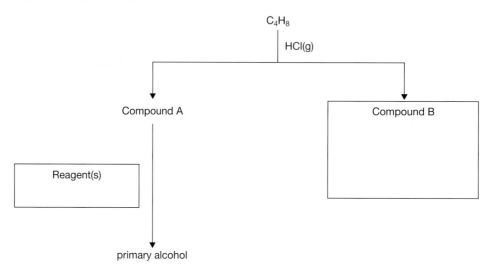

C_4H_8 reacts with HCl(g) to form two unbranched isomers — Compound A and Compound B. Only Compound A can react to produce a primary alcohol.

a. Identify the type of reaction that converts C_4H_8 into Compound A. **(1 mark)**

b. Write the semi-structural formula for Compound B in the box provided. **(1 mark)**

c. State the reagent(s) required to convert Compound A into a primary alcohol in the box provided. **(1 mark)**

d. Propan-1-ol can react with methanoic acid to produce an organic molecule.

    i. Identify the catalyst for this reaction. **(1 mark)**

    ii. Write a balanced chemical equation for the reaction. **(2 marks)**

    iii. Write the systematic IUPAC name for the organic molecule produced. **(1 mark)**

▶ Question 12 (5 marks)

Source: VCE 2019 Chemistry Exam, Section B, Q.2.b; © VCAA

2-chloropropane can be reacted with ammonia to produce an uncharged organic molecule, Compound R.

a. Write the equation for the reaction that occurs. **(1 mark)**

b. Give the IUPAC name of Compound R. **(1 mark)**

c. Name the type of reaction that produces Compound R. **(1 mark)**

d. Calculate the percentage atom economy for the production of Compound R. **(2 marks)**

Source: VCE 2021 Chemistry Exam, Section B, Q.6.b–d; © VCAA

A reaction pathway beginning with 1-bromopentane is shown below.

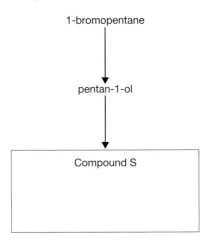

a. i. Write a balanced equation for the reaction that will produce pentan-1-ol from 1-bromopentane and a
        sodium salt.   **(2 marks)**
    ii. Calculate the atom economy in the production of pentan-1-ol from 1-bromopentane and a sodium
        salt.   **(3 marks)**
b. Pentan-1-ol is fully oxidised to Compound S.
    Write the IUPAC name of Compound S.   **(1 mark)**
c. In an alternative reaction pathway, pentanamine can be formed from 1-bromopentane.
    Draw the skeletal formula for pentanamine.   **(1 mark)**

Source: Adapted from VCE 2019 Chemistry Exam, Section B, Q.1.a,b,d; © VCAA

A commercial chocolate spread is commonly used in sandwiches and desserts. This food contains high amounts
of proteins, triglycerides and sucrose.

Proteins are an important part of food. Proteins are broken down into smaller molecules during digestion.

a. Proteins can be hydrolysed to produce alpha (α-) amino acids.

    Identify **one** structural feature common to all alpha (α-) amino acids.   **(1 mark)**
b. i. What is the process by which amino acids are obtained from the chocolate spread?   **(1 mark)**
    ii. Identify the chemical process in which amino acids are predominantly used in the body.   **(1 mark)**
    iii. Two of the amino acids in the chocolate spread are aspartic acid and cysteine.
        Draw the chemical structure of the dipeptide Cys-Asp.   **(2 marks)**
c. Draw the structure of a triglyceride that contains only palmitoleic acid using semi-structural formulas.
    Circle and label the triglyceride functional group.   **(2 marks)**

Question 15 (4 marks)

Source: Adapted from VCE 2020 Chemistry Exam, Section B, Q.3.a–d; © VCAA

Below is a reaction pathway beginning with hex-3-ene.

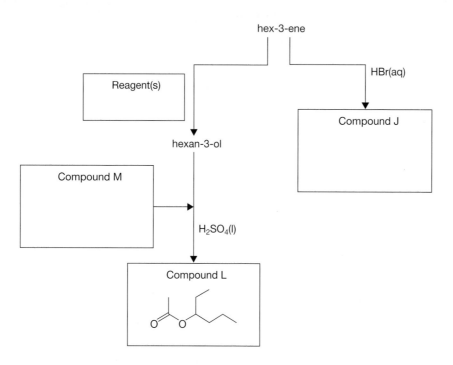

a. Write the IUPAC name of Compound J in the box provided. **(1 mark)**
b. State the reagent(s) required to convert hex-3-ene to hexan-3-ol in the box provided. **(1 mark)**
c. Draw the structural formula for a tertiary alcohol that is an isomer of hexan-3-ol. **(1 mark)**
d. Hexan-3-ol is reacted with Compound M under acidic conditions to produce Compound L.
   Draw the semi-structural formula for Compound M in the box provided. **(1 mark)**

AREA OF STUDY 1 How are organic compounds categorised and synthesised?

OUTCOME 1

Analyse the general structures and reactions of the major organic families of compounds, design reaction pathways for organic synthesis, and evaluate the sustainability of the manufacture of organic compounds used in society.

PRACTICE EXAMINATION

STRUCTURE OF PRACTICE EXAMINATION		
Section	Number of questions	Number of marks
A	20	20
B	4	30
Total		50

Duration: 50 minutes

Information:

- This practice examination consists of two parts. You must answer all question sections.
- Pens, pencils, highlighters, erasers, rulers and a scientific calculator are permitted.
- You may use the VCE Chemistry Data Book for this task.

  Resources

🔗 **Weblink** VCE Chemistry Data Book

SECTION A — Multiple choice questions

All correct answers are worth 1 mark each; an incorrect answer is worth 0.

1. The formulas of four organic compounds are as follows:
   $CH_2CHCHCH_2$
   $CH_2CHCH_2CH_3$
   $CH_3CHCHCH_3$
   CH_3CHCH_3
   Which of the following is demonstrated within these formulas?
   A. Unsaturation only
   B. Isomerism only
   C. Unsaturation and isomerism
   D. Saturation, unsaturation and isomerism

2. What is the correct semi-structural formula for the following molecule?

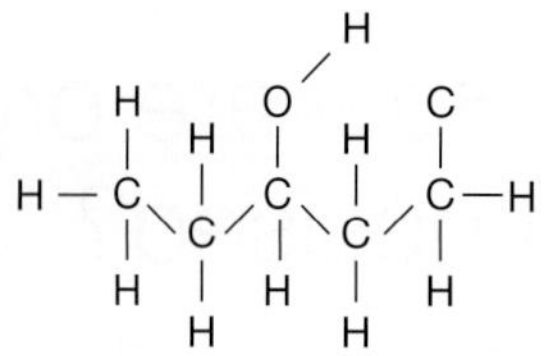

A. $CH_3CH_2C(HOH)CH_2CH_3$
B. $CH_3CH_2CH(OH)CH_2CH_3$
C. $CH_3CH_2HCOHCH_2CH_3$
D. $CH_3CH_2HC(OH)CH_2CH_3$

3. Which of the following pairs are isomers?
A. CH_3CH_2OH, CH_3OCH_3
B. CH_3CH_2OH, $CH_3CH_2CH_2OH$
C. CH_3CH_2OH, CH_3COOH
D. CH_3CH_2OH, CH_3CHO

4. Which of the following is a tertiary alcohol?
A. Pentan-3-ol
B. Pent-2-en-1-ol
C. Propan-2-ol
D. 2-methylpropan-2-ol

5. A hydrocarbon has the formula C_6H_{12}.
Which of the following statements is *not* true of the hydrocarbon?
A. It could be cyclohexane.
B. It has the empirical formula CH_2.
C. It will burn in a plentiful supply of oxygen to produce carbon dioxide and water.
D. It is definitely an alkene.

6.

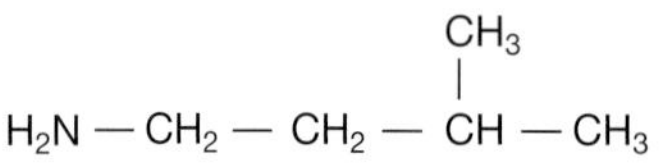

The compound shown is
A. an amino acid.
B. an amide.
C. an amine.
D. an ammonium salt.

7. What is the name of the following molecule?

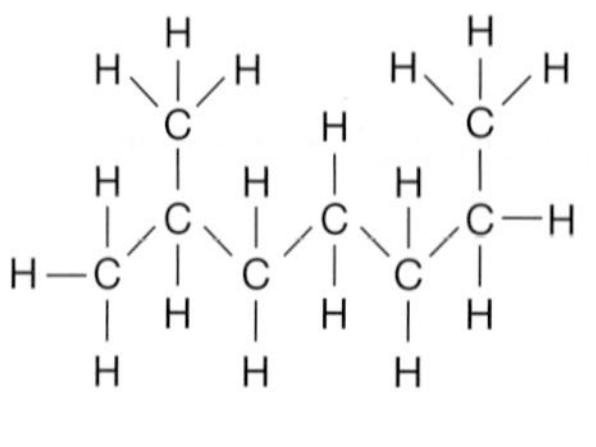

A. 1,6-dimethylhexane
B. 2,6-dimethylhexane
C. 2-methylheptane
D. 6-methylheptane

8. What is the correct semi-structural formula of 2,4-dimethylpentane?

   A. $CH_3CH(CH_3)CH_2CH_2CH(CH_3)CH_3$
   B. $CH_2(CH_3)CH_2CH(CH_3)CH_3$
   C. $CH_3CH(CH_3)CH_2CH(CH_3)CH_3$
   D. $CH_3CH(CH_3)CHCH(CH_3)CH_3$

9. What is the correct name of the molecule with the semi-structural formula $CH_3CH_2CH_2CHO$?

   A. Butan-1-one
   B. Butanoic acid
   C. Butan-1-ol
   D. Butanal

10. What is the correct name for the following molecule?

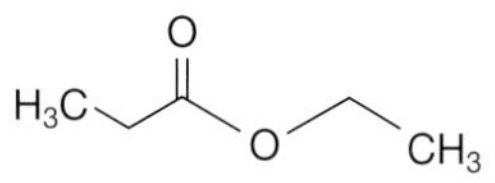

   A. Methyl methanoate
   B. Methyl ethanoate
   C. Ethyl ethanoate
   D. Ethyl propanoate

11. A particular organic compound is represented by the following skeletal structure:

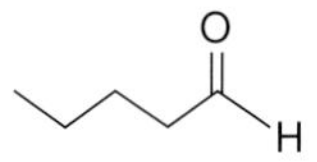

   The name of this compound is
   A. butanal.
   B. pentan-1-ol.
   C. pentanoic acid.
   D. pentanal.

12. The following table shows the boiling points of the first six members of the alcohol homologous series (primary alcohols).

Primary alcohol	Boiling point (°C)
Methanol	65
Ethanol	79
Propan-1-ol	97
Butan-1-ol	117
Pentan-1-ol	138
Hexan-1-ol	157

   The trend in boiling points from methanol to hexan-1-ol is best explained by the increasing strength of
   A. dispersion forces.
   B. covalent bonding forces.
   C. hydrogen bonding forces.
   D. hydroxyl bonding forces.

13.

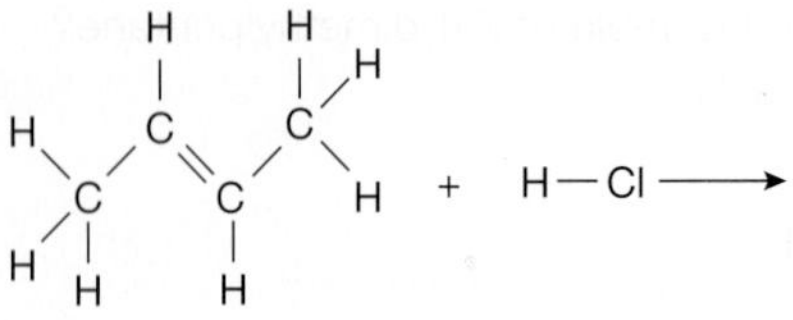

For the reaction shown, the reaction type and products are respectively

A. addition, 2,3-dichlorobutane.

B. addition, 2-chlorobutane.

C. substitution, 2-chlorobutane.

D. substitution, 2,3-dichlorobutane.

14. Two compounds with formulas CH_3CH_2OH and HCOOH react together under suitable conditions to form a new compound or compounds.
The name(s) of the new compound(s) is/are

A. methyl ethanoate.

B. ethyl methanoate.

C. methyl ethanoate and water.

D. ethyl methanoate and water.

15. Which of the following is true of the formation of one molecule of fat?

A. One water molecule is produced in the reaction.

B. The fat produced is always a solid.

C. The reaction is between one fatty acid molecule and one glycerol molecule.

D. Ester linkages form between the hydroxyl groups and carboxylic acid groups.

16. Starch and glycogen are large and biologically important molecules.
These are synthesised from

A. glucose molecules in a process that removes water.

B. glucose molecules in a process that adds water.

C. glycerol molecules in a process that removes water.

D. glycerol molecules in a process that adds water.

17. A protein chain is made up of 146 amino acids.
What mass of water is released when 1.00×10^{-3} mol of this protein is formed?

A. 0.018 g

B. 0.146 g

C. 2.61 g

D. 2.63 g

18. Chloroethane can be prepared from ethane according to the following equation:

$$CH_3CH_3 + Cl_2 \rightarrow CH_3CH_2Cl + HCl$$

In a particular experiment it was predicted that the maximum possible amount of chloroethane would be 9.4 g. However, only 1.4 g was obtained.
The percentage yield and the atom economy of this preparation, respectively, are

A. 14% and 100%.

B. 15% and 64%.

C. 32% and 64%.

D. 32% and 100%.

19. The molar mass of a starch molecule made from 5000 glucose units is

A. 9.90×10^5 g mol^{-1}

B. 9.00×10^5 g mol^{-1}

C. 8.10×10^5 g mol^{-1}

D. 4.50×10^5 g mol^{-1}

20. Short-chain alcohols such as methanol and ethanol are important in the production of biodiesel. The table shows information about how these alcohols may be produced.

Method	Alcohol	Details
I	Methanol	Using catalysts and low to moderate temperatures to convert methane from natural gas into methanol
II	Methanol	Capture of carbon dioxide from current processes or removal of carbon dioxide from the atmosphere, followed by reaction with water using suitable catalysts
III	Ethanol	Addition reaction of water to ethene, using steam and performed at moderate industrial temperatures. Ethene is obtained from the refining of crude oil.
IV	Ethanol	Fermentation of glucose derived from plant feedstocks followed by distillation

The two most sustainable methods are

A. I and III.

B. I and IV.

C. II and III.

D. II and IV.

SECTION B — Short answer questions

Question 21 (6 marks)

The incredible number and variety of carbon compounds can be explained by some characteristics of the carbon atom. Two of these are:

- its ability to form multiple carbon-to-carbon bonds
- isomerism.

a. How many valence electrons are present in a carbon atom? **(1 mark)**

b. Consider a hydrocarbon that contains five carbon atoms.

   i. Draw structural formulas for two possible saturated isomers. **(2 marks)**

   ii. Draw a structural formula for an unsaturated molecule with a degree of unsaturation of one. **(1 mark)**

   iii. Draw a structural formula for an unsaturated molecule with a degree of unsaturation of two. **(1 mark)**

c. Besides the two reasons mentioned, give one other reason for carbon's ability to form so many different compounds. **(1 mark)**

Question 22 (5 marks)

Biodiesel can be produced by reacting plant-based oils with methanol. One such oil is canola oil. A typical biodiesel molecule derived from canola oil has the chemical formula $C_{16}H_{32}O_2$.

a. What is the name of the process used to convert canola oil to biodiesel? **(1 mark)**

b. Write the semi-structural formula for this molecule and then circle and name the functional group that it contains. **(2 marks)**

c. This process produces a useful by-product. Name this by-product. **(1 mark)**

d. Give one disadvantage for using canola oil to make biodiesel. **(1 mark)**

Question 23 (11 marks)

Consider the following reaction pathway.

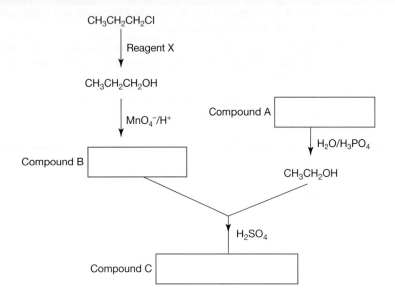

CH$_3$CH$_2$CH$_2$Cl

| Reagent X

CH$_3$CH$_2$CH$_2$OH

| MnO$_4^-$/H$^+$

Compound B

Compound A

| H$_2$O/H$_3$PO$_4$

CH$_3$CH$_2$OH

H$_2$SO$_4$

Compound C

a. **i.** Identify reagent X. (1 mark)
    ii. Identify the type of reaction occurring to convert CH$_3$CH$_2$CH$_2$Cl into CH$_3$CH$_2$CH$_2$OH. (1 mark)
    iii. Identify the by-product of the reaction forming CH$_3$CH$_2$CH$_2$OH. (1 mark)
    iv. Calculate the atom economy for this reaction. (1 mark)
b. **i.** Write the semi-structural formula of compound A. (1 mark)
    ii. Identify the type of reaction occurring to convert compound A into CH$_3$CH$_2$OH. (1 mark)
c. **i.** Name the type of reaction that produces compound B. (1 mark)
    ii. Name compound B. (1 mark)
d. **i.** Draw the skeletal formula of compound C. (1 mark)
    ii. Calculate the atom economy for the production of compound C. (1 mark)
    iii. Draw the structural formula of the other product formed when compound C is formed. (1 mark)

Question 24 (8 marks)

Butanone is an organic compound that can be prepared from the alcohol butan-2-ol.

a. Is butan-2-ol a primary, secondary or tertiary alcohol? (1 mark)
b. **i.** What type of organic compound is butanone? (1 mark)
    ii. Draw the structural formula for butanone. (1 mark)
c. **i.** What is the type of reaction that converts butan-2-ol into butanone? (1 mark)
    ii. Give an example of the other reactant that must be present for this to occur. (1 mark)
d. If 18.6 g of butan-2-one is obtained from 22.5 g of butan-2-ol, calculate the percentage yield. (3 marks)

PRACTICE SCHOOL-ASSESSED COURSEWORK

ASSESSMENT TASK — PROBLEM-SOLVING IN REAL-WORLD CONTEXTS

For this task, you will utilise chemistry principles and techniques, including calculations, to solve problems within a real-world context.

- Pens, pencils, highlighters, erasers, rulers and a scientific calculator are permitted.
- You may use your practical logbook and the VCE Chemistry Data Book to complete this task.

Total time: 45 minutes + 5 minutes reading time
Total marks: 40 marks

SYNTHESISING FRAGRANCES AND FLAVOURS

Esters are utilised in perfumes and food for their aromatic properties. In perfumes, esters contribute to the overall scent composition by providing various fruity, floral or other desirable notes. They add depth, complexity and a unique character to fragrances.

In food, esters are used as flavouring agents to enhance and recreate natural flavours. They can imitate the taste and aroma of fruits, flowers and other food ingredients. Esters are particularly valuable in creating artificial fruit flavours, as they can mimic the specific scent and taste profiles of different fruits.

1. The following list presents different fruits and their corresponding esters that play a significant role in creating their distinct aromas. These esters can be found naturally or produced synthetically in the industry.

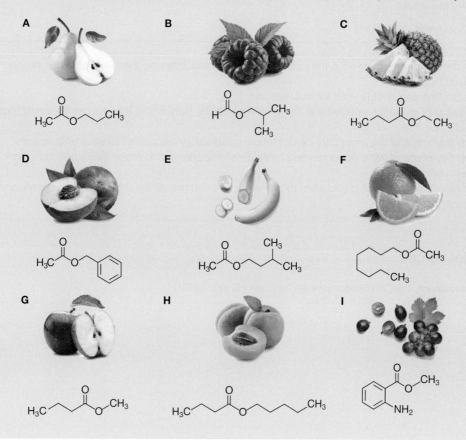

a. Give the IUPAC name for two of the esters shown. **(2 marks)**
b. Choose one ester and use structural formulas to show the pathway to synthesise it from an appropriate alkane and alkene. Your method should:
   i. include seven different structural formulas **(7 marks)**
   ii. show the reagents required for each reaction **(4 marks)**
   iii. name and label at least three different reaction types. **(3 marks)**
c. Suggest why artificial flavours and aromas are less expensive than natural ones. **(2 marks)**
d. Propyl butanoate is an ester with a fruity odour and taste that can be used to make flavoured ice-cream.
   i. Name the alcohol and the carboxylic acid from which propyl butanoate is made. **(2 marks)**
   ii. Write an equation to show how this acid can be made from a suitable alcohol. **(2 marks)**
   iii. Draw the full structural formula for the ester, and label the functional group. **(2 marks)**
2. The following experiment was undertaken as an investigation.

Procedure
   1. Set up the hotplate.
   2. Obtain about 4.1 g of carboxylic acid; add to the 100 mL beaker.
   3. Weigh out 2.5 g of alcohol into the beaker.
   4. Add five drops of sulfuric acid to the mixture.
   5. Heat for 10 minutes.
   6. Remove the beaker from the hotplate; waft the beaker.
   7. Allow to cool.

The teacher then took the beaker and gave the following instructions.
   8. Wash the crude ester product twice with 2 mL of 10% sodium carbonate solution and once with distilled water.
   9. Add sodium carbonate slowly to prevent the mixture from bubbling out due to the generation of CO_2.
   10. Remove as much of the lower aqueous product as possible using a plastic pipette.
   11. Dry the ester.

The final weight was 3.8 g.

---

a. What products are expected in the beaker at the end, and why was the lower aqueous product removed? **(2 marks)**
b. Calculate the percentage yield for this reaction. **(5 marks)**
c. One group of students reported that the yield was 100%. Why would the teacher be suspicious of such a result? **(1 mark)**
d. Which two steps of the procedure do not follow good safety protocols? Explain your answer. **(2 marks)**
e. Which two steps of the procedure need more details or changes to make the experiment reproducible? Explain your answer. **(4 marks)**
f. What laboratory test would you need to complete to confirm that the product was an ester, and that it was pure? **(2 marks)**

on Resources

   Digital document U4AOS1 School-assessed coursework (doc-39701)

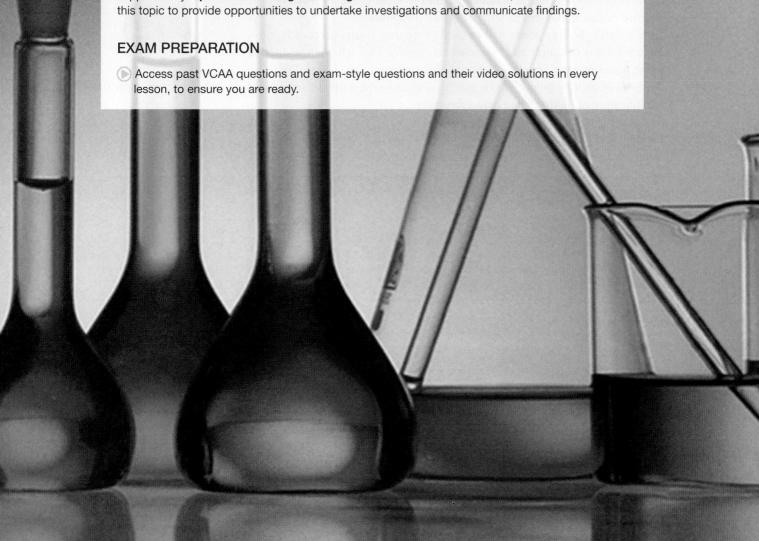

9 Laboratory analysis of organic compounds

KEY KNOWLEDGE

In this topic you will investigate:

Laboratory analysis of organic compounds

- qualitative tests for the presence of carbon-carbon double bonds, hydroxyl and carboxyl functional groups
- applications and principles of laboratory analysis techniques in verifying components and purity of consumer products, including melting point determination and distillation (simple and fractional)
- measurement of the degree of unsaturation of compounds using iodine
- volumetric analysis, including calculations of excess and limiting reactants using redox titrations (excluding back titrations).

Source: VCE Chemistry Study Design (2024–2027) extracts © VCAA; reproduced by permission.

PRACTICAL WORK AND INVESTIGATIONS

Practical work is a central component of VCE Chemistry. Experiments and investigations, supported by a **practical investigation eLogbook** and **teacher-led videos**, are included in this topic to provide opportunities to undertake investigations and communicate findings.

EXAM PREPARATION

▶ Access past VCAA questions and exam-style questions and their video solutions in every lesson, to ensure you are ready.

9.1 Overview

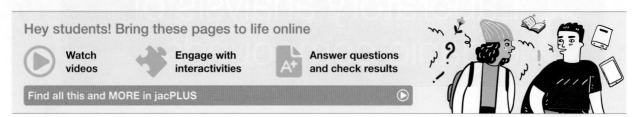
9.1.1 Introduction

In topic 8 we learned that the structure of an organic compound determines its properties. Functional groups are of particular importance, as they both underpin chemical reactivity as well as greatly affecting the formation of intermolecular forces and subsequent physical properties.

Differences in the chemical and physical properties of organic molecules may be utilised by a variety of laboratory techniques to perform both qualitative and quantitative analysis, as well as to separate mixtures.

This topic will draw on an understanding of intermolecular forces and organic reactions to introduce a suite of laboratory techniques for the analysis of organic compounds. These techniques may be applied in a range of contexts, including the analysis of consumer products.

- Qualitative tests will utilise organic reactions to identify the presence or absence of functional groups within organic compounds.
- Differences in boiling points will be used to separate components of liquid mixtures via both simple and fractional distillation. This will then be applied to the fractionation of crude oil into useful components.
- Analysis of melting points will be used to determine the purity of solid organic substances, and allow for identification of mixtures and pure substances.
- The number of carbon-to-carbon double bonds in organic molecules will be quantified by addition of iodine, I_2, to calculate iodine number.
- Volumetric analysis will be applied to redox reactions.

FIGURE 9.1 A redox titration

LEARNING SEQUENCE

---

on Resources

☑ **Solutions**	Solutions — Topic 9 (sol-0836)
Practical investigation eLogbook	Practical investigation eLogbook — Topic 9 (elog-1708)
📄 **Digital documents**	Key science skills — VCE Chemistry Units 1–4 (doc-37066) Key terms glossary — Topic 9 (doc-37297) Key ideas summary — Topic 9 (doc-37298)
Exam question booklet	Exam question booklet — Topic 9 (eqb-0120)

9.2 Tests for functional groups

KEY KNOWLEDGE

- Qualitative tests for the presence of carbon-carbon double bonds, hydroxyl and carboxyl functional groups
- Measurement of the degree of unsaturation of compounds using iodine

Families of organic compounds are classified based on shared **functional groups**. These atoms or groups of atoms undergo structural change during chemical reactions and largely determine the chemical properties of each family. They may also affect physical properties by contributing to polarity and determining the intermolecular forces acting between molecules.

Given the importance of functional groups, tests to detect their presence within organic molecules are of value. Tests for the following functional groups will be introduced:

- Carbon-to-carbon double bonds
- Hydroxyl groups
- Carboxyl groups.

9.2.1 Carbon-to-carbon double bonds

The carbon-to-carbon double bonds (C=C) present in alkenes and other unsaturated compounds allow such compounds to readily undergo addition reactions with a variety of substances, including halogens. By contrast, reaction of a halogen with an alkane or other saturated compound occurs via substitution and requires UV light or heat.

Bromine water test

Bromine water is an orange-brown aqueous solution of Br_2 that can be mixed with an organic compound to detect the presence of at least one C=C. Figure 9.2 illustrates how the bromine water test works.

functional group an atom or group of atoms that is attached to or part of a hydrocarbon chain, and influences the physical and chemical properties of the molecule

FIGURE 9.2 a. Decolouration of bromine water indicates the presence of at least one C=C. **b.** Saturated compounds do not undergo an addition reaction, so the orange-brown colour of the bromine persists.

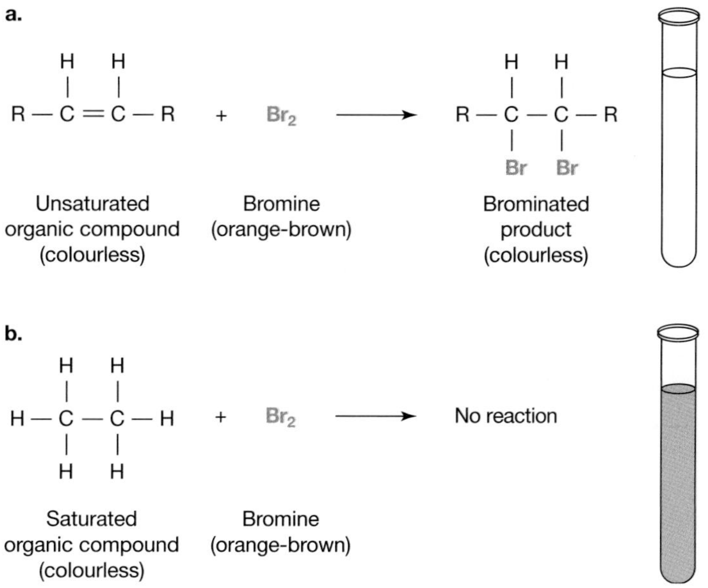

The addition of Br_2 to the unsaturated compound generates a colourless product, so the colour of the solution fades (figure 9.2a). However, saturated compounds will not react with bromine (in the absence of UV light), so the orange-brown colour persists (figure 9.2b).

The bromine water test is routinely used qualitatively to distinguish between alkanes and alkenes. However, the test can also be used quantitatively, as the number of C=C in a molecule can be determined using stoichiometry, given one Br_2 molecule reacts with each C=C. Application of this principle to calculate iodine number will be discussed in the next section.

TIP: Don't forget about substitution reactions. A saturated hydrocarbon will react with Br_2 in the presence of UV light (or heat). Therefore, it is important to avoid these conditions and observe an immediate colour change when performing a bromine water test.

 Resources

 Video eLesson Bromination of ethene (eles-1674)

Iodine number

The number of grams of iodine that react with 100 grams of a chemical substance is defined as the **iodine number**. Given one mole of I_2 reacts with one mole of C=C, iodine numbers provide an indication of the degree of unsaturation of a compound. This technique is routinely used to determine the degree of unsaturation in fats and oils that commonly contain one or more C=C. Higher iodine numbers indicate more C=C and a greater degree of unsaturation.

FIGURE 9.3 Each C=C in an unsaturated molecule will react with one molecule of I_2. Molecules with more C=C will react with a greater mass of I_2, which can be used to determine iodine number.

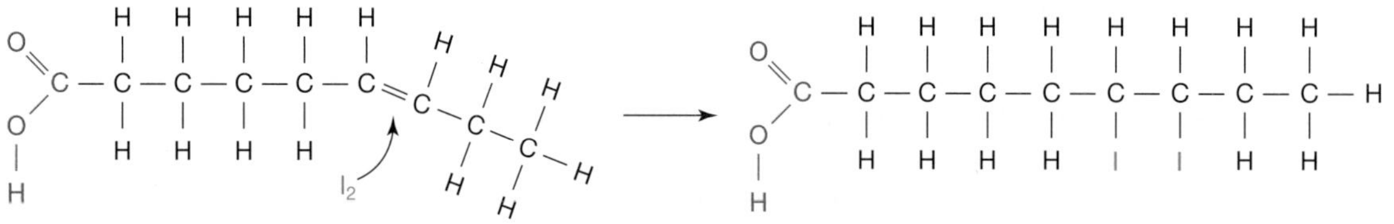

To determine the iodine number:
1. 100 g of a compound is fully reacted with I_2.
2. The mass of iodine added is determined from the increase in mass of an unsaturated molecule as it undergoes an addition reaction with iodine.

TABLE 9.1 Iodine number for some common fats and oils

Fat or oil	Iodine number
Canola oil	110–126
Olive oil	75–94
Butter	25–42
Coconut oil	6–11

iodine number the mass of iodine that reacts with 100 g of a compound

SAMPLE PROBLEM 1 Determining the degree of unsaturation of a compound using the iodine number

To determine the degree of unsaturation, a 100 g sample of a fat was reacted with iodine. The final mass of the sample was 601 g. $M(\text{fat}) = 304.0\,\text{g mol}^{-1}$
a. Calculate the iodine number of the fat.
b. Calculate the amount of I_2, in mol, that reacted with the fat.
c. Calculate the amount of C=C, in mol, present in the fat sample.
d. Calculate the amount, in mol, of fat in the 100 g sample.
e. Determine the number of C=C present (degree of unsaturation) in each fat molecule.

THINK	WRITE
a. The iodine number is the mass of iodine that reacts with 100 g of fat.	a. Iodine number $= m(\text{fat after reaction with } I_2) - m(\text{fat})$ $= 601 - 100 = 501$
b. Divide the mass of I_2 that reacted by the molar mass of I_2.	b. $n(I_2) = \dfrac{m}{M}$ $= \dfrac{501}{253.8}$ $= 1.97\,\text{mol}$
c. One I_2 molecule reacts with each C=C, so the mole ratio is 1 : 1.	c. $n(\text{C=C}) = n(I_2)$ $= 1.97\,\text{mol}$
d. Divide the mass of fat used for the reaction (100 g) by the molar mass provided in the stem of the question.	d. $n(\text{fat}) = \dfrac{m}{M}$ $= \dfrac{100}{304.0}$ $= 0.329\,\text{mol}$
e. The number of C=C in each molecule can be determined by dividing the total amount of C=C in the sample by the amount of fat in the sample.	e. Number of C=C per fat molecule $= \dfrac{n(\text{C=C})}{n(\text{fat})}$ $= \dfrac{1.97}{0.329}$ $= 6$

PRACTICE PROBLEM 1

To determine the degree of unsaturation, a 100 g sample of a fat was reacted with iodine. The final mass of the sample was 200 g. $M(\text{fat}) = 254.0\,\text{g mol}^{-1}$
a. Calculate the iodine number of the fat.
b. Calculate the amount of I_2, in mol, that reacted with the fat.
c. Calculate the amount of C=C, in mol, present in the fat sample.
d. Calculate the amount, in mol, of fat in the 100 g sample.
e. Determine the number of C=C present (degree of unsaturation) in each fat molecule.

9.2.2 Hydroxyl groups

Compounds in which a hydroxyl group (OH) is the parent functional group are called **alcohols**.

Two laboratory tests to positively identify alcohols based on the presence of a hydroxyl group, and another to distinguish between primary, secondary, and tertiary alcohols, will be discussed.

FIGURE 9.4 Hydroxyl groups in ethanol

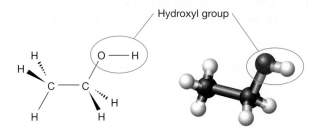

Sodium metal test

Alcohols have a similar acidity to water, so do not undergo many of the familiar reactions of acids, such as reacting with carbonates and metal hydroxides. However, they do react with active metals, including sodium.

For example, the reaction of methanol and sodium is as follows:

$$CH_3OH(l) + Na(s) \rightarrow CH_3ONa(l) + \frac{1}{2}H_2(g)$$

You may recall that water will undergo a similar reaction with sodium, so for the test to be informative, alcohol samples must be pure liquids and not aqueous solutions.

Sodium metal test:
1. A piece of sodium metal is added to a tube containing the sample to be analysed.
2. Hydrogen gas bubbles are produced if the test substance is an alcohol.

Carboxylic acids will also react with sodium metal to produce hydrogen gas. When investigating functional groups, considering which test(s) to use and in which order is frequently important; this will be considered in section 9.2.4.

FIGURE 9.5 Reaction of an alcohol with sodium metal produces hydrogen gas. If desired, the identity of the hydrogen gas could be confirmed with a pop test, taking care not to light the flammable alcohol.

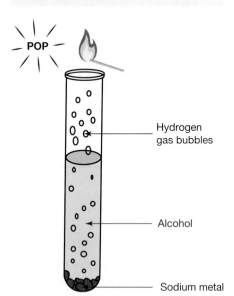

Oxidation test

Once a compound has been identified as an alcohol, additional information about the type of alcohol can be obtained via an oxidation test. Recall that alcohols are classified as primary, secondary or tertiary depending on how many alkyl (R) groups are attached to the same carbon atom as the hydroxyl group.

alcohols compounds in which a hydroxyl group (–OH) is the parent functional group

Oxidation test:
1. A sample of the alcohol to be analysed is added to a tube containing potassium dichromate and dilute sulfuric acid.
2. The tube is heated to observe any colour change.

Figure 9.6 illustrates the different capacity of these alcohols to be oxidised by acidified dichromate, allowing for the identification of tertiary alcohols. To distinguish between primary and secondary alcohols, the reaction with dichromate can be performed under reflux conditions. The primary alcohol will be further oxidised to form a carboxylic acid, which can be detected using the tests discussed in section 9.2.3.

FIGURE 9.6 Tertiary alcohols are not oxidised; therefore, the orange dichromate ions persist, allowing for positive identification of tertiary alcohols. Primary and secondary alcohols do react with dichromate ions, producing a green solution containing Cr^{3+} ions. Under reflux, primary alcohols will be further oxidised to form carboxylic acids.

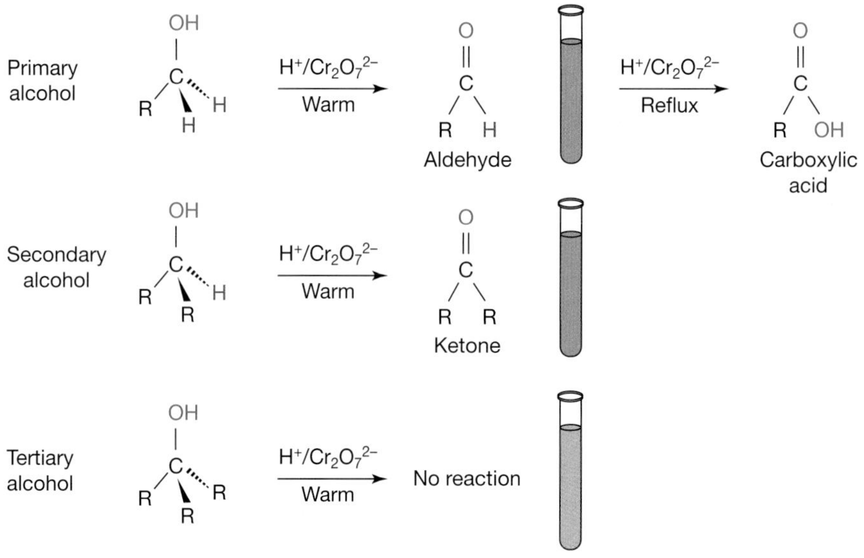

TABLE 9.2 Oxidation test with acidified dichromate ions to identify tertiary alcohols

Type of alcohol	Organic product	Solution colour
Primary	Aldehyde (or carboxylic acid under reflux)	Green
Secondary	Ketone	Green
Tertiary	No reaction	Orange

Permanganate ions (MnO_4^-) act as another oxidising agent that could be used in this test. The purple permanganate ions are reduced to colourless Mn^{2+} ions.

Describing a colour change observation

A colour *change* needs to include both the initial and final colours. For example, 'orange → green' is correct as it describes a change in colour from orange to green. 'Green', however, is insufficient, as it does not state what the colour changed from.

Esterification test

Alcohols react with carboxylic acids to form esters, which are recognisable by their fruity aroma.

Esterification test:
1. The sample to be tested, a carboxylic acid and a concentrated sulfuric acid catalyst are added to a tube and heated.
2. The production of an ester can be detected by a fruity smell, confirming that the test substance is an alcohol.

The esterification reaction of methanol and ethanoic acid is as follows:

$$\underset{\text{Methanol}}{CH_3OH(l)} + \underset{\text{Ethanoic acid}}{CH_3COOH(l)} \underset{}{\overset{\text{Conc. } H_2SO_4}{\rightleftharpoons}} \underset{\text{Methyl ethanoate}}{CH_3COOCH_3(aq)} + \underset{\text{Water}}{H_2O(l)}$$

9.2.3 Carboxyl groups

Carboxyl groups (COOH) are the parent functional groups in carboxylic acids. Tests for carboxyl groups can utilise the acidic nature or chemical reactivity of compounds containing these groups.

FIGURE 9.7 Carboxyl groups in ethanoic acid

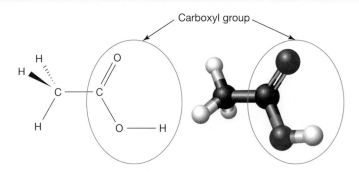

pH test

Carboxylic acids are classified as weak acids because they partially ionise in aqueous solutions. The ionisation reaction for ethanoic acid in water is as follows:

$$CH_3COOH(aq) + H_2O(l) \rightleftharpoons CH_3COO^-(aq) + H_3O^+(aq)$$

The production of H^+ (or H_3O^+) ions increases the acidity and lowers the pH of the solution. Acid–base indicators or pH probes can be used to measure the pH and detect the presence of a carboxylic acid.

FIGURE 9.8 A universal indicator strip turns red in an acidic solution.

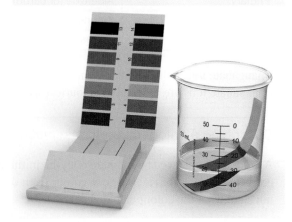

Hydrogen carbonate test

Carboxylic acids react with hydrogen carbonates to produce carbon dioxide gas. The reaction of ethanoic acid with sodium hydrogen carbonate is as follows:

$$CH_3COOH(aq) + NaHCO_3(s) \rightarrow CH_3COONa(aq) + CO_2(g) + H_2O(l)$$

Hydrogen carbonate test:
1. Sodium hydrogen carbonate powder is added to a tube containing the sample to be tested.
2. The rapid formation of carbon dioxide is observed as effervescence, confirming that the test substance is a carboxylic acid.
3. If desired, the gas may be confirmed as carbon dioxide if it turns limewater (calcium hydroxide) cloudy due to the formation of insoluble calcium carbonate.

FIGURE 9.9 A reaction between a carboxylic acid and sodium hydrogen carbonate will produce carbon dioxide gas bubbles. Bubbling the gas through limewater will result in a cloudy appearance if the gas is carbon dioxide.

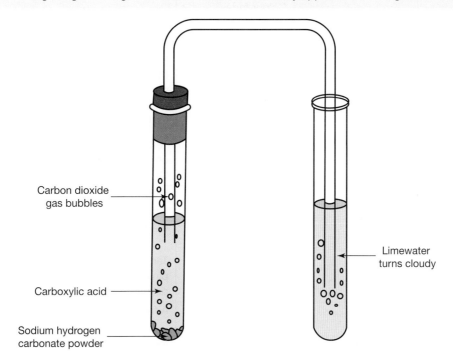

Carboxylic acids will react with metal hydrogen carbonates and metal carbonates, so similar tests could be performed with a variety of compounds.

Esterification test

A similar esterification test to that used for alcohol identification can be employed to detect carboxylic acids.

Esterification test:
1. The sample to be tested, an alcohol and a concentrated sulfuric acid catalyst are added to a tube and heated.
2. The production of an ester can be observed by a fruity smell, confirming that the test substance is a carboxylic acid.

The esterification reaction of ethanol and methanoic acid is as follows:

$$\underset{\text{Ethanol}}{CH_3CH_2OH(l)} + \underset{\text{Methanoic acid}}{HCOOH(l)} \underset{}{\overset{\text{Conc. } H_2SO_4}{\rightleftharpoons}} \underset{\text{Ethyl ethanoate}}{HCOOCH_2CH_3(aq)} + \underset{\text{Water}}{H_2O(l)}$$

9.2.4 Using qualitative laboratory tests to identify an unknown organic substance

The qualitative tests for carbon-to-carbon double bonds, hydroxyl groups and carboxyl groups introduced in this subtopic are summarised in table 9.3. This information is valuable when selecting a test or designing a series of tests to identify these functional groups in unknown organic compounds.

TABLE 9.3 A summary of qualitative test observations for C–C, C=C, hydroxyl groups and carboxyl groups. X indicates that no changes are observed.

	Observations			
Test	**C–C (saturated compounds)**	**C=C (unsaturated compounds)**	**Hydroxyl groups (alcohols)**	**Carboxyl groups (carboxylic acids)**
Bromine water	Orange-brown colour of Br_2 persists	Orange-brown → colourless Colour fades	X	X
Sodium metal	X	X	Hydrogen gas bubbles (confirm with pop test)	Hydrogen gas bubbles (confirm with pop test)
Oxidation with acidified dichromate ions ($H^+/Cr_2O_7^{2-}$)	X	X	Primary and secondary alcohols: orange $Cr_2O_7^{2-}$ → green Cr^{3+} Tertiary alcohols: orange colour of $Cr_2O_7^{2-}$ persists	X
Oxidation with acidified permanganate ions (H^+/MnO_4^-)	X	X	Primary and secondary alcohols: pink MnO_4^- → colourless Mn^{2+} Tertiary alcohols: pink colour of MnO_4^- persists	X
Esterification with a carboxylic acid	X	X	A fruity smell	X
pH	Neutral pH	Neutral pH	Neutral pH	Acidic pH
Hydrogen carbonate	X	X	X	Carbon dioxide gas bubbles (confirm with limewater test)
Esterification with an alcohol	X		X	A fruity smell

tlvd-9655

An unknown organic compound was known to be a hydrocarbon, or to contain either a hydroxyl or carboxyl group. A sample was tested to identify the presence of functional groups and observations were recorded in the following table.

Test	Observations
The sample was mixed with bromine water.	The orange-brown colour disappeared, leaving a colourless solution.
A probe was used to measure the pH of the sample.	The solution had a pH of 4.
A piece of sodium metal was added to the sample.	Bubbles were observed.

a. Explain whether the unknown substance was saturated or unsaturated.
b. Which test(s) produced observations consistent with the presence of a hydroxyl group?
c. Which test observation could only be produced by a carboxylic acid?
d. Identify the functional group(s) present in the compound.

THINK

a. The bromine water test is used to test for saturation. Br_2 reacts with C=C in unsaturated compounds, producing a colourless solution.

b. The sodium metal test will produce hydrogen gas if a hydroxyl group is present.

c. Carboxyl groups ionise to produce solutions with low pH.

d. The acidic pH confirms the molecule contains a carboxyl group.
The bromine water test confirms the molecule is unsaturated.

WRITE

a. The Br_2 reacted with C=C in the unsaturated compound, turning the orange-brown solution colourless.

b. Reaction with a piece of sodium metal to produce hydrogen gas is consistent with the presence of a hydroxyl group.

c. pH = 4

d. The molecule contains a carboxyl group and a C=C.

PRACTICE PROBLEM 2

An unknown organic compound known to be a hydrocarbon, alcohol or carboxylic acid was tested to identify the presence of functional groups. The following observations were recorded.

Test	Observations
1. The sample was mixed with bromine water.	The orange-brown colour of the solution persisted.
2. The sample was heated with acidified potassium dichromate.	The orange colour of the solution persisted.
3. The sample was reacted with an alcohol in the presence of concentrated sulfuric acid.	A fruity smell was not detected.
4. The sample was reacted with a carboxylic acid in the presence of concentrated sulfuric acid.	A fruity smell was produced.

a. Deduce whether the unknown substance was saturated or unsaturated. Identify the test result used to make your decision.
b. Which test result(s) were consistent with the presence of a carboxyl group?
c. Which test observation could only be produced by an alcohol?
d. Identify the functional group(s) present in the compound.

9.2 Activities

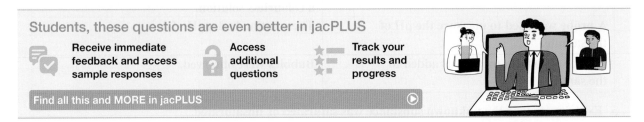

| 9.2 Quick quiz on | 9.2 Exercise | 9.2 Exam questions |

9.2 Exercise

1. Bacon contains approximately 30% fat. Saturated fats have negative health effects, so it is valuable to measure the degree of unsaturation of fats in foods such as bacon.
   a. Identify a qualitative test to determine whether bacon contains any unsaturated fats.
   b. Identify a quantitative test to determine the degree of unsaturation of fats in bacon.
2. Linolenic acid has the formula $CH_3CH_2(CH=CHCH_2)_3(CH_2)_6COOH$. M(linolenic acid) = 278.0 g mol^{-1}
   a. Deduce the number of C=C in linolenic acid.
   b. What amount, in mol, of linolenic acid is in a 76.8 g sample?
   c. What amount, in mol, of I_2 is required to fully react with the sample of linolenic acid?
   d. Calculate the mass of I_2, in g, from part c.
   e. Define the term *iodine number* and determine the iodine number for linolenic acid.
3. Propose two chemical tests that could be used to distinguish between cyclohexene and methanol. Complete the following table with your predicted observations for each test.

Test	Cyclohexene observation	Methanol observation

4. The following tests were performed on a primary alcohol:
   Test 1: The alcohol was heated under reflux with acidified potassium dichromate.
   Test 2: The product of test 1 was added to sodium carbonate powder.
   a. Identify the functional group in the organic molecule present after test 1.
   b. Predict observations for the two tests.
   c. Explain your observations from part b.
5. Explain how a student could use esterification to identify whether an unknown compound is an alcohol or a carboxylic acid.

6. When tidying the laboratory after a practical, Dr Bunty found a bottle with a damaged label that read 'prop'. She knew that the chemical was propan-1-ol, propen-1-ol or propanoic acid. She ran some tests to identify the compound and recorded her results in the following table.

Test	Observation
Add sodium hydrogen carbonate powder	No reaction
Add a piece of sodium metal	Rapid effervescence
Mix with bromine water	No change in orange-brown colour
Add a few drops of universal indicator	Green (neutral pH)

a. Identify the chemical.
b. Justify your answer to part a, referring to the test results.
c. Identify one test that was not required to identify the unknown compound. Justify your answer.

9.2 Exam questions

Question 1 (1 mark)

MC α-hydroxy acids contain a hydroxyl group on the carbon neighbouring the carboxyl group. They have gained prominence as an ingredient in cosmetics intended to improve the appearance of skin.

Which combination of laboratory tests could distinguish between a sample of an α-hydroxy acid and similar organic molecules containing either a carboxyl group or a hydroxyl group?
A. Sodium metal and bromine water
B. Sodium hydrogen carbonate and sodium metal
C. pH and esterification with an alcohol
D. Esterification with an alcohol and esterification with a carboxylic acid

Question 2 (2 marks)
Source: Adapted from VCE 2020 Chemistry Exam, Section B, Q.10.a; © VCAA

Analytical chemistry deals with methods for determining the chemical composition of samples of matter. A qualitative method yields information about the identity of atomic or molecular species or the functional groups in the sample …
Analytical methods are often classified as being either *classical* or *instrumental*.

Source: DA Skoog, FJ Holler and SR Crouch, *Principles of Instrumental Analysis*, 6th edition, Thomson Brooks/Cole, Belmont (CA), 2007, p. 1

Classical methods include qualitative analysis, such as treating a compound with reagents to observe any reaction, and quantitative methods, such as volumetric analysis, where the amount of a compound is determined by its reaction with a standard reagent.

Instrumental methods include a variety of spectroscopy, such as IR spectroscopy and NMR spectroscopy.

Explain how the classical methods of analytical chemistry can be used to determine information about alcohols. In your answer, refer to qualitative analysis and how it can be used to determine whether a compound is an alcohol and, if it is, the type of alcohol.

Question 3 (1 mark)
Source: VCE 2012 Chemistry Exam 1, Section A, Q.18; © VCAA

MC 2.1 g of an alkene that contains only one double bond per molecule reacted completely with 8.0 g of bromine, Br_2. The molar mass of bromine, Br_2, is 160 g mol^{-1}.

Which one of the following is the molecular formula of the alkene?
A. C_5H_{10}
B. C_4H_8
C. C_3H_6
D. C_2H_4

▶ **Question 4 (1 mark)**

Source: VCE 2008 Chemistry Exam 1, Section A, Q.17; © VCAA

MC A student was given the task of identifying a liquid organic compound that contains only carbon, hydrogen and oxygen. The following tests were carried out.

	Procedure	Result
Test 1	Some brown Br_2(aq) was added to a sample of the compound.	A reaction occurred and a colourless product formed.
Test 2	Some Na_2CO_3(s) was added to a sample of the compound.	A reaction occurred and a colourless gas was evolved.

Based on the above test results, the compound could be

A.

$$H-\underset{\underset{H}{|}}{\overset{\overset{H}{|}}{C}}-\underset{\underset{H}{|}}{\overset{\overset{H}{|}}{C}}-\underset{\underset{H}{|}}{\overset{\overset{H}{|}}{C}}-\overset{\overset{O}{\|}}{C}-O-H$$

B.

$$H-\underset{\underset{H}{|}}{\overset{\overset{H}{|}}{C}}-\overset{\overset{H}{|}}{C}=\underset{\underset{H}{|}}{C}-\overset{\overset{O}{\|}}{C}-O-H$$

C.

$$H-\underset{\underset{H}{|}}{\overset{\overset{H}{|}}{C}}-\underset{\underset{H}{|}}{\overset{\overset{H}{|}}{C}}-O-\underset{\overset{O}{\|}}{C}-\underset{\underset{H}{|}}{\overset{\overset{H}{|}}{C}}-H$$

D.

$$\underset{\overset{O}{\|}}{\overset{H}{C}}-O-\underset{\underset{H}{|}}{C}=\underset{\underset{H}{|}}{C}-\underset{\underset{H}{|}}{\overset{\overset{H}{|}}{C}}-H$$

▶ **Question 5 (4 marks)**

Source: VCE 2006 Chemistry Exam 1, Section B, Q.1; © VCAA

A student was given four colourless liquids that were labelled **A**, **B**, **C** and **D**. They were known to be ethanol, ethanoic acid, pentane and hexene, but the exact identity of each liquid was unknown.

The student tested the properties of each liquid and obtained the following results.

	A	B	C	D
Solubility in water	insoluble	soluble	soluble	insoluble
Addition of red-coloured bromine (Br_2) solution	colour disappears	no immediate reaction	no immediate reaction	no immediate reaction
Addition of sodium carbonate (Na_2CO_3) powder	no reaction	gas evolved	no reaction	no reaction

Identify each of the liquids **A**, **B**, **C** and **D**.

More exam questions are available in your learnON title.

9.3 Laboratory techniques for analysis of consumer products

KEY KNOWLEDGE

- Applications and principles of laboratory analysis techniques in verifying components and purity of consumer products, including melting point determination and distillation (simple and fractional)

Source: VCE Chemistry Study Design (2024–2027) extracts © VCAA; reproduced by permission.

Traditional laboratory techniques have long been used to analyse pure substances and mixtures. The physical properties of organic compounds differ based on their structure and the resultant forces between molecules. Both melting point and boiling point are determined by intermolecular force strength and are characteristic for each substance.

In the laboratory, melting point may be measured to gain insights into the identity and purity of a sample. For example, melting point determination may be used for quality control in the pharmaceutical industry to ensure newly synthesised batches are pure samples of the desired compound.

Differences in boiling point allow the separation of liquid mixtures by distillation. This is useful on a small scale in the laboratory and also industrially — for example, in the separation of crude oil into petroleum products.

9.3.1 Melting point determination

The **melting point** of a substance is the temperature at which the phase change between solid and liquid occurs. At this temperature, sufficient energy is present to overcome the intermolecular forces within the crystalline lattice of solid covalent molecular substances. As illustrated in figure 9.10, melting can be considered as a three-step process:

1. As a solid substance is heated, particles within the lattice absorb the energy and vibrate faster, which is observed as rising temperature.
2. At the melting point, energy is absorbed as the latent heat of fusion to weaken the intermolecular bonds within the solid lattice, although with no increase in temperature. During this process, both solid and liquid phases are present in the sample.
3. Once the entire lattice has been melted, the energy absorbed increases vibration of particles in the liquid phase, observed as increased temperature.

FIGURE 9.10 The effect of adding energy to a crystalline substance

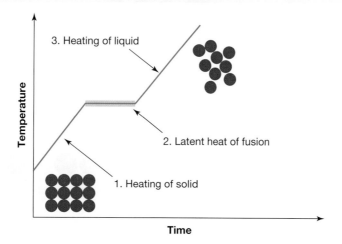

Each crystalline organic substance has a characteristic melting point. This is determined by the strength of the intermolecular forces present within the lattice, which can be affected by a number of factors, including:

- the type(s) of intermolecular forces present; for example, hydrogen bonding, dipole–dipole attraction and dispersion forces. Stronger bonds will require more energy to overcome them, resulting in a higher melting point.
- the number of each type of intermolecular bond; for example, larger molecules may form multiple hydrogen bonds and/or many dispersion forces, resulting in a higher melting point.

Many of the organic compounds with low molecular masses introduced in Unit 4 are gases (C1–4 hydrocarbons) or liquids (alcohols, amines, carboxylic acids, esters) at room temperature. However, members of these organic families with larger molecular masses are solids at room temperature, due to the increased strength of the dispersion forces contributed by the longer hydrocarbon chain. It is these larger molecules that may undergo melting point determination.

> **melting point** the temperature at which a substance changes between the solid and liquid states

Melting point determination in the laboratory allows for:
- identification of a compound, by comparison of an experimental melting point value to a literature value
- the purity of a sample to be determined.

TABLE 9.4 The effect of sample purity on the observed melting point

Sample purity	Observation	Explanation
Pure	Narrow melting point range (0.5–2 °C)	The regular lattice structure means all intermolecular forces are of similar strength and are broken within a narrow temperature range.
Impure	Lower melting point value than for a pure sample (melting point depression)	Intermolecular forces are weaker in the disrupted lattice structure of an impure sample, so require less energy to break than for a pure sample.
	Wide melting point range (5+ °C)	Uneven distribution of impurities within the lattice leads to areas with different degrees of melting point depression. As a result, the melting point extends over a wider temperature range.

FIGURE 9.11 Impurities disrupt the lattice structure in crystalline solids, leading to a wider temperature range for melting point.

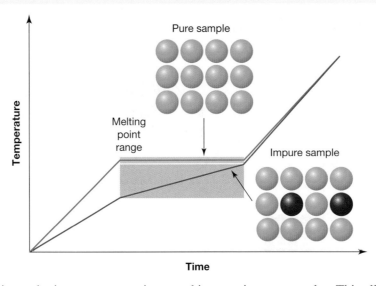

The effects of impurities on lattice structure are increased in more impure samples. This allows melting point determination to be used to identify the relative impurity of multiple samples of the same substance.

Measuring melting point in the laboratory

Modern methods of melting point determination use specialised equipment, although operate using the same principles as simpler methods available in a high-school laboratory.

Figure 9.12 shows a chemist inserting a capillary tube into equipment used for melting point determination. The temperature is slowly increased until the sample melts. A beam of light shone on the solid sample in the capillary tube will not pass through until the moment the sample melts, allowing for identification of the melting point.

FIGURE 9.12 A specialised melting point system

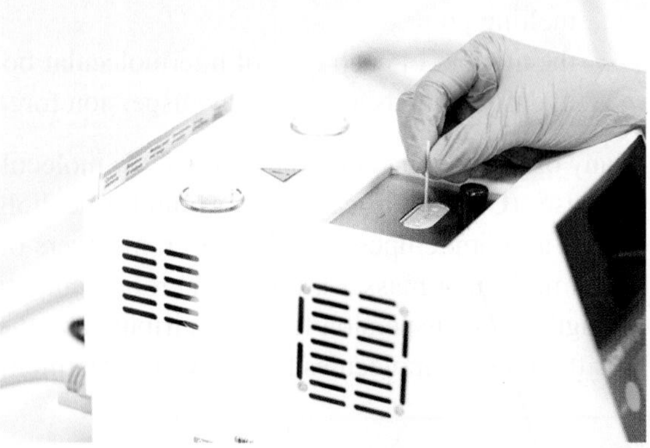

Figure 9.13 illustrates a common laboratory apparatus for melting point determination.

1. A small sample of powder is added to a glass capillary that has been sealed at one end.
2. The capillary is attached to a thermometer to accurately measure the melting temperature.
3. The sample is heated slowly. In this example, a Thiele tube is filled with mineral oil that is heated by a Bunsen burner. Alternative heating equipment includes a metal block, hot plate or specialist device (figure 9.12).
4. The sample within the capillary is observed (often using a magnifying lens) for the first signs of melting (figure 9.14). The temperature at which melting begins is recorded.
5. The sample continues to be heated and the temperature at which the entire sample has melted is recorded (figure 9.14).

Experimental generation of accurate melting point values is dependent on several factors, including:

- preparation of the sample as a dry, finely ground powder
- the use of thin, glass capillary tubes of the required dimensions, filled with the sample to a set height (2–3 mm)
- a slow rate of temperature increase (1 °C min^{-1}).

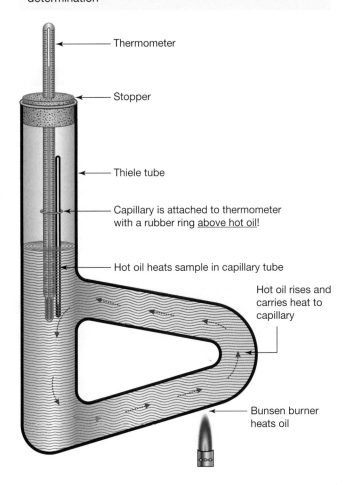

FIGURE 9.13 A laboratory apparatus for melting point determination

Thermometer

Stopper

Thiele tube

Capillary is attached to thermometer with a rubber ring above hot oil!

Hot oil heats sample in capillary tube

Hot oil rises and carries heat to capillary

Bunsen burner heats oil

FIGURE 9.14 Changes in crystal appearance during the melting process. The initial temperature measurement is made when the crystals first begin to move away from the edge of the capillary. The final measurement is made when the sample has fully melted.

First sign of melting   Fully melted sample

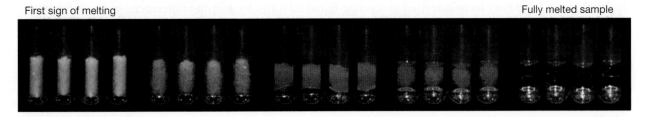

Rate of temperature increase is important, as faster changes (5–10 °C min^{-1}) will produce melting temperatures higher than the literature values. Nevertheless, this can be useful to quickly estimate the melting temperature of a sample prior to repeat measurements using a much lower and more accurate rate of temperature increase (1 °C min^{-1}).

When comparing experimental melting points with literature values, it is important to use the same conditions. Moreover, calibration of equipment with a pure sample of a known compound may be used to detect any differences between the experimental and literature values. The equipment can then be adjusted prior to taking further measurements.

Analysing melting point results

Once an accurate melting point determination has been conducted, the results can be analysed in several ways to either identify the compound or determine its purity.

TABLE 9.5 Analysing melting point determination results

Objective	Method
Sample purity	A wide temperature range that is lower than the literature value is evidence for an impure sample. The greater the effects, the less pure the sample.
Compound identification	The experimental melting point is compared to a pure compound run under the same conditions, or a literature value. A match is inconclusive evidence for the compound identity.
	To confirm the identify of an unknown compound, the experimental sample can be mixed with a known substance in a 1 : 1 ratio. 1. If the melting point of this mixture is the same as the unknown compound, then the two compounds are identical. 2. If the melting point of the mixture is lower and wider, then they are different compounds. This process can be repeated until the unknown compound is identified.

elog-1920

EXPERIMENT 9.1

online only

Melting point determination for samples of covalent solids

Aim

To determine the melting point of a variety of samples containing covalent solids

9.3.2 Distillation

The **boiling point** of an organic compound is determined by the strength of the intermolecular forces between its molecules. In a similar manner to melting point, larger molecules and those containing polar groups require more energy to overcome the stronger intermolecular forces present, so have higher boiling points. **Volatility** is an associated concept, describing how readily a substance will form a vapour. Substances with lower boiling points have weaker intermolecular forces and are more volatile.

Distillation is a laboratory technique used to separate liquid mixtures by utilising differences in the ability of the components to form a gas (boiling points).

Simple distillation

Simple distillation uses the apparatus illustrated in figure 9.15 to separate one component from a mixture.

1. The mixture is heated to a target temperature to vaporise the component to be separated.
2. The vapour passes into the condenser where it is cooled by cold, flowing water.
3. The resultant liquid (termed *distillate*) is collected.
4. Steps 1–3 may be repeated to further enrich the desired component in the distillate.

> **boiling point** the temperature at which a substance changes between the liquid and gas states
>
> **volatility** describes how readily a liquid substance will form a vapour
>
> **distillation** a process for separating components in a mixture that is dependent on the differing boiling points of the components

FIGURE 9.15 Simple distillation apparatus

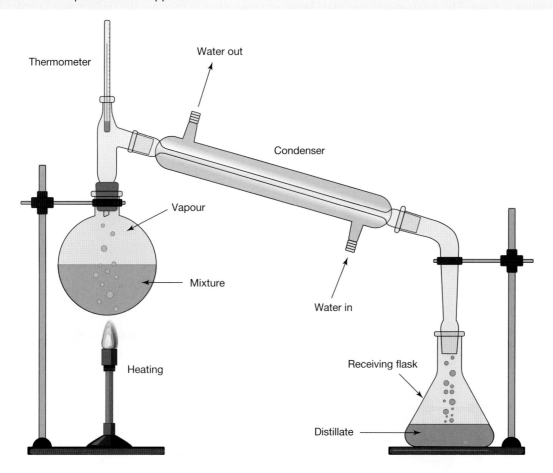

A familiar example of distillation is separation of the aldehyde produced by oxidation of a primary alcohol before it can be further oxidised to a carboxylic acid. Unlike alcohols and carboxylic acids, hydrogen bonds do not form between aldehyde molecules, so they have a lower boiling temperature. A target temperature above the boiling point of the aldehyde but below the boiling point of the other components would allow the aldehyde to be vapourised, separated and collected.

Distillation, or similar methods, may also be used to determine the boiling point of a substance by heating the pure substance and recording the highest vapour temperature reached. The compound may then be identified by comparing this temperature with literature values. However, unlike melting point determination, boiling point is rarely used to analyse purity.

Simple distillation is most effective at enriching the more volatile compound when the boiling points of the components differ by at least 100 °C. This is sufficient to increase the ethanol content of wine (13%) and beer (5%) to produce spirits (40%), and for the desalination of water. However, organic mixtures in which the components have similar boiling points cannot be effectively separated by simple distillation and require a more powerful technique.

Fractional distillation

Fractional distillation is a modified form of simple distillation that effectively allows multiple rounds of distillation to occur without having to move the distillate back to the distilling flask. This allows much greater enrichment of the more volatile component in the distillate.

The apparatus illustrated in figure 9.16 is similar to that used for simple distillation, with the addition of a fractionating column. This column contains glass beads or projections to increase the surface area in order to maximise condensation. There is also a temperature gradient caused by the ascending vapour, moving from the hotter base to the cooler top of the column. It is these two features of fractionating columns that allow the repeated boil–condense cycles, which have a similar effect to multiple rounds of distillation.

FIGURE 9.16 Fractional distillation apparatus

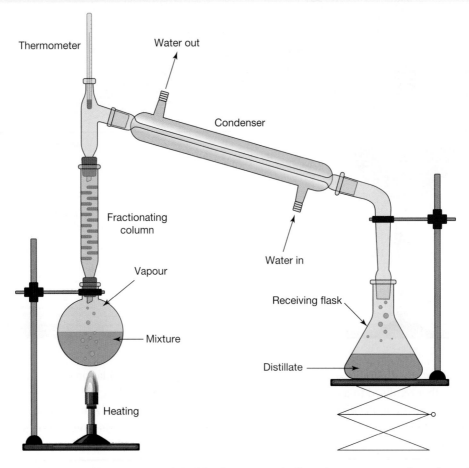

1. The mixture is heated, and a vapour enriched in the more volatile substance moves into the column.
2. The vapour condenses once it reaches a point on the column with a temperature equal to its boiling point.
3. This liquid then flows a short distance down the column until it reaches a point hot enough to vaporise it again.
4. After each boil–condense cycle (steps 2–3) the percentage of the more volatile substance in the mixture increases. This lowers its boiling point, allowing it to condense in cooler areas further up the column.
5. After multiple cycles the highly enriched vapour exits the top of the column to be condensed and collected.

elog-1921

tlvd-9729

EXPERIMENT 9.2

online only

Fractional distillation of an ethanol/water mixture

Aim

To separate a mixture of ethanol and water by fractional distillation

Fractional distillation of crude oil

As introduced in topic 1, the refining of crude oil into a variety of petroleum products is a major application of fractional distillation. Crude oil is typically a viscous mixture composed of many different hydrocarbon compounds. It was formed from the remains of marine organisms, such as bacteria, algae and plankton. These marine organisms were altered by the combined effects of pressure, temperature, moisture and decay.

The fractional distillation of crude oil is termed *refining* and occurs in purpose-designed towers. These separate components into fractions containing compounds with similar boiling points, which can then be collected from trays positioned at different heights along the tower. Figure 9.17 shows a simplified outline of this process.

1. The crude oil is heated and then introduced to the base of the tower.
2. Many of its components vaporise and these vapours rise up the tower, being cooled as they do so.
3. When the vapours reach a point at which the tower's temperature equals their boiling temperature, condensation occurs.
4. Specially designed trays are placed inside the tower at strategic intervals. These are designed to allow the vapours to continue rising, but stop condensed fractions from dripping back down to lower levels in the tower.
5. The condensed fractions may then be removed from these trays to undergo further processing.

FIGURE 9.17 A schematic of fractional distillation of crude oil, showing levels of the fractionating column

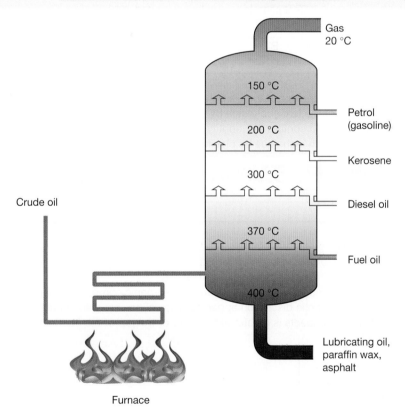

The hydrocarbon fractions separated by refining are used for many purposes, including fuels for transport and heating, construction materials, industrial lubricants, and in the manufacture of plastics, paints, synthetic fibres, medicines and pesticides.

9.3 Quick quiz on	9.3 Exercise	9.3 Exam questions

9.3 Exercise

1. Identify the two effects that impurities will have on an experimentally determined melting point compared to a pure sample.
2. A chemist performed a melting point determination and recorded their results in the following table with literature values for selected compounds.

Substance	Melting point (°C)
Test sample	111–112
A	112
B	106
C	168

   a. State and justify whether the test substance is pure.
   b. Deduce the likely identity of the test substance by referring to the table.
   c. Explain how the chemist could confirm the identity of the test substance.
3. Indicate the applications for each technique by placing ticks in the correct locations in the following table.

Application	Melting point determination	Distillation
Compound identification		
Purity analysis		

4. An aqueous solution containing ethanol, ethanoic acid and propanoic acid can be separated by distillation.
   a. Identify the intermolecular forces present in the mixture.
   b. List the organic compounds in order of increasing volatility.
   c. Which laboratory technique could be used to separate the organic compounds from the mixture?
5. The use of a fractionating column in distillation apparatus allows increased separation of components. Explain whether each of the following aspects is greater at the base or the top of the fractionating column.
   a. Column temperature
   b. Boiling point of the mixture
   c. Purity of volatile substance

9.3 Exam questions

▶ **Question 1 (1 mark)**

MC Which of the following statements about melting point determination are correct?
 I The melting point of a mixture is the average melting points of the components.
 II Fast temperature changes will often result in an overestimated melting temperature.
 III Wide melting point ranges may be caused by uneven distribution of impurities within the crystal lattice of the sample.
A. I and II only
B. I and III only
C. II and III only
D. I, II and III

Question 2 (4 marks)

A pharmaceutical chemist synthesised drug X from reactants A and B.

Substance	Melting point (°C)	Boiling point (°C)
A	98	318
B	110	420
X	31	268

a. Identify techniques that the chemist could use to perform the following functions.
   i. Separate X from excess A and B in the reaction mixture   **(1 mark)**
   ii. Test the purity of the sample of X   **(1 mark)**
b. X has a higher molar mass than A and B.

   Identify a reason that the melting and boiling points of X are lower than those of A and B.   **(2 marks)**

Question 3 (4 marks)

A mixture of propanoic acid, propanal and propan-1-ol is fractionally distilled.
a. List the order that these molecules will leave the fractionating column, from fastest to slowest.   **(1 mark)**
b. Explain your answer to part **a**.   **(1 mark)**
c. Identify two features of a fractionating column that increase the separation of components.   **(2 marks)**

Question 4 (5 marks)

A student prepared two batches of aspirin by reacting salicylic acid with ethanoic anhydride. The aspirin was crystallised from the reaction mixture, and its purity determined by melting point analysis.

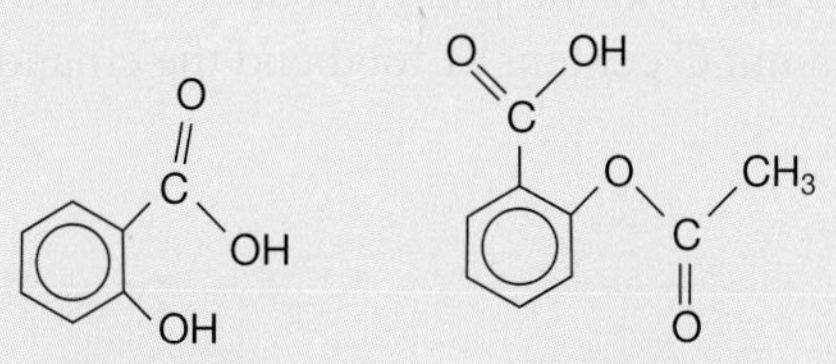

Salicylic acid   Aspirin

Substance	Melting point (°C)
Batch 1	125–129
Batch 2	123–127
Pure aspirin	139–140
Pure salicylic acid	159–160

a. Explain why salicylic acid has a higher melting point than aspirin.   **(2 marks)**
b. Deduce the purity of batch 1 of the aspirin by referring to the table.   **(2 marks)**
c. Which batch of aspirin contains the lowest yield?   **(1 mark)**

Question 5 (3 marks)

Explain how the hydrocarbons in crude oil are separated.

More exam questions are available in your learnON title.

9.4 Volumetric analysis by redox titration

KEY KNOWLEDGE

- Volumetric analysis, including calculations of excess and limiting reactants using redox titrations (excluding back titrations)

Source: VCE Chemistry Study Design (2024–2027) extracts © VCAA; reproduced by permission.

9.4.1 Volumetric analysis procedure

Volumetric analysis is a quantitative technique that involves reactions in solution. The concentration of a solution can be determined using accurately measured volumes and reacting it with a **standard solution** with an accurately known concentration.

Volumetric analysis versus titration

Titration is a type of volumetric analysis using a **burette**. The volume delivered by a burette is called a **titre**, which gives the technique its name.

Titrations were introduced in Unit 2 for the analysis of acidic and basic solutions, although the same principles apply to redox reactions.

Redox titrations are commonly used in the food and pharmaceutical industries. Examples include determining the vitamin C content of food and the ethanol concentration in alcoholic drinks.

volumetric analysis determination of the concentration, by volume, of a substance in a solution, such as by titration

standard solution a solution that has an accurately known concentration

titration a type of volumetric analysis used to determine the concentration of a substance; a pipette is used to deliver one substance and a burette is used to deliver another substance until they have reacted exactly in the reaction equation mole ratios

burette a graduated glass tube used for delivering known volumes of a liquid, especially in titrations

titre the volume delivered by a burette

primary standard a substance used in volumetric analysis that is of such high purity and stability that it can be used to prepare a solution of accurately known concentration

---

BACKGROUND KNOWLEDGE: Practical aspects that are common to both acid–base and redox titrations

Standard solutions

A standard solution is one whose concentration is accurately known. There are usually two methods by which a solution can have its concentration determined accurately. These are:
1. by performing volumetric analysis (titration) using a solution with an accurately known concentration. This is called standardisation.
2. by taking a substance called a **primary standard** and dissolving it in a known volume of water. Primary standards are pure substances that satisfy a special list of criteria.

To qualify as a primary standard, a substance must have a number of the following properties:
- It must have a high state of purity.
- It must have an accurately known formula.
- It must be stable (its composition or formula must not change over time, which may happen as a result of storage or reaction with the atmosphere).
- It should be cheap and readily available.
- It should have a relatively high molar mass so that weighing uncertainties are minimised.

Not all substances are suitable for use as primary standards.

To prepare a primary standard, chemists use special flasks called *volumetric flasks*. These are filled to a previously calibrated etched line on their necks, so that the volume of their contents is accurately known.

Glassware and rinsing

FIGURE 9.18 Basic equipment required for titration

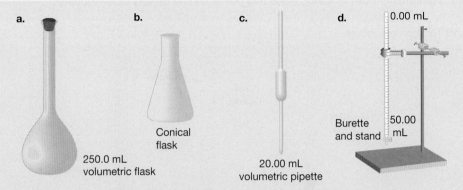

a. 250.0 mL volumetric flask

b. Conical flask

c. 20.00 mL volumetric pipette

d. Burette and stand

0.00 mL

50.00 mL

Table 9.6 summarises the different glassware, their use, their precision and the rinsing methods used for titration.

TABLE 9.6 Summary of glassware used for titration

Glassware	Use	Precision (number of decimal places)	Rinsed with
Burette	Delivers the standard solution or unknown solution. Readings from a burette should be taken from the bottom of the meniscus formed by the solution.	Two decimal places, with the second decimal place estimated (e.g. 25.68 mL). Note that the volume of a titre is commonly estimated to the nearest 0.02 mL.	The solution it will deliver. A few mL of the solution is added, and the burette carefully rotated to ensure thorough rinsing, before the liquid is expelled using the tap.
Volumetric pipette	Used to deliver an aliquot of standard solution or unknown solution directly to a conical flask or into a volumetric flask for dilution	Two decimal places (e.g. 20.00 mL)	The solution it will deliver. A partial or full aliquot can be used to rinse a pipette as long as all of its inner surface has been rinsed.
Volumetric flask	Used to dilute aliquots Used to prepare standard solutions	One decimal place. Commonly used volumes are 50.0, 100.0 and 250.0 mL.	Distilled water. The aliquots added to a volumetric flask will be diluted with distilled water, so the number of moles from the pipette is unchanged.
Conical flask	Used to hold the aliquot of solution that will then be titrated against the solution in the burette. If it is a titration that needs an indicator, this will also be added.	Not used to measure volumes	Distilled water. This does not change the moles of standard solution or unknown solution delivered by the pipette and burette.

Performing a titration

In volumetric analysis, the calculations require that a titration be stopped when one substance has *just finished* reacting with the other one. This point is called the **equivalence point**. Detection of this point is critical to the success of a volumetric procedure.

equivalence point the point at which two reactants have reacted in their correct mole proportions in a titration

In acid–base titrations the **end point** is reached when the indicator first undergoes a permanent colour change. Because this occurs only after a slight excess is added, we often do not have the true equivalence point.

Thus, we can say that *the end point approximates the equivalence point*. However, in a carefully designed procedure with a carefully chosen indicator, these two points should be very close together.

FIGURE 9.19 Titration procedure

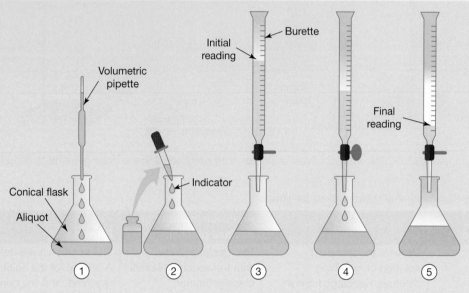

A titration is performed as follows:

1. An accurate volume of one of the solutions is transferred to a conical flask with a volumetric **pipette** (this volume is called an **aliquot**).
2. A small volume of an **indicator** is added to the conical flask in order to generate an evident colour change in step 4.
3. The other solution is added to a burette and the initial reading is taken.
4. The solution in the burette is then added, carefully.
5. When the first permanent colour change is noted in the conical flask, the end point has been reached. The volume delivered from the burette at this point is known as the titre.
6. By knowing the volumes involved, the concentration of one of the solutions and the mole ratio of the reactants (from the reaction equation), the concentration of the other solution can be determined.

Concordant titres

Repeat titrations are routinely performed to reduce the effect of random errors.

When performing repeat titrations, one usually aims for **concordant** titres. These are titres that are within a defined volume of each other, with 0.10 and 0.05 mL being commonly accepted values. 0.05 mL is an exacting standard, and requires very careful attention to detail and excellent technique because it represents approximately one drop. However, in many situations, including school laboratories, titres within 0.10 mL of each other is a more realistic standard. On this basis, if a titration produces results of 19.26, 19.20, 19.40 and 19.20 mL, all except the third value can be considered concordant.

end point the point at which the indicator changes colour in a titration

pipette a piece of glassware used for transferring accurate volumes of liquid

aliquot the liquid from a pipette

indicator a chemical compound that changes colour and structure when exposed to certain conditions and is therefore useful for chemical tests

concordant describes titres that are within a defined volume of each other, such as 0.10 mL

TABLE 9.7 Differences between acid–base and redox titrations

Acid–base titration	Redox titration
The reaction involves the transfer of protons from an acid to a base.	The reaction involves the transfer of electrons from a reducing agent to an oxidising agent.
The end point is identified by the change in colour of an acid–base indicator.	Many redox reactions involve a colour change, so an indicator may not be required to identify the end point.

9.4.2 Analysis of organic compounds using redox titration

Volumetric analysis can be used to analyse substances with redox properties, including a variety of organic compounds. Variations to the basic method of titration have been developed, all aimed at taking into account the properties of the substances involved. Here we will consider simple (or direct) titration — the process in which one reactant is added to the other until the correct stoichiometric proportions are present. However, analysis of organic substances by redox titration is frequently performed using more sophisticated techniques that are outside the VCE curriculum.

Performing redox titrations

Several of the standard solutions used to analyse organic compounds by redox titration are strong oxidising agents. Recall that certain organic compounds, including alcohols, have the capacity to be oxidised.

Table 9.8 lists some common standard solutions used in redox titrations, each of which undergoes a colour change during the reaction. As a result, some redox titrations can be self-indicating. This is fortunate, as redox indicators are much less common than acid–base indicators.

TABLE 9.8 Redox standard solutions used to analyse organic substances undergo colour changes.

Standard solution	Oxidised form	Reduced form
Iodine	I_2 is brown (dark blue when bound to starch).	I^- is colourless.
Permanganate ion	MnO_4^- is purple/pink.	Mn^{2+} is colourless.
Dichromate ion	$Cr_2O_7^{2-}$ is orange.	Cr^{3+} is green.

An example of an indicator used in redox titrations is starch. Starch is used to detect the presence of iodine, I_2, which is formed in titrations from the oxidation of iodide ions, I^-. Starch is dark blue in the presence of iodine. The appearance, or disappearance, of the blue colour formed in the presence of iodine with starch signals the end point of the titration. Another indicator suitable for redox titrations is methylene blue, which is blue in the presence of an oxidising agent and colourless in the presence of a reducing agent.

FIGURE 9.20 The presence of a starch–iodine complex indicates the end point of this redox titration.

Of the substances listed in table 9.8, only compounds containing the dichromate ion — for example, potassium dichromate — are suitable as primary standards.
- Iodine is unsuitable as a primary standard as iodide ions are added to improve its low solubility in water.
- Standard solutions containing permanganate ions cannot be produced by dissolving known amounts of permanganate-containing compounds in water, as once mixed with water, the permanganate ions react to produce manganese(IV) oxide.

As such, solutions of these substances must be standardised via titration with another standard solution.

A variety of organic compounds may be quantified by redox titration, allowing analysis of food, drinks, pharmaceuticals and other consumer products.

TABLE 9.9 Analysis of organic compounds by redox titration

Substance	Reactant	Standard solution(s)
Food and supplements	Vitamin C	Iodine solution
Alcoholic drinks	Ethanol	Acidified permanganate solution
Hand sanitiser		Acidified dichromate solution
Leafy green vegetables	Oxalic acid	Acidified permanganate solution

---

 Resources

⬥ **Interactivity** Simulation of a redox titration (int-1225)

Redox titration calculations

Once concordant titres have been obtained, the average volume of the titres (in L) is ultimately used to calculate the concentration of the unknown solution. Figure 9.21 shows how a thumbnail sketch of the apparatus can be a useful tool to organise the numbers and set out the sequence of steps in titration calculations.

Steps in a titration calculation

1. Quickly draw a burette and conical flask. Two vertical lines of different length will suffice.
2. Identify where the standard solution and unknown solution are located. This will vary so must always be checked. In this example, the standard solution is in the burette and the unknown solution is in the flask.
3. List all the information that is known about each solution. This comprises the volumes in the burette (titre) and conical flask (aliquot), and the concentration of the standard solution.
4. Begin by calculating the moles of standard solution as both the volume and concentration are known.
5. Use the mole ratio from the reaction equation to calculate the moles of the unknown substance.
6. Finally, calculate the concentration of the unknown solution.

FIGURE 9.21 Using a thumbnail sketch of the apparatus to complete a titration calculation

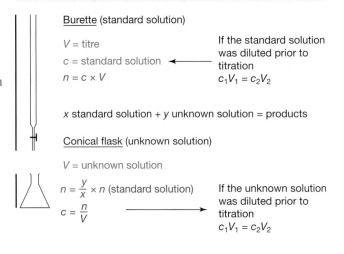

Burette (standard solution)

V = titre

c = standard solution ← If the standard solution was diluted prior to titration

$n = c \times V$ $c_1V_1 = c_2V_2$

x standard solution + y unknown solution = products

Conical flask (unknown solution)

V = unknown solution

$n = \dfrac{y}{x} \times n$ (standard solution) → If the unknown solution was diluted prior to titration

$c = \dfrac{n}{V}$ $c_1V_1 = c_2V_2$

Including dilutions

Many titrations include dilutions, so require use of the formula $c_1V_1 = c_2V_2$. This step may be performed before or after the main titration calculation, depending on which solution was diluted.

- *If the standard solution was diluted,* then it's necessary to determine the concentration of the diluted standard solution used in the titration. As such, $c_1V_1 = c_2V_2$ is used *before* performing the main titration calculation.
- *If the unknown solution was diluted,* this concentration must first be determined to then calculate the concentration of the undiluted unknown solution. As such, $c_1V_1 = c_2V_2$ is used *after* performing the main titration calculation.

Ionic equations for redox titrations

When performing redox titrations, use the half-equation method to write ionic equations for the reaction between the standard solution and the unknown solution. This will remove spectator ions; for example, K^+ ions from $KMnO_4$.

tlvd-9656

SAMPLE PROBLEM 3 Calculating the concentration of ethanol using the results of a redox titration

The ethanol content in wine can be determined by a redox titration using acidified potassium dichromate. The ethanol in the wine is oxidised to ethanoic acid, while the orange dichromate ions, $Cr_2O_7^{2-}$, are reduced to green Cr^{3+} ions.

In a particular analysis, a 25.00 mL sample of wine was pipetted into a volumetric flask and carefully diluted to 250.0 mL. 20.00 mL aliquots were then titrated against 0.1500 M acidified potassium dichromate solution. The average titre obtained was 17.50 mL.

Calculate the concentration of ethanol, in mol L^{-1}, in the wine.

THINK	WRITE
1. Write the balanced ionic equation for the reaction, using the half-equation method, to obtain the mole ratios of the reactants.	$CH_3CH_2OH(aq) + H_2O(l)$ $\quad \rightarrow CH_3COOH(aq) + 4H^+(aq) + 4e^- \times 3$ $Cr_2O_7^{2-}(aq) + 14H^+(aq) + 6e^-$ $\quad \rightarrow 2Cr^{3+}(aq) + 7H_2O(l) \times 2$ $2Cr_2O_7^{2-}(aq) + 3CH_3CH_2OH(aq) + 16H^+(aq)$ $\quad \rightarrow 4Cr^{3+}(aq) + 3CH_3COOH(aq) + 11H_2O(l)$
2. Calculate the amount of the substance with the known concentration, $Cr_2O_7^{2-}$, by first identifying the variables and checking the units required. Volume is required in litres; convert mL to L.	$c(K_2Cr_2O_7) = 0.1500 \text{ M}$ $V(K_2Cr_2O_7) = \dfrac{17.50 \text{ mL}}{1000}$ $\quad = 0.01750 \text{ L}$ $n(K_2Cr_2O_7)_{reacted} = c \times V$ $\quad = 0.1500 \times 0.01750$ $\quad = 0.002\,625 \text{ mol}$ $n\left(Cr_2O_7^{2-}\right)_{reacted} = n(K_2Cr_2O_7)_{reacted}$ $\quad = 0.002\,625 \text{ mol}$
3. Use the mole ratio from the equation to calculate the amount of the unknown substance, CH_3CH_2OH, in the aliquot. To avoid rounding errors, an additional significant digit is retained in intermediate steps of the calculation. Only round the final answer to the appropriate number of significant figures.	$\dfrac{n(CH_3CH_2OH)_{diluted}}{n\left(Cr_2O_7^{2-}\right)_{reacted}} = \dfrac{3}{2}$ $\therefore n(CH_3CH_2OH)_{diluted} = \dfrac{3}{2} \times n(Cr_2O_7^{2-})_{reacted}$ $\quad = \dfrac{3}{2} \times 0.002\,625$ $\quad = 0.003\,937\,5 \text{ mol}$
4. The concentration of ethanol can be calculated using the amount from step 3 and the volume of the aliquot.	$c(CH_3CH_2OH)_{diluted} = \dfrac{n}{V}$ $\quad = \dfrac{0.003\,937\,5}{0.02000}$ $\quad = 0.19688 \text{ mol L}^{-1}$
5. The concentration of ethanol in the undiluted sample can be calculated using $c_1 V_1 = c_2 V_2$, in which the original undiluted sample is c_1.	$c_1 = \dfrac{c_2 V_2}{V_1}$ $\quad = \dfrac{0.19688 \times 250.0}{25.00}$ $\quad = 1.9688 \text{ mol L}^{-1}$ $\therefore c(CH_3CH_2OH)_{undiluted} = 1.969 \text{ mol L}^{-1}$

PRACTICE PROBLEM 3

The concentration of oxalic acid, $(COOH)_2$, in a vegetable smoothie can be determined by titration with acidified potassium permanganate. The oxalic acid is oxidised to carbon dioxide, while the purple/pink permanganate ions, MnO_4^-, are reduced to colourless Mn^{2+} ions.

A 25.00 mL sample of smoothie was diluted in a 100.0 mL volumetric flask. 20.00 mL aliquots of the diluted solution were then titrated with a 0.1015 M solution of acidified potassium permanganate. The average of the concordant titres obtained was 11.22 mL.

Calculate the concentration, in mol L^{-1}, of oxalic acid in the smoothie.

elog-2119

tlvd-9730

EXPERIMENT 9.3	online only

Standardisation of vitamin C

Aim

To determine the concentration of a vitamin C solution by titration with a standard solution of iodine

Determining purity

Titrations are frequently used to determine the concentration of an unknown solution, expressed as mass or moles of solute per volume of solvent; for example, mol L^{-1}. However, titrations can also be used to determine **percentage purity**. Purity is commonly expressed as %(m/m), although %(m/v) and %(v/v) may be encountered.

Many substances are impure mixtures containing a percentage of a particular substance. Examples include a mineral containing calcium carbonate and a tablet containing vitamin C.

FIGURE 9.22 a. The purity of a substance relates to the component of interest. **b.** Vitamin C tablets may contain additional substances and not be 100% pure.

a.

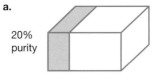
20% purity

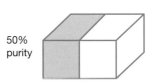
50% purity

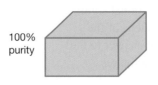
100% purity

b.

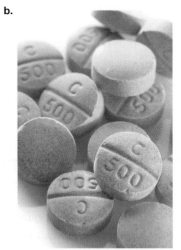

Percentage purity expressed as %(m/m) is calculated using the following formula:

$$\text{Percentage purity (\%)} = \frac{\text{mass of particular substance}}{\text{mass of sample}} \times 100$$

There are a few additional steps when using a titration to determine percentage purity as %(m/m).
- Weigh the sample, in g.
- Dissolve the sample in water to prepare a solution with an accurately known volume.
- Perform the titration.
- Calculate the mass, in g, of the particular substance in the sample.
- Calculate the percentage purity using the given formula.

percentage purity the percentage of a sample that is the desired substance

tlvd-9657

SAMPLE PROBLEM 4 Calculating the percentage purity of vitamin C from titration

Vitamin C (ascorbic acid) is found in many foods and supplements. It may be analysed by titration, in which it is oxidised to form dehydroascorbic acid.

HO
HO
O
O
HO
OH
Oxidation
Reduction
HO
HO
O
O
O
O

Ascorbic acid — Dehydroascorbic acid

A 0.7385 g vitamin C tablet was crushed and used to prepare a solution in a 200.0 mL volumetric flask. 20.00 mL aliquots of the solution were transferred to a conical flask and starch indicator added. This was then titrated against a 0.01833 M iodine (I_2) solution according to the following reaction equation. The average of the concordant titres obtained was 15.78 mL.

$$C_6H_8O_6(aq) + I_2(aq) \rightarrow C_6H_6O_6(aq) + 2I^-(aq) + 2H^+(aq)$$

Calculate the percentage purity %(m/m) of vitamin C in the tablet.

THINK

1. Calculate the amount, in mol, of the substance with the known concentration, I_2, by first identifying the variables and checking the units required. Volume is required in litres; convert mL to L.

WRITE

$$c(I_2) = 0.01833 \text{ mol L}^{-1}$$

$$V(I_2) = \frac{15.78 \text{ mL}}{1000} = 0.01578 \text{ L}$$

$$n(I_2)_{\text{reacted}} = c \times V = 0.01833 \times 0.01578 = 0.000\,289\,25 \text{ mol}$$

2. Use the mole ratio from the equation to calculate the number of moles of the unknown substance, $C_6H_8O_6$, in the aliquot.

$$\frac{n(C_6H_8O_6)_{\text{aliquot}}}{n(I_2)_{\text{reacted}}} = \frac{1}{1}$$

$$\therefore n(C_6H_8O_6)_{\text{aliquot}} = n(I_2)_{\text{reacted}} = 0.000\,289\,25$$

3. Calculate the amount, in mol, of vitamin C in the volumetric flask as it contains all the vitamin C from the tablet.

$$n(C_6H_8O_6)_{\text{vol. flask}} = \frac{V(\text{vol. flask})}{V(\text{aliquot})} \times n(C_6H_8O_6)_{\text{aliquot}} = \frac{200.0}{20.00} \times 0.000\,289\,25 = 0.002\,892\,5 \text{ mol}$$

4. To calculate the percentage purity %(m/m) of vitamin C in the tablet, the mass of vitamin C must first be determined by applying the formula $n = \frac{m}{M}$.

$$m(C_6H_8O_6) = n \times M = 0.002\,892 \times 176.0 = 5.0908 \text{ g}$$

5. The percentage purity %(m/m) of vitamin C can be calculated using the mass from step 4 and the mass of the tablet.
To avoid rounding errors, an additional significant digit is retained in intermediate steps of the calculation. Only round the final answer to the appropriate number of significant figures.

$$\%(m/m) = \frac{m(C_6H_8O_6)}{m(\text{tablet})} \times 100 = \frac{0.50908}{0.7385} \times 100 = 68.93\ \%(m/m)$$

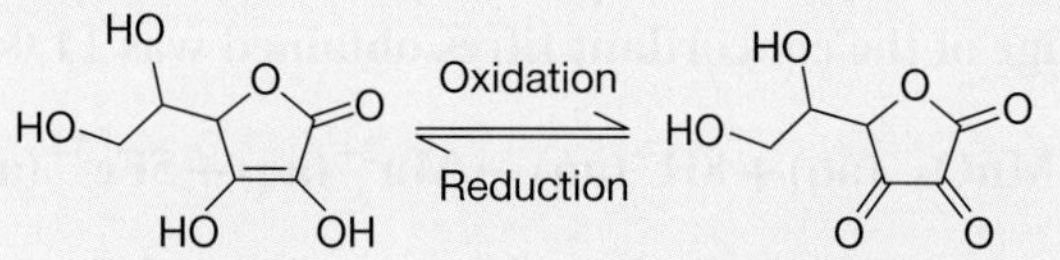

A 7.875 g sample of an iron-supplement powder was used to prepare a solution in a 100.0 mL volumetric flask. 20.00 mL aliquots of the solution were transferred to a conical flask and reacted with a 0.1258 M solution of acidified potassium permanganate, $KMnO_4$, according to the following reaction equation. The average of the concordant titres obtained was 11.04 mL.

$$5Fe^{2+}(aq) + MnO_4^-(aq) + 8H^+(aq) \rightarrow Mn^{2+}(aq) + 5Fe^{3+}(aq) + 4H_2O(l)$$

Calculate the percentage purity %(m/m) of $FeSO_4$ in the tablet.

9.4.3 Error analysis

Volumetric analysis is a laboratory technique that is capable of high accuracy (the correct result) and precision (repeatable results). However, errors can and do occur, so it is important to understand their effect on the final calculated value and how they could be reduced in the future. This process is called error analysis. Table 9.10 contains a brief reminder of the different error types.

TABLE 9.10 Types of error

Type of error	Effect on the calculated value	How to minimise impact
Systematic	Always biased in one direction; that is, the **erroneous value** is either higher or lower than the **true value** (if the error had not occurred)	Improve the design and/or implementation of the methodology
Random	More varied; that is, repeated measurements will be scattered over a greater range, both higher and lower than the true value	Use more precise equipment and/or methodology Take repeated measurements and calculate an average
Mistakes	Avoidable errors that mean the affected data cannot be used for calculations	Repeat the measurement, avoiding the mistake

The goal of error analysis

Error analysis is frequently performed following an experiment and any associated calculations. At this time, it is recognised that an error has occurred at an earlier step, meaning the calculated value is erroneous.

In this situation, the goal of error analysis is to identify which of the following is the case:
- The *erroneous* value is higher than the *true* value.
- The *erroneous* value is lower than the *true* value.
- The error has led to repeated measurements being scattered over a wider range than would otherwise occur.

Alternatively, if an experimental value is shown to be erroneous, possibly by comparison to a literature value, error analysis could also be used to identify a potential source of error.

Random errors

Random errors will lead to an increase in the range over which repeated measurements are scattered; that is, the data points are 'more varied'. Examples of random errors in titrations include:
- using equipment with low precision (measure to fewer decimal places) to make measurements; for example, delivering aliquots with a measuring cylinder rather than a volumetric pipette
- failing to read measurements to the appropriate level of precision; for example, taking burette readings to only one decimal place
- changes in the surroundings affecting measurements; for example, fluctuations in lighting affecting perception of colour and end point identification.

erroneous value an inaccurate value resulting from an error
true value an accurate value

Systematic errors

The effect of simple systematic errors can be obvious; for example, dilution of the unknown solution by residual water in a burette or pipette will lead to the calculated concentration being lower than the true value.

More complex scenarios may benefit from a structured approach.
1. Identify the point at which the error impacts the calculations; that is, where the erroneous value is first used.
2. Determine whether this *erroneous* value used in the calculation is higher or lower than the *true* value.
3. Follow the effect of the error through the calculations to determine whether the final value (generally the concentration of the unknown solution) is higher or lower than the true value. This may also be phrased as 'overestimated' or 'underestimated'.

This approach is illustrated in sample problem 5.

tlvd-9658

SAMPLE PROBLEM 5 Analysing experimental errors

Experiment: **The concentration of a vitamin C solution is determined by titration with an I_2 standard solution. The vitamin C solution is in the conical flask and the I_2 is in the burette.**

Error: **After the vitamin C concentration is calculated, the experimenter realises the burette was only rinsed with deionised water prior to titrating.**

Goal: **The experimenter now wishes to identify the effect of the error by determining whether the calculated concentration is higher or lower than the true value.**

THINK	WRITE
1. Identify the goal of the question.	To determine the effect of the error on the calculated c(vitamin C)
2. Identify the: i. quantity affected by the error ii. erroneous value that impacted the calculations iii. true value that would be used in an accurate calculation.	i. The I_2 solution will be diluted by the residual water in the burette, so the $c(I_2)$ will be affected. ii. The erroneous $c(I_2)$ value was the one previously used to calculate c(vitamin C), as it incorrectly assumed dilution had not occurred. iii. The true $c(I_2)$ value for this titration would be the actual concentration of the diluted I_2.
3. Identify whether the erroneous $c(I_2)$ is higher or lower than the true value.	The erroneous $c(I_2)$ used in the calculation was higher than the true value, as it did not take into account the dilution that had occurred.
4. Identify where erroneous $c(I_2)$ has been used and follow the calculation logic through to determine the effect on the c(vitamin C). *Note:* Arrows next to a quantity indicate that the value is higher or lower than the true value; for example, $\uparrow V(I_2)$ indicates that the volume of I_2 is higher than the true value.	1. $\uparrow n(I_2) = \uparrow c \times V$ 2. $\uparrow n(\text{vitamin C}) \propto \uparrow n(I_2)^*$ 3. $\uparrow c(\text{vitamin C}) = \dfrac{\uparrow n}{V}$ The calculated concentration of vitamin C is higher than the true value. *The mole ratio is not required to show the direction of the effect on n(vitamin C); that is, the number of moles of vitamin C is proportional to the number of moles of iodine.

PRACTICE PROBLEM 5

The concentration of a vitamin C solution is determined by titration with an I_2 standard solution. The vitamin C solution is in the burette and the I_2 is in the conical flask. After the vitamin C concentration is calculated, the experimenter realises the end point was overshot.

State whether the calculated vitamin C concentration is higher or lower than the true value. Justify your answer.

9.4 Activities

learn on

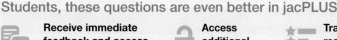

Students, these questions are even better in jacPLUS

 Receive immediate feedback and access sample responses

 Access additional questions

Track your results and progress

Find all this and MORE in jacPLUS ▶

| 9.4 Quick quiz on | 9.4 Exercise | 9.4 Exam questions |

9.4 Exercise

1. State what you would rinse the following items with before a titration.
   a. Burette
   b. Pipette
   c. Conical flask
   d. Volumetric flask
2. A new batch of iodine, I_2, was standardised with sodium thiosulfate, $Na_2S_2O_3$, according to the following reaction equation:

$$I_2(aq) + 2S_2O_3^{2-}(aq) \rightarrow S_4O_6^{2-} + 2I^-(aq)$$

   A 20.00 mL sample of iodine was diluted to 250.0 mL in a volumetric flask. A volumetric pipette was then used to transfer 20.00 mL aliquots of the diluted iodine into a conical flask. Starch indicator was added prior to titration with a 0.1500 mol L^{-1} solution of sodium thiosulfate. The average titre obtained was 21.58 mL.
   a. Calculate the amount, in mol, of $S_2O_3^{2-}$ in the average titre.
   b. Calculate the amount, in mol, of iodine in the aliquot.
   c. Calculate the concentration, in mol L^{-1}, of iodine in the aliquot.
   d. Calculate the concentration, in mol L^{-1}, of iodine in the undiluted batch.
3. Sulfur dioxide, SO_2, is used as a preservative in wine. The sulfur dioxide concentration in a bottle of wine was determined by titration with iodine, I_2, with the reaction producing sulfate ions, SO_4^{2-}, and iodide ions, I^-.

   A 20.00-mL aliquot of wine was transferred into a conical flask and starch solution was added.
   The 0.09021 mol L^{-1} iodine solution produced an average titre of 18.68 mL.
   a. Why was the starch solution added to the conical flask?
   b. Write the oxidation half-equation for SO_2.
   c. Write the reduction half-equation for I_2.
   d. Write the ionic equation for the overall reaction occurring during this titration, including states.
   e. Calculate the concentration, in mol L^{-1}, of SO_2 in the wine.

4. The active ingredient in hand sanitiser is ethanol. A bottle of hand sanitiser was analysed by titration to ensure it met the minimum 80 %(m/v) ethanol content.

A 10.00 mL aliquot of hand sanitiser was titrated against a 1.988 mol L^{-1} solution of acidified potassium permanganate. The reaction equation is as follows:

$$4MnO_4^-(aq) + 5CH_3CH_2OH(aq) + 12H^+(aq) \rightarrow 4Mn^{2+}(aq) + 5CH_3COOH(aq) + 11H_2O(l)$$

The average titre obtained was 45.74 mL.
   a. Calculate the amount, in mol, of ethanol in the aliquot.
   b. Determine the percentage purity of the hand sanitiser, expressed as %(m/v).
   c. Does the hand sanitiser meet the required standards?
5. A titration was performed with the unknown solution in a burette and the standard solution in a conical flask. State the effect of the following errors on the calculated concentration of the unknown solution. Justify your answer.
   a. Some of the unknown solution was spilt when filling the burette.
   b. The pipette used to transfer the aliquot of standard solution to the conical flask was rinsed with water.

9.4 Exam questions

▶ Question 1 (1 mark)
Source: VCE 2021 Chemistry Exam, Section A, Q.4; © VCAA

MC A titration was performed to determine the concentration of an ethanoic acid, $C_2H_4O_2$, solution using the following procedure:
1. 25.00 mL of the $C_2H_4O_2$ solution was pipetted into a conical flask.
2. A few drops of indicator were added to the flask.
3. A burette was filled with standard sodium hydroxide, NaOH, solution.
4. The $C_2H_4O_2$ solution was then titrated with the NaOH solution.
5. Steps 1–4 were repeated until three concordant titres were obtained.

A systematic error could result if the
   A. burette tap leaked during one of the titrations.
   B. burette readings were recorded to the nearest 0.1 mL.
   C. number of drops of indicator was not consistent for each titration.
   D. actual concentration of the standard NaOH solution was lower than the stated concentration.

▶ Question 2 (2 marks)
Source: VCE 2021 Chemistry Exam, Section B, Q.5.c; © VCAA

The loss of vitamin C, $C_6H_8O_6$, in sweet potato after heating can be determined in a titration by reacting vitamin C with iodine, I_2, solution. The balanced titration equation is shown below.

$$C_6H_8O_6(aq) + I_2(aq) \rightarrow 2HI(aq) + C_6H_6O_6(aq)$$

A sample of sweet potato was blended with water and filtered. The filtrate was titrated against 0.0500 M of I_2(aq). The average of three concordant titres was 21.81 mL.

Calculate the mass of vitamin C in the sweet potato sample.

Question 3 (1 mark)

Source: VCE 2021 Chemistry Exam, Section A, Q.23; © VCAA

MC A student titrated 25 mL aliquots of three different concentrations of an organic acid against a standardised potassium hydroxide, KOH, solution. The student's results are shown in the table below.

	KOH titre for Sample 1 (mL)	KOH titre for Sample 2 (mL)	KOH titre for Sample 3 (mL)
Titration 1	20.35	19.85	21.55
Titration 2	20.45	19.65	21.45
Titration 3	20.30	20.45	21.65
Average titre	20.37	19.98	21.55

Which one of the following statements is consistent with the results shown in the table?
A. Sample 2 is the most concentrated acid.
B. Sample 3 is the most concentrated acid.
C. There is not enough information to draw a valid conclusion.
D. The averages in the table are correct as the results are concordant.

Question 4 (1 mark)

Source: VCE 2020 Chemistry Exam, Section A, Q.24; © VCAA

MC A solution of citric acid, $C_3H_5O(COOH)_3$, was analysed by titration.

25.0 mL aliquots of the $C_3H_5O(COOH)_3$ solution were titrated against a standardised solution of 0.0250 M sodium hydroxide, NaOH. Phenolphthalein indicator was used and the average titre was found to be 24.0 mL.

Which one of the following would have resulted in a concentration that is higher than the actual concentration?
A. The pipette was rinsed with NaOH solution.
B. The pipette was rinsed with $C_3H_5O(COOH)_3$ solution.
C. The conical flask was rinsed with NaOH solution.
D. The conical flask was rinsed with $C_3H_5O(COOH)_3$ solution.

Question 5 (1 mark)

Source: VCE 2019 Chemistry Exam, Section A, Q.29; © VCAA

MC The concentration of vitamin C in a filtered sample of grapefruit juice was determined by titrating the juice with 9.367×10^{-4} M iodine, I_2, solution using starch solution as an indicator. The molar mass of vitamin C is 176.0 g mol^{-1}. The reaction can be represented by the following equation.

$$C_6H_8O_6(aq) + I_2(aq) \rightarrow C_6H_6O_6(aq) + 2H^+(aq) + 2I^-(aq)$$

The following method was used:
1. Weigh a clean 250 mL conical flask.
2. Use a 10 mL measuring cylinder to measure 5 mL of grapefruit juice into the conical flask and reweigh it.
3. Add 20 mL of deionised water to the conical flask.
4. Add a drop of starch solution to the conical flask.
5. Titrate the diluted grapefruit juice against the I_2 solution.

Which one of the following errors would result in an underestimation of the concentration of vitamin C in grapefruit juice?
A. 19 mL of deionised water was added to the conical flask.
B. The concentration of the I_2 solution was actually 9.178×10^{-4} M.
C. The initial volume of the I_2 solution in the burette was 1.50 mL, but it was read as 2.50 mL.
D. The balance was faulty and the measured mass of grapefruit juice was lower than the actual mass.

More exam questions are available in your learnON title.

9.5 Review

9.5.1 Topic summary

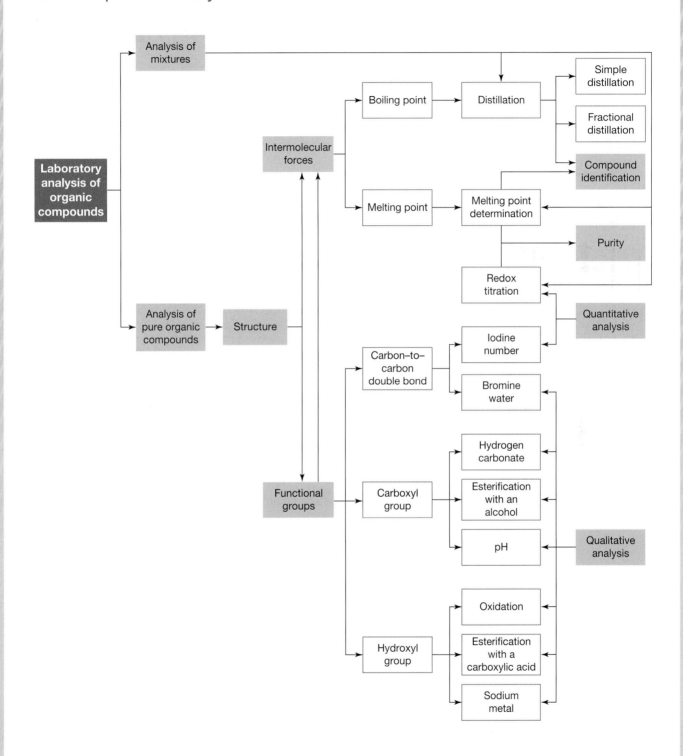

Hey students! Now that it's time to revise this topic, go online to:

Access the topic summary | Review your results | Watch teacher-led videos | Practise past VCAA exam questions

Find all this and MORE in jacPLUS

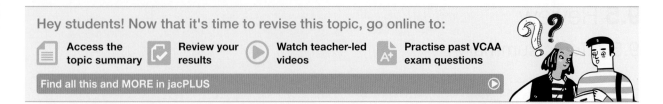

9.5.2 Key ideas summary

online only

9.5.3 Key terms glossary

online only

on Resources

Solutions Solutions — Topic 9 (sol-0836)

Practical investigation eLogbook Practical investigation eLogbook — Topic 9 (elog-1708)

Digital documents Key science skills — VCE Chemistry Units 1–4 (doc-37066)
Key terms glossary — Topic 9 (doc-37297)
Key ideas summary — Topic 9 (doc-37298)

Exam question booklet Exam question booklet — Topic 9 (eqb-0120)

9.5 Activities

learnon

Students, these questions are even better in jacPLUS

Receive immediate feedback and access sample responses | Access additional questions | Track your results and progress

Find all this and MORE in jacPLUS

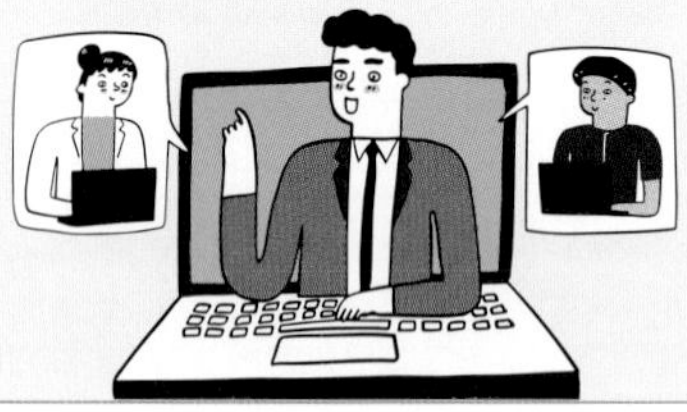

9.5 Review questions

1. Maleic and malic acid are closely related compounds that can be difficult to distinguish using qualitative tests.

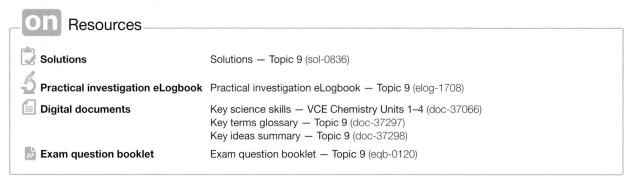

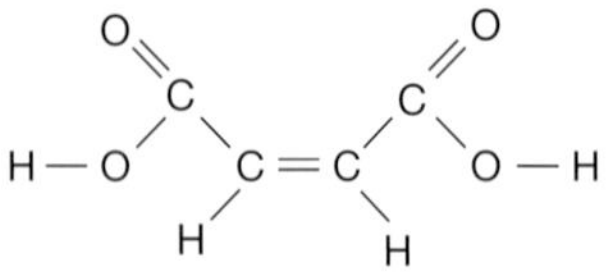

Maleic acid

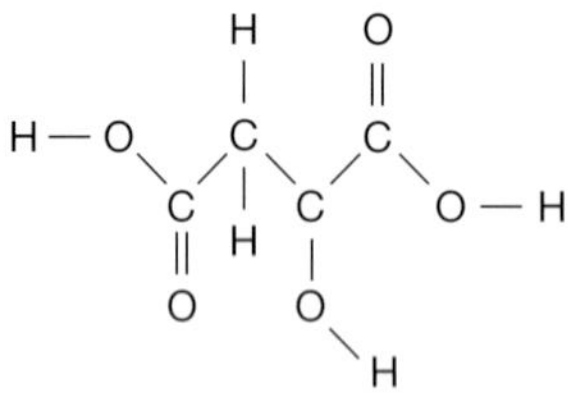

Malic acid

Identify two tests and the observations for each compound in the following scenarios.

a. Observations will be the same for both compounds.
b. Observations will be different for each compound.

2. Soap is produced by reacting a fat or oil with a metal hydroxide solution to produce a compound with both polar and non-polar regions. The degree of unsaturation of the fat or oil affects the hardness of the soap. Softer soaps produced from less saturated oils are preferred by customers.

a. Describe how the iodine number could be determined for a new batch of oil.
b. Calculate the minimum number of carbon-to-carbon double bonds required in each molecule for the iodine number to be at least 110. The oil has a molar mass of 876.6 g mol^{-1}.

3. Explain how the hydrocarbons in crude oil separate during fractional distillation.

4. The level of vitamin C (ascorbic acid) in citrus fruits can be determined by titration. The reaction involved produces dehydroascorbic acid and iodide ions as its products, and starch is used as an indicator. The equation for this reaction is as follows:

$$C_6H_8O_6(aq) + I_2(aq) \rightarrow C_6H_6O_6(aq) + 2I^-(aq) + 2H^+(aq)$$

In an experiment to determine the level of vitamin C in oranges, the juice from a 210 g orange was diluted with water up to the calibration line in a 100.0 mL volumetric flask. 20.00 mL of this solution was then transferred to another volumetric flask and water added to a final volume of 200.0 mL. 25.00 mL aliquots of this solution were titrated against a standardised 0.000 500 mol L^{-1} iodine solution. The average titre required was 24.80 mL.

a. Calculate the concentration, in mol L^{-1}, of vitamin C in the diluted 200.00 mL orange juice solution.
b. Calculate the mass, in g, of vitamin C in the 100.0 mL volumetric flask.
c. Calculate the concentration, in %(m/m), of vitamin C in the orange tested.

5. Malic acid, $C_4H_6O_5$, is used to impart the sour taste in candy. A student dissolved a 10.0 g sample of candy in distilled water, then transferred this solution to a 250.0 mL volumetric flask and made up to the mark with more distilled water. A 25.00 mL aliquot of the candy solution was placed in a conical flask.

The student then filled a burette with a 0.0100 mol L^{-1} acidified KMnO$_4$ solution and titrated it against the candy solution, with the reaction producing Mn^{2+} ions and $C_4H_4O_5$. This was repeated three times with an average titre of 15.50 mL.

a. Write the oxidation half-equation (states are not required).
b. Write the reduction half-equation (states are not required).
c. Write the ionic equation for the overall reaction occurring during this titration, including states.
d. Calculate the average amount, in mol, of malic acid in the 25.00 mL aliquot.
e. Calculate the %(m/m) of malic acid in the 10.0 g sample.
f. The student later realised that they had been using a 0.0200 mol L^{-1} solution of potassium permanganate. State the effect of this error on the calculated %(m/m) of malic acid in the candy. Justify your answer.

9.5 Exam questions

Section A — Multiple choice questions

All correct answers are worth 1 mark each; an incorrect answer is worth 0.

▶ Question 1

MC A series of tests were performed to determine the degree of unsaturation of two organic substances. Results were recorded in the following table.

Compound	Mixing with bromine water	Reaction with iodine
A	Colour change from orange-brown to colourless	The mass of a 10.0 g sample increased by 3.8 g.
B	Colour change from orange-brown to colourless	The mass of a 5.0 g sample increased by 2.3 g.

Consider the following statements:

I An addition reaction occurred between compound A and iodine.
II Both tests provided quantitative data.
III Compound B has a greater degree of unsaturation than compound A.

Which combination of statements is a correct interpretation of the results?

A. I and II only   **B.** II and III only   **C.** I and III only   **D.** I, II and III

Question 2

MC Which of the following combinations correctly describes the column temperature and mixture purity at the top of a fractionating column?

	Column temperature	Purity of mixture
A.	Lower	Higher
B.	Higher	Higher
C.	Lower	Lower
D.	Higher	Lower

Question 3

Source: VCE 2019 Chemistry Exam, Section A, Q.30; © VCAA

MC The concentration of vitamin C in a filtered sample of grapefruit juice was determined by titrating the juice with 9.367×10^{-4} M iodine, I_2, solution using starch solution as an indicator. The molar mass of vitamin C is 176.0 g mol^{-1}. The reaction can be represented by the following equation.

$$C_6H_8O_6(aq) + I_2(aq) \rightarrow C_6H_6O_6(aq) + 2H^+(aq) + 2I^-(aq)$$

The following method was used:
1. Weigh a clean 250 mL conical flask.
2. Use a 10 mL measuring cylinder to measure 5 mL of grapefruit juice into the conical flask and reweigh it.
3. Add 20 mL of deionised water to the conical flask.
4. Add a drop of starch solution to the conical flask.
5. Titrate the diluted grapefruit juice against the I_2 solution.

If the measured mass of grapefruit juice was 4.90 g and the titre was 21.50 mL, what was the measured percentage mass/mass %(m/m) concentration of vitamin C in the grapefruit juice?

A. 0.00987   **B.** 0.0723   **C.** 0.354   **D.** 3.36

Use the following information to answer Questions 4 and 5.

A clear, colourless liquid extract of the rhubarb plant was analysed for the concentration of oxalic acid, $H_2C_2O_4$, by direct titration with a recently standardised and acidified potassium permanganate solution, $KMnO_4(aq)$. The balanced equation for this titration is shown below.

$$2MnO_4^-(aq) + 5C_2O_4^{2-}(aq) + 16H^+(aq) \rightarrow 2Mn^{2+}(aq) + 10CO_2(g) + 8H_2O(l)$$
$$\text{purple} \qquad \text{colourless} \qquad\qquad\qquad \text{colourless}$$

The steps in the titration were as follows:

Step 1 – A 20.00 mL aliquot of the rhubarb extract was placed in a 200 mL conical flask.

Step 2 – The burette was filled with acidified 0.0200 M $KMnO_4$ solution.

Step 3 – The acidified 0.0200 M $KMnO_4$ solution was titrated into the rhubarb extract in the conical flask. The titration was considered to have reached the end point when the solution in the conical flask showed a permanent change in colour to pink. The volume of the titre was recorded.

Step 4 – The titration was repeated until three concordant results were obtained. The average of the concordant titres was 21.7 mL.

Question 4

Source: VCE 2018 Chemistry Exam, Section A, Q.18; © VCAA

MC Which of the following rinses is **least** likely to affect the accuracy of the results?

	Item	Rinse solution
A.	burette	distilled water
B.	burette	rhubarb extract
C.	pipette	$KMnO_4(aq)$
D.	conical flask	distilled water

Question 5

Source: VCE 2018 Chemistry Exam, Section A, Q.17; © VCAA

MC The concentration of $H_2C_2O_4$ in the rhubarb extract is closest to

A. 5.43×10^{-2} M

B. 5.00×10^{-2} M

C. 2.17×10^{-2} M

D. 7.40×10^{-4} M

Use the following information to answer Questions 6 and 7.

A group of students was required to determine the concentration of a solution of hydrochloric acid, HCl, provided for a titration competition. In each titration, a 25.00 mL aliquot of a freshly standardised solution of 0.2450 M sodium hydroxide, NaOH, was pipetted into a conical flask and titrated against the HCl solution. An appropriate indicator was added. The experiment was repeated until three concordant results were obtained.

The data for these titrations is shown in the following table.

volume of aliquot of NaOH	25.00 mL
concentration of NaOH solution	0.2450 M
mean titre of HCl solution	13.49 mL

Question 6

Source: VCE 2016 Chemistry Exam, Section A, Q.7; © VCAA

MC Based on these results, the concentration of HCl is

A. 0.1322 M   **B.** 0.4540 M   **C.** 1.322 M   **D.** 2.202 M

Question 7

Source: VCE 2016 Chemistry Exam, Section A, Q.8; © VCAA

MC The experimental value of the concentration of HCl obtained from these titrations was less than the actual value.

Which one of these actions by the students most likely accounts for the lower than expected result?

A. rinsing the burette with water

B. rinsing the pipette with water

C. rinsing the conical flask with water

D. leaving the funnel in the top of the burette

Question 8

Source: VCE 2013 Chemistry Exam, Section A, Q.3; © VCAA

MC In a titration, a 25.00 mL titre of 1.00 M hydrochloric acid neutralised a 20.00 mL aliquot of sodium hydroxide solution.

If, in repeating the titration, a student failed to rinse one of the pieces of glassware with the appropriate solution, the titre would be

A. equal to 25.00 mL if water was left in the titration flask after final rinsing.
B. less than 25.00 mL if the final rinsing of the burette is with water rather than the acid.
C. greater than 25.00 mL if the final rinsing of the 20.00 mL pipette is with water rather than the base.
D. greater than 25.00 mL if the titration flask had been rinsed with the acid prior to the addition of the aliquot.

Question 9

Source: VCE 2012 Chemistry Exam 1, Section A, Q.13; © VCAA

MC 15.0 mL of 10.0 M HCl is added to 60.0 mL of deionised water.

The concentration of the diluted acid is

A. 3.33 M
B. 2.50 M
C. 2.00 M
D. 0.500 M

Question 10

Source: VCE 2008 Chemistry Exam 1, Section A, Q.1; © VCAA

MC The diagram shows a section of a 50.00 mL burette containing a colourless solution.

The reading indicated on the burette is closest to

A. 14.50
B. 14.58
C. 15.42
D. 15.50

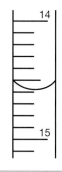

Section B — Short answer questions

Question 11 (3 marks)

Source: VCE 2020 Chemistry Exam, Section B, Q.7.c.ii; © VCAA

Calculate the mass of iodine, I_2, in grams, that reacts in an addition reaction with 100.0 g of the triglyceride formed from three palmitoleic acid molecules.

Source: VCE 2016 Chemistry Exam, Section B, Q.9.b; © VCAA

Standard solutions of sodium hydroxide, NaOH, must be kept in airtight containers. This is because NaOH is a strong base and absorbs acidic oxides, such as carbon dioxide, CO_2, from the air and reacts with them. As a result, the concentration of NaOH is changed to an unknown extent.

A 10.00 L container is completely filled with a freshly made 0.1000 M NaOH solution.

During a Chemistry class, 9.90 L of the solution is used and air enters the empty space above the remaining solution before the container is completely sealed off from the outside air.

The container is then opened. Air enters the container at 101.3 kPa and 21.5 °C. Assume that the concentration of CO_2 in the air is 0.0400 %(v/v).

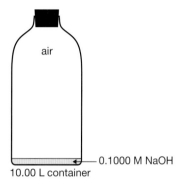

a. Calculate the amount of CO_2, in mol, that entered the container. **(2 marks)**

b. Calculate the amount of NaOH, in mol, that would be present in the solution that remains in the container. Assume that the NaOH did not react with the CO_2 in the air that entered when the container was opened. **(1 mark)**

c. The container is then shaken thoroughly, ensuring that all the CO_2 in the air is absorbed into the solution.

Calculate the resulting concentration of NaOH in the solution in the container. **(3 marks)**

▶ **Question 13 (3 marks)**

Source: VCE 2009 Chemistry Exam 1, Section B, Q.3.b,c; © VCAA

A student is to accurately determine the concentration of a solution of sodium hydrogencarbonate in a titration against a standard solution of hydrochloric acid, HCl.

The first step in this experiment is to accurately dilute 100.0 mL of a 1.00 M HCl stock solution to a 0.100 M solution using a 1.00 L volumetric flask.

However, instead of using distilled water in the dilution, the student mistakenly adds 900.0 mL of 0.0222 M sodium hydroxide, NaOH, solution.

a. Calculate the concentration of the hydrochloric acid in the 1.00 L volumetric flask after the student added the sodium hydroxide solution. Give your answer to correct significant figures. **(2 marks)**

b. The student then uses this contaminated hydrochloric acid solution to determine the accurate concentration of the unknown sodium hydrogencarbonate solution.

Will the calculated concentration of sodium hydrogencarbonate solution be greater or smaller than the true value? Justify your answer. **(1 mark)**

Source: VCE 2010 Chemistry Exam 1, Section B, Q.1; © VCAA

The amount of iron in a newly developed, heat-resistant aluminium alloy is to be determined.

An 80.50 g sample of alloy is dissolved in concentrated hydrochloric acid and the iron atoms are converted to Fe^{2+}(aq) ions.

This solution is accurately transferred to a 250.0 mL volumetric flask and made up to the mark.

20.00 mL aliquots of this solution are then titrated against a standard 0.0400 M potassium permanganate solution.

$$5Fe^{2+}(aq) + MnO_4^-(aq) + 8H^+(aq) \rightarrow 5Fe^{3+}(aq) + Mn^{2+}(aq) + 4H_2O(l)$$

Four titrations were carried out and the volumes of potassium permanganate solution used were recorded in the table below.

Titration number	1	2	3	4
Volume of $KMnO_4$ (mL)	22.03	20.25	21.97	21.99

a. Write a balanced half-equation, including states, for the conversion of MnO_4^- ions, in an acidic solution, to Mn^{2+} ions. **(2 marks)**

b. Calculate the average volume, in mL, of the concordant titres of the potassium permanganate solution. **(1 mark)**

c. Use your answer to **part b.** to calculate the amount, in mol, of MnO_4^-(aq) ions used in this titration. **(1 mark)**

d. Calculate the amount, in mol, of Fe^{2+}(aq) ions present in the 250.0 mL volumetric flask. **(2 marks)**

e. Calculate the percentage, by mass, of iron in the 80.50 g sample of alloy. Express your answer to the correct number of significant figures. **(3 marks)**

▶ **Question 15 (4 marks)**

Source: VCE 2013 Chemistry Exam, Section B, Q.5.b,a; © VCAA

A 20.00 mL aliquot of 0.200 M CH_3COOH (ethanoic acid) is titrated with 0.150 M NaOH.

The equation for the reaction between the ethanoic acid and NaOH solution is represented as

$$OH^-(aq) + CH_3COOH(aq) \rightarrow H_2O(l) + CH_3COO^-(aq)$$

a. Define the terms 'equivalence point' and 'end point'. **(2 marks)**

b. What volume of the NaOH solution is required to completely react with the ethanoic acid? **(2 marks)**

Hey students! Access past VCAA examinations in learnON

 Sit past VCAA examinations

 Receive immediate feedback

 Identify strengths and weaknesses

Find all this and MORE in jacPLUS

Hey teachers! Create custom assignments for this topic

 Create and assign unique tests and exams

 Access quarantined tests and assessments

 Track your students' results

Find all this and MORE in jacPLUS

10 Instrumental analysis of organic compounds

KEY KNOWLEDGE

In this topic you will investigate:

Instrumental analysis of organic compounds

- applications of mass spectrometry (excluding features of instrumentation and operation) and interpretation of qualitative and quantitative data, including identification of molecular ion peak, determination of molecular mass and identification of simple fragments
- identification of bond types by qualitative infrared spectroscopy (IR) data analysis using characteristic absorption bands
- structural determination of organic compounds by low resolution carbon-13 nuclear magnetic resonance (^{13}C-NMR) spectral analysis, using chemical shift values to deduce the number and nature of different carbon environments
- structural determination of organic compounds by low and high resolution proton nuclear magnetic resonance (^{1}H-NMR) spectral analysis, using chemical shift values, integration curves (where the height is proportional to the area underneath a peak) and peak splitting patterns (excluding coupling constants), and application of the $n + 1$ rule (where n is the number of neighbouring protons) to deduce the number and nature of different proton environments
- the principles of chromatography, including high performance liquid chromatography (HPLC) and the use of retention times and the construction of a calibration curve to determine the concentration of an organic compound in a solution (excluding features of instrumentation and operation)
- deduction of the structures of simple organic compounds using a combination of mass spectrometry (MS), infrared spectroscopy (IR), proton nuclear magnetic resonance (^{1}H-NMR) and carbon-13 nuclear magnetic resonance (^{13}C-NMR) (limited to data analysis)
- the roles and applications of laboratory and instrumental analysis, with reference to product purity and the identification of organic compounds or functional groups in isolation or within a mixture.

Source: VCE Chemistry Study Design (2024–2027) extracts © VCAA; reproduced by permission.

PRACTICAL WORK AND INVESTIGATIONS

Practical work is a central component of VCE Chemistry. Experiments and investigations, supported by a **practical investigation eLogbook** and **teacher-led videos**, are included in this topic to provide opportunities to undertake investigations and communicate findings.

EXAM PREPARATION

▶ Access past VCAA questions and exam-style questions and their video solutions in every lesson, to ensure you are ready.

10.1 Overview

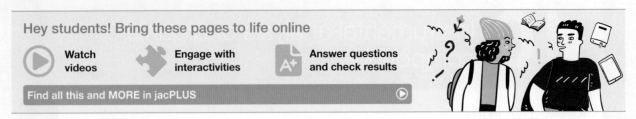
10.1.1 Introduction

Advances in the development of instrumentation to qualitatively and quantitatively analyse chemical compounds has allowed research and technology to progress at an extraordinary pace. These instruments take advantage of the interactions between electromagnetic radiation and matter, as well as the different interactions between matter itself.

This topic covers the analysis of organic compounds by qualitative spectroscopic techniques and introduces a quantitative version of chromatography.

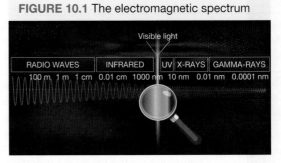

FIGURE 10.1 The electromagnetic spectrum

Visible light

RADIO WAVES | INFRARED | UV | X-RAYS | GAMMA-RAYS
100 m 1 m 1 cm 0.01 cm 1000 nm 10 nm 0.01 nm 0.0001 nm

In this topic you will learn how mass spectrometry (MS), and infrared (IR) and nuclear magnetic resonance (NMR) spectroscopy, are used to identify the structures of organic molecules. You will also learn how high-performance liquid chromatography (HPLC) is used to separate mixtures into their components for qualitative and quantitative analysis.

In combination, this powerful suite of techniques allows separation, identification and quantitation of the components in a mixture. As a consequence, such instrumentation is routinely used in the medical, food and pharmaceutical industries, as well as forensics.

You will learn how to use MS, and IR and NMR spectroscopy, to deduce the structure of simple organic compounds containing amino, hydroxyl, carbonyl and other common functional groups from spectra, and investigate how matter responds to different types of electromagnetic radiation. Using HPLC, you will utilise your understanding of intermolecular forces to identify and quantify organic compounds present in a mixture.

📲 Resources

📋 **Solutions**	Solutions — Topic 10 (sol-0837)
🔬 **Practical investigation eLogbook**	Practical investigation eLogbook — Topic 10 (elog-1709)
📄 **Digital documents**	Key science skills — VCE Chemistry Units 1–4 (doc-37066)
	Key terms glossary — Topic 10 (doc-37299)
	Key ideas summary — Topic 10 (doc-37300)
A+ **Exam question booklet**	Exam question booklet — Topic 10 (eqb-0121)

10.2 Mass spectrometry

10.2.1 Principles of mass spectroscopy

Technique overview

Mass spectrometry measures the **mass-to-charge ratio (*m/z*)** of particles, from which particle mass can easily be determined. The technique was introduced in Unit 1, topic 6 to determine atomic mass and the relative intensity (or abundance) of isotopes within a sample of an element. The mass spectrum in figure 10.2 shows that boron has two isotopes and that the heavier isotope is more abundant.

Similarly, analysis of organic compounds by mass spectrometry can:
- identify the molar mass of a compound
- provide information about the arrangement of atoms within a molecule.

When combined with physical separation of individual components by chromatography (subtopic 10.5), identification of molecules within complex mixtures can be performed using a single instrument. Analysis can be performed with as little as one or two milligrams of a sample, although the sample is destroyed in the process.

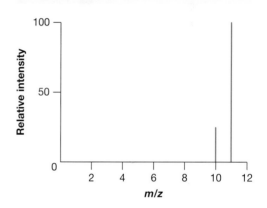

FIGURE 10.2 A mass spectrometer readout for boron

Ionisation and fragmentation

Ions are formed by two processes: ionisation and fragmentation.

1. Ionisation generates the parent or **molecular ion (M⁺)**, according to the following reaction equation, by bombarding the sample with high-energy electrons. Only positively charged ions are detected by the mass spectrometer.

$$M(g) + e^- \rightarrow M^+(g) + 2e^-$$

2. Fragmentation of the unstable molecular ion produces a variety of smaller ions as different bonds are broken. The specific fragmentation pattern can help determine the structure of the original molecule.

A mass spectrometer detects all the positive ions formed to determine their *m/z* ratio. Different positive ions are deflected to varying degrees by a magnetic field, allowing each ion to be detected.

mass spectrometry the investigation and measurement of the masses of isotopes, molecules and molecular fragments by ionising samples and separating the fragments produced, using a combination of electric and magnetic fields

mass-to-charge ratio (*m/z*) the mass of a particle divided by its overall charge; when the charge is +1 the *m/z* and mass have the same numerical value

molecular ion the positive ion produced by ionisation of a whole molecule

EXTENSION: Mass spectrometer

The details of the instrumentation are not required in this course. Nevertheless, an awareness of the process provides useful context when analysing mass spectra data.

FIGURE 10.3 Operation of a mass spectrometer

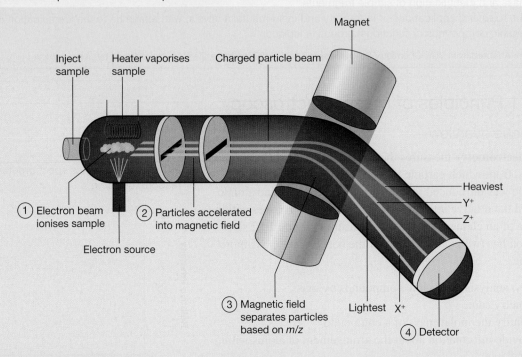

Key steps:
1. Solid and liquid samples are vaporised and an electron beam is used to ionise the gaseous sample.
2. Positive ions are accelerated through the instrument by an electric field.
3. Ions are deflected by a magnetic field according to the mass-to-charge ratio (m/z). The lighter the ion, the greater the deflection. In the example shown in figure 10.3, which comprises three ions of like charge, the mass of $X^+ < Y^+ < Z^+$, as X^+ is deflected the most and Z^+ the least. Ions with a larger positive charge will be deflected to a greater degree, although are not considered in the VCE curriculum.
4. A detector measures the m/z and relative intensity of each ion to generate a mass spectrum.

Spectrometry or spectroscopy?

Infrared (IR) spectroscopy and nuclear magnetic resonance (NMR) spectroscopy are related techniques that will be introduced in subtopics 10.3 and 10.4, and used in conjunction with mass spectrometry to determine the structure of simple organic molecules in subtopic 10.5. However, unlike IR and NMR spectroscopy, mass spectrometry is not strictly **spectroscopy** because it does not use electromagnetic radiation.

spectroscopy the investigation and measurement of spectra produced when matter interacts with or emits electromagnetic radiation

10.2.2 Interpreting mass spectra

Mass spectra plot the *m/z* versus the abundance of each fragment.

Consider the mass spectrum of propane (see figure 10.4a). Observe how each fragment ion produces a specific peak.

FIGURE 10.4 a. Mass spectrum for propane **b.** Formation of fragments of propane

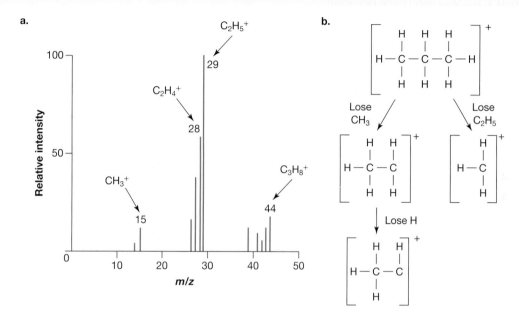

The peak with the largest *m/z* is due to the molecular ion. The *m/z* of this peak is 44, which can be used to infer the molar mass of the compound as 44 g mol⁻¹.

The most abundant ion generates the **base peak** and is usually assigned a height of 100. This is the most common fragment, either because it is the most stable or because it can be formed in different ways.

Reading the *x*-axis of mass spectra

The *x*-axis represents mass/charge, although the only ions considered in this course are unipositive (+1), so it is effectively a mass scale. Nevertheless, when reading values directly from the graph they should be stated as *m/z* rather than with units involving mass; for example, at *m/z* 16, not 16 g mol⁻¹.

Fragmentation may involve breakage of multiple bonds and lead to the generation of a large number of fragments. For example, in figure 10.4b the molecular ion undergoes breakage of a C–C and C–H bond to produce $C_2H_4^+$. The fragmentation pattern provides useful structural clues. However, it may be complex and it is not always necessary to identify every peak in the spectrum.

FIGURE 10.5 Fragmentation of the propane molecular ion at a C–C bond may result in two different signals. Only the positively charged (blue) fragments are detected at the *m/z* shown.

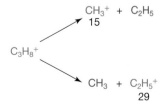

> **base peak** identifies the most abundant ion in a mass spectrum

Table 10.1 shows *m/z* ratios for a selection of small ions that may be observed in a mass spectrum. Sometimes it is easier to determine the fragment causing a peak by subtracting the fragment that is lost. For example, if a peak is present with an *m/z* value 15 less than the *m/z* of the molecular ion, this suggests that a methyl group, CH_3, was part of the molecule.

TABLE 10.1 *m/z* values for small ions

m/z	Positively charged fragment
15	CH_3^+
28	CO^+
29	$C_2H_5^+$, CHO^+
31	CH_3O^+
35 (37)	$^{35}Cl^+$ ($^{37}Cl^+$)
44	$C_3H_8^+$
45	$COOH^+$

Notation for ions

Include the positive charge on ions detected in mass spectra. Square brackets — for example, $[C_3H_8]^+$ — are frequently used for larger fragments, given that the location of the electron removed and hence, the location of the positive charge centre, can vary between ions. Use of such brackets is optional.

Ions are gaseous

When writing ionisation equations, ensure the states are gaseous (g).

SAMPLE PROBLEM 1 Interpreting the mass spectra of a carboxylic acid

tlvd-9659

The mass spectrum for methanoic acid, HCOOH, is shown.

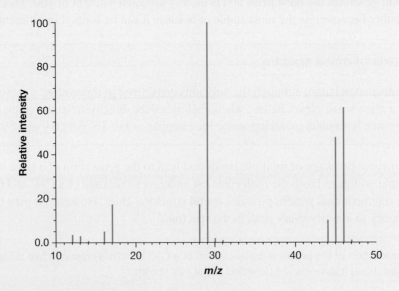

a. Write the formula for the molecular ion.
b. Identify the *m/z* of the molecular ion and compare this to the molar mass of methanoic acid.
c. Write the ionisation equation for the formation of the molecular ion, including states.
d. Identify the *m/z* of the base peak and propose a formula for the ion responsible for this peak.
e. What particle must have been lost from the molecular ion to produce the ion responsible for the base peak?

THINK	WRITE
a. The formula of the molecular ion is that of the molecule with a +1 charge.	a. $HCOOH^+$
b. Observe the peak with the largest m/z. Calculate the $M(HCOOH)$ and compare to the m/z.	b. The molecular ion has an m/z of 46. $M(HCOOH) = 12.0 + (2 \times 1.0) + (2 \times 16.0)$ $= 46.0\,g\,mol^{-1}$ They are the same.
c. HCOOH is bombarded with an electron to form the molecular ion. *Note:* The states are gaseous.	c. $HCOOH(g) + e^- \rightarrow HCOOH^+(g) + 2e^-$
d. The base peak is the tallest peak. Identify a fragment of the molecular ion with the same molar mass as the m/z of this peak.	d. The base peak has an m/z of 29. $M(CHO^+) = 12.0 + 1.0 + 16.0$ $= 29.0\,g\,mol^{-1}$
e. Remove the fragment responsible for the base peak from the molecular ion. *Note:* This fragment is not charged so will not be detected.	e. Removal of CHO from $HCOOH^+$ results in OH.

PRACTICE PROBLEM 1

The mass spectrum for ethanol, CH_3CH_2OH, is shown.

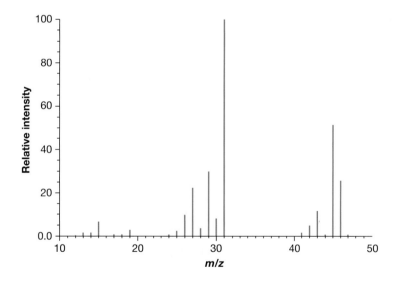

a. Write the formula for the molecular ion.
b. Identify the m/z of the molecular ion and compare this to the molar mass of ethanol.
c. Write the ionisation equation for the formation of the molecular ion, including states.
d. Identify the m/z of the base peak and propose a formula for the ion responsible for this peak.
e. What particle must have been lost from the molecular ion to produce the ion responsible for the base peak?

The isotope effect

Close observation of mass spectra will frequently reveal a very small peak with an m/z ratio slightly higher than that of the molecular ion. These peaks are due to the **isotope effect** and indicate that the molecular ion contains an atom of a heavier isotope. The molecular ion in sample problem 1, $HCOOH^+$, contains an atom of ^{13}C in place of the much more common ^{12}C. Given approximately one per cent of carbon on Earth is ^{13}C, these peaks are approximately $\dfrac{1}{100}$ as tall as the molecular ion peak.

isotope effect the generation of multiple peaks for fragments with the same formula due to the presence of isotopes of the constituent elements

The isotope effect is most evident when analysing compounds containing chlorine and bromine. As detailed in table 10.2, each of these elements has two abundant isotopes, resulting in the molecular ion being represented by two clear peaks with an m/z difference of two.

TABLE 10.2 Common isotopes

Element	Isotope 1		Isotope 2	
	A_r	Abundance (%)	A_r	Abundance (%)
Carbon	12	99	13	1
Chlorine	35	76	37	24
Bromine	79	51	81	49

SAMPLE PROBLEM 2 Determining ions in a mass spectrum diagram

tlvd-9704

The following diagram shows the mass spectrum for chloroethane, C_2H_5Cl. What ions are responsible for the peaks at $m/z = 66$ and 64, 49 and 51, 29 and 28?

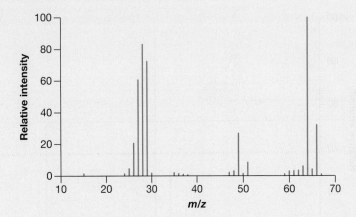

THINK

1. Determine the molecular mass of chloroethane, remembering chorine has two isotopes: ^{35}Cl and ^{37}Cl.

2. Check for isotope peaks (e.g. for carbon and chlorine).

WRITE

$M(C_2H_5{}^{35}Cl) = (2 \times 12.0) + (5 \times 1.0) + 35$
$\qquad = 64.0$

$M(C_2H_5{}^{37}Cl) = (2 \times 12.0) + (5 \times 1.0) + 37$
$\qquad = 66.0$

The peak at $m/z = 64$ corresponds to $[C_2H_5{}^{35}Cl]^+$.
The peak at $m/z = 66$ corresponds to $[C_2H_5{}^{37}Cl]^+$.

3. Look for a difference of 15 lost from the molecular ion, showing the loss of a methyl group, CH_3, to identify fragments.

$64 - 15 = 49$ corresponds to:

$[C_2H_5{}^{35}Cl]^+ - CH_3 = [CH_2{}^{35}Cl]^+$

Hence, $[CH_2{}^{35}Cl]^+$ shows a peak at $m/z = 49$.

$66 - 15 = 51$ corresponds to:

$[C_2H_5{}^{37}Cl]^+ - CH_3 = [CH_2{}^{37}Cl]^+$

Hence, $[CH_2{}^{37}Cl]^+$ shows a peak at $m/z = 51$.

4. Identify any other fragments (table 10.1 may assist) and calculate the masses to confirm the peaks observed.

$[C_2H_5]^+$ occurs at $m/z = 29$, and the peak for $[C_2H_4]^+$ occurs at $m/z = 28$.

5. Write a concluding statement.

$m/z = 64$ corresponds to $[C_2H_5{}^{35}Cl]^+$.

$m/z = 66$ corresponds to $[C_2H_5{}^{37}Cl]^+$.

$m/z = 49$ corresponds to $[CH_2{}^{35}Cl]^+$.

$m/z = 51$ corresponds to $[CH_2{}^{37}Cl]^+$.

$m/z = 29$ corresponds to $[C_2H_5]^+$.

$m/z = 28$ corresponds to $[C_2H_4]^+$.

PRACTICE PROBLEM 2

The following spectrum is for chloromethane, CH_3Cl. Which ions are responsible for the peaks at $m/z = 52, 50$ and 15?

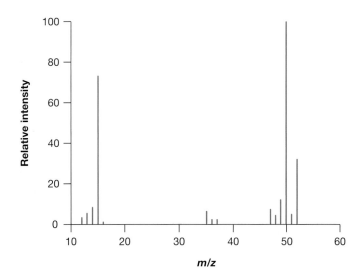

on Resources

Interactivity Interpreting a mass spectrum (int-1230)

10.2 Activities

learnon

Students, these questions are even better in jacPLUS

- Receive immediate feedback and access sample responses
- Access additional questions
- Track your results and progress

Find all this and MORE in jacPLUS

10.2 Quick quiz	10.2 Exercise	10.2 Exam questions

10.2 Exercise

1. How can the *m/z* ratio on a mass spectrum be thought of as a relative mass scale?
2. **MC** In the spectrum shown, which peak corresponds to the $[CH_2F]^+$ fragment?
   A. 1
   B. 2
   C. 3
   D. 4

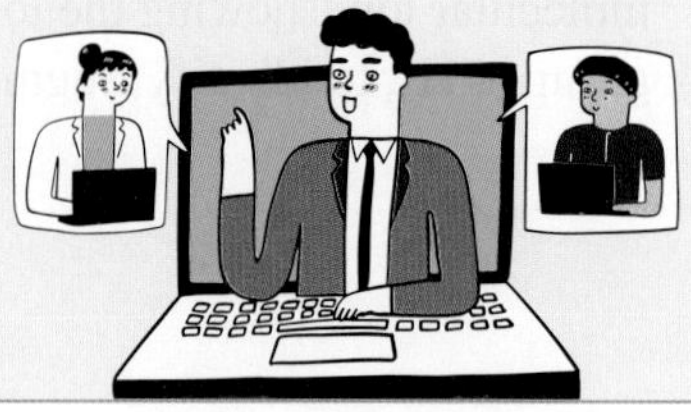

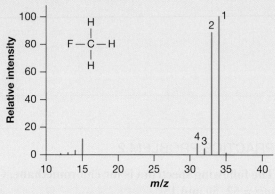

3. Write an equation for the molecular ion formation of ethanoic acid in a mass spectrometer.
4. Propanone, commonly called acetone, CH_3COCH_3, is an important solvent in industry. In a mass spectrometer, propanone breaks down into a series of fragment ions as shown in the figure.
   a. Which peak corresponds to $CH_3COCH_3{}^+$?
   b. Identify which fragment ions correspond to the other labelled peaks.

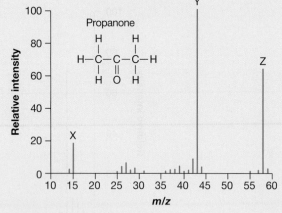

5. A molecule with the formula C_3H_9N produced the mass spectrum shown.
   a. What is the *m/z* ratio of the molecular ion?
   b. What is the *m/z* ratio of the ion responsible for the base peak?
   c. Write the formula of the molecular ion.
   d. Write the formula of the fragment that produced the base peak.

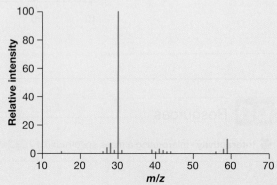

6. The mass spectrum shown is produced by an aldehyde.
   a. Write the m/z that would be caused by the aldehyde functional group.
   b. Write the formula for a fragment that may be responsible for the peak at m/z = 43.
   c. What is the peak at m/z = 43 called?
   d. Identify the molecular formula of the compound using the molar mass.
   e. Name the aldehyde that produced the spectrum.

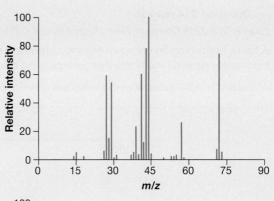

7. A compound with the empirical formula CH_2Cl produced the spectrum shown. Deduce the molecular formula of the compound.

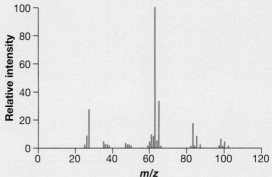

10.2 Exam questions

▶ Question 1 (4 marks)
Source: VCE 2017 Chemistry Exam, Section B, Q.5.b; © VCAA

There are a number of structural isomers for the molecular formula C_3H_6O. Three of these are propanal, propanone and prop-2-en-1-ol.

The skeletal structure for the aldehyde propanal is as follows.

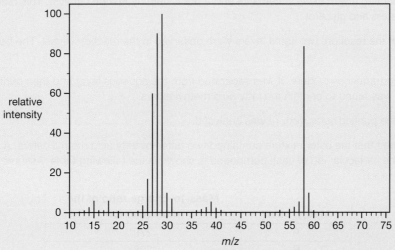

The mass spectrum below was produced by one of the three named isomers of C_3H_6O.

Data: SDBS Web, <http://sdbs.db.aist.go.jp>,
National Institute of Advanced Industrial Science and Technology

a. Identify the fragment at 29 m/z.   (1 mark)
b. Name the isomer of C_3H_6O that produced this spectrum and justify your answer.   (3 marks)

▶

Question 2 (4 marks)

Source: VCE 2016 Chemistry Exam, Section B, Q.4.a; © VCAA

A bottle containing an unknown organic compound was examined in a university laboratory. There was an incomplete label on the bottle that gave only the empirical formula for the contents: CH_4N.

A chemist hypothesised that the unknown compound was 1,2-ethanediamine, $NH_2CH_2CH_2NH_2$.

Mass spectrometry produced the following spectral data.

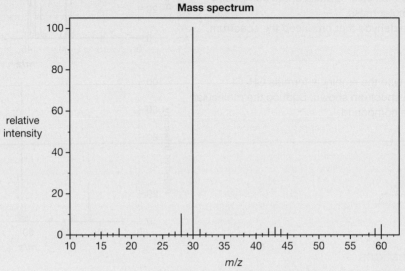

Data: SDBS Web, http://sdbs.db.aist.go.jp,
National Institute of Advanced Industrial Science and Technology

a. On the diagram, circle the base peak. **(1 mark)**

b. At what *m/z* ratio is the principal peak that supports the chemist's hypothesis that the unknown compound has the formula $NH_2CH_2CH_2NH_2$? Justify your answer. **(2 marks)**

c. Write the semi-structural formula of the species that produces the peak at 30 *m/z*. **(1 mark)**

Question 3 (1 mark)

Source: Adapted from VCE 2015 Chemistry Exam, Section B, Q.9.c; © VCAA

Biodiesel is a mixture of fatty acid methyl esters. A particular triglyceride used in the manufacture of biodiesel was analysed by reacting it with excess methanol and a potassium hydroxide catalyst. This reaction produced fatty acid methyl esters and glycerol.

At the conclusion of the reaction, two liquid layers were observed in the reaction vessel. The bottom layer was an aqueous solution.

The top layer is a non-aqueous mixture. It was separated from the aqueous layer and then purified. The non-aqueous layer was found to contain the fatty acid methyl esters.

A small sample of the purified ester mixture was analysed.

The analysis indicated that the ester mixture contained two different fatty acid methyl esters, A and B. The mass-to-charge ratio of the molecular ion of each compound is shown in the following table. Assume that the charge on each molecular ion is +1.

Methyl ester	Mass-to-charge ratio of the molecular ion
A	270
B	298

The mass spectrum of methyl ester A corresponds to that of methyl palmitate, $CH_3(CH_2)_{14}COOCH_3$.

What is the semi-structural formula of methyl ester B? (Refer to the VCE Chemistry Data Book.)

▶ Question 4 (1 mark)

Source: Adapted from VCE 2014 Chemistry Exam, Section A, Q.13; © VCAA

MC Four straight chain alcohols, S, T, U, V, with a general formula ROH, were analysed using a mass spectrometer.

The mass spectrum of alcohol T is provided below.

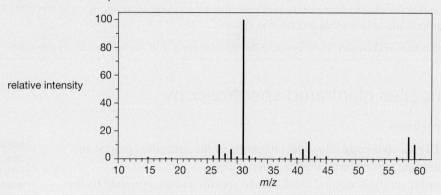

Source: National Institute of Advanced Industrial Science and Technology

What is alcohol T?

A. butan-1-ol

B. ethanol

C. methanol

D. propan-1-ol

▶ Question 5 (1 mark)

Source: VCE 2013 Chemistry Sample Exam for Units 3 and 4, Section A, Q.15; © VCAA

MC The mass spectrum of an unknown compound is given below. The empirical formula of this compound is CH_4N.

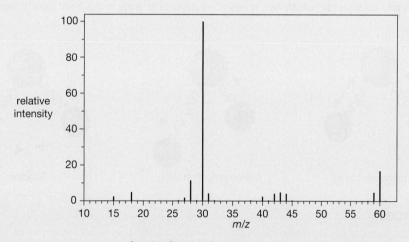

Source: Spectral Database for Organic Compounds SDBS

Which of the following correctly identifies the relative molecular mass and the formula of the molecular ion of this unknown compound?

	Relative molecular mass	Formula of the molecular ion
A.	60	$C_2H_8N_2^+$
B.	60	$C_2H_8N_2$
C.	30	CH_4N^+
D.	30	CH_4N

More exam questions are available in your learnON title.

10.3 Infrared spectroscopy

10.3.1 Principles of infrared spectroscopy

Technique overview

- **Infrared (IR) spectroscopy** allows identification of the functional groups and single, double and triple bonds in organic molecules.
- This **qualitative analysis** method identifies the specific energy absorbed by the various covalent bonds present in a molecule when exposed to radiation in the IR portion of the electromagnetic spectrum.
- Once the bonds are identified, the functional groups present in the molecule may be inferred.

infrared (IR) spectroscopy describes spectroscopy that deals with the infrared region of the electromagnetic spectrum

qualitative analysis the determination of non-numerical information, such as the presence or absence of elements, ions, functional groups or molecules in a sample

dipole moment a way to describe the asymmetrical charge distribution in a polar molecule

Vibration of covalent bonds

Covalent bonds can be likened to vibrating springs. When IR radiation is absorbed it can change the vibration of covalent bonds, leading to bending and stretching.

FIGURE 10.6 Infrared radiation can cause two types of vibrational change in covalent bonds: bending and stretching.

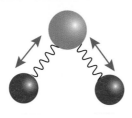
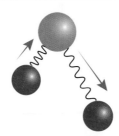
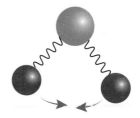

Symmetric stretching   Asymmetric stretching   Bending

 **Resources**

▶ **Video eLesson** Bending and stretching of covalent bonds caused by infrared radiation (eles-6040)

Just as the electrons in atoms have a number of possible electronic energy levels, bond vibrations have a number of possible vibrational energy levels. Therefore, it is possible to talk about 'ground-state' vibrational energy levels and 'excited-state' vibrational energy levels. A molecule can move from a lower to a higher vibrational energy level if it absorbs an amount of energy equal to the difference between levels. The region of the electromagnetic spectrum corresponding to such amounts of energy is the IR region.

The specific energy required for these transitions in molecular vibration, and therefore the frequency (or wavelength), can give clues about the types of covalent bonds present.

Not all molecules absorb infrared radiation. Only molecules that have a change in **dipole moment** when they absorb IR energy will generate a signal in an IR spectrum. For example, molecules consisting of two atoms of the same element, such as N_2 and O_2, do not absorb infrared radiation, so will not generate a signal.

Change in vibration

The covalent bonds analysed by IR spectroscopy are already vibrating before they are analysed. Absorbed infrared radiation *changes* the vibration of a covalent bond in a manner that can be detected, providing clues about the type of bond present.

Infrared spectra

An IR spectrum looks upside-down compared with a UV–visible or atomic absorption spectroscopy (AAS) spectrum. This is because it measures transmittance, which is the opposite of absorbance, on the vertical (y) axis.

FIGURE 10.7 Those wavelengths of IR radiation that are not absorbed by covalent bonds in a molecule are transmitted.

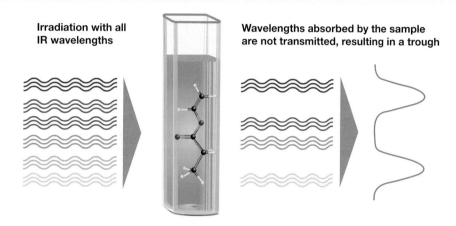

The IR spectrum has a base line of 100 per cent transmittance running along the top of the graph, meaning that no energy/wavelengths from the IR region have been absorbed by the sample. A peak occurs in the UV–visible spectrum when energy is absorbed, whereas a trough appears in the IR spectrum when energy is absorbed. However, it is reasonable to refer to IR absorbances as either troughs or peaks.

FIGURE 10.8 An example of an IR spectrum with wave number (cm^{-1}) on the x-axis and transmittance (%) on the y-axis. The peaks are characteristic of specific covalent bonds.

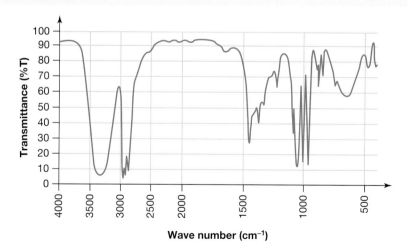

The IR spectrum measures **wave number**, which is the inverse of wavelength, on the horizontal (x) axis.

wave number the number of waves per centimetre

Calculating wave number

$$\frac{1}{\text{wavelength (cm)}} = \text{wave number (cm}^{-1})$$

The relationship between the energy required to change the vibrational energy of the covalent bond and the wave number is shown in table 10.3.

TABLE 10.3 The relationship between the energy, frequency and wavelength of IR radiation required to change the vibration of a covalent bond, and the wave number observed in IR spectra

Energy required to change bond vibration	Frequency	Wavelength	Wave number
Low	Low	Long	Low
High	High	Short	High

The region with wave numbers above 1500 cm^{-1} can be used to identify the functional groups present. Table 10.6 contains information allowing identification of peaks in this region and attribution of them to certain types of bonds.

The region with wave numbers below 1500 cm^{-1} is known as the **fingerprint region**, as the pattern of peaks is unique for each molecule. It can be challenging to identify the bonds associated with individual peaks in this region. However, the fingerprint region in the IR spectrum of an unknown compound may be compared to those of known compounds to find a match, which is almost certain evidence for the identity of the compound being analysed. This is illustrated in figure 10.9, in which two hydrocarbons that only differ by a CH$_3$ group have unique peak patterns in the fingerprint region.

FIGURE 10.9 Small differences in molecular structure are observed in the fingerprint region.

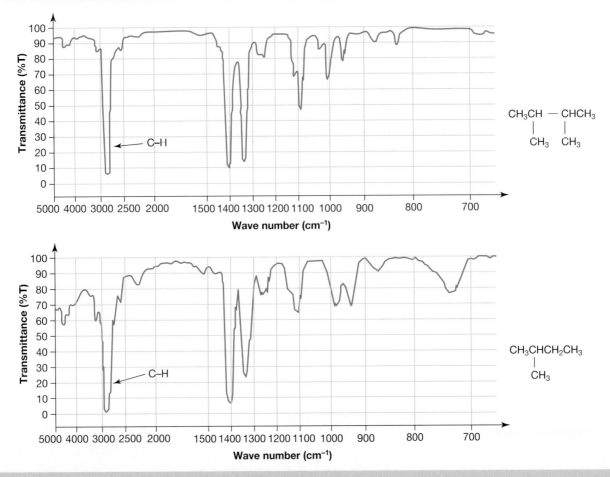

Factors affecting bond vibration energy

Why do different bonds absorb different wavelengths of infrared radiation? Recall that covalent bonds can be likened to springs (bonds) connecting weights (atoms).

FIGURE 10.10 Covalent bonds are similar to springs connecting weights.

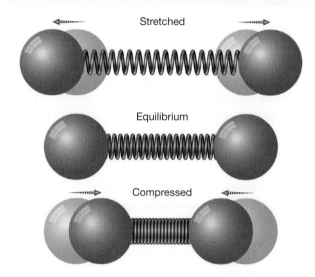

Variation in both of the following factors affects the energy of the IR energy absorbed:
- The strength of the bonds
- The mass of the atoms.

Comparison of similar bonds demonstrates some general trends in the energy required to change their vibration:
- Seen in table 10.4: Stronger bonds require more energy to change their vibration, will vibrate at a higher frequency and have a higher wave number than weaker bonds. Picture a short, stiff spring vibrating more rapidly than a long, flexible spring.
- Seen in table 10.5: Bonds between lighter atoms require more energy to change their vibration, will vibrate at a higher frequency and have a higher wave number than bonds between heavier atoms. Picture lighter weights vibrating more rapidly than heavier weights.

TABLE 10.4 The effect of bond strength (mean bond enthalpy) on bond length and wave number (cm^{-1})

Bond	Mean bond enthalpy (kJ mol^{-1})	Length (pm)	Wave number ranges (cm^{-1})
C–O (alcohols, esters, ethers)	358	143	1050–1410
C=O (aldehydes)	804	122	1660–1745

Source: Adapted from VCE Chemistry Written Examination Data Book (2022) extracts © VCAA; reproduced by permission.

TABLE 10.5 The effect of relative atomic mass (A_r) on wave number (cm^{-1})

Bond	A_r	Wave number ranges (cm^{-1})
C–H (alkanes, alkenes, arenes)	C 12.0	2850–3090
N–H (amines and amides)	N 14.0	3300–3500
O–H (alcohols)	O 16.0	3200–3600

Source: Adapted from VCE Chemistry Written Examination Data Book (2022) extracts © VCAA; reproduced by permission.

10.3.2 Interpreting infrared spectra

Wave number ranges for various bonds are shown in table 10.6. The variety of molecular environments in which the bonds listed are found affect the wave number in a variety of ways. This leads to wave numbers across a characteristic range of values for each bond. Also, the shape (or intensity) of the peaks can provide useful information when identifying bonds.

The IR spectrum for methanol in figure 10.12 shows characteristic peaks at 3200–3600 cm^{-1} for –O–H (alcohols) and 2850–3090 cm^{-1} for –C–H (alkanes, alkenes, arenes). The C–H peak is almost always present in organic molecules and is less helpful because it is not a characteristic identifier.

FIGURE 10.11 Common peak intensities observed in IR spectra.

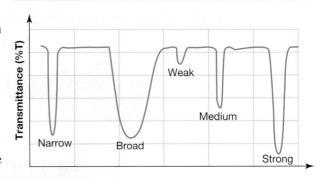

TABLE 10.6 Characteristic ranges for IR absorption and peak intensity

Bond	Wave number (cm^{-1})	Intensity
C–Cl (chloroalkanes)	600–800	Strong
C–O (alcohols, esters, ethers)	1050–1410	Strong
C=C (alkenes)	1620–1680	Medium-weak
C=O (amides)	1630–1680	Strong
C=O (aldehydes)	1660–1745	Strong
C=O (acids)	1680–1740	Strong
C=O (ketones)	1680–1850	Strong
C=O (esters)	1720–1840	Strong
C–H (alkanes, alkenes, arenes)	2850–3090	Strong
O–H (acids)	2500–3500	Strong, very broad
O–H (alcohols)	3200–3600	Strong, broad (narrower than acids)
N–H (primary amines)	3350–3500	Medium, two bands

Source: Adapted from VCE Chemistry Written Examination Data Book (2022) extracts © VCAA; reproduced by permission.

FIGURE 10.12 The IR spectrum for methanol, CH_3OH, with characteristic peak wave number and intensity for O–H and C–H. The C–O peak within the fingerprint region is only identifiable in very simple molecules.

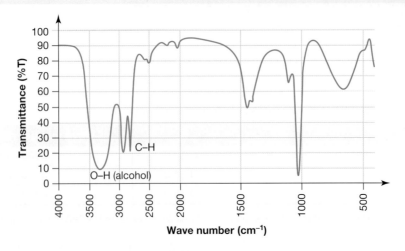

Identifying peaks in IR spectra

When identifying peaks in IR spectra, refer specifically to the bonds and wave number ranges in the VCE Chemistry Data Book.

on Resources

Video eLesson Infrared spectroscopy (eles-3523)

Interactivity Interpreting IR spectra (int-1229)

tlvd-9705

SAMPLE PROBLEM 3 Identifying major peaks in an IR spectrum to identify the molecule

Identify the major peaks in this IR spectrum for a molecule that has only one carbon atom in its molecular structure, and then identify the molecule.

THINK	WRITE
1. Identify characteristic peaks caused by functional groups listed in the VCE Chemistry Data Book.	The peak at approximately 1600–1750 cm^{-1} corresponds to C–O. The peak at approximately 2500–3200 cm^{-1} corresponds to O–H acids.
2. Deduce a structure that must have both the C–O and O–H acid groups, and contains one carbon atom only (as stated in the question).	The functional group must be –COOH.
3. Identify the molecule. The C atom in –COOH has one unbonded electron so it could be bonded to an H. The C–H peak at 2950 cm^{-1} is mostly hidden by the broad O–H peak. The semi-structural formula is HCOOH.	The molecule must be HCOOH, methanoic acid.

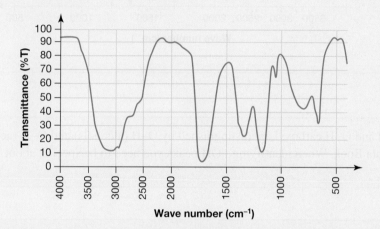

PRACTICE PROBLEM 3

Identify the major peaks in this IR spectrum and deduce a structure for a molecule with the formula C_3H_6O.

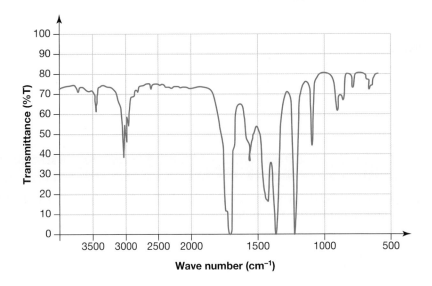

TIP: O–H hydroxyl and O–H carboxyl are distinguished by O–H (alcohols) and O–H (acids) in the VCE Chemistry Data Book. When identifying –OH peaks you need to be specific about which one you are referring to.

CASE STUDY: Infrared astronomy

The infrared photograph of the Trifid Nebula in figure 10.13 was taken by the Spitzer Space Telescope. The nebula is 5400 light-years away from Earth in the Sagittarius constellation. Visible-light telescopes cannot see into the nebula, but infrared cameras can detect infrared radiation coming from its interior, allowing us to 'see' what's inside it.

Infrared cameras take pictures using the infrared part of the electromagnetic spectrum. The differences in infrared wavelengths between parts of an object or between objects can be used to show different colours.

FIGURE 10.13 Infrared photograph of the Trifid Nebula

10.3 Activities

10.3 Quick quiz on	10.3 Exercise	10.3 Exam questions

10.3 Exercise

1. A sample of propene is analysed by infrared spectroscopy.
   a. Describe the effect of infrared radiation on the covalent bonds within the sample.
   b. Explain whether the C–C or C=C bond will produce a peak with a higher wave number.
2. The following infrared spectrum was produced by ethyl ethanoate.

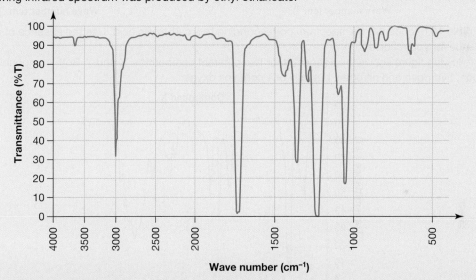

Identify the peaks outside the fingerprint region (>1500 cm^{-1}).
3. A molecule with the molecular formula $C_3H_6O_2$ was analysed by infrared spectroscopy.

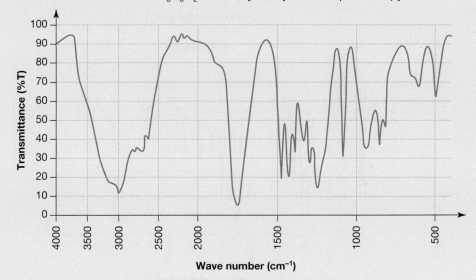

a. Use table 10.6 to identify the bonds responsible for the major peaks in the spectrum.
b. Name the molecule.

4. The following figure shows two infrared spectra for two different compounds, X and Y. Only one is a carboxylic acid; the other is an alcohol.

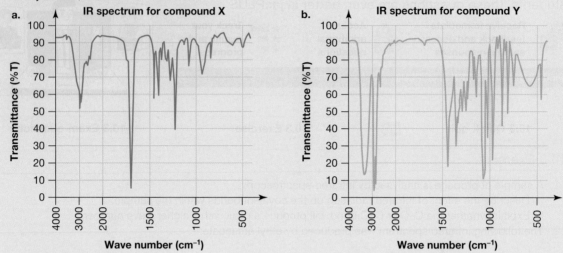

a. IR spectrum for compound X

b. IR spectrum for compound Y

Use table 10.6 to identify the spectrum corresponding to a carboxylic acid. Use the absence of one peak as evidence for your choice.

5. The following figure shows the IR spectra for two different compounds, X and Y. Both compounds contain carbon, hydrogen and nitrogen, but only one contains oxygen.

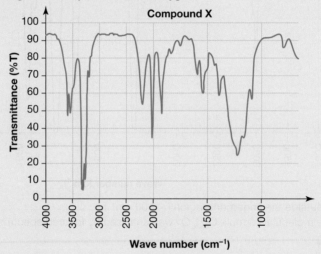

Compound X

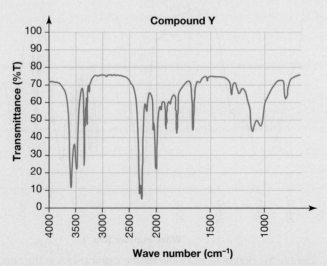

Compound Y

Use table 10.6 to identify the homologous series that compounds X and Y belong to.

10.3 Exam questions

▶ **Question 1 (1 mark)**

Source: VCE 2021 Chemistry Exam, Section A, Q.16; © VCAA

MC Which one of the following statements about IR spectroscopy is correct?
A. IR radiation changes the spin state of electrons.
B. Bond wave number is influenced only by bond strength.
C. An IR spectrum can be used to determine the purity of a sample.
D. In an IR spectrum, high transmittance corresponds to high absorption.

▶ **Question 2 (1 mark)**

Source: VCE 2020 Chemistry Exam, Section A, Q.21; © VCAA

MC The infrared (IR) spectrum of an organic compound is shown below.

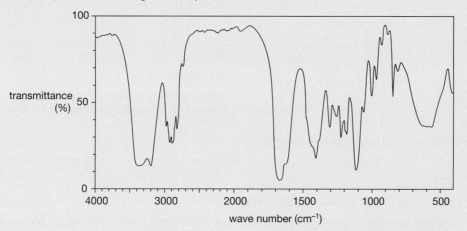

Data: SDBS Web, <https://sdbs.db.aist.go.jp>,
National Institute of Advanced Industrial Science and Technology

Referring to the IR spectrum above, the compound could be

A. $CH_3CH_2COOCH_3$ B. $CH_3CH_2CH_2CHO$ C. $NH_2CH_2CH_2CONH_2$ D. $NH_2CH_2CH_2CHOHCH_3$

▶ **Question 3 (2 marks)**

Source: VCE 2020 Chemistry Exam, Section B, Q.8.a; © VCAA

An unknown organic compound has a molecular formula of C_4H_8O.

The compound is **non-cyclic** and contains a **double bond**.

The infrared (IR) spectrum of the molecule is shown below.

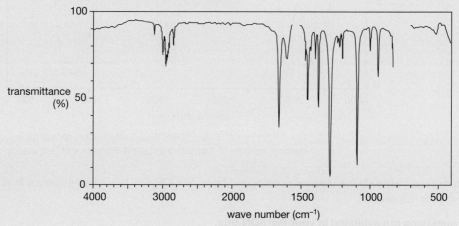

Data: SDBS Web, <https://sdbs.db.aist.go.jp>,
National Institute of Advanced Industrial Science and Technology

What does the region 3100–4000 cm^{-1} indicate about the bonds in C_4H_8O? Give your reasoning.

Question 4 (1 mark)

Source: VCE 2017 Chemistry Exam, Section A, Q.17; © VCAA

MC Shown below is the infrared spectrum of an organic compound.

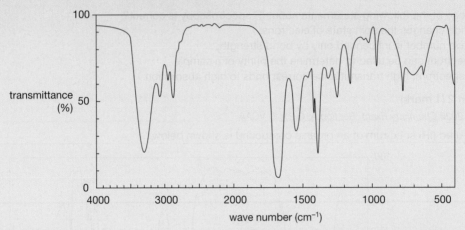

Data: SDBS Web, <http://sdbs.db.aist.go.jp>,
National Institute of Advanced Industrial Science and Technology

The organic compound that produces this spectrum is an

A. aldehyde.   **B.** alcohol.   **C.** amide.   **D.** ester.

Question 5 (2 marks)

Source: VCE 2016 Chemistry Exam, Section B, Q.4.b; © VCAA

A bottle containing an unknown organic compound was examined in a university laboratory. There was an incomplete label on the bottle that gave only the empirical formula for the contents: CH_4N.

A chemist hypothesised that the unknown compound was 1,2-ethanediamine, $NH_2CH_2CH_2NH_2$.

Infrared (IR) spectroscopy was used to analyse the sample. The spectrum is shown below.

IR spectrum

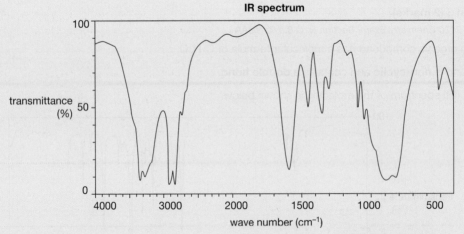

Data: SDBS Web, <http://sdbd.db.aist.go.jp>,
National Institute of Advanced Industrial Science and Technology

Is this spectrum consistent with the unknown compound being $NH_2CH_2CH_2NH_2$? Use evidence from the IR spectrum in your response.

More exam questions are available in your learnON title.

10.4 NMR spectroscopy

10.4.1 Principles of NMR spectroscopy

CASE STUDY: MRI scans

Nuclear magnetic resonance (NMR) images are called MRI scans in the medical field. MRI stands for magnetic resonance imaging. These images provide doctors with pictures of the soft tissues of the body. When NMR was introduced, many patients refused to have NMR scans because they thought the process had something to do with being bombarded with radiation from a nuclear reactor. However, the word 'nuclear' in this case refers to the nucleus of an atom and how it interacts with a magnetic field. To alleviate patients' fears, NMR scanning is now called MRI.

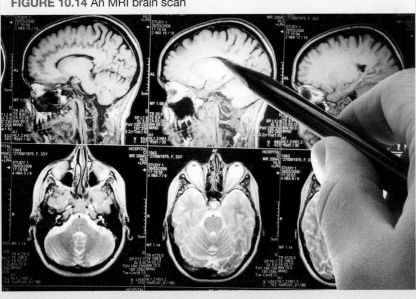

FIGURE 10.14 An MRI brain scan

Technique overview

NMR is a qualitative analysis method used to determine molecular structure. It utilises a property of certain nuclei called 'spin' to identify the location of atoms within a molecule.

The two particles most commonly used in NMR analysis are carbon-13 atoms, ^{13}C, and protons, ^{1}H. Other atoms within an organic molecule can be analysed, but examining the environments of the carbon and hydrogen atoms reveals valuable information about the structure of the molecule under investigation.

There are three main types of NMR spectra, each of which are based on similar principles:
- Carbon-13 NMR (^{13}C)
- Low-resolution proton NMR (^{1}H)
- High-resolution proton NMR (^{1}H)

Fundamentals of NMR

Nuclei with an odd number of nucleons can be detected by NMR, as they have two overall spin states and behave as if they are magnets. ^{1}H and ^{13}C are two such nuclei, although ^{12}C is not.

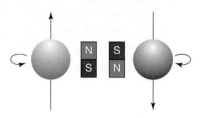

FIGURE 10.15 Some nuclei have spin and can be thought of as small bar magnets.

1. When placed in an external magnetic field, most nuclei will line up with the field (termed parallel), although some will have enough energy to line up against the field (termed antiparallel).
2. Radio waves are provided to change the spin state or 'flip' the nuclei from the low-energy (with the field) to the high-energy (against the field) alignment.
3. When a nucleus moves back to the low-energy alignment it releases the specific energy difference between the two states, which can be detected.

FIGURE 10.16 The magnetic field generated by spinning nuclei can be aligned with (low energy) or against (high energy) an external magnetic field.

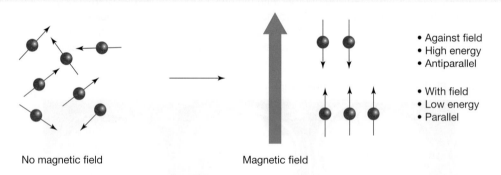

No magnetic field

Magnetic field

- Against field
- High energy
- Antiparallel

- With field
- Low energy
- Parallel

FIGURE 10.17 Radio-wave energy can be provided to flip nuclei from the low-energy to the high-energy alignment. These high-energy nuclei will spontaneously flip back to the low-energy state and release the specific energy they initially absorbed. This is called resonance.

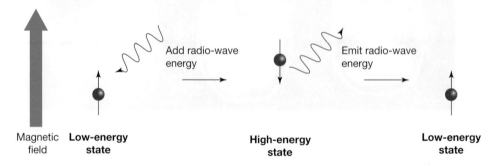

Magnetic field **Low-energy state**

Add radio-wave energy

High-energy state

Emit radio-wave energy

Low-energy state

The energy difference between these two spin states depends on the strength of the external magnetic field that is 'felt' by the nucleus. This is not always the same as the external magnetic field because other atoms that surround a given nucleus can modify it via a process called *shielding*.

EXTENSION: Operation of an NMR spectrometer

NMR spectrometers apply an external magnetic field and radio-wave pulses to a sample. The radio frequencies required to flip nuclei are detected and output as spectra.

FIGURE 10.18 A schematic diagram of an NMR spectrometer

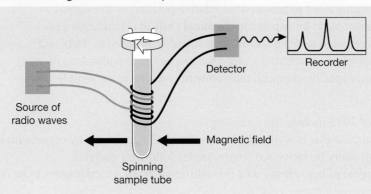

Shielding and chemical environments

Electrons surrounding a nucleus will somewhat shield it from the applied magnetic field of an NMR spectrometer. As such, the radio-wave energy required to flip the spin state of nuclei differs depending on the extent of shielding. This property allows nuclei in different chemical environments to produce separate signals.

Nuclei that are connected to the same atoms are in the same chemical environment and will produce one NMR signal. This is true for both hydrogen and carbon environments in ^{1}H-NMR and ^{13}C-NMR, respectively, although hydrogen environments will be the focus of this discussion.

- All the hydrogen atoms bonded to the same carbon atom are in the same hydrogen environment.
- However, hydrogen atoms bonded to different carbon atoms may be in the same hydrogen environment if the molecule is symmetrical.

Figure 10.19 shows the carbon and hydrogen environments for butane and methylbutane.

- The two blue CH_3 groups in butane are each connected to a $CH_2CH_2CH_3$ group, so are in the same chemical environment.
- Similarly, the two red CH_2 groups in butane are connected to a CH_3 on one side and a CH_2CH_3 on the other, so are also in the same chemical environment.
- Methylbutane has four hydrogen environments, as the two blue CH_3 groups present are each connected to $CH(CH_3)CH_2CH_3$ and so share a chemical environment.
- By contrast, the purple methyl group is attached to a different set of atoms, $CH_2CH(CH_3)_2$, and is in a different chemical environment.
- Methylbutane only contains one CH (pink) group and one CH_2 (green) group, so these must be the only hydrogen atoms in each of these environments.

FIGURE 10.19 Hydrogen and carbon chemical environments in butane and methylbutane

Butane

Methylbutane

These examples highlight that structural symmetry in molecules is a clue that groups of atoms are likely to share a chemical environment.

on Resources

Interactivity Predicting carbon and hydrogen environments in different compounds (int-1227)

Chemical shift

All NMR signals produced by a sample are compared to that produced by a standard, tetramethylsilane (TMS). This inert molecule is added to samples prior to analysis and produces a single peak for both ^{1}H- and ^{13}C-NMR due to symmetry, resulting in equivalent environments for all twelve H and all four C atoms.

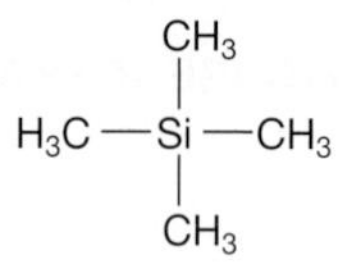

The value for the TMS signal is set to zero and the relative position of the signal generated by nuclei in a sample is known as the **chemical shift** (δ). This represents the difference in energy required to flip a nucleus in a sample compared to TMS and is specific for each chemical environment. Comparison can be made with tables of literature values to identify specific chemical environments present in the sample being analysed.

FIGURE 10.20
Structure of TMS

CH$_3$
|
H$_3$C —Si —CH$_3$
|
CH$_3$

Additional advantages of TMS include the following:
- It produces a signal peak that is well away from other peaks generated by organic molecules.
- It is volatile, so can easily be recovered from samples following analysis.
- Setting the TMS signal to zero allows data from different NMR spectrometers to be compared.

Low-resolution proton NMR

There are two different types of NMR spectra: high resolution and low resolution. Low-resolution spectra include a variety of information.
- The number of peaks indicates the number of unique hydrogen environments present in the molecule.
- The ratio of the areas under the peaks shows the ratio of hydrogen atoms in that environment.
- The chemical shift provides information about the specific hydrogen environment.

The ratio of peak areas is frequently written above the peak on the spectrum or provided in a table. Alternatively, a line called an integration trace may be superimposed on the spectrum. Each time the trace crosses a peak, it gains height proportional to the area under the peak. To determine the ratio of peak areas, the integration trace heights are measured and the simplest whole-number ratio determined; for example, for integration trace height increases of 1.2 cm and 0.4 cm, the ratio of hydrogen atoms in the two different environments is 3 : 1.

The low-resolution ^{1}H-NMR spectrum of ethanol in figure 10.21 provides the following information:
- Three peaks indicates three hydrogen environments.
- The peak areas, 2 : 1 : 3, are proportional to the ratio of hydrogen nuclei in each environment.
- The chemical shift value for each peak provides clues to the specific hydrogen environment.

chemical shift the horizontal scale on an NMR spectrum

FIGURE 10.21 Low-resolution ^{1}H-NMR spectrum of ethanol

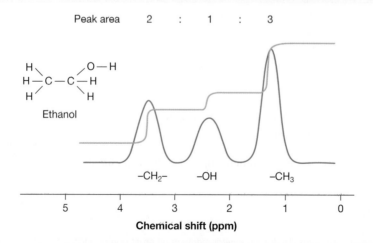

High-resolution proton NMR and peak splitting

A high-resolution spectrum provides the same information as the low-resolution spectrum, but signals in the low-resolution spectrum may each be split into two or more peaks.

^{1}H nuclei can interact with other ^{1}H nuclei near them. If the neighbours are in a different chemical environment, that interaction may cause the original peak to split into multiple peaks. This happens because neighbouring nuclei have a small magnetic effect on each other, further impacting on the level of shielding that may be experienced by the specific nuclei.

The splitting pattern or number of peaks is related to the number of adjacent hydrogen atoms by the **$n + 1$ rule**. For simple molecules, the number of peaks is one more than the number of hydrogen atoms on the neighbouring carbon atom(s). Hydrogen atoms further away than this are not considered neighbouring and will not cause splitting. It is important to note that all the neighbouring hydrogen atoms must be in the same hydrogen environment; if not, they will cause splitting independently, resulting in the overlay of two splitting patterns. This is known as a *multiplet*.

TABLE 10.7 The $n + 1$ rule and splitting patterns

Number of hydrogen atoms on neighbouring carbon atom(s)	Peak splitting pattern: $n + 1$	Pattern name
0	$0 + 1 = 1$	Singlet
1	$1 + 1 = 2$	Doublet
2	$2 + 1 = 3$	Triplet
3	$3 + 1 = 4$	Quartet
2+ different hydrogen environments	Overlay of 2+ splitting patterns	Multiplet

In the high-resolution ^{1}H-NMR spectrum of ethanol in figure 10.22, the signal of three peaks (triplet) indicates that there are two hydrogen atoms attached to the neighbouring carbon atom in the molecule. The set of four peaks (quartet) indicates that there are three neighbouring hydrogen atoms. O–H groups always present as a single peak (singlet) in high-resolution spectra.

$n + 1$ **rule** a rule used for simple molecules; the number of peaks is one more than the number of equivalent hydrogen atoms on the neighbouring carbon atom(s)

FIGURE 10.22 The high-resolution ^{1}H-NMR spectrum of ethanol, CH_3CH_2OH, has peak areas proportional to the number of protons producing the signal and peak splitting according to the $n + 1$ rule. O–H always presents as a single peak.

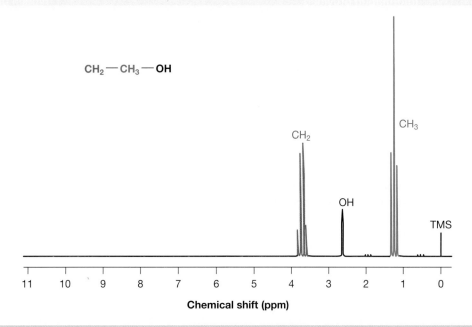

Neighbouring hydrogen environments that are equivalent do not split each other. For example, 1,2-ethanediol, $HOCH_2CH_2OH$, produces two singlets because the CH_2 groups have equivalent hydrogen atoms; therefore, no splitting occurs (see figure 10.23).

FIGURE 10.23 ^{1}H-NMR spectrum for 1,2-ethanediol

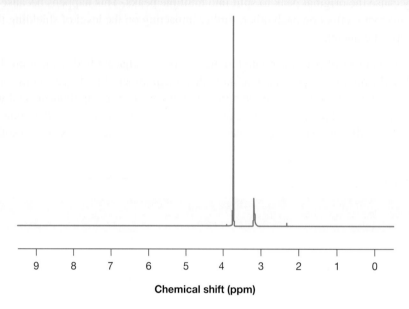

Chemical shift (ppm)

 Resources

▶ **Video eLesson** High-resolution proton NMR (eles-3254)

10.4.2 Interpreting NMR spectra

All NMR techniques produce spectra that show the chemical shift, in parts per million (ppm), of all peaks produced and can be analysed using a similar approach.

TABLE 10.8 A summary of the key information in NMR spectra

Feature	Information provided	Type(s) of NMR
The number of unique sets of peaks	This indicates the number of H or C environments present in the molecule.	All
Peak area	The signal intensity (area under peaks) indicates the ratio of equivalent H or C atoms responsible for a peak. This may be determined from an integration trace.	All
Peak splitting	In high-resolution ^{1}H-NMR spectra the number of peaks within a set can be used to identify the number of equivalent hydrogens on neighbouring carbon atom(s) using the $n + 1$ rule.	Only high-resolution ^{1}H-NMR
Chemical shift	The chemical shift is stated relative to the TMS standard and is affected by the extent of shielding experienced by the H or C atom. These values can be compared to tabulated literature values to identify the H or C environment. Chemical shift data is frequently useful to confirm other information about a structure rather than as a starting point, given there may be multiple possible chemical environments for a peak with a particular chemical shift.	All

Interpreting ^{1}H-NMR spectra

The chemical shifts for ^{1}H-NMR spectra are summarised in table 10.9.

TABLE 10.9 ^{1}H-NMR data

Type of proton	Chemical shift (ppm)	Type of proton	Chemical shift (ppm)
$R–CH_3$	0.9–1.0	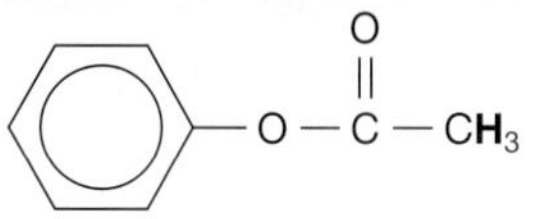 (phenyl)$–O–C(=O)–CH_3$	2.3
$R–CH_2–R$	1.3–1.4	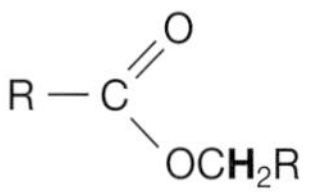 $R–C(=O)–OCH_2R$	3.7–4.8
$RCH=CH–CH_3$	1.6–1.9	$R–O–\mathbf{H}$	1–6 (varies under different conditions)
$R_3–CH$	1.5	$R–NH_2$	1–5
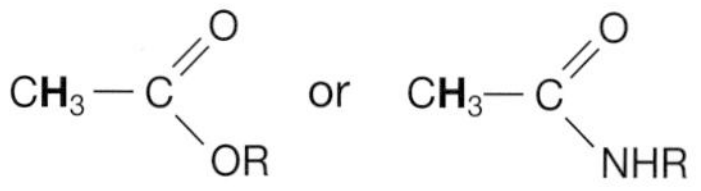 $CH_3–C(=O)–OR$ or $CH_3–C(=O)–NHR$	2.0	$RHC=CHR$	4.5–7.0
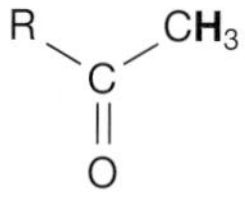 $R–C(CH_3)=O$	2.1–2.7	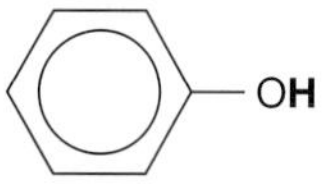 (phenyl)–OH	4.0–12.0
$R–CH_2–X$ (X = F, Cl, Br or I)	3.0–4.5	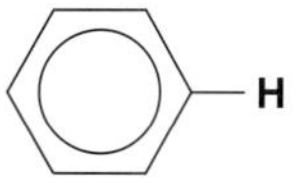 (phenyl)–$\mathbf{H}$	6.9–9.0
$R–CH_2–OH$, $R_2–\mathbf{CH}–OH$	3.3–4.5	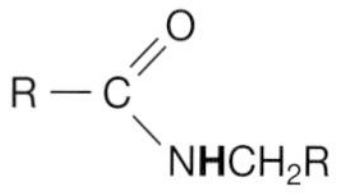 $R–C(=O)–NHCH_2R$	8.1
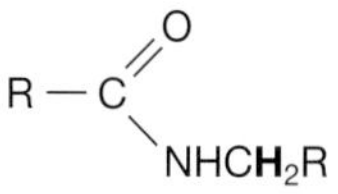 $R–C(=O)–NHCH_2R$	3.2	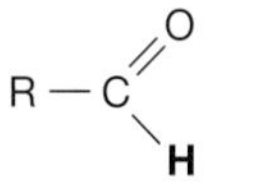 $R–C(=O)–\mathbf{H}$	9.4–10.0
$R–O–CH_3$ or $R–O–CH_2R$	3.3–3.7	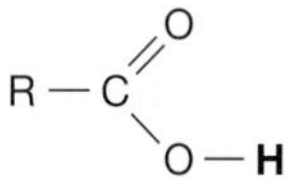 $R–C(=O)–O–\mathbf{H}$	9.0–13.0

Source: VCE Chemistry Written Examination Data Book (2022) extracts © VCAA; reproduced by permission.

Given that multiple proton environments could have been responsible for a particular peak, chemical shifts are frequently useful in support of other information provided in ^{1}H-NMR spectra.

Consider the ^{1}H-NMR spectrum of a sample with the molecular formula C_2H_5Br shown in figure 10.24. Quickly 'eye-balling' the spectra reveals that there are two sets of peaks and so two different hydrogen environments, and that the ratio of their peak areas is 2 : 3. The splitting pattern of a quartet (blue) and triplet (red) is frequently observed in spectra. It is generated by a $–CH_2CH_3$ group, in which a quartet is produced by a neighbouring CH_3 group and the triplet is produced by a neighbouring CH_2 group. This is sufficient information to draw a structure for the molecular formula, with the chemical shift data confirming this interpretation. The chemical shift of 3.7 ppm is in the correct range for $R–CH_2X$ (3.0–4.5 ppm) and is adjacent to a CH_3 group at 1.7 ppm.

FIGURE 10.24 ^{1}H-NMR spectrum of C_2H_5Br

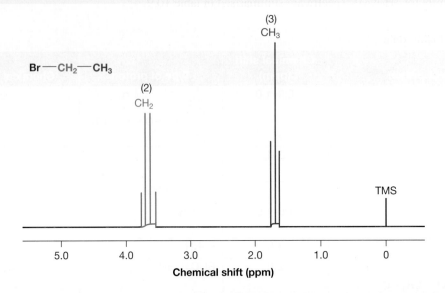

SAMPLE PROBLEM 4 Using ^{1}H-NMR spectroscopy to identify the structure of a molecule

tlvd-9660

Analyse the following ^{1}H-NMR spectrum and use table 10.9 to identify the structure of the molecule. The molecular formula for the molecule is $C_4H_8O_2$.

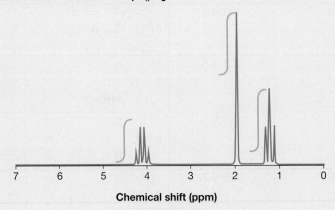

THINK	WRITE
1. Identify the number of different hydrogen environments.	There are three sets of peaks, so three different hydrogen environments.
2. Identify the ratio of hydrogen atoms in each environment. Measure the height of the integration trace for each peak and determine the simplest whole-number ratio.	The ratio is 2 : 3 : 3. As there are eight hydrogen atoms in the molecular formula, this is the number of hydrogen atoms in each environment.
3. Identify groups according to the splitting pattern.	The peak at 4.1 ppm is a CH_2 group (relative peak area of 2) and is neighbouring a CH_3 group (as its peak is split into a quartet). The peak at 1.9 ppm is a CH_3 group (relative peak area of 3) and it has no neighbouring hydrogen atoms (a singlet due to no splitting).

The peak at 1.2 ppm is a CH_3 group (relative peak area of 2) and is neighbouring a CH_2 group (as its peak is split into a triplet).

Recall that a triplet and quartet splitting pattern is suggestive of a $-CH_2CH_3$ group.

4. Assemble the molecule that matches the number of peak sets and splitting patterns.

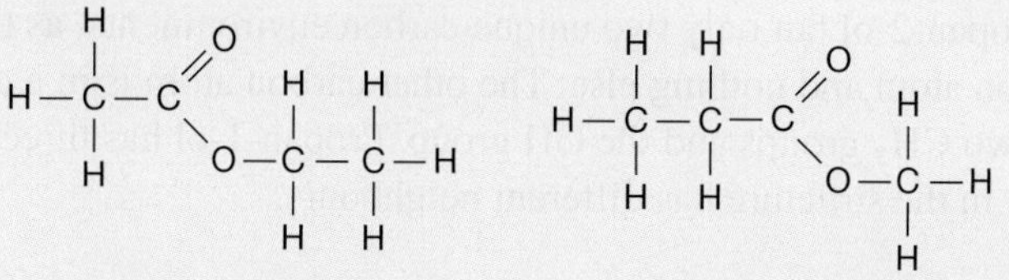

Ethyl ethanoate   Methyl propanoate

There are two possible structures using peak number, area and splitting, so chemical shift is required.

5. Use chemical shift information to identify the correct structure.

The chemical shift of the CH_3 group that produces the singlet will vary between these two structures.

Singlet peak (ppm)	Ethyl ethanoate		Methyl propanoate	
2.0	CH_3-C with $=O$ and OR 2.0 ppm	✓	$R-C$ with $=O$ and OCH_2R 3.7–4.8 ppm	✗

The correct structure is:

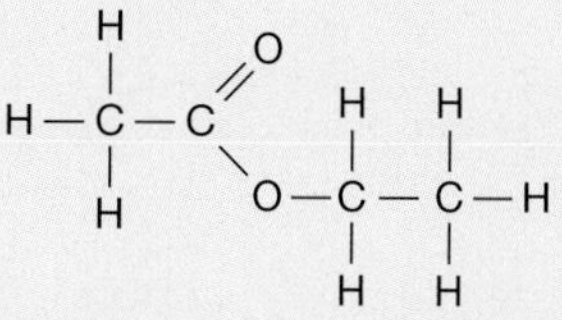

PRACTICE PROBLEM 4

Draw the structure of an isomer of $C_2H_4Cl_2$ that produced the following spectrum.

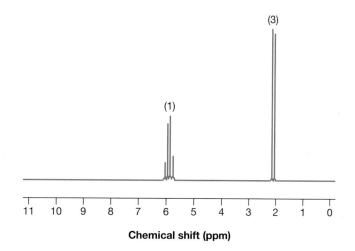

Interpreting ^{13}C-NMR spectra

Analysis of ^{13}C-NMR spectra is similar to that for ^{1}H-NMR, although peak splitting does not occur.

Recall that carbon environments can be identified in the same manner as hydrogen environments and that molecules with structural symmetry are likely to contain multiple carbon atoms in the same environment.

Consider the carbon environments of propan-2-ol and propan-1-ol in figure 10.25. There are three carbon atoms in propan-2-ol but only two unique carbon environments, as the CH_3 groups are both connected to the central carbon atom and nothing else. The other carbon atom is in a different environment due to it being connected to the two CH_3 groups and the OH group. Propan-1-ol has three unique carbon environments because each carbon atom in the structure has different neighbours.

FIGURE 10.25 Structures of **a.** propan-2-ol and **b.** propan-1-ol

a.

$$CH_3$$
$$|$$
$$H_3C-CH$$
$$|$$
$$OH$$

b.

$$H_3C-CH_2-CH_2-OH$$

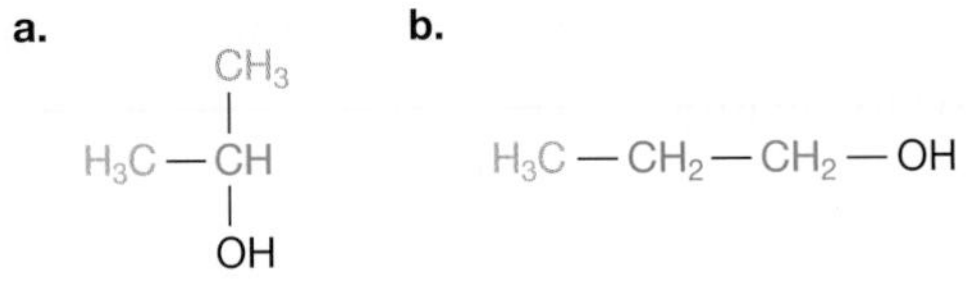

If we examine the spectra of propan-2-ol and propan-1-ol (figure 10.26) we can see the difference in the chemical shift and the number of peaks visible on the spectra. The hydrogens in the CH_3 groups in propan-2-ol have the same chemical shift and thus produce a single peak with a relative area twice that of the other peak. When peaks are extremely narrow, peak height can be used as a rough approximation of peak area.

FIGURE 10.26 The spectra of **a.** propan-2-ol and **b.** propan-1-ol

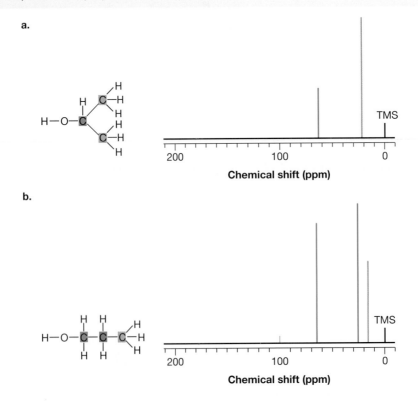

The ^{13}C-NMR chemical shifts are summarised in table 10.10.

TABLE 10.10 ^{13}C-NMR chemical shift data relative to TMS = 0

Type of carbon	Chemical shift (ppm)
R–CH$_3$	8–25
R–CH$_2$–R	20–45
R$_3$–CH	40–60
R$_4$–C	36–45
R–CH$_2$–X	15–80
R$_3$C–NH$_2$, R$_3$C–NR	35–70
R–CH$_2$–OH	50–90
RC≡CR	75–95
R$_2$C–CR$_2$	110–150
RCOOH	160–185
$\begin{array}{c}R \\ \!\diagdown \\ C=O \\ \!\diagup \\ RO\end{array}$	165–175
$\begin{array}{c}R \\ \!\diagdown \\ C=O \\ \!\diagup \\ H\end{array}$	190–200
R$_2$C–O	205–220

Source: VCE Chemistry Written Examination Data Book (2022) extracts © VCAA; reproduced by permission.

10.4 Activities

learn**on**

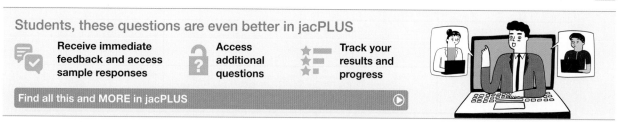

Students, these questions are even better in jacPLUS

- Receive immediate feedback and access sample responses
- Access additional questions
- Track your results and progress

Find all this and MORE in jacPLUS ▶

10.4 Quick quiz on	10.4 Exercise	10.4 Exam questions

10.4 Exercise

1. **MC** Which of the following molecules will produce a ^{1}H-NMR spectrum with a peak area ratio of 3 : 1?
   I CH$_3$CH$_2$CH$_3$
   II CH$_3$CHCl$_2$
   III CH$_3$OH

   A. I and II only
   B. I and III only
   C. II and III only
   D. I, II and III

2. **MC** How many signals does the carboxylic acid $(CH_3)_2CHCOOH$ have in its ^1H-NMR and ^{13}C-NMR spectra?
   A. Three ^1H signals and three ^{13}C signals
   B. Three ^1H signals and four ^{13}C signals
   C. Four ^1H signals and four ^{13}C signals
   D. Five ^1H signals and three ^{13}C signals
3. Identify and explain two reasons that TMS is used as the NMR reference standard.
4. List possible chemical shifts, in ppm, observed in a ^1H-NMR spectrum for CH_3CH_2Cl.
5. The molecule C_4H_{10} has two isomers. Sketch the isomers and examine the chemical environment of each carbon atom. Decide how many signals each isomer would produce in a ^{13}C-NMR spectrum.
6. Draw the structural formulas of the two isomers of C_3H_7Br and explain how ^{13}C-NMR spectroscopy could be used to identify each.
7. A sample of methylpropan-2-ol is analysed by ^{13}C-NMR spectroscopy.
   a. Identify how many different carbon environments are present in the compound.
   b. Identify the ratio of signals produced by these environments.
   c. Use table 10.10 to identify the chemical shifts of the peaks.
8. Propanoic acid is used as a preservative and anti-mould agent for animal feed, as well as in packaged food for human consumption. Complete the following table. Include the hydrogen environment, splitting pattern, relative peak height and the chemical shift for each type of hydrogen atom in this molecule. The first hydrogen environment has been done for you.

Hydrogen set or atom	Splitting pattern	Relative peak area	Chemical shift (ppm)
CH_3CH_2COOH			

9. Other than chemical shift values, identify and explain two differences you would expect to see in high-resolution ^1H-NMR spectra for 1,1-dichloroethane and 1,2-dichloroethane.
10. Draw the peak splitting pattern for the methyl ethanoate, $CH_3CH_2COOCH_3$, high-resolution ^1H-NMR spectrum. The approximate chemical shift values may be obtained from table 10.9.

10.4 Exam questions

▶ Question 1 (1 mark)

Source: VCE 2021 Chemistry Exam, Section A, Q.30; © VCAA

MC The ^1H-NMR spectrum of an organic compound has three unique sets of peaks: a single peak, seven peaks (septet) and two peaks (doublet).

The compound is
A. 3-methylbutanoic acid.
B. 2-methylpropanoic acid.
C. 2-chloro-2-methylpropane.
D. 1,2-dichloro-2-methylpropane.

Source: VCE 2020 Chemistry Exam, Section B, Q.8.b,c; © VCAA

An unknown organic compound has a molecular formula of C_4H_8O.

The compound is **non-cyclic** and contains a **double bond**.

The infrared (IR) spectrum of the molecule is shown below.

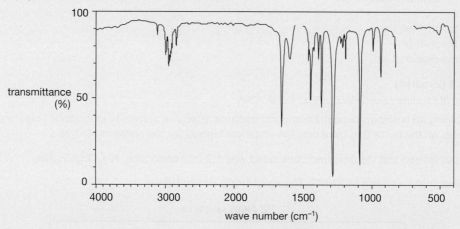

Data: SDBS Web, <https://sdbs.db.aist.go.jp>,
National Institute of Advanced Industrial Science and Technology

a. The ^{13}C-NMR spectrum of the unknown compound has four distinct peaks.

Draw **two** possible structural formulas of the unknown compound using the information provided.   **(2 marks)**

b. The high-resolution ^{1}H-NMR spectrum of the unknown compound has three single peaks, as shown below.

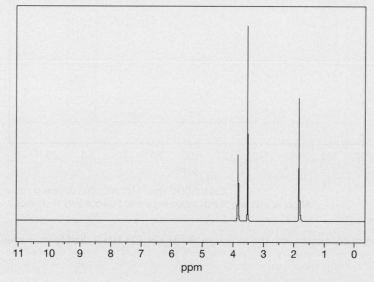

Data: SDBS Web, <https://sdbs.db.aist.go.jp>,
National Institute of Advanced Industrial Science and Technology

Chemical shift (ppm)	Relative peak area
1.82	3
3.53	3
3.85	2

Refer to the ^{1}H-NMR spectrum and the table of spectrum information provided.

Identify three pieces of information about the unknown compound and indicate how each would assist in determining its structure.   **(3 marks)**

Question 3 (1 mark)

Source: VCE 2019 Chemistry Exam, Section A, Q.27; © VCAA

MC An organic compound has a molar mass of 88 g mol⁻¹.

The ^{13}C-NMR spectrum of the organic compound shows four distinct peaks.

The organic compound is **most** likely
A. butan-1-ol.
B. 2-methyl-butan-1-ol.
C. 2-methyl-butan-2-ol.
D. 2,2-dimethyl-propan-1-ol.

Question 4 (2 marks)

Source: VCE 2016 Chemistry Exam, Section B, Q.4.c; © VCAA

A bottle containing an unknown organic compound was examined in a university laboratory. There was an incomplete label on the bottle that gave only the empirical formula for the contents: CH_4N.

A chemist hypothesised that the unknown compound was 1,2-ethanediamine, $NH_2CH_2CH_2NH_2$.

The sample was analysed using ^{13}C-NMR. The spectrum is shown below.

^{13}C-NMR spectrum

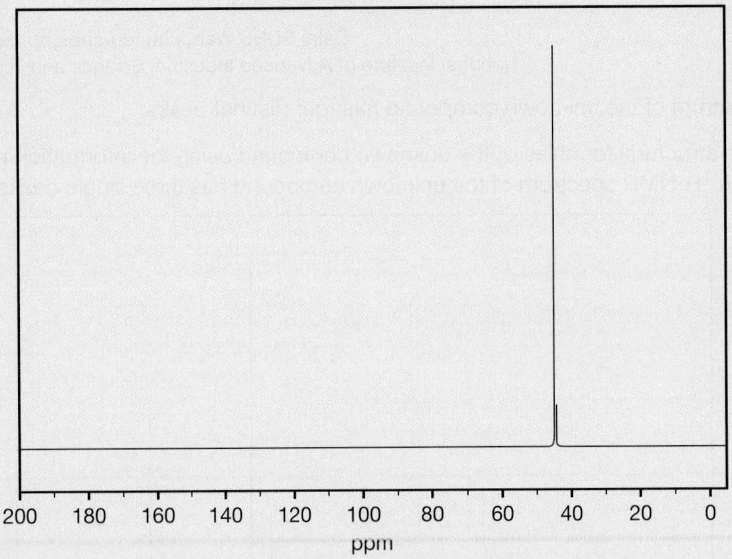

Data: SDBS Web, <https://sdbs.db.aist.go.jp>,
National Institute of Advanced Industrial Science and Technology

Is the ^{13}C-NMR spectrum consistent with the structure of $NH_2CH_2CH_2NH_2$? Justify your answer.

▶ Question 5 (3 marks)

Source: VCE 2017 Chemistry Exam, Section B, Q.5.c; © VCAA

There are a number of structural isomers for the molecular formula C_3H_6O. Three of these are propanal, propanone and prop-2-en-1-ol.

The skeletal structure for the aldehyde propanal is as follows.

Consider the ^{13}C-NMR and ^{1}H-NMR spectra below.

^{13}C-NMR spectrum

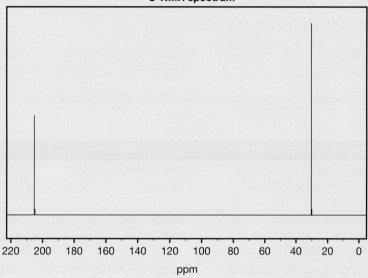

Data: SDBS Web, <https://sdbs.db.aist.go.jp>,
National Institute of Advanced Industrial Science and Technology

^{1}H-NMR spectrum

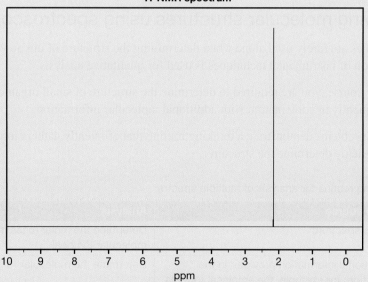

Data: SDBS Web, <https://sdbs.db.aist.go.jp>,
National Institute of Advanced Industrial Science and Technology

Identify which one of the three named isomers of C_3H_6O produced the NMR spectra shown. Justify your answer by referencing both spectra.

More exam questions are available in your learnON title.

10.5 Combining spectroscopic techniques

10.5.1 Summary of techniques

Each of the techniques introduced in the preceding subtopics provide a stimulus to a substance of interest in order to generate detectable signals that can be interpreted to provide insights into molecular structure.

It can be useful to summarise this information in a table to clarify the relationships between these aspects.

TABLE 10.11 Summary of MS and IR and NMR spectroscopy

Technique	Stimulus applied to sample	Signal detected	Interpretation of signal
MS	Electron beam ionisation of molecule and subsequent fragmentation	*m/z* of positively charged fragments	Mass of fragments, including molar mass
IR spectroscopy	Vibration of covalent bonds changed by absorbance of infrared radiation	Absorbances for individual bonds	Identification of presence (and absence) of bonds
NMR (^{1}H and ^{13}C) spectroscopy	Change in nuclear spin states (flipping) within a magnetic field by absorbance of radio waves	Energy released when nuclei 'flip' back to lower-energy spin state	Information about the location and abundance of H and C atoms

10.5.2 Identifying molecular structures using spectroscopic data

Spectroscopic techniques are rarely used alone when determining the structure of unknown organic compounds. Typically, a combination of instrumental techniques is used for qualitative analysis.

In the VCE Chemistry course, you are required to determine the structure of small organic molecules from a variety of spectra, frequently in combination with additional molecular information.

The following sample problems demonstrate a thinking routine that efficiently gathers information about the molecule in order to quickly determine the structure.

TABLE 10.12 A thinking routine for analysis of multiple spectra

Step number	Data source	Analysis
1	Mass spectrum	Determine the molar mass from the *m/z* of the molecular ion peak
2	Additional information provided in the question; for example, the empirical formula or percentage composition by mass	Determine the molecular formula
3	IR spectrum	Identify (bonds and infer) functional groups present
4	NMR spectra (^{1}H and ^{13}C)	This is the most complex data so is best used to confirm the identity/structure of the molecule once a short list of possible candidates has been generated.

Analysis of an unknown compound has revealed that it has an empirical formula of C_2H_4O. The mass, IR and NMR spectra are shown. Identify and name the compound.

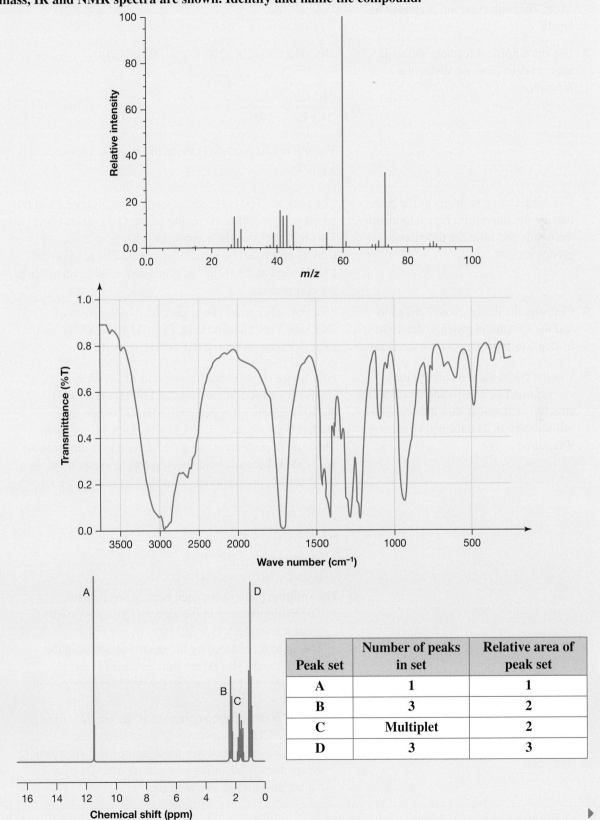

Peak set	Number of peaks in set	Relative area of peak set
A	1	1
B	3	2
C	Multiplet	2
D	3	3

THINK	WRITE
1. Identify the molar mass from the m/z of the molecular ion peak on the IR spectrum. *Note*: The molecular ion peak may be small.	As the m/z of the molecular ion peak is 88, the molar mass = 88 g mol^{-1}.
2. Use the empirical formula and molar mass to determine the molecular formula.	$M(C_2H_4O) = (2 \times 12.0) + (4 \times 1.0) + 16.0$ $= 44.0$ g mol^{-1} $\dfrac{M(\text{MF})}{M(\text{EF})} = \dfrac{88}{44.0}$ $= 2$ The molecular formula is twice the empirical formula: $C_4H_8O_2$.
3. Use the IR wave numbers of the peaks outside the fingerprint region to identify the bonds and infer the functional groups present.	The peak at 1700 cm^{-1} corresponds to a C=O bond and the broad peak at 3000 cm^{-1} could be the O–H (acid) bond. This indicates that the compound is likely to contain a COOH group and be a carboxylic acid. The presence of a C–H bond at 2900 cm^{-1} is of limited value in identifying the compound.
4. Consider the molar mass of 88 g mol^{-1} and the functional group(s) determined in step **3** to propose possible structures.	Two possible carboxylic acids with a molar mass of 88 g mol^{-1} are butanoic acid, $CH_3CH_2CH_2COOH$, and methylpropanoic acid, $CH_3CH(CH_3)COOH$.
5. Use the NMR data to confirm the correct order of groups and therefore the structure. Chemical shift data could be utilised here in a confirmatory manner, if required.	Four unique hydrogen environments are evident in the NMR spectrum. Butanoic acid, $CH_3CH_2CH_2COOH$, has four unique hydrogen environments, whereas methylpropanoic acid, $CH_3CH(CH_3)COOH$, has three. The ratio of peak areas, 3 : 2 : 2 : 1, also indicates that the molecule is butanoic acid because they correspond to the number of hydrogen atoms contributing to each signal.

The splitting patterns for each peak are as follows:
- The hydrogen atom in the carboxyl group (A) will present a singlet.
- The signals produced by the hydrogen atoms in the CH_2 (B) and CH_3 (D) are split into triplets as each group has two hydrogens on the neighbouring carbon atom.
- Peak C is the most complicated as the neighbouring CH_3 and CH_2 groups have different hydrogen environments. Recall that these groups will split peak C independently, resulting in a quartet overlaid on a triplet, also known as a multiplet.

PRACTICE PROBLEM 5

An organic compound has the empirical formula $C_3H_6O_2$. When sodium carbonate is added to this compound, bubbling is observed. The mass, IR and ^{1}H-NMR spectra of the compound are shown.

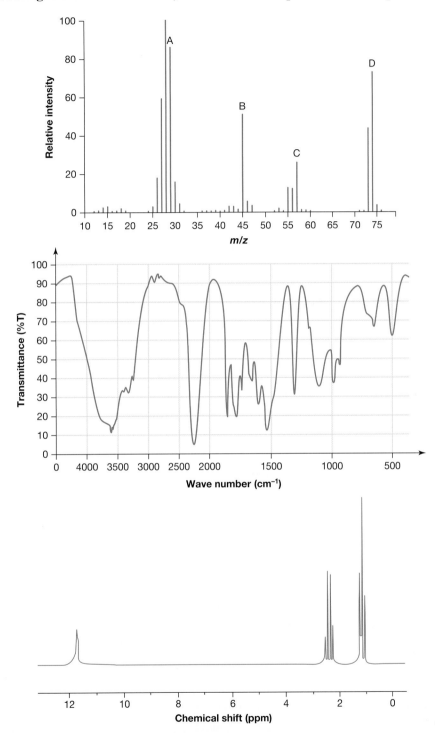

a. Identify the molar mass from the mass spectrum.
b. Deduce the molecular formula.
c. Use the IR spectrum to identify the bonds responsible for the peaks at 1720 cm^{-1} and 3000 cm^{-1}.
d. Identify the organic family to which this compound belongs, citing two pieces of evidence.
e. Deduce the structures indicated by the splitting pattern in the ^{1}H-NMR spectrum.
f. Name and draw the structural formula for the compound.

SAMPLE PROBLEM 6 Using percentage composition by mass, IR peaks, mass spectrometry and ^{1}H-NMR spectroscopy to identify a compound

A compound was found to contain 38.4% C, 4.8% H and 56.8% Cl, and generate narrow peaks at 1650 cm^{-1} and 3000 cm^{-1} on the IR spectrum. The mass and ^{1}H-NMR spectra are shown. Identify and name the compound.

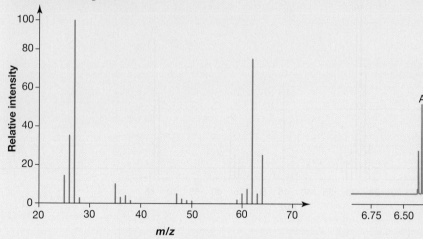

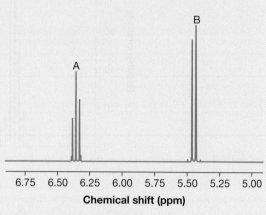

Peak set	Number of peaks in set	Relative area of peak set
A	3	1
B	2	2

THINK

1. Use the % by mass values to determine the empirical formula.

2. Identify the molar mass from the m/z of the molecular ion peak on the IR spectrum.

3. Use the empirical formula and molar mass to determine the molecular formula.

WRITE

Assume the sample mass is 100 g.

	C	H	Cl
m (g)	38.4	4.8	56.8
$n = \dfrac{m}{M}$ (mol)	$\dfrac{38.4}{12.0} = 3.20$	$\dfrac{4.8}{1.0} = 4.8$	$\dfrac{56.8}{35.5} = 1.60$
Divide by least-abundant element	$\dfrac{3.20}{3.20} = 1.00$	$\dfrac{4.8}{3.20} = 1.5$	$\dfrac{1.60}{3.20} = 0.500$
Multiply by 2 to achieve integer ratio	$1.00 \times 2 = 2.00$	$1.5 \times 2 = 3.0$	$0.500 \times 2 = 1.00$

The empirical formula is C_2H_3Cl.

As the m/z of the molecular ion peak is 62, the molar mass = 62 g mol^{-1}.
Note: This molecular ion must contain the ^{35}Cl isotope since:
$m/z = (2 \times 12) + (3 \times 1) + 35$
$\quad = 62$

$M(C_2H_3Cl) = (2 \times 12.0) + (3 \times 1.0) + 35.5$
$\qquad\qquad = 62.5$ g mol^{-1}

$\dfrac{M(\text{MF})}{M(\text{EF})} = \dfrac{62.5}{62.5}$
$\qquad\quad = 1$

The molecular formula is the same as the empirical formula: C_2H_3Cl.

4. Use the IR wave numbers of the peaks outside the fingerprint region to identify the bonds and infer the functional groups present.

The narrow peak at 3000 cm⁻¹ could be C–H bonds. This is not particularly informative as C–H bonds are present in almost all organic compounds.
The narrow peak at 1650 cm⁻¹ could be a C=O bond; however, as there is no O in the compound, a C=C bond is likely.

5. Use the ¹H-NMR data to confirm the correct order of groups and therefore the structure.

There is only one possible structure, so the molecule is chloroethene.

$$\begin{array}{c} \text{H} \qquad\qquad \text{H} \\ \diagdown \qquad\qquad \diagup \\ \text{C} = \text{C} \\ \diagup \qquad\qquad \diagdown \\ \text{H} \qquad\qquad \text{Cl} \end{array}$$

The ¹H-NMR data is confirmatory.
Two unique hydrogen environments are evident in the ¹H-NMR spectrum. The CH₂ signal is split into a doublet due to the neighbouring CH group and the CH signal is split into a triplet due to the neighbouring CH₂ group.
The ratio of peak areas, 2 : 1, corresponds to the number of hydrogen atoms contributing to each signal.

PRACTICE PROBLEM 6

A compound was found to contain 36.4% C, 6.1% H and 57.5% F, and generate a narrow peak at 3000 cm⁻¹ on the IR spectrum. The mass and ¹H-NMR spectra are shown.

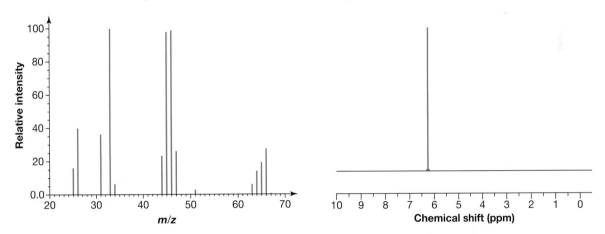

a. Use the percentage composition by mass to determine the empirical formula.
b. Identify the molar mass from the mass spectrum.
c. Deduce the molecular formula.
d. Use the IR spectrum to identify the bonds present.
e. Deduce the possible structures.
f. Use the ¹H-NMR spectrum to identify and name the compound.

elog-1926

Spectroscopy

Aim

To use spectroscopy to investigate the compounds in a chemical reaction

10.5 Activities   learn on

Students, these questions are even better in jacPLUS

 Receive immediate feedback and access sample responses

 Access additional questions

 Track your results and progress

Find all this and MORE in jacPLUS ▶

| 10.5 Quick quiz on | 10.5 Exercise | 10.5 Exam questions |

10.5 Exercise

1. ^{1}H- and ^{13}C-NMR spectroscopy can be used to differentiate between ethanol and ethanal.
   a. Identify one similarity and one difference in the ^{13}C spectra of these two molecules.
   b. Identify two differences in the ^{1}H spectra of these two molecules.
2. IR and low-resolution ^{1}H-NMR spectroscopy were used to analyse samples of propan-1-ol and propan-2-ol.
   a. Identify the signals, including wave number ranges, common to the IR spectra of both molecules.
   b. Identify the following for the low-resolution ^{1}H-NMR spectra of each molecule:
      i. Number of sets of peaks
      ii. Ratio of hydrogen atoms in each environment
   c. Justify whether IR or low-resolution ^{1}H-NMR spectroscopy is more informative when distinguishing between these two molecules.
3. A compound with the molecular formula $C_4H_8O_2$ produced the following spectra.

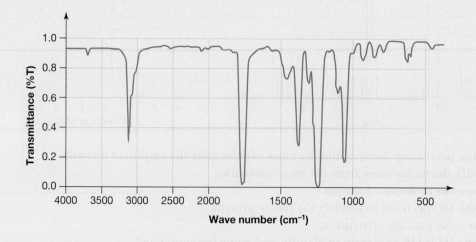

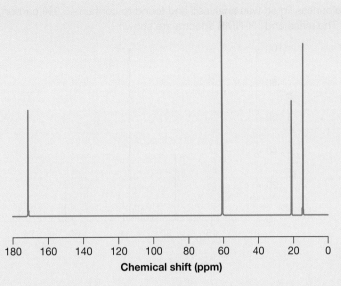

a. Identify the bonds responsible for the two peaks outside the fingerprint region of the IR spectrum.
b. How many unique carbon environments are present?
c. Draw a structure for this compound.

4. Two isomers of C_3H_9N produced the following spectra.

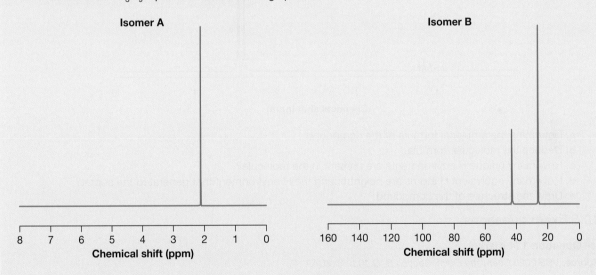

Isomer A

Isomer B

a. i. Name the technique used to produce the spectrum for isomer A.
   ii. Draw the structure of isomer A.
b. i. Name the technique used to produce the spectrum for isomer B.
   ii. Draw the structure of isomer B.

5. A sample of a colourless liquid was analysed and found to contain 45.9% carbon, 8.8% hydrogen and 45.2% chlorine. The mass and ^{1}H-NMR spectra are shown.

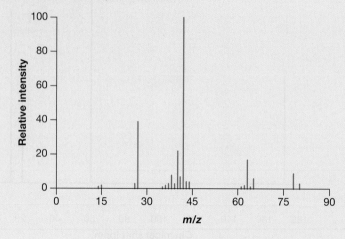

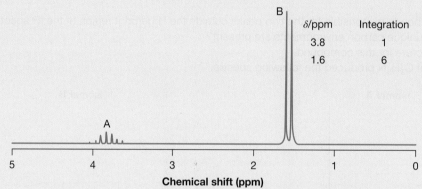

δ/ppm	Integration
3.8	1
1.6	6

a. Determine the empirical formula of the compound.
b. Deduce the molecular formula.
c. How many unique H environments are present in the molecule?
d. How many equivalent H atoms are neighbouring the H environment that generated the septet?
e. Draw the structure of the compound.

10.5 Exam questions

Question 1 (4 marks)

Source: VCE 2020 Chemistry Exam, Section B, Q.10.b; © VCAA

> Analytical chemistry deals with methods for determining the chemical composition of samples of matter. A qualitative method yields information about the identity of atomic or molecular species or the functional groups in the sample …
>
> Analytical methods are often classified as being either *classical* or *instrumental*.

Source: DA Skoog, FJ Holler and SR Crouch, *Principles of Instrumental Analysis*, 6th edition, Thomson Brooks/Cole, Belmont (CA), 2007, p. 1

Classical methods include qualitative analysis, such as treating a compound with reagents to observe any reaction, and quantitative methods, such as volumetric analysis, where the amount of a compound is determined by its reaction with a standard reagent.

Instrumental methods include a variety of spectroscopy, such as IR spectroscopy and NMR spectroscopy.

C_3H_6O can exist as a ketone or as a primary alcohol.

Explain how the principles of IR spectroscopy and ^{1}H-NMR spectroscopy lead to different spectra for the ketone and primary alcohol isomers of C_3H_6O, which can then be used to differentiate between the two molecules.

Question 2 (9 marks)

Source: VCE 2019 Chemistry Exam, Section B, Q.8; © VCAA

An unknown organic compound contains carbon, hydrogen and oxygen.

It is known that:

- the compound does not contain carbon-to-carbon double bonds (C=C)
- the molecular ion peak is found at a mass-to-charge ratio (m/z) of 74
- the ^{13}C-NMR has three distinct peaks.

a. A small peak in the mass spectrum can be identified at $m/z = 75$.

   Explain the presence of this peak. **(1 mark)**

b. i. Use the information provided to give **two** possible molecular formulas for this compound. **(2 marks)**

   ii. The ^{1}H-NMR spectrum of the compound shows three sets of peaks with a peak area ratio of 3 : 2 : 1.

    What does this information tell you about the structure of the compound and its molecular formula? Justify your answer by referring to the information given about the peaks in the ^{1}H-NMR spectrum. **(2 marks)**

c. There are many structural isomers of this compound.

   Draw the structural formulas of **two** possible isomers. **(2 marks)**

d. The infrared (IR) spectrum of the compound is shown below.

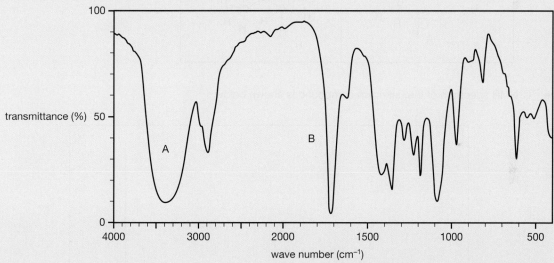

Data: SDBS Web, <http://sdbs.db.aist.go.jp>,
National Institute of Advanced Industrial Science and Technology

   i. Identify the functional groups responsible for the absorption peaks labelled A and B in the IR spectrum. **(1 mark)**

   A _____

   B _____

   ii. Using the ^{1}H-NMR information given in part **b.ii.** and the IR spectrum provided above, draw the structural formula of the compound. **(1 mark)**

Source: VCE 2021 Chemistry Exam, Section B, Q.7; © VCAA

Two students are given a homework assignment that involves analysing a set of spectra and identifying an unknown compound. The unknown compound is one of the molecules shown below.

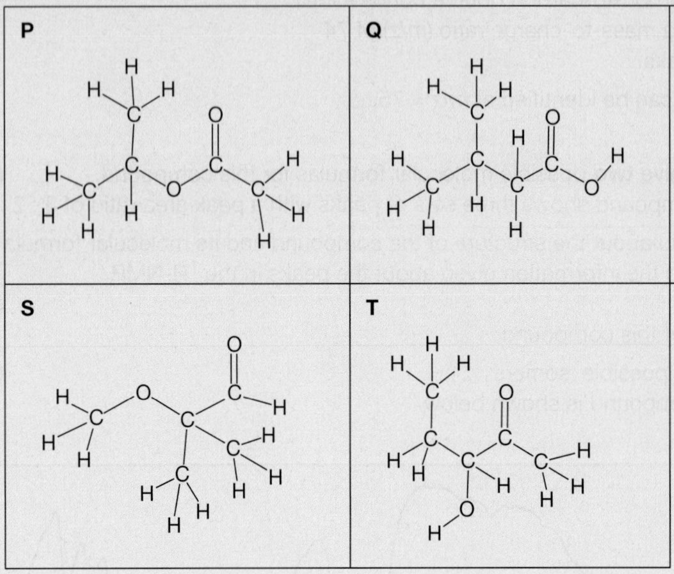

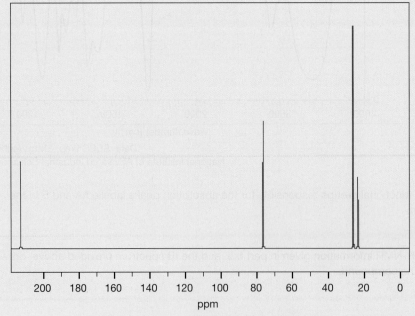

The ^{13}C-NMR spectrum of the unknown compound is shown below.

Data: SDBS Web, <https://sdbs.db.aist.go.jp>,
National Institute of Advanced Industrial Science and Technology

a. Based on the number of peaks in the ^{13}C-NMR spectrum above, which compound — P, Q, R, S or T — could be eliminated as the unknown compound? **(1 mark)**

b. The infrared (IR) spectrum of the unknown compound is shown below.

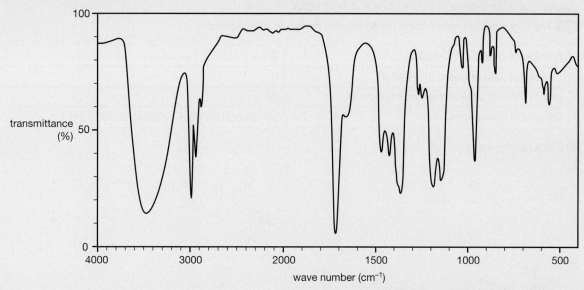

Data: SDBS Web, <https://sdbs.db.aist.go.jp>,
National Institute of Advanced Industrial Science and Technology

Identify which of the five compounds can be eliminated on the basis of the IR spectrum. Justify your answer using data from the IR spectrum. **(3 marks)**

c. The mass spectrum of the unknown compound is shown below.

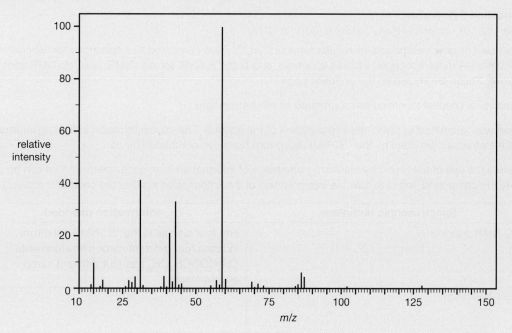

Data: SDBS Web, <https://sdbs.db.aist.go.jp>,
National Institute of Advanced Industrial Science and Technology

  i. Write the chemical formula of the species that produces a peak at $m/z = 43$. **(1 mark)**
 ii. Define m/z as used in mass spectroscopy. **(1 mark)**
iii. Explain why one molecule can produce multiple peaks on a mass spectrum. **(2 marks)**

Source: VCE 2021 Chemistry Exam, Section A, Q.11; © VCAA

MC The spectroscopy information for an organic molecule is given below.

number of peaks in ^{13}C-NMR	2
number of sets of peaks in ^{1}H-NMR	3
m/z of the last peak in the mass spectrum	60
infrared (IR) spectrum	an absorption peak appears at 3350 cm^{-1}

The organic molecule is

A.

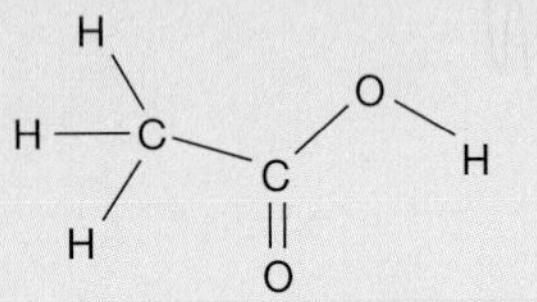

B.

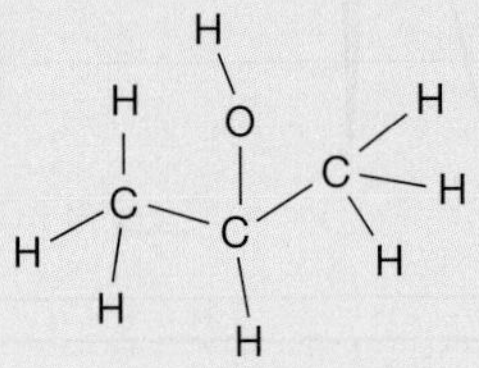

C.

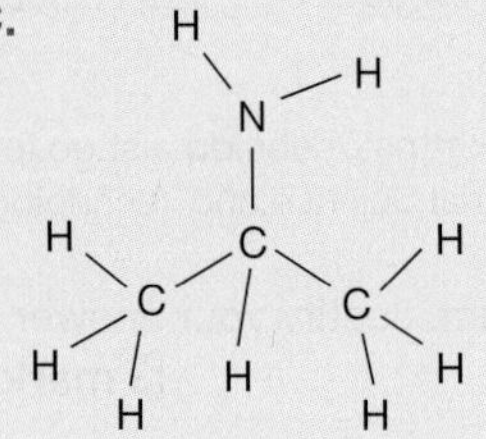

D.

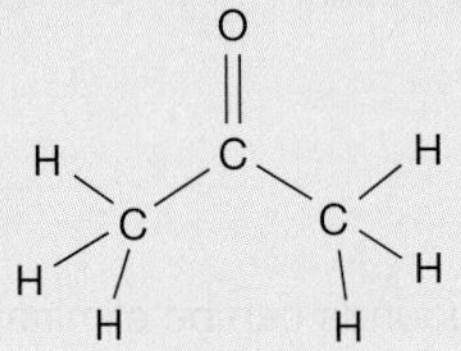

▶ **Question 5 (6 marks)**

Source: VCE 2013 Chemistry Exam, Section B, Q.9.a; © VCAA

An unknown organic compound, molecular formula $C_4H_8O_2$, was presented to a spectroscopy laboratory for identification. A mass spectrum, infrared spectrum, and both ^{1}H-NMR (proton NMR) and ^{13}C-NMR spectra were produced. These are shown on the opposite page.

The analytical chemist identified the compound as ethyl ethanoate.

A report was submitted to justify the interpretation of the spectra. The chemist's report indicating information about the structure provided by the ^{13}C-NMR spectrum has been completed for you.

Complete the rest of the report by identifying **one** piece of information from each spectrum that can be used to identify the compound. Indicate how the interpretation of this information justifies the chemist's analysis.

Spectroscopic technique	Information provided
^{13}C-NMR spectrum	The four signals in the ^{13}C-NMR spectrum indicate four different carbon environments. $CH_3COOCH_2CH_3$ has four different carbon environments.
mass spectrum	
infrared spectrum	
^{1}H-NMR spectrum	

Mass spectrum

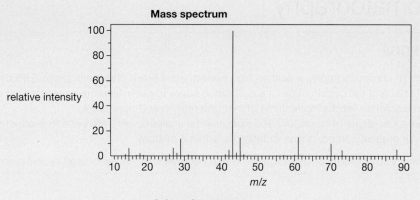

Infrared spectrum

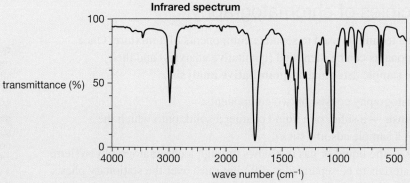

Proton NMR spectrum

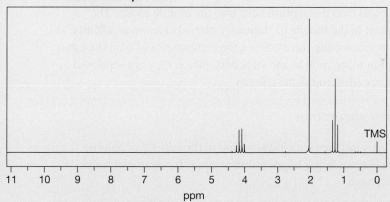

¹³C-NMR spectrum

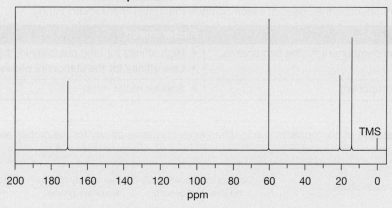

Source: National Institute of Advanced Industrial Science and Technology;
http://sdbs.riodb.aist.go.jp/sdbs/cgi-bin/direct_frame_top.cgi

More exam questions are available in your learnON title.

10.6 Chromatography

10.6.1. Principles of chromatography

Chromatography is a technique used to separate components of a mixture. Once separated, the components can be identified (qualitative analysis) and their concentration in the sample determined (**quantitative analysis**).

All forms of chromatography consist of two components:
- **Stationary phase** — a solid, or a liquid coating a solid, onto which the components of a sample adsorb (stick)
- **Mobile phase** — the liquid or gas that flows through a chromatography system, moving the materials to be separated at different rates over the stationary phase.

As seen in figure 10.27, the components undergo a continual process of **adsorption** to the stationary phase and then **desorption** back into the mobile phase. The attraction of a component to the mobile or stationary phase is known as **affinity** and is dependent on intermolecular forces. Since the components of a mixture vary in their relative attraction to the mobile and stationary phase, they are separated as they travel at different speeds through the column.

Adsorption versus absorption

Adsorption refers to a substance sticking/adhering to the surface of another. Absorption is one substance being taken within another; for example, water being drawn into a sponge.

quantitative analysis the determination of numerical information, such as the amount of a given element or compound in a known mass or volume of a sample

stationary phase a solid with a high surface area, or a finely divided solid coated with liquid; it shows different affinities for various components of a sample mixture when separating them by chromatography

mobile phase the liquid or gas that flows through a chromatography system, moving the materials to be separated at different rates over the stationary phase

adsorption the adhesion of atoms, ions or molecules from a gas, liquid or dissolved solid to a surface

desorption the removal of a substance from a surface; the opposite of adsorption

affinity the attraction of a component to a phase, either mobile or stationary

TABLE 10.13 Factors affecting the speed at which components travel in chromatography

Factor	Faster speed
Relative affinity of the component for the two phases	• High affinity for (and solubility in) the mobile phase • Low affinity for the stationary phase
Molar mass of the component	• Smaller molar mass

FIGURE 10.27 Separation of components due to differences in relative affinity for the mobile and stationary phases

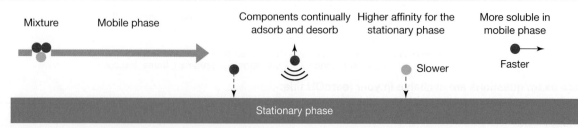

Intermolecular forces and affinity

Intermolecular forces are fundamental to chromatography as they determine the relative affinity of components for the two phases. More specifically, they determine how readily a component will dissolve in the mobile phase and adsorb to the stationary phase. Substances of like polarity tend to have higher affinity and be soluble, while those of unlike polarity tend to have lower affinity and be insoluble. The degree of polarity, and hence affinity, is affected by:

- the number and type of functional groups/atoms; for example, polar functional groups
- the proportion of the molecule that is non-polar; for example, length of hydrocarbon chains.

FIGURE 10.28 The proportion of polar and non-polar components determines the relative polarity of molecules.

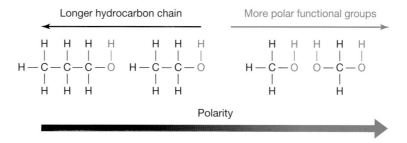

When comparing similar molecules, the proportion of the polar and non-polar components can be used to estimate relative polarity. In figure 10.28, the molecules with the shortest hydrocarbon chain and the largest number of hydroxyl groups are the most polar. Small differences in polarity may be sufficient to allow separation of components depending on the type of chromatography used.

Types of chromatography

In Unit 1, thin-layer chromatography (TLC) was introduced. Although it is a simple technique using easily obtained reagents and equipment, it operates based on the same principles as more sophisticated techniques, including column chromatography, **high-performance liquid chromatography (HPLC)** and gas chromatography (GC).

high-performance liquid chromatography (HPLC) a method used to separate the components of a mixture

---

BACKGROUND KNOWLEDGE: Thin-layer and paper chromatography

In thin-layer chromatography (TLC), a finely divided adsorbent material is coated onto either a glass, plastic or aluminium sheet to form the stationary phase. The mobile phase can be any of a wide range of mixtures of solvents (including water). Paper chromatography comprises a paper stationary phase and a polar mobile phase, which is frequently water.

FIGURE 10.29 Thin-layer chromatography — we can see that the red and yellow dyes are more strongly attracted to the stationary phase than the blue dye because they have not travelled as far up the paper (stationary phase)

For both TLC and paper chromatography, retardation factor (R_f) may be calculated and used to compare components to each other and to a database of known substances run under the same conditions.

$$R_f = \frac{\text{distance travelled by component from the origin}}{\text{distance travelled by solvent from the origin}}$$

10.6.2 Column chromatography

Column (or liquid) chromatography operates based on the same principles as paper chromatography and TLC, although has the advantage that the separated components may be collected. The stationary phase is composed of tiny particles (resin) tightly packed into a column through which the mobile phase moves by gravity. The apparatus and operation are shown in figure 10.30 and are as follows:

1. The mixture to be analysed (referred to as the **analyte**) is loaded at the top of the column.
2. The mobile phase (referred to as the **eluent**) is added continuously, carrying the components through the resin to the base of the column.
3. The components exit the column at different times depending on their relative affinity for the mobile and stationary phases.
4. The various components can be collected for qualitative (identity) and quantitative (concentration) analysis.

FIGURE 10.30 Column chromatography can be used to collect separated mixture components for further analysis.

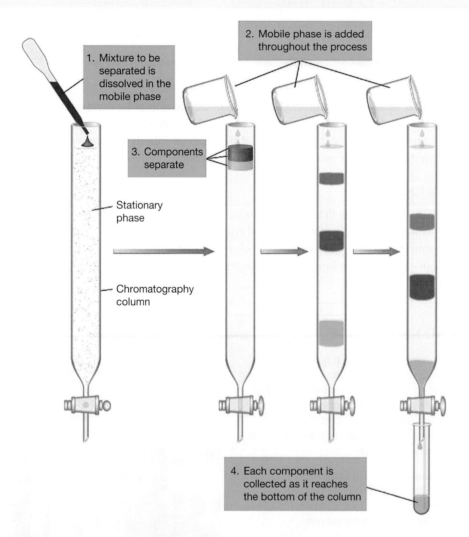

1. Mixture to be separated is dissolved in the mobile phase

2. Mobile phase is added throughout the process

3. Components separate

Stationary phase

Chromatography column

4. Each component is collected as it reaches the bottom of the column

In figure 10.30 the green dye has travelled the fastest, so has a higher relative affinity for the mobile phase than the red or blue dye. Conversely, the blue dye has the highest relative affinity for the stationary phase, so moves more slowly and will elute last.

analyte the sample undergoing analysis

eluent a substance used as a solvent in separating materials; for example, the mobile phase in chromatography

10.6.3 High-performance liquid chromatography (HPLC)

High-performance liquid chromatography (HPLC) is a form of column chromatography that is highly sensitive and able to separate complex mixtures.

- The most common stationary phase is a tiny diameter resin, packed into a narrow column. The large surface area of the resin provides improved separation of components, although increases resistance to the mobile phase flow.
- The mobile phase is a liquid solvent (eluent) that is pumped through the column under high pressure.

The components and operation of a modern HPLC instrument are shown in figure 10.31.

1. The **sample** is injected onto the start of the column as a liquid.
2. The eluent is then pumped through the column, carrying the sample with it.
3. As the mobile phase moves through the column, the process of adsorption and desorption results in the components of the sample moving at different speeds and thus being separated from each other.
4. As the components exit the column they are usually detected by measuring the absorbance of ultraviolet (UV) light. Unlike the dyes in figure 10.29, many organic molecules are colourless, although do absorb UV light. This is recorded as a series of peaks on a chart called a **chromatogram**, including both the **retention time** (R_t) and area of each peak. Analysis of this data will be discussed later in this section.

sample a substance to be analysed

chromatogram a chart that shows the results from analysis by chromatography

retention time the time taken for a component in a sample to travel from the injection port to the end of the column

FIGURE 10.31 A schematic diagram of a high-performance liquid chromatography instrument

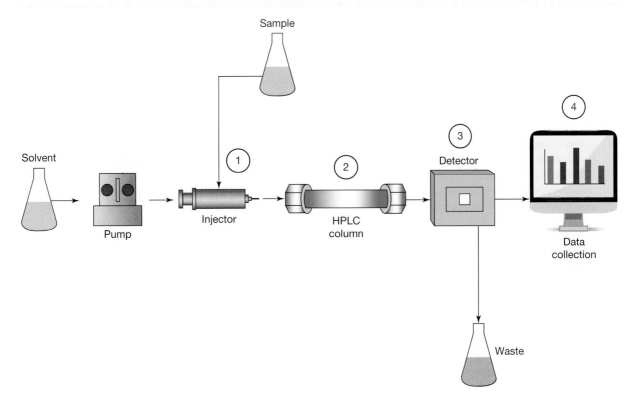

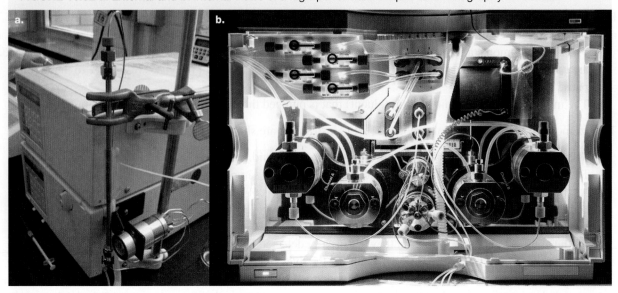

FIGURE 10.32 a. External and **b.** internal views of a high-performance liquid chromatography instrument

Uses of HPLC

HPLC is an extremely sensitive and widely used technique. Detection of concentrations in parts per million (or mg L^{-1}) and parts per billion levels is routine, with some instruments capable of detecting parts per trillion. Applications of HPLC include research, medicine, pharmaceutical science, forensic analysis, food analysis, drug detection in sport and environmental monitoring.

HPLC is often categorised according to the nature of the stationary phase (resin) and mobile phase (eluent) used.

EXTENSION: Types of HPLC

The two most common types of HPLC are:
1. *normal-phase liquid chromatography (NPLC).* In this form of HPLC, the resin is more polar than the eluent. Because of this, the more polar components in the sample adsorb more strongly to the resin and move more slowly through the column. Therefore, they have a longer retention time.
2. *reverse-phase liquid chromatography (RPLC).* This is the opposite of NPLC, as the resin is less polar than the eluent being pumped through it. The columns used often contain silica particles that have been coated with long hydrocarbon chains (C$_8$ and C$_{18}$ are commonly used) to achieve a level of 'non-polarity'. This has the opposite effect on retention times to NPLC. More polar molecules in the sample are not as strongly adsorbed to the resin and therefore move through it more quickly, thus displaying shortened retention times. RPLC is the most commonly used form of HPLC.

---

on Resources

▷ **Video eLesson**   High-performance liquid chromatography (HPLC) (eles-3251)

🔗 **Weblink**   HPLC

Qualitative analysis

The time taken for each component of a sample to travel from the injection port to the end of the column where it is detected is referred to as its retention time, R_t. This corresponds to the position of the peak on the chromatogram. Retention time can be used to identify a component, by comparing the retention time for an unknown substance with those for known substances under the same operating conditions.

Retention time and retardation factor

Retention time (R_t) measured by HPLC should not be confused with retardation factor (R_f), which is determined by TLC.
- In HPLC, a component with high relative affinity for the mobile phase will elute quickly from the column. This will result in a low R_t.
- In TLC, a component with a high relative affinity for the mobile phase will produce a dot close to the solvent front and hence will have a high R_f.

Figure 10.33 shows results obtained from testing a brand of decaffeinated coffee. Note that a caffeine standard has been run through the instrument so that the caffeine peak on the chromatogram of the sample can be identified from the retention time (four minutes). Therefore, the reduction in concentration of caffeine in decaffeinated coffee when compared to normal coffee becomes obvious by noting the decrease in the area of the peak due to caffeine.

FIGURE 10.33 HPLC chromatograms for **a.** caffeine, **b.** normal coffee and **c.** decaffeinated coffee

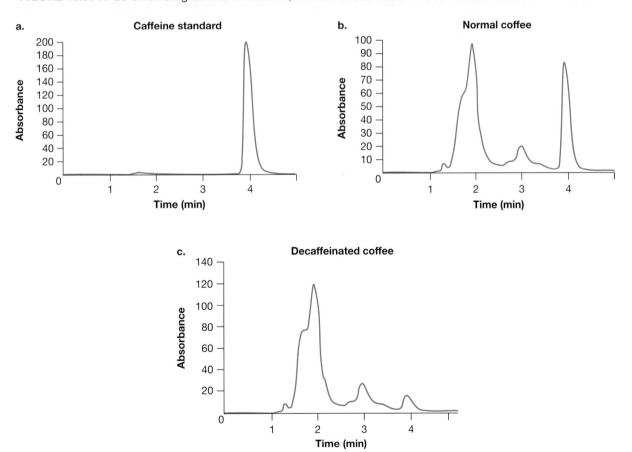

Quantitative analysis

The concentration of a component is proportional to the area under the corresponding peak on the chromatogram. As such, it is possible to perform quantitative analysis of a substance by using a **calibration curve** (also known as a standard curve) to convert peak area into concentration.

1. Prepare a series of **standard solutions** of the same compound.
2. Run the standard solutions through the HPLC under identical conditions to the original sample.
3. A calibration curve (concentration versus peak area) is plotted and a line of best fit drawn.
4. The concentration of the component can be determined using the peak area and the calibration curve.

Returning to the previous example, quantitative measurement of the concentration of caffeine in the sample could be performed using the calibration curve in figure 10.34 and a peak area of 17 000.

FIGURE 10.34 Using the calibration curve, the concentration of the unknown sample (shown by the pink arrow) can be estimated as approximately 17 mg L^{-1}.

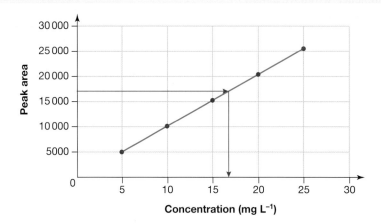

Resources

▶ **Video eLesson** Calibration curves (eles-3252)

Comparing peak areas

Peak areas are only comparable for the same substance run under the same conditions. As such, it is *not* appropriate to compare the area under peaks produced:
- by the same compound under different condition
- by different compounds under the same conditions.

When plotting a calibration curve:
- only include a data point at the origin (0, 0) when there is evidence to do so. Such evidence could be either mention of the instrument having been **calibrated**, or if a peak area of 0 has been recorded for a standard solution with zero concentration.
- the line of best fit should be a smooth line (not dot-to-dot) ending at the lowest and highest data points, as the relationship with peak area is only known within this concentration range
- values for peak height may be used when peaks are very narrow
- peak area is frequently expressed with 'arbitrary units' or without units.

calibration curve a graph of concentration versus peak area; also known as a standard curve

standard solution a solution that has an accurately known concentration

calibrate adjusting an instrument using standards of known measurements to ensure the instrument's accuracy

Most modern instruments are programmed to automatically calculate concentration from peak area. They utilise statistical algorithms to determine the *equation of the line of best fit* using the calibration data. The test result for the unknown is then fed into this equation and the resulting concentration displayed.

tlvd-9663

SAMPLE PROBLEM 7 Using HPLC to determine the concentration of ethanol in a solution

Although it is usually performed using a related technique called gas chromatography (GC), the level of ethanol in alcoholic drinks can be determined using HPLC.

In one such analysis using HPLC, a set of six ethanol solutions of known concentration were run through the instrument for the purpose of calibration. A sample of a vintage brandy was then diluted 1 in 5 and analysed under exactly the same conditions as the standard solutions.

Chromatograms for one of the standard ethanol solutions (figure a) and the brandy sample (figure b), as well as the data table, are shown.

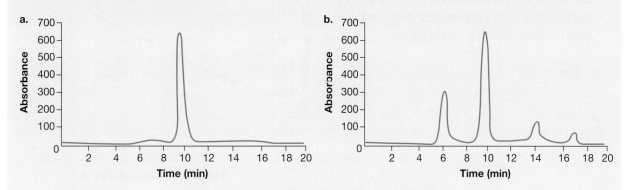

Standard concentration, %(v/v)	Peak area
7	342 401
8	391 318
9	440 230
10	489 136
11	538 058
12	586 970
Sample	450 012

a. Explain why only one peak is produced in the chromatogram for the standard ethanol solution.
b. Identify the R_t for the ethanol peak and justify your response.
c. Using these results, plot a calibration curve of concentration versus peak area.
d. Use the calibration curve to deduce the ethanol content in the diluted sample of brandy.
e. Calculate the concentration of ethanol in the original, undiluted bottle of brandy.
f. Explain why it was necessary to dilute the sample of brandy before analysing the sample.
g. Convert the concentration of ethanol in %(v/v) in the undiluted brandy into mL L^{-1}.

THINK

a. In a chromatogram, each peak corresponds to a particular substance.

b. It is the R_t of the peak in the chromatogram for the reference standard.

WRITE

a. Ethanol is the only substance present, so only one peak was produced.

b. 9.5 minutes

c. To plot the graph, consider the scale required. For each percentage the peak area differs by approximately 50 000. Include a line of best fit.

Note: In this example the scales on both the *x*- and *y*-axes have been broken to position the curve in the centre of the graph. However, you are advised to use unbroken scales when plotting calibration curves.

c.

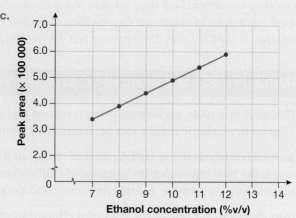

d. Locate the peak area of the sample (450 012) and rule a straight line from it until you reach the line of best fit. Drop straight down and read the corresponding concentration.

d.

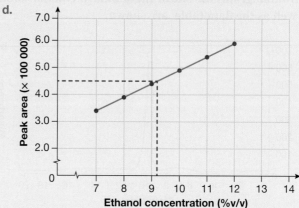

Reading from the calibration curve, the sample of brandy gives an ethanol concentration of 9.2 %(v/v).

e. The concentration of the diluted brandy was determined in part **d.** As the original (undiluted) sample of brandy was diluted 1 in 5 before being analysed, it will be fives times more concentrated than the diluted sample.

e. $c(\text{ethanol})_{\text{undiluted}} = \text{dilution factor} \times c(\text{ethanol})_{\text{diluted}}$
$= 5 \times 9.2\%$
$= 46\,\%(\text{v/v})$

f. Consider the absorbance of an undiluted sample of brandy and the scale of the line of best fit on the calibration curve.

f. The absorbance of an undiluted sample of brandy would be five times that of the undiluted sample. Dilution allowed the peak area to fall within the range of the line of best fit on the calibration curve and allowed the concentration to be determined.

g. 1 %(v/v) is 1 mL in 100 mL. 1 L = 1000 mL.

g. $V(\text{ethanol})_{\text{in 1000 mL}} = 46\% \times 1000 \text{ mL}$
$= \dfrac{46}{100} \times 1000 \text{ mL}$
$= 460 \text{ mL}$
$\therefore c(\text{ethanol}) = 4.6 \times 10^2 \text{ mL L}^{-1}$

PRACTICE PROBLEM 7

The ester methyl butanoate, $CH_3CH_2CH_2COOCH_3$, is used as a flavour additive and in perfumes. It has both a pleasant odour and taste. However, butanoic acid, from which it is made, has an extremely unpleasant odour. Therefore, it is desirable that residual butanoic acid levels be kept to a minimum in methyl butanoate preparations that are used for these purposes.

HPLC was used to measure the level of butanoic acid in a sample of food-grade methyl butanoate. A number of standards were run through the instrument, together with a sample of the methyl butanoate, which had been diluted 1 in 10. The results are shown in the following table.

Concentration of butanoic acid ($mg\ L^{-1}$)	Peak area
4.0	640
6.0	958
8.0	1280
10.0	1605
Diluted sample	1150

a. Using these results, plot a calibration curve of concentration versus peak area.
b. Use the calibration curve to deduce the butanoic acid content in the diluted sample.
c. Calculate the concentration of butanoic acid in the original, undiluted sample.
d. Calculate the concentration of butanoic acid in the undiluted sample expressed as $mol\ L^{-1}$.
   M(butanoic acid) = 88.0 $g\ mol^{-1}$

Optimising HPLC

The conditions of HPLC can be adjusted to improve performance in two ways:
- To change the retention time for components
- To improve the separation of components.

TABLE 10.14 Factors affecting retention time

Factor	Explanation of effect on retention time
Polarity of the mobile and stationary phases	Changing these will affect the relative affinity of a component for the mobile and stationary phases, and thus the R_t for that component. For example, greater similarity in the polarity of the component and the mobile phase will increase affinity and solubility, reducing R_t.
Temperature	Higher temperature increases the time spent by a component in the mobile phase, reducing R_t. It does so by: • reducing the strength of intermolecular forces between the component and the stationary phase • increasing solubility of all components in the mobile phase. Higher temperature will also reduce the viscosity of the mobile phase.
Viscosity of the mobile phase	Lower viscosity will increase the rate at which the mobile phase flows through the column, reducing R_t.
Diameter of the resin	Resin with a smaller diameter has a larger surface area, which slows the rate of mobile phase flow, increasing R_t.
Packing of the resin	Packing the resin more tightly slows the rate of mobile phase flow, increasing R_t.
Pressure applied by the pump	Higher pressure will increase the rate of mobile phase flow, decreasing R_t.
Length of the column	A shorter column will reduce the time required for a component to exit, reducing R_t.

Effective separation of components within a sample (referred to as **resolution**) is essential for HPLC to operate effectively. Figure 10.35 demonstrates how varying resolution between two components in a column is shown on a chromatogram. As the top of the peaks are still evident in figure 10.35c, it may still be possible to determine the R_t for these components; however, peak area could not be calculated as the peaks are overlapping.

Improving resolution is more complex than changing retention time. Nevertheless, chemists can increase the column length, and adjust the polarity of the mobile and stationary phases, in order to maximise component separation.

Determining product purity

The detection of impurities in products such as pharmaceuticals, food and water is vital to ensure quality and safety. HPLC has a number of capabilities that make it an ideal technique to determine product purity:

- Mixture separation
- Identification of components — comparison of R_t to known compounds
- Quantification of components — calibration curves

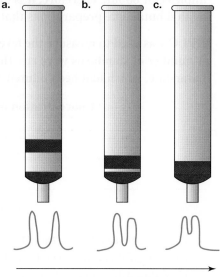

FIGURE 10.35 The effect of **a.** high, **b.** medium and **c.** low component separation on chromatograms

Lower resolution

The three capabilities listed are illustrated in figure 10.36, which shows HPLC analysis of a paracetamol tablet. The presence of two peaks indicates that an unknown impurity was present.

Components were identified by matching the R_t of each peak to known standards analysed under the same conditions. Paracetamol produced a peak at 2.6 minutes, with the contaminant peak at 3.5 minutes identified as caffeine.

Purity could be quantified by using calibration curves of paracetamol and caffeine to determine the concentration of each substance and therefore, the percentage of paracetamol and caffeine in the tablet.

resolution (with reference to chromatography) the degree of component separation

FIGURE 10.36 HPLC analysis identifying a caffeine impurity in a paracetamol tablet

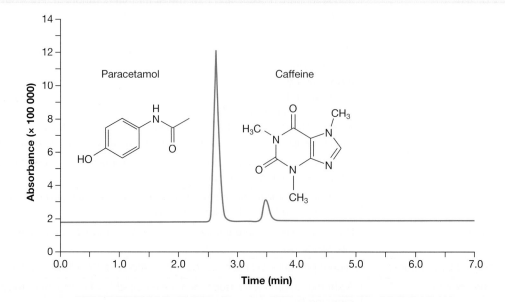

EXPERIMENT 10.2

Separating mixtures using column chromatography

Aim

To separate a mixture of dyes using column chromatography

10.6.4 Liquid chromatography—mass spectrometry (LC–MS)

What do I need to know about LC–MS?

This section explores some of the principles underpinning the modern application of instrumental analysis techniques. The details of LC–MS do not need to be memorised.

HPLC can be used to separate mixtures and provide qualitative and quantitative information about the components. However, identification of components using only R_t comparison to standards is less effective than some other instrumental techniques.

Much like the fingerprint region of IR spectra, the fragmentation pattern of mass spectra is characteristic for each compound and can be compared to spectra from known compounds until a match is found. These control spectra may be produced by analysing standards on the same instrument, or accessed in a spectral database containing many thousands of compounds. Given the complexity of most mass spectra, matching is performed computationally rather than manually (figure 10.37).

FIGURE 10.37 Identification of a compound by comparing the fragmentation pattern with known compounds

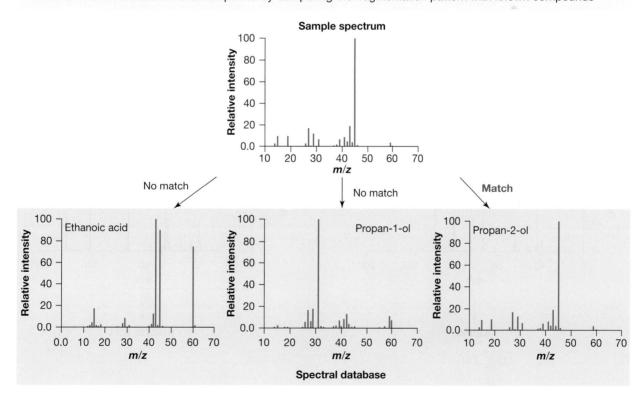

Modern instruments containing both liquid chromatography and mass spectrometry (LC–MS) capabilities are able to perform both the separation and identification steps.

1. The sample is injected into the HPLC.
2. As the separated components leave the column, they are analysed by MS.
3. Mass spectra are matched to identify the component.

LC–MS instruments are particularly powerful tools for the analysis of mixtures and are routinely used in many fields, including the pharmaceutical industry. Aspirin is a common pain-relief medication that is manufactured by chemical modification of salicylic acid. In figure 10.38, the results are shown for a batch of aspirin that was analysed for purity. The sample was first separated by chromatography and then the two components were identified by MS, confirming the presence of the impurity, salicylic acid.

FIGURE 10.38 LC–MS analysis identifies a salicylic acid impurity in a batch of aspirin.

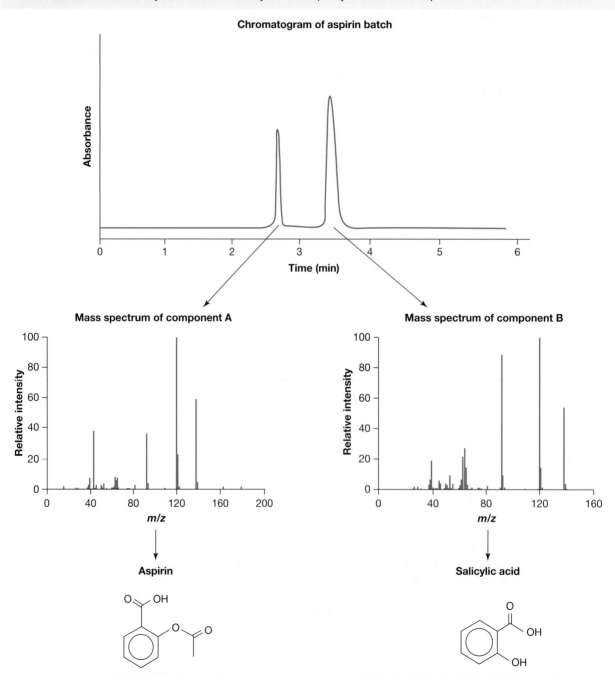

10.6 Activities

| 10.6 Quick quiz   on | 10.6 Exercise | 10.6 Exam questions |

10.6 Exercise

1. **MC** Which of the following factors would increase the retention time of an analyte?
   I Increasing the length of the column
   II Increasing the pressure used to pump the mobile phase through the column
   III Decreasing the particle size of the resin

   A. I and II only
   B. I and III only
   C. II and III only
   D. I, II and III

2. **MC** An analyte was analysed by HPLC using a polar mobile phase and non-polar stationary phase.

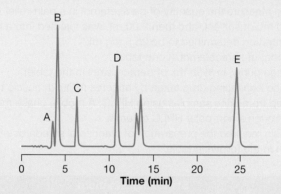

   Which one of the following statements is correct?
   A. E is more polar than B.
   B. B has a higher R_t than D.
   C. C is more soluble in the mobile phase than B.
   D. B has a higher affinity for the stationary phase than A.

3. In chromatography, why is it important that the stationary phase has a large surface area?

4. The first four members of the carboxylic acid homologous series were analysed by HPLC, using a polar mobile phase and a non-polar stationary phase.
   a. In what sequence would the acids elute?
   b. Identify two effects to support your answer.

5. Analysis of a multivitamin supplement was performed by LC–MS, using a methanol mobile phase and a stationary phase of particles coated in long-chain hydrocarbons.
   a. Justify why methanol is a more appropriate mobile phase than hexane when using this column.
   b. Identify the type(s) of intermolecular forces formed between methanol and the hydrocarbon stationary phase.
   c. How could the mass spectra be used to identify the vitamins in the supplement?

6. The following chromatogram of amino acids was generated by HPLC analysis with a polar stationary phase. The mixture was thought to consist of leucine, isoleucine, phenylalanine and serine. Amino acid structures can be found in the VCE Chemistry Data Book.

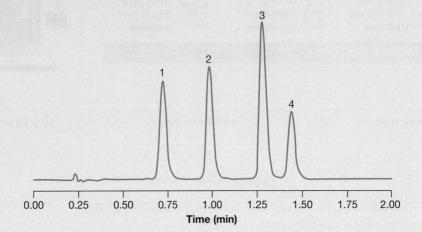

a. Which of the four peaks (1–4) is most likely to be serine? Explain your answer.
b. Explain how you could confirm that the R_t for the peak you identified in part a is caused by serine.
c. Explain whether threonine would have a higher or lower R_t than serine.
d. A student proposed that comparing the peak areas would be a way to accurately determine the relative concentration of each amino acid in the mixture. Justify whether this idea will work.

7. Analysis was performed to measure the quantity of paracetamol in a pain-relief tablet. A 500 mg tablet was crushed, dissolved in 10.0 mL of solvent, and then 2.00 mL was injected into a HPLC. A calibration curve was generated and the concentration determined to be 25.5 mg mL^{-1}.
a. Calculate the mass, in mg, of paracetamol in one tablet.
b. Determine the percentage purity, in %(m/m), of paracetamol in the tablet.

8. Stanozolol is a performance-enhancing drug taken by athletes to build muscle tissue and increase power. Stanozolol can be analysed from urine samples using HPLC. A mobile phase mixture of methanol (90%) and water (10%) is pumped through a non-polar HPLC column.

One particular urine analysis required the preparation of stanozolol standards of 1.0, 2.0, 3.0 and 4.0 mg L^{-1}. The peak areas are shown in the following table.

Retention time (min)	Stanozolol standard (mg L^{-1})	Peak area (× 10 000)
4.1	1.0	5.0
4.1	2.0	9.8
4.1	3.0	15.2
4.1	4.0	20.0

A 20 μL sample of undiluted urine was run through the chromatograph under the same conditions.
a. Explain how this procedure can be used for the qualitative analysis of stanozolol in urine.
b. Explain how it can be used to determine how much stanozolol is present in the urine.
c. Using the approach you proposed in part b, determine the stanozolol concentration in a urine sample that returned a peak at $R_t = 4.1$ minutes with an area of 125 000.
d. Explain why a sample of urine from an athlete with a suspected stanozolol concentration of 5.0 mg L^{-1} could not be reliably tested using this method.
e. Suggest one alteration to the procedure that would allow an athlete suspected of having a stanozolol concentration above 5.0 mg per litre of urine to be tested.

10.6 Exam questions

▶ Question 1 (2 marks)

Source: VCE 2021 Chemistry Exam, Section B, Q.9.c; © VCAA

Aspartame is an ingredient in some soft drinks. Aspartame is unstable in some conditions and reacts to form four main products. One of the products of aspartame breakdown is 5-benzyl-3,6-dioxo-2-piperazineacetic acid (DKP). It is thought that DKP may be harmful to humans.

A student, Kim, investigates the effect of storage temperature on the rate of production of DKP from aspartame in lemonade. Experimental data is obtained using high-performance liquid chromatography (HPLC) to analyse the aspartame and DKP content in lemonade samples.

HPLC calibration

Kim first calibrated the HPLC using the following method:
1. Prepare and refrigerate a standard solution of pure aspartame with a concentration of 1000 mg L^{-1}.
2. Transfer a 10.00 mL aliquot of the pure aspartame solution into a 1.000 L volumetric flask.
3. Fill the volumetric flask up to the 1.000 L mark with deionised water and shake the flask.
4. Inject a sample of the diluted aspartame solution into the HPLC to obtain a chromatogram.
5. Repeat steps 1–4 with DKP.

The following two calibration chromatograms were obtained.

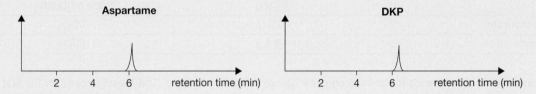

Analysis of lemonade samples

Kim then followed the method given in steps 6–14 to investigate the rate of production of DKP from aspartame in lemonade at different storage temperatures.
6. Open a can of lemonade.
7. Transfer a 10.00 mL aliquot of lemonade from the can into a 1.000 L volumetric flask.
8. Fill the volumetric flask up to the 1.000 L mark with deionised water and shake the flask.
9. Inject a sample of the diluted lemonade into the HPLC using the same operating conditions used during calibration.
10. Set up three water baths at temperatures of 15 °C, 25 °C and 35 °C.
11. Put three unopened cans of lemonade into each of the three water baths.
12. After one day, take one can from each water bath and follow steps 6–9.
13. After two days, take one can from each water bath and follow steps 6–9.
14. After three days, take one can from each water bath and follow steps 6–9.

One of the chromatograms from the diluted lemonade is given below.

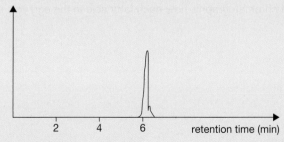

a. State a change to the operating conditions of the HPLC that could be made to reduce the errors in measuring the concentrations of aspartame and DKP. **(1 mark)**

b. State how this change would reduce the measurement errors. **(1 mark)**

▶

Question 2 (1 mark)

Source: VCE 2020 Chemistry Exam, Section A, Q.20; © VCAA

MC Consider the following changes that could be applied to the operating parameters for a chromatogram set up to carry out high-performance liquid chromatography (HPLC) with a polar stationary phase and a non-polar mobile phase:

I decreasing the viscosity of the mobile phase
II using a more tightly packed stationary phase
III using a mobile phase that is more polar than the stationary phase

Which of the changes would be most likely to reduce the retention time of a sugar in the HPLC?

A. I only   B. I and III only   C. III only   D. II and III only

Question 3 (3 marks)

Source: VCE 2020 Chemistry Exam, Section B, Q.7.b.i; © VCAA

Inside the shell of an egg is egg white that encircles egg yolk. The nutrition information for egg yolk and egg white is given in Table 1.

Table 1

Nutrient	Per 100 g of egg yolk	Per 100 g of egg white
energy	1437 kJ	184 kJ
fat	27.0 g	trace amounts
carbohydrate	0.0 g	0.0 g
protein	16.4 g	10.8 g

The composition of fatty acids found in an egg yolk sample is given in Table 2. The melting points for the first three fatty acids are provided.

Table 2

Fatty acid	Percentage (%)	Melting point (°C)
palmitic	25.9	63
stearic	9.1	69
palmitoleic	3.4	0
oleic	40.9	
linoleic	16.3	
linolenic	2.9	
arachidonic	1.5	

The composition of fatty acids in an egg yolk was determined by reacting the fatty acids with methanol to produce methyl esters and then analysing the methyl esters using chromatography.

Explain, using the principles of chromatography, how each fatty acid in the egg yolk sample can be identified and the percentage determined.

Use the following information to answer Questions 4 and 5.

The mass of caffeine in a particular coffee drink was determined by high-performance liquid chromatography (HPLC).

The calibration curve produced from running standard solutions of caffeine through an HPLC column is shown below.

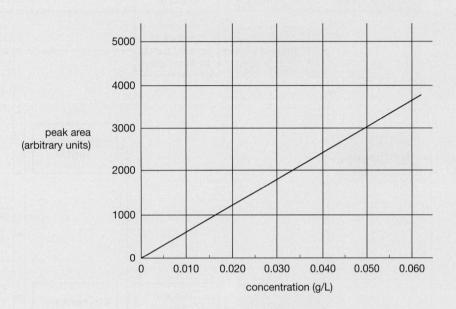

A 5.0 mL aliquot of the coffee drink was diluted to 50.0 mL with de-ionised water. A sample of the diluted coffee drink was run through the HPLC column under identical conditions to those used to obtain the calibration curve.

The peak area obtained for this diluted sample was 2400 arbitrary units.

▶ Question 4 (1 mark)
Source: VCE 2017 Chemistry Exam, Section A, Q.21; © VCAA

MC The HPLC column used has a non-polar stationary phase.

The most suitable solvent for determining the concentration of caffeine in the sample is
A. carbon tetrachloride, CCl_4
B. methanol, CH_3OH
C. octanol, $C_8H_{17}OH$
D. hexane, C_6H_{14}

▶ Question 5 (1 mark)
Source: VCE 2017 Chemistry Exam, Section A, Q.22; © VCAA

MC The mass of caffeine, in grams, in 350 mL of the undiluted coffee drink is closest to
A. 0.014
B. 0.070
C. 0.14
D. 0.40

More exam questions are available in your learnON title.

10.7 Review

10.7.1 Topic summary

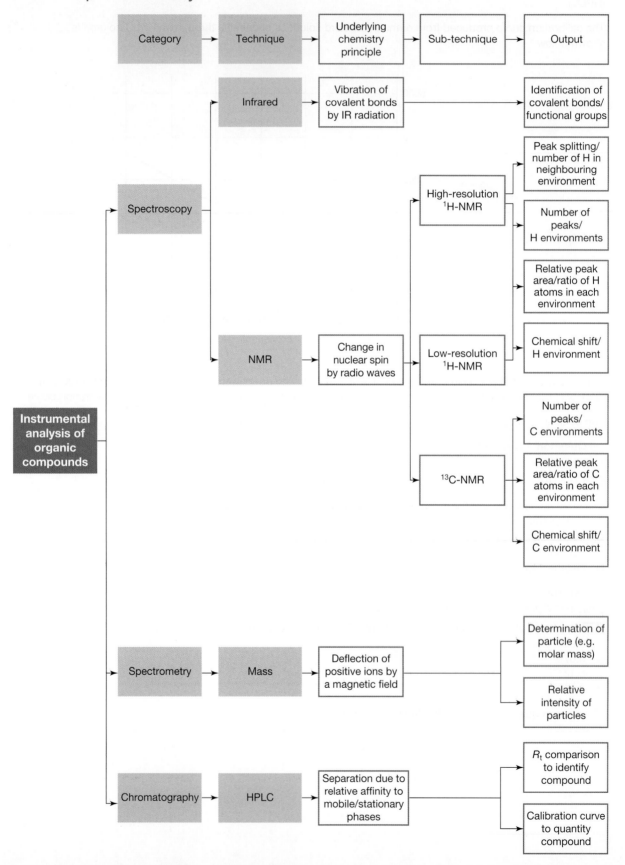

Hey students! Now that it's time to revise this topic, go online to:

Access the topic summary | Review your results | Watch teacher-led videos | Practise past VCAA exam questions

Find all this and MORE in jacPLUS

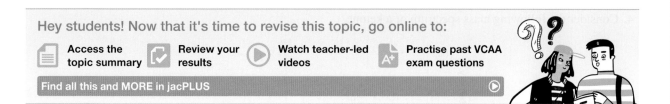

10.7.2 Key ideas summary

online only

10.7.3 Key terms glossary

online only

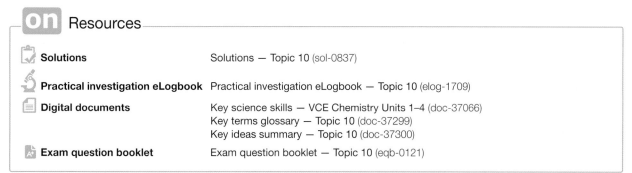

on Resources

Solutions	Solutions — Topic 10 (sol-0837)
Practical investigation eLogbook	Practical investigation eLogbook — Topic 10 (elog-1709)
Digital documents	Key science skills — VCE Chemistry Units 1–4 (doc-37066) Key terms glossary — Topic 10 (doc-37299) Key ideas summary — Topic 10 (doc-37300)
Exam question booklet	Exam question booklet — Topic 10 (eqb-0121)

10.7 Activities

learnon

Students, these questions are even better in jacPLUS

Receive immediate feedback and access sample responses | Access additional questions | Track your results and progress

Find all this and MORE in jacPLUS

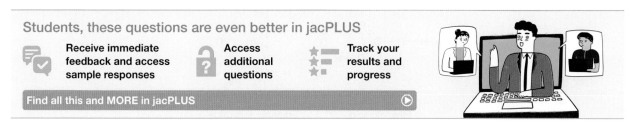

10.7 Review questions

1. **MC** Which of the following would be an example of quantitative analysis?
   - **A.** Comparing a mass spectrum fragmentation pattern to a database of compounds
   - **B.** Using chromatography to identify the number of different compounds present in pen ink
   - **C.** Identifying the amino acids present in orange juice
   - **D.** Using a calibration curve to analyse HPLC data

2. **MC** Ethanoic acid and methyl methanoate both have the same molecular formula, $C_2H_4O_2$.

   If IR spectroscopy was used to distinguish between these compounds, which of the following wave number ranges would be most useful?
   - **A.** 1000–1300 cm^{-1}
   - **B.** 1680–1740 cm^{-1}
   - **C.** 2500–3500 cm^{-1}
   - **D.** 2850–3090 cm^{-1}

3. **MC** The ethyl group, CH_3CH_2–, is easily identified using high-resolution ^{1}H-NMR spectroscopy because its splitting pattern is
   - **A.** a triplet and a doublet.
   - **B.** a triplet and a quartet.
   - **C.** a singlet and a doublet.
   - **D.** two triplets.

4. Consider the following mass spectrum of a ketone.

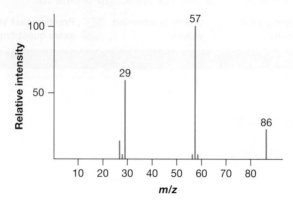

a. What is the m/z value for the parent ion?
b. What is the m/z value for the base peak?
c. The peak at $m/z = 57$ represents the loss of what possible fragment from the molecule?
d. Suggest a possible structure for the compound.

5. The painkiller aspirin may be formed by reacting salicylic acid with either ethanoic acid or ethanoic anhydride. The reaction with ethanoic anhydride is preferred as it is much faster. The structures of ethanoic acid and ethanoic anhydride are shown in the following figure.

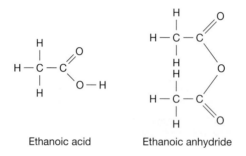

Ethanoic acid   Ethanoic anhydride

Explain how these two molecules could be distinguished using IR spectroscopy.

6. A ^{1}H-NMR spectrum of a molecule with the molecular formula $C_3H_6O_2$ is shown in the following figure.

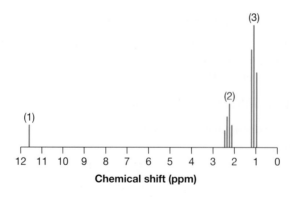

a. Identify the peaks using the table of chemical shifts (table 10.9).
b. How many peaks would you most likely find in the set of peaks for the CH_2 group?
c. Sketch the structure of the molecule.

7. Compound A is an alcohol with the molecular formula C_3H_8O.

   a. Draw and name the possible structural isomers represented by this formula.

   Compound A reacts with acidified potassium dichromate solution to form compound B, which has the molecular formula C_3H_6O. The ^{1}H-NMR spectrum of compound B shows only one peak, and its mass spectrum is shown in the following figure.

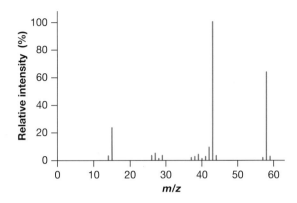

   b. Name and draw the structure of compound B. Justify your answer by referring to the ^{1}H-NMR information.
   c. Identify the fragment with $m/z = 43$ and comment on whether this supports your answer to part b.
   d. How many peaks would you expect in the ^{13}C-NMR spectrum of compound B?

8. A new brand of throat lozenges called 'Throat Eze' makes the claim that each lozenge contains 1.2 mg of dichlorobenzyl alcohol.

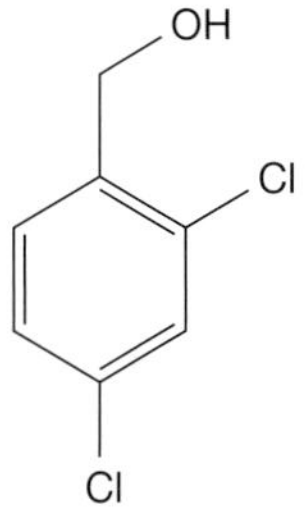

   To test this claim, a government analyst dissolved the lozenge in a solvent made from water and ethanol and made it up to 500 mL. A small sample was then injected into a high-performance liquid chromatograph. A chromatogram containing a large number of peaks was obtained.

   The operator then ran a series of dichlorobenzyl alcohol standards of known concentration through the instrument. Chromatograms for each standard were obtained, as well as a measure of the area under each of the reference peaks.

   Dichlorobenzyl alcohol

   a. Explain how the standards would allow the dichlorobenzyl alcohol peak from the original chromatogram to be identified.
   b. What is the purpose of using a set of standards as described, and subsequently obtaining the area under their peaks?
   c. The following table shows the results from the standards, together with a measurement for the area under the peak that was identified as dichlorobenzyl alcohol from the original chromatogram.

Concentration of standard (mg L^{-1})	Area under peak (arbitrary units)
1.0	83
2.0	160
3.0	241
4.0	315
Lozenge extract	193

   Is the claim made by the manufacturer true?

9. The painkiller phenacetin was the world's first synthetic pharmaceutical drug. It was developed by an American chemist and began distribution in 1887. Phenacetin was often accompanied by aspirin and caffeine in what were called 'APC' pills. These pills were widely distributed during and after World War II. The use of phenacetin in the United States was discontinued in the 1980s because of links to cancer and other adverse side effects, but it remains available in some countries.

An APC pill was dissolved in a suitable solvent and analysed using HPLC. The results are shown in the following chromatogram.

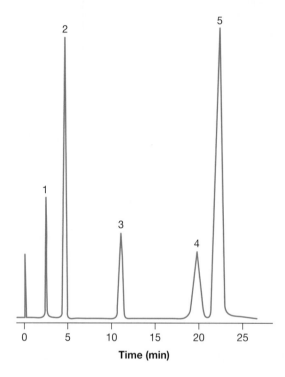

a. Which of the peaks (1–5) in the chromatogram of the APC adsorbed least to the stationary phase?
b. Describe an experiment that would enable you to determine which of the peaks in the chromatogram of APC was due to the presence of phenacetin.
c. To determine the amount of phenacetin in an APC tablet using HPLC, a set of five reference standards of known phenacetin concentration were run through the instrument for the purpose of calibration. A 0.0336 g sample of an APC tablet was then dissolved in a suitable solvent and made up to 100.0 mL in a volumetric flask. This APC sample solution was then analysed under exactly the same conditions as the reference samples. The results obtained are shown in the following table.

Concentration of phenacetin (mg L^{-1})	Peak area
0.00	0
1.00	154
2.00	298
3.00	448
4.00	606
5.00	750
Sample	375

Using these results, plot a calibration curve of concentration, on the x-axis, against peak area.
d. Use the graph to deduce the concentration of the phenacetin solution, in mg L^{-1}, to one decimal place.
e. Determine the concentration of phenacetin, in %(m/m), in the tablet.

10. Compound X was found to contain 48.63% carbon, 8.18% hydrogen and 43.19% oxygen. The mass spectrum of X showed a molecular ion peak at $m/z = 74$, along with fragments at $m/z = 59$, 45, 29 and 15.

The information in the following table was obtained from its ^{1}H-NMR.

Peak	Number of H atoms	Chemical shift (δ)	Splitting
1	3	0.9–1.1	Triplet
2	2	4.1–4.3	Quartet
3	1	8.1	Singlet

The IR spectrum is shown in the following figure.

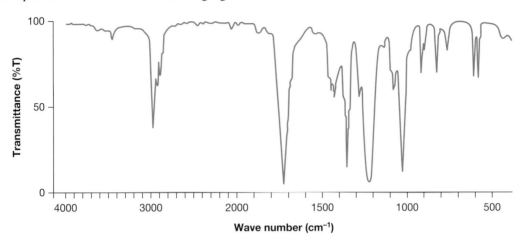

a. Determine the empirical formula of X.
b. Use information from the mass spectrum to determine the molecular formula of X.
c. The molecular formula suggests that X is either a carboxylic acid or an ester. Draw the structural formulae of *three* carboxylic acid and/or ester molecules that have this molecular formula.
d. Use the information from the IR spectrum to complete the following table.

Wave number (cm^{-1})	Bond	Present or absent
	C–H	
1750		
	O–H (acids)	

e. Draw the structure of X.
f. Show that the fragments at $m/z = 59$, 45, 29 and 15 are consistent with the structure in part **e** by providing their semi-structural formulae.

10.7 Exam questions

Section A — Multiple choice questions

All correct answers are worth 1 mark each; an incorrect answer is worth 0.

Question 1

MC Which one of the following best describes what occurs when a substance absorbs infrared radiation?

A. Some of its bonds begin to vibrate.
B. Some of its bonds begin to spin.
C. An electron jumps to a higher energy level.
D. Vibrating bonds increase the intensity of their vibration.

Use the following information to answer Questions 2 and 3.

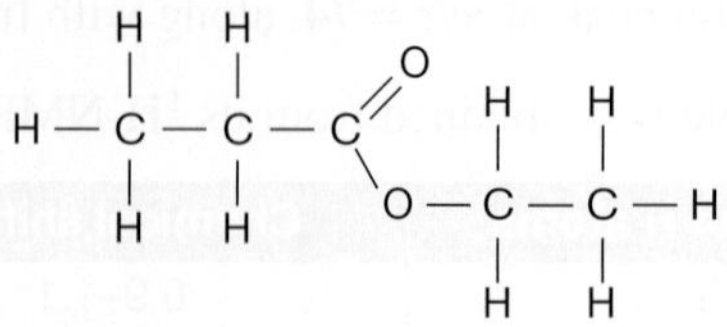

$$H-\underset{\underset{H}{|}}{\overset{\overset{H}{|}}{C}}-\underset{\underset{H}{|}}{\overset{\overset{H}{|}}{C}}-\overset{\overset{O}{\|}}{C}\underset{O-\underset{\underset{H}{|}}{\overset{\overset{H}{|}}{C}}-\underset{\underset{H}{|}}{\overset{\overset{H}{|}}{C}}-H}{}$$

▶ **Question 2**

Source: VCE 2012 Chemistry Exam 1, Section A, Q.10; © VCAA

MC The species that produces the molecular ion peak in the mass spectrum of this compound is

A. $[CH_3CH_2COOCH_2CH_3]^+$

B. $[CH_3CH_2COOCH_2CH_3]^{2+}$

C. $[CH_3CH_2COOCH_2CH_3]^-$

D. $CH_3CH_2COOCH_2CH_3$

▶ **Question 3**

Source: VCE 2012 Chemistry Exam 1, Section A, Q.11; © VCAA

MC Which one of the following infrared (IR) spectra is consistent with the structure of this compound?

A.

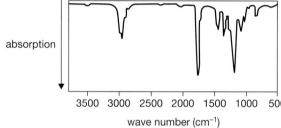

B.

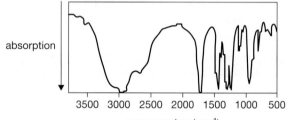

C.

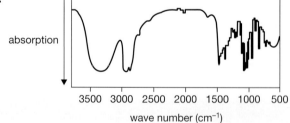

D.

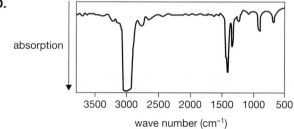

Question 4

Source: VCE 2015 Chemistry Exam, Section A, Q.10; © VCAA

MC The high-resolution proton NMR spectrum of chloroethane has two sets of peaks. Both peaks are split.

Which of the following correctly describes the splitting pattern?

A. a singlet and a doublet
B. a doublet and a doublet
C. a doublet and a triplet
D. a triplet and a quartet

Question 5

Source: VCE 2015 Chemistry Exam, Section A, Q.9; © VCAA

MC Which two isomers of $C_3H_6Br_2$ have two peaks (other than the TMS peak) in their ^{13}C-NMR spectrum?

A. $CH_3CBr_2CH_3$ and $CHBr_2CH_2CH_3$
B. $CHBr_2CH_2CH_3$ and $CH_2BrCHBrCH_3$
C. $CH_2BrCHBrCH_3$ and $CH_2BrCH_2CH_2Br$
D. $CH_2BrCH_2CH_2Br$ and $CH_3CBr_2CH_3$

Question 6

Source: VCE 2009 Chemistry Exam 1, Section A, Q.20; © VCAA

MC The separation and identification of proteins that can be used as disease markers is an exciting area of research.

Researchers must separate and identify proteins that could be used as disease markers from the many thousands of proteins that exist in our bodies.

Which of the following sequence of techniques could be used to
  i. separate these molecules, then
 ii. accurately determine their molecular mass, and then
iii. determine their molecular structure?

A. NMR spectroscopy, followed by mass spectrometry, followed by high-performance liquid chromatography
B. high-performance liquid chromatography, followed by mass spectrometry, followed by NMR spectroscopy
C. high-performance liquid chromatography, followed by infrared spectroscopy, followed by mass spectrometry
D. mass spectrometry, followed by high-performance liquid chromatography, followed by infrared spectroscopy

Question 7

MC An unknown compound was analysed by several spectroscopic techniques. The mass spectrum included a molecular ion peak at $m/z = 58$. The IR spectrum showed a strong band at 1700 cm^{-1} but none near 3400 cm^{-1}. The ^{1}H-NMR spectrum showed only one peak at 2.1 ppm.

Which of the following compounds could match this evidence?

A. Butane
B. Acetone, CH_3COCH_3
C. Propanol
D. Propanoic acid

Question 8

MC An unknown compound was analysed by several spectroscopic techniques. The mass spectrum included a molecular ion peak at $m/z = 58$. The IR spectrum did not show a strong band at 1700 cm^{-1}, nor one in the 3000–3500 cm^{-1} region. The low-resolution ^{1}H-NMR spectrum showed only two peaks (at 0.9 ppm (intensity = 3) and at 1.3 ppm (intensity = 2)).

Which of the following compounds could match this evidence?

A. Butane

B. Aminopropane

C. Acetone, CH_3COCH_3

D. Propanol

Question 9

Source: VCE 2015 Chemistry Exam, Section A, Q.8; © VCAA

MC Consider the following statements about a high-performance liquid chromatography (HPLC) column that uses a polar solvent and a non-polar stationary phase to analyse a solution:

Statement I — Polar molecules in the solution will be attracted to the solvent particles by dipole–dipole attraction.

Statement II — Non-polar molecules in the solution will be attracted to the stationary phase by dispersion forces.

Statement III — Polar molecules in the solution will travel through the HPLC column more rapidly than non-polar molecules.

Which of these statements are true?

A. I and II only

B. I and III only

C. II and III only

D. I, II and III

Question 10

Source: VCE 2007 Chemistry Exam 1, Section A, Q.5; © VCAA

MC Chromatogram 1 was obtained by analysis of a sample of a mixture of two sugars, A and B, using high-performance liquid chromatography (HPLC). Chromatogram 2 was obtained by analysing another sample of the same mixture by HPLC under different conditions.

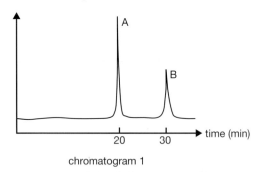

chromatogram 1

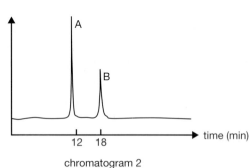

chromatogram 2

Consider the following changes which could be made to the operating conditions for HPLC.

I decreasing the pressure of the mobile phase

II decreasing the temperature

III using a less tightly packed column

Which of the changes would be most likely to produce chromatogram 2?

A. I only   **B.** II only   **C.** III only   **D.** I and II only

Section B — Short answer questions

Question 11 (5 marks)

Source: *VCE 2009 Chemistry Exam 1, Section B, Q.5; © VCAA*

The structure of an organic molecule, with empirical formula CH_2O, is determined using spectroscopic techniques.

The mass spectrum, infrared spectrum and ^{1}H-NMR spectrum for this molecule are given below.

mass spectrum

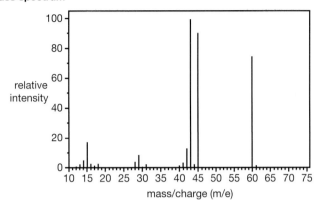

^{1}H-NMR spectrum

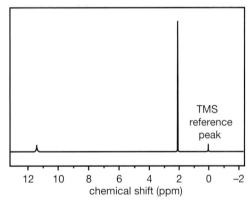

infrared spectrum

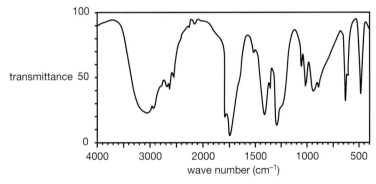

Use the information provided by these spectra to answer the following questions.

a. What is the molecular formula of this molecule? **(1 mark)**

b. How many different proton environments are there in this molecule? **(1 mark)**

c. Draw the structure of the unknown molecule, clearly showing all bonds. **(1 mark)**

d. Explain how the structure of the compound you have drawn in **part c.** is consistent with its IR spectrum. **(1 mark)**

e. Name the compound you have drawn in **part c.** **(1 mark)**

Source: VCE 2011 Chemistry Exam 1, Section B, Q.3; © VCAA

Caffeine is a stimulant drug that is found in coffee, tea, energy drinks and some soft drinks.

The concentration of caffeine in drinks can be determined using HPLC.

Four caffeine standard solutions containing 50 ppm, 100 ppm, 150 ppm and 200 ppm were prepared. 25 μL of each sample was injected into the HPLC column. The peak areas were measured and used to construct the calibration graph below. The chromatograms of the standard solutions each produced a single peak at a retention time of 96 seconds.

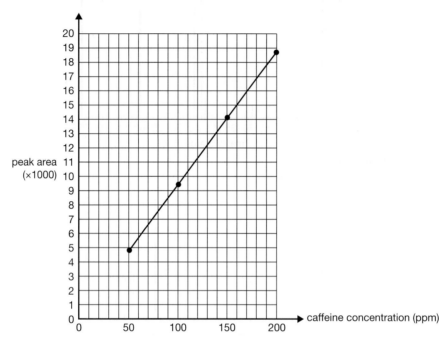

Peak area of caffeine standard solutions: retention time = 96 seconds

25 μL samples of various drinks thought to contain caffeine were then separately passed through the HPLC column. The results are summarised below.

Sample	Retention time of major peak (seconds)	Peak area of largest peak
Soft drink A	96	12 000
Soft drink B	32	8 500
Espresso coffee	96	211 000

a. Determine the caffeine content, in ppm, of soft drink A. **(1 mark)**

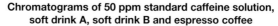

Chromatograms of 50 ppm standard caffeine solution,
soft drink A, soft drink B and espresso coffee

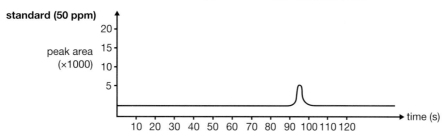

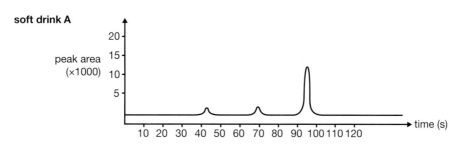

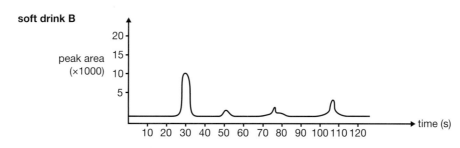

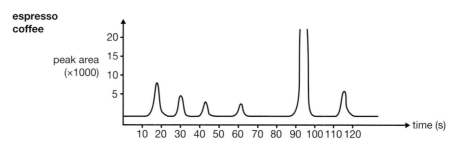

b. What evidence is presented in the chromatogram that supports the conclusion that soft drink B does not contain any caffeine? **(1 mark)**

c. i. Explain why the caffeine content of the espresso coffee sample cannot be reliably determined using the information provided. **(1 mark)**

ii. Describe what could be done to the espresso coffee sample so that its caffeine content can be reliably determined using the information provided. **(1 mark)**

Source: VCE 2010 Chemistry Exam 1, Section B, Q.2; © VCAA

The molecular formula of an unknown compound, X, is $C_3H_6O_2$.

The infrared ^{13}C-NMR and ^1H-NMR spectra of this compound are shown below.

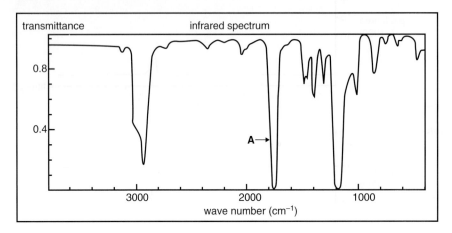

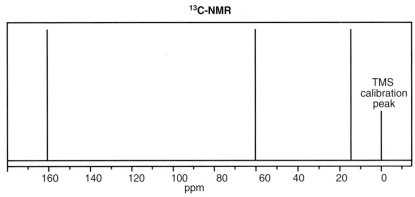

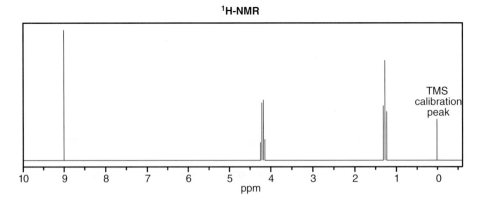

The ^1H-NMR spectrum data is summarised in the following table.

Chemical shift (ppm)	Relative peak area	Peak splitting
1.3	3	triplet (3)
4.2	2	quartet (4)
9.0	1	singlet (1)

a. Using the **Infrared absorption data** in the VCE Chemistry Data Book, identify the atoms and the bonds between them that are associated with the absorption labelled A on the infrared spectrum. **(1 mark)**

b. How many different carbon environments are present in compound X? **(1 mark)**

c. How many different hydrogen environments are present in compound X? **(1 mark)**

d. i. The signal at 1.3 ppm is split into a triplet. What is the number of equivalent protons bonded to the adjacent carbon atom? **(1 mark)**

ii. Draw the grouping of atoms that would give rise to the triplet and quartet splitting patterns. **(1 mark)**

e. A chemical test showed that compound X does **not** react with a base.

Propose a structure for compound X that is consistent with all the evidence provided. **(2 marks)**

▶ Question 14 (8 marks)

Source: VCE 2011 Chemistry Exam 1, Section B, Q.2 © VCAA

a. Bromine exists as two isotopes, ^{79}Br and ^{81}Br.

The mass spectrum of bromoethane, C_2H_5Br, with two molecular ion peaks at *m/z* 108 and 110, is shown below.

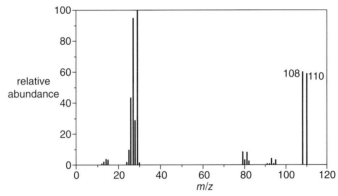

i. Identify the species that produces the peak at *m/z* = 29. **(1 mark)**

ii. What do the two molecular ion peaks indicate about the relative intensity of ^{79}Br and ^{81}Br? Give a reason for your answer. **(2 marks)**

b. There are two compounds that have the molecular formula $C_2H_4Br_2$.

The ^1H-NMR spectrum of **one** of these compounds is provided below.

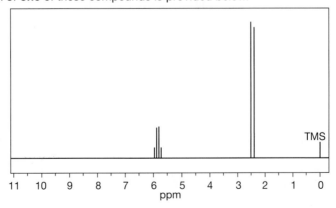

i. Draw the structural formula of each of the two compounds that have the molecular formula $C_2H_4Br_2$. **(2 marks)**

ii. Circle the structure from **part b.i.** that corresponds to the ^1H-NMR spectrum provided. Justify your selection by referring to both the ^1H-NMR spectrum and to the structure of the compound. **(3 marks)**

▶ **Question 15 (7 marks)**

Source: Adapted from VCE 2008 Chemistry Exam 1, Section B, Q.4.b.ii,iii,c; © VCAA

A mixture contains several different organic liquids, all of which boil at temperatures greater than 50 °C.

The compounds present in the mixture are separated and analysed.

Compound Y is an alkanol of molecular formula of $C_4H_{10}O$.

a. **i.** Compound Y shows 3 lines in the ^{13}C-NMR spectrum and undergoes reaction with $Cr_2O_7^{2-}$(aq) in acid to produce a carboxylic acid.
- What evidence about the structure of Y can be gained from the ^{13}C-NMR spectrum?
- What evidence about the structure of Y can be gained from the reaction with $Cr_2O_7^{2-}$(aq) in acid solution? **(2 marks)**

ii. Based on the evidence gained from the ^{13}C-NMR spectrum and reaction with $Cr_2O_7^{2-}$(aq) in acid solution:
- draw, showing all bonds, the structural formula of compound Y
- write the systematic name of compound Y. **(2 marks)**

Compound Z has the molecular formula $C_5H_{10}O$ and shows a strong band in the infrared spectrum at about 1700 cm^{-1}. The ^1H-NMR spectrum of compound Z is given below.

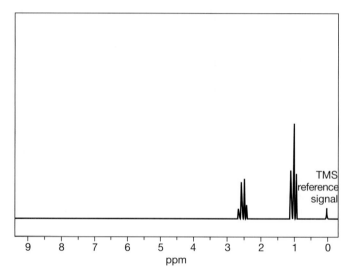

b. **i.** What information about the structure of Z can be deduced from the
- IR data
- ^1H-NMR spectrum? **(2 marks)**

ii. Draw a structure for compound Z that is consistent with the spectral data. **(1 mark)**

11 Medicinal chemistry

KEY KNOWLEDGE

In this topic you will investigate:

Medicinal chemistry

- extraction and purification of natural plant compounds as possible active ingredients for medicines, using solvent extraction and distillation
- identification of the structure and functional groups of organic molecules that are medicines
- significance of isomers and the identification of chiral centres (carbon atom surrounded by four different groups) in the effectiveness of medicines
- enzymes as protein-based catalysts in living systems: primary, secondary, tertiary and quaternary structures and changes in enzyme function in terms of structure and bonding as a result of increased temperature (denaturation), decreased temperature (lowered activity), or changes in pH (formation of zwitterions and denaturation)
- medicines that function as competitive enzyme inhibitors: organic molecules that bind through lock-and-key mechanism to an active site preventing binding of the actual substrate.

Source: VCE Chemistry Study Design (2024–2027) extracts © VCAA; reproduced by permission.

PRACTICAL WORK AND INVESTIGATIONS

Practical work is a central component of VCE Chemistry. Experiments and investigations, supported by a **practical investigation eLogbook** and **teacher-led videos**, are included in this topic to provide opportunities to undertake investigations and communicate findings.

EXAM PREPARATION

▶ Access past VCAA questions and exam-style questions and their video solutions in every lesson, to ensure you are ready.

11.1 Overview

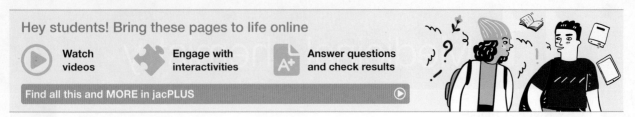
11.1.1 Introduction

Traditional cultures have used a variety of substances from their environment to treat illness for thousands of years. Modern medicine has seen these and other precursor compounds used to synthesise sophisticated medicines with both improved safety and efficacy, for a myriad of illnesses. The development of vaccines to prevent infection, and therapies for heart disease, cancer, and diabetes, among others, has greatly improved the quality of life and life expectancy of millions of people. These advances can rightly be considered one of humanity's greatest achievements.

In this topic we will be considering medicines that are organic molecules. More specifically, they are drugs that provide a beneficial health effect when administered to a patient, to either prevent disease or to treat an existing disease. Organic compounds comprise a remarkably large and diverse class of medicines, which is unsurprising given the enormous variety of possible structures.

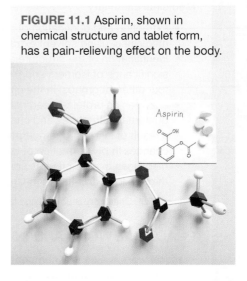

FIGURE 11.1 Aspirin, shown in chemical structure and tablet form, has a pain-relieving effect on the body.

Most drugs exert their effect by binding to a target molecule, which then affects the function of the molecule (often activating or inhibiting a process), leading to a beneficial health outcome. These binding interactions may be reversible or irreversible, and occur via covalent, ionic and/or weaker intermolecular forces. Whatever the nature of the interaction, a 'chemical fit' is required to facilitate binding of the drug to the intended target. This highlights the key role that chemical structures play in the effect of medicines.

The structure of medicinal organic molecules will be considered from several perspectives in this topic:
- The chirality or 'handedness' of certain organic compounds
- Extraction and purification of natural plant compounds for synthesis into effective medicines
- Competitive enzyme inhibitors as a class of therapeutics

LEARNING SEQUENCE

on Resources

📋 **Solutions**	Solutions — Topic 11 (sol-0838)
🔬 **Practical investigation eLogbook**	Practical investigation eLogbook — Topic 11 (elog-1710)
📄 **Digital documents**	Key science skills — VCE Chemistry Units 1–4 (doc-37066)
	Key terms glossary — Topic 11 (doc-37301)
	Key ideas summary — Topic 11 (doc-37302)
📝 **Exam question booklet**	Exam question booklet — Topic 11 (eqb-0122)

11.2 Structures and isolation of organic medicines

In this subtopic we will consider medicines from a variety of perspectives, beginning with an examination of the structures that underpin their physical and chemical properties, as well as looking at their beneficial health effects. This will provide an opportunity to identify those structural features that may be utilised when extracting and purifying natural compounds from plant sources.

11.2.1 Structures and functional groups of organic medicines

The chemical and physical properties of organic compounds are determined by their structures. This includes features such as:
- type and number of functional groups
- hydrocarbon chain length
- molecular size
- overall arrangement of atoms.

In topics 7 and 9, these features were used to determine the polarity of molecules and the effects on physical properties, including solubility, volatility, melting points and boiling points. In topic 8 the key role that functional groups play in chemical reactivity was discussed.

As introduced in this topic's overview, an additional consideration for most medicines is the ability of the drug to bind to a target molecule. This binding requires a 'chemical fit' between the two molecules, which is dependent on complementary structures, to facilitate a variety of interactions that may be covalent, ionic or intermolecular.

We will begin our discussion of medicines by identifying the functional groups present, as summarised in table 11.1.

TABLE 11.1 Functional groups. Open bonds (–) represent any group of atoms connected to the structure shown.

Functional group structure	Functional group name *Homologous series*	Functional group structure	Functional group name *Homologous series*	Functional group structure	Functional group name *Homologous series*
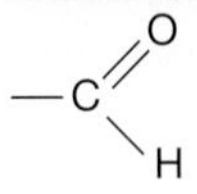 ![aldehyde]	Aldehyde *Aldehyde*	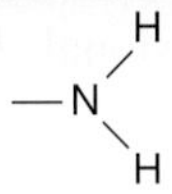 ![amine]	Amino (primary) *Amine*	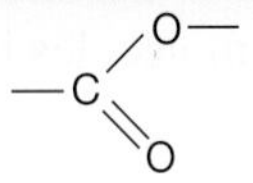 ![ester]	Ester *Ester*
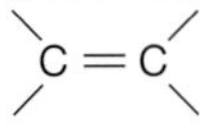 ![alkene]	Alkenyl *Alkene*	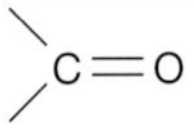 ![ketone]	Carbonyl *Ketone*	–O–H	Hydroxyl *Alcohol*
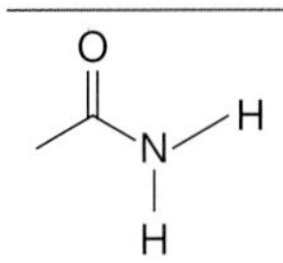 ![amide]	Amide (primary) *Primary amide*	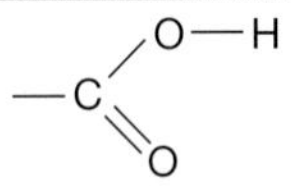 ![carboxyl]	Carboxyl *Carboxylic acid*	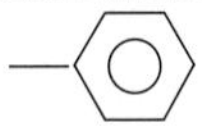 ![phenyl]	Phenyl *Arene*

Salicylic acid and oseltamivir (Tamiflu) are two organic medicines (table 11.2).
- Salicylic acid has long been used to relieve pain and fever. The process used to isolate salicylic acid from the bark of willow trees will be discussed later in this subtopic.
- Tamiflu is an organic compound used for the treatment of influenza. It works by inhibiting a key process required for the infection to spread. The viral target and design process for Tamiflu and a related drug, Relenza, will be discussed in subtopic 11.3.

TABLE 11.2 Identifying functional groups in the structures of medicines

Compound	Salicylic acid	Tamiflu
Structure	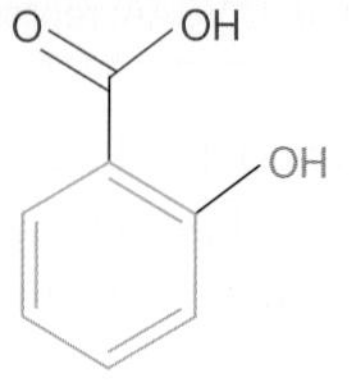	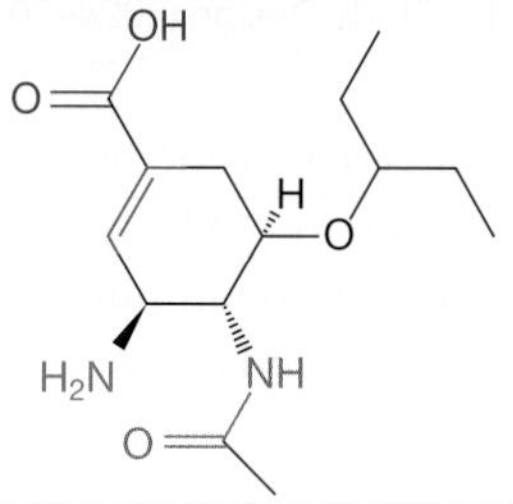
Functional groups	Carboxyl Hydroxyl Phenyl	Carboxyl Amide Amino Alkenyl

Salicylic acid has a relatively simple structure; however, most drugs are larger and more complicated. Tamiflu is more complex than salicylic acid, while codeine and Taxol are more complicated again (figure 11.2).
- Codeine is used for pain relief.
- Taxol is a chemotherapy drug used to treat a variety of cancers.

The relationship between these drugs and similar, yet inactive, isomers will be discussed later in this subtopic.

FIGURE 11.2 The structures of **a.** codeine and **b.** Taxol are complicated.

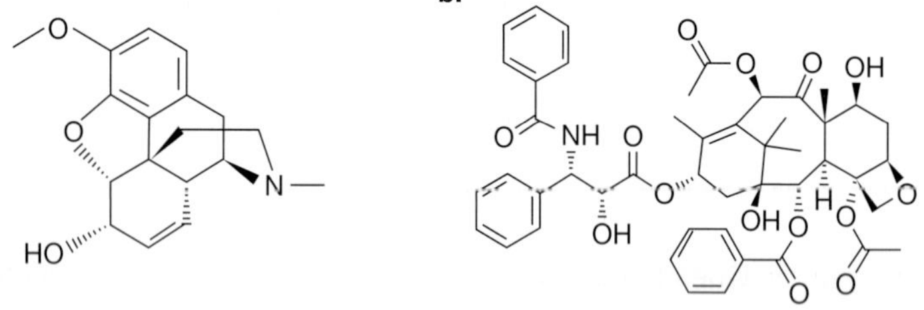

11.2.2 Isomers

Isomers are two or more compounds with the same molecular formula but different arrangements of atoms. The effect on the properties of the substances depends on the type of isomerism present. One requirement for the binding of drugs to target molecules is a match in the three-dimensional shapes, so all types of isomers may affect the activity of a drug.

There are two main types of isomers; these are:
- structural isomers
- stereoisomers.

Structural isomers differ in the order in which their atoms are arranged. The chain, positional and functional subtypes of structural isomers were introduced in topic 7.

structural isomers molecules that have the same molecular formula but different structural formulas

In **stereoisomers** the atoms are connected in the same order but are oriented differently in space. As a result, stereoisomers tend to have different chemical properties. There are a variety of subtypes of stereoisomers, although in this topic we will only consider optical isomers.

Optical isomers and chirality

Optical isomers are non-superimposable mirror images of each other, known as **enantiomers**. Molecules that form enantiomers do not contain a plane of symmetry.

A simple way to demonstrate the difference between symmetrical and asymmetrical (not symmetrical) objects is by considering our hands, feet or household objects. Symmetry is observed when a line cuts through an object and one half is the mirror image of the other half, and this can be extended to organic molecules. Organic molecules with a plane of symmetry are achiral (not chiral), whereas asymmetrical molecules are **chiral**.

> **stereoisomers** two or more compounds differing only in the spatial arrangement of their atoms
>
> **optical isomers** *see* **enantiomers**
>
> **enantiomers** chiral molecules that are non-superimposable mirror images of one another
>
> **chiral** describes compounds that contain an asymmetric carbon atom or chiral centre; the molecule cannot be superimposed upon its mirror image
>
> **chiral centre** an asymmetric carbon atom; a carbon atom bonded to four different groups of atoms

FIGURE 11.3 Object **a.** and molecule **c.** are chiral as they have a plane of symmetry. Object **b.** and molecule **d.** have no line of symmetry and are thus achiral.

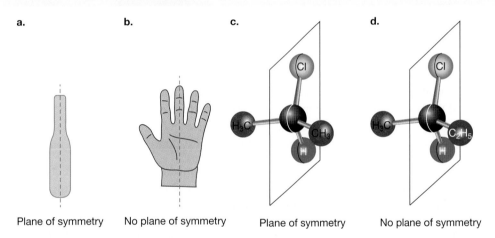

a.	b.	c.	d.
Plane of symmetry	No plane of symmetry	Plane of symmetry	No plane of symmetry

We can recognise whether a molecule will form enantiomers by the presence of a **chiral centre**, which is a carbon atom bonded to four different atoms or groups of atoms.

Figure 11.4 shows a pair of enantiomers (I and II) with the same four atoms bonded to the chiral carbon. Note that W and Y have swapped places, so the molecules are mirror images of each other. We can tell the atoms are now arranged differently in space because we cannot place or rotate the two molecules so that they sit exactly on top of one another; that is, the mirror images cannot be superimposed, so they must be different molecules.

Properties of enantiomers

Physical properties including density, solubility and melting point are identical for enantiomers. However, enantiomers interact differently with plane polarised light.

Figure 11.5 demonstrates how a device called a polarimeter is used to detect the presence of different enantiomers.

FIGURE 11.4 Chiral objects are non-superimposable mirror images of each other.

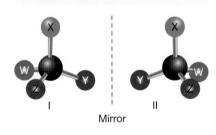

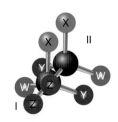

1. Normally, an unpolarised light wave is made up of a mixture of waves vibrating in every direction perpendicular to its direction of movement. A polarising filter (which is present in sunglasses) results in a single, polarised beam.
2. Enantiomers are 'optically active', as when polarised light is passed through a sample containing the chiral compound it is rotated.
3. If the polarised light is rotated clockwise, it is the (+) enantiomer; if it is rotated anticlockwise, it is the (−) enantiomer.

FIGURE 11.5 A polarimeter is used to distinguish between optical isomers.

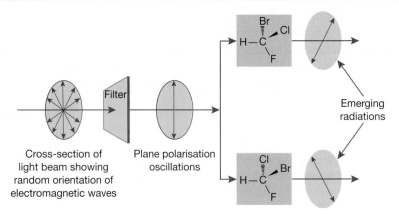

Medicines and chirality

It is this interaction with polarised light that led to the name *optical isomers*.

When optically active substances are synthesised in the laboratory, they often produce a 50 : 50 mixture of the two enantiomers, known as a **racemic mixture** or **racemate**. These mixtures rotate light equally in both directions so do not have an overall effect on polarised light. By contrast, stereoisomers formed in biological systems consist of only one enantiomer.

> **racemic mixture** *see* **racemate**
> **racemate** a 50 : 50 mixture of two enantiomers; often occurs when optically active substances are synthesised in the laboratory

As a result of this, many of the natural and synthetic drugs used in medicine have different effects on the body. This is because the enantiomer in the body has a unique three-dimensional shape, so the drug that interacts (binds) with it must have a matching three-dimensional shape.

FIGURE 11.6 The active enantiomer can bind to the target molecule as the binding sites all match. The inactive enantiomer is unable to bind as the binding sites do not align even if the molecule is rotated.

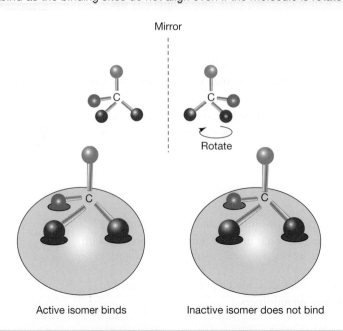

Modern synthetic chemists are looking to develop pathways that only produce the desired medicinal enantiomer for two key reasons:

- Some enantiomers may be harmful if administered to patients.
- It is more cost effective to produce only the desired enantiomer.

Frequently, one of the enantiomers is simply inactive. For example, codeine sourced from opium poppies consists entirely of the active (–) isomer (figure 11.2), whereas 50 per cent of the synthetic racemic mixture is the inactive (+) form. If this mixture were administered to a patient, it would require a higher dosage to achieve the same concentration of the active form.

However, synthesis of some medicines can produce both beneficial and harmful enantiomers. For example, between 1957 and 1962 the drug thalidomide was used to treat morning sickness in pregnant women. More than 10 000 babies were born with birth defects as a result of the use of this drug. Eventually, investigations found that thalidomide was a racemic mix of two enantiomers due to the presence of one chiral carbon atom. While one enantiomer did indeed cure morning sickness, the other enantiomer caused deformities in organs and limbs.

FIGURE 11.7 a. Thalidomide has two enantiomers, one of which causes birth defects. **b.** The hands of a person affected by thalidomide.

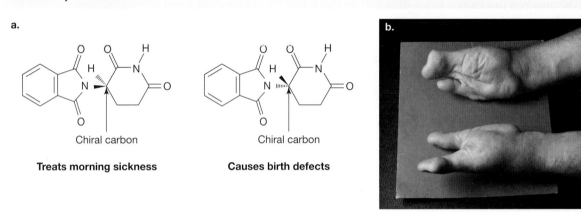

Taxol is a chemotherapy drug used for the treatment of cancer. It is a naturally occurring compound that was first extracted from the bark of Pacific yew trees in the 1970s. Due to the low concentration of Taxol in yew bark, it is now synthesised. However, Taxol contains 11 chiral carbons, so a synthetic mixture will contain many different stereoisomers and only a very small proportion of the desired active stereoisomer. As such, it is preferable to use methods of synthesis that only produce the desired stereoisomer.

FIGURE 11.8 a. Taxol is a chemotherapy drug with multiple chiral centres. **b.** It is present in the bark of Pacific yew trees.

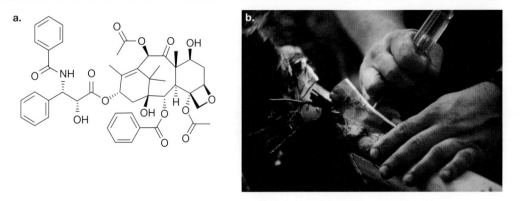

tlvd-9706

SAMPLE PROBLEM 1 Determining if different organic molecules will produce two enantiomers

Determine whether each of the following organic molecules would produce two enantiomers.

a. CH_4 b. CH_2Cl_2 c. $CHBrF_2$ d. $CHBrClF$

THINK

A carbon atom must be bonded to four different atoms or groups of atoms to be classified as chiral and produce enantiomers.

a. CH_4 has four atoms of the same type attached to a carbon atom.

b. CH_2Cl_2 has two sets of two atoms of the same type attached to a carbon atom. This will produce molecules that are symmetrical. Therefore, it is achiral.

c. $CHBrF_2$ has one set of two atoms of the same type attached to a carbon atom. This will produce molecules that are symmetrical. Therefore, it is achiral.

d. $CHBrClF$ has four different atoms attached and therefore has a chiral centre. It will produce two enantiomers; that is, mirror images of each other that are not superimposable.

WRITE

a. CH_4 does not have a chiral centre and will not produce enantiomers.

b. CH_2Cl_2 does not have a chiral centre and will not produce enantiomers.

c. $CHBrF_2$ does not have a chiral centre and will not produce enantiomers.

d. $CHBrClF$ has a chiral centre and will therefore produce two enantiomers.

PRACTICE PROBLEM 1

Which of the following would produce two enantiomers?

A. CH_2ClF

B. CH_3CHBrF

C. $(CH_3)_2CHF$

D. $CHBr_2F$

11.2.3 Extraction and purification of medicines sourced from plants

Whether drugs are synthesised using modern approaches or extracted from plant sources, they must be isolated from impurities before they can be used. In this section we will consider techniques that exploit differences in the physical properties of substances to achieve separation of these mixtures.

Structure and physical properties

As discussed earlier, the structure of organic molecules determines physical properties, including volatility and solubility. The polarity of functional groups affects the types of intermolecular forces that form in a pure sample of an organic substance, and with other substances in a mixture. This, in turn, affects volatility and solubility.

TABLE 11.3 Structural features determine the types of interactions formed by a substance.

Structural features	Example(s)	Interactions
Polar functional groups	Hydroxyl, –OH Amino, –NH$_2$ Carboxyl, –COOH Carbonyl (ketone, aldehyde, ester, amide), –C=O	Hydrogen bonding Dipole–dipole interactions Ion–dipole interactions Dispersion forces
Ionic groups	Conjugate bases (acidic salts), –COO$^-$ Conjugate acids (basic salts), –NH$_3{}^+$	Ionic Ion–dipole interactions Dispersion forces
Long carbon chains Aromatic rings	Hydrocarbons, C$_x$H$_y$	Dispersion forces

Volatility and boiling point are associated concepts. Stronger intermolecular forces require more energy to overcome, so increase the boiling point and lower the volatility of molecular substances.

Solubility of organic molecules requires stable interactions to form with the solvent. This is dependent on the two substances having similar polarity.
- Polar molecules tend to dissolve well in polar solvents such as water and ethanol due to the presence of polar functional groups that can form hydrogen bonds, and/or dipole–dipole interactions.
- Relatively strong ion–dipole bonds may also form when charged groups are present, such as in salts, increasing solubility in polar solvents.
- Non-polar organic molecules dissolve well in non-polar solvents such as hexane through the formation of dispersion forces.

Isolation techniques

In this section we will consider two techniques that utilise differences in physical properties:
- Distillation separates substances based on differences in volatility (see topic 9).
- **Solvent extraction** separates substances based on differences in solubility.

Distillation

The principles and technical details for simple and fractional distillation were introduced in topic 9. Both techniques separate volatile compounds from solutions by heating until vapours are formed. The most volatile compounds readily escape from the solution and can then be collected as a liquid via condensation. Subsequent fractions contain compounds that are progressively less volatile.

Solvent extraction

Extraction is the process of moving a substance from one phase to another, generally to isolate a specific component (analyte) from a mixture. For example, preparing a cup of tea or an espresso coffee selectively dissolves or extracts certain compounds, leaving others behind. Chromatography (topic 10) is a type of extraction as the component of interest (analyte) will bind more readily to either the stationary phase or mobile phase, allowing its separation from the mixture.

Solvent extraction (figure 11.9) is a related laboratory technique in which two **immiscible** solvents are used to separate components of a mixture. Solvents are selected based on their differing polarities.

> **volatility** describes how readily a liquid substance will form a vapour
>
> **solvent extraction** a technique used to separate solutes based on their relative solubility in two solvents with different polarity
>
> **immiscible** refers to liquids that do not form a homogeneous mixture when mixed with another liquid

1. A solution containing the mixture of interest is prepared in a solvent and placed in a piece of glassware called a separation funnel.
2. An immiscible solvent (of the opposite polarity) is then added to the funnel, generating two layers.
3. The funnel is sealed and shaken to increase the surface area between the layers. This allows most of the solute molecules to move to the layer with similar polarity. Polar molecules will move to the polar solvent and vice versa.
4. The layers are allowed to settle and separate before being collected from the funnel via the stopcock.
5. Solutes may be collected via evaporation of the solvent or by distillation.

FIGURE 11.9 Solvent extraction of a polar aqueous solution using a non-polar organic solvent. **a.** The organic compound is dissolved in water. **b.** A non-polar solvent is added to the separation funnel, which is shaken. **c.** This results in most of the organic compound moving to the non-polar organic solvent.

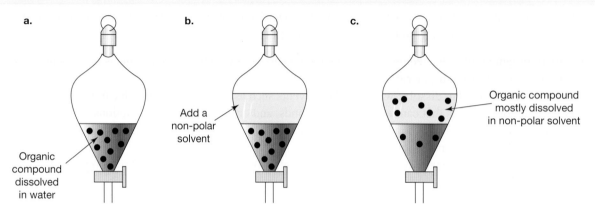

Several modifications can be made to improve the yield and purity of analyte generated by simple extractions.

TABLE 11.4 Modifications can be made to simple extractions to improve yield and purity.

Improvement	Technique	Description
Yield	Multiple extractions	Each extraction will transfer only a percentage of the analyte molecules to the added solvent. To improve yield, the original solution can be repeatedly extracted with fresh solvent, and the resultant fractions combined.
Purity	Distillation	Following exaction, the analyte may be retrieved by evaporation of the solvent.
		If the solution contains multiple compounds, then distillation may be used to provide additional separation and increase purity.
		Fractional distillation may be required to separate mixtures containing components with similar boiling points.
	Acid–base extractions	Acid–base extractions allow uncharged molecules to be converted into ions. Increased solubility facilitates movement of ions into the polar phase, improving purification.
		Despite the presence of a polar carboxyl group, larger carboxylic acids have low solubility in water due to the presence of significant non-polar hydrocarbon chains and/or aromatic rings. However, their corresponding salt is much more soluble due to the charged –COO⁻ group. Addition of a strong base such as NaOH will convert the acid into a salt form and increase its solubility in polar solvents.
		The application of this technique for the extraction of salicylic acid will be discussed in the next section.

Extraction and purification of plant compounds

The techniques of solvent extraction and distillation can be applied to the separation of medicinal compounds from plants. However, plants contain a complex mixture of compounds, some of which may have similar physical properties to the desired medicinal molecule. As a result, modifications to these techniques may be required to optimise separation (table 11.4).

Salicylic acid (table 11.2) has been used for the relief of pain and fever since ancient times, and is a precursor reagent in the production of aspirin (figure 11.10).

FIGURE 11.10 Synthesis of aspirin from salicylic acid

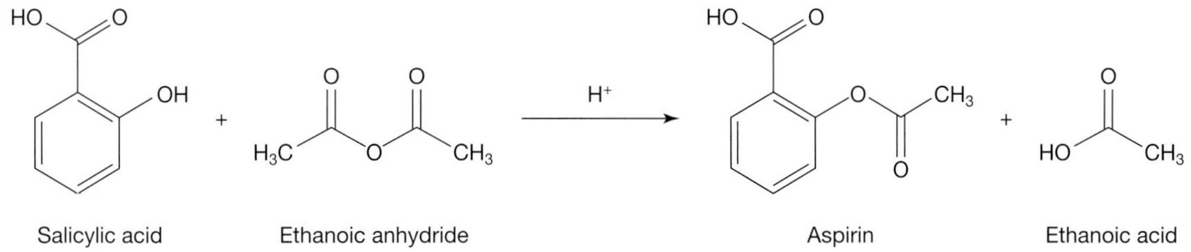

Salicylic acid   Ethanoic anhydride   Aspirin   Ethanoic acid

Salicylic acid can be found in a variety of plants, such as in the bark of willow trees and in the herb meadowsweet. Extraction of salicylic acid from plants may be performed by acid–base solvent extraction (table 11.4) and evaporation.

1. A non-polar organic solvent (e.g. hexane) is added to dried plant powder. This dissolves salicylic acid and many other non-polar compounds. Despite the presence of polar carboxyl and hydroxyl groups, salicylic acid has low solubility in water due to the non-polar aromatic ring.
2. Undissolved substances, including any polar compounds, are removed from the mixture by filtration. The mixture is then transferred to a separation funnel.
3. Aqueous NaOH solution is added, which reacts with the salicylic acid to produce a salicylate salt (figure 11.11). As salicylate salt contains a charged $-COO^-$ group, it will readily dissolve in the polar aqueous layer. However, other non-polar substances remain in the organic layer.
4. An aqueous solution of HCl may be added to convert the salicylate back into salicylic acid, which may be recovered via evaporation.

FIGURE 11.11 Production of sodium salicylate by reaction of salicylic acid with sodium hydroxide

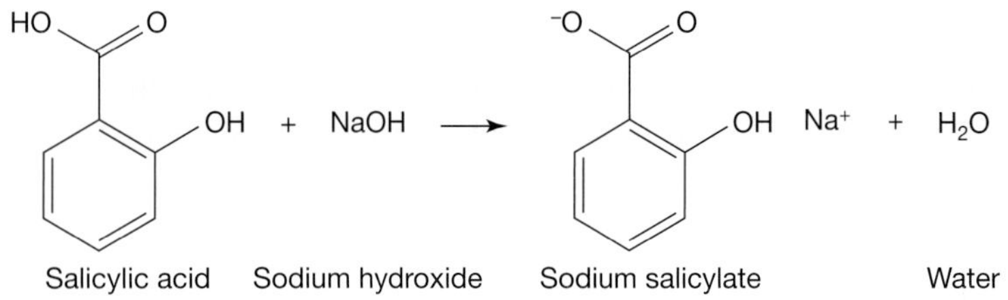

Salicylic acid   Sodium hydroxide   Sodium salicylate   Water

FIGURE 11.12 Acid–base extraction of salicylic acid. **a.** A solution of NaOH is added to a non-polar organic solvent containing salicylic acid and other dissolved non-polar compounds. **b.** The NaOH reacts with the salicylic acid to produce salicylate, which readily dissolves in the polar aqueous layer. Non-polar compounds remain in the organic layer.

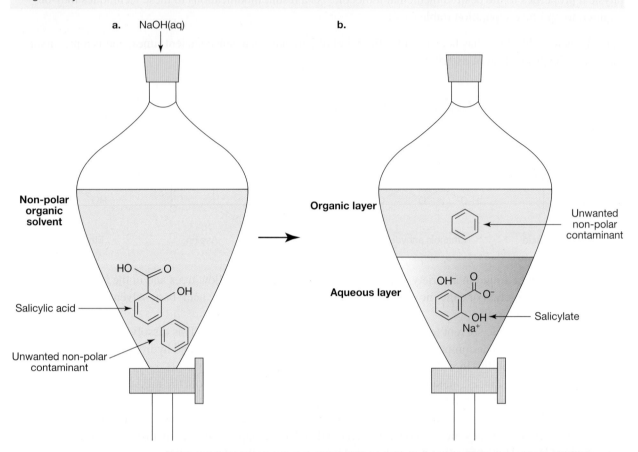

tlvd-9664

SAMPLE PROBLEM 2 Using solvent extraction to separate an organic mixture

A mixture of propan-1-ol, butan-1-ol, propanal and butanal in a non-polar solvent is to be separated using the following solvent extraction and distillation method.
1. Add water to the mixture, shake, and wait for the layers to separate.
2. Collect each layer.
3. Distil each layer and collect the fractions.

a. Identify which two compounds are most likely to dissolve in the water layer. Justify your answer.
b. Identify which of the four compounds in the initial mixture will be the last separated from the polar solvent during distillation.

THINK	WRITE
a. Water is a polar solvent, so molecules that form the strongest bonds with polar solvents will be most soluble.	**a.** Propan-1-ol and butan-1-ol Due to the presence of the highly polar hydroxyl group, propan-1-ol and butan-1-ol can form stronger interactions with water than the propanal and butanal, which contain a carbonyl group.
b. Distillation separates compounds in order from lowest to highest boiling point. Butan-1-ol has a higher boiling point than propan-1-ol as it is a larger molecule and will form stronger dispersion forces.	**b.** Butan-1-ol

---

PRACTICE PROBLEM 2

A mixture of ethyl ethanoate, methyl ethanoate, propanoic acid and butanoic acid is prepared in water, and separated using the following solvent extraction and distillation method.

1. Add hexane to the mixture, shake, and wait for the layers to separate.
2. Collect each layer.
3. Distil each layer and collect the fractions.

a. Identify which two compounds are most likely to dissolve in the hexane layer. Justify your answer.

b. Identify which of the four compounds in the initial mixture will be the first separated from the non-polar solvent during distillation. Justify your answer.

11.2 Activities

learn on

| 11.2 Quick quiz | on | 11.2 Exercise | 11.2 Exam questions |

11.2 Exercise

1. What is the difference between chiral and achiral carbon atoms?
2. Why are enantiomers called optical isomers?
3. Draw the enantiomers of CH_3CHBrF.
4. Shikimic acid is a precursor chemical used for the synthesis of oseltamivir (Tamiflu). It may be sourced from star anise by extraction. The structure of shikimic acid is shown in the figure. 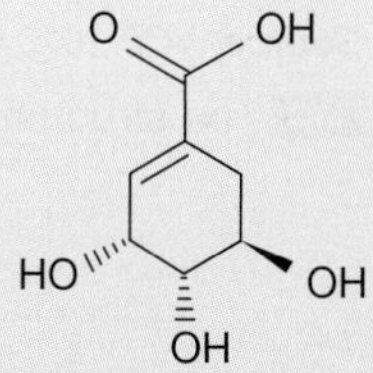
   a. Identify a suitable solvent for the extraction of shikimic acid and explain your choice.
   b. Suggest a method to recover the shikimic acid from the solvent assuming that there are impurities remaining in the solution.

5. A solid mixture contains two compounds, A and B. Describe how solvent extraction could be used to separate the two compounds, given the following properties:
   • A and B are both non-polar and have similar polarity.
   • A has acidic properties and B has no acid–base properties.

11.2 Exam questions

▶ **Question 1 (1 mark)**

Source: VCE 2022 Chemistry Exam, Section A, Q.19; © VCAA

MC Which one of the following chemical compounds contains a chiral carbon centre?

A. glycine
C. butan-2-o1

B. glycerol
D. 1,1-dichloropropane

▶ **Question 2 (1 mark)**

Source: VCE 2019 Chemistry Exam, Section A, Q.15; © VCAA

MC Aspartame has only

A. one chiral centre.
C. four optical isomers.

B. two stereoisomers.
D. three structural isomers.

Question 3 (2 marks)

Source: VCE 2015 Chemistry Exam, Section B, Q.5.c.i; © VCAA

A student mixed salicylic acid with ethanoic anhydride (acetic anhydride) in the presence of concentrated sulfuric acid. The products of this reaction were the painkilling drug aspirin (acetyl salicylic acid) and ethanoic acid.

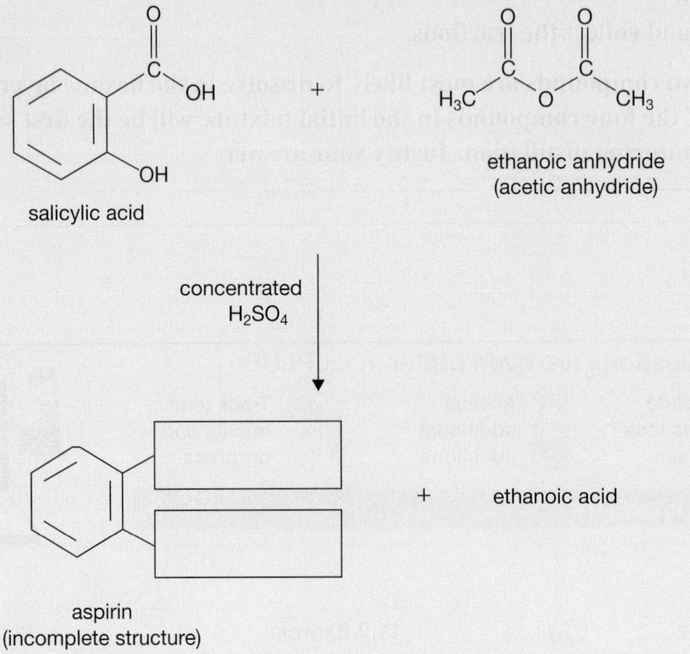

salicylic acid

ethanoic anhydride
(acetic anhydride)

concentrated
H$_2$SO$_4$

+ ethanoic acid

aspirin
(incomplete structure)

An incomplete structure of the aspirin molecule is shown above.

Complete the structure by filling in the two boxes provided in the diagram.

Question 4 (1 mark)

Source: VCE 2010 Chemistry Exam 1, Section A, Q.19; © VCAA

MC The structure of Tamiflu®, an antiflu drug, is shown below.

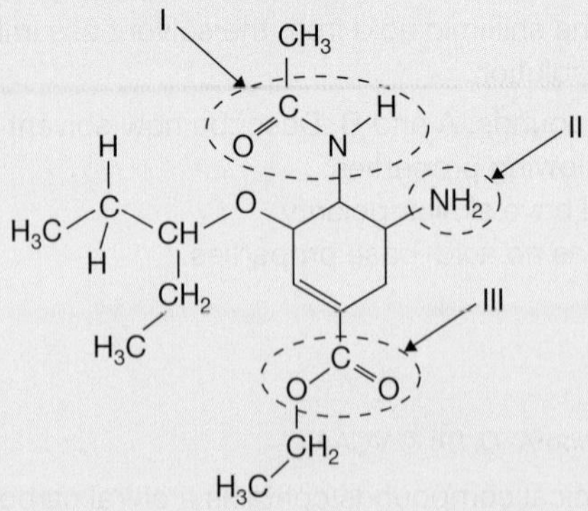

The names of the functional groups labelled I, II and III are

	I	II	III
A.	amide	amino	carboxylic acid
B.	amino	amide	ester
C.	amide	amino	ester
D.	amino	amide	carboxylic acid

11.3 Enzymes and inhibitors

KEY KNOWLEDGE

- Enzymes as protein-based catalysts in living systems: primary, secondary, tertiary and quaternary structures and changes in enzyme function in terms of structure and bonding as a result of increased temperature (denaturation), decreased temperature (lowered activity), or changes in pH (formation of zwitterions and denaturation)
- Medicines that function as competitive enzyme inhibitors: organic molecules that bind through lock-and-key mechanism to an active site preventing binding of the actual substrate

Source: VCE Chemistry Study Design (2024–2027) extracts © VCAA; reproduced by permission.

This subtopic builds on understanding developed in:
- topic 4 — catalysts increase the rate of chemical reactions
- topics 7 and 8 — polypeptides are polymers of amino acids.

Here, we will introduce enzymes as protein-based catalysts in living systems and look at factors affecting the rate of enzymatic reactions. Finally, we will examine competitive enzyme inhibitors and consider how they are utilised as medicines in the treatment of disease.

11.3.1 Amino acids and zwitterions

Proteins are polymers built from monomers called 2-amino acids. Figure 11.13 shows the general structure of a 2-amino acid in which the central carbon atom (C2) is attached to a carboxyl group, an amino group, a hydrogen atom and a group of atoms known as 'R'.

Amino acids used to make proteins are named 2-amino acids because the amino group is attached to the second carbon counting from the carboxyl group (see figure 11.13). They are also called α-amino acids ('alpha amino acids'). α-amino acids contain an α-carbon, which is the first carbon atom directly bonded to a functional group; in this case, a carboxyl group.

FIGURE 11.13 The general structure of a 2-amino acid

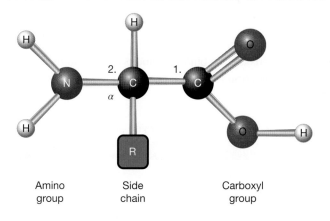

Amino group   Side chain   Carboxyl group

proteins large molecules composed of one or more long chains of amino acids

Differences between amino acids

Amino acids differ in the structure of the R group (also termed the side chain), and it is the nature of the R group that forms the classification of 2-amino acids. The R groups can be one or more of the following.

TABLE 11.5 R group properties determine the amino acid classification.

Amino acid classification	R group contains	Example(s)	Structure
Non-polar (hydrophobic)	Alkyl	Alanine	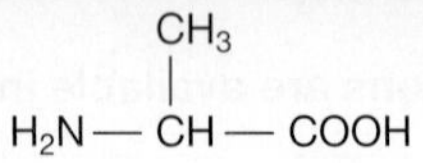 CH_3 | $H_2N — CH — COOH$
	Phenyl/aromatic	Phenylalanine	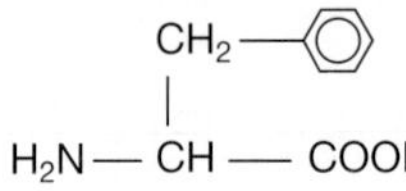 CH_2—⬡ | $H_2N — CH — COOH$
Polar — neutral	Hydroxyl, –OH Amide, –$CONH_2$ Sulfhydryl, –SH	Serine	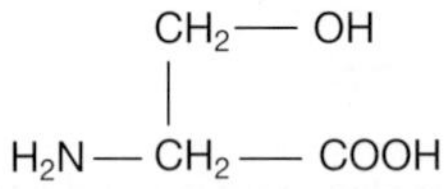 $CH_2— OH$ | $H_2N — CH_2 — COOH$
Polar — acidic (proton donors)	Carboxyl, –COOH	Glutamic acid	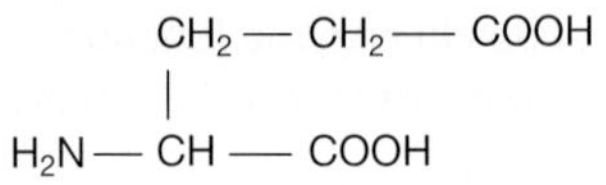 $CH_2 — CH_2 — COOH$ | $H_2N — CH — COOH$
Polar — basic (proton acceptors)	Amino, –NH_2 (and N or NH)	Lysine	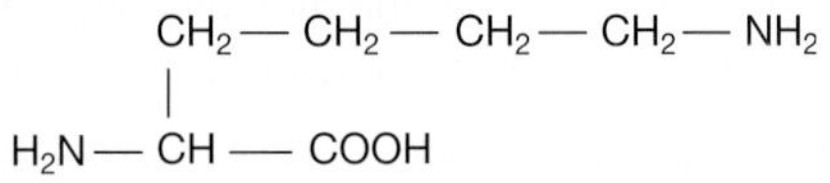 $CH_2 — CH_2 — CH_2 — CH_2 — NH_2$ | $H_2N — CH — COOH$

The simplest 2-amino acids are glycine, in which the R group is just a hydrogen atom, and alanine, in which the R group is a methyl group (see figure 11.14).

FIGURE 11.14 Structures of glycine and alanine

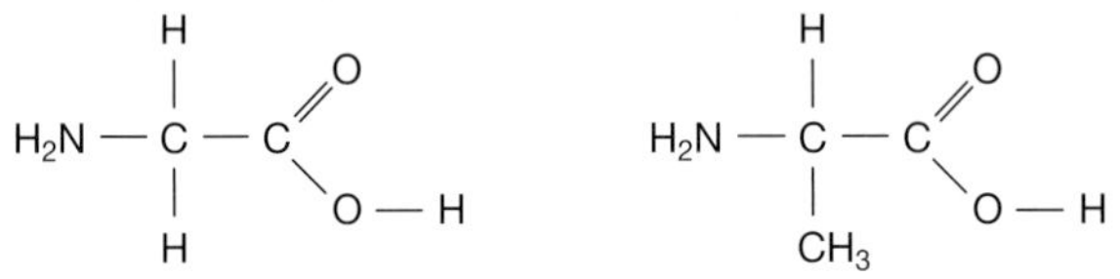

Glycine (Gly)
Non-polar
$C_2H_5NO_2$

Alanine (Ala)
Non-polar
$C_3H_7NO_2$

Hundreds of amino acids are known, but only 20 have been found in proteins in the human body. These 2-amino acids are the building blocks of thousands of proteins and are shown in table 11.6, with the three-letter abbreviation for their names. Amino acids can also be identified by a single letter, but this is not used in VCE Chemistry.

TABLE 11.6 2-amino acids (α-amino acids)

Name	Symbol	Structure
alanine	Ala	CH_3 | $H_2N — CH — COOH$
arginine	Arg	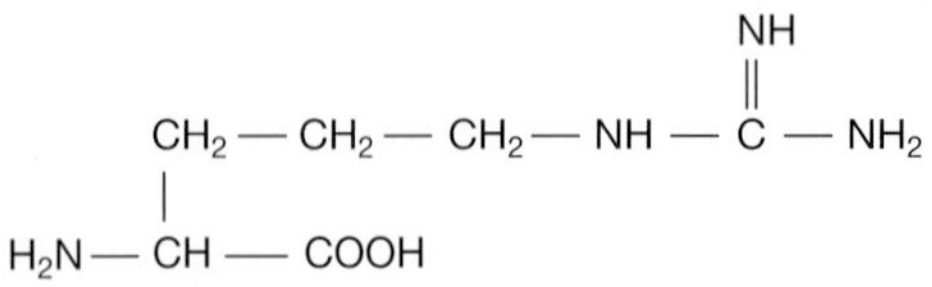 NH ‖ $CH_2 — CH_2 — CH_2 — NH — C — NH_2$ | $H_2N — CH — COOH$

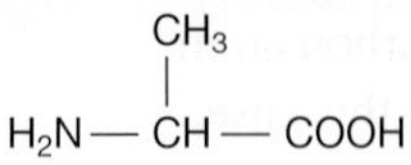

Name	Symbol	Structure
asparagine	Asn	$H_2N - CH - COOH$ with side chain $CH_2 - C(=O) - NH_2$
aspartic acid	Asp	$CH_2 - COOH$ $H_2N - CH - COOH$
cysteine	Cys	$CH_2 - SH$ $H_2N - CH - COOH$
glutamic acid	Glu	$CH_2 - CH_2 - COOH$ $H_2N - CH - COOH$
glutamine	Gln	$CH_2 - CH_2 - C(=O) - NH_2$ $H_2N - CH - COOH$
glycine	Gly	$H_2N - CH_2 - COOH$
histidine	His	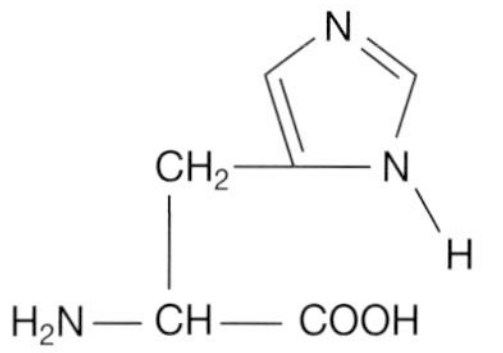 imidazole ring — CH_2 $H_2N - CH - COOH$
isoleucine	Ile	$CH_3 - CH - CH_2 - CH_3$ $H_2N - CH - COOH$
leucine	Leu	$CH_3 - CH - CH_3$ CH_2 $H_2N - CH - COOH$
lysine	Lys	$CH_2 - CH_2 - CH_2 - CH_2 - NH_2$ $H_2N - CH - COOH$
methionine	Met	$CH_2 - CH_2 - S - CH_3$ $H_2N - CH - COOH$
phenylalanine	Phe	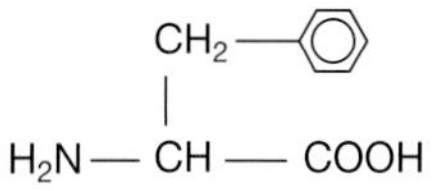 $CH_2 -$ benzene ring $H_2N - CH - COOH$
proline	Pro	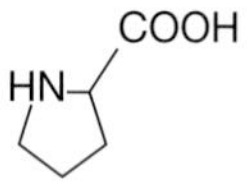 pyrrolidine ring, HN, $COOH$
serine	Ser	$CH_2 - OH$ $H_2N - CH_2 - COOH$

(continued)

TABLE 11.6 2-amino acids (α-amino acids) *(continued)*

Name	Symbol	Structure
threonine	Thr	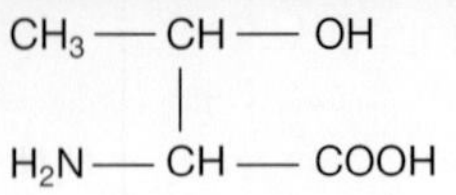 CH_3—CH—OH | H_2N—CH—COOH
trytophan	Trp	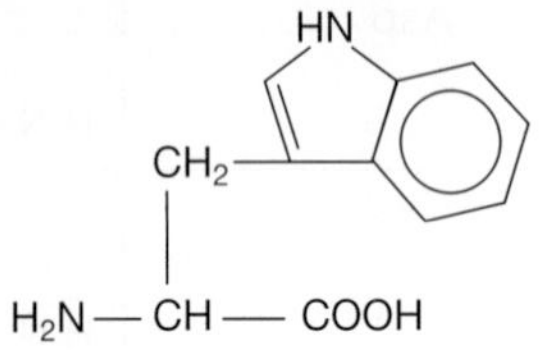 
tyrosine	Tyr	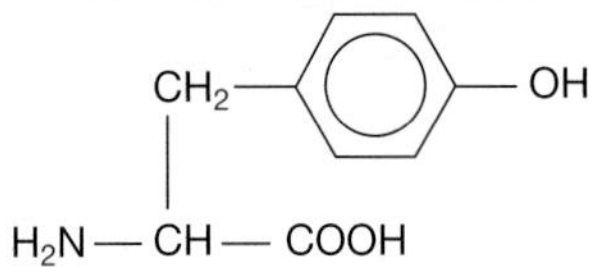 CH_2—⬡—OH | H_2N—CH—COOH
valine	Val	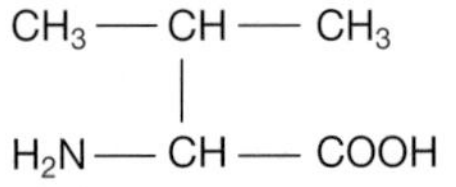 CH_3—CH—CH_3 | H_2N—CH—COOH

Source: VCE Chemistry Written Examination Data Book (2022) extracts © VCAA; reproduced by permission.

The names, symbols and structures of the 2-amino acids can be found in the VCE Chemistry Data Book.

---

on Resources

⬥ **Interactivity** Classifying properties of amino acids (int-1237)

Zwitterions

Although 2-amino acids are commonly shown as containing an amino group ($-NH_2$) and a carboxyl group ($-COOH$), certain physical and chemical properties — including melting points, solubilities and acid–base properties — are not consistent with this structure.

The acid–base properties of carboxyl and amino groups have an effect on amino acid structure. The weakly acidic proton of the carboxyl group easily transfers to the weakly basic amino group, forming a zwitterion.

FIGURE 11.15 An amino acid and its corresponding zwitterion

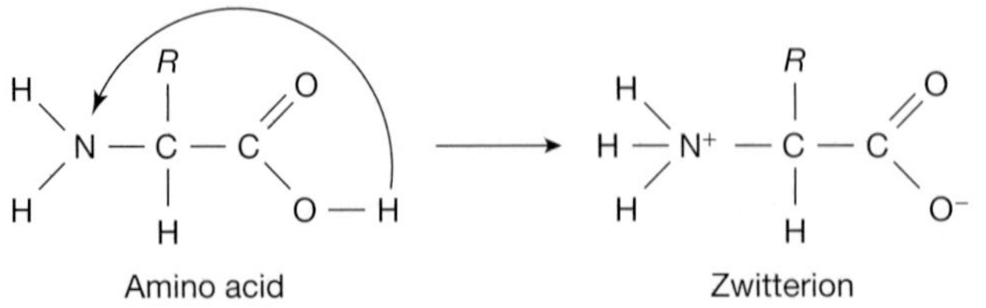

A zwitterion is a molecule with a net charge of zero, but negative and positive charges on individual atoms in its structure. In the pure solid state and in aqueous solutions with an approximately neutral pH, the amino acids exist almost completely as zwitterions.

Amino acids can behave as both acids and bases and can exist in several forms, depending on the pH of the solution. It can be useful to consider these effects with Le Chatelier's principle.

TABLE 11.7 Using Le Chatelier's principle to consider the acid and base behaviour of amino acids in solutions with different pH

Solution pH	Le Chatelier's principle	Reaction	Acting as an acid or a base
Low	The high concentration of H^+ ions will be partially opposed by favouring the protonation of the $-COO^-$ group.	The forward reaction will be favoured. $-COO^- + H^+ \rightleftharpoons -COOH$	Accepting an H^+ ion so acting as a base
High	The low concentration of H^+ ions will be partially opposed by favouring the deprotonation of the $-NH_3^+$ group.	The forward reaction will be favoured. $-NH_3^+ \rightleftharpoons -NH_2 + H^+$	Donating an H^+ ion so acting as an acid

FIGURE 11.16 The effect of solution pH on the protonation of amino acids

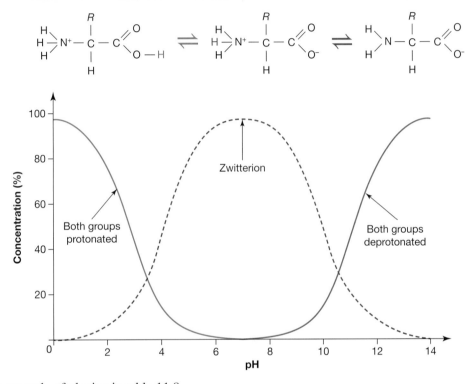

Consider the example of glycine in table 11.8.

TABLE 11.8 The effect of pH on the ionic form of glycine

Solution pH	Effect on carboxyl and amino groups	Structure	Ionic form
Neutral	Both the $-COO^-$ and $-NH_3^+$ will be present in the charged forms.	$H_3N^+ - CH_2 - COO^-$	Zwitterion
Low	The $-COO^-$ will accept an H^+ ion to form a $-COOH$ group.	$H_3N^+ - CH_2 - COO^- + H^+ \rightleftharpoons H_3N^+ - CH_2 - COOH$	Cation
High	The $-NH_3^+$ will donate an H^+ ion to form an $-NH_2$ group.	$H_3N^+ - CH_2 - COO^- \rightleftharpoons H_2N - CH_2 - COO^- + H^+$	Anion

 Resources

⊳ **Video eLesson** Glycine — amphoteric behaviour (eles-2593)

✦ **Interactivity** Classifying the effect of pH on amino acids (int-1238)

11.3.2 Protein structure

The structure of a protein is critical to its function. There are four levels of structure that contribute to a protein's overall structure: primary, secondary, tertiary and quaternary.

Primary structure

The simplest level of protein structure is the order of amino acids in a polypeptide chain and is referred to as its **primary structure**.

- This is composed of the amino acid residues covalently bonded with peptide links.
- Recall that the polypeptide ends are referred to as the C-terminus and N-terminus due to the presence of an unbonded –COOH or –NH$_2$ group in the final amino acid residue.
- The amino acid sequence is unique for each protein and determines how the chain will be arranged in later levels of structure.

For example, insulin consists of two amino acid chains: chain A has 21 amino acids and chain B has 30 amino acids. Insulin is a hormone produced by the pancreas that helps the cells in the body receive glucose from the blood and use it for providing energy.

> **primary structure** the order of amino acids in a protein molecule

FIGURE 11.17 The primary structure of human insulin, showing the order of amino acid residues

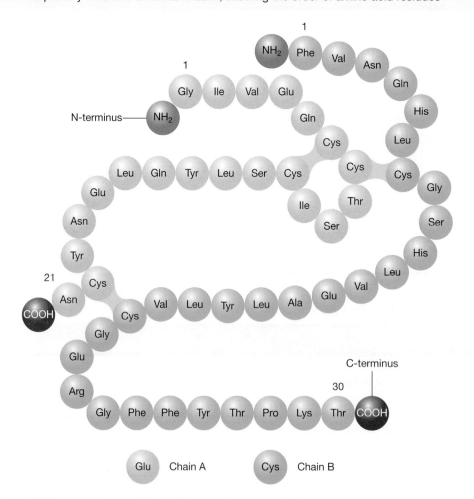

Primary structure

The primary structure of proteins is the order of the amino acid residues covalently bonded in a polypeptide chain. The amino acid residues are covalently bonded by peptide links.

Secondary structure of proteins

Folding of a polypeptide chain due to hydrogen bonding between peptide bonds is known as the **secondary structure** of a protein. The hydrogen bonds form between the slightly positive hydrogen atom in an –NH group and the slightly negative oxygen atom in a –C=O group of two peptide bonds at different positions along the chain. Note that hydrogen bonding between R groups is *not* involved in the secondary structure.

Two main folded arrangements maximise the number of hydrogen bonds formed and hence, the stability of the secondary structure. These are **α-helices** and **β-pleated sheets**.

- In α-helices, hydrogen bonds are formed between peptide bonds that are four amino acids apart on the same chain (see figure 11.18), resulting in a rigid, stable coiled structure.
- In β-pleated sheets, two sections of the same peptide chain line up and are held together in a sheet-like structure. In figure 11.18, the arrows at the ends of the ribbons point towards the carboxyl end of the chain.

Some protein structures consist of more than a thousand atoms in complex arrangements with a multitude of intermolecular interactions, so they are sometimes represented by computer-generated models, such as that shown in figure 11.18. The sections of α-helices and β-pleated sheets are linked by thin sections, showing random coils and loops.

> **secondary structure** the structure formed from hydrogen bonding between carboxyl and amino groups in peptide links at different positions in a protein molecule
>
> **α-helices** refers to when hydrogen bonds are formed between an oxygen atom of a –C=O bond and a hydrogen atom of an –NH bond that is four amino acids away on the same chain
>
> **β-pleated sheets** refers to when two sections of the peptide chain line up and are held together in a sheet-like structure by hydrogen bonds between one oxygen atom of a –C=O bond and a hydrogen atom of an –NH bond in the parallel or anti-parallel sheet

FIGURE 11.18 The secondary structure of a protein shown in a computer representation

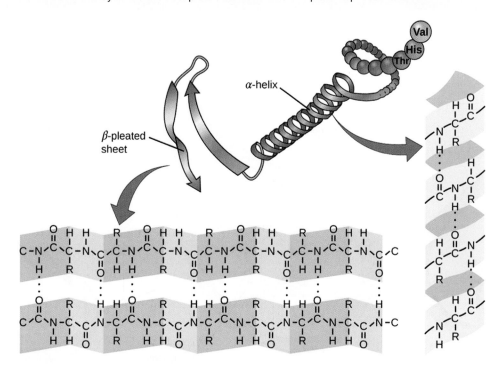

Secondary structure

The secondary structure of proteins is the arrangement of the amino acid chains into α-helices and β-sheets. These structures result from hydrogen bonding between peptide bonds (not R groups).

Tertiary structure

The three-dimensional structure of a protein, referred to as its **conformation**, is critical to its function. A wide variety of interactions form between R groups (side chains) of the amino acid residues, resulting in a complex three-dimensional shape called the **tertiary structure**.

The conformation is also affected by the aqueous environment, as non-polar R groups are **hydrophobic** and gather on the inside of the protein, leaving **hydrophilic** R groups on the surface.

conformation the three-dimensional structure of a protein

tertiary structure the structure formed in a protein molecule from side-group interactions, including hydrogen bonding, ionic bonding, dipole–dipole interactions and disulfide bridges

hydrophobic describes non-polar molecules that repel water molecules

hydrophilic describes molecules more likely to interact with water and other polar substances

TABLE 11.9 Types of interactions between R groups in a protein's tertiary structure

Interaction	Description
Dispersion forces	Form between non-polar R groups
Hydrogen bonding	Form between R groups containing C=O, –NH or –OH groups
Ionic bonding	Can occur between oppositely charged $-NH_3^+$ and $-COO^-$ groups
Disulfide bridge	A covalent bond formed between two cysteine residues. These strong bonds stabilise protein structures and help to maintain their conformation.

FIGURE 11.19 Types of bonding present in the tertiary structure of proteins

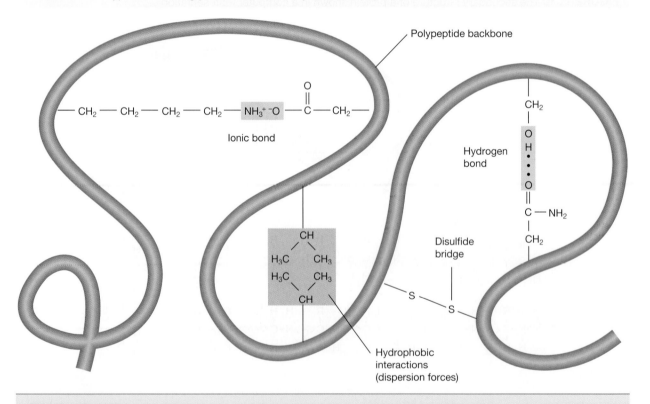

Tertiary structure

A variety of interactions between R groups contribute to the tertiary structure of proteins. In the tertiary structure, hydrogen bonding occurs between R groups, whereas in the secondary structure hydrogen bonds are formed between peptide groups.

Quaternary structure

Some proteins consist of two or more folded peptide chains combined. When individual protein molecules link together in a particular spatial arrangement, a **quaternary structure** is formed.

These proteins can be the same or different and form a larger group of proteins known as a complex. A complex is stabilised by interactions similar to the tertiary structure, including hydrogen bonding, ionic attractions and dispersion forces.

An example is haemoglobin, the oxygen transport protein, which has four chains: two identical α-chains and two identical β-sheets (see figure 11.20a). Collagen, the main structural protein found in skin and tendons, has three polypeptide chains wound around each other (see figure 11.20b).

> **quaternary structure** the structure formed when individual protein molecules link together in a particular spatial arrangement

FIGURE 11.20 Models of **a.** human haemoglobin and **b.** collagen

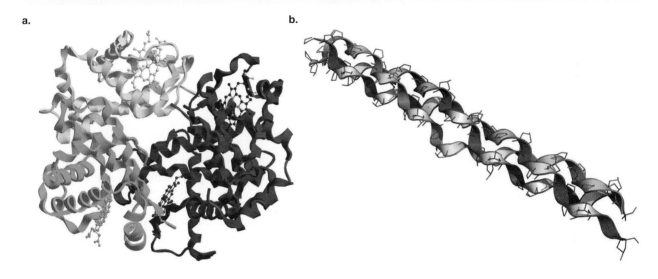

a.

b.

Quaternary structure

The quaternary structure of proteins refers to interactions between peptide chains.

Summary of protein structures

Proteins contain four levels of structure, as presented in figure 11.21.
- The primary structure is the sequence of amino acids in the molecule.
- The secondary structure is formed by hydrogen bonding between the hydrogen atom in an –NH group and the oxygen atom in a –C=O group of two peptide bonds at different positions along the chain.
- In the tertiary structure, all of the main types of bonding between R groups may be involved in forming the three-dimensional arrangement of the protein.
- The quaternary structure comprises the interactions between multiple folded polypeptide chains forming a protein complex.

 Resources

🔗 **Weblink** Video: Protein folding

FIGURE 11.21 The four levels of protein structure

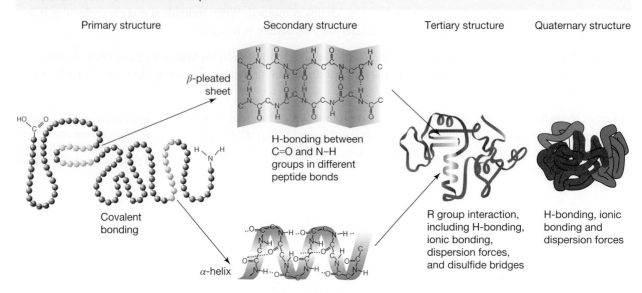

Primary structure   Secondary structure   Tertiary structure   Quaternary structure

β-pleated sheet

H-bonding between C=O and N–H groups in different peptide bonds

Covalent bonding

α-helix

R group interaction, including H-bonding, ionic bonding, dispersion forces, and disulfide bridges

H-bonding, ionic bonding and dispersion forces

11.3.3 Enzymes as protein-based catalysts

What are enzymes?

Enzymes are protein-based catalysts that increase the rate of reactions in biological systems. Amazingly, reactions with enzymes occur up to millions of times faster than reactions without enzymes. As a result, life could not be maintained without enzymes.

There are an extraordinary number of different processes required for cells to function normally, and enzymes play a key role in many of them. Familiar examples from topic 8 include the condensation and hydrolysis reactions of various macromolecules.

In many respects enzymes are similar to the inorganic catalysts studied in topic 4, although there are some crucial differences. This comparison is summarised in table 11.10.

TABLE 11.10 Comparison of enzymes and inorganic catalysts

Property	Inorganic catalysts	Enzymes
Required in small amounts	✓	✓
Not consumed in the reaction	✓	✓
Do not alter the position of equilibrium	✓	✓
Provide an alternative pathway with a lower activation energy	✓	✓
The range of reactions able to be catalysed by each catalyst (specificity)	A wide range, so low specificity	A single reaction or reactions with one functional group, so highly specific
The conditions required for optimal function	Various conditions are acceptable	A narrow range of conditions (e.g. pH and temperature)

For two molecules to react in a chemical reaction, they must collide with one another with the correct orientation and sufficient energy to overcome the energy barrier to the reaction. This is called the **activation energy** (E_a). A key similarity between inorganic catalysts and enzymes is their effect to lower activation energy by providing an alternative reaction pathway (figure 11.22).

enzyme a protein that acts as a biological catalyst

activation energy (E_a) the minimum energy required by reactants in order to react

FIGURE 11.22 An activation energy profile for enzyme-catalysed and uncatalysed reactions

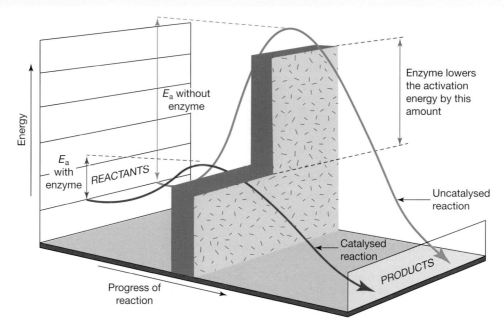

Two crucial differences between enzymes and inorganic catalysts are specificity and the conditions required to function optimally.

- Most inorganic catalysts are not selective and can speed up many different chemical reactions, but enzymes are very specific. Each enzyme generally catalyses one specific reaction or a series of closely related reactions sharing a functional group, so there are many different enzymes working in our bodies. The specificity lies in the shapes and chemical composition of the enzyme molecules.
- Unlike inorganic catalysts, enzymes only operate effectively in the narrow temperature and pH range of their cellular environment. For human enzymes this is typically body temperature (about 37.5 °C) and at a specific pH depending on the part of the body in which they function.

> ### Action of enzymes
>
> Enzymes lower the activation energy by providing an alternative pathway.

The lock-and-key model

The reactants in enzymatic reactions are termed substrates, and temporarily bind to a zone on the enzyme called the active site. The active site is a uniquely shaped indentation on the surface of the enzyme. It is lined with amino acid residues, which will form a variety of weak, non-covalent interactions with the substrate that has a complementary shape. It is this interaction between the active site (lock) and substrate (key) that provides enzyme specificity and gave the lock-and-key model its name.

The process of enzymatic catalysis consists of three steps, which are illustrated in figure 11.24.

1. The substrate collides with the enzyme and binds to the active site, forming an enzyme–substrate complex.
2. The enzyme assists in the breaking and/or forming of bonds to generate the product from the substrate.
3. The product no longer has the required conformation to bind to the active site so is released, allowing the enzyme to repeat the process.

FIGURE 11.23 Effect of the active site conformation on enzyme specificity

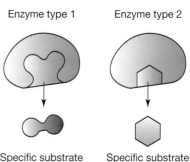

Enzyme type 1   Enzyme type 2

Specific substrate of enzyme 1   Specific substrate of enzyme 2

FIGURE 11.24 The lock-and-key-model of enzyme action

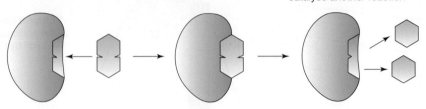

1. Substrate binds to the active site
2. The enzyme catalyses the reaction
3. The product moves out of the active site, allowing the enzyme to catalyse another reaction

It is important to remember that enzymes are three-dimensional and chiral, so they are highly selective as to the substrate they will interact with. For substrates with chiral centres, it is likely that only one of the enantiomers will fit into the active site of the enzyme and react.

FIGURE 11.25 Substrate chirality affects binding to the active site.

a. One enantiomer binds to the active site.

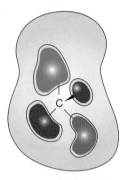

b. The other enantiomer does not bind to the active site.

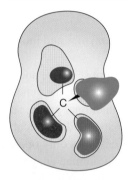

Enzyme-catalysed reactions are stereospecific

Only substrates with the correct shape will fit into the active site.

---

elog-1958

tlvd-9732

EXPERIMENT 11.1 online only

Enzymes as catalysts

Aim

To investigate the effect of enzyme activity

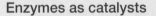

Environmental factors that affect enzyme activity

The rate at which an enzyme converts reactants into products is known as its activity. This is dependent on effective binding of the substate to the active site, so any factors that affect this binding will change the activity of the enzyme.

Enzymes have evolved to operate optimally under the temperature and pH conditions in their cellular environment. These conditions allow the proteins to fold appropriately and generate an active site with the conformation required for substrate binding.

Temperature

The temperature at which an enzyme has maximum activity is known as the optimum temperature. As seen in figure 11.26, the optimum temperature for a human enzyme is approximately 37.5 °C, which is body temperature, but different organisms have different optimum temperatures depending on the environment in which they live. If the temperature is lower or higher than the optimum, the activity of the enzyme is reduced, with inactivation occurring at extreme temperatures.

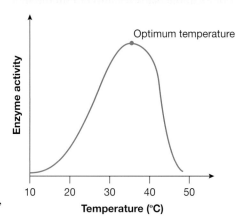

FIGURE 11.26 The effect of temperature on enzyme activity

The shape of the curve in figure 11.26 can be explained by two effects:

- Below body temperature, the enzyme activity is very slow, although as the temperature rises the average kinetic energy of the particles is increased. This leads to an increase in both the frequency of collisions and proportion of collisions, with $E \geq E_a$, and in turn an increase in the frequency of successful collisions.
- However, beyond the optimum temperature, the stronger vibrational energy increases the strain on the interactions holding the enzyme in its shape. When this happens, weak interactions in the secondary, tertiary and quaternary structures of the protein are disrupted, affecting the conformation of the active site.

At sufficiently high temperatures, the enzyme loses its three-dimensional shape entirely and is said to be denatured (figure 11.27). Under these conditions, only the strong covalent bonds in the primary structure are maintained. The variety of weak interactions required for all other levels of protein structure are disrupted and the protein unfolds.

During **denaturation**, **coagulation** of the protein commonly occurs. You may have noticed this effect when cooking an egg. The 'white' of a raw egg is clear and runny, but after heating it becomes white and solid. This transformation shows that it has been denatured and the protein has formed different interactions when coagulating. This example helps illustrate that denaturation is frequently an irreversible process.

denaturation a change in the structure or function of a large molecule, such as a protein

coagulation the process of turning a liquid into a solid or a thicker liquid

FIGURE 11.27 Denaturation disrupts the secondary, tertiary and quaternary structures of a protein while the primary structure remains intact.

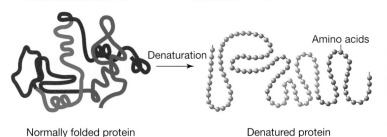

Normally folded protein   Denaturation   Amino acids   Denatured protein

FIGURE 11.28 The protein in egg white is denatured when the egg is heated.

Temperature has two effects on enzyme activity

Low temperatures slow reaction rate, as explained by collision theory. High temperatures change the shape of the active site, affecting substrate binding, and can denature the enzyme.

pH

Enzymes have an optimum pH, which is determined by the cellular environment in which they normally function. For example, salivary amylase in the mouth operates best at a neutral pH, whereas pepsin is optimised for the acidic conditions of the stomach. A larger deviation in pH from the optimum results in greater reduction in enzyme activity, with denaturation occurring at both high and low pH extremes. Indeed, instead of heating, it is possible to 'cook' an egg with acid.

The pH of the environment affects the ionisation of acidic and basic groups, in a similar manner to zwitterions.
- At low pH, high H^+ ion concentration leads to protonation of acidic and basic groups, forming $-COOH$ and $-NH_3^+$.
- At high pH, low H^+ ion concentration leads to deprotonation of acidic and basic groups, forming $-COO^-$ and $-NH_2$.

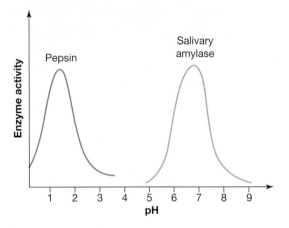

FIGURE 11.29 Different enzymes have different optimum pH levels.

The charge of these groups affects the types of interactions they form. Two charged groups form strong ionic bonds (e.g. $-COO^- \cdots H_3N^+-$); whereas uncharged groups may participate in weaker ion–dipole interactions (e.g. $-COO^- \cdots H_2N-$) or hydrogen bonds (e.g. $-COOH \cdots H_2N-$).

Changes in ionisation of R groups can affect substrate binding and formation of the enzyme–substrate complex in two ways:
- An altered tertiary structure can change confirmation of the active site.
- Amino acid residues within the active site may no longer be able to bind to the substrate.

Consider the example shown in figure 11.30. If the pH is optimal (figure 11.30a), ionic bonding will occur between the charged groups in the enzyme and the substrate. However, if the conditions become more acidic (figure 11.30b), the $-COO^-$ groups will become protonated, thus preventing formation of ionic bonds.

FIGURE 11.30 a. At optimal pH, ionic bonding occurs between the $-COO^-$ and the $-NH_3^+$ groups. b. In more acidic conditions, ionic bonds cannot form between the substrate and the enzyme.

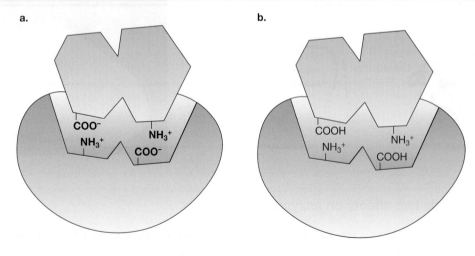

Discussing effects on enzyme activity

Make sure you refer to the different levels by name when discussing protein structure.

tlvd-3054

SAMPLE PROBLEM 3 Determining optimal conditions for enzyme activity

An experiment was conducted to investigate the activity of salivary amylase. Samples of starch were mixed with the enzyme at different pH values (2, 7 and 12) and at different temperatures (5, 35 and 80 °C). A glucose meter was used to measure the rate at which the starch was converted into glucose by this enzyme.

Which combination of conditions would produce the fastest rate of reaction and why?

THINK	WRITE
1. Amylase has an optimum pH of close to 7, as seen in figure 11.29. This will therefore need to be one of the conditions. As salivary amylase works in the mouth, this pH is consistent with that environment.	pH = 7
2. Amylase works in the body, so its optimum temperature would be body temperature. 35 °C is closest to this value.	Temperature = 35 °C
3. The optimum conditions for producing the fastest rate of reaction are at a pH of 7 and a temperature of 35 °C. Provide a reason for your answer.	The fastest rate of reaction will be at 35 °C and at a pH of 7. These conditions are consistent with the conditions found in the mouth and represent the optimum pH and temperature for the functioning of this enzyme. These conditions allow the fastest rates of reaction to occur.

PRACTICE PROBLEM 3

A similar experiment to that described in sample problem 3 was conducted to investigate the effectiveness of pepsin, an enzyme that works in the stomach to break down proteins (see figure 11.29).

Samples of protein were mixed with the enzyme at different pH values (2, 7 and 12) and at different temperatures (5, 35 and 80 °C). Each sample was tested to see if any breakdown had occurred.

Which combination of conditions would produce the fastest rate of reaction and why?

elog-1960

EXPERIMENT 11.2   online only

Investigating proteins

Aim

To investigate factors affecting denaturation of proteins

11.3.4 Competitive inhibition of enzymes

Temperature and pH are conditions that will affect all enzymes within a cell; however, normal cellular functioning requires specific enzymatic reactions to operate at appropriate rates. This can be achieved with the use of substances that reduce the activity of specific enzymes, called inhibitors.

Competitive inhibitors are a class of inhibitor that bind to the active site of an enzyme in place of its substrate. The inhibitor and substrate tend to have similar shapes, so there is a 'competition' between the two molecules for active site binding.
- If the inhibitor 'wins' it will bind and prevent the reaction from proceeding.
- If the substrate 'wins' it will bind and the reaction will proceed.

The extent of inhibition depends on the following factors:
- Concentration of the substrate
- Concentration of the inhibitor
- Relative binding affinity of the substrate and inhibitor for the active site.

For many reactions, higher reactant concentration will increase the frequency of collisions, and the frequency of successful collisions, resulting in a faster rate. In this way, higher substrate concentration will increase the rate of an enzymatic reaction, although the maximum rate is also determined by the concentration of enzyme present. Once all the available enzyme is actively catalysing reactions, further increasing of the substrate concentration will have no effect.

Competition between the inhibitor and substrate is illustrated in figure 11.32. Adding an inhibitor reduces the rate of the reaction. However, as the substrate concentration rises, it will outcompete the inhibitor, bind to the active site of more enzyme molecules, and therefore increase the reaction rate. Once the substrate is bound to all available active sites, the maximum rate has been reached, although the substrate concentration required will be higher than in the absence of an inhibitor.

The reaction rate may be affected by varying inhibitor concentration or by using an inhibitor with a different binding affinity for the active site. Increasing the concentration of the inhibitor or using an inhibitor that binds more strongly to the active site will both decrease reaction rate.

Competitive enzyme inhibitors as medicines

Modern medicine has seen the development of a wide array of drugs for both the prevention and treatment of disease. How these drugs work, termed the *mechanism of action*, varies greatly between different classes of medicines. Enzyme inhibitors are a promising class of drugs, given the key role that enzymes play in most cellular processes. As a result, competitive enzyme inhibitors have been developed to treat a variety of conditions, some of which are listed in table 11.11.

FIGURE 11.31 A competitive inhibitor binds to the active site, preventing the substrate from binding.

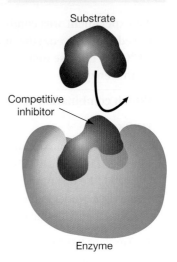

FIGURE 11.32 The effect of a competitive inhibitor on the rate of an enzymatic reaction

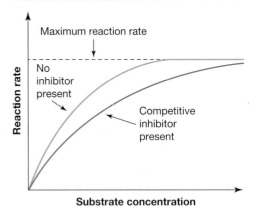

competitive inhibitors molecules that bind to the active site of an enzyme and prevent the substrate from binding

TABLE 11.11 Examples of competitive inhibitors used in medicine

Medical condition	Inhibitor(s)	Enzyme	Substrate	Effect
Methanol poisoning	Ethanol	Alcohol dehydrogenase	Methanol	Ethanol is preferentially metabolised, slowing conversion of methanol to toxic methanal
Influenza	Zanamivir (Relenza) Oseltamivir (Tamiflu)	Neuraminidase	Sialic acid	Inhibits cleavage of sialic acid on the host cell surface, and release of new viral particles
Excess stomach acid	Omeprazole (Prilosec) Esomeprazole (Nexium)	H^+/K^+ ATPase (gastric proton pump)	ATP	Inhibits H^+ ions being pumped into the stomach, reducing acidity
High cholesterol	Statins (lovastatin, atorvastatin (Lipitor))	HMG-CoA reductase	HMG-CoA	Lowers cholesterol to treat atherosclerosis (narrowing of arteries) and heart disease
Cancer	Amethopterin (methotrexate)	Dihydrofolate reductase	Folate	Inhibits production of folic acid, which is required for DNA and RNA synthesis by rapidly proliferating cancer cells

The following examples demonstrate some structural features commonly found in therapeutic competitive inhibitors:

- Inhibitors frequently mimic the shape of the substrate, thus facilitating binding to the active site.
- Small structural differences between the inhibitor and substrate may act to increase the binding affinity of the inhibitor to the active site and/or prevent the inhibitor undergoing an enzymatic reaction, thus increasing the duration of inhibition.

A simple example is the administration of ethanol to patients who have ingested methanol.

1. Methanol is broken down into methanal (formaldehyde) by the enzyme alcohol dehydrogenase.
2. Methanal is highly damaging to tissues, particularly those in the eyes, and may result in blindness.
3. Ethanol is also a substrate for alcohol dehydrogenase and will outcompete methanol, thus slowing the rate of methanal formation.
4. This delay ensures less methanal is produced over a specific time and can be excreted harmlessly before dangerous levels are reached.

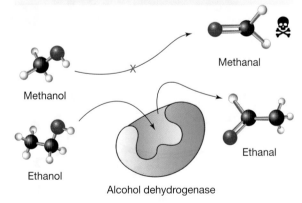

FIGURE 11.33 Ethanol slows the conversion of methanol to the toxic product methanal by competitive inhibition of the enzyme alcohol dehydrogenase.

Methanol

Methanal

Ethanol

Ethanal

Alcohol dehydrogenase

Influenza is a common viral disease that is estimated to kill approximately half a million people each year. The life cycle of the virus requires two proteins on the surface of the viral particle to perform specific functions.

- Hemagglutinin allows the viral particles to attach to a host cell.
- Neuraminidase allows new viral particles to leave the host cell and continue the infection. It does so by enzymatically cleaving a molecule on the surface of the cell called sialic acid.

FIGURE 11.34 a. Hemagglutinin (red) and neuraminidase (blue) proteins on the surface of an influenza viral particle **b.** The structure of neuraminidase

Scientists in Australia studied the shape of the neuraminidase enzyme and designed competitive inhibitors to bind to the active site. This led to the development of zanamivir (Relenza), which was closely followed by oseltamivir (Tamiflu). Both drugs are claimed to reduce the duration of influenza if taken soon after onset of symptoms by disrupting the release of new viral particles required to spread infection.

FIGURE 11.35 Schematic of neuraminidase inhibition, to disrupt the release of new viral particles

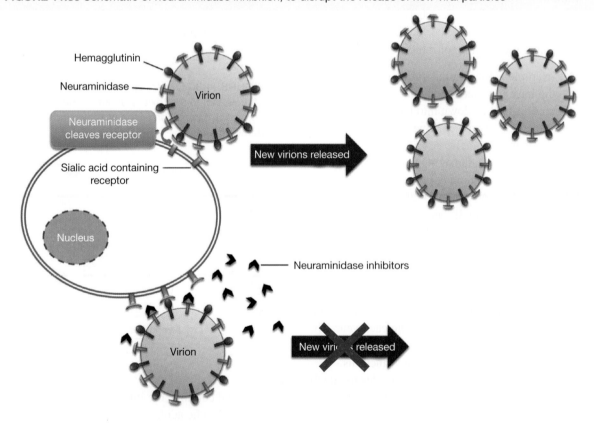

FIGURE 11.36 Comparison of the structures of the substrate **a.** sialic acid, and the competitive inhibitors **b.** zanamivir (Relenza) and **c.** oseltamivir (Tamiflu), allows identification of similarities and differences.

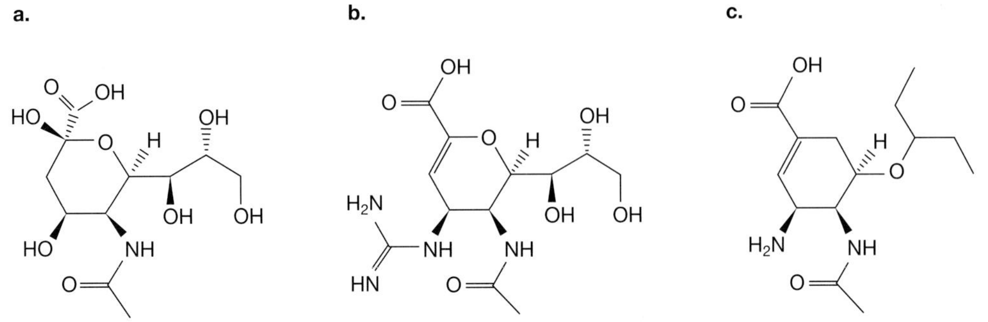

a. Sialic acid

b. Zanamivir (Relenza)

c. Oseltamivir (Tamiflu)

EXTENSION: Use of reaction rates to measure enzyme activity

Investigating enzyme activity in the laboratory can help provide useful information about how enzymes function. The rate of an enzyme-catalysed reaction can be observed by measuring the rate of disappearance of the substrate or the appearance of the product over time. Laboratory investigations examining the factors previously discussed could involve changing the pH or temperature of the reaction as the independent variable. The dependent variable could include measuring the volume of gas produced over a given time, or using simple observations, colour changes or colorimetry to measure a colour change over time. A glucose monitor that allows measurement of glucose levels within the blood is useful to obtain quantitative results. Some possible experiments are shown in table 11.12.

TABLE 11.12 Possible experiments to measure enzyme activity

Enzyme	Substrate	Products	Possible measurement
Amylase	Starch	Maltose and some glucose	Iodine (for starch) Benedict's solution Glucose test strips Glucose meter
Catalase	Hydrogen peroxide (H_2O_2): byproduct of respiration	Oxygen (O_2) and water (H_2O)	Count bubbles Gas syringe
Lactase	Lactose	Glucose and galactose	Benedict's solution Glucose test strips Glucose meter
Lipase	Lipid in milk	Fatty acids and glycerol	Phenolphthalein
Pectinase	Pectin	Simple sugars	Volume of liquid
Pepsin	Protein	Short polypeptides	Observation of egg white
Sucrase	Sucrose	Glucose and fructose	Benedict's solution Glucose test strips Glucose meter

Measuring enzyme activity and the effect of pH and temperature

To measure the effect of pH on enzyme activity, an experiment might measure the reaction times of amylase and starch to determine the optimum pH of the reaction. The pH of the solution is varied by using two different concentrations of sodium carbonate solution (base) and ethanoic acid (acid). The presence of starch can be observed by using iodine, which changes to a blue-black colour in the presence of starch.
- Independent variable: pH for colour change
- Dependent variables: Time
- Constants: Temperature, volumes and concentrations of enzyme (amalyse) and substrate

To begin the experiment, place two drops of iodine in each well of a tray. Place the 5 mL of starch in a test tube and add 2 mL of sodium carbonate solution. Start timing as 1 mL of amylase is added. Every 15 seconds, remove a sample of the mixture using a clean dropping pipette and place in order in the tray.

The iodine will change colour to blue-black in the initial wells, but at a particular time there will not be a change in colour. Note this time. If possible, repeat this experiment twice more, record the results in a table and determine an average time.

Repeat the procedure using clean equipment and solutions of different pH (e.g. ethanoic acid), including pure water, as shown in figure 11.37. Observe the pH at which the reaction took place in the least amount of time to find out the optimum pH for this reaction to occur at the fastest rate.

FIGURE 11.37 Experiment to investigate the effect of pH on an enzyme reaction

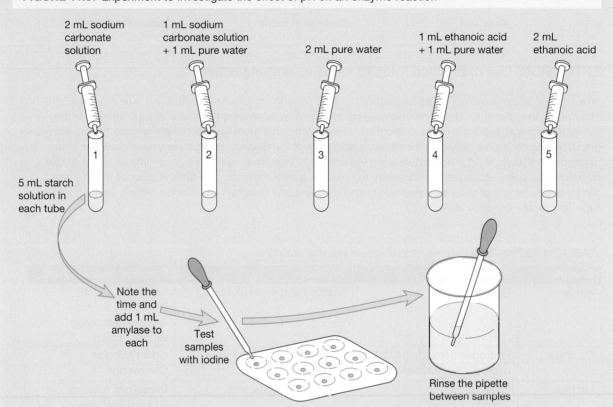

This experiment could be repeated to find the effect of temperature on enzyme activity. By using the optimum pH as determined previously, all solutions would be placed in a water bath for 10 minutes before mixing. The experiment would be carried out at different temperatures: at very low temperatures, room temperature and then at higher temperatures. This variation would find the optimum temperature for enzyme effectiveness. All other factors must be kept constant.

Another experiment could measure the activity of catalase. Catalase reacts with hydrogen peroxide and produces oxygen bubbles, which can then be measured. This experiment would also be carried out using water baths at different temperatures. The rate of reaction would be monitored by the rate of appearance of the product oxygen. Hydrogen peroxide is a by-product of many chemical reactions in the body, but is toxic if it builds up in cells and so must be removed.

$$2H_2O_2 \xrightarrow{\text{Catalase}} 2H_2O(l) + O_2(g)$$

11.3 Activities

11.3 Quick quiz `on`	11.3 Exercise	11.3 Exam questions

11.3 Exercise

1. State two examples of amino acids with basic side chains.
2. Which type of bonding is found in both the secondary and tertiary structures of proteins?
3. Identify the strongest type of interaction that can form between the R groups of two glutamic acid residues.
4. Enzymes affect the activation energy of a reaction. On the diagram shown, label the activation energy with an enzyme and the activation energy without an enzyme.

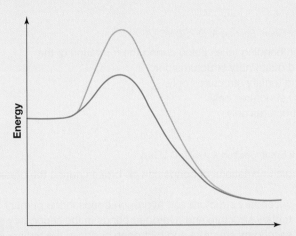

5. On the diagram shown, label the products, enzyme–substrate complex, active site, enzyme and substrate.

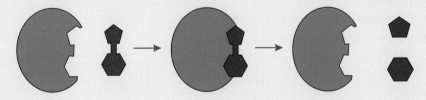

6. What type of interaction might occur between a –CH$_3$ group in a substrate and a functional group in an enzyme's active site?
7. Explain why glycine does not have an optical isomer.
8. Draw the structure of aspartic acid in a highly alkaline solution.

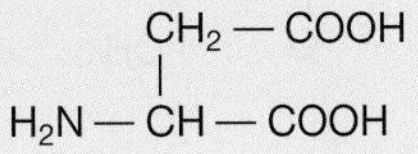

$$CH_2 - COOH$$
$$|$$
$$H_2N - CH - COOH$$

9. **a.** What does the word 'optimum' mean when referring to enzyme temperature?

   b. State the optimum temperature and pH of the enzyme shown in the following graphs.

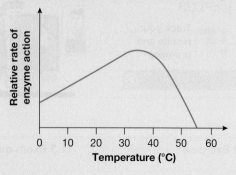 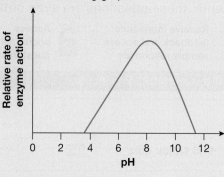

10. Statins are a class of drugs used to treat high cholesterol by competitive inhibition of the enzyme HMG-CoA reductase. Explain how the following factors will affect the activity of HMG-CoA reductase:

   a. Increasing the concentration of the statin for a fixed concentration of substrate

   b. Increasing the concentration of the substrate, HMG-CoA, for a fixed concentration of statin

   c. Using a statin with a more similar shape to the HMG-CoA substrate than an alternative statin.

11.3 Exam questions

Question 1 (1 mark)

Source: VCE 2022 Chemistry Exam, Section A, Q.7; © VCAA

MC In a protein, hydrogen bonding takes place during the formation of the
A. secondary, tertiary and quaternary structures only.
B. primary, secondary and tertiary structures only.
C. tertiary and quaternary structures only.
D. primary and tertiary structures only.

Question 2 (1 mark)

Source: VCE 2020 Chemistry Exam, Section A, Q.29; © VCAA

MC Which of the following combinations of bonds can be broken during the breakdown of a protein that a person has eaten?
A. covalent bonds in the secondary structure and hydrogen bonds in the primary structure
B. covalent bonds in the tertiary structure and hydrogen bonds in the secondary structure
C. covalent bonds in the secondary structure and hydrogen bonds in the tertiary structure
D. covalent bonds in the quaternary structure and hydrogen bonds in the primary structure

Question 3 (1 mark)

Source: VCE 2020 Chemistry Exam, Section A, Q.12; © VCAA

MC The diagram below represents a section of an enzyme.

The diagram can be described as a
A. secondary structure consisting of glutamine, glycine and lysine.
B. primary structure consisting of asparagine, glycine and lysine.
C. secondary structure consisting of asparagine, alanine and lysine.
D. primary structure consisting of glutamine, alanine and lysine.

Question 4 (1 mark)

Source: Adapted from VCE 2019 Chemistry Exam, Section A, Q.6; © VCAA

MC Which one of the following statements about enzymes is correct?

A. The lock-and-key model suggests that the shape of the active site changes when it binds to a competitive inhibitor.
B. Enzymes may have their tertiary structure altered during a catalysed reaction.
C. Enzymes can catalyse most reactions over a broad range of temperatures.
D. Enzymes may change the equilibrium constant of a catalysed reaction.

Question 5 (4 marks)

Source: VCE 2015 Chemistry Exam, Section B, Q.6.a; © VCAA

After a murder had been committed, a forensic chemist obtained crime scene blood samples and immediately placed them in two sterile containers labelled Sample I and Sample II.

The chemist discovered that Sample I contained a particular protein, which was analysed to reveal the following sequence of amino acid residues.

-ser-gly-tyr

a. Referring to the VCE Chemistry Data Book, draw the structure of this sequence of amino acid residues and circle one amide link/peptide bond in your drawing. **(3 marks)**
b. The protein was hydrolysed in the presence of a suitable enzyme and the amino acid glycine was isolated. The glycine sample was then dissolved in a 0.1 M solution of sodium hydroxide. Draw the structure of glycine in this solution. **(1 mark)**

More exam questions are available in your learnON title.

11.4 Review

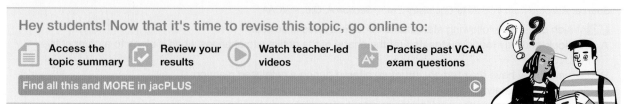

11.4.1 Topic summary

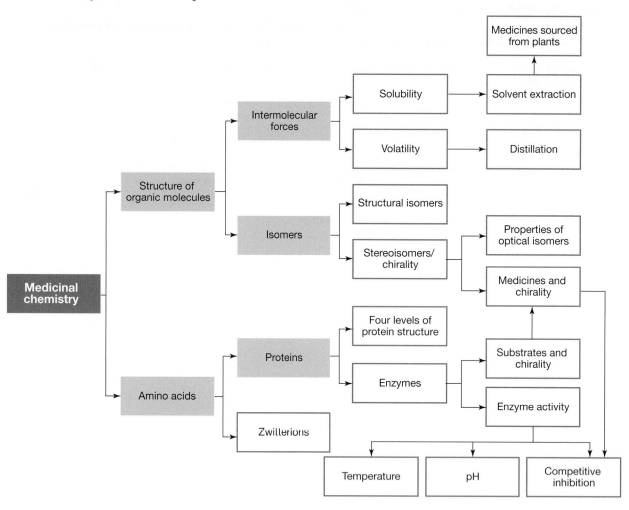

11.4.2 Key ideas summary

online only

11.4.3 Key terms glossary

online only

Resources

Solutions	Solutions — Topic 11 (sol-0838)
Practical investigation eLogbook	Practical investigation eLogbook — Topic 11 (elog-1710)
Digital documents	Key science skills — VCE Chemistry Units 1–4 (doc-37066)
	Key terms glossary — Topic 11 (doc-37301)
	Key ideas summary — Topic 11 (doc-37302)
Exam question booklet	Exam question booklet — Topic 11 (eqb-0122)

11.4 Review questions

1. **MC** Which of the following contains a chiral carbon?

   A. Prop-1-ene, CH_2CHCH_3
   B. 2,2-dichloropropane, $CH_2Cl_2CH_2CH_3$
   C. 2-chloropropane, $CH_3CHClCH_3$
   D. 2-bromobutane, $CH_3CHBrCH_2CH_3$

2. What are the structural features of all amino acids found in naturally occurring proteins?

3. Describe the differences in structure between glutamine and glutamic acid.

4. When many amino acid molecules react together, a protein is formed. The four levels of protein structure are primary, secondary, tertiary and quaternary.

   a. Describe the group interactions required for the secondary structure to be maintained.
   b. State three ways in which the tertiary structure of a protein is maintained.

5. Enzymes are organic catalysts that operate in living things to facilitate chemical reactions essential to life. They are often referred to as 'biological catalysts'. List two differences between enzymes and inorganic catalysts.

6. Newborn babies are tested for phenylketonuria (PKU). PKU is a genetic disorder that prevents the breakdown of the amino acid phenylalanine, which then builds up in the body. Brain development can be limited if treatment is not provided. The structure of the phenylalanine molecule is shown.

   a. Draw the structure of phenylalanine in a solution of pH 3.
   b. Draw the structure of phenylalanine in a solution of pH 10.
   c. At a pH of about 5.9, phenylalanine is ionised but is not attracted to either the positive or negative electrodes of an electrolytic cell. Draw the structure of phenylalanine at pH 5.9.

7. a. Explain why the mechanism of action of an enzyme is sometimes referred to as a 'lock-and-key' mode of operation.
   b. Imagine you want to use a catalyst to speed up the reaction $A + B \rightarrow C + D$. Draw a diagram that shows how an enzyme can facilitate this reaction. Label the substrate and active site.

8. **MC** Two optical isomers of dopamine, L-DOPA and D-DOPA, have very different effects in the body. L-DOPA is effective in treating Parkinson's disease, whereas D-DOPA has no effect.

   The reason for this is that

   A. the functional groups of L-DOPA and D-DOPA differ, causing different biological activity.
   B. D-DOPA is a smaller molecule than L-DOPA and will only bond through dispersion forces.
   C. the shape of the L-DOPA molecule forms a better fit with the relevant enzyme than D-DOPA does.
   D. L-DOPA can be ionised, whereas D-DOPA cannot.

9. The enzyme succinate dehydrogenase catalyses the conversion of succinate to fumarate. This reaction is inhibited by malonate.

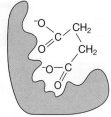 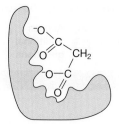

Succinate–enzyme complex   Malonate–enzyme complex

   a. Identify the functional group present in both succinate and malonate.

   b. Explain why malonate is an effective inhibitor of succinate dehydrogenase.

10. Morphine is a naturally occurring pain medication found in opium poppies. It can be chemically converted into codeine.

Morphine   Codeine

   a. Identify a polar functional group present in morphine

   b. Justify why morphine is more soluble in a non-polar solvent, such as cyclohexane, than a polar solvent, such as water.

   c. A student placed a mixture of codeine and morphine dissolved in cyclohexane in a separation funnel. They then added an equal volume of ethanol and shook the funnel. Explain whether morphine or codeine would more readily dissolve in the ethanol.

11.4 Exam questions

Section A — Multiple choice questions

All correct answers are worth 1 mark each; an incorrect answer is worth 0.

Question 1

Source: VCE 2021 Chemistry Exam, Section A, Q.6; © VCAA

MC Which of the following correctly identifies the bonds that break in a protein when it undergoes denaturation and when it undergoes hydrolysis?

	Denaturation	Hydrolysis
A.	covalent	hydrogen
B.	covalent	covalent
C.	hydrogen and ionic	hydrogen
D.	hydrogen and ionic	covalent

Question 2

Source: VCE 2022 Chemistry Exam, Section A, Q.12; © VCAA

MC Enzymes are commonly not effective in acidic conditions because acids

A. change the charges on the enzymes.
B. react with the enzymes to form zwitterions.
C. esterify the enzymes into smaller molecules.
D. react with the carboxyl groups on the enzymes' amino acid residues.

Question 3

Source: VCE 2018 Chemistry Exam, Section A, Q.4; © VCAA

MC At the molecular level, Protein P is shaped like a coil. When a solution of Protein P is mixed with citric acid, solid lumps form.

The change in the structure of Protein P is due to

A. hydrolysis.
B. denaturation.
C. polymerisation.
D. the formation of peptide bonds.

Question 4

Source: VCE 2017 Chemistry Exam, Section A, Q.4; © VCAA

MC Which of the following contains a chiral carbon?

	Name	Semi-structural formula
A.	2-methylbut-1-ene	$CH_2C(CH_3)CH_2CH_3$
B.	2-chlorobutane	$CH_3CHClCH_2CH_3$
C.	propanoic acid	CH_3CH_2COOH
D.	1,2-dichloroethene	$ClCHCHCl$

Question 5

Source: VCE 2017 Chemistry Exam, Section A, Q.15; © VCAA

MC Which one of the following is a correct statement about the denaturation of a protein?

A. Denaturation is characterised by the release of peptides.
B. Alcohol denatures proteins by disrupting the hydrogen bonding.
C. Denaturation involves disruption of all bonds in the tertiary structure.
D. The primary and secondary structures are disrupted when denaturation occurs.

Question 6

Source: VCE 2015 Chemistry Exam, Section A, Q.14; © VCAA

MC Which one of the following is **not** true of protein denaturation?

A. It could result from a temperature change.
B. It may be caused by a pH change.
C. It alters the primary structure.
D. It results in a change in the shape of the protein.

Question 7

Source: *VCE 2017 Chemistry Exam, Section A, Q.10; © VCAA*

MC Which one of the following structures represents a zwitterion of a 2-amino acid?

A.

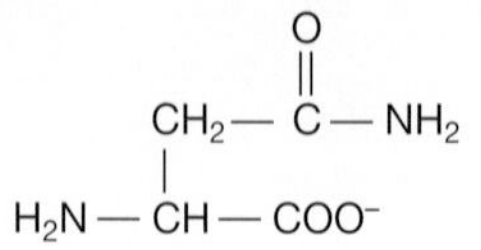
O
CH2— C— NH2
H2N— CH— COO−

B.

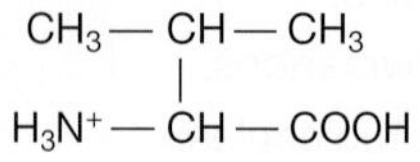
CH3— CH— CH3
H3N+— CH— COOH

C.

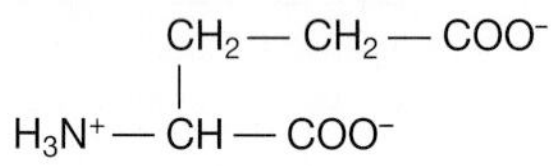
CH2— CH2— COO−
H3N+— CH— COO−

D.

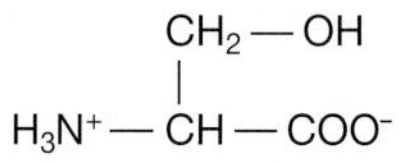
CH2— OH
H3N+— CH— COO−

Question 8

Source: *VCE 2013 Chemistry Exam, Section A, Q.11; © VCAA*

MC Australian jellyfish venom is a mixture of proteins for which there is no antivenom. Jellyfish stings are painful, can leave scars and, in some circumstances, can cause death.

Some commercially available remedies disrupt ionic interactions between the side chains on amino acid residues.

These products most likely disrupt the protein's

A. primary structure only.
B. secondary structure only.
C. tertiary structure only.
D. primary, secondary and tertiary structures.

Question 9

Source: *VCE 2013 Chemistry Exam, Section A, Q.12; © VCAA*

MC Which figure best represents the bonding between adenine and thymine in the structure of DNA?

A.

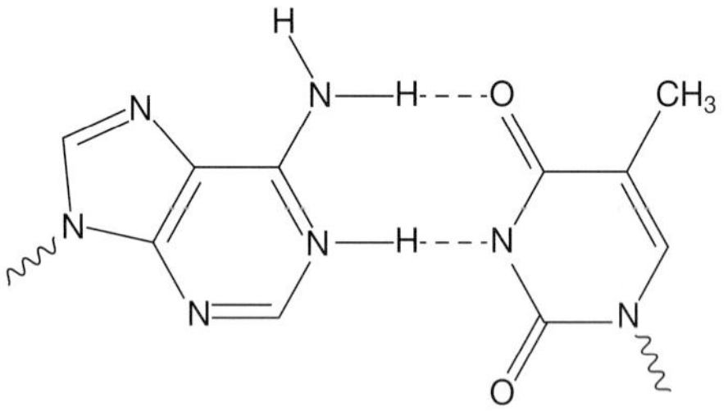

B.

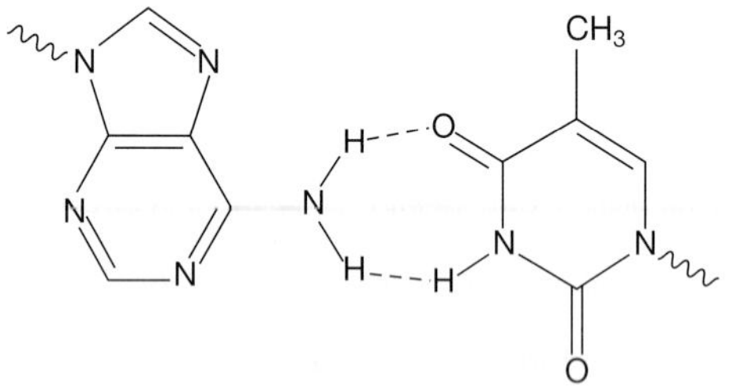

C.

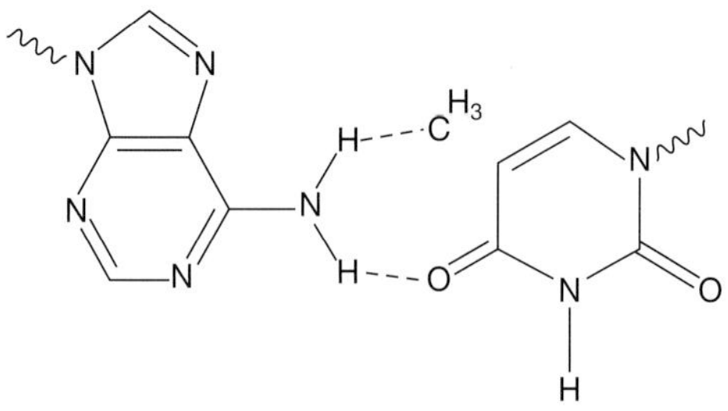

D.

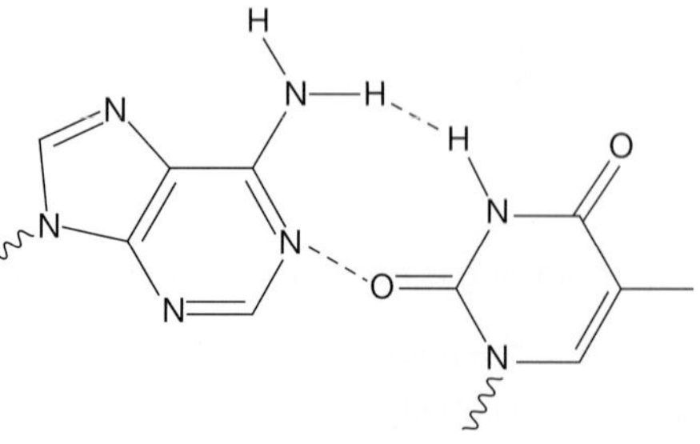

Source: VCE 2013 Chemistry Sample Exam for Units 3 and 4, Section A, Q.9; © VCAA

MC Enzymes, which are composed mostly of protein, catalyse many chemical reactions. The structure of a portion of an enzyme, with some of its constituent atoms shown, is represented below.

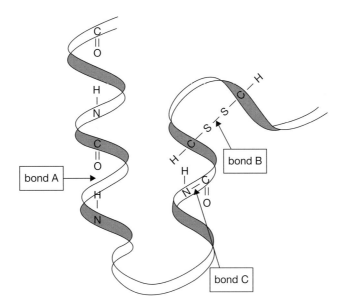

Which level of protein structure is each of the chemical bonds labelled involved in?

	Bond A	**Bond B**	**Bond C**
A.	primary	tertiary	secondary
B.	secondary	tertiary	primary
C.	tertiary	primary	secondary
D.	primary	secondary	tertiary

Section B — Short answer questions

Question 11 (3 marks)

Source: VCE 2017 Chemistry Exam, Section B, Q.3.b.ii; © VCAA

Glucagon is a peptide hormone that works with insulin to help regulate blood glucose levels. Glucagon acts to increase blood glucose levels through targeted action on the polysaccharide stored in the liver.

Glucagon consists of a chain of 29 amino acids, the sequence of which is given below, and folds to form a short alpha-helix.

H₂N-His-Ser-Gln-Gly-Thr-Phe-Thr-Ser-Asp-Tyr-Ser-Lys-Tyr-Leu-Asp-Ser-Arg-Arg-
Ala-Gln-Asp-Phe-Val-Gln-Trp-Leu-Met-Asn-Thr-COOH

Describe the bonding that is found in the primary and secondary structures of the glucagon molecule.

Question 12 (2 marks)

Source: VCE 2016 Chemistry Exam, Section B, Q.7.b; © VCAA

An incomplete reaction pathway for the synthesis of aspirin is given below.

a. Draw the structural formula of salicylic acid in the box provided. **(1 mark)**

b. The structural formula of the other reactant is provided.
State its systematic name in the box provided. **(1 mark)**

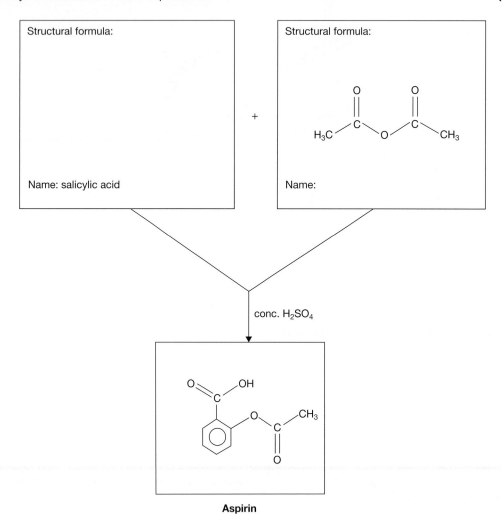

Structural formula:

Name: salicylic acid

+

Structural formula:

Name:

conc. H_2SO_4

Aspirin

Source: VCE 2013 Chemistry Exam, Section B, Q.3.b,d,e; © VCAA

Spider webs are very strong and elastic. Spider web silk is a protein that mainly consists of glycine and alanine residues.

a. What is the name of the bond between each amino acid residue? **(1 mark)**

Glycine forms an ion at a pH of 6 that has both a positive and negative charge.

b. Draw the structure of a glycine ion at a pH of less than 4. **(1 mark)**

c. Describe the bonds that contribute to the spiral secondary structure of this protein. **(2 marks)**

Question 14 (7 marks)

Source: VCE 2014 Chemistry Exam, Section B, Q.7; © VCAA

Amino acids can be classified according to the nature of their side chains (Z groups). These may be polar, non-polar, acidic or basic.

a. Referring to the VCE Chemistry Data Book, name one amino acid that has a non-polar side chain and one amino acid that has an acidic side chain. **(2 marks)**

The table below provides examples of different categories of side chains at a pH of 7.

Name of amino acid	Structure of side chain of pH 7
alanine (Ala)	$-CH_3$
asparagine (Asn)	$-CH_2-CO-NH_2$
aspartic acid (Asp)	$-CH_2COO^-$
cysteine (Cys)	$-CH-SH$
lysine (Lys)	$-CH_2-CH_2-CH_2-CH_2-NH_3^+$
serine (Ser)	$-CH_2OH$

b. The tertiary structure of proteins is a result of the bonding interactions between side chains of amino acid residues.
Use the information provided in the table above to

  i. identify the amino acid that is involved in the formation of disulfide bonds (sulfur bridges) **(1 mark)**

  ii. give an example of **two** amino acid side chains that may form hydrogen bonds between each other **(1 mark)**

  iii. give an example of amino acid side chains that may form ionic bonds (salt bridges) between each other **(1 mark)**

  iv. identify the type of bonding that exists between the side chains of two alanine residues. **(1 mark)**

c. The enzyme trypsin catalyses the breaking of peptide bonds in proteins. Trypsin is active in the upper part of the small intestine, where the pH is between 7.5 and 8.5.
Trypsin is not effective in the stomach, where the pH is 4.
Suggest a reason why. **(1 mark)**

Question 15 (7 marks)

Source: VCE 2018 Chemistry NHT Exam, Section B, Q.9; © VCAA

Enzymes are crucial for the reactions involved in the metabolism of food in the human body. Even when conditions vary in the human body, there are enzymes that function to ensure the chemical reactions needed to sustain life take place.

In the digestive tract, there is a variation in pH. The stomach can have a pH in the range of 1 to 4, while in the intestines, the pH can vary from 5 to 7.

a. Describe the tertiary structure of enzymes and explain the chemistry that enables enzymes to function in different parts of the digestive tract. Your response should:
  - describe the chemical bonding that enables the tertiary structure to be maintained
  - comment on the significance of chemical bonding to the correct functioning of the enzyme
  - explain how the enzyme chemically interacts with the substrate.

Diagrams may be used to support your answer. **(4 marks)**

b. State **one** factor, other than pH, that would affect the activity of an enzyme. Outline the effect this factor would have on the rate of reaction of the enzyme and explain why. **(3 marks)**

AREA OF STUDY 2 How are organic compounds analysed and used?

OUTCOME 2

Apply qualitative and quantitative tests to analyse organic compounds and their structural characteristics, deduce structures of organic compounds using instrumental analysis data, explain how some medicines function, and experimentally analyse how some natural medicines can be extracted and purified.

PRACTICE EXAMINATION

STRUCTURE OF PRACTICE EXAMINATION		
Section	Number of questions	Number of marks
A	20	20
B	4	30
Total		50

Duration: 50 minutes

Information:

- This practice examination consists of two parts. You must answer all question sections.
- Pens, pencils, highlighters, erasers, rulers and a scientific calculator are permitted.
- You may use the VCE Chemistry Data Book for this task.

 Resources

🔗 **Weblink** VCE Chemistry Data Book

SECTION A — Multiple choice questions

All correct answers are worth 1 mark each; an incorrect answer is worth 0.

1. An unknown carbon-containing compound was found to turn litmus paper red and produced vigorous effervescence upon the addition of sodium carbonate.
This compound is most likely to be
   A. 1-chlorohexane.
   B. methanoic acid.
   C. methanol.
   D. methanal.

2. Four different substances (labelled W, X, Y and Z) have been isolated in a mixture from a recently discovered plant in the Amazon rainforest. The following table shows the melting and boiling points of these four substances.

Substance	Melting point (°C)	Boiling point (°C)
W	−15	75
X	−3	85
Y	55	230
Z	−12	91

It is desired to isolate substance Z for further study.
Based on this information, which of the following would be the most appropriate technique to use?
A. Solvent extraction using water
B. Evaporation
C. Simple distillation
D. Fractional distillation

3. A chemist attempted to prepare a sample of acetylsalicylic acid (aspirin). The substance produced was then tested in a melting-point apparatus. The melting point of acetylsalicylic acid is 135 °C.
The sample tested was observed to melt over the range 121–123 °C.
Which of the following is true?
A. A pure sample of acetylsalicylic acid has been prepared.
B. An impure sample of acetylsalicylic acid has been prepared.
C. A pure sample has been prepared but it is not acetylsalicylic acid.
D. An impure sample has been prepared and it is not acetylsalicylic acid.

4. Which of the following fatty acids would form a triglyceride with the lowest iodine number (mass of iodine that reacts with 100 g of a substance)?

A. Palmitoleic
B. Stearic
C. Oleic
D. Linoleic

5. A student carried out a redox titration to determine the amount of vitamin C in lemon juice. They standardised an iodine solution and then titrated this against a diluted lemon juice sample.
If the student rinsed their burette with deionised water, the implication would be

A. the amount of lemon juice calculated would be too high.
B. the amount of lemon juice calculated would be too low.
C. there would be no effect on the calculated amount of lemon juice.
D. 5.0 mL more lemon juice would be needed to rectify the error.

6. A mass spectrum is shown.

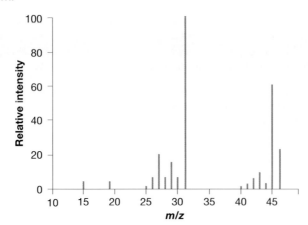

Which molecule is likely to have produced the mass spectrum shown?
A. Methanoic acid
B. Ethanol
C. Ethanal
D. Propane

7. A molecule has the molecular formula $C_3H_6O_2$. Information about the ^{1}H-NMR spectrum for the molecule is shown in the table.

Chemical shift (ppm)	Relative peak area	Peak splitting
1.3	3	Triplet (3)
4.2	2	Quartet (4)
9.0	1	Single (1)

Based on this information, the molecule is likely to be

A. methyl ethanoate.

B. ethyl methanoate.

C. propanoic acid.

D. propan-1,2-diol.

8. How many peaks would be seen on a ^{13}C-NMR spectrum for 1-chlorobutane?

A. 1

B. 2

C. 3

D. 4

9. The infrared spectrum for a molecule is shown.

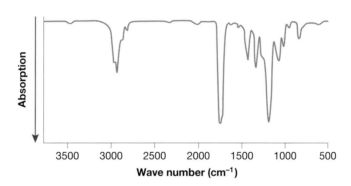

The molecule is most likely to be

A. an alcohol.

B. an ester.

C. an amine.

D. a carboxylic acid.

10. The peak area in a high-performance liquid chromatography (HPLC) chromatogram can be used to determine

A. the retention time.

B. the concentration of the analyte.

C. the amount of analyte.

D. the polarity of the analyte.

11. Which of the following would be used to determine the concentration of an organic compound using HPLC?

A. The chromatogram only

B. The chromatogram and a set of standards of known concentration only

C. The calibration curve and a set of standards of known concentration only

D. The chromatogram, a calibration curve and a set of standards of known concentration

12. A colourless organic liquid is tested in a number of ways. The results of these tests are shown in the following table.

Test	Results
Mass spectrometry	Base peak at m/z 59
	Parent peak at m/z 74
Reaction with acidified dichromate solution	No reaction
Infrared spectroscopy	No peak at 1680–1740 cm^{-1}
	Large peak at 3200–3600 cm^{-1}
Elemental analysis	Consists of C, H, O
Reaction with potassium carbonate	No reaction

Based on this information, the unknown compound is

A. $CH_3CH_2CH_2COOH$

B. $(CH_3)_3COH$

C. $CH_3C(OH)CH_2CH_3$

D. $CH_3CH_2OCH_2CH_3$

13. Currently there are moves to legalise cannabis-based products for therapeutic use. Cannabis contains a substance called THC, which is responsible for its effects on the nervous system.
The structure of THC is shown.

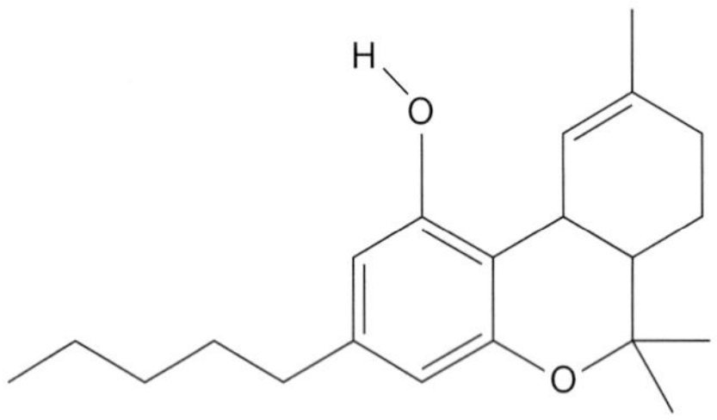

A company is examining the use of solvent extraction to remove THC from cannabis extract. Which of the following would be the most effective for performing this task?

A. Cyclohexane

B. Ethanol

C. Water

D. Propanoic acid

14. The structural formula of aspartame is shown.

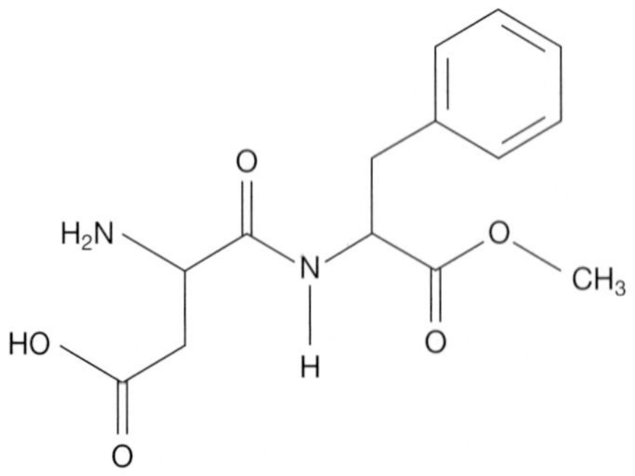

What are the functional groups present in aspartame?

A. Carboxyl, amine, carbonyl, ester

B. Carboxyl, amine, amide, ester

C. Carboxyl, amine, amide, ester, methyl

D. Carbonyl, hydroxyl, carbonyl, amine, ester

15. Which amino acid does *not* form an optical isomer?

    A. Alanine

    C. Glutamine

    B. Phenylalanine

    D. Glycine

16. Diagrams of amino acid structures can be found in the VCE Chemistry Data Book.
In the pure solid state with an approximately neutral pH, the amino acids exist almost completely

    A. as uncharged molecules.

    C. in cationic form.

    B. as zwitterions.

    D. in anionic form.

17. Which of the following statements about enzymes are correct?

    I Changing pH can change the shape of an enzyme and its active site, reducing its activity.

    II Enzymes cannot be used again; once they have been used they are denatured.

    III Hydrogen bonding contributes to an enzyme's secondary and tertiary structures.

    A. I and II

    C. II and III

    B. I and III

    D. I, II and III

18. The following graph shows the activity of an enzyme plotted against temperature.

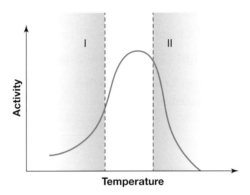

Which of the following statements is *not* true?

    A. In region I, the increase in activity is due to substrate molecules binding and unbinding from the active site more often.

    B. In region II, the decrease in activity is caused by denaturation.

    C. The decrease in activity as temperature decreases in region I is due to denaturation.

    D. The graph illustrates that there is an optimal temperature for enzyme activity.

19. One of the essential growth compounds for bacteria is folic acid. This is manufactured from a substance called *p-aminobenzoic acid* in a stepwise pathway, using enzymes for each step.

One of the first classes of antibacterial drugs developed was the *sulfonamides*. Sulfonamide molecules all have a shape and charge distribution that is very similar to p-aminobenzoic acid. Sulfonamides work because they block the active site of the enzyme responsible for the first step in the production of folic acid.
This is an example of

    A. isomerism.

    C. denaturation.

    B. chirality.

    D. enzyme inhibiting.

20. The column in a HPLC consists of fine silica particles coated with octadecane, $C_{18}H_{38}$, to provide the stationary phase.
Which of the following would have the longest retention time when passed through this instrument under identical conditions?

    A. Cyclohexane

    C. Ethanoic acid

    B. Ethanol

    D. Water

Question 21 (6 marks)

Commercial bleach contains hypochlorite ions, OCl^-, as its active ingredient. Typically, the concentration of this is about 5.25 %(m/m), which equates to a hypochlorite ion concentration of approximately 0.7 M. However, this concentration decreases over time due to decomposition.

A redox titration can be used to accurately measure this level.

In one such analysis, the following steps were used.

1. A pipette was used to accurately dilute 10.0 mL of bleach solution to 100 mL using a volumetric flask.
2. A standardised solution of sodium thiosulfate ($Na_2S_2O_3$) was placed in a burette.
3. 20 mL of the diluted bleach solution from step 1 was pipetted into a conical flask. An acidified solution containing an excess amount of iodide ions was then added.
4. The contents of the flask were then titrated using starch as an indicator. Steps 3 and 4 were repeated until three concordant titres were obtained.

The equations for the reactions occurring are as follows:

Reaction I (in step 3): $OCl^-(aq) + 2I^-(aq) + 2H^+(aq) \rightarrow I_2(s) + Cl^-(aq) + H_2O(l)$

Reaction II (in step 4): $I_2(s) + 2S_2O_3^{2-}(aq) \rightarrow 2I^-(aq) + S_4O_6^{2-}(aq)$

a. Explain why the iodide ions added in step 3 must be in excess. **(1 mark)**

b. What is the value of x in the following equation? **(1 mark)**

$$n(OCl^-) = \frac{n(S_2O_3^{2-})}{x}$$

c. In this experiment, the following results were obtained:

Concentration of thiosulfate solution: 0.192 M

Average of concordant titres: 10.20 mL

   i. Calculate the average number of moles of thiosulfate solution added for the titrations. **(1 mark)**

   ii. Calculate the number of moles of hypochlorite ions present during each titration. **(1 mark)**

   iii. Calculate the concentration (in mol L^{-1}) of hypochlorite ions in the original bleach. **(2 marks)**

Question 22 (12 marks)

Propanone and propan-2-ol both contain three carbon atoms and one oxygen atom.

a. Draw the structural formula of:

   i. propanone **(1 mark)**

   ii. propan-2-ol. **(1 mark)**

b. How many peaks would you expect to see on a ^{13}C-NMR spectrum for:

   i. propanone **(1 mark)**

   ii. propan-2-ol? **(1 mark)**

c. Explain how many peaks (the splitting pattern) you would expect to see on a ^1H-NMR spectrum for:

   i. the $R–CH_3$ in propanone **(2 marks)**

   ii. the $R–CH_3$ in propan-2-ol. **(2 marks)**

d. Describe the key difference expected in the infrared spectrum of propanone compared to that of propan-2-ol. **(2 marks)**

e. Identify the peak at $m/z = 43$ for propanone. **(1 mark)**

f. A student has a bottle labelled 'P' that contains either propanone or propan-2-ol. When they react their unknown compound with acidified dichromate ions, they note a change in odour. Identify the compound in the bottle labelled 'P'. **(1 mark)**

Question 23 (4 marks)

Amlodipine is a drug used to control high blood pressure. The structure of amlodipine is shown.

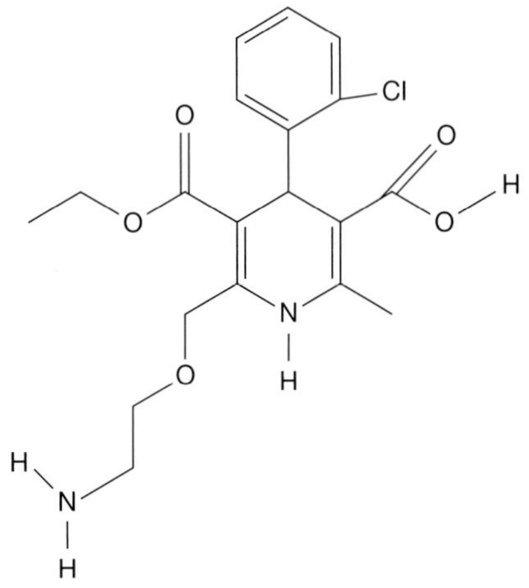

On the diagram, circle and name four different functional groups.

Question 24 (8 marks)

A section of a protein chain from an enzyme is shown.

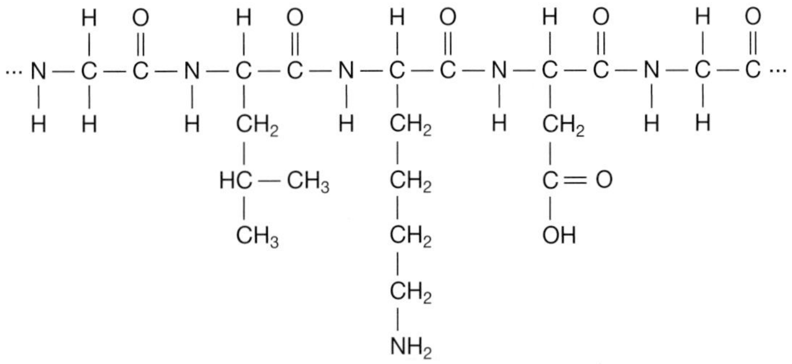

a. On the diagram, circle a peptide bond. **(1 mark)**

b. Using their symbols, identify the amino acids in the chain from left to right. **(1 mark)**

c. Identify the functional groups responsible for the secondary structure of the enzyme. **(2 marks)**

d. Based on the amino acids present, what type of bonding would be expected in the tertiary structure of this enzyme? **(2 marks)**

e. Explain what would happen if the enzyme were subjected to high temperatures. **(2 marks)**

PRACTICE SCHOOL-ASSESSED COURSEWORK

ASSESSMENT TASK — ANALYSIS AND EVALUATION OF SECONDARY DATA

In this task you will analyse and evaluate secondary data, including identified assumptions or data limitations, and conclusions.
- Pens, pencils, highlighters, erasers, rulers and a scientific calculator are permitted.
- You may use the VCE Chemistry Data Book to complete this task.

Total time: 50 minutes
Total marks: 39 marks

IDENTIFYING ORGANIC ACIDS FROM LC–MS

Vinegar is a popular condiment produced from the double fermentation of fruit. The first fermentation converts sugars into ethanol and the further fermentation produces the ethanoic acid that gives vinegar its sour taste. Spirit vinegars are typically between 5.0 and 20.0 %(v/v) ethanoic acid.

Other organic acids can be formed during the fermentation process and these impart their own characteristics on the overall flavour, depending on the concentrations. Possible organic acids in spirit vinegars are shown below.

| 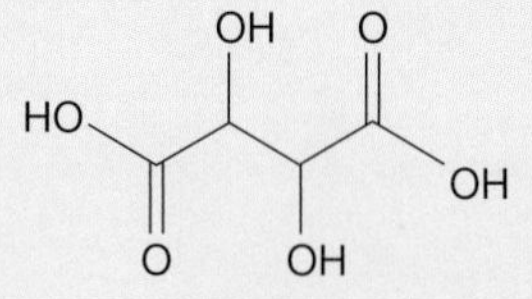 Tartaric acid | 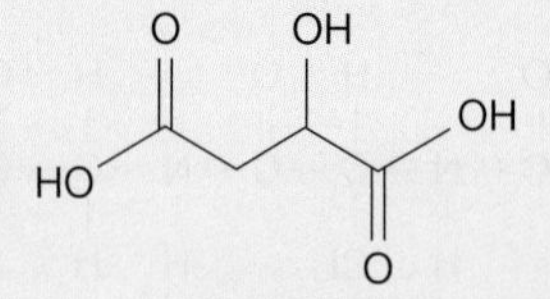 Malic acid | 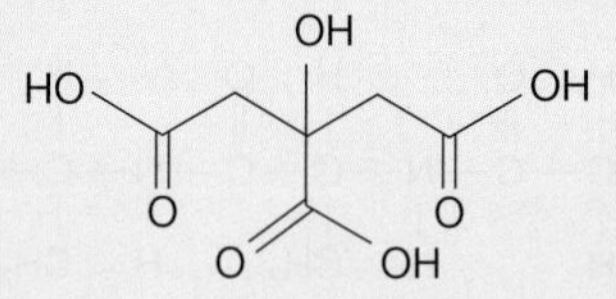 Citric acid | 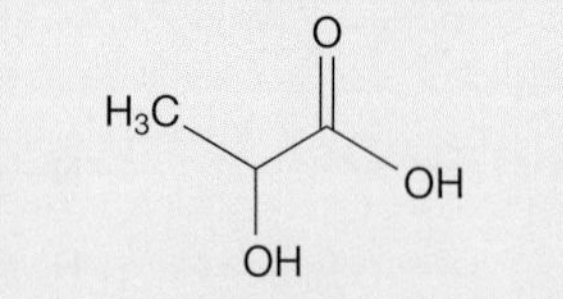 Lactic acid |
| 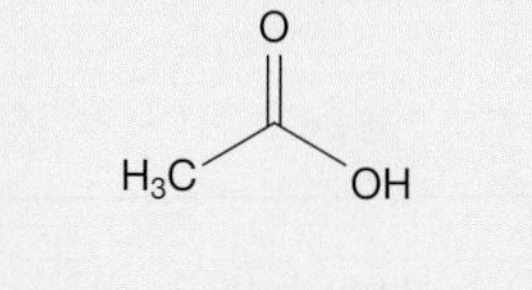 Ethanoic acid | 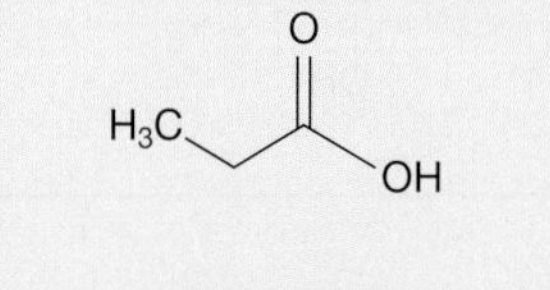 Propanoic acid | 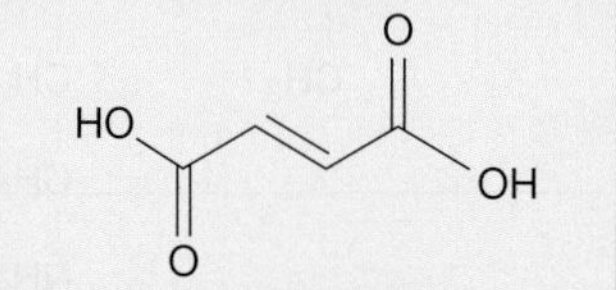 Fumaric acid | 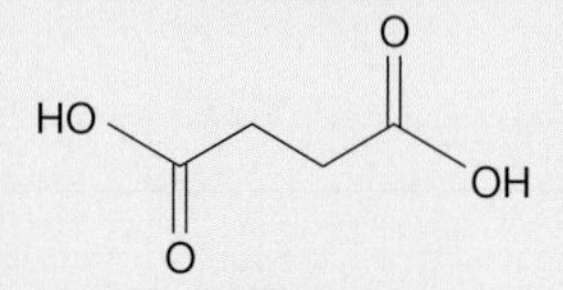 Succinic acid |

Vinegar producers send their spirit vinegar samples to laboratories for qualitative and quantitative analysis that uses a technique known as LC–MS (liquid chromatography–mass spectrometry).

LC–MS is a technique used to determine what is in a mixture when the components are similar. LC works in the same way as high-performance liquid chromatography (HPLC), except the mobile phase is not pumped through under high pressure. This slower flow rate of the mobile phase allows the mass spectrometer to function with high accuracy. The chromatograph separates the acids in the vinegar, while the mass spectrometer acts as the detector and identifies each component as it passes through.

One particular vinegar analysis used 20 μL vinegar samples injected onto a HPLC column at 50 °C. The mobile phase was a mixture of water and methanol with a flow rate of 1.0 mL/min. The following chromatogram, peak area data and mass spectra were produced.

FIGURE 1 Vinegar sample chromatogram

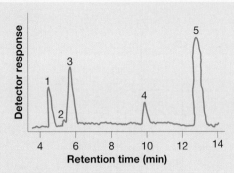

TABLE 1 Vinegar sample peak area data

Peak	Peak area ×10 000	Retention time (min)
1	5.50	4.7
2	0.68	5.6
3	9.95	5.9
4	3.65	9.7
5	24.40	13.0

FIGURE 2 Vinegar sample Peak 1 mass spectrum

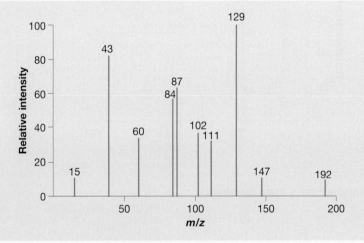

FIGURE 3 Vinegar sample Peak 2 mass spectrum

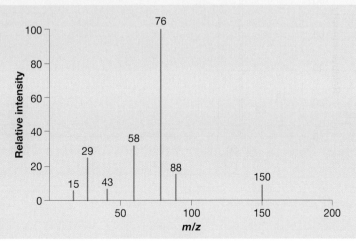

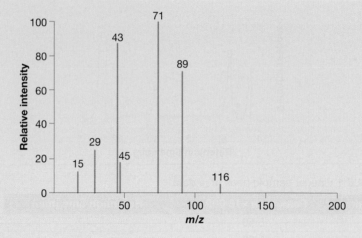

FIGURE 4 Vinegar sample Peak 3 mass spectrum

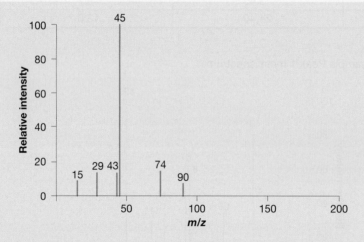

FIGURE 5 Vinegar sample Peak 4 mass spectrum

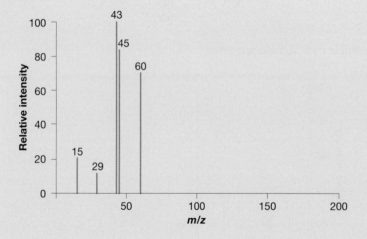

FIGURE 6 Vinegar sample Peak 5 mass spectrum

Analyse the data given to answer the following questions.

1. Using the principles of chromatography, explain how the mixture of organic acids was able to be separated in this chromatograph. **(5 marks)**

2. Explain how mass spectrometry can be used as a qualitative technique and how the peaks on a mass spectrum are produced. **(4 marks)**

3. How could you change the experimental conditions to:
   a. separate Peaks 2 and 3 on the chromatogram **(2 marks)**
   b. accurately determine the identity of Peak 4? **(2 marks)**

4. What information does the molecular ion peak provide? Does it confirm the identity of the acid present in each spectrum? Justify your answer. **(3 marks)**

5. Circle the molecular ion peak in figure 6. Write an equation to show the formation of this species. **(2 marks)**

6. Three spectra have a peak at 45 *m/z*.
   a. Suggest what could be responsible for the observed peak. Use an equation to show the formation of this. **(2 marks)**
   b. Does the equation you gave in part **a** help to identify the acids present? Justify your answer. **(1 mark)**

7. Consider the five mass spectra and the set of possible organic acids given in table 1.
   a. For each spectrum, identify one likely organic acid. **(4 marks)**
   b. Provide at least one piece of evidence from the spectrum to justify your answers to part **a**. **(4 marks)**

8. Suggest which of the organic acids could be distinguished from the others using infrared spectroscopy. Justify your answer. **(2 marks)**

9. Suggest which of the organic acids would produce a 1-H-NMR spectrum that has four peaks with a relative ratio of 1 : 1 : 1 : 3. Justify your answer. **(3 marks)**

10. Outline changes and additions you would make to this HPLC procedure to determine the percentage by volume, %(v/v), of ethanoic acid in this sample of spirit vinegar. Include any changes you would make to the sample. Comment on the data to be collected and show how you would use it to calculate the percentage by volume, %(v/v), of ethanoic acid in the spirit vinegar. **(5 marks)**

 Resources

Digital document   U4AOS2 School-assessed coursework (doc-39704)

12 Scientific investigations

online only

KEY KNOWLEDGE

In this area of study, you will adapt or design and then conduct a scientific investigation related to the production of energy and/or chemicals and/or the analysis or synthesis of organic compounds, which must include the generation of primary data. You will organise and interpret the data and reach a conclusion in response to your research question and present it as a scientific poster.

Investigation design

- Chemical concepts specific to the selected scientific investigation and their significance, including definitions of key terms
- Characteristics of the selected scientific methodology and method, and appropriateness of the use of independent, dependent and controlled variables in the selected scientific investigation
- Techniques of primary quantitative data generation relevant to the selected scientific investigation
- The accuracy, precision, repeatability, reproducibility, resolution and validity of measurements
- The health, safety and ethical guidelines relevant to the selected scientific investigation

Scientific evidence

- The nature of evidence that supports or refutes a hypothesis, model or theory
- Ways of organising, analysing and evaluating primary data to identify patterns and relationships, including sources of error and uncertainty
- Authentication of generated primary data through the use of a logbook
- Assumptions and limitations of investigation methodology and/or data generation and/or analysis methods

Science communication

- Conventions of science communication: scientific terminology and representations, symbols, formulas, standard abbreviations and units of measurement
- Conventions of scientific poster presentation, including succinct communication of the selected scientific investigation, and acknowledgements and references
- The key findings and implications of the selected scientific investigation

KEY SCIENCE SKILLS

- Develop aims and questions, formulate hypotheses and make predictions
- Plan and conduct investigations
- Comply with safety and ethical guidelines
- Generate, collate and record data
- Analyse and evaluate data and investigation methods
- Construct evidence-based arguments and draw conclusions
- Analyse, evaluate and communicate scientific ideas

Source: VCE Chemistry Study Design (2024–2027) extracts © VCAA; reproduced by permission.

This topic is available online at **www.jacplus.com.au**.

12.1 Scientific investigations

Hey students! Bring these pages to life online

Watch videos

Engage with interactivities

Answer questions and check results

Find all this and MORE in jacPLUS

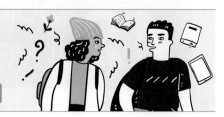

LEARNING SEQUENCE

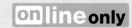

online only

UNIT 4 | AREA OF STUDY 3 REVIEW

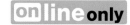

online only

AREA OF STUDY 3 How is scientific inquiry used to investigate the sustainable production of energy and/or materials?

OUTCOME 3

Design and conduct a scientific investigation related to the production of energy and/or chemicals and/or the analysis or synthesis of organic compounds, and present an aim, methodology and method, results, discussion and conclusion in a scientific poster.

The key science skills are a core component of the study of VCE Chemistry and apply across Units 1 to 4 in all Areas of Study.

- Develop aims and questions, formulate hypotheses and make predictions
- Plan and conduct investigations
- Comply with safety and ethical guidelines
- Generate, collate and record data
- Analyse and evaluate data and investigation methods
- Construct evidence-based arguments and draw conclusions
- Analyse, evaluate and communicate scientific ideas

Information:

- This Area of Study Review is a collation of past VCAA exam questions that focus on key science skills.
- You may use the VCE Chemistry Data Book and a scientific calculator for this task.

on Resources

Weblink VCE Chemistry Data Book

Answers

1 Carbon-based fuels

1.2 What are fuels?

Practice problem 1

Advantages:
- Renewable — crops can be grown in a short period of time
- Fewer pollutants emitted when combusted (e.g. NO_x, SO_2, C, CO)
- Close to carbon neutral as CO_2 is required to grow the crops.

Disadvantages:
- Less energy-dense/less energy per gram
- Engines need modification to use high ethanol concentrations
- Land used to grow crops for fuel could be used to grow food
- Crops require large amounts of water.

1.2 Exercise

1. a. A fuel is a substance that releases energy, commonly in the form of heat.

   b. A fossil fuel is a carbon-based energy source that was formed extremely slowly from the decaying remains of plants and animals that accumulated millions of years ago. They are non-renewable due to the amount of time required for their formation. Although biofuels are also carbon-based energy sources, their formation time is significantly shorter. They are formed primarily from plant matter over the course of months or years.

2. Any two of:
   - Higher energy content
   - Higher efficiency
   - Less carbon dioxide emissions
   - Less pollution, such as SO_2 and soot.

3. There are several brown coal mines in Victoria (including Loy Yang and Yallourn). The economics of using the lower quality fuel for on-site power generation outweighs the cost of transporting higher grade coal from interstate.

4. a. Renewable fuels can be replaced by natural processes within a relatively short period of time. Non-renewable fuels are used more quickly than they can be produced.

   b. Yes, all biofuels are renewable because the rate of production can exceed the rate of consumption, but advancements still need to be made in order to improve the efficiency of their manufacture.

5. The combustion of fossil fuels puts carbon back into the environment (as carbon dioxide) that has been locked underground for millions of years. By comparison, biofuels are considered to be carbon neutral because the carbon released into the atmosphere was absorbed relatively recently via photosynthesis.

6. a. Any three of:
   - Fossil fuels are a major contributing factor in the enhanced greenhouse effect, which is warming Earth. Changes in Earth's temperature can affect the climate and environment.
   - Fossil fuels contribute to the emission of unpleasant gases that cause smog in cities, leading to poor air quality and the health risks that come with it.
   - Fossil fuels are a non-renewable resource — if they continue to be used at the current rate they will not be available to future generations.
   - Methane from coal seam gas (CSG) extraction involves fracking, which can potentially pollute ground water.
   - Oxides of sulfur and nitrogen from burning fossil fuels can cause acid rain.
   - Biofuels are renewable and can be replenished at rates equal to or greater than consumption.

   b. Any three of:
   - Fossil fuels are cheap sources of energy.
   - Australia has a large supply of fossil fuels, and the extraction of them is a large and important economical industry.
   - Fossil fuels provide large amounts of energy relative to their mass, so are very reliable sources of energy.
   - Infrastructure for extraction, production and supply of fossil fuels is already in place.
   - Biofuels require land and fresh water that can be used to grow food, or is the existing habitat of plants and animals.
   - Many combustion engines for transport vehicles would require modification to use high percentages of bioethanol.
   - Stopping production of fossil fuels would impact the existing export markets and jobs of people in many countries.

1.2 Exam questions

1. Biogas is considered renewable because its production-and-use cycle is continuous so that it is constantly replenished, whereas coal seam gas is used at a faster rate than it can be replenished.

2. Any two of:

   Similarities — methane from both sources
   - Both produce atmospheric carbon dioxide through combustion.
   - Methane from both sources contains small amounts of nitrogen and sulfur; combustion of natural gas leads to the formation of acidic oxides, such as SO_x and NO_x.

   Differences — landfill versus natural gas
   - Methane from landfill can be produced renewably, whereas methane from natural gas releases stored carbon.
   - Methane from landfill is more carbon neutral, whereas methane from natural gas increases atmospheric CO_2 levels.
   - Obtaining methane from natural gas via fracking causes additional significant environmental damage, whereas when obtaining methane from a landfill the damage has already been done in the formation of the landfill.
   - Landfill gases contain less methane and release more CO_2 (for the same amount of energy generated), whereas natural gas contains more methane and releases comparatively less CO_2.

- Methane captured from landfill and used as a source of energy may have a positive impact as it is a more potent greenhouse gas than CO_2.
- CH_4 from landfill is more easily collected compared to fracking/sourcing methane from fossil fuels.

3. C

4. Valid discussion points include:

Carbon neutrality

For:
- CO_2 is absorbed/used by the crops/plants (used to produce the biodiesel).
- Biodiesel is more carbon neutral as it produces less new CO_2 than other fuels.

Against:
- Petroleum diesel (or other fuels) is used to produce biodiesel — a large amount of energy is required to produce biodiesel fuel from soy crops, as energy is needed for sowing, fertilising, harvesting, transporting and processing crops.
- Clearing land for crops by burning trees releases CO_2 and destroys habitats.
- There is less photosynthesis when land is cleared.
- Burning biomass directly emits a bit more carbon dioxide than fossil fuels for the same amount of generated energy.

Sustainability of using biodiesel as a fuel

For:
- Plants can be produced/grown in a short period of time.
- It can be made from waste vegetable oils, animal fats or restaurant grease.
- It releases fewer toxic chemicals if spilled or released to the environment/many by-products are biodegradable.
- Biodiesel produces less soot (particulate matter)/carbon monoxide/unburned hydrocarbons/sulfur dioxide.
- Crops (that produce oil) can be grown in many places.
- Second-generation technologies can be used to convert material such as crop residues into bioenergy and avoid competition for land.

Against:
- Some regions are not suitable for oil-producing crops.
- Crops and land are used that could be used for food/food production.
- The excess use of fertilisers can result in soil erosion and land pollution.
- Nitrous oxide released from fertilisers could have a greater (300 times more) global warming effect than carbon dioxide.
- The use of water to produce more crops can put pressure on local water resources.

Using biodiesel as a fuel for transport

For:
- It produces less toxic pollutants and greenhouse gases than petroleum diesel.
- It reduces dependence on foreign oil reserves as it is domestically produced.
- It can be used in any diesel engine with little or no modification to the engine or the fuel system.
- The lubricating property of the biodiesel may lengthen the lifetime of engines.

Against:
- It has a higher viscosity/it is not suitable for use in low temperatures.

- Biodiesel fuel is more expensive than petroleum diesel fuel.

5. A

Practice problem 2

a. Exothermic

b. 100 kJ mol^{-1}

Practice problem 3

$2C_2H_6(g) + 5O_2(g) \rightarrow 4CO(g) + 6H_2O(l)$

1.3 Thermochemical reactions

1.3 Exercise

1. a. Endothermic

  b. Endothermic

  c. Exothermic

  d. Exothermic

  e. Exothermic

  f. Endothermic

2. a. The energy released is 1.13×10^3 kJ.

  b. The energy absorbed is 282 kJ.

3. a. $CH_4(g) + 2O_2(g) \rightarrow CO_2(g) + 2H_2O(l)$
$\Delta H = -890 \text{ kJ mol}^{-1}$

  b. $C_3H_8(g) + 5O_2(g) \rightarrow 3CO_2(g) + 4H_2O(l)$
$\Delta H = -2220 \text{ kJ mol}^{-1}$

  c. $CH_3OH(l) + \frac{3}{2}O_2(g) \rightarrow CO_2(g) + 2H_2O(l)$
$\Delta H = -726 \text{ kJ mol}^{-1}$

$2CH_3OH(l) + 3O_2(g) \rightarrow 2CO_2(g) + 4H_2O(l)$
$\Delta H = -1452 \text{ kJ mol}^{-1}$

  d. $C_2H_5OH(l) + 3O_2(g) \rightarrow 2CO_2(g) + 3H_2O(l)$
$\Delta H = -1360 \text{ kJ mol}^{-1}$

4. a. i. $CH_4(g) + \frac{3}{2}O_2(g) \rightarrow CO(g) + 2H_2O(l)$

  ii. $CH_4(g) + O_2(g) \rightarrow C(s) + 2H_2O(l)$

  b. i. $C_3H_8(g) + \frac{7}{2}O_2(g) \rightarrow 3CO(g) + 4H_2O(l)$

  ii. $C_3H_8(g) + 2O_2(g) \rightarrow 3C(s) + 4H_2O(l)$

  c. i. $CH_3OH(l) + O_2(g) \rightarrow CO(g) + 2H_2O(l)$

  ii. $CH_3OH(l) + \frac{1}{2}O_2(g) \rightarrow C(s) + 2H_2O(l)$

  d. i. $C_2H_5OH(l) + 2O_2(g) \rightarrow 2CO(g) + 3H_2O(l)$

  ii. $C_2H_5OH(l) + O_2(g) \rightarrow 2C(s) + 3H_2O(l)$

5. a. Exothermic

  b. i. $\Delta H = -30 \text{ kJ mol}^{-1}$

  ii. $E_a = 20 \text{ kJ mol}^{-1}$

  iii. $E_a = 20 \text{ kJ mol}^{-1}$

  iv. Energy released = 50 kJ mol^{-1}

6. a. $\Delta H = -425 \text{ kJ mol}^{-1}$

  b. $E_a(\text{reverse}) = 575 \text{ kJ mol}^{-1}$

7. See figure at the bottom of the page*

1.3 Exam questions

1. A
2. C
3. Either of:
   - $2CH_4(g) + 3O_2(g) \rightarrow 2CO(g) + 4H_2O(l)$
   - $CH_4(g) + \dfrac{1}{2}O_2(g) \rightarrow CO(g) + 2H_2O(l)$
4. B
5. B

1.4 Fuel sources for plants and animals

Practice problem 4

Photosynthesis reduces the number of C–O bonds and increases the number of C–H bonds, producing glucose; energy is needed to form these bonds. Respiration breaks the C–H bonds in glucose and forms C–O bonds; these bonds require less energy, so energy is released.
The total energy of the bonds in the reactants is less than the energy of the bonds of the products in respiration, while the reverse is true for photosynthesis.

1.4 Exercise

1. Large molecules of starch, protein and fat must be converted into smaller molecules so that they can be easily absorbed into the bloodstream.
2. a. Carbohydrates → (hydrolysis) → disaccharides → (hydrolysis) → monosaccharides → glucose → (condensation polymerisation/respiration) → glycogen/energy
   b. Fats and oils → (hydrolysis) → glycerol and fatty acids → (condensation reaction) → triglycerides → stored energy

c. Proteins → (hydrolysis) → amino acids → (condensation reaction) → proteins → growth and repair
3. A hydrolysis reaction is the chemical breakdown of a compound due to reaction with water. For example, maltose is hydrolysed to two molecules of glucose.
4. Glucose is broken down to carbon dioxide via an oxidation reaction, which is exothermic.
5. The cramps could be due to the build-up of lactic acid because there is a limited supply of oxygen, resulting in anaerobic respiration. The reaction for the breakdown of glucose in the absence of oxygen producing lactic acid is: $C_6H_{12}O_6(aq) \rightarrow 2CH_3CH(OH)COOH(aq)$
6. a. $C_6H_{12}O_6(aq) \rightarrow 2CH_3CH_2OH(aq) + 2CO_2(g)$
   b. Fermentation
   c. The carbon dioxide produced causes bubbles in dough, which helps the bread to rise.

1.4 Exam questions

1. D
2. B
3. Triglycerides/fats/lipids
4. a. i. Glucose
      ii. $C_2H_{12}O_6 \rightarrow 2CH_3CH_2OH(aq) + 2CO_2(g)$
          Note: C_2H_6O and C_2H_5OH are also acceptable formulas for ethanol.
   b. i. Either of:
        - $3CH_3(CH_2)_{16}COOH + 1C_3H_8O_3$
        - $3CH_3(CH_2)_{16}COOH + \_C_3H_8O_3$ (i.e. no coefficient for glycerol)
      ii. Stearic acid or octadecanoic acid
      iii. $CH_3(CH_2)_{16}COOH + 26O_2 \rightarrow 18CO_2 + 18H_2O$
           Note: $C_{17}H_{35}COOH$ and $C_{18}H_{36}O_2$ are also acceptable formulas for stearic acid.
5. C

---

*7.

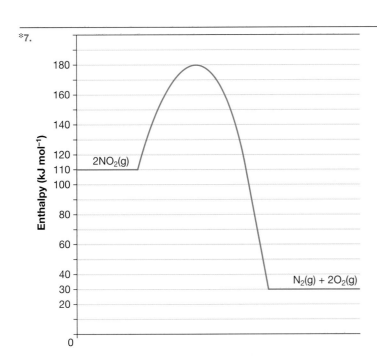

1.5 Review

1.5 Review questions

1. B

2. A

3. a. A fossil fuel is a fuel formed from the remains of living organisms, such as animals, trees and smaller plants that lived millions of years ago.

   b. Examples of fossil fuels include coal, petroleum and natural gas.

4. a. A biofuel is a fuel that is produced from renewable, organic resources, especially biomass (organic material).

   b. Examples of biofuels include bioethanol, biodiesel and biogas.

5. a. $CH_3CH_2OH(l) + 3O_2(g) \rightarrow 2CO_2(g) + 3H_2O(l)$
      $\Delta H = -1364 \text{ kJ mol}^{-1}$

   b. $CH_3CH_2OH(l) + 2O_2(g) \rightarrow 2CO(g) + 3H_2O(l)$
      $\Delta H = -1192 \text{ kJ mol}^{-1}$

   c. Carbon dioxide forms in plentiful air ($3O_2$) while carbon monoxide forms in limited air ($2O_2$).

6. $2C_2H_6(g) + 7O_2(g) \rightarrow 4CO_2(g) + 6H_2O(l)$
   $\Delta H = -3.12 \times 10^3 \text{ kJ mol}^{-1}$
   $C_2H_4(g) + 3O_2(g) \rightarrow 2CO_2(g) + 2H_2O(l)$
   $\Delta H = -1.41 \times 10^3 \text{ kJ mol}^{-1}$

7. a.

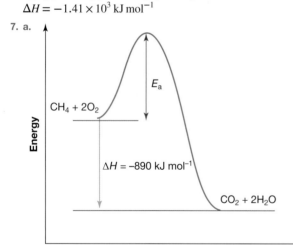

   b.

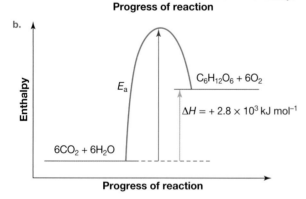

1.5 Exam questions

Section A — Multiple choice questions

1. C

2. D

3. B

4. D

5. D

6. A

7. B

8. D

9. D

10. D

Section B — Short answer questions

11. $m(CO_2) = 9.85 \text{ kg}$

12. For example:
    Canola oil (biodiesel) is a mixture, and hence does not have a specific formula or molar mass, so the number of mole in a sample cannot be determined.

13. Step 1, because the energy (enthalpy) of the products, $[(CH_3)_3C^+, Br^-]$, is higher than that of the reactant, $(CH_3)_3CBr$

14. $C_{12}H_{22}O_{11}(s) + 12O_2(g) \rightarrow 12CO_2(g) + 11H_2O(g \text{ or } l)$

15. Explanation may include the following (or similar):
    - Because it is produced from a biological source/biomass
    - Glucose is produced by plants.

2 Measuring changes in chemical reactions

2.2 Fuel calculations

Practice problem 1

a. $\Delta T = -52.1 \,°C$

b. %(ethanol) = 9.17%

Practice problem 2

$m(CO_2) = 30.9 \text{ g}$

Practice problem 3

$m(C_2H_5OH) = 9.3 \text{ g}$

Practice problem 4

a. $V(O_2) = 50 \text{ mL}$

b. $V(CO_2) = 25 \text{ mL}$

c. $V(H_2O) = 50 \text{ mL}$

Practice problem 5

$V(C_2H_5OH) = 1.76 \times 10^4 \text{ L}$

2.2 Exercise

1. a. $n(O_2) = 0.60 \text{ mol}$

   b. $n(Cl_2) = 1.0 \text{ mol}$

2. a. $V(H_2) = 32 \text{ L}$

   b. $V(CH_4) = 5.6 \text{ L}$

   c. $V(Ar) = 0.22 \text{ L}$

3. **a.** $m(Ne) = 13.4\,g$

   b. $m(SO_2) = 2.71\,g$

4. $m(CO) = 0.960\,kg$

5. $M(gas) = 83.3\,g\,mol^{-1}$

   The gas is most likely krypton (Kr).

6. **a.** Increase in mass $= 224\,g$

   b. % increase $= 175\%$

   c. If pressure and temperature are kept constant, $V(CH_4)$ consumed $= V(CO_2)$ produced; that is, there is no percentage change in volume. This contrasts significantly with the 175% increase when comparing masses.

7. $V(CO_2)_{SLC} = 42.5\,L$

8. **a.** $V(O_2) = 40\,mL$

   b. $V(reactants) = 60\,mL$

   c. $V(NO_2) = 40\,mL$

   d. There is a decrease in volume of 20 mL.

2.2 Exam questions

1. D

2. $P(CH_4) = 6.2 \times 10^2\,kPa$

3. D

4. $m(C_2H_5OH) = 673\,g$

5. **a.** Any one of the following:
   - $C_6H_{12}O_6(aq) \rightarrow 2C_2H_6O(aq) + 2CO_2(g/aq)$
   - $C_6H_{12}O_6(aq) \rightarrow 2CH_3CH_2OH(aq) + 2CO_2(g/aq)$
   - $C_6H_{12}O_6(aq) \rightarrow 2C_2H_5OH(aq) + 2CO_2(g/aq)$

   b. $m(C_2H_{12}O_6)$ required $= 1.54 \times 10^3\,g/1.54\,kg$

2.3 Energy from food and fuels

Practice problem 6

a. $\Delta H = -723\,kJ\,mol^{-1}$

b. $2CH_3OH(l) + 3O_2(g) \rightarrow 2H_2O(g) + CO_2(g)$

$\Delta H = -1.45 \times 10^3\,kJ\,mol^{-1}$

c. Amount of energy evolved $= 179\,MJ$

Practice problem 7

$V(CO_2) = 35.2\,L$

Practice problem 8

$x = 1.7 \times 10^3\,MJ$

2.3 Exercise

1. $m(CO_2) = 64.7\,g$

2. **a.** $m(CO_2) = 60.7\,g$

   b. $V(CO_2) = 34.2\,L\,MJ^{-1}$

   c. 1 MJ of energy produced by burning ethanol produces 64.5 g of carbon dioxide, while 1 MJ of energy released from burning methanol produces 60.7 g of carbon dioxide. Methanol has a slightly lower greenhouse gas emission per MJ of energy produced.

3. **a.** $m(CO_2) = 180\,g\,MJ^{-1}$

   b. $V(CO_2) = 101.3\,L\,MJ^{-1}$

4. $2CH_3OH(l) + 3O_2(g) \rightarrow 2CO_2(g) + 4H_2O(g)$

   $\Delta H = -1450\,kJ\,mol^{-1}$

5. Heat energy released $= 607\,kJ$

6. **a.** Heat evolved $= 4.97 \times 10^3\,kJ$

   b. $m(octane) = 103\,g$

   c. It is important to show symbols of state in thermochemical equations because the ΔH values would be different. For example, more energy would need to be removed if liquid water is produced instead of water vapour in a combustion reaction.

7. Energy from 15 g of tomato sauce $= 82\,kJ$

8. Number of minutes ≈ 63

2.3 Exam questions

1. D

2. Number of bananas $= 0.66$ (accept 0.65–0.67)

3. Energy $= 4.4 \times 10^2\,kJ$ (444 kJ)

4. **a.** $CH_4(g) + 2O_2(g) \rightarrow CO_2(g) + 2H_2O(l)$

   $\Delta H = -890\,kJ\,mol^{-1}$

   b. Energy $= 1.07 \times 10^4\,kJ$ (10 680) kJ or 10.7 MJ

5. D

2.4 Calorimetry

Practice problem 9

Energy per gram of almonds $= 7.55\,kJ\,g^{-1}$

Practice problem 10

$CF = 332\,J\,°C^{-1}$

Practice problem 11

$CF = 0.466\,kJ\,°C^{-1}$ ($466\,J\,°C^{-1}$)

Practice problem 12

$KNO_3(s) \rightarrow K^+(aq) + NO_3^-(aq)$ $\Delta H = +2.80\,kJ\,mol^{-1}$

Practice problem 13

$NaOH(s) \rightarrow Na^+(aq) + OH^-(aq)$ $\Delta H = -37.8\,kJ\,mol^{-1}$

Practice problem 14

Energy per gram $= 8.95\,kJ\,g^{-1}$

2.4 Exercise

1. Energy required $= 3.13 \times 10^2\,kJ$

2. **a.** Energy per gram $= 21\,kJ\,g^{-1}$

   b. To prevent heat escape and improve the accuracy of the experiment, use a metal shield to help contain the heat and direct it to the water.

   c. A corn chip is a mixture and so will not have a molar mass. The appropriate units are grams.

3. **a.** Calorimetry is a method used to determine the changes in energy of a system by measuring heat exchanges with the surroundings.

   b. When carrying out a reaction in a calorimeter not all of the heat produced goes into the water, due to imperfect insulation or absorbance by parts of the calorimeter. By calibrating the instrument, these factors are taken into account and the relationship of temperature to heat can be more accurately calculated.

c. There would be no difference to the calculations because 1 °C is the same as 1K and it is the difference of two temperatures being used.

d. If 50.0 mL was used in a calibration instead of 100.0 mL, the change in temperature of the water would be greater. The calibration factor would be lower (this can be determined by using the equation $CF = \dfrac{VIt}{\Delta T_c}$).

4. $\Delta T_r = 0.492 \,°C$

5. a. The student would have used an electric heater, connected into an electric circuit, to pass a known quantity of energy through 100.0 mL of water in the same calorimeter used for the experiment. Then, by calculating the change in temperature of the water in the calorimeter, the calibration factor (CF) could be calculated from the relationship

$CF = \dfrac{VIt}{\Delta T}$, where ΔT = change in temperature of the 100 mL of water.

b. $HCl(aq) + NaOH(aq) \rightarrow NaCl(aq) + H_2O(l)$
$\Delta H = -54.9 \,kJ\,mol^{-1}$

c. A higher concentration of base than acid was used to ensure that all of the acid had reacted.

d. There would be no change because the amount in moles of the limiting reactant did not change.

e. Random errors might involve measurement of the solution volume and temperature, and not stirring the reaction mixture during the reaction. Random errors can be minimised by repeating experiments. Another error is heat losses due to insufficient insulation.

6. a.

Property	Ethanol	Candle wax	Butane
Mass of 'burner' before heating (g)	23.77	32.72	43.94
Mass of 'burner' after heating (g)	22.54	32.50	43.71
Mass of fuel used (g)	1.23	0.22	0.23
Mass of water (g)	200	200	200
Initial temperature of water (°C)	20.0	20.0	20.0
Highest temperature of water (°C)	35.0	30.0	29.0
Temperature rise (°C)	15.0	10.0	9.0
Molar mass (g mol^{-1})	46.0	282	58

b. Energy per g (ethanol) = $10.2 \,kJ\,g^{-1}$
Energy per g (candle wax) = $38 \,kJ\,g^{-1}$
Energy per g (butane) = $32.6\,(33)\,kJ\,g^{-1}$

c. Ethanol: $\Delta H = -469 \,kJ\,mol^{-1}$
Candle wax: $\Delta H = -1.1 \times 10^4 \,kJ\,mol^{-1}$
Butane: $\Delta H = -1.9 \times 10^3 \,kJ\,mol^{-1}$

d. i. Exothermic

ii. Percentage accuracy = 34.5%

iii. Sources of error: heat loss from the candle flame to the surrounding air; heat loss from the copper can; heat loss from the water in the can; incomplete combustion of the fuel.

How to minimise errors: totally surround the flame with a heat-reflecting barrier; cover the side of the can with insulating material; place a lid on the can.

2.4 Exam questions

1. B

2. a. $T_{max} = 28.9 \,°C\,(or\,302\,K)$

b. For example, any one of:
- loss of heat/energy to the atmosphere
- heat/energy loss in the combustion chamber
- heat/energy loss since the tank material also is heated
- heat/energy loss from the piping
- faulty insulation.

3. C

4. % energy lost = 33.3%

5. B

2.5 Review

2.5 Review questions

1. a. $n(O_2) = 0.060 \,mol$

b. $n(H_2) = 0.103 \,mol$

c. $n(N_2) = 0.0101 \,mol$

2. a. $V(H_2) = 37.9 \,L$

b. $V(CH_4) = 21.1 \,L$

c. $V(N_2) = 1.0 \times 10^8 \,L$

3. a. $m(H_2O) = 9.9 \,g$

b. $V(CO_2) = 11 \,L$

c. $V(C_4H_{10}O) = 0.092 \,mL$

4. Energy released = $1.57 \times 10^3 \,kJ\,kg^{-1}$

5. a. $m(propane) = 19.8 \,g$

b. $m(CO_2) = 59.4 \,g\,per\,MJ$

c. $m(octane) = 20.9 \,g$

d. $m(CO_2) = 64.4 \,g$

e. Net reduction of CO_2 emissions = $5.0 \,g\,MJ^{-1}$

f. $m(butane) = 20.2 \,g$
$m(CO_2) = 61.2 \,g\,per\,MJ$
Net reduction of CO_2 emissions = $3.2 \,g\,MJ^{-1}$

g. $V(propane) = 1.3 \,L$

h. According to the calculations, LPG has a net reduction of CO_2 emission of $5.0 \,g\,MJ^{-1}$, whereas petrol has a reduction of only $3.0 \,g\,MJ^{-1}$. Therefore, LPG is the better fuel on this basis.

6. a. Heat produced per gram = $18.4 \,kJ$

b. $\Delta H(CH_3CH_2OH) = -846 \,kJ\,mol^{-1}$

c. % accuracy = 62.0%

d. Sources of error include heat loss from the flame, heat absorbed by the can, heat loss from the water in the can, and incomplete combustion of the ethanol. An improved design could include a barrier surrounding the apparatus to minimise radiated heat loss from the flame, using insulating material surrounding the can and placing a lid on the can.

7. Heat is progressively lost at each energy transformation stage.

8. a. The heat is measured in $kJ\ kg^{-1}$ rather than $kJ\ mol^{-1}$ because kerosene is measured in industry in kilograms rather than moles.

   b. Number of cups = 5.52

9. Energy = $9.4 \times 10^2\ kJ$

10. $CF = 587\ J\,^\circ C^{-1}$

2.5 Exam questions

Section A — Multiple choice questions

1. D
2. C
3. B
4. D
5. B
6. B
7. C
8. C
9. C
10. B

Section B — Short answer questions

11. a. Calibration method

    b. For example, any one of:
    - error associated with calibration
    - error associated with inaccuracy of voltmeter/ammeter/power supply (voltage)
    - error with electrical connections.

    c. Possible limitations that affect accuracy but not reliability include:
    - temperature not stabilised before KNO_3 added
    - no indication of how long to record temperature
    - purity of KNO_3
    - not all KNO_3 may have been transferred
    - limitations that affect reliability
    - only one set of data collected
    - the exercise was not repeated.

    Reliability would be improved by repeating the exercise or sharing data.

    d. For example, any one of the following differences:
    - Student A (25 °C) reached a higher maximum temperature than Student B (24 °C).
    - Student A reached maximum temperature later (450 s) than Student B (390 s).
    - Student A heated water for longer than Student B.

    For example, any one of the following variations:
    - Student A turned the power/current/voltage off later than Student B, so reached higher temperature.
    - Student B turned the power/current/voltage off earlier than Student A, so reached lower temperature.
    - Different calorimeters were used so there were different levels of energy absorption by components.

    e. $CF = 6.5 \times 10^2\ J\,^\circ C^{-1}$

12. a. $m(fat) = 15\ g$

    b. Energy in 15 g of fat = $5.6 \times 10^2\ kJ$

    c. Mass of protein = 40.6 g

    d. Remaining mass = 44.4 g, which is probably mainly water.

13. $V(H_2) = 6.2$ (L)

14. a. $CH_3(CH_2)_{14}COOH + 23O_2 \rightarrow 16CO_2 + 16H_2O$

    b. Energy from 20.00 g = $7.80 \times 10^2\ kJ$

15. a. Energy = $2.5 \times 10^2\ kJ$ (247 kJ)

    b. i. Mass of bread = 0.5 g

    ii. % efficiency = $4 \times 10^1\%$

3 Primary galvanic cells and fuel cells as sources of energy

3.2 Redox reactions

Practice problem 1

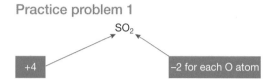

Practice problem 2

This is a redox reaction. WO_3 is the oxidising agent and H_2 is the reducing agent.

Practice problem 3

Na(s)	+	H$_2$O(l)	→	NaOH(aq)	+	H$_2$(g)
Na(s) gets oxidised. Acts as a reducing agent Forms a conjugate oxidising agent		H$_2$O(l) gets reduced. Acts as an oxidising agent Forms a conjugate reducing agent		Conjugate oxidising agent		Conjugate reducing agent

3.2 Exercise

1. a. $H = +1,\ Br = -1$ b. $Na = +1,\ O = -2$
   c. $C = -4,\ H = +1$ d. $Na = +1,\ Cl = +5,\ O = -2$
   e. $Al = +3,\ O = -2$ f. $H = +1,\ P = +5,\ O = -2$

2. a. $N = -3,\ H = +1$ b. $Mn = +7,\ O = -2$
   c. $H = +1,\ S = -2$ d. $V = +4,\ O = -2$
   e. $I = +5,\ O = -2$ f. $P = +5,\ O = -2$

3. a. This is a redox reaction. Fe is oxidised and Cl_2 is reduced.

   b. This is not a redox reaction as there has been no change in oxidation numbers.

   c. This is a redox reaction. N (in NO) is oxidised and O_2 is reduced.

   d. This is not a redox reaction as there has been no change in oxidation numbers.

   e. This not a redox reaction as there has been no change in oxidation numbers.

   f. This is not a redox reaction as there has been no change in oxidation numbers.

   g. This is a redox reaction. C (in CO) is oxidised and O_2 is reduced.

h. This is a redox reaction. H_2 is oxidised and C (in C_2H_4) is reduced.

4. a. $Mg(s) + H_2O(l) \rightarrow MgO(s) + 2H^+(aq) + 2e^-$
   b. $Mg(s) + 2OH^-(aq) \rightarrow MgO(s) + H_2O(l) + 2e^-$
5. a. O_2
   b. $CH_3OH(aq) + H_2O(l) \rightarrow CO_2(g) + 6H^+(aq) + 6e^-$

3.2 Exam questions

1. C
2. D
3. A
4. A
5. C

3.3 Galvanic cells and the electrochemical series

Practice problem 4

a., d., e. See figure at the bottom of the page*
b. Oxidation: $Zn(s) \rightarrow Zn^{2+}(aq) + 2e^-$
   Reduction: $Ag^+(aq) + e^- \rightarrow Ag(s)$
c. $Zn(s) + 2Ag^+(aq) \rightarrow Zn^{2+}(aq) + 2Ag(s)$

Practice problem 5

a. i. MnO_4^- **ii.** Al   **iii.** Al^{3+} **iv.** Mn^{2+}

b. $Cl_2(g) + 2e^- \rightleftharpoons 2Cl^-(aq)$

$I_2(s) + 2e^- \rightleftharpoons 2I^-(aq)$

$Al^{3+}(aq) + 3e^- \rightleftharpoons Al(s)$

$MnO_4^- + 8H^+(aq) + 5e^- \rightleftharpoons Mn^{2+}(aq) + 4H_2O(l)$

$Pb^{2+}(aq) + 2e^- \rightleftharpoons Pb(s)$

Practice problem 6

In this cell, $Fe^{3+}(aq)$ is reduced and $Ni(s)$ is oxidised. Because reduction occurs in the $Fe^{3+}(aq)/Fe^{2+}(aq)$ half-cell, the electrode in this half-cell acts as the cathode. The nickel electrode in the $Ni^{2+}(aq)/Ni(s)$ half-cell acts as the anode, and undergoes oxidation.

Overall equation: $Ni(s) + 2Fe^{3+}(aq) \rightarrow Ni^{2+}(aq) + 2Fe^{2+}(aq)$

$E^0_{cell} = +1.02$ V

3.3 Exercise

1. A salt bridge or porous barrier is needed to connect the two half-cells because electrons cannot flow in the external circuit unless ions can move in the internal circuit. The circuit must be complete and the half-cells must be separated while still allowing the flow of charge through the salt bridge or porous barrier.

2. **a.** $Zn(s) + 2Ag^+(aq) \rightarrow Zn^{2+}(aq) + 2Ag(s)$
   b. See figure at the bottom of the page**

---

*a., d., e.

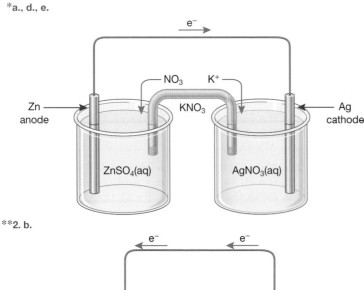

**2. b.

c. Cathode: $Ag^+(aq) + e^- \rightarrow Ag(s)$

   Anode: $Zn(s) \rightarrow Zn^{2+}(aq) + 2e^-$

d. Not all of the chemical energy is transformed into electrical energy. Some of the chemical energy stored in the reactants is released as heat energy.

3. a.–e. See figure at the bottom of the page*

   f. Oxidation (at anode): $Mg(s) \rightarrow Mg^{2+}(aq) + 2e^-$

      Reduction (at cathode): $Zn^{2+}(aq) + 2e^- \rightarrow Zn(s)$

   g. Oxidation (at anode): $Mg(s) \rightarrow Mg^{2+}(aq) + 2e^-$

      Reduction (at cathode): $Zn^{2+}(aq) + 2e^- \rightarrow Zn(s)$

   h. Overall reaction: $Mg(s) + Zn^{2+}(aq) \rightarrow Mg^{2+}(aq) + Zn(s)$

   i. Zn^{2+} is the oxidising agent; Mg is the reducing agent.

4. a. $Cl_2(g) + Ni(s) \rightarrow 2Cl^-(aq) + Ni^{2+}(aq)$

   b. $2Al^{3+}(aq) + 3Mg(s) \rightarrow 2Al(s) + 3Mg^{2+}(aq)$

   c. $2MnO_4^- + 6H^+(aq) + 5ClO_3^-(aq)$

      $\rightarrow 2Mn^{2+}(aq) + 3H_2O(l) + 5ClO_4^-(aq)$

   d. $2MnO_4^- + 16H^+(aq) + 5Fe(s)$

      $\rightarrow 2Mn^{2+}(aq) + 8H_2O(l) + 5Fe^{2+}(aq)$

5. Predicted spontaneous redox reactions may not occur because the rate of reaction may be too slow to be observed initially or non-standard conditions may have been used, making the reaction less favourable/observable.

3.3 Exam questions

1. C
2. C
3. C
4. C
5. C

3.4 Energy from primary cells and fuel cells

3.4 Exercise

1. • Reactants are supplied from an external source in fuel cells but are stored in a primary cell.
   • Excess reactants and products of reactions in a fuel cell are removed.

• Fuel cells don't go flat. They produce electricity as long as reactants are supplied.

2. a. The term 'renewable' is used to describe an energy source that can be produced at a rate equal to or greater than the rate of consumption. These are not fossil fuels. A feedstock is simply a chemical used in a large-scale production/an ongoing chemical reaction.

   b. • Electrolysis of water using electricity generated from renewable means, such as solar and wind
      • Catalytic conversion of alcohol produced from fermentation of biomass
      • Thermochemical splitting of water

   c. Storage of hydrogen is difficult as high pressures and low temperatures are needed to store it as a liquid. Producing hydrogen from the thermal or electrical decomposition of water is expensive, and producing it from alcohol is a challenge in terms of the yield and efficient use of biomass being low.

3. a.

	Equation
Anode (−) (**oxidation**)	$H_2(g) + 2OH^-(aq) \rightarrow 2H_2O(l) + 2e^-$ or $2H_2(g) + 4OH^-(aq) \rightarrow 4H_2O(l) + 4e^-$
Cathode (+) (**reduction**)	$\frac{1}{2}O_2(g) + H_2O(l) + 2e^- \rightarrow 2OH^-(aq)$ or $O_2(g) + 2H_2O(l) + 4e^- \rightarrow 4OH^-(aq)$
Overall	$H_2(g) + \frac{1}{2}O_2(g) \rightarrow H_2O(l)$ or $2H_2(g) + O_2(g) \rightarrow 2H_2O(l)$

b.

	Equation
Anode (−) (**oxidation**)	$CH_4(g) + 2H_2O(l)$ $\rightarrow CO_2(g) + 8H^+(aq) + 8e^-$
Cathode (+) (**reduction**)	$O_2(g) + 4H^+(aq) + 4e^- \rightarrow 2H_2O(l)$
Overall	$CH_4(g) + 2O_2(g)$ $\rightarrow 2H_2O(l) + CO_2(g)$

*3. a.–e.

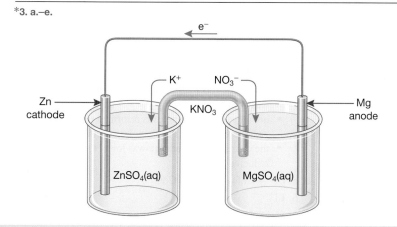

	Equation
Anode (−) (oxidation)	$C_2H_5OH(l) + 3H_2O(l)$ $\rightarrow 2CO_2(g) + 12H^+(aq) + 12e^-$
Cathode (+) (reduction)	$O_2(g) + 4H^-(aq) + 4e^- \rightarrow 2H_2O(l)$
Overall	$C_2H_5OH(l) + 3O_2(g)$ $\rightarrow 3H_2O(l) + 2CO_2(g)$

4. • As a compressed gas in high-pressure tanks
   • As a liquid in tanks (stored at −253 °C)
   • As a solid by either absorbing or reacting with metals or chemical compounds, or storing in an alternative solid hydride compound

5. Advantages include:
   • water is the only by-product (non-polluting and no carbon emissions)
   • greater efficiency.
   Disadvantages include:
   • hydrogen is difficult to store
   • hydrogen is difficult to initially manufacture and has a limited life cycle of H_2.

3.4 Exam questions

1. a. Fuel cells are a type of galvanic cell or a cell that converts chemical energy into electrical energy. They require a constant supply of reactants to be added.
   b. i. Any one of the following is acceptable:
      • Hydrogen ions
      • H^+
      • Protons
      • Hydronium ions
      • H_3O^+.
      ii. The arrow must clearly be shown pointing to the right-hand side and inside the box as directed.
   c. Either of the following
      • $C_2H_6O(l) + 3H_2O(l)$
         $\rightarrow 2CO_2(aq) + 12H^{1}(aq) + 12e^-$
      • $CH_3CH_2OH(l) + 3H_2O(l)$
         $\rightarrow 2CO_2(aq) + 12H^+(aq) + 12e^-$
      Note: States are not required for this question.
   d. Energy = 30 kJ (29.6 kJ)
   e. The electrodes should be highly porous or have a high surface area.
      The electrodes could either be a catalyst or incorporate a catalyst.
   f. Any one of the following:
      • Source the ethanol from a renewable/sustainable source, so the CO_2 released is balanced out.
      • There is a net reduction in the greenhouse gas CO_2 emission when produced in a biologically sourced ethanol fuel cell operating at 100% efficiency.
      • Obtain electrodes from renewable/sustainable sources and reuse electrode materials.
      • Manage end-of-life waste streams, in particular acidic electrolyte.
      • Capture CO_2 gas through CCS (carbon capture sequestration).

2. B
3. a. Energy released $= 3.4 \times 10^3$ kJ (3410 kJ) (3411 kJ)
   b. i. Anode
      ii. $CH_3CH_2OH(l) + 3H_2O(l)$
         $\rightarrow 2CO_2(g) + 12H^+(aq) + 12e^-$
      iii. Any two advantages similar to the following:
         • Fuel cells are inherently more efficient than internal combustion engines and so produce less CO_2.
         • Ethanol can be produced renewably.
         • Bioethanol is closer to carbon neutral than fossil fuels are.
         • Ethanol reacts more cleanly than fossil fuels (less carbon monoxide, other gases and particulates).
         • Fuel cells produce less total greenhouse gases/pollutants.
         • Fuel cells are quieter.

4. C
5. A

3.5 Calculations involved in producing electricity from galvanic cells and fuel cells

Practice problem 7
$m(Zn) = 5.4$ g

Practice problem 8
$n(W) = 0.027$ mol

Practice problem 9
$m(Zn) = 1.24$ g

Practice problem 10
$m(H_2) = 0.27$ g

Practice problem 11
$t = 4.19$ hours

3.5 Exercise
1. $n(Cr) = 0.0073$ mol
2. 0 g
3. $t = 27$ hours
4. a. $H_2(g) + O^{2-}(s) \rightarrow H_2O(l) + 2e^-$
   b. $O_2(g) + 4e^- \rightarrow 2O^{2-}(s)$
   c. $Q = 1.30 \times 10^6$ C
   d. $m(CH_3OH) = 69.4$ g
   e. The actual mass used was greater than the theoretical mass required to deliver the current for the 24.0-hour period. Possible reasons include:
      • it was not 100% efficient — chemical energy was converted to heat
      • not all of the methanol was reformed into hydrogen gas; therefore, more methanol was required
      • the operating temperature was lower than ideal, reducing efficiency.
5. a. $Li^+ + CoO_2 + e^- \rightarrow LiCoO_2$
   b. Li^+
   c. $m = 8.4$ g

1. A

2. C

3. D

4. 1.8 hours

5. $m(Zn)$ reacting = 0.029 g

3.6 Review

3.6 Review questions

1. a. H^+ has an initial oxidation number of +1 and a final oxidation number of 0. Therefore, HCl has been reduced and Zn has been oxidised.

  b. O_2 has an initial oxidation number of 0 and a final oxidation number of −2. Therefore, O_2 has been reduced and NO has been oxidised.

  c. Mg has an initial oxidation number of 0 and a final oxidation number of +2. Therefore, Mg has been oxidised and H_2SO_4 has been reduced.

  d. Al has an initial oxidation number of 0 and a final oxidation number of +3. Therefore, Al has been oxidised and Cl_2 has been reduced.

2. a. $2Br^- \rightarrow Br_2 + 2e^-$

  $SO_4^{2-} + 4H^+ + 2e^- \rightarrow H_2SO_3 + H_2O$

  Therefore:

  $2Br^-(aq) + SO_4^{2-} + 4H^+(aq)$
    $\rightarrow Br_2(l) + SO_2(g) + 2H_2O(l)$

  b. $I_2 + 2e^- \rightarrow 2I^-$

  $H_2S \rightarrow 2H^+ + S + 2e^-$

  Therefore:

  $I_2(s) + H_2S(g) \rightarrow 2I^-(aq) + S(s) + 2H^+(aq)$

  c. $Cu \rightarrow Cu^{2+} + 2e^-$ [1]

  $2NO_3^- + 8H^+ + 5e^- \rightarrow 2NO + 4H_2O$ [2]

  [1] × 5

Therefore:

$5Cu(s) + 2NO_3^-(aq) + 8H^+(aq)$
  $\rightarrow 5Cu^{2+}(aq) + 2NO(g) + 4H_2O(l)$

  d. $Cu \rightarrow Cu^{2+} + 2e^-$

  $2NO_3^- + 4H^+ + 2e^- \rightarrow 2NO_2 + 2H_2O$

  Therefore:

  $Cu(s) + 2NO_3^-(aq) + 4H^+(aq)$
    $\rightarrow Cu^{2+}(aq) + 2NO_2(g) + 2H_2O(l)$

3. a. $Zn(s) \rightarrow Zn^{2+}(aq) + 2e^-$

  b. $Fe^{2+}(aq) + 2e^- \rightarrow Fe(s)$

  c. $Zn(s) + Fe^{2+}(aq) \rightarrow Zn^{2+}(aq) + Fe(s)$

4. a. Yes

  b. No

  c. Yes

  d. No

5. a. See figure at the bottom of the page*
  Reduction: $Cu^{2+}(aq) + 2e^- \rightarrow Cu(s)$

  Oxidation: $Al(s) \rightarrow Al^{3+}(aq) + 3e^-$

  Overall: $3Cu^{2+}(aq) + 2Al(s) \rightarrow 3Cu(s) + 2Al^{3+}(aq)$

  b. The intensity of the blue colour would slowly fade to become colourless as the Cu^{2+} ions react at the cathode.

  c. There would be no electron flow — the transformation from chemical energy to electrical energy would stop.

6. a. A fuel cell is an electrochemical device that converts chemical energy into electricity without combustion as an intermediary step — that is, a fuel cell converts chemical energy directly into electrical energy rather than into heat energy. There is also a continual supply of reactants.

  b. In fuel cells, operating and maintenance costs are lower because they can be used continually rather than being discarded when depleted. Fuel cells often need the supply of only one reagent, as oxygen can be readily sourced from the atmosphere.

***5. a.**

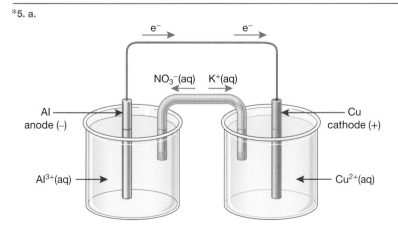

7. a. $CH_4(g) + 2H_2O(l) \rightarrow CO_2(g) + 8H^+(aq) + 8e^-$

    b. $O_2(g) + 4H^+(aq) + 4e^- \rightarrow 2H_2O(l)$

    c. See figure at the bottom of the page*

9. a. $Q = 289\,C$

    b. $n(e^-) = 0.00300$

    c. $n(Cr) = 0.00100\,mol$

    d. The charge on the chromium ion is 3+.

10. a. 2.0 faradays

    b. 5.0 faradays

    c. 1.7 faradays

    d. 0.23 faradays

    e. 0.014 faradays

3.6 Exam questions

Section A — Multiple choice questions

    1. B

    2. B

    3. B

    4. B

    5. A

    6. C

    7. C

    8. D

    9. C

    10. C

Section B — Short answer questions

11. a. Fe(s)/Fe/iron

b. i. R = Fe(s)/Fe/iron

    S = Ni(s)/Ni/nickel/Pt/C or any other inert electrode, such as Sn, Pb, Cu, Ag or Au (i.e. any metal higher than Ni on the electrochemical series)

  ii. T = $Ni(NO_3)_2$/$NiSO_4$/$NiCl_2$ or any other Ni^{2+} solution

    Note: Not Ni^{2+} alone or Ni^{2+}(aq).

c.

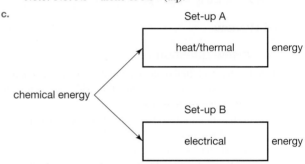

d. Anode

e. $O_2 + 4e^- \rightarrow 2O^{2-}$

f. SOFCs involve direct conversion of chemical to electrical energy, so are more efficient than a conventional power station.

Both reactions will produce the greenhouse gases carbon dioxide and water vapour.

An SOFC is more efficient and will hence produce less CO_2/greenhouse gas(es) for a given amount of energy produced.

Note: Other suitable statements of comparison are also acceptable.

---

*7. c.

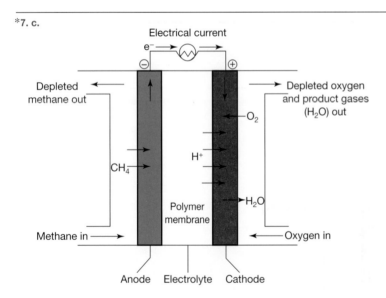

12. a. i. See figure at the bottom of the page*

ii. $O_2(g) + 4H^+(aq) + 4e^- \rightarrow 2H_2O(g)$

b. $m(NiOOH) = 3.9\,g$

13. a. $Zn(s) + H_2O(l) \rightarrow ZnO(s) + H_2(g)$

b. $H_2(g) + 2OH^-(aq) \rightarrow 2H_2O(l) + 2e^-$

c. The arrow should be pointing to the left.

d. The amount of H_2 produced in the generator cell is equal to the amount of H_2 consumed in the fuel cell.
The half-equations show that the $n(e^-)$ released (2 per mol H_2 produced) in the generator cell is equal to the $n(e^-)$ used (2 per mol H_2 consumed) in the fuel cell. Two electrons are transferred in the production of each mole of hydrogen in the Zn–H_2 generator cell and two electrons are transferred in the reaction of each mole of hydrogen consumed in the hydrogen fuel cell.

e. Oxygen is a stronger oxidising agent than H_2O, so would be reduced at the cathode.
$O_2(g) + 2H_2O(l) + 4e^- \rightarrow 4OH^-(aq)$ would occur at the cathode and no H_2 would be produced.
The reduction of O_2 is higher on the electrochemical series, so water will not be reduced and hydrogen will not be produced.

f. The Zn–H_2 generator cell converts chemical energy of zinc to electrical energy and chemical energy in the form of hydrogen.
The H_2 fuel cell converts chemical energy of hydrogen to electrical energy

14. a. $C_3H_8(g) + 6H_2O(l) \rightarrow 3CO_2(g) + 20H^+(aq) + 20e^-$

b. $O_2(g) + 4H^+(aq) + 4e^- \rightarrow 2H_2O(l)$

c. Using propane as a fuel in a fuel cell rather than combusting propane results in more efficient energy transformation and less CO_2 is evolved per unit of transformed energy.

d. The electrodes of a fuel cell must be porous, and they must catalyse the half-cell reactions. Electrodes in primary cells are not catalysts, as the areas of oxidation and reduction are separated.

15. a. Cars powered by hydrogen would not be truly carbon neutral because the hydrogen gas to power the cars would need to be made in power stations (and potentially by burning fossil fuels), resulting in carbon dioxide formation/global warming/greenhouse gas emission/acid rain production; and because manufacturing the hydrogen cars produces polluting gas.

b. An advantage of using hydrogen instead of petrol is that combusting hydrogen will not produce carbon dioxide or sulfur dioxide pollutants.

c. Advantages of storing hydrogen as a solid hydride rather than as a gas include:
- Hydrogen gas is very flammable, and storing hydrogen as a solid hydride (MgH_2) is safer because it is not flammable.
- A solid will occupy a much smaller volume than a gas, saving weight and therefore increasing overall efficiency.

Unit 3 | Area of Study 1 review
Section A — Multiple choice questions

1. C
2. D
3. A
4. A
5. C
6. C
7. A
8. C

---

*12. a. i.

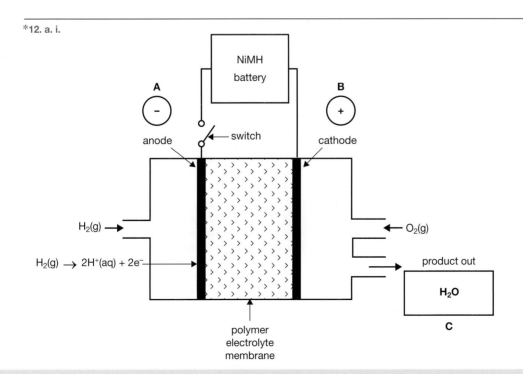

9. C
10. A
11. D
12. B
13. C
14. A
15. A
16. B
17. C
18. C
19. C
20. D

Section B — Short answer questions

21. a. V(ethanol) = 95.7 mL

    b. $C_6H_{12}O_6(aq) \rightarrow 2CH_3CH_2OH(l) + 2CO_2(g)$

22. a. m(ethanol) = 1.17 g

    b. The actual value is expected to be higher because there will be significant heat loss to the surroundings.

    c. The reliability is low because the experiment was not repeated.

23. a. Energy = 7.8×10^4 J

    b. Energy content = 397 kJ

    c. 22 slices

24. a. Anode: $Zn(s) \rightarrow Zn^{2+}(aq) + 2e^-$
    Cathode: $Cu^{2+}(aq) + 2e^- \rightarrow Cu(s)$

    b. Maximum voltage = 1.10 V

    c. m(Cu) = 0.95 g (increase)

    d. Any reasonable answer is acceptable. For example:
    - The zinc rod will need replacing periodically.
    - The copper can may need cleaning periodically due to copper build-up.
    - The design is cumbersome (it is easy to spill liquids when moving).

25. a. The Ni/Ni^{2+} half-cell

    b. Positive

    c. It will decrease as the Cd gets oxidised.

26. a. $Zn(s) + 2OH^-(aq) \rightarrow Zn(OH)_2(s) + 2e^-$

    b. Graphite (carbon, C) or platinum (Pt)

    c. $x = -1.0$ V

4 Rates of chemical reactions

4.2 Factors affecting the rate of a chemical reaction

Practice problem 1

Set A

4.2 Exercise

1. Reacting particles must collide, the collisions must have correct orientation, and collisions must possess the minimum amount of energy required to break bonds.

2. a. Collisions occur at random. Some of these will slow particles down, while others will speed particles up. This results in a range of velocities and energies.

   b. There will always be some particles in the tail of the Maxwell–Boltzmann distribution curve, at or above the activation energy of the reaction, which possess enough energy to break bonds upon collision. However, this number may sometimes be so small that it is effectively negligible.

3. Increase temperature, increase concentration, increase pressure, increase surface area, use of a catalyst

4. Fireworks need their powder to react quickly with oxygen. A large surface area (powder) facilitates this.

5. The rate will increase due to a higher frequency of collisions caused by the particles being closer together.

6. a. The concentration of H_2CO_3 decreases as it is used up, which decreases the rate of CO_2 production.

   b. The rate of carbon dioxide production (and hence the rate of the forward reaction) would be higher if the drink was warm. This is due to more of the colliding particles having energy greater than the activation energy, which leads to a higher proportion of successful collisions.

7. Increasing the concentration of acid meant that more acid molecules collided with the surface of the solid calcium carbonate in any given time period. This increased frequency of collisions also led to an increased frequency of 'successful' collisions, thereby increasing the rate of reaction.

8. a. Powder, 2 M HCl and a hot water bath

   b. Strip, 0.5 M HCl and a cool water bath

9. a. The molecules in the gas/air mixture may not initially have enough energy to overcome the activation energy barrier and hence do not react. However, a spark provides enough energy for some of the molecules to do so and so the reaction starts.

   b. Following on from part a, energy is given out and this provides the energy for further molecules to overcome the activation energy barrier. This enables the reaction (explosion) to build up and continue.

4.2 Exam questions

1. A

2. C

3. Increasing temperature (constant volume):
   Increasing pressure (constant temperature):
   - Effect: Increase
   - Reasoning: Increasing temperature increases the (average) kinetic energy of reactant molecules so:
     - more collisions have energy greater than the activation energy
     - the proportion of collisions that are successful (fruitful) increases.

   Increasing pressure (constant temperature):
   - Effect: Increase
   - Reasoning: Increasing the pressure increases the closeness (concentration) of the reactant molecules and so the frequency (number) of collisions increases.

4. a. Sample response:
   - As temperature increases, the rate of reaction increases/the time taken for the balloon to reach 10 cm decreases.
   - This is because increased temperature increases the speed of particles, which increases the frequency of collisions (number of collisions per unit time) between reactants, thereby increasing the rate of the reaction.
   - Increased temperature also increases the energy of the particles, which means that a greater proportion of particles have enough energy to overcome the activation energy barrier, resulting in a greater proportion of collisions that are successful, which also increases the reaction rate.

   b. Any one of:
   - No. There is not a consistent trend; time for reaction is lower/rate is higher at 15 °C than at 25 °C.
   - Yes. If the data was presented using a line of best fit the overall trend is consistent with the prediction.
   - Yes, if the outlier (data point showing 25 °C) is ignored.

5. D

4.3 Catalysts and reaction rates

Practice problem 2

The profile labelled Y corresponds to catalyst B.

4.3 Exercise

1. D
2. D
3. a. Exothermic (ΔH is negative)
   b. i. $\Delta H = -30 \text{ kJ mol}^{-1}$
      ii. $E_a = 20 \text{ kJ mol}^{-1}$
      iii. $E_a = 20 \text{ kJ mol}^{-1}$. This is the same as the minimum energy required to break the reactant bonds.
      iv. Energy evolved $= 50 \text{ kJ mol}^{-1}$
4. The activation energy must be low.

5. This is not possible because the products would need to have an enthalpy higher than that of the activated complex (transition state).
6. a. The E_a for the forward reaction is lower than for the backward reaction.
   b. The E_a for the forward reaction is higher than for the backward reaction.
7. No. A catalyst lowers the activation energy — it does not change the enthalpy of the reactants or the products. Therefore, ΔH will remain unchanged.
8. See figure at the bottom of the page*
9.

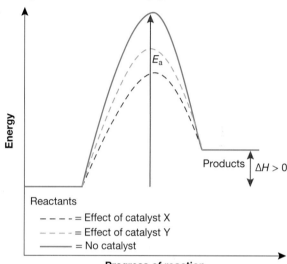

10. a. Powdered form has more surface area.
    b. The catalyst is improving the orientation for collisions to be successful.
    c. The temporary bonds between X and AB weaken the AB bond.
    d. Once formed, molecules of CA leave the surface, thus freeing it up to repeat the process with more molecules of AB.

*8.

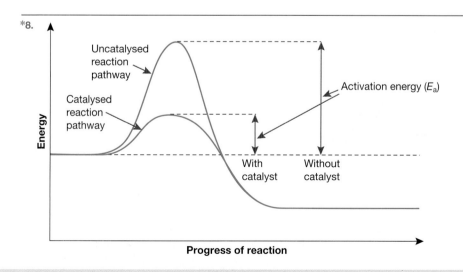

4.3 Exam questions

1. D
2. B
3. A
4. a. See figure at the bottom of the page*
   b. Tungsten; with it the reaction has a lower activation energy, which means the proportion of collisions that are successful between NH_3 molecules will be higher in a given time period.
5. See figure at the bottom of the page**

4.4 Review

4.4 Review questions

1. A catalyst lowers the activation energy of a reaction; that is, it makes it easier to break the bonds in reactants. This is described as 'providing an alternative reaction pathway'. Hence, catalysts speed up the rate of a chemical reaction.

---

*4. a.

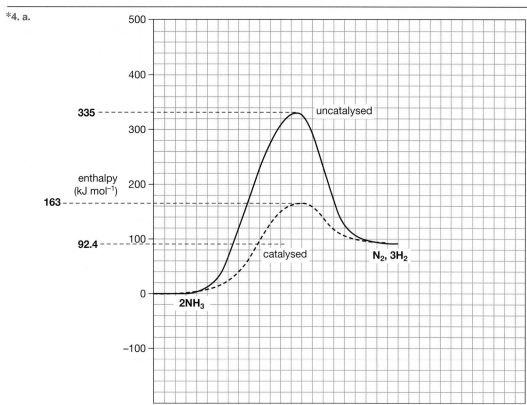

**5.

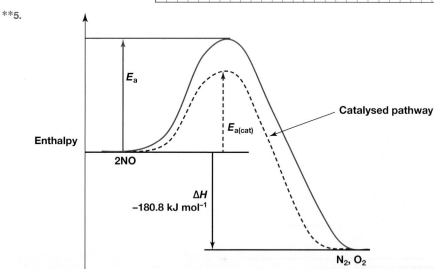

2. Caster sugar is finely ground and has a much greater surface area than granulated sugar. Consequently, more caster sugar particles are exposed to the sulfuric acid molecules, resulting in an increased frequency of collision between the reacting particles.

3. a. The reactant molecules are consumed, resulting in a progressive decrease in the frequency of successful collisions of reactant molecules.

b. Four reactant gas molecules are being converted into three product gas molecules. The net volume of gas will decrease as the reaction progresses, so the total gas pressure will decrease continuously.

4. a. The necessary reactions happen at a quicker rate as the average energy of the colliding particles is greater and thus able to overcome the activation energy barrier.

b. The steel wool has a greater surface area and hence is more exposed to the oxygen in the air.

c. The manganese dioxide is acting as a catalyst, allowing peroxide to decompose faster, producing oxygen continually.

d. Increased temperature causes a faster reaction due to particles colliding with more energy.

e. The reaction slows because the concentration of reactants decreases as they are used up.

f. The chemicals in the infrared film are sensitive to heat. With cooling during storage, the rate at which these chemicals react is lowered.

5. a. A catalyst is a substance that increases the rate of a chemical reaction without being consumed. It provides an alternative reaction pathway with a lower activation energy.

b.

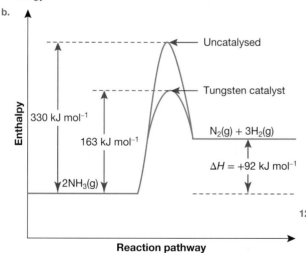

Reaction pathway

c. i. $E_a = 238$ kJ mol^{-1}
Note: Question is about reverse reaction
ii. $E_a = 71$ kJ mol^{-1}
Note: Question is about reverse reaction

6. a. Lipase, an enzyme, acts as a catalyst, thereby lowering the activation energy.

b. It is the same because catalysts are not consumed in chemical reactions.

7. Reaction rates can also be increased by an increase in surface area for solid reactants, an increase in pressure for gas reactions, an increase in concentration for reactions occurring in solutions and using an appropriate catalyst.

8. a. The rate of production of C will be higher during the initial stages in both reactions because the reactant concentration is greatest at this stage in both scenarios. This is shown by the steeper initial gradients on both graphs.

b. Equal amounts of A and B are used, leading to equal amounts of C produced. A catalyst does not affect the nature of the products.

c. Y is a catalyst. The rate of reaction is increased and the concentration of Y stays the same.

d. Repeat the experiment under identical conditions (except) without X present.

4.4 Exam questions

Section A — Multiple choice questions

1. A
2. D
3. B
4. D
5. B
6. D
7. C
8. C
9. C
10. D

Section B — Short answer questions

11.

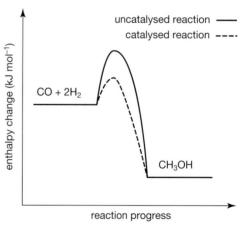

12. Answers will vary. Some acceptable responses are given below.

Acceptable improvements/modifications include:
- use the same volume of H_2O_2 in both trials
- use the same concentration of H_2O_2 in both trials
- use the same mole amounts of both catalysts
- use a device with better insulation such as a calorimeter to minimise heat losses
- investigate more than two different catalysts
- repeat the trials a number of times
- use a mechanical stirrer to ensure thorough mixing of reactants
- use catalysts of the same state.

Acceptable responses for identification of the expected outcome of the trials include the following:

- Different catalysts will have no effect on the molar heat of reaction since the molar enthalpy of the decomposition reaction of hydrogen peroxide is independent of the catalyst used.
- The same molar heat of reaction should be determined irrespective of the catalyst used, assuming all other variables are controlled.
- A catalyst has no effect on the relative enthalpies of the reactants and products, hence does not affect the molar heat of reaction.
- The same molar enthalpy should be determined for the decomposition of H_2O_2, irrespective of the catalyst used.

Acceptable responses for the application of chemistry ideas in the context of the question include the following:

- If investigating the effect of different catalysts on the molar enthalpy of decomposition, the only variable should be the catalyst. The results presented were inconclusive because independent variables, such as the concentration of H_2O_2, were not kept constant.
- The conclusion that the different temperatures for the two trials verify that the molar enthalpy of decomposition depends on the catalyst used is invalid because $n(H_2O_2)$ reacting in Trial 2 is four times the $n(H_2O_2)$ reacting in Trial 1 and this will cause a greater temperature change.
- Different catalysts will decrease the activation energy for the decomposition by different amounts but will have no effect on the molar enthalpy of decomposition. (Some students used energy profiles to emphasise this point.) However, since one catalyst may increase the rate of reaction more than the other, and because the faster reaction will reach completion in a shorter time, this will produce a larger temperature change in that time. If the catalyst is the only variable, the overall temperature change, in well-insulated reaction vessels, will be the same for both catalysts, consistent with the fact that a catalyst does not affect the molar enthalpy of decomposition.

13. a. Increasing $S_2O_8{}^{2-}$(aq) concentration produces a faster rate in reaction 1.
    Comparing trials 1 and 2 shows that keeping I^-(aq) volume constant and doubling $S_2O_8{}^{2-}$(aq) volume results in a decrease in the time taken for the blue colour to appear; that is, the rate of reaction 1 is increased.

   b. Comparing trials 1 and 3 shows that keeping $S_2O_8{}^{2-}$(aq) volume constant and doubling the I^-(aq) volume results in a decrease in the time taken for the blue colour to appear; that is, the rate of reaction 1 is increased.

   c. The reaction times would increase.

   d. Adding water as indicated ensures concentration does not become another variable in the experiment, as this would vary the frequency of collisions between reacting particles and thus affect the rate of the reaction.

e. To identify when reaction 1 reaches the same stoichiometric point for each trial, the appearance of the blue colour is used to indicate when the same amount of $S_2O_3{}^{3-}$(aq) is used up, leaving I_2 in excess to react with starch.

14. a. For example, any one of the following:
    - Measuring the mass (loss of mass) of the beaker and contents over a time interval
    - Measuring the mass/volume of CO_2 produced over a time interval
    - Plotting the mass loss of beaker and contents, or mass/volume of CO_2 produced against time.

   b. i. For example, one of:
       - surface area
       - particle size of $CaCO_3$.

      ii. Beaker A, because of the higher concentration of HCl. This will increase the frequency (number) of collisions between reactant particles/successful collisions/fruitful collisions/collisions with energy greater than activation energy.

   c. Collision theory states that only collisions with energy equal to or greater than the activation energy for the reaction can result in reaction. Chemical reaction requires breaking of bonds in the reactant particles; this requires collisions to have energy greater than or equal to the activation energy, or the reactant particles must collide with the correct orientation.

15. a. Endothermic

   b. The new method would produce substance C quicker.

   c. Because the temperature and concentration is the same across both methods, these factors will not influence collision rate and hence rate of reaction. In the new method, the greater surface area of B (and thus of contact between the two immiscible liquids) would lead to more collisions between A and B, and hence more successful collisions in a given period of time and a faster rate of reaction.

   d. Sample response:
      As liquid B now interacts with liquid A for a longer time, the droplets of B in the spray will become coated with C as their reaction is a surface reaction. The collected C will therefore not be pure. This could be minimised by making the droplets in the spray as small as possible.

   e. Increased temperature will result in more frequent collisions between molecules of A and B. The contact time between the two liquids can be reduced.

5 Extent of chemical reactions

5.2 Reversible and irreversible reactions

5.2 Exercise

1. Irreversible reaction: the reactants form products that cannot be converted back into reactants. Only the forward reaction is possible.
   Reversible reaction: the reactants react to form products, the products can react to form reactants. Both the forward and backward reactions can occur.

2. The rate of a reaction measures how quickly reactants are converted into products. The extent of a reaction measures the quantity of reactants that are converted into products.

3. The rate at which the hydrogen peroxide decomposes (under normal conditions) is very slow.

4. No. The amount of product will always be less than the stoichiometrically predicted amount. This is because there will be some product that is being converted back into reactants.

5. a. Not obvious and difficult to detect
   b. Not obvious and difficult to detect
   c. Not obvious and difficult to detect

6. a. Reaction I: type 4
   Reaction II: type 1, type 2, type 3
   Reaction III: type 2
   Sample responses can be found in your digital formats. It is possible that your answers may have been slightly different due to your interpretation of the qualitative terms *slow*, *fast*, *small* and *large*.
   b. The provision of more time or the addition of a catalyst would provide evidence for the isolation of type 2. Distinguishing between types 1 and 3 would prove more difficult. A more sensitive method for detecting any product may work.

5.2 Exam questions

1. Both acids only react to a very small degree with water. This is because their K_a values are so small.

2. B

3. a. Rate is affected by concentration. At the start, the concentration of NO is high so the rate of the backward reaction will be high. Conversely, the concentrations of N_2 and O_2 are low, resulting in the forward reaction being slow.
   b. The rate of the forward reaction will increase and the rate of the backward reaction will decrease.
   c. The two rates will eventually become equal.

4. In a closed system the products that are formed are not able to escape, making them available for collisions that will revert them to reactants.

5. a. The graph shows that at least one reactant and one product are present together and have reached a point at which their concentrations do not change. This indicates a reversible (or equilibrium) reaction (see figure 5.3).
   b. The rate of the forward reaction is decreasing because the concentration of methane is decreasing. The rate of the backward reaction is increasing because the concentration of carbon monoxide is increasing.

5.3 Homogeneous equilibria

Practice problem 1

a. The concentration of N_2O_4 at t_1 will be equal to its concentration at t_2.

b. N_2O_4 is increasing but at an ever slower rate until the reaction reaches equilibrium. After this its value stays constant.

5.3 Exercise

1. The dynamic nature of equilibrium refers to the fact that reaction is still taking place; it has not stopped. The rate of the forward reaction is equal to the rate of the reverse reaction. The result is that all amounts therefore stay constant.

2. A reaction depends on successful collisions (see topic 4). Statistically, the more particles there are, the more successful collisions there will be in a given period of time. Reactions slow down as their reactants are being used up. When this occurs to a reaction, its reverse reaction will speed up as the number of its reactants have increased by the first reaction.

3. a. Radioactive iodine would be found only in I_2.
   b. Radioactive iodine will be distributed between I_2, I^- and I_3^-.

4. a. Forward reaction > reverse reaction
   b. Forward reaction > reverse reaction
   c. Forward reaction = reverse reaction
   d. Forward reaction = reverse reaction.

5. a. There would be no further changes to any of the concentrations.
   b. Z increases because there is a net forward reaction. It increases at three times the rate that Y decreases and at $\frac{3}{2}$ times the rate that X decreases (due to the stoichiometry of the reaction).
   c. The concentration of X = 2.5, Y = 1.5 and Z = 1.

5.3 Exam questions

1. D

2.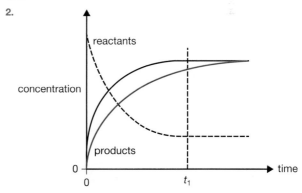

3. D

4. a. See figure at the bottom of the page*

   b. See figure at the bottom of the page**

5. D

5.4 Calculations involving equilibrium systems

Practice problem 2

$$K = \frac{[CO_2][H_2O]^2}{[CH_4][O_2]^2}$$

Practice problem 3

$K = 31\,M^{-1}$

Practice problem 4

$K(2) = 0.00400\,M^{-1}$

Practice problem 5

$K = 1.2\,M$

Practice problem 6

$[NH_3] = 0.089\,M$

5.4 Exercise

1. a. The very high value of $K = 10^{140}$ suggests the reactants are completely converted into products.

   b. No. The high equilibrium constant means this can be considered a complete reaction.

2. a. i. $K = \dfrac{[ClF_3]^2}{[Cl_2][F_2]^3}$ ii. $K = \dfrac{[NO]^2}{[N_2][O_2]}$

   iii. $K = \dfrac{[H_2O]^2[Cl_2]^2}{[HCl]^4[O_2]}$ iv. $K = \dfrac{[CF_4][CO_2]}{[COF_2]^2}$

   v. $K = \dfrac{[PF_5]^4}{[P_4][F_2]^{10}}$

   b. i. K units $= M^{-2}$ ii. K units $=$ no unit
   iii. K units $= M^{-1}$ iv. K units $=$ no unit
   v. K units $= M^{-7}$

3. $2A + B \rightleftharpoons C + 2D$

4. $K = 4.0$
   Note: There are no units.

5. a. $m(Z) = 17\,g$ (to two sig. figs.)

   b. $K(\text{new}) = 3.57\,M^2$

6. a. $K = 0.617$ b. $K = 2.62$

7. $K = 16\,M$

8. a. There are no units (concentration units cancel out).

   b. $K = 0.16$

---

*4. a.

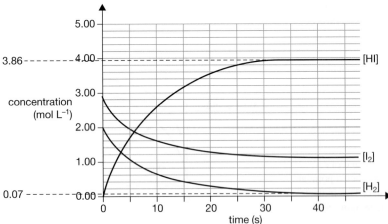

**4. b.

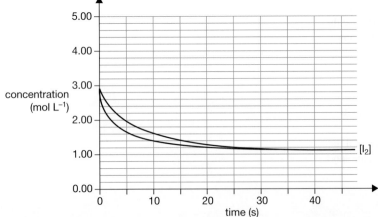

5.4 Exam questions

1. $0.38\,M^{-1}$
2. A
3. B
4. D
5. i. Any one of:
   - $K = \dfrac{[HI]^2}{[H_2][I_2]}$
   - $\dfrac{[HI]^2}{[H_2][I_2]}$

   ii. $[H_2] = 0.07\,M$
   $[HI] = 3.86\,M$
   $K = 1.99 \times 10^2$ or 1.99

5.5 The reaction quotient (Q)

Practice problem 7

The reaction is not at equilibrium. As $Q > K$ there is a net backward reaction. The rate of the reverse reaction is faster than the rate of the forward reaction.

5.5 Exercise

1. Measure the concentrations that are present. Use these to evaluate the value for Q and compare this to the value of K. If the two values are equal, the reaction is at equilibrium.
2. The units are the same because Q and K are both derived from the same mathematical expression.
3. B
4. After evaluating the reaction quotient (Q) for each experiment, experiment 4 is the only one where $Q = K$. Therefore, experiment 4 is at equilibrium.
   Sample responses can be found in your digital formats.
5. a. The reaction is not at equilibrium.
   b. The rate of the forward reaction is greater than the rate of the reverse reaction.

5.5 Exam questions

1. $Q = \dfrac{[NOBr]^2}{[NO]^2\,[Br_2]}$
2. C
3. $Q > K$
4. D
5. $Q = \dfrac{[CH_3OCH_3][H_2O]}{[CH_3OH]^2}$

5.6 Changes to equilibrium and Le Chatelier's principle

Practice problem 8

The reaction would partially oppose the removal of the hydrogen by attempting to replace some of it. This is achieved by the forward reaction becoming temporarily faster than the backward reaction. The 'position' of the equilibrium would, therefore, shift to the right. The forward reaction is favoured.

Practice problem 9

a. The addition of water immediately lowers the total concentration. The system responds by increasing the number of moles in an effort to compensate. The amount of I_3^- present would decrease.
b. If the volume increases the concentration decreases. The equilibrium will try to produce more moles and shift to the right. If the volume decreases (concentration increases) there will be a shift to the left.

Practice problem 10

The reaction is exothermic.

Practice problem 11

a. i. Decrease   ii. Increase
b. i. Decrease   ii. Decrease
c. i. Remain unaltered   ii. Increase

5.6 Exercise

1. Temperature is the only variable for which a change affects the position of the equilibrium. Exothermic and endothermic reactions are affected differently by temperature.
2. a. Produces a net forward reaction to partially replace the removed product
   b. Produces a net forward reaction to partially use up the added reactant
   c. Produces a net forward reaction that turns four molecules into two, thus partially countering the effect of the increased pressure
   d. A net backward reaction
   e. No change
3. a. Produces net backward reaction
   b. Reaction is unchanged
   c. Produces net backward reaction
   d. Produces net forward reaction
4. The student is correct in saying that an increase in pressure will cause the system to move to the right. However, assuming the temperature is kept constant, there will be no change in the value of the equilibrium constant.
5. a. The new equilibrium is to the right (there was a net forward reaction).
   b. From top to bottom, the graphs represent substances C, B and A.
6. a. At time t_1 there was a decrease in temperature.
   b. Between times t_1 and t_2 there was a net backward reaction.
   c. If substance D was added at t_3 there would be no effect.
7. a. Endothermic
   b. $K(2) = 160$
8. a. It is a reaction between ammonia and oxygen to make nitrogen monoxide and water.

b. The reaction is exothermic due to the negative energy value given in kJ mol^{-1}.

c. To obtain a high value for K (i.e. favour the forward reaction) the reaction should be carried out at low temperatures.

d. The formation of products would be favoured by low pressure.

9. a. Increased pressure will maximise the yield of methanol.

b. Lowering the temperature will increase the yield of methanol.

c. If carbon monoxide (CO) is added the yield of methanol will increase.

d. Changing the temperature is the only way to change the value of the equilibrium constant.

10. a. The lines indicate a system at equilibrium.

b. Increase in volume

c. Additional substance A was added.

d. If a catalyst was added at t_3 the time between t_3 and t_4 would be decreased.

e. All results would be the same.

5.6 Exam questions

1. D

2. D

3. A

4. A

5. a. The system is at equilibrium, so the rates of the forward and reverse reactions are equal; hence, the rate of formation of NOBr is constant. Frequency of successful collisions (between NO and Br$_2$) is also constant.

b. A volume decrease has the immediate effect of increasing all concentrations, and therefore increasing the frequency (number) of collisions between NO and Br$_2$, so the rate of formation of NOBr increases.

c. The rate of formation of NOBr decreases because:
• the forward reaction — formation of NOBr — is favoured, so as the reactant concentrations [NO] and [Br$_2$] decrease, the frequency of collisions decreases and the rate of this reaction decreases
• as the [NOBr] increases due to the forward reaction being favoured, the rate of decomposition of NOBr (reverse reaction) increases, so the net effect is a decrease in the rate of formation of NOBr.

5.7 Review

5.7 Review questions

1. Isla is correct.
Sample responses can be found in your digital formats.

2. a. $[I_2] = 0.067$ M

b. The value of K is temperature dependent.

c. $K = 40.8$ (no units)

3. $K = 4.2$ M^{-1}

4. a.

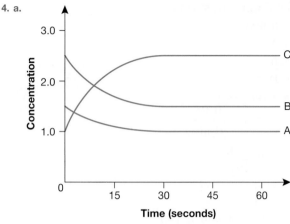

b. Rate of forward reaction > rate of backward reaction

c. Rate of forward reaction = rate of backward reaction

d. $y = 2$; $z = 3$

e. $K = 6.9$

5. a. Equilibrium has been achieved in experiments 2 and 5, as $Q = K = 4$.

b. A net forward reaction will occur in experiments 1, 3 and 6 as $Q < K$. A net backward reaction will occur for experiment 4 as $Q > K$.

6. a. Exothermic reaction

b. Fewer particles on the right-hand (product) side

7. a. $K = 1.85$ M^{-2}

b. $p = 42.9T$ kPa

c. The amount of NH$_3$ would increase.

8. a. The system was in the process of reaching equilibrium.

b. The system was at equilibrium.

c. Some HI being removed

d. The spike in the graph indicates that some HI was added.

e. The volume was lowered.

f. No, because there are equal numbers of moles on each side of the equation.

9. $K = 1.75$

10. a. Change 1: darken
Change 2: stay the same
Change 3: lighten

b. Le Chatelier's principle states that if a change is made to a system at equilibrium it will partially oppose it (if possible).
Change 1: Adding KBr adds Br$^-$ ions to the right-hand side. The system opposes this change with a net backward reaction and more Br$_2$ is produced, making it darker.
Change 2: As NaCl is not involved in the reaction there would be no change.
Change 3: As the reaction is endothermic, increasing the temperature would favour the forward reaction, consuming the brown Br$_2$(g) and resulting in the solution becoming lighter in colour.

5.7 Exam questions

Section A— Multiple choice questions

1. C
2. C
3. A
4. B
5. A
6. A
7. C
8. A
9. B
10. D

Section B — Short answer questions

11. a. i. $K = \dfrac{[N_2]^2[H_2O]^6}{[NH_3]^4[O_2]^3}$ ii. M or mol L^{-1}

    b. i. The forward reaction is exothermic, so the equilibrium constant decreases.
       The increase in temperature causes the system to partially oppose this stress by causing a shift in the reverse direction. The increase in temperature increases the rate of the backward reaction more than the rate of the forward reaction.
       The [reactants] increases and the [products] decreases.

       ii. The rate of the forward reaction will be greater at the new equilibrium after the temperature increase.
       The average energy/speed of all particles will increase, hence the frequency of collisions increases. The reactant concentration increases due to the backward reaction being favoured, which in turn means a greater number of collisions per unit time.

There are more particles with energy above the activation energy (E_a), and therefore an increased probability/proportion/ratio of collisions that are successful.

12. a. Exothermic
    Any one of the following reasons:
    • Lower absorbance at 80 °C indicates there is a smaller amount of Co^{2+}(aq), a product, at the higher temperature, and a larger amount of Co^{2+}(aq), a product, at the lower temperature, so the forward reaction is exothermic.   .
    • Higher absorbance at 30 °C indicates there is more Co^{2+}(aq) at the lower temperature, so forward reaction is favoured at the lower temperature and is exothermic.
    • Absorbance is proportional to the concentration of Co^{2+} and the graph indicates that a smaller amount of products are formed at high temperatures, and a larger amount of products at lower temperatures, hence the forward reaction is exothermic.

    b. The final colour is darker than just after the water was added but lighter than the colour at the initial equilibrium, and not quite as pink as before the change (between the diluted colour and original colour). Adding water decreases the concentration of all species. The system partially opposes this change, favouring the reaction producing more particles in aqueous solution, the forward reaction (1 mol → 5 mol). The concentration of Co^{2+}(aq) at the new equilibrium is less than prior to the addition of water but greater than after dilution.

    c. See figure at the bottom of the page*

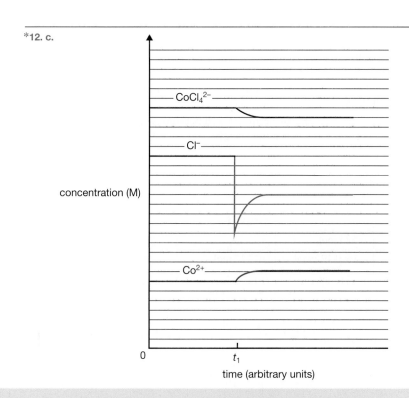

*12. c.

concentration (M)

CoCl$_4$$^{2-}$

Cl$^-$

Co^{2+}

0   t_1

time (arbitrary units)

13. a. i. Either one of:
- The yield is higher at low temperatures.
- The yield is lower at high temperatures.

And, either one of:
- The forward reaction is exothermic so, according to Le Chatelier's principle, if the temperature is lowered the system will move to raise the temperature (partially compensate) by favouring the exothermic forward reaction.
- The forward reaction is exothermic so, according to Le Chatelier's principle, if the temperature is increased the system will move to lower the temperature (partially compensate) by favouring the endothermic backward (reverse) reaction.

ii. Either one of:
- The yield is greater at high pressure.
- The yield is lower at low pressure.

And, either one of:
- According to Le Chatelier's principle, at high pressure, the system moves to decrease pressure by favouring the reaction that produces fewer particles in a given volume (forward reaction, 3 mol → 1 mol), thus increasing the yield of methanol.
- According to Le Chatelier's principle, at low pressure, the system moves to increase pressure by favouring the reaction that produces more particles in a given volume (backward reaction, 1 mol → 3 mol), thus decreasing the yield of methanol.

b. $K = \dfrac{[CH_3OH]}{[CO][H_2]^2}$

c. See table at the bottom of the page*
$K = 0.605 \, M^{-2}$

14. a. $K = 81.4 \, M^{-1} \, (L \, mol^{-1})$

b. Temperature should be decreased. The forward reaction is exothermic and is favoured at lower temperature, hence the yield of SO_3 will increase.
Volume should be decreased. Smaller volume causes the concentrations/pressure to increase. As the system moves to partially compensate for this change, the side of the equilibrium with fewer molecules is favoured so the yield of SO_3 will increase.

15. a. i. No change
The system cannot compensate for (partially oppose) the increase in pressure because there are the same number of particles (moles) on both sides of the equilibrium. The reaction quotient has not changed and the system remains at equilibrium.

ii. Increase
The system moves to compensate for (partially oppose) the removal of product by favouring the forward reaction (production of CH_3Br) or removal of product reduces the rate of the reverse reaction, and so the forward reaction is favoured.

b.

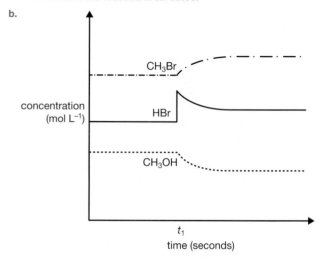

6 Production of chemicals using electrolysis

6.2 What is electrolysis?

Practice problem 1

a. Cathode: $K^+(l) + e^- \rightarrow K(l)$
Anode: $2I^-(l) \rightarrow I_2(g) + 2e^-$

b. $2K^+(l) + 2I^-(l) \rightarrow 2K(l) + I_2(g)$

6.2 Exercise

1. a. A galvanic cell transforms chemical energy into electrical energy.

b. An electrolytic cell transforms electrical energy into chemical energy.

---

*13. c.

	CO	H₂	CH₃OH
$n_{initial}$	0.760 mol	0.525 mol	0
n_{change}	−0.122 mol	−2 × 0.122 mol	+0.122 mol
$n_{equilibrium}$	0.638 mol	0.281 mol	0.122 mol
Final concentration	$\dfrac{0.638}{0.500} = 1.276 \, M$	$\dfrac{0.281}{0.500} = 0.562 \, M$	$\dfrac{0.122}{0.500} = 0.244 \, M$

2. Cathode: Lead metal. The lead ions gain electrons (are reduced) at the cathode.

$$Pb^{2+} + 2e^- \rightarrow Pb(s)$$

Anode: Bromine. Bromide ions will be oxidised at the anode, forming bromine.

$$2Br^- \rightarrow Br_2(g) + 2e^-$$

(The overall reaction would be:

$$Pb^{2+} + 2Br^- \rightarrow Pb(s) + Br_2(g).)$$

3. Liquid lithium will form around the cathode due to the reduction of Li^+ ions to Li. Gaseous I_2 will form around the anode due to the oxidation of I^- ions to I_2.

4. Pure water is a non-conductor because it contains almost no mobile ions. Adding a small amount of KNO_3 provides some mobile $K^+(aq)$ ions and some mobile $NO_3^-(aq)$ ions. Since there are charged particles that are free to move, the solution conducts electricity and the circuit is completed, allowing oxidation and reduction to happen at the anode and cathode.

5. Oxidation occurs at the anode. When water undergoes oxidation, oxygen gas is formed. Therefore, oxygen will collect around the positively charged (left) side of the apparatus. Reduction occurs at the cathode. When water undergoes reduction, hydrogen gas is formed. Therefore, hydrogen gas will collect around the negatively charged (right) side of the apparatus. Also, the equation indicates that there is a 2 : 1 mole/volume ratio between the hydrogen produced and the oxygen produced. This further supports that hydrogen is on the negative (right) side and oxygen is on the positive (left) side in this diagram.

6. a. Cathode: $2H_2O(l) + 2e^- \rightarrow H_2(g) + 2OH^-(aq)$

   Anode: $2H_2O(l) \rightarrow O_2(g) + 4H^+(aq) + 4e^-$

   b. Cathode: Phenolphthalein would turn pink. This is due to OH^- ions making the region basic.
   Anode: Phenolphthalein would remain clear. This is due to H^+ ions making the region acidic.

7. Electrodes must conduct electricity from their external circuit connections to their surfaces. Glass is a non-conductor.

8. a. The cathode has a negative charge because it is connected to the negative terminal of the power supply. It receives electrons from the external circuit. Electrons are removed from its surface by the process of reduction.

   b. The anode has a positive charge because it is connected to the positive terminal of the power supply. Electrons are added to its surface by the process of oxidation, and these flow back into the external circuit, towards the cathode.

9. Reactants in a galvanic cell would react spontaneously if they were placed in the same compartment. Reactants in an electrolytic cell do not react until a critical amount of energy is supplied, so a single compartment cell can be used.

10. a. Reduction occurs at the cathode: $K^+(l) + e^- \rightarrow K(l)$.

    b. The cathode is negative (−).

    c. Oxidation occurs at the anode: $2Br^-(l) \rightarrow Br_2(g) + 2e^-$. At the temperature of molten KBr, Br_2 is a gas.

    d. The anode is positive (+).

    e. Combining the half-equations, the overall equation is:
    $$2K^+(l) + 2Br^-(l) \rightarrow 2K(l) + Br_2(g).$$

6.2 Exam questions

1. A

2. a. (−) iron electrode; (+) carbon electrode

   b. Either of:
   - $2F^-_{(HF)} \rightarrow F_2(g) + 2e^-$
   - $2HF_{(HF)} \rightarrow F_2(g) + 2H^+_{(HF)} + 2e^-$

3. A

4. See figure at the bottom of the page*

5. $2Br^-(l) \rightarrow Br_2(g) + 2e^-$

6.3 Using the electrochemical series in electrolysis

Practice problem 2

Cathode: K^+ ions and H_2O molecules

Anode: NO_3^- ions and H_2O molecules. The NO_3^- ions can be ignored because they are inert, as they cannot be further oxidised.

*4.

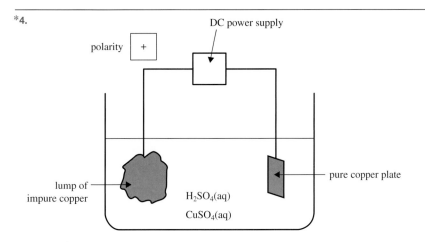

Practice problem 3

Cathode: $Ag^+(aq) + e^- \rightarrow Ag(s)$. Silver metal will form.

Anode: $2H_2O(l) \rightarrow O_2(g) + 4H^+(aq) + 4e^-$. Oxygen gas will form and the pH will decrease.

Overall: $4Ag^+(aq) + 2H_2O(l) \rightarrow 4Ag(s) + O_2(g) + 4H^+(aq)$

Minimum voltage required > 0.43 V

Practice problem 4

Cathode: $Pb^{2+}(aq) + 2e^- \rightarrow Pb(s)$. Lead metal will form.

Anode: $Cu(s) \rightarrow Cu^{2+}(aq) + 2e^-$. Cu^{2+} will form.

Overall: $Cu(s) + Pb^{2+}(aq) \rightarrow Pb(s) + Cu^{2+}(aq)$

Minimum voltage required > 0.47 V

6.3 Exercise

1. Sample responses include:
   - hydrogen gas (bubbles) produced around the cathode
   - pH increases around the cathode (due to OH^- production)
   - blue colour around the anode (due to Cu^{2+} ions).

2. **a.** Cathode: $2H^+(aq) + 2e^- \rightarrow H_2(g)$
   (H^+ is a stronger oxidising agent than H_2O.)
   Anode: $Ag(s) \rightarrow Ag^+(aq) + e^-$
   (Ag is a stronger reducing agent than H_2O and Cl^-.)
   b. Overall: $2Ag(s) + 2H^+(aq) \rightarrow 2Ag^+(aq) + H_2(g)$

3. Cathode: $2H_2O(l) + 2e^- \rightarrow H_2(g) + 2OH^-(aq)$
   Anode: $2H_2O(l) \rightarrow O_2(g) + 4H^+(aq) + 4e^-$

4. Cathode: Water reacts in preference to Al^{3+}.
   Anode: At 1 M, both chloride ions and water molecules will be oxidised at the anode.
   Therefore, the products are hydrogen, oxygen and chlorine.
   Minimum cell voltage required: >2.06 V

5. **a.** Cathode: Copper metal, $Cu(s)$
   Anode: Fluorine gas, $Fl_2(g)$
   b. Cathode: Copper metal, $Cu(s)$
   Anode: $O_2(g)$ and $H^+(aq)$

6. **a.** Copper metal
   b. Lead metal

c. Hydrogen gas and hydroxide ions
Sample responses can be found in your digital formats.

7. **a.** Iron can be used at the cathode because it cannot be reduced (iron does not accept electrons to produce negative ions). It is not used at the anode because it would be oxidised in preference to the chloride ions (Fe is a stronger reducing agent than Cl^- ions).
   b. See figure at the bottom of the page*

8. **a.** Cathode: $Cu^{2+}(aq) + 2e^- \rightarrow Cu(s)$
   Anode: $2H_2O(l) \rightarrow O_2(g) + 4H^+(aq) + 4e^-$
   b. The concentration of Cu^{2+} ions will decrease.
   c. Cathode: $Cu^{2+}(aq) + 2e^- \rightarrow Cu(s)$
   Anode: $Cu(s) \rightarrow Cu^{2+}(aq) + 2e^-$
   d. The concentration of Cu^{2+} ions will stay constant.

9. The electrolysis is not being performed at standard conditions.

10. This solution will react spontaneously to produce Sn^{4+} and Fe^{2+}.

6.3 Exam questions

1. D

2. A

3. Either one of the following approaches is acceptable:
   Galvanic cell approach
   Li(s) is a very strong reducing agent and will reduce water to produce $H_2(g)$.
   $2Li(s) + 2H_2O(l) \rightarrow H_2(g) + 2Li^+(aq) + 2OH^-(aq)$
   For the product that causes the hazard and the safety risk, either of:
   - $H_2(g)$ is explosive/flammable/builds up pressure in the battery.
   - The reaction is highly exothermic and may cause the battery to catch fire.

   Electrolytic cell approach
   Water will be reduced in preference to Li^+ ions, producing $H_2(g)$.
   Overall: $2H_2O(l) \rightarrow O_2(g) + 2H_2(g)$
   There would be a build-up of H_2/O_2, leading to a potential explosion due to either pressure or spontaneous combustion.

*7. b.

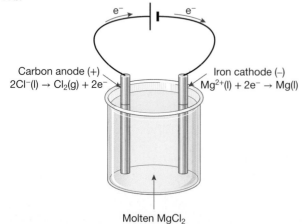

Carbon anode (+)
$2Cl^-(l) \rightarrow Cl_2(g) + 2e^-$

Iron cathode (−)
$Mg^{2+}(l) + 2e^- \rightarrow Mg(l)$

Molten $MgCl_2$

4. Explanation should include the following:
   - Iron is a stronger reducing agent than $F^-_{(HF)}$ and would be preferentially oxidised at the anode.
   - No F_2 would be produced.
   - $Fe(s) \rightarrow Fe^{2+}_{(HF)} + 2e^-$

5. B

6.4 Commercial electrolytic cells

6.4 Exercise

1. Calcium ions are harder to reduce than sodium ions. The sodium ions will be reduced in preference to the calcium ions.

2. Sodium hydroxide, chlorine gas and hydrogen gas

3. OH^- ions are formed at the cathode and will be attracted to the anode. Membranes impervious to OH^- ions ensure that they stay in the cathode compartment so that they can form sodium hydroxide, and prevent the reaction of OH^- with the chlorine being formed at the anode.

4. a. Aluminium can be produced at a lower temperature and therefore saves energy.

   b. Alumina does not dissolve in water, so cryolite is used as a solvent instead. Even if it alumina did dissolve in water, water would react preferentially at the cathode.

5. The electrolysis of water would cause a mixture of chlorine and oxygen to be produced at the anode. If the dilution continued, eventually only oxygen would be produced.

6. Na^+ and F^- are extremely weak oxidising agents and reducing agents respectively and will therefore not interfere with the desired reactions at the cathode and anode.

7. Cathode: $Al^{3+}(l) + 3e^- \rightarrow Al(l)$
   Anode: $C(s) + 2O^{2-}(l) \rightarrow CO_2(g) + 4e^-$
   The carbon cathode acts as an inert electrode and therefore remains intact, whereas the carbon anode is consumed in the oxidation reaction that takes place.

8.

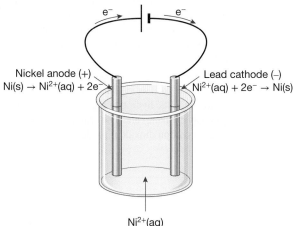

Nickel anode (+)
$Ni(s) \rightarrow Ni^{2+}(aq) + 2e^-$

Lead cathode (−)
$Ni^{2+}(aq) + 2e^- \rightarrow Ni(s)$

$Ni^{2+}(aq)$

Nickel will be oxidised at the anode and then will be reduced at the cathode to plate the lead cathode. $Ni(s)$ is a stronger reducing agent than water, and $Ni^{2+}(aq)$ is a stronger oxidising agent than water.

9. Reduction always takes place at the cathode according to the reaction: $M^{x+}(aq) + xe^- \rightarrow M(s)$.

10. a. Anode: silver

    b. Electrolyte: a solution containing Ag^+ ions

6.4 Exam questions

1. A

2. Reponses include:
   - If the products H_2 and F_2 can mix they will react explosively.
   - The diaphragm keeps the products of the electrolysis, $H_2(g)$ and $F_2(g)$, from coming in contact.

3. A

4. If water is present, it will be reduced in preference to sodium since Na^+ is a weaker oxidising agent.

5. a. i. Cathode: $Mg^{2+}(l) + 2e^- \rightarrow Mg(l)$
   ii. Anode: $2Cl^-(l) \rightarrow Cl_2(g) + 2e^-$

   b. Reponses include:
   - to prevent molten Mg reacting with oxygen in the air
   - to prevent contact between Mg and air/oxygen.

   c. According to the electrochemical series:
   - both Na^+ and Ca^{2+} are weaker oxidising agents than Mg^{2+} and so are unlikely to interfere with the production of Mg at the cathode
   - Zn^{2+} is a stronger oxidising agent than $Mg^{2+}(aq)$ and could be reduced to Zn, thus either preventing the production of Mg or contaminating the Mg produced.

   d. According to the electrochemical series, Fe is a stronger reducing agent than Cl^-.
   At the anode, Fe would be oxidised instead of Cl^-/ Fe^{2+} would be produced rather than Cl_2.
   Half-equation: $Fe(s) \rightarrow Fe^{2+}(l) + 2e^-$
   The cations $Fe^{2+}(l)$ would migrate to the cathode/ Fe^{2+} is a stronger oxidising agent than Mg^{2+}. Hence, Fe could be produced/the cathode half-equation would be $Fe^{2+}(l) + 2e^- \rightarrow Fe(s)$.

6.5 Rechargeable batteries (secondary cells)

Practice problem 5

$VO^{2+}(aq) + H_2O(l) \rightarrow VO_2^+(aq) + 2H^+(aq) + e^-$

6.5 Exercise

1. The pH will fall. During the recharging process, H^+ ions are produced.

2. The positive terminal of the battery is positive because electrons flow *into* it during discharge. To recharge the battery, the direction of electron flow needs to be reversed. To achieve this, the positive terminal must be connected to the positive terminal of the charging device. This means that electrons will be *removed* (i.e. move in the opposite direction).

3. a. Cathode: $Cd(OH)_2(s) + 2e^- \rightarrow Cd(s) + 2OH^-(aq)$
   b. The negative terminal
   c. During recharging:
   $Cd(OH)_2(s) + 2Ni(OH)_2(s)$
   $\rightarrow Cd(s) + 2NiO(OH)(s) + 2H_2O(l)$
   d. Anode (during recharging):
   $Ni(OH)_2(s) + OH^-(aq) \rightarrow NiO(OH)(s) + H_2O(l) + e^-$

4. a. $V^{2+}(aq) + VO_2^+(aq) + 2H^+(aq)$
   $\rightarrow V^{3+}(aq) + VO^{2+}(aq) + H_2O(l)$

b. $V^{3+}(aq) + VO^{2+}(aq) + H_2O(l)$
$\rightarrow V^{2+}(aq) + VO_2{}^+(aq) + 2H^+(aq)$

c. Cathode: $V^{3+}(aq) + e^- \rightarrow V^{2+}(aq)$
Anode: $VO^{2+}(aq) + H_2O(l)$
$\rightarrow VO_2{}^+(aq) + 2H^+(aq) + e^-$

5. a. Cathode: $2H_2O(l) + 2e^- \rightarrow H_2(g) + 2OH^-(aq)$
Anode: $2H_2O(l) \rightarrow O_2(g) + 4H^+(aq) + 4e^-$
Overall: $2H_2O(l) \rightarrow 2H_2(g) + O_2(g)$

b. Water is used up from each cell — it must be topped up with water.

c. Because hydrogen is released and is explosive, safety precautions would involve the removal of all possible ignition sources.

6.5 Exam questions

1. C

2. a. i. See figure at the bottom of the page*
ii. $Mg \rightarrow Mg^{2+} + 2e^-$

b. Responses include:
- content of device/electrolyte is toxic/harmful/corrosive to the body if it leaks
- battery may overheat
- Mg is relatively reactive so may react in the body
- leakage would cause an imbalance in natural Mg^{2+}/Na^+ ion levels in the body.

3. D

4. a. $2Li_2CO_3 + C \rightarrow 4Li + 3CO_2$

b. This would affect the performance of the battery by reducing battery life, reducing performance, limiting the extent of recharging/the number of recharges, and reducing the ability to hold full charge.
As lithium carbonate breaks away from the cathode, this reduces the amount of Li_2CO_3 available for recharging
OR
the products of electrolysis need to stay in contact with electrodes for effective recharge.

c. No. The $CO_2(g)$ absorbed during discharge will be released during recharge.

5. A

6.6 Contemporary responses to meeting society's energy needs

6.6 Exercise

1. a. Possible responses include:
- They both require an energy input
- In both cases, catalysts are important to electrode function
- They both decompose water into hydrogen and oxygen.

b. Different types of energy input are used (electrical in one case/light in the other).

2. a. Large-scale operation. This is because:
- small-scale operation is more likely to produce a fluctuating load
- more opportunities exist to heat the cell to its required temperature on a large scale than on a small scale
- any other suitable response.

b. $O^{2-} \rightarrow O_2 + 2e^-$

c. The cathode is negative and the anode is positive.

d. It could be co-located with another process that produces heat that would otherwise be wasted.

3. a. A is positive, B is negative.

b. H^+ ions

c. $2H^+(aq) + 2e^- \rightarrow H_2(g)$

d. $2H_2O(l) \rightarrow O_2(g) + 4H^+(aq) + 4e^-$

e. Possible responses include:
- high cost of catalysts
- membrane is expensive.

4. Possible responses include:
- lack of infrastructure
- difficulty of storage
- expense of fuel and electrolysis cells
- hydrogen is a flammable gas.

5. a. Both methods use methane to produce hydrogen and also form carbon dioxide. For grey hydrogen, this carbon dioxide is allowed to enter the atmosphere, whereas for blue hydrogen, it is captured in some way and prevented from entering the atmosphere.

b. Green hydrogen does not rely on the use of a non-renewable input (such as methane) and does not produce carbon dioxide. Blue hydrogen requires a non-renewable input (such as methane) and also requires a means of capturing the carbon dioxide it produces.

---

*2. a. i.

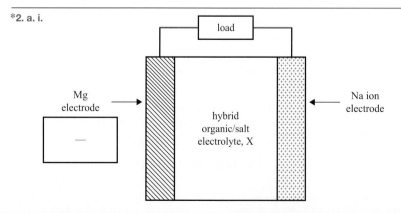

6.6 Exam questions

1. A sample response is as follows:

   I PEMEC. The wind can be used to generate electricity to power the cell and would be reasonably reliable. An extra source of energy would be required to raise an SOEC to its operating temperature.

   II PEMEC. The cells are adaptable to small-scale operation and could be powered by solar power.

   III SOEC. These could be co-located with the processes that generate heat, which could then be used to raise the cell to its operating temperature. These cells are very efficient at these temperatures. The electricity could be used from the grid or from a dedicated renewable supply method.
Note: PEMEC could also be acceptable.

   IV PEMEC. Hydroelectricity is renewable. This means that it could be used to produce green hydrogen in a PEMEC. An issue would be that it would not take electricity away from other uses and would therefore require the import of electricity made from non-renewable resources.
Note: SOEC could also be acceptable.

2. a. Light is used as the energy source. It does not require a means of generating electricity (such as, for example, solar arrays or wind turbines).

  b. Possible responses include:
- it produces hydrogen as a replacement for fossil fuels
- it can make further fuels using carbon dioxide from the atmosphere
- it reduces reliance on large-scale electricity generation.

  c. A catalyst increases the rate of a chemical reaction without itself being consumed.

  d. Either of:
- Yes, because light could be used to produce fuels from hydrogen and carbon dioxide.
- No, because synthesis is not the production of simpler substances, such as hydrogen, from more complicated substances, such as water.

  e. Oxygen

3. a. A: Cathode
    B: Membrane
    C: Anode
    D: H_2
    E: H^+ ions

  b. $2H^+(aq) + 2e^- \rightarrow H_2(g)$

  c. $2H_2O(l) \rightarrow O_2(g) + 4H^+(aq) + 4e^-$

  d. A conductor is required to permit the electrons that are produced or consumed to move to and from the external electrical circuit. A catalyst is required to increase the rate of the reactions that produce or consume these electrons.

  e. It keeps the products of the reaction separated from each other.

4. a. Water is oxidised.
    $2H_2O(l) \rightarrow O_2(g) + 4H^+(aq) + 4e^-$

  b. Hydrogen ions (protons) are reduced.
    $2H^+(aq) + 2e^- \rightarrow H_2(g)$

  c. The reactions at each electrode are different and will therefore require different catalysts for optimal efficiency.

5. a. Hydrogen is produced at the cathode (or negative electrode).
    Oxygen is produced at the anode (or positive electrode).

  b. $2H_2O(l) + 2e^- \rightarrow H_2(g) + 2OH^-(aq)$
    $4OH^-(aq) \rightarrow O_2(g) + 2H_2O(l) + 4e^-$

  c. The electricity required currently generates greenhouse gases in its production.

  d. Diaphragms are required to allow the flow of OH^- ions from the cathode to the anode, and to keep the product gases separated (prevent crossover).

6.7 Applications of Faraday's Laws

Practice problem 6
$n(X) = 0.0032$ mol

Practice problem 7
a. $m(Cu) = 0.173$ g
b. $V(O_2)_{SLC} = 0.0337$ L or 33.7 mL

Practice problem 8
2.98 minutes

Practice problem 9
The charge on the aluminium ion is $3+ (Al^{3+})$.

Practice problem 10
$m(Mg) = 109$ kg

6.7 Exercise
1. a. 13.2 C   **b.** 7.6×10^2 C   **c.** 6.6×10^3 C
  d. 1.9×10^4 C

2. a. 5.33×10^{-3} mol
  b. 0.482 mol
  c. 2.58 mol

3. a. $m(Ag) = 0.805$ g
  b. $V(O_2)_{SLC} = 0.0463$ L or 46.3 mL

4. $t = 936$ s

5. a. $m(Al) = 19.3$ g   **b.** $V(CO_2) = 8.83$ L

6. 36.3 minutes

7. $I = 45$ A

8. The charge on the gold ion is $1+ (Au^+)$.

9. 0 g

10. $t = 39.6$ seconds

6.7 Exam questions
1. A
2. A
3. D
4. B
5. $V(F_2) = 1.39$ L

6.8 Review

6.8 Review questions

1. See table at the bottom of the page*

2. a. In the solid state, $Na^+(s)$ and $Cl^-(s)$ ions are held firmly in the ionic lattice.

   b. It is only at the electrodes that electrons can be accepted in a reduction reaction (at the cathode) and lost in an oxidation reaction (at the anode), because electrons move freely though the metal/graphite but not through the molten liquid.

   c. The external power supply drives the current flow in the external circuit, which in turn drives the ion flow in the internal circuit.

3. a. Products at the cathode: hydrogen gas and hydroxide ions
   Products at the anode: oxygen gas and water
   Minimum cell voltage: > 1.23 V
   Overall equation: $2H_2O(l) \rightarrow 2H_2(g) + O_2(g)$

   b. Products at the cathode: hydrogen gas and hydroxide ions
   Product at the anode: iodine
   Minimum cell voltage: > 1.37 V
   Overall equation:
   $2H_2O(l) + 2I^-(aq) \rightleftharpoons H_2(g) + 2OH^-(aq) + I_2(s)$

   c. Product at the cathode: zinc
   Product at the anode: bromine
   Minimum cell voltage: > 1.85 V
   Overall equation: $Zn^{2+}(aq) + 2Br^-(aq) \rightarrow Zn(s) + Br_2(l)$

   d. Product at the cathode: zinc
   Product at the anode: bromine
   Minimum cell voltage: > 1.85 V
   Overall equation: $Zn^{2+}(aq) + 2Br^-(aq) \rightarrow Zn(s) + Br_2(l)$

   e. Product at the cathode: zinc
   Product at the anode: silver ions
   Minimum cell voltage: > 1.56 V
   Overall equation:
   $Zn^{2+}(aq) + 2Ag(s) \rightarrow Zn(s) + 2Ag^+(aq)$

   f. Products at the cathode: hydrogen gas and hydroxide ions
   Product at the anode: iron(II) ions
   Minimum cell voltage: > 0.39 V
   Overall equation:
   $2H_2O(l) + Fe(s) \rightarrow H_2(g) + 2OH^-(aq) + Fe^{2+}(aq)$

4. a.

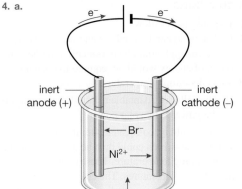

   NiBr$_2$(aq)

   b. Anode: $2Br^-(aq) \rightarrow Br_2(l) + 2e^-$
   Cathode: $Ni^{2+}(aq) + 2e^- \rightarrow Ni(s)$
   Overall equation: $2Br^-(aq) + Ni^{2+}(aq) \rightarrow Br_2(l) + Ni(s)$

   c. The minimum cell voltage is > 1.34 V.

   d. If nickel electrodes were used, nickel(II) ions would be the strongest oxidising agent and nickel metal would be the strongest reducing agent.
   Anode: $Ni(s) \rightarrow Ni^{2+}(aq) + 2e^-$
   Cathode: $Ni^{2+}(aq) + 2e^- \rightarrow Ni(s)$

5. a. The iron cathode is the negative electrode and the carbon anode is the positive electrode.

   b. Anode: $2Cl^-(l) \rightarrow Cl_2(g) + 2e^-$
   Cathode: $Na^+(l) + e^- \rightarrow Na(l)$

   c. The perforated iron plate is necessary to ensure that the products of electrolysis do not mix. If they did, a dangerous and spontaneous reaction between sodium and chlorine would occur.

   d. If an iron anode was used, the possibility of Fe being oxidised would need to be considered. Because Fe is a stronger reducing agent than Cl^- ions, it will be preferentially oxidised to Fe^{2+} ions, and Cl_2 would not be produced.

   e. Ca^{2+} ions are a weaker oxidising agent than Na^+ ions. Therefore, Na^+ ions are the preferred reactant. They are still able to undergo reduction to Na.

   f. $V(Cl_2)_{SLC} = 13.9\,L$

*1.

Electrolyte type	Electrodes	Reaction at	
		Anode(+)	Cathode(−)
Molten salt	Inert	$2Cl^-(l) \rightarrow Cl_2(g) + 2e^-$	$Na^+(l) + e^- \rightarrow Na(l)$
0.1 M aqueous salt solution	Inert	$2H_2O(l) \rightarrow O_2(g) + 4H^+(aq) + 4e^-$	$2H_2O(l) + 2e^- \rightarrow H_2(g) + 2OH^-(aq)$
6 M aqueous salt solution	Inert	$2Cl^-(aq) \rightarrow Cl_2(g) + 2e^-$	$2H_2O(l) + 2e^- \rightarrow H_2(g) + 2OH^-(aq)$

6. See figure at the bottom of the page*

7. **a.** $Q = 289$ C

   b. $n(e^-) = 0.00300$ mol

   c. $n(Cr) = 0.00100$ mol

   d. The charge on the chromium ion is 3+.

8. **a.** $m(Sn) = 1.10$ g

   b. No magnesium will be deposited

9. **a.** **i.** Anode (−):

      $Fe(s) + 2OH^-(aq) \rightarrow Fe(OH)_2(s) + 2e^-$

      Cathode (+):

      $NiO(OH)(s) + H_2O(l) + e^- \rightarrow Ni(OH)_2(s) + OH^-(aq)$

   ii. Cathode (−):

      $Fe(OH)_2(s) + 2e^- \rightarrow Fe(s) + 2OH^-(aq)$

      Anode (+):

      $Ni(OH)_2(s) + OH^-(aq) \rightarrow NiO(OH)(s) + H_2O(l) + e^-$

   b. Negative electrode: Iron in contact with iron(II) hydroxide. This ensures that the reaction can go in either direction depending on whether the cell is discharging or recharging.

     Positive electrode: Inert electrode in contact with a mixture of nickel(III) oxyhydroxide and nickel(II) hydroxide. This ensures that the reaction can go in either direction depending on whether the cell is discharging or recharging. An inert electrode material is used so as to not interfere with these desired reactions.

   c. Fe is oxidised during discharge so this electrode is the anode. The NiO(OH) electrode undergoes reduction as the oxidation state of Ni decreases from +3 to +2, so this electrode is the cathode.

   d. During recharge, $Ni(OH)_2$ is oxidised to NiO(OH). The oxidation state of nickel increases from +2 to +3. Therefore, this is the anode. At the other electrode, $Fe(OH)_2$ is reduced to Fe, so this electrode is the cathode.

10. **a.** $Q = 304$ C   **b.** $n(Cu) = 0.00157$ mol

   c. $n(e^-) = 0.00315$ mol   **d.** 96.5×10^3 C

   e. $N_A = 6.0 \times 10^{23}$

6.8 Exam questions

Section A — Multiple choice questions

1. D

2. B

3. C

4. D

5. D

6. A

7. B

8. B

9. B

10. D

Section B — Short answer questions

11. **a.** **i.**

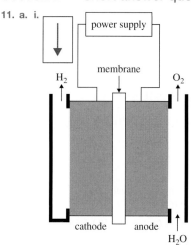

   Any one of the following justifications:
- Electrons move/are pushed/forced towards the negative electrode/cathode.
- To produce the H_2 from H^+, electrons are gained, hence electrons must move towards the cathode/ $2H^+ + 2e^- \rightarrow H_2$ at the cathode.
- In all cells, the anode produces electrons and the cathode accepts electrons.
- In all cells, electrons flow from the anode to the cathode.

   ii. The membrane allowed H^+ ions/protons to pass through in order to complete the circuit.

     The membrane kept the O_2 gas and H_2 gas separated

     OR

     the membrane prevented a spontaneous reaction between the products.

   b. **i.** $2H_2O(l) \rightarrow 2H_2(g) + O_2(g)$

     ii. $n(H_2) = 37.9$ mol

---

*6.

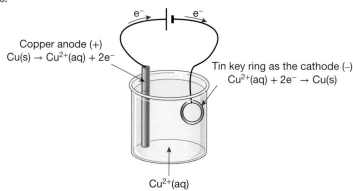

Copper anode (+)
$Cu(s) \rightarrow Cu^{2+}(aq) + 2e^-$

Tin key ring as the cathode (−)
$Cu^{2+}(aq) + 2e^- \rightarrow Cu(s)$

$Cu^{2+}(aq)$

12. a. Cathode

b. > 0.73 V

c. $2CO_2 + 3H_2O \rightarrow C_2H_5OH + 3O_2$

d. CO_2

e. 37 or 37.3 minutes

13. a. Either of:
- $Zn(s) + 2Ce(CH_3SO_3)_4(aq)$

   $\rightarrow 2Ce(CH_3SO_3)_3(aq) + Zn(CH_3SO_3)_2(aq)$
- $Zn(s) + 2Ce^{4+}(aq) \rightarrow Zn^{2+}(aq) + 2Ce^{3+}(aq)$

b. The oxidising agent is $Ce(CH_3SO_3)_4/Ce^{4+}$.

Either of the following justifications:
- The oxidation number of Ce decreases from +4 (in $Ce(CH_3SO_3)_4$) to +3 (in $Ce(CH_3SO_3)_3$).
- $Ce(CH_3SO_3)_4$ is reduced according to:

   $Ce(CH_3SO_3)_4(aq) + H^+(aq) + e^-$

   $\quad\quad +4 \quad\quad\quad\quad +3$

   $\rightarrow Ce(CH_3SO_3)_3(aq) + CH_3SO_3H(aq)$

c. $E_{cell} = E^0_{(oxidising\ agent)} - E^0_{(reducing\ agent)}$

   $= 1.64 - (-0.76)$

   $= 2.40\ V$

d. Either of the following half-equations:
- $Ce(CH_3SO_3)_3(aq)$

   $\rightarrow Ce(CH_3SO_3)_4(aq) + CH_3SO_3H(aq) + H^+(aq) + e^-$
- $Ce^{3+}(aq) \rightarrow Ce^{4+}(aq) + e^-$

e. Any one of the following:
- To prevent the oxidising agent and reducing agent from coming into direct contact
- To prevent a spontaneous redox reaction occurring when the reducing agent and oxidising agent come into contact with each other
- To separate the two half-cells
- To prevent the excessive release of thermal energy in the cell.

f. Any one of the following
- Loss/breakdown/oxidation/corrosion of the Zn electrode
- Side reactions at the electrodes
- Electrolysis of water during recharging
- $(-)$: $2H^+(aq) + 2e^- \rightarrow H_2(g)$

   $(+)$: $2H_2O(l) \rightarrow O_2(g) + 4H^+(aq) + 4e^-$
- Significant temperature change
- Build-up of gases around electrode.

g. Fuel cell: supply of reactants $(Zn^{2+}/Ce^{4+}/Ce^{3+})$ from outside the cell

Secondary cell: rechargeable/discharge reaction can be reversed

14. Acceptable strengths of the experimental design include the following:

See table at the bottom of the page*

Suggested improvements or modifications could include any two of the following:

See table at the bottom of the next page**

Comments on the validity of a conclusion (referencing directly or indirectly one of Faraday's Laws) include the following:
- Results show limited consistency with Faraday's First Law — m(Ni) deposited should be proportional to charge passed and mass of Ni deposited is higher for the higher current.
- Trials Y and Z reflect Faraday's First Law — show mass is proportional to charge.

   For Y:

   $\dfrac{m}{current} = \dfrac{0.201}{0.54}$

   $= 0.372$

   For Z:

   $\dfrac{m}{current} = \dfrac{0.188}{0.50}$

   $= 0.376$

   For X:

   $\dfrac{m}{current} = 0.256$

   Hence, more trials are needed to confirm the relationship.
- Although X and Y support this conclusion (for Faraday's First Law), to some degree, more trials in a more tightly controlled experimental set-up should be considered.
- To test the validity of Faraday's First Law — constant voltage — the same starting concentration and consistent electrode properties are needed. Then the effect of time and current on mass deposited needs to be tested.
- The experiment only attempts to test Faraday's First Law. To test his Second Law, electrolytes with ions of different charges need to be investigated. As this has not been done, a conclusion about Faraday's Laws in general is not valid.
- Giving a logical consideration showing some appreciation of the link (or lack thereof) between the data collected and Faraday's Laws.

*14.

Strength	Explanation
Time was the same in all trials	Reduced the number of variables
Each brass key was sanded before weighing	Removed possible contaminants from the key's surface
Three trials were carried out	Allowed for better verification of results/to test whether $m \propto Q$
Nickel anode was used	Maintained electrolyte concentration

15. a. Negative (–)

b. $Li^+ + e^- + C_6 \rightarrow LiC_6$

c. Li is a strong reducing agent that reacts readily with water.
- $Li(s) \rightarrow Li^+(aq) + e^-$ /
  $2H_2O(l) + 2e^- \rightarrow H_2(g) + 2OH^-(aq)$
- $2Li(s) + 2H_2O(l) \rightarrow H_2(g) + 2LiOH(aq)$

Any one of the following consequences of lithium making contact with water:
- Hydrogen gas is explosive.
- Heat is generated.
- LiOH is a strong base.
- Current does not flow.

d. The movement of lithium ions into and out of the electrodes enables the reactions at the electrodes to be reversed.

e. Responses include:
- A secondary cell is more convenient for on–off usage since it does not need a continuous external supply of reactants.

- A secondary cell can be recharged (electrically), whereas a fuel cell continuously needs fresh reactants.
- Secondary cells are usually cheaper than fuel cells.
- Secondary cells are more suitable for most of today's electronic devices.
- Secondary cells are storage devices.

Unit 3 | Area of Study 2 review

Section A — Multiple choice questions

1. C
2. D
3. B
4. A
5. C
6. C
7. D

**14.

Improvement or modification	Justification
A suitable concentration of $Ni^{2+}(aq)$ is used	The products of the electrolysis are dependent on the concentration of $Ni^{2+}(aq)$.
Ensure that the $Ni^{2+}(aq)$ solution is pure/free of other metal ions	Prevents other metals being deposited on the key.
Ni electrode should be weighed before and after each experiment	This will enable a check of whether the amount of Ni oxidised at the anode is the same as the amount plated onto the key/give more accurate results.
Ensure that voltages are below 2.06 V	High voltages may lead to water reacting depending on solution concentration. Results are unreliable due to possible O_2 and/or H_2 production.
Maintain a constant voltage (without stating constant current)	Possible side reactions if voltage is too high/ eliminates another variable
Use constant voltage and current	Multiple trials for averaging (assuming first interpretation of aim, as mentioned above)
Current should be measured at regular intervals during the plating, not just at the start	Variations in current make it difficult to accurately calculate the amount of charge passing through the cell. By taking an average of current readings over the 20 minutes, a more accurate value is obtained.
Patting dry with paper towel is inaccurate	The keys need to be dried to constant mass to ensure that they are dry.
Keys should have same shape and surface area	Reduces variables that could affect results
Use a pure Cu electrode sheet rather than brass keys	Reduces possibility of side reactions. Other components of the brass key may react with Ni^{2+} in solution. The mass of the key would be less than predicted.
Carry out more trials at different currents	Get a more consistent set of results from which to draw conclusions/to obtain concordant results
Extend the time for electroplating	Allows for greater mass to be deposited and hence, weighing errors are less significant
Keep current constant and vary the time of electroplating	To investigate further the relationship between mass deposited and the amount of charge passed
Wash key after sanding/before weighing	Removes fine particles/soluble impurities
Connect circuit before key is added	Avoid reaction between Zn in brass and $Ni^{2+}(aq)$

8. B
9. D
10. D
11. D
12. D
13. B
14. D
15. C
16. B
17. B
18. C
19. D
20. C

Section B — Short answer questions

21. a. A catalyst increases the rate of a reaction by lowering the activation energy, which results in more particles having sufficient energy for successful collisions to occur in a given time.

    b. A catalyst has no effect on the extent of a reaction because it is not involved in the reaction.

22. a. $K = 9.5 \times 10^{-4}$ M

    b. The equilibrium position lies to the left, favouring the products.
    This is evident because the concentration of the reactants increased as the concentration of the products decreased. Also, the value for K is very low, which is an indicator that equilibrium lies to the left.

    c. i. An increase in temperature would increase the rate of reaction because the increased average kinetic energy of particles would result in an increased frequency of successful collisions.

       ii. Equilibrium will be shifted to the left, increasing the concentration of the reactants and decreasing the yield.

    d. i. A decrease in pressure will decrease the rate of reaction because the particles will be more spread out and successful collisions will therefore be less likely to occur.

       ii. The equation has more moles of gas on the products side than the reactants side. To counteract the pressure decrease, the system will act to increase pressure, shifting right where there are more moles, favouring the products.

23. a. Oxidation: $2H_2O(l) \rightarrow O_2(g) + 4H^+(aq) + 4e^-$
    Reduction: $2H_2O(l) + 2e^- \rightarrow H_2(g) + 2OH^-(g)$

    b. $2H_2O(l) \rightarrow 2H_2(g) + O_2(g)$

    c. > 2.06 V

    d. The product at the anode would be $Cl_2(g)$.

    e. $Mg^{2+}(l) + 2e^- \rightarrow Mg(l)$

24. The charge on the ion is $+3\,(Cr^{3+})$.

25. a. i. Electrode 2

       ii. Electrode 2

       iii. The mass will decrease.

    b. $Zn(s) + 2Ce^{3+}(aq) \rightarrow Zn^{2+}(aq) + 2Ce^{2+}(aq)$

7 Structure, nomenclature and properties of organic compounds

7.2 Characteristics of the carbon atom

Practice problem 1
Energy $= 6.6 \times 10^3$ kJ

Practice problem 2
181 g/100 g

7.2 Exercise
1. B
2. B
3. C
4. B
5. Carbon is able to form four bonds, which can include single, double or triple bonds. Carbon compounds are able to form rings or long chain compounds.
6. Difference in bond angles $= 60°$
7. Energy $= 1134$ kJ
8. Total energy $= 1812$ kJ mol^{-1}
9. C–C: 346 kJ mol^{-1}
   C–O: 358 kJ mol^{-1}
   H–C: 414 kJ mol^{-1}
   O–H: 463 kJ mol^{-1}
   C=C: 614 kJ mol^{-1}
10. Ethane

7.2 Exam questions
1. B
2. B
3. C=C bonds are stronger than C–C bonds because they have twice the number of electrons between the nuclei. This causes a shorter, stronger and more stable bond.
4. Energy $= 2.03 \times 10^3$ kJ
5. a. i. Energy $= 967$ kJ mol^{-1}

      ii. To vaporise the water the hydrogen bonds between the water molecules must be broken; however, to separate the hydrogen atoms the much stronger covalent bonds must be broken, which requires considerably more energy.

   b. O–O bond

   c. The hydrogen peroxide molecule would be less stable because the O–O bond is easier to break.

7.3 Structure and systematic naming

Practice problem 3
Semi-structural formula: $CH_3CH_2COCH_3$
Skeletal structure:

Practice problem 4

4-ethyl-2,3-dimethylhexane

7.3 Exercise

1. D
2. B
3. B
4. $C_{18}H_{38}$
5. a. Molecular formula: C_3H_8
   b. Empirical formula: C_3H_8
   c. Semi-structural formula: $CH_3CH_2CH_3$
   d. Skeletal structure:

   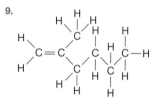

6. a. 2,3,3-trimethylpentane
   b. Cyclohexane
   c. Cyclopropane
7. a. There are two methyl (CH_3) groups and one ethyl ($-CH_2CH_3$) group.
   b. 5-ethyl-3,3-dimethyloctane
8. Unsaturated means that there is at least one multiple bond between carbons in the molecule.
9.

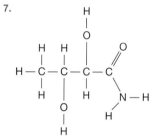

10. Hex-3-yne

7.3 Exam questions

1. D
2. C
3. B
4. D
5. a.

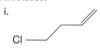

   b. The longest chain has five carbons and there is a methyl group attached to the third carbon. The correct name is 3-methylpentane.
   c. C_6H_{14}
   d. Alkanes
   e. C_nH_{2n+2}

7.4 Functional groups

Practice problem 5

a. Functional group: Carboxyl
   Homologous series: Carboxylic acids
b. 2,4-dimethylpentanoic acid

Practice problem 6

3-amino-2-methylpropanal

7.4 Exercise

1. D
2. B
3. B
4. $CH_3CHNH_2(CH_2)_3OH$
5. No. The IUPAC name is 5-chlorohexan-2-amine because the highest priority functional group present is the amino group, so the suffix '*-amine*' is used.
6. a. 4,5-dibromohex-2-ene
   b. $CH_3CHBrCHBrCHCHCH_3$ or $CH_3(CHBr)_2(CH)_2CH_3$
   c. $C_6H_{10}Br_2$
   d. C_3H_5Br
7.

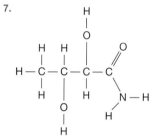

   The semi-structural formula is $CH_3(CHOH)_2CONH_2$ or $H_2NCO(CHOH)_2CH_3$.
8. 1, 2-dichloropropane
9. a. i. 4-chlorobut-1-ene
       ii. 2-bromo-3-chloro-4-methylpentane
       iii. 4-hydroxy-3-methylbutanoic acid
       iv. 2-bromopentan-3-one or 2-bromo-3-pentanone
       v. 3-aminopropanal
   b. *Condense* the structural formulas to obtain the semi-structural formulas.
       i. $H_2CH(CH_2)_2Cl$ or $Cl(CH_2)_2CHCH_2$
       ii. $CH_3CHBrCHClCH(CH_3)_2$ or $(CH_3)_2CHCHClCHBrCH_3$
       iii. $CH_2OHCH(CH_3)CH_2COOH$ or $HOOCCH_2CH(CH_3)CH_2OH$
       iv. $CH_3CHBrCOCH_2CH_3$ or $CH_3CH_2COCHBrCH_3$
       v. $H_2N(CH_2)_2CHO$ or $CHO(CH_2)_2NH_2$
   c. Remove the carbon and hydrogen atoms from the structural formula and join the bonds to form the skeletal structures.
       i.

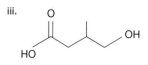

       ii.

       iii.

iv.

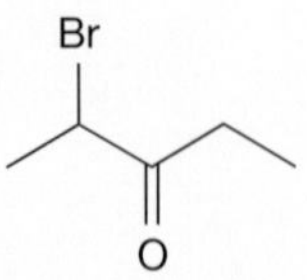

v.

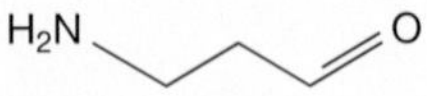

10.

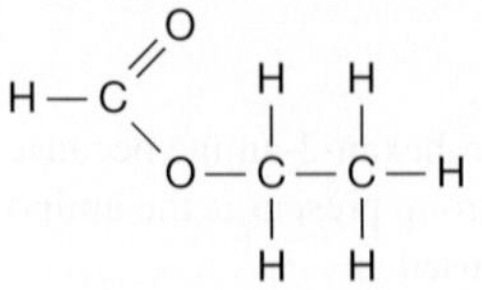

7.4 Exam questions

1.

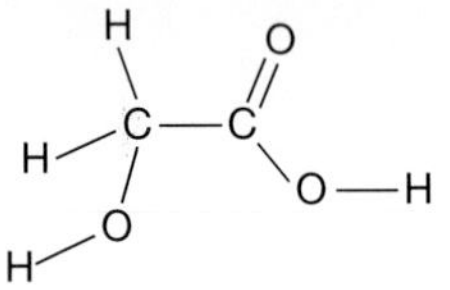

2. C

3. B

4. B

5. a.

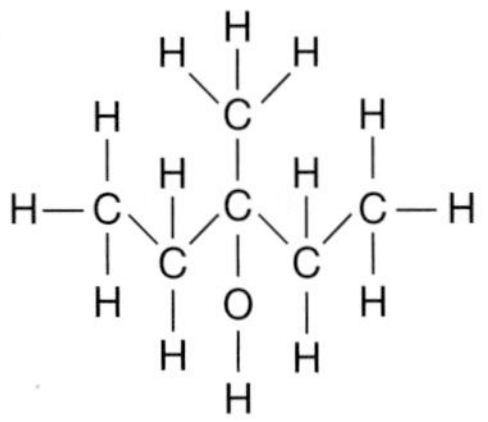

b. $C_6H_{14}O$

c. $CH_3CH_2C(CH_3)OHCH_2CH_3$

7.5 Isomers

Practice problem 7

There are two butanols possible: butan-1-ol and butan-2-ol.

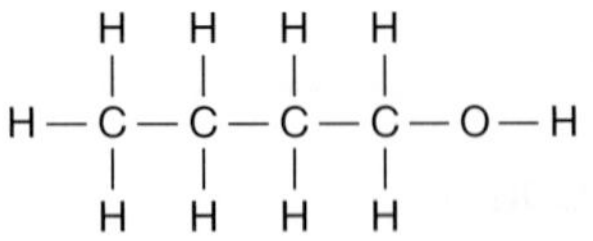

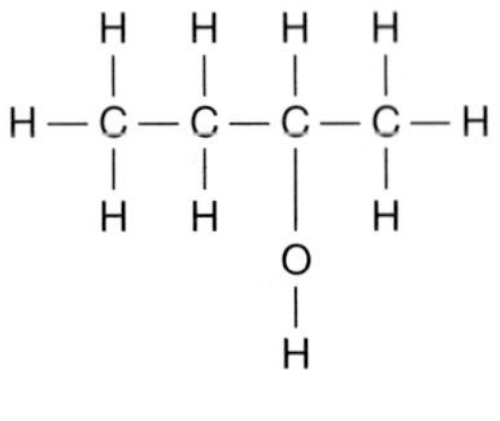

There are two propanols possible: 2-methylpropan-1-ol and 2-methylpropan-2-ol.

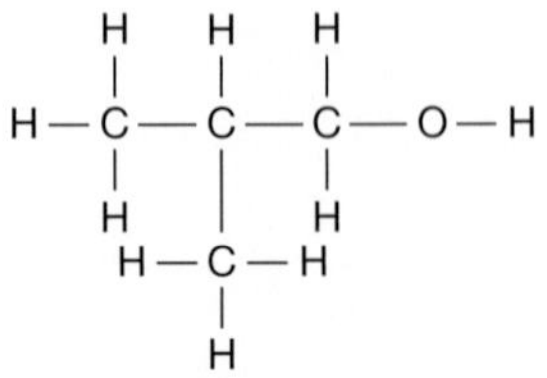

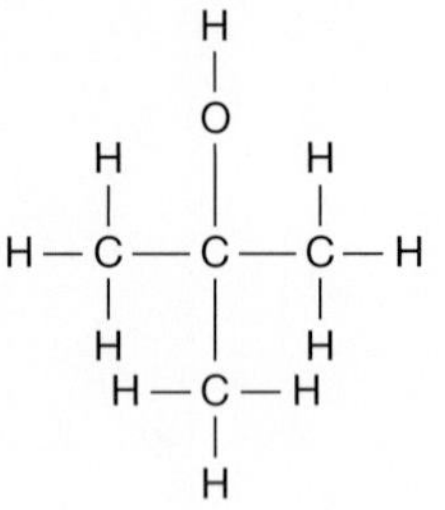

7.5 Exercise

1. C

2.

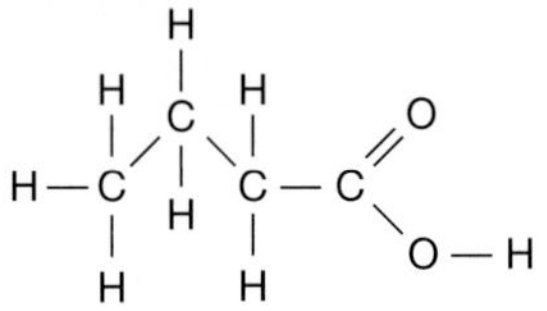

and

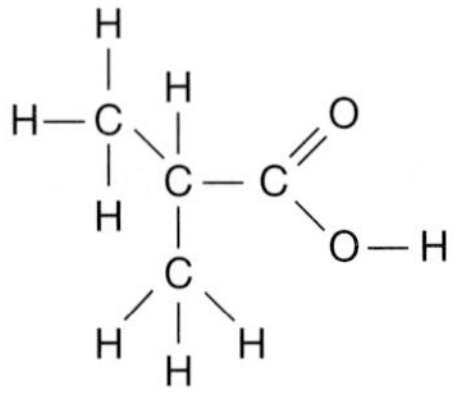

3.

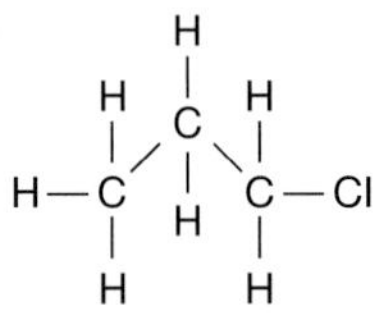

1-chloropropane 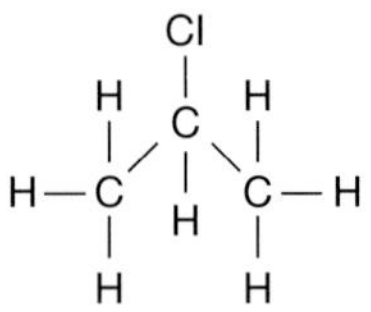   2-chloropropane

4.

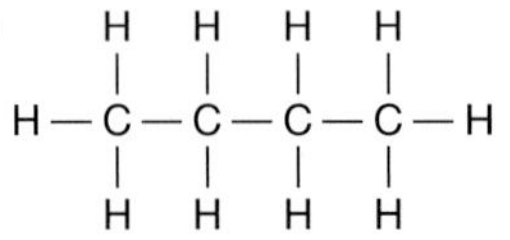

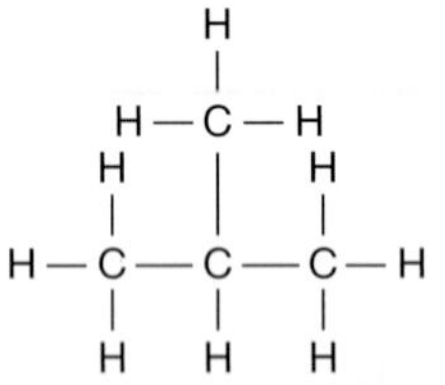

5.

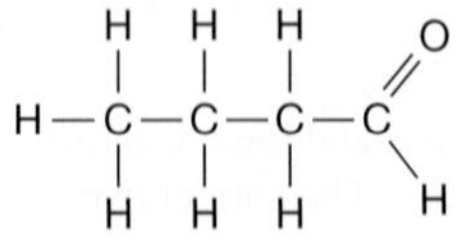

Butanal

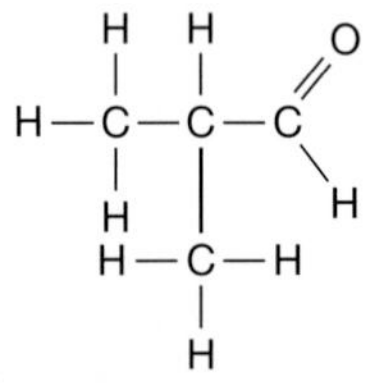

2-methylpropanal

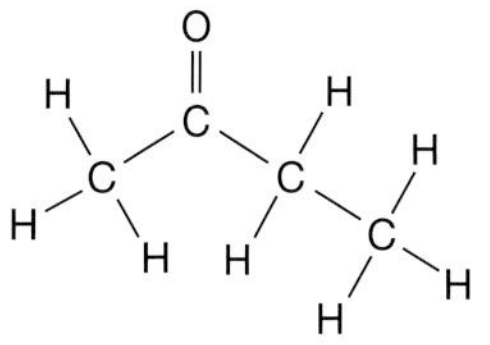

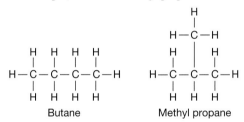

Butanone

6. a. Positional

b. Not isomers — same molecule

c. Functional

d. Functional

e. Chain

7. a.

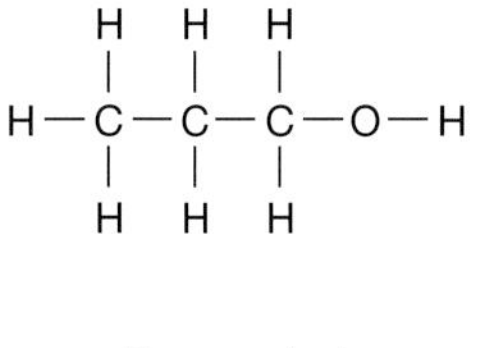

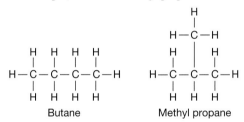

Propan-1-ol 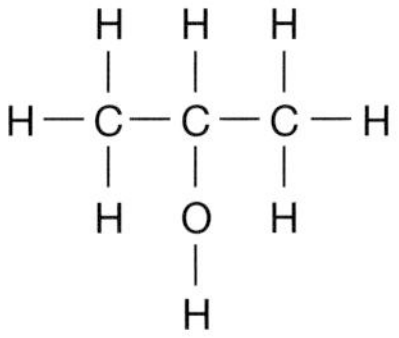 Propan-2-ol

b. Propan-1-ol is a primary alcohol. Propan-2-ol is a secondary alcohol.

c. Positional isomerism

8. a. i. Chain isomers involve a rearrangement of branches and a chain.

ii. For example, butane and methyl propane:

Butane Methyl propane

b. i. Positional isomers have the same basic structure of carbon atoms, but the functional group is attached to a different carbon atom.

ii. For example, 1-chlorobutane and 2-chlorobutane:

1-chlorobutane 2-chlorobutane

7.5 Exam questions

1. C

2. B

3. a. CH_3COCH_3

b.

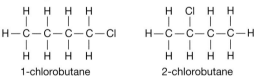

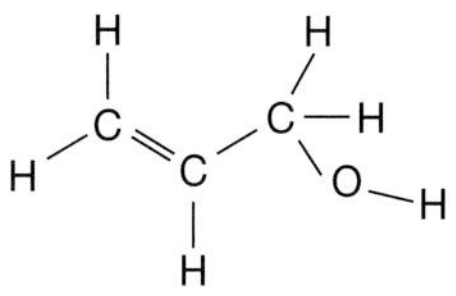

4. A

---

5. Acceptable structures include:

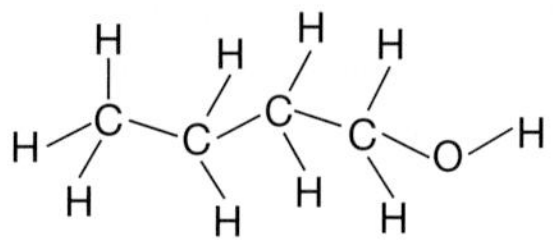

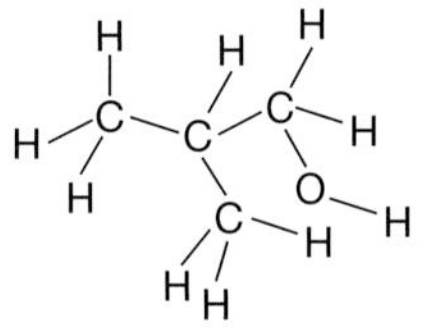

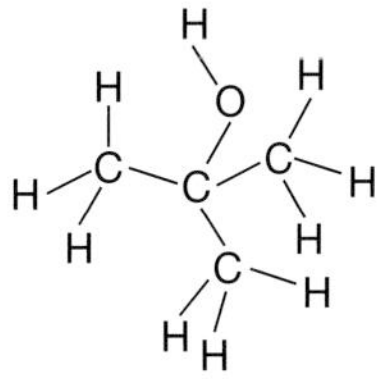

7.6 Trends in physical properties

Practice problem 8

Propane:

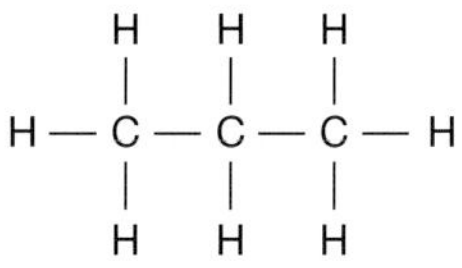

H — C — C — C — H

1-chloropropane:

H — C — C — C — H

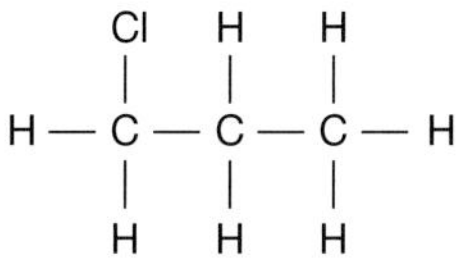

Propane is a symmetrical hydrocarbon with weak dispersion forces only operating between molecules. This results in it boiling at a relatively low temperature.

1-chloropropane has the electronegative Cl atom that causes the molecule to become polar. Permanent dipole–dipole attractions also exist between molecules, which are much stronger than the dispersion forces. This explains why its boiling point is considerably higher than that of propane.

7.6 Exercise

1. The intermolecular forces that exist between hydrocarbons are dispersion forces, also known as van der Waals forces. They are temporary dipole moments that form as electrons move around, creating attractions between neighbouring molecules with opposing temporary charges.
The intramolecular forces are the forces within the molecule that hold it together. These are very strong covalent bonds.

2. a. Dispersion forces and hydrogen bonding

b. Dispersion forces only

c. Dispersion forces and dipole–dipole attractions

d. Dispersion forces and hydrogen bonding

3. Both methanol and ethanol are soluble in water because they can form hydrogen bonds with water molecules. Methane and ethane, as non-polar molecules, are unable to form hydrogen bonds with water molecules and are therefore insoluble.

4. Butanamide has a higher boiling point than ethyl ethanoate because it has hydrogen bonding between its molecules, while ethyl ethanoate has weaker dipole–dipole attractions between its molecules.

5. a. The fact that the candle wax is a solid at room temperature and melts at temperatures in a range of 50–60 °C suggests the alkane chains are reasonably long compared to those that exist in the gas and liquid states at SLC.

   b. Dispersion forces between the wax and oils allow miscibility.
   Sample responses can be found in your digital formats.

6. a. Hexane is not suitable for removing salt from water.
   Sample responses can be found in your digital formats.

   b. Hexane will be able to separate oil from soy beans.
   Sample responses can be found in your digital formats.

   c. Hexane will be able to remove oil contaminants in water.
   Sample responses can be found in your digital formats.

7. Carbon tetrachloride is non-polar, reducing its ability to dissolve in water. Dichloromethane has permanent, unsymmetrical dipoles, resulting in higher solubility. However, the smaller dichloromethane, collectively, has fewer intermolecular forces acting, resulting in a lower boiling point.

8. Ethyl ethanoate and ethyl butanoate both have dipole–dipole attractions with water. Ethyl butanoate molecules, with their longer hydrocarbon sections, are far more likely to form dispersion forces between themselves or with non-polar molecules. As a result, it is easier for the smaller ethyl ethanoate to dissolve in water. The boiling point also increases as the size of the ester increases due to the number of intermolecular forces acting. We observe this trend in a homologous series.

9. a. Homologous series

   b. The boiling points of alcohols increase with molecular size because the non-polar sections of the molecules increase, and so the dispersion forces increase. The solubility decreases because for smaller alcohol molecules, the polarity of the –OH group allows hydrogen bonding with water, but as the non-polar section of the molecules increase in size, the solubility decreases.

   c.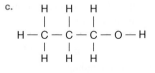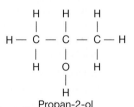

   Propan-1-ol   Propan-2-ol

10. a. Octane, C_8H_{18}

    b. Octane, C_8H_{18}

    c. Methylbutane, C_5H_{12}

7.6 Exam questions

1. A

2. B

3. A

4. Responses may include:
   - correctly describing the bonding (dispersion force attraction) between butane molecules
   - correctly describing the bonding (hydrogen bonding) between propan-1-ol molecules
   - indicating that the higher boiling point of propan-1-ol is due to the hydrogen bonding between propan-1-ol molecules being stronger than the dispersion forces bonding between butane molecules.

5.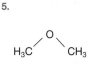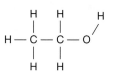

   Compound A   Compound B

7.7 Review

7.7 Review questions

1. a. –OH, *–ol*

   b. –COOH, *–oic* or *–oic acid*

2. Bond energy = 168 kJ

3. a. Pent-2-ene

   b. Chloroethene

   c. Dichloro-difluoromethane

   d. Ethanol

   e. 2,2-dimethylpropane

   f. Hexan-2-ol

   g. 1,1-dibromo-2,2-dichloroethane

4. a.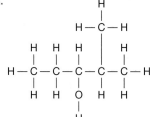

   b.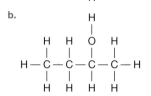

   c.

d.

```
                     H
                     |
                 H — C — H
     H   H   H       |        H
     |   |   |       |        |
 H — C — C — C       C — C — H
     |   |   |       |   |
     H   H   |       O   H
         H — C — H       |
             |           H
             H
```

e.

```
     H   H   H   H       O
     |   |   |   |      //
 H — C — C — C — C — C
     |   |   |   |      \
     H   H   H   H       O — H
```

f.

```
         H
         |
     H — C — H
         |
     H   |       O
     |   |      //
 H — C — C — C
     |   |      \
     H   H       O — H
```

g.

```
     H   H   H       O
     |   |   |      //
 H — C — C — C — C
     |   |   |      \
     H   H   H       O — H
```

h.

```
     H   H   H   H   O
     |   |   |   |   ||
 H — C — C — C — C — C — H
     |   |   |   |
     H   H   H   H
```

i.

```
     H   Cl  H           H
     |   |   |          /
 H — C — C — C — N
     |   |   |          \
     H   H   H           H
```

j.

```
       O
      //
 H — C         H   H   H   H
      \        |   |   |   |
       O — C — C — C — C — H
               |   |   |   |
               H   H   H   H
```

k.

```
     H   H   O   H
     |   |   ||  |
 H — C — C — C — C — H
     |   |       |
     H   H       H
```

l.

```
     H           H   H
     |           |   |
 H — C — C ≡ C — C — C — H
     |           |   |
     H           H   H
```

5. 3-methylbutan-2-amine

6.

Compound	Intermolecular forces
$CH_3CH_2CH_2OH$	Hydrogen bonds, dispersion forces
$CH_3CH_2OCH_2CH_3$	Dispersion forces and dipole–dipole attractions
$CH_3CH_2CH_2F$	Dispersion forces and dipole–dipole attractions
$CH_3CH_2N(CH_3)_2$	Dispersion forces and dipole–dipole attractions

7. CH_3CH_2OH, $CH_3CH_2CH_2OH$, CH_3CH_3Cl, $CH_3CH_2CH_3$

8. Propan-1-ol contains the polar O–H bond, which will allow it to hydrogen bond to other propanol molecules. This is a stronger form of intermolecular bonding, resulting in a higher boiling point.

9. CH_4, C_2H_2, C_6H_6

10. a.

```
     H   H   H   H            H   Cl  H   H
     |   |   |   |            |   |   |   |
 H — C — C — C — C — Cl    H — C — C — C — C — H
     |   |   |   |            |   |   |   |
     H   H   H   H            H   H   H   H
      1-chloropropane          2-chloropropane
```

```
     H   CH₃  H              H   CH₃  H
     |   |    |              |   |    |
 H — C — C —  C — H      H — C — C —  C — H
     |   |    |              |   |    |
     H   Cl   H              Cl  H    H
  2-chloro-2-methylpropane   1-chloro-2-methylpropane
```

b.

1-chlorobutane

```
         Cl
         |
    (2-chloro-2-methylpropane skeletal structure)
```
2-chloro-2-methylpropane

```
    (2-chlorobutane skeletal structure)
    Cl
```
2-chlorobutane

```
    (1-chloro-2-methylpropane skeletal structure)
                      Cl
```
1-chloro-2-methylpropane

c. 1-chlorobutane: $CH_3CH_2CH_2CH_2Cl$
   2-chlorobutane: $CH_3CHClCH_2CH_3$
   2-chloro-2-methylpropane: $CH_3CCl(CH_3)_2$
   1-chloro-2-methylpropane: $CH_2ClCH(CH_3)_2$

7.7 Exam questions

Section A — Multiple choice questions

1. B
2. D
3. D
4. A
5. A
6. C
7. C
8. C
9. B
10. D

Section B — Short answer questions

11. a. i.

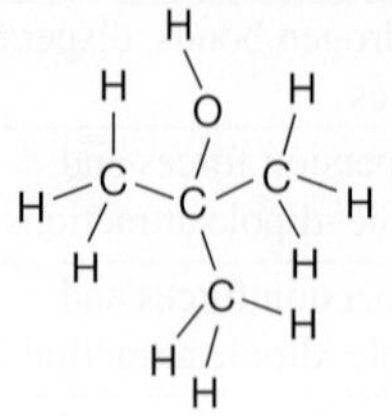

ii. C_4H_6

iii. 2,3-dibromo-4-methylhexane

12. a. Either of:

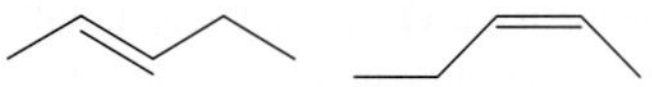

b. i. Pentan-2-ol/2-pentanol and pentan-3-ol/3-pentanol

ii. $CH_3COCH_2CH_2CH_3$

13. a. Heptanoic and pentanoic acids are both straight-chain carboxylic acids, but pentanoic acid molecules $(CH_3(CH_2)_3COOH)$ have shorter hydrocarbon chains than heptanoic acid molecules $(CH_3(CH_2)_5COOH)$. Thus, pentanoic acid has weaker dispersion force attraction between its molecules and hence a lower melting point

OR

heptanoic acid has a higher molecular mass than pentanoic acid and hence stronger intermolecular dispersion forces and a higher melting point.

b. Heptanoic and 2-methylhexanoic molecules have the same molecular formula $(C_8H_{13}COOH)$, but heptanoic acid molecules are straight chains and 2-methylhexanoic acid molecules are branched.

The branch in the 2-methylhexanoic acid hydrocarbon chain prevents the 2-methylhexanoic acid molecules from packing together as effectively as heptanoic acid molecules, hence the intermolecular forces are much weaker than in heptanoic acid, and 2-methylhexanoic acid has a lower melting point.

14. a.

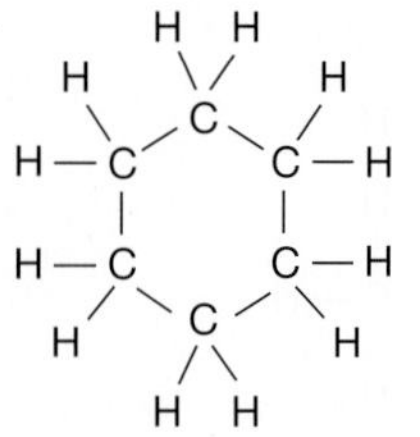

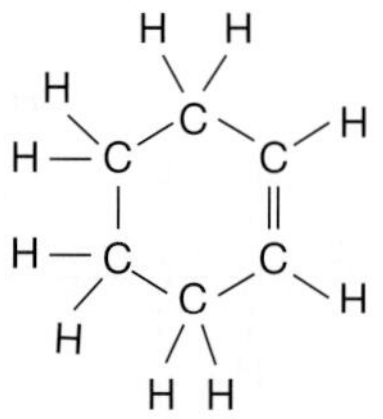

b. Any straight-chain isomer of hexene *and* any isomer of hexadiene or hexyne. For example:

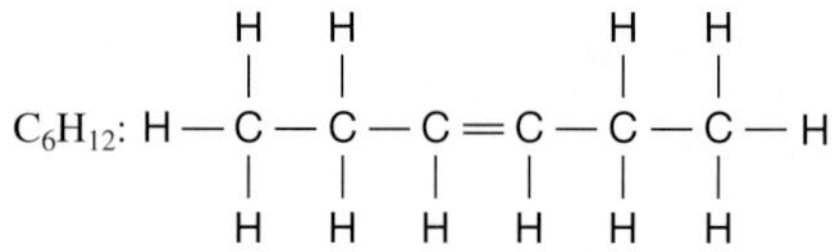

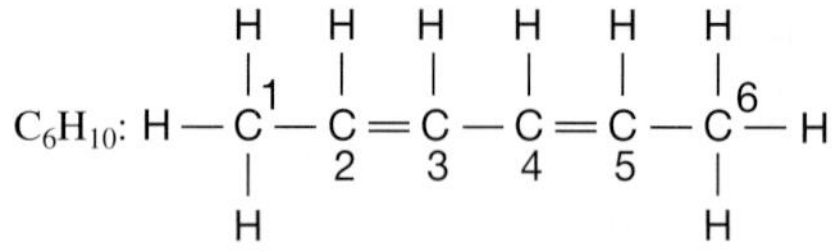

c. Alkene, alkyne

15. a.

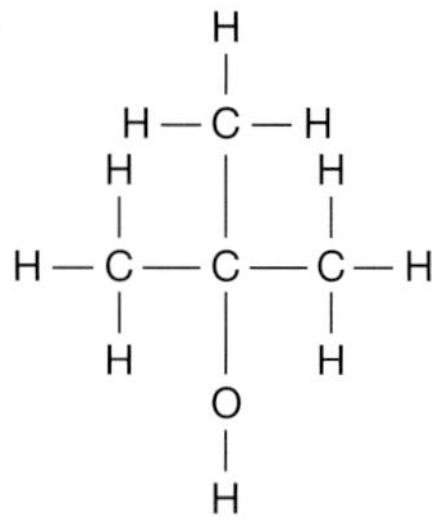

b. Hydroxyl group

c. This is a tertiary alcohol because the carbon bonded to the –OH group is attached to three other carbon atoms (alkyl groups).

d. $CH_3C(CH_3)OHCH_3$

e.

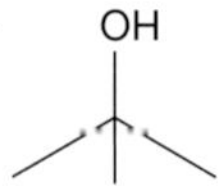

8 Reactions of organic compounds

8.2 Substitution, addition and oxidation reactions

Practice problem 1

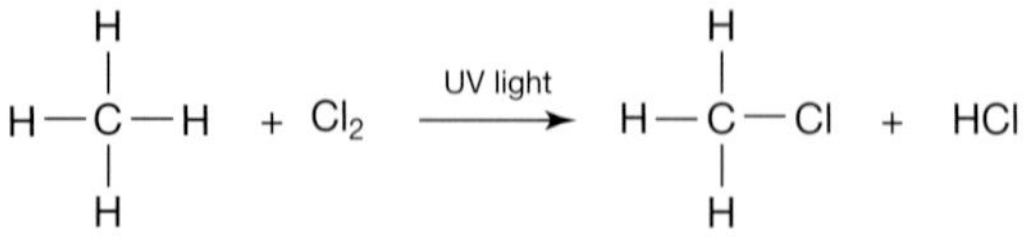

Separate chloromethane by fractional distillation.

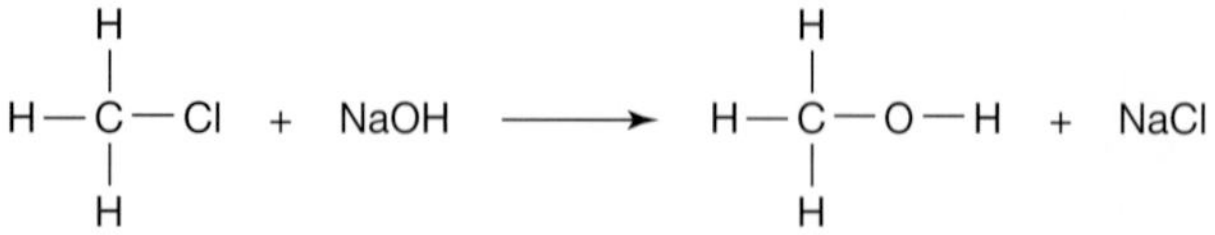

Practice problem 2

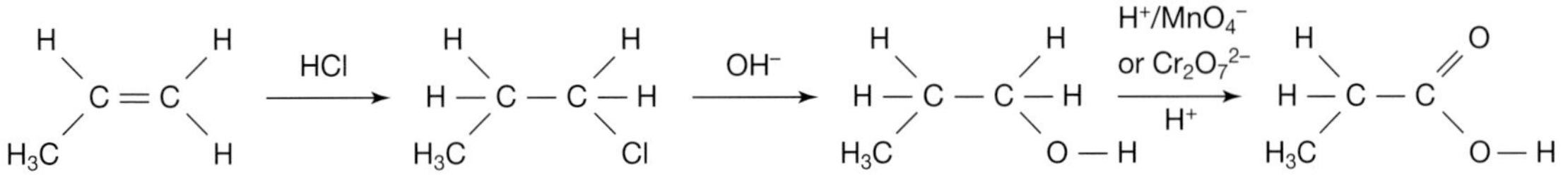

8.2 Exercise

1. 1,1-dichloroethane would be made in two steps using chlorine gas and UV light to catalyse the reaction.

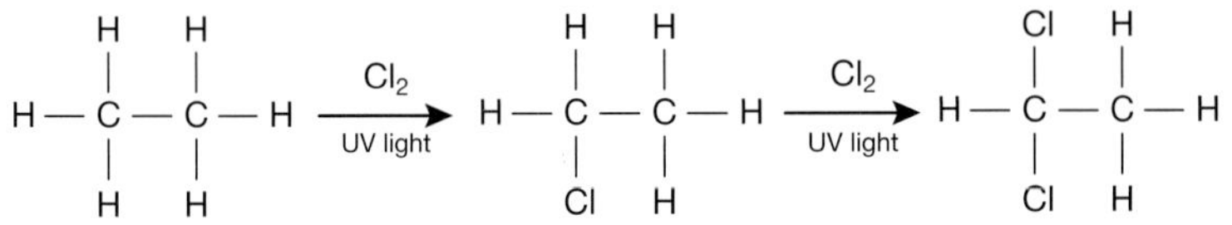

2.

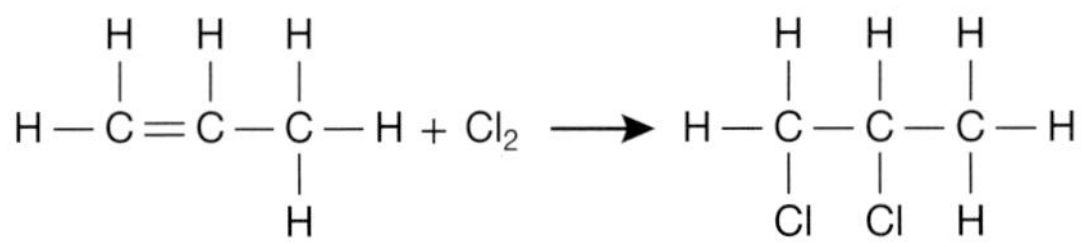

1,2-dichloropropane

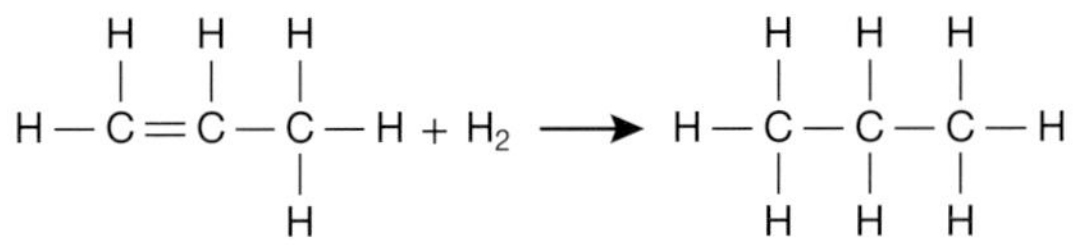

Propane

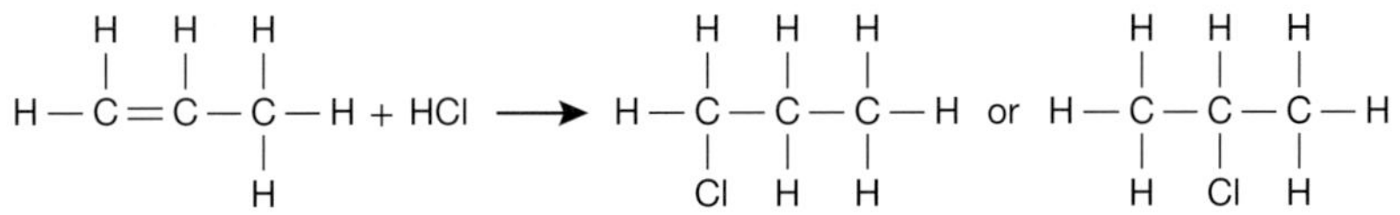

1-chloropropane   2-chloropropane

3. If bromine was added to ethane, the red colour of the bromine would remain because there would be no reaction. However, if bromine was added to ethene, the bromine would undergo an addition reaction with the ethene and the reaction mixture would turn colourless. Only unsaturated molecules react with the bromide.

4. a. 3-methylbutan-1-ol

   b. 3-methylbutanoic acid. Sample responses can be found in your digital formats.

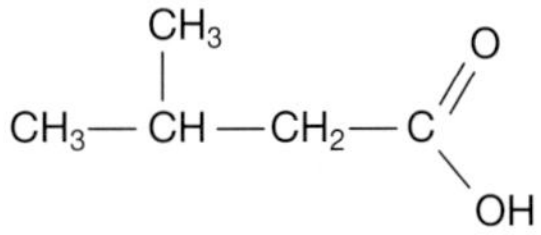

5. The O–H group needs to be on the end of a chain to be able to form the three covalent bonds with the two oxygen atoms in a carboxyl functional group.

6. C

7. a. Hydrogen gas $H_2(g)$ needs to react with ethene in the presence of a metal catalyst, such as Ni or Pt.

   b. Steam $H_2O(g)$, heat and an acid catalyst (e.g. H_3PO_4) are required.

   c. HCl(g) is required.

   d. First add HCl(g) to produce chloroethane, then add $NH_3(g)$ with an aluminium catalyst.

8. a. $CH_3CH{=}CHCH_3 + HCl \rightarrow CH_3CH_2CHClCH_3$

   b.

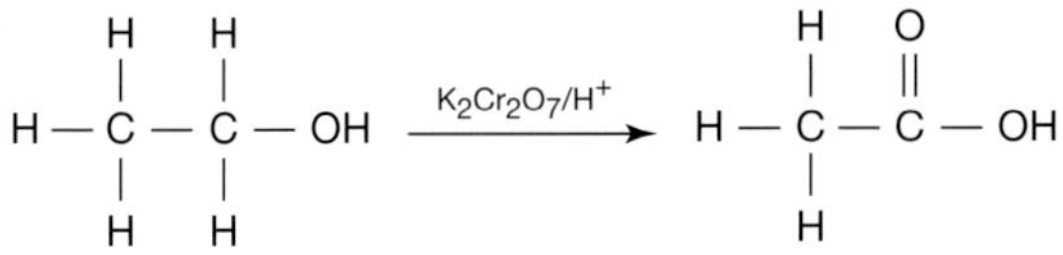

   c.

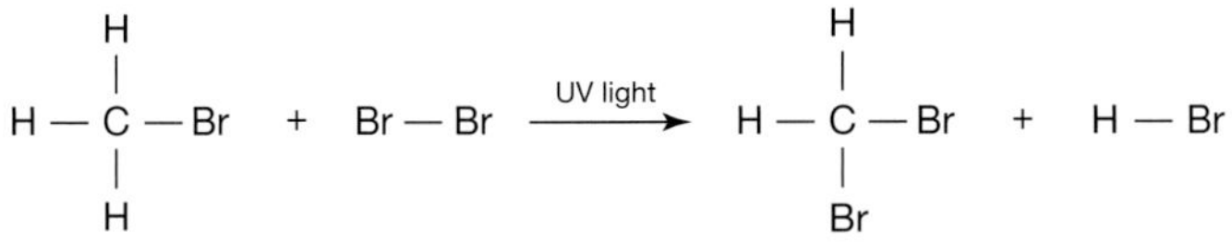

d.

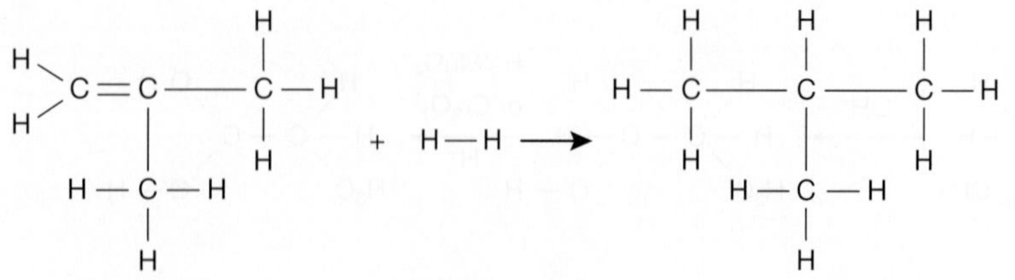

9. $CH_4 \xrightarrow{\text{UV light and } Cl_2} CH_3Cl \xrightarrow{\text{Concentrated } NH_3} CH_3NH_2$

8.2 Exam questions

1. A
2. C
3. a.

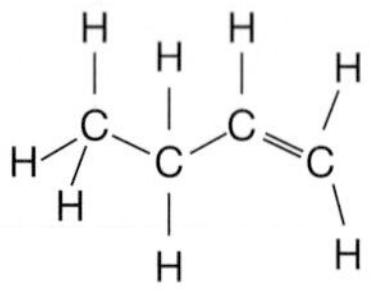

   b. H_2O and H_3PO_4 (acid catalyst)

   c. Butan-1-ol/1-butanol

   d. $CH_3CH_2CH_2COOH$

4. a. i. HCl

    ii. Addition reaction

   b. i. NH_3

    ii. Substitution reaction

5.

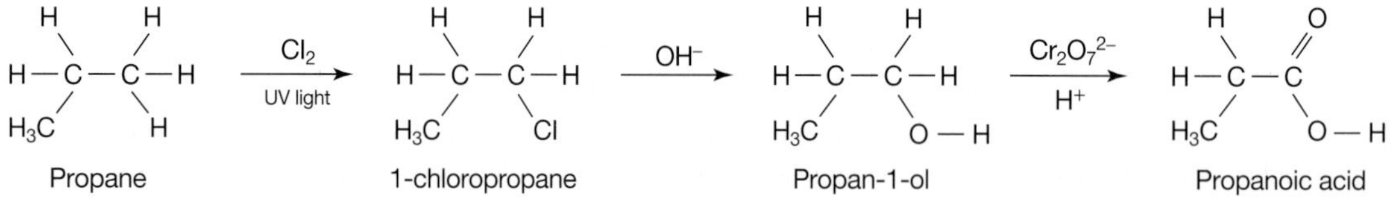

Propane   1-chloropropane   Propan-1-ol   Propanoic acid

8.3 Condensation and hydrolytic reactions of esters

8.3 Exercise

1. Condensation and hydrolysis of esters both involve water molecules and require sulfuric acid as a catalyst, but condensation uses concentrated acid and hydrolysis involves dilute acid. Condensation involves building up an ester molecule, whereas hydrolysis involves breaking down an ester molecule. Water is consumed in a hydrolytic reaction and produced in a condensation reaction. A peptide link is formed in a condensation reaction and broken in a hydrolytic reaction.

2. To convert ethene into propyl ethanoate, first add $H_2O(g)$, heat and H_3PO_4 to produce ethanol. Then add acidified $K_2Cr_2O_7$ to produce ethanoic acid. React this with propan-1-ol in the presence of concentrated H_2SO_4 to produce the ester.

3. a. Propanol and ethanoic acid

   b. Ethanol and butanoic acid

   c. Ethanol and propanoic acid

4. a.

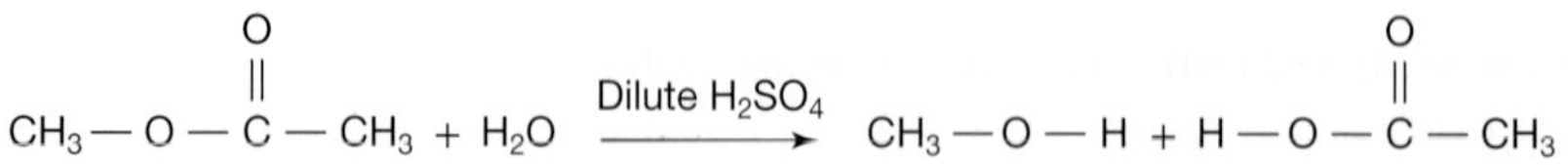

   b.

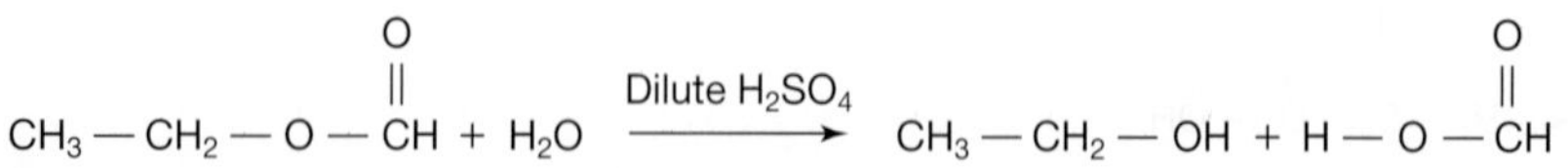

5. a. Ethyl butanoate

   b. Methyl butanoate

6. Biodiesel is produced from a triglyceride and an alcohol such as methanol. The triglyceride can come from plant or animal fats. The methanol can be made from natural gas or using glycerol produced in the manufacture of biodiesel.

7. *Condensation* is a general term referring to the combination of simpler molecules to form a more complex molecule with the loss of a small molecule.

Esterification relates to condensation reactions that result in the formation of an ester.

Transesterification relates to converting one ester into another ester.

8. $CH_3(CH_2)_{14}COOCH_3$

9. Advantages: Using biodiesel limits our dependence on fossil fuels and waste oil can be used. Biodiesel is renewable and carbon dioxide is consumed by the organic material used for production.

Disadvantages: Using plant materials as raw material requires land and water that may be needed to produce food. Biodiesel is currently more expensive to produce in bulk compared to diesel from fossil fuels.

8.3 Exam questions

1. a. i. $H_2O(g)$

   ii. Addition reaction

   b. $Cr_2O_7{}^{2-}(aq)/H^+(aq)$; $MnO_4{}^-(aq)/H^+(aq)$; $K_2Cr_2O_7$, $KMnO_4$

   c. i. $CH_3COOH(l) + CH_3CH_2CH_2OH(l) \rightarrow CH_3COOCH_2CH_2CH_3(l) + H_2O(l)$ (Hint: Don't forget the water molecule.)

   ii. Propyl ethanoate

2.

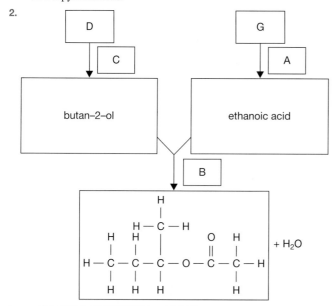

3. • $CH_3(CH_2)_7CHCH(CH_2)_{11}COOCH_3$

   • $CH_3(CH_2)_7CH = CH(CH_2)_{11}COOCH_3$

   • $H_3COOC(CH_2)_{11}CH = CH(CH_2)_7CH_3$

4. A

5. a.

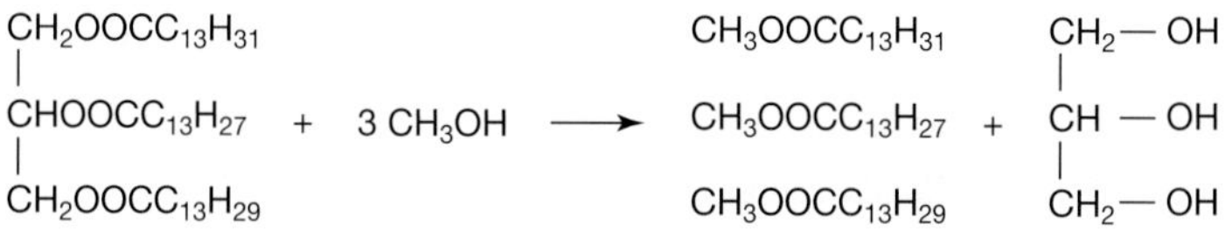

$$CH_2OOCC_{13}H_{31} \qquad\qquad CH_3OOCC_{13}H_{31} \qquad CH_2-OH$$
$$|\qquad\qquad\qquad\qquad\qquad\qquad\qquad\qquad |$$
$$CHOOCC_{13}H_{27} \; + \; 3\,CH_3OH \longrightarrow CH_3OOCC_{13}H_{27} \; + \; CH-OH$$
$$|\qquad\qquad\qquad\qquad\qquad\qquad\qquad\qquad |$$
$$CH_2OOCC_{13}H_{29} \qquad\qquad CH_3OOCC_{13}H_{29} \qquad CH_2-OH$$

   b. 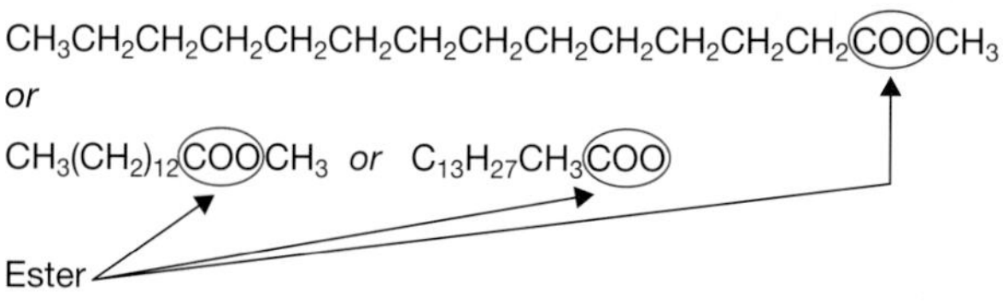 $CH_3CH_2CH_2CH_2CH_2CH_2CH_2CH_2CH_2CH_2CH_2CH_2CH_2COOCH_3$

   or

   $CH_3(CH_2)_{12}COOCH_3$ or $C_{13}H_{27}CH_3COO$

   Ester

   c. $M(CH_3OOCC_{13}H_{27}) = 242\ \text{g mol}^{-1}$

8.4 Hydrolytic and condensation reactions of biomolecules

Practice problem 3

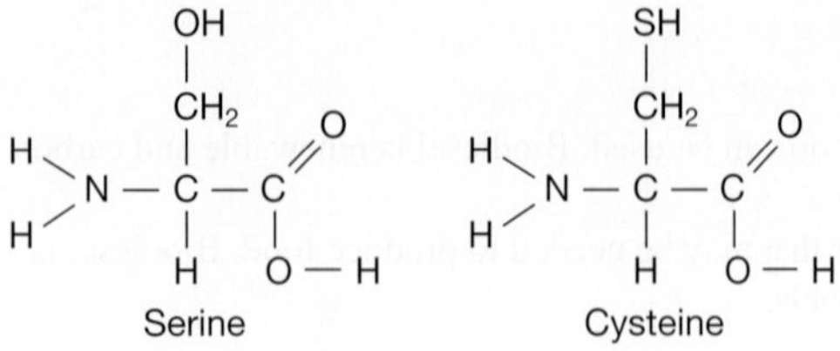

Serine

Cysteine

Practice problem 4

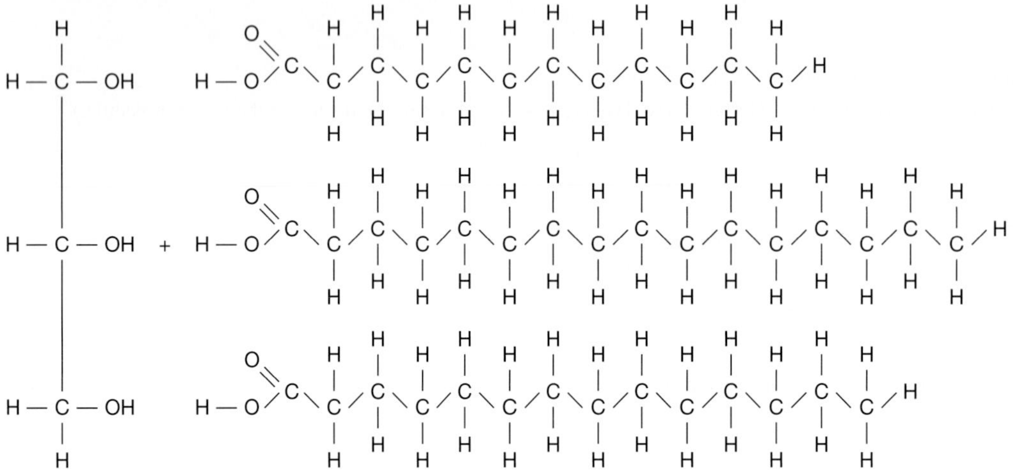

Practice problem 5

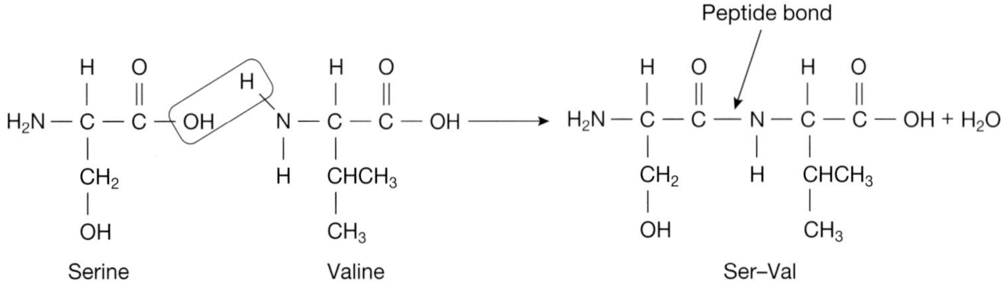

Serine

Valine

Ser–Val

Peptide bond

Practice problem 6

Molar mass of starch molecule $= 1.05 \times 10^5 \text{ g mol}^{-1}$

8.4 Exercise

1. A

2. a. Condensation
   c. Hydrolysis
   e. Hydrolysis
   g. Hydrolysis
   b. Condensation
   d. Condensation
   f. Hydrolysis

3. a. Glycerol and saturated fatty acids
   c. Amino acids
   e. Glycerol and unsaturated fatty acids
   b. Glucose
   d. Glucose
   f. Glucose

4.

Amino group

Carboxyl group

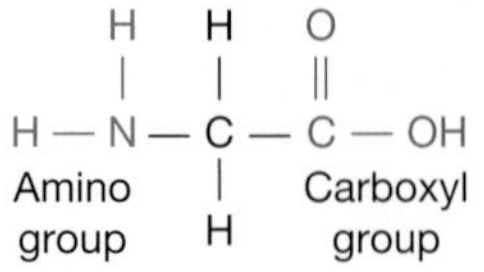

5. a. and **b.**

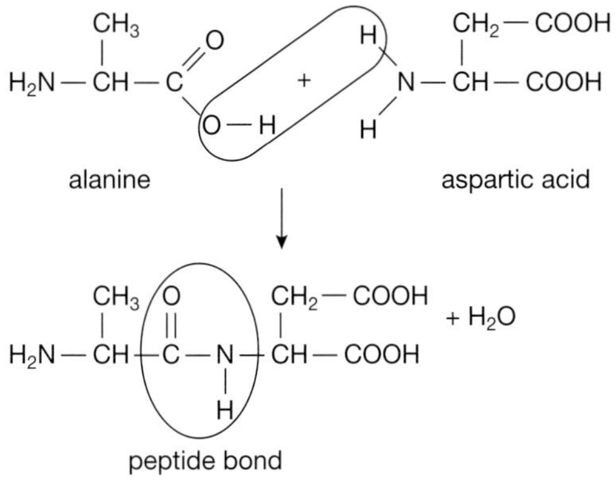
alanine
aspartic acid
peptide bond

c. Condensation

d. 204 g mol^{-1}

6. a.

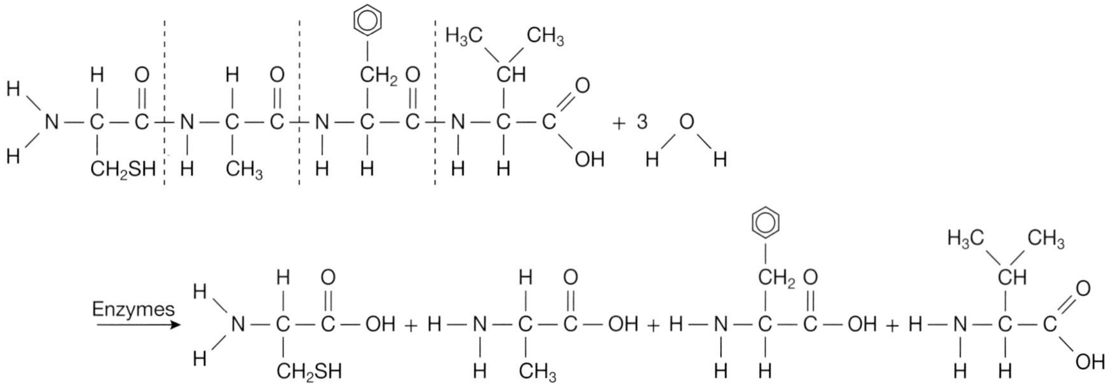
Enzymes

b. Cysteine, alanine, phenylalanine and valine

7. a. A peptide, or amide, link is lost.

b. A carboxyl group and an amino group are formed.

8. a. $C_3H_8O_3$

$CH_2(OH)CH(OH)CH_2OH$

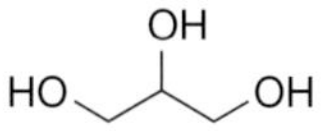

b. It is an alcohol.

9. a.

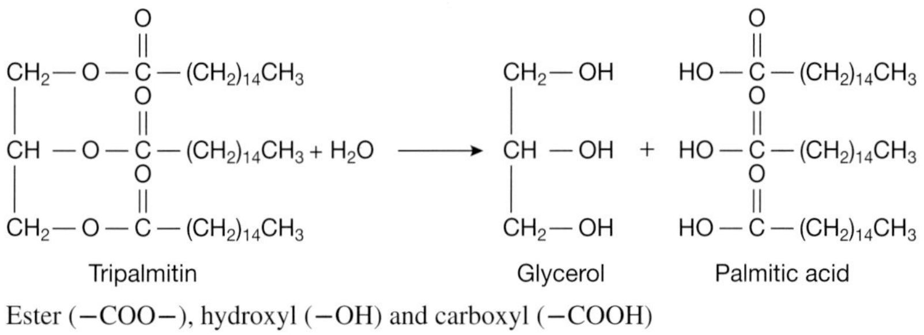
Tripalmitin
Glycerol
Palmitic acid

b. Ester (−COO−), hydroxyl (−OH) and carboxyl (−COOH)

8.4 Exam questions

1. a. i.

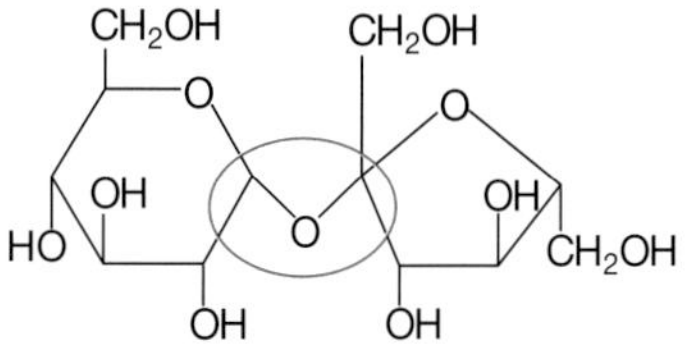

Ether link or glycosidic link

ii. Glucose and fructose

iii. Humans lack the enzyme (cellulase) necessary for the digestion of cellulose.

b. Hydrolysis

2.

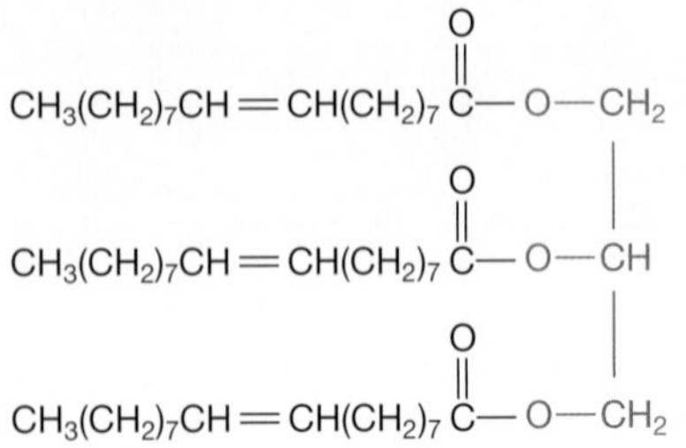

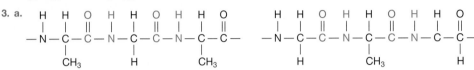

$CH_3(CH_2)_7CH = CH(CH_2)_7 C - O - CH_2$

$CH_3(CH_2)_7CH = CH(CH_2)_7 C - O - CH$

$CH_3(CH_2)_7CH = CH(CH_2)_7 C - O - CH_2$

3. a.

b. Names may include:
- peptide bond/group
- amide bond
- covalent bond.

c. Condensation (polymerisation)

d. Functional groups are:
- carboxyl or COOH
- amino, NH_2 or $NH_3{}^+$.

4.

Characteristic	Biomolecule letter(s) (A.–G.)
contains a glycosidic linkage	A
can produce an ester when reacted with an alcohol in the presence of a concentrated acid	E or G
is soluble in water (give letters for **two** examples)	Two of A, B, D, E, F
contains an ester linkage (give letters for **two** examples)	Two of B, C, D
can be a key constituent of biodiesel	C
has phenylalanine as a component	D

5. C

8.5 The production of chemicals and green chemistry

Practice problem 7

% yield = 91%

Practice problem 8

% atom economy = 80.4%

8.5 Exercise

1. % yield = 86.3%

2. a. % atom economy = 18%

   b. 100% (there will be no waste products — all atoms are used to make desired products)

3. % yield = 89.7%

4. $m(CO_2) = 181$ g

5. B

6. Method 1 has the higher atom economy for the production of hydrogen gas.

   $\%_{\text{Method 1}} = 18\%$

   $\%_{\text{Method 2}} = 6.7\%$

7. D

8. Factors involved in designing a new chemical could include the following:

Chemicals:
- What raw materials are required?
- Are they hazardous?
- How readily available are they?
- Are they biodegradable?
- Do the raw materials or product have a detrimental impact on the environment?
- What is the cost of the raw materials?
- Is the product effective?

Process:
- Is it fast?
- Does it have a minimum of steps?
- Does it have a good yield?
- Does it have a good atom economy?
- Does it require complex technology?
- What waste is produced?
- Is it economical?

9. No

10. Preferred catalysts are plentiful, not hazardous to health or the environment and are heterogenous, so they are easily separated from the reactants and products.

Heavy metals, for example, are often toxic, in limited supply and cause considerable damage to the environment during extraction.

8.5 Exam questions

1. C

2. a. n(salicylic acid) $= 0.0159$ mol

   b. m(aspirin) $= 2.86$ g

   c. % yield $= 78.4\%$

3. A

4. a. Atom economy for fermentation $= 100\%$

   Atom economy for hydration $= 51.1\%$

   b.

	Hydration	Fermentation
Advantages	• Fast rate • 100% atom economy • Requires one step	• Can use renewable biomass as raw material • Does not require technology to contain gas and for heating • Uses carbon dioxide in crop growing
Disadvantages	• Raw material is non-renewable crude oil • Uses acid catalyst, which could be hazardous for workers • Uses more energy • Requires technology	• Slow process • Produces carbon dioxide (greenhouse gas) • Lower atom economy • Land use may be required for food/oxygen production in some countries • Requires distillation equipment • Requires two steps

   c. The atom economy of 100% is a theoretical calculated quantity; in practice, 100% will not be achieved because it is a reversible reaction and so there will always be reactants and products present.

   d. Answers will vary but must use scientific reasons; for example, those given in part **b**.

5. a. There is only one step in method 1, whereas method 2 has two steps, so method 1 is preferable from this perspective.

   In method 1 the waste is HCl, which is a corrosive chemical, so method 2 is preferable because ethanoic acid is a weak acid and potentially less harmful.

   In addition, the ethanoic acid produced can be reused in step 1 of method 2, so it can be recycled immediately, and the atom economy is 100%. The atom economy of method 1 will be less.

   Method 2 is preferable because of the high atom economy and minimal hazardous waste.

   b. Further useful information could include:
   - if the reaction is an equilibrium reaction, to know which direction is favoured
   - the conditions of temperature and pressure, so that the energy requirements are known
   - whether a catalyst is used for one or both reactions, as a catalyst would save energy
   - if a particular type of catalyst is used, as some catalysts have limited availability/are toxic

- the technology that is required for the process to take place, and the energy and construction requirements to conserve energy and cost
- if another chemical can be used to manufacture the same products, because prop-2-en-1-ol is a toxic liquid
- whether it is difficult to separate the prop-2-en-1-ol from the ethanoic acid in method 2, as this may consume more energy
- if there is a potential use for the HCl in method 1 to reduce waste
- if there is adequate containment of any hazardous gases and liquids to prevent damage to workers
- if there is adequate containment of any hazardous gases and liquids to prevent environmental damage.

8.6 Review

8.6 Review questions

1. a. $H_2C=CH_2 + HI \rightarrow CH_3CH_2I$
   b. $H_2C=CHCH_3 + H_2 \rightarrow CH_3CH_2CH_3$
   c. $H_2C=CH_2 + Br_2 \rightarrow CH_2BrCH_2Br$
   d. $H_2C=CHCH_2CH_3 + Cl_2 \rightarrow CH_2ClCHClCH_2CH_3$
   e. $CH_4 + 2O_2 \rightarrow CO_2 + 2H_2O$
   f. $CH_3CH_3 + Cl_2 \rightarrow CH_3CH_2Cl + HCl$
   g. $H_2C=CH_2 + H_2O \rightarrow CH_3CH_2OH$
   h. $CH_3CH=CHCH_3 + H_2 \rightarrow CH_3CH_2CH_2CH_3$

2. $X = C_2H_4$, $Y = C_2H_5OH$ and $Z = Br_2$

3. Esterification is a condensation reaction. The reactants are carboxylic acid and alcohol, and the products are an ester and water.

4.

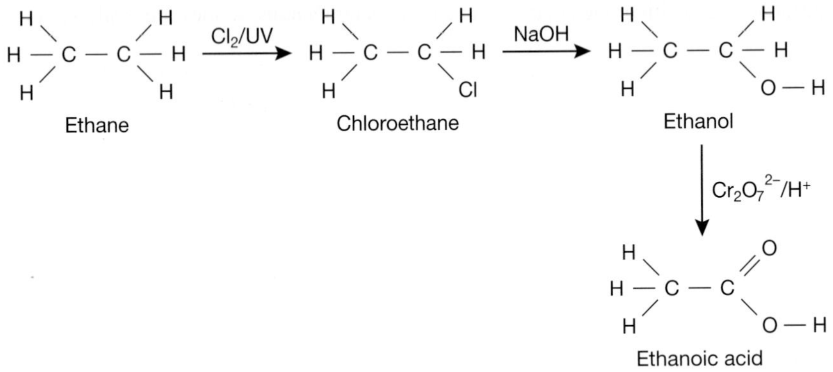

5. % yield = 80.0%

6. % atom economy = 75.7%

7. Atom economy is a measure of how much of the mass of the reactants transfers into the desired product. As all of the atoms in the reactants are directly present in the product, the atom economy is 100%.

8. a.

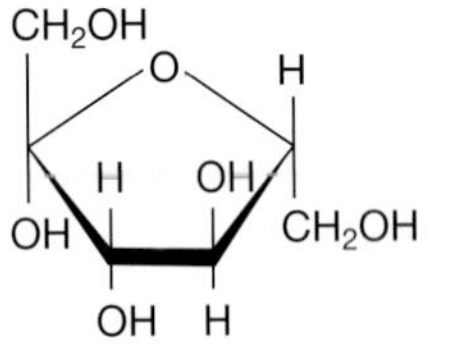

Fructose

   b. Condensation reaction
   c. Water
   d. Hydroxyl and glycosidic
   e. $C_{12}H_{22}O_{11}$

9. $C_8H_{15}O_5N_3 + 2H_2O \rightarrow C_2H_5O_2N + C_3H_7O_2N + C_3H_7O_3N$

10.

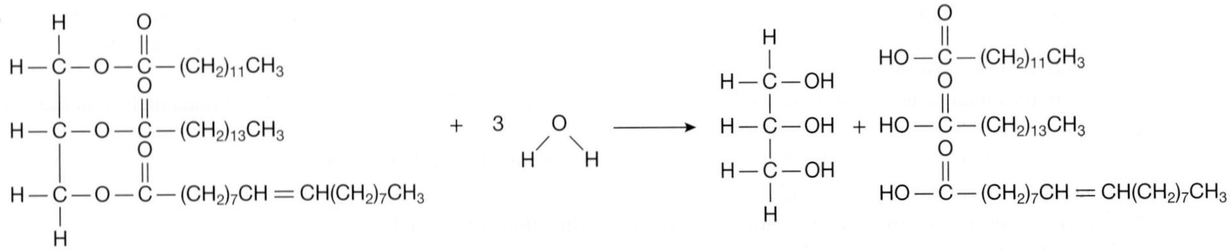

8.6 Exam questions

Section A — Multiple choice questions

1. D
2. D
3. C
4. B
5. C
6. C
7. D
8. B
9. A
10. C

Section B — Short answer questions

11. a. Addition

    b. $CH_3CH_2CHClCH_3$

    c. Dilute sodium hydroxide, $NaOH(aq)$

    d. i. Concentrated sulfuric acid

    ii. $CH_3CH_2CH_2OH + HCOOH \rightarrow HCOOCH_2CH_2CH_3 + H_2O$

    iii. Propyl methanoate

12. a. Any one of the following:
    - $CH_3CHClCH_3 + NH_3 \rightarrow CH_3CHNH_2CH_3 + HCl$
    - $(CH_3)_2CHCl + NH_3 \rightarrow (CH_3)_2CHNH_2 + HCl$
    - $C_3H_7Cl + NH_3 \rightarrow C_3H_9N + HCl$

    b. Propan-2-amine/2-propanamine

    c. Substitution reaction

    d. % atom economy = 61.8% (62%)

13. a. i. Any one of the following:
    - $CH_3(CH_2)_3CH_2Br + NaOH \rightarrow CH_3(CH_2)_3CH_2OH + NaBr$
    - $C_5H_{11}Br + NaOH \rightarrow C_5H_{11}OH + NaBr$
    - $C_5H_{11}Br + NaOH \rightarrow C_5H_{12}O + NaBr$

    ii. % atom economy = 46.1%

    b. Pentanoic acid

    c.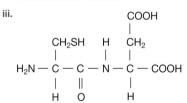

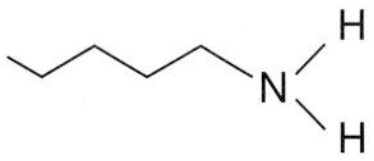

14. a. Any one of the following:
    - Carboxyl and amino functional groups attached to the same C atom
    - $-COOH$ and $-NH_2/-NH$ attached to the same C atom
    - $H_2NCHXCOOH$ or diagram of the same
    - Amino ($-NH_2$) functional group attached to second C atom (C2).

    b. i. Hydrolysis reaction

    ii. Condensation reaction

    iii.

```
                           COOH
                            |
        CH₂SH     H        CH₂
         |        |         |
  H₂N — C —— C —— N —— C — COOH
         |      ‖          |
         H      O          H
```

c. Ester group

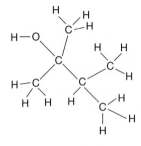

CH$_2$ — O — C — (CH$_2$)$_7$ — CH = CH — (CH$_2$)$_5$ — CH$_3$

CH — O — C — (CH$_2$)$_7$ — CH = CH — (CH$_2$)$_5$ — CH$_3$

CH$_2$ — O — C — (CH$_2$)$_7$ — CH = CH — (CH$_2$)$_5$ — CH$_3$

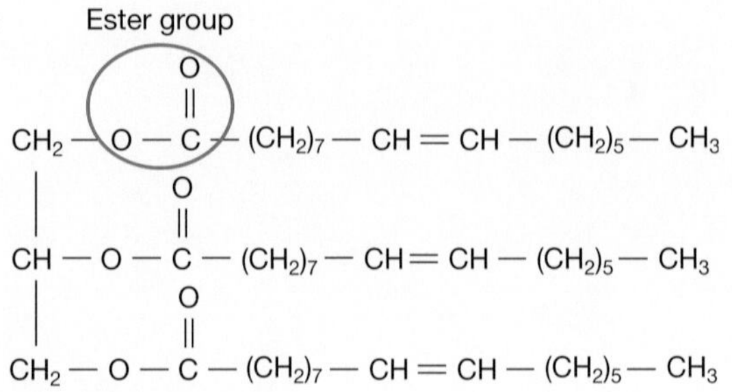

15. a. 3-bromohexane

b. H$_2$O/H$^+$

c. Any one of the following:

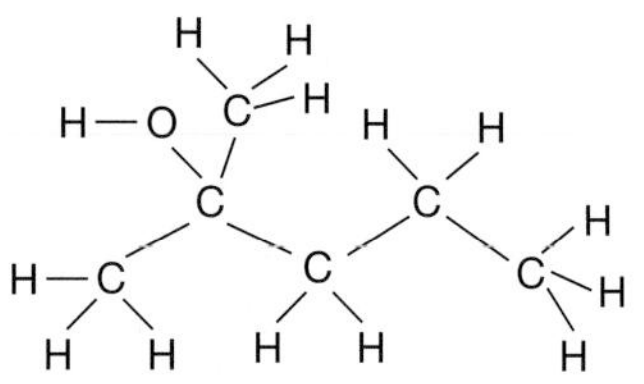

2–methylpentan–2–ol 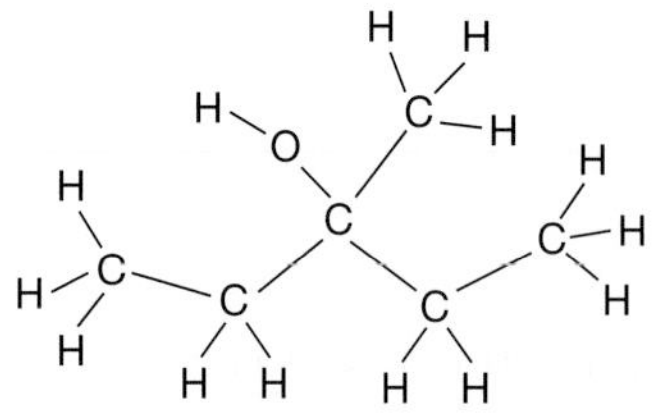   3–methylpentan–3–ol   2,3–methylpentan–2–ol

d. Either of:
- CH$_3$COOH
- HOOCCH$_3$.

Unit 4 | Area of Study 1 review
Section A — Multiple choice questions

1. C
2. B
3. A
4. D
5. D
6. C
7. C
8. C
9. D
10. D
11. D
12. A
13. B
14. D
15. D
16. A
17. C
18. B
19. C
20. D

Section B — Short answer questions

21. a. Carbon has four valence electrons.

b. i. Any two of the following:

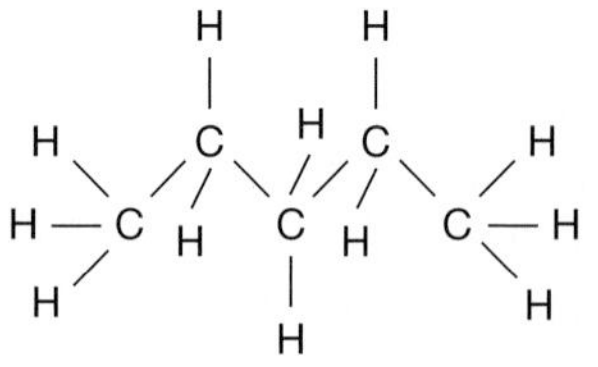

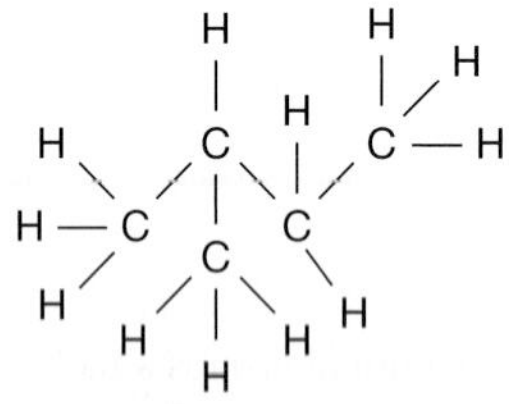

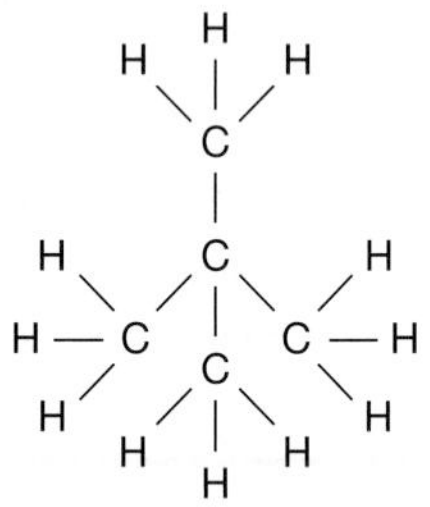

ii. Sample response:

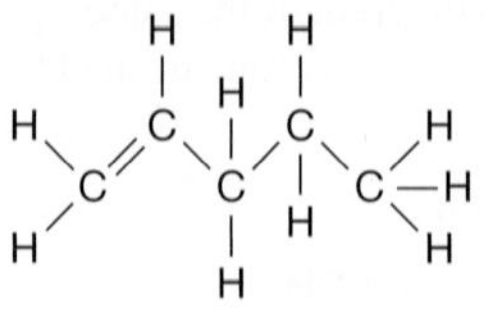

Other answers are possible; the double bond can be placed anywhere along the molecule. Ensure the carbon either side has exactly four bonds in total.

iii. Sample response:

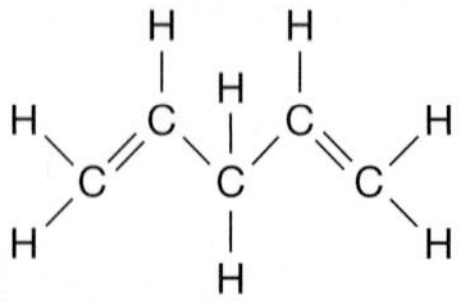

Other answers are possible; the two double bonds can be placed anywhere along the molecule. Ensure the carbons either side have exactly four bonds in total.

c. A number of answers are possible. Accept any correct answer; for example:
  - Bond strength relative to other elements
  - Forms stable bonds with other elements

22. a. Transesterification
   b. See figure at the bottom of the page*
      Ester group
   c. Glycerol
   d. A number of answers are possible. Sample response: Canola is an important food crop. Land used for its production would be diverted from food production to fuel production

23. a. i. Reagent X is NaOH or KOH.
      ii. Substitution
      iii. NaCl or KCl
      iv. 50.5% if reagent X is NaOH, or 44.6% if reagent X is KOH
   b. i. CH_2CH_2
      ii. Addition
   c. i. Oxidation
      ii. Propanoic acid
   d. i.

      ii. 85%
      iii.

24. a. Secondary
   b. i. Ketone
      ii.

c. i. Oxidation
   ii. Either acified dichromate $(H^+/Cr_2O_7{}^{2-})$ or acidified permanganate $(H^+/MnO_4{}^-)$
d. 85.0%

9 Laboratory analysis of organic compounds

9.2 Tests for functional groups

Practice problem 1

a. Iodine number $= 100$
b. $n(I_2) = 0.394\,mol$
c. $n(C{=}C) = 0.394\,mol$
d. $n(fat) = 0.394\,mol$
e. Number of C=C per fat molecule $= 1$

Practice problem 2

a. Test 1 identified that the substance was saturated, as the bromine water remained an orange-brown colour.
b. Tests 1 and 2
c. Test 4
d. Hydroxyl group

9.2 Exercise

1. a. Bromine water
   b. Iodine number
2. a. Three
   b. n(linolenic acid) $= 0.276\,mol$
   c. $n(I_2) = 0.829\,mol$
   d. $m(I_2) = 210\,g$
   e. Iodine number is the mass of I_2 required to react with 100 g of a substance.
      Iodine number for linolenic acid $= 273$

3. Any two of the following:
   See table at the bottom of the page**

---

*22. b.

$CH_3CH_2\,CH_2\,CH_2\,CH_2\,CH_2\,CH_2\,CH_2\,CH_2\,CH_2\,CH_2\,CH_2\,CH_2\,CH_2\boxed{COO}CH_3$

or

$CH_3(CH_2)_{13}\boxed{COO}CH_3$ or $C_{14}H_{29}\boxed{COO}CH_3$

**3.

Test	Cyclohexene observation	Methanol observation
Bromine water	Orange-brown → colourless Colour fades	No change Orange-brown colour persists
Sodium metal	No reaction	Bubbles of (hydrogen) gas produced/effervescence
Esterification with a carboxylic acid (heat with concentrated sulfuric acid catalyst)	No reaction	A fruity-smelling ester will be produced.

4. a. Carboxyl (group)

b. Test 1: Colour change from orange to green
Test 2: Bubbles (of carbon dioxide)/effervescence

c. Test 1: Reduction of $Cr_2O_7^{2-}$ to form Cr^{3+}
Test 2: The carboxylic acid reacts to form carbon dioxide gas.

5. • Heat two tubes containing the unknown compound, concentrated sulfuric acid and either an alcohol or a carboxylic acid.
• If a fruity smell is observed following the reaction, then an ester has been produced.
• Reaction with an alcohol indicates the unknown compound is a carboxylic acid. Reaction with a carboxylic acid indicates the unknown compound is an alcohol.

6. a. Propan-1-ol

b. The bromine water test indicated the chemical is saturated (Br_2 did not react with C=C and become decolourised).
Any one of the following justifications:
• No reaction with sodium hydrogen carbonate indicated the chemical could not be an acid (so must be an alcohol).
• Neutral pH indicates the chemical is not an acid (so must be an alcohol).

c. Any one of the following:
• Sodium metal, as both alcohols and carboxylic acids will react to form hydrogen gas
• Sodium hydrogen carbonate, as pH will determine if the unknown compound is an alcohol or carboxylic acid
• Universal indicator to test for pH, as sodium hydrogen carbonate will determine if the unknown compound is an alcohol or carboxylic acid.

9.2 Exam questions

1. D

2. Any two of the following or similar:
• Alcohols react with carboxylic acids to produce esters, which can be identified by smell.
• No colour change with $Cr_2O_7^{2-}/H^+$: not a primary or secondary alcohol, but could be tertiary
• Colour change with $Cr_2O_7^{2-}/H^+$: either primary or secondary alcohol
• If the product of reaction with $Cr_2O_7^{2-}/H^+$ reacts with $NaHCO_3$: primary alcohol
• If the product of reaction with $Cr_2O_7^{2-}/H^+$ does not react with $NaHCO_3$: secondary alcohol
• Addition of Br_2: decolourisation implies a possible alkenol
• If a pH test shows neutral: could be an alcohol.

3. C

4. B

5. **A** = hexene
B = ethanoic acid
C = ethanol
D = pentane

9.3 Laboratory techniques for analysis of consumer products

9.3 Exercise

1. • Depressed/reduced melting point
• Wider melting point range

2. a. It is pure, as the melting point range is narrow.

b. The test sample is likely to be substance A as they have similar melting points.

c. Mix the test substance with A in a 1 : 1 ratio.
If the melting point range of the mixture is 111–112 °C, the test sample is substance A. Alternatively, if the melting point range of the mixture is wider and/or lower than this, the test sample is not substance A.

3.

Application	Melting point determination	Distillation
Compound identification	✓	✓
Purity analysis	✓	

4. a. Hydrogen bonding, dispersion forces

b. Propanoic acid < ethanoic acid < ethanol

c. Fractional distillation

5. a. The base, as it is nearest the heater and the hottest ascending vapour

b. The base, as it contains more impurities with higher boiling point

c. The top, after the mixture has undergone multiple boil–condense cycles

9.3 Exam questions

1. C

2. a. Fractional distillation

b. Melting point determination

c. Melting point and boiling point are determined by the energy required to overcome the intermolecular forces between molecules. X may be less polar than A and B, so have weaker intermolecular forces.

3. a. Propanal < propan-1-ol < propanoic acid

b. Compounds with the weakest intermolecular forces have lower boiling points so leave the column first, and vice versa.

c. Any two of the following:
• High surface area
• Increased condensation
• Temperature gradient
• Multiple boil–condense cycles.

4. a. Salicylic contains two hydrogen atoms capable of forming hydrogen bonds, whereas aspirin only contains one hydrogen atom capable of forming a hydrogen bond. More energy is required to overcome stronger intermolecular forces between salicylic acid molecules.

b. The sample is impure, as the melting point range is lower and wider than for pure aspirin.

c. Batch 2

5. The components have different molar masses/different intermolecular force strengths.
The boiling points are different.
Components with the lowest boiling point rise highest up the tower.

9.4 Volumetric analysis by redox titration

Practice problem 3

$c((COOH)_2)_{undiluted} = 0.5694 \text{ mol L}^{-1}$

Practice problem 4

$\%(m/m) = 24.60\%$

Practice problem 5

The titre of vitamin C is erroneous as the end point was overshot. The erroneous titre will be higher than the true value.

$n(I_2) = c \times V$

$n(\text{vitamin C}) \propto n(I_2)$

$\uparrow V(\text{vitamin C})_{\text{erroneous titre}}$*

$\downarrow c(\text{vitamin C}) = \dfrac{n}{\uparrow V}$

The calculated concentration of vitamin C is lower than the true value.

*The point at which the error feeds into the calculation. The values in the earlier steps of the calculation are correct.

9.4 Exercise

1. a. The solution it will deliver
   b. The aliquot solution
   c. Distilled water
   d. Distilled water
2. a. $n(S_2O_3^{2-})_{reacted} = 0.003\,237 \text{ mol}$
   b. $n(I_2)_{diluted} = 0.001\,619 \text{ mol}$
   c. $c(I_2)_{diluted} = 0.08093 \text{ mol L}^{-1}$
   d. $c(I_2)_{undiluted} = 1.012 \text{ mol L}^{-1}$
3. a. It is an indicator to help identify the end point.
   b. $SO_2 + 2H_2O \rightarrow SO_4^{2-} + 4H^+ + 2e^-$
   c. $I_2 + 2e^- \rightarrow 2I^-$
   d. $SO_2(aq) + I_2(aq) + 2H_2O(l)$
      $\rightarrow SO_4^{2-}(aq) + 4H^+(aq) + 2I^-(aq)$
   e. $c(SO_2) = 0.08426 \text{ mol L}^{-1}$
4. a. $n(CH_3CH_2OH) = 0.1137 \text{ mol}$
   b. $\%(m/v) = 52.29\%$
   c. No, as its ethanol concentration of 52.29% is below the 80% requirement
5. a. No effect, as the concentration of the standard solution and hence the titre will not be affected by spillage while filling the burette
   b. The standard solution will be diluted. The erroneous c(standard solution) used in the calculations was higher than the true value as it did not take the dilution into account.

$\uparrow c(\text{standard solution})_{\text{erroneous aliquot}}$
$\uparrow n(\text{standard solution}) = \uparrow c \times V$
$\uparrow n(\text{unknown solution}) \propto \uparrow n(\text{standard solution})$

$\uparrow c(\text{unknown solution}) = \dfrac{\uparrow n}{V}$

The concentration of the unknown solution will be higher than the true value.

9.4 Exam questions

1. D
2. 0.192 g/192 mg
3. B
4. D
5. C

9.5 Review

9.5 Review questions

1. a. Any two of the following:
   - Sodium metal — (hydrogen) bubbles
   - Hydrogen carbonate/carbonate — (carbon dioxide) bubbles
   - Esterification with an alcohol — fruity smell
   - pH — acidic/low pH.
   b. Any two of the following:
   - Bromine water:
     Maleic acid — orange-brown to colourless/colour fades
     Malic acid — orange-brown colour persists
   - Esterification with a carboxylic acid:
     Maleic acid — no observation/fruity smell
     Malic acid — fruity smell
   - Oxidation with acidified dichromate/permanganate:
     Maleic acid — orange/pink colour persists
     Malic acid — orange to green/pink to colourless.
2. a. React a known mass of the oil with I_2.
   The iodine number is the mass, in g, of I_2 that reacts with 100 g of the oil.
   b. The molecule must contain a minimum of $4 \times C=C$.
3. Components have different molar masses/different intermolecular force strengths.
   The boiling points are different.
   Components with the lowest boiling point rise highest up the tower.
4. a. $c(C_6H_8O_6)_{diluted} = 0.000\,620 \text{ mol L}^{-1}$
   b. $m(C_6H_8O_6)_{undiluted} = 0.1091 \text{ g}$
   c. $c(C_6H_8O_6) = 0.00520 \%(m/m)$
5. a. $C_4H_6O_5 \rightarrow C_4H_4O_5 + 2H^+ + 2e^-$
   b. $MnO_4^- + 8H^+ + 5e^- \rightarrow Mn^{2+} + 4H_2O$
   c. $5C_4H_6O_5(aq) + 2MnO_4^-(aq) + 6H^+(aq)$
      $\rightarrow 5C_4H_4O_5(aq) + 2Mn^{2+}(aq) + 8H_2O(l)$
   d. $3.88 \times 10^{-4} \text{ mol}$
   e. $5.19 \%(m/m)$

f. The concentration of the potassium permanganate standard solution will be affected. The erroneous $c(KMnO_4)$ used in the calculations (0.0100 mol L^{-1}) was lower than the true value (0.0200 mol L^{-1}).

$\downarrow c(KMnO_4)_{burette}$

$\downarrow n(KMnO_4) = \downarrow \times V$

$\downarrow n(C_4H_6O_5) \propto \downarrow n(KMnO_4)$

$\downarrow m(C_4H_6O_5) = \downarrow n \times M$

$\downarrow \%(m/m) = \dfrac{\downarrow m(C_4H_6O_5)}{m(candy)} \times 100$

The calculated %(m/m) malic acid in the candy will be lower than the true value/underestimated.

9.5 Exam questions

Section A — Multiple choice questions

1. C
2. A
3. B
4. D
5. A
6. B
7. A
8. A
9. C
10. B

Section B — Short answer questions

11. $m(I_2) = 95.2$ g

12. **a.** $n(CO_2)$ entering container $= 1.64 \times 10^{-4}$ mol

    b. $n(NaOH) = 0.010$ mol

    c. $c(NaOH) = 0.097$ M (0.097 mol L^{-1})

13. **a.** $c(HCl) = 0.080$ (M)

    b. The calculated concentration of sodium hydrogencarbonate will be greater because a larger volume of the HCl solution will be required.

14. **a.** Either of:
- $MnO_4^-(aq) + 8H^+(aq) + 5e^-$
  $\rightarrow Mn^{2+}(aq) + 4H_2O(l)$
- $MnO_4^-(aq) + 8H_3O^+(aq) + 5e^-$
  $\rightarrow Mn^{2+}(aq) + 12H_2O(l)$

    b. 22.00 mL

    c. $n(MnO_4^-) = 8.80 \times 10^{-4}$ mol

    d. $n(Fe^{2+})$ in 250 mL flask $= 5.5 \times 10^{-2}$ mol

    e. % Fe in alloy $= 3.81\%$

15. **a.** Equivalence point: The point when the reactants have combined (reacted) in the exact stoichiometric ratio (mole ratio) indicated in the equation/the point in the reaction where no reactant is in excess
End point: The point at which the indicator used in the reaction changes colour

    b. $V(NaOH) = 26.7$ mL

10 Instrumental analysis of organic compounds

10.2 Mass spectrometry

Practice problem 1

a. $CH_3CH_2OH^+/C_2H_5OH^+/C_2H_6O^+$

b. The molecular ion has an m/z of 46.
$M(CH_3CH_2OH) = 46.0$ g mol^{-1}
They are the same.

c. $CH_3CH_2OH(g) + e^- \rightarrow CH_3CH_2OH^+(g) + 2e^-$

d. The base peak has an m/z of 31.
CH_2OH^+

e. The particle lost has $m/z = 15$, thus the particle lost is CH_3.

Practice problem 2

The peak at $m/z = 50$ corresponds to $[CH_3{}^{35}Cl]^+$.

The peak at $m/z = 52$ corresponds to $[CH_3{}^{37}Cl]^+$.

The peak with $m/z = 15$ is a fragment of the molecular ion, $CH_3{}^+$.

10.2 Exercise

1. The m/z ratio can be thought of as a mass scale because usually only one electron is removed per molecule in the spectrometer. There are examples where two electrons are removed and in this instance the ratio would be half the molar mass of the molecule.

2. B

3. Any one of:
$C_2H_4O_2(g) + e^- \rightarrow C_2H_4O_2{}^+(g) + 2e^-$
$C_2H_4O_2(g) + e^- \rightarrow [C_2H_4O_2]^+(g) + 2e^-$
$C_2H_4O_2(g) \rightarrow C_2H_4O_2{}^+(g) + e^-$

4. **a.** Peak Z

    b. Peak Y $= CH_3CO^+$
Peak X $= CH_3{}^+$

5. **a.** 59

    b. 30

    c. $C_3H_9N^+$ or $[C_3H_9N]^+$

    d. CH_4N^+ or $[CH_4N]^+$ or $[CH_2NH_2]^+$

6. **a.** $m/z = 29$

    b. CH_2CHO^+ and/or $C_3H_7{}^+$. It is probably both, which explains it being the highest peak on the spectrum.

    c. The base peak

    d. C_4H_8O

    e. Butanal

7. $C_2H_4Cl_2$

10.2 Exam questions

1. **a.** $[CH_3CH_2]^+/[CHO]^+$
Note: $[COH]^+$ is also acceptable.

    b. Propanal
Only CH_3CH_2CHO can readily be fragmented to produce $[CH_3CH_2]^+$ and/or $[CHO]^+$. Neither the alkenol nor the ketone can readily be fragmented to produce $[CH_3CH_2]^+$ or $[CHO]^+$. The two fragments $[CH_3CH_2]^+/[CHO]^+$ are consistent with the base peak at $m/z = 29$.

2. a. Peak at $m/z = 30$

   b. $m/z = 60$

     Acceptable justifications include:
- $M_r(NH_2CH_2CH_2NH_2) = 60$
- $M(NH_2CH_2CH_2NH_2) = 60$ g mol^{-1}
- Peak at $m/z = 60$ is caused by its molecular ion $[NH_2CH_2CH_2NH_2]^+$.

   c. $[NH_2CH_2]^+$

3. $CH_3(CH_2)_{16}COOCH_3$

4. D

5. A

10.3 Infrared spectroscopy

Practice problem 3

The peak at 1670–1750 cm^{-1} corresponds to C=O.

Deduce a structure that has C=O and C–H bonds, with a molecular formula C_3H_6O. The molecule has the structural formula CH_3COCH_3.

10.3 Exercise

1. a. The covalent bonds will absorb the infrared energy and undergo a change in vibration.

   b. The C=C bond is stronger than the C–C bond, so the change in vibration will occur at a higher frequency and produce a peak with a higher wave number.

2. C=O (esters) at 1750 cm^{-1}
C–H at 2800–3150 cm^{-1}

3. a. C=O (acids) at 1680–1740 cm^{-1}
O–H (acids) at 2500–3500 cm^{-1}

   b. Propanoic acid

4. Compound X is the carboxylic acid. Compound Y is the alcohol.

5. Compound X is an amine.
Compound Y is an amide.

10.3 Exam questions

1. C

2. C

3. The compound is not an alcohol, since there is no O–H bond in the molecule.
This is shown by a lack of a peak in the O–H (alcohols) range (3200–3600 cm^{-1}).

4. C

5. Yes, a peak in the absorption band characteristic of an N–H bond — that is, 3350–3500 cm^{-1} — is evident.

10.4 NMR spectroscopy

Practice problem 4

10.4 Exercise

1. D

2. A

3. Any two of the following:
- It is inert, so will not react with the sample being analysed.
- It produces a signal peak that is well way from other peaks generated by organic molecules.
- It is volatile, so can easily be recovered from samples following analysis.
- Setting the TMS signal to zero allows data from different NMR spectrometers to be compared.

4. CH_3CH_2Cl consists of R–CH$_3$, R–CH$_2$–X.
The CH$_3$ group will produce a signal in the 0.9–1.0 ppm range for a ^{1}H-NMR spectrum and the R–CH$_2$–X will produce a signal in the 3.0–4.5 ppm range.

5. Methyl propane:

There would be two peaks in the ratio of 3 : 1.
Butane:

Butane will also produce two peaks in a 1 : 1 ratio.

6. 1-bromopropane has three different carbon environments, therefore there will be three peaks on the spectrum.
2-bromopropane only has two environments due to the CH$_3$ groups being in identical environments and only producing one peak.

2-bromopropane, (CH$_3$)$_2$CHBr (two peaks)

1-bromopropane, CH$_3$CH$_2$CH$_2$Br (three peaks)

7. a. Two

   b. 3 : 1

   c. The R–CH$_3$ peak is seen at 25–30 ppm while the R$_3$C–OH peak is observed at approximately 70 ppm.

8.

Hydrogen set or atom	Splitting pattern	Relative peak height	Chemical shift (ppm)
CH₃CH₂COOH	Triplet	3	0.8–1.0
CH₃**CH₂**COOH	Quartet	2	2.1–2.7
CH₃CH₂COO**H**	Singlet	1	9–13

9. Any two of the following:

Observation	CH₂ClCH₂Cl	CH₃CHCl₂
Number of peaks	A single peak, because the molecule is symmetrical and the hydrogen atoms are in exactly the same environment	Two peaks since there are two unique hydrogen environments
Ratio of hydrogen environments	No ratio as there is a single peak	3 : 1
Peak splitting	No splitting as there is one hydrogen environment	The R–CH₃ signal will be split into a doublet as there is one hydrogen atom bound to the neighbouring carbon atom. The other hydrogen environment will be split into a quartet of peaks as there are three hydrogen atoms bound to a neighbouring carbon atom.
Chemical shift	R–CH₂–X at 3.0–4.5 ppm	R–CH₃ at 0.9–1.0 ppm and R–CH at 5.9 ppm. *Note:* the R–CH peak is not in table 10.9 so not required for a correct answer.

1,2-dichloroethane, CH₂ClCH₂Cl   1,1-dichloroethane, CH₃CHCl₂

10.

10.4 Exam questions

1. B

2. a. Any two of the following:

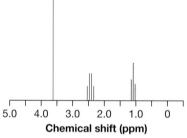

Correct structure for the compound being analysed

b. Any three of the following:
- Three sets of peaks on the ^{1}H-NMR spectrum indicates three (hydrogen) environments.
- No signal splitting (single peaks) on the ^{1}H-NMR spectrum indicates there are no regions in the molecule where H atoms are attached to adjacent C atoms/there are no neighbouring H atoms.
- The peak area ratio 3 : 3 : 2 indicates the three different hydrogen environments contain three, three and two hydrogen atoms respectively.
- The type of environment is indicated by the chemical shift; for example, δ = 1.82 indicates –CH=CH–**CH$_3$**; δ = 3.53 indicates –O**CH$_3$**.

3. C

4. Yes, one peak on the spectrum is consistent with only one carbon environment.
$NH_2CH_2CH_2NH_2$ is a symmetrical molecule/both C atoms have identical bonding environments.

5. Propanone/CH_3COCH_3
It is a symmetrical molecule and end C atoms are equivalent, therefore it has two C environments and two peaks on the ^{13}C-NMR spectrum, or the peak at chemical shift of 205 ppm on the ^{13}C-NMR spectrum is consistent with a ketone (R_2CO).
It has one H environment since all H atoms are equivalent, therefore it has one peak on the ^{1}H-NMR spectrum, or the peak at chemical shift of 2.2 ppm is consistent with a ketone ($RCOCH_3$).
The spectra showing two C environments and one H environment cannot be of the alkenol or the aldehyde since they both have three C environments and multiple H environments.

10.5 Combining spectroscopic techniques

Practice problem 5

a. Molar mass = 74 g mol^{-1}

b. The molecular formula is $C_3H_6O_2$.

c. 1720 cm^{-1} corresponds to C=O (acids) and 3000 cm^{-1} corresponds to O–H (acids).

d. The presence of O–H (acids) and C=O (acids) bonds indicates this is a carboxylic acid. The observation of bubbles in an acid carbonate reaction confirms the identification.

e. The triplet (2.2 ppm) and quartet (1.6 ppm) indicate the presence of an ethyl (–CH$_2$CH$_3$) group. The singlet (11.9 ppm) is suggestive of an O–H bond.

f. Propanoic acid

$$
\begin{array}{c}
\mathsf{H\ \ H}\\
\mathsf{|\ \ \ |}\\
\mathsf{H-C-C-C}\overset{\displaystyle O}{\diagdown}\\
\mathsf{|\ \ \ |\ \ \ \diagdown}\\
\mathsf{H\ \ H\ \ \ O-H}
\end{array}
$$

Practice problem 6

a. The empirical formula is CH$_2$F.

b. Molar mass = 66 g mol^{-1}

c. The molecular formula is $C_2H_4F_2$.

d. The narrow peak at 3000 cm^{-1} in the IR spectrum is likely to be C–H bonds.

e. There are two possible structures: 1,2-difluoroethane and 1,1-difluoroethane.

f. The ^{1}H-NMR spectrum contains one peak, indicating that all H atoms are in the same environment. Therefore, the molecule is 1,2-difluoroethane.

$$
\begin{array}{c}
\mathsf{F\ \ \ F}\\
\mathsf{|\ \ \ |}\\
\mathsf{H-C-C-H}\\
\mathsf{|\ \ \ |}\\
\mathsf{H\ \ \ H}
\end{array}
$$

10.5 Exercise

1. a. Ethanol, CH_3CH_2OH, and ethanal, CH_3CHO, both have two unique carbon environments and the same relative peak areas.
However, with different functional groups, the peaks will be seen at different chemical shifts. The RCH_2OH peak produced in ethanol will be observed in the 50–90 ppm range of a ^{13}C-NMR spectrum, while the RCHO peak in ethanal will be seen in the 190–200 ppm range characteristic of aldehydes.

b. Any two of the following:
- The ^{1}H-NMR spectra will differ in the number of unique hydrogen environments. Ethanol will produce three sets of peaks, whereas ethanal will only produce two.
- The relative peak areas will differ.
- The H atoms in the OH (1–6 ppm) and CHO (9.4–10.0 ppm) produce signals at different chemical shifts.
- Ethanol will also produce a set of four peaks (quartet), a set of three peaks (triplet) and a singlet due to the CH_3CH_2OH sequence. The signal at 9.4–10.0 ppm in ethanal will be split into a quartet due to the neighbouring CH_3 group. The other signal at approximately 2 ppm will be split into a doublet due to the neighbouring CHO group.

2. a. O–H (alcohols): 3200–3600 cm^{-1}
C–H (alkanes, alkenes, arenes): 2850–3090 cm^{-1}

b. i. Propan-1-ol = four peaks
Propan-2-ol = three peaks
ii. Propan-1-ol = 3 : 2 : 2 : 1
Propan-2-ol = 6 : 1 : 1

c. Low-resolution ^{1}H-NMR spectroscopy shows differences in peak number and the ratio of peak areas, while there are no differences between the IR spectra.

3. a. C=O at 1700 cm^{-1} and C–H at 3000 cm^{-1}

b. Four

c.

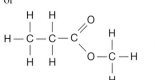

Ethyl ethanoate

or

Methyl propanoate

4. a. i. ^{1}H-NMR or proton NMR

ii.

b. i. ^{13}C-NMR

ii.

5. a. The empirical formula is C_3H_7Cl.
   b. The molecular formula is C_3H_7Cl.
   c. Two
   d. Six
   e.

10.5 Exam questions

1. **IR spectroscopy — Principles**
   Either of the following:
   - IR spectroscopy measures the vibrations of atoms in a molecule and can give information about the type of bonds (functional groups) present in a molecule.
   - IR spectroscopy involves the absorption of infrared radiation, the wave number of which alters with different bond types.

   ^{1}H-NMR spectroscopy — Principles
   Any one of the following:
   - ^{1}H-NMR spectroscopy measures the change in the spin state of nuclei and can give information about the H environment present in the molecule.
   - ^{1}H-NMR spectroscopy involves the absorption of radio waves; the chemical shift alters with the neighbouring environment.
   - Neighbouring hydrogens will induce a splitting in the peak of a hydrogen atom, leading to the 'n + 1' rule. The peak area can indicate the number of hydrogens in the environment.

Any two of the following:
Interpretation of spectra — IR
- In primary alcohols the O–H bond will show a distinct peak between 3200–3600 cm^{-1}.
- In ketones the C=O bond will show a distinct peak between 1680–1850 cm^{-1}.
- The lack of a peak in a region of the spectra can indicate the absence of a type of bond in a molecule. For example, the lack of a peak between 1680–1850 cm^{-1} due to C=O indicates that the compound is unlikely to be a ketone.
- Since an alcohol C_3H_6O is unsaturated it must have a C=C bond and it would have a peak at 1620–1680 cm^{-1}.

Interpretation of spectra — NMR
- Primary alcohols show a chemical shift of 3.3–4.5 ppm for H atoms adjacent to the –OH functional group and a singlet between 1–6 ppm for functional group H.
- Ketones show a chemical shift of 2.1–2.7 ppm for H atoms adjacent to the C=O functional group.
- The symmetry of the C_3H_6O ketone means that there is only one ^{1}H-NMR peak in the spectrum, whereas there are four peaks in the spectrum of the C_3H_6O primary alcohol.
- The ratio of peak areas on the alcohol (CH_2=$CHCH_2OH$) spectrum will be 2 : 1 : 2 : 1; HO–CH=CH–CH_3 is 1 : 1 : 1 : 3.

2. a. The peak at m/z = 75 could be due to the presence of a one-unit heavier isotope, such as ^{13}C, ^{2}H or ^{17}O.
   b. i. $C_3H_6O_2$
         $C_4H_{10}O$
      ii. There are three hydrogen environments (from three sets of peaks)
          OR
          there are six hydrogen atoms (or a multiple of six H atoms) from the peak area ratio 3 : 2 : 1. Three environments contain 3, 2 and 1 H atoms respectively.
          The molecular formula is $C_3H_6O_2$
   c. Possible structural isomers include:
      See figure at the bottom of the page*

---

*2. c.

d. **i.** A: –OH (hydroxyl)

 B: C=O (carbonyl/ketone)

ii.

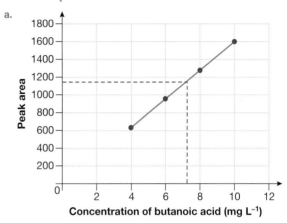

3. a. Compound T

 b. Compounds eliminated: P, Q and S.

 Compound P has no –O–H group present, whereas the IR spectrum shows a clear –O–H alcohol with an absorbance at 3500 cm^{-1}.

 Compound Q has an –O–H group of an acid, whereas the IR spectrum shows a clear –O–H alcohol with an absorbance at 3500 cm^{-1}. There is no evidence of a large broad –O–H of an acid showing up between 2500 cm^{-1} and 3500 cm^{-1}.

 Compound S has no –O–H group present, whereas the IR spectrum shows a clear –O–H alcohol with an absorbance at 3500 cm^{-1}.

 c. **i.** Any one of:

 • $[CH_3CO]^+$

 • $[(CH_3)_2CH]^+$

 • $[C_2H_3O]^+$

 • $[C_3H_7]^+$.

 ii. Mass-to-charge ratio

 iii. Ions (cations) need to be formed in order for any peak to appear on a mass spectrum.

 Multiple ions form because of either:

 • the fragmentation pattern that occurs because of the initial parent ion being unstable

 • the molar masses of the different isotopes present in the parent ion.

4. B

5. Mass spectrum

Information may include the following:

 • The parent (molecular) ion peak at $m/z = 88$ is consistent with the molar mass (relative molecular) mass of ethyl ethanoate ($CH_3COOCH_2CH_3$).

 • Other peaks on the spectrum are consistent with fragments that may result from ethyl ethanoate ($CH_3COOCH_2CH_3$); for example:

 • $m/z = 15$: $[CH_3]^+$

 • $m/z = 29$: $[CH_3CH_2]^+$

 • $m/z = 43$: $[CH_3CO]^+$

 • $m/z = 45$: $[CH_3CH_2O]^+$

 • $m/z = 70$: $[CH_2COOC]^+$

 • $m/z = 73$: $[CH_3COOCH_2]^+$.

Infrared spectrum

Information may include the following:

 • A strong absorption around 1750 cm^{-1} (in the 1670–1750 absorption band) is consistent with the presence of C=O. This bond is present in ethyl ethanoate, $CH_3COOCH_2CH_3$.

 • The absence of a broad peak in the 2500–3300 cm^{-1} absorption band shows that the O–H (acids) bond is not present and so the compound cannot be a carboxylic acid with four C atoms.

^{1}H-NMR spectrum

Information may include the following:

 • The three different sets of peaks mean/indicate that $CH_3COOCH_2CH_3$ has three different hydrogen environments.

 • The triplet and quartet on the spectrum are consistent with the presence of an ethyl group, $-CH_2CH_3$, in $CH_3COOCH_2CH_3$.

 • The chemical shifts $\delta = 4.1$ ppm ($RCOOCH_2R$) or $\delta = 2.0$ ppm (CH_3COOR) are characteristic of an ester.

10.6 Chromatography

Practice problem 7

a.

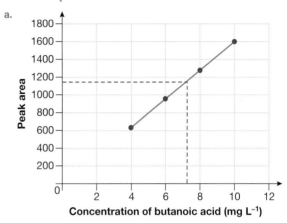

b. c(butanoic acid)$_{diluted} = 7.2$ mg L^{-1}

c. c(butanoic acid)$_{undiluted} = 72$ mg L^{-1}

d. c(butanoic acid) $= 8.2 \times 10^{-4}$ mol L^{-1}

10.6 Exercise

1. B

2. D

3. Separation occurs through the processes of adsorption and desorption onto and from the surface of the stationary phase. The larger the surface area, the greater the degree of adsorption and desorption, producing better separation.

4. a. R_t of methanoic acid < ethanoic acid < propanoic acid < butanoic acid

 b. • Increasing the length of the hydrocarbon chain will increase the non-polar component of the carboxylic acid, increasing affinity for the stationary phase.

 • Increasing the molar mass

5. a. Methanol is polar and would have a different polarity to the non-polar stationary phase. Hexane is non-polar so would have very similar polarity to the stationary phase.

 b. Dispersion forces

 c. The mass spectra of the components could be identified by comparison to those of known compounds (vitamins).

6. a. Peak 4 is likely to be serine. All amino acids have both the COOH and NH$_2$ functional groups. However, serine is the most polar of the four with its hydroxyl group and short chain. Serine will have a greater affinity for the polar stationary phase and have the highest R_t.

 b. To confirm the R_t of peak 4 is caused by the detection of serine, a pure standard could be run through the column under the same conditions to see if the retention times match.

c. Threonine would have a higher R_t than serine. Threonine has the same polar functional groups as serine, although it also contains an additional CH_3 group. This will increase molar mass and, hence, the R_t.

d. This idea will not work. It is only possible to compare the areas of peaks for the same substance generated by HPLC performed under identical HPLC conditions.

7. a. $m = 255$ mg

b. % purity = 51.0 %(m/m)

8. a. Pure stanozolol can be run through HPLC and its retention time recorded. A urine sample would be analysed and if a unique retention time of stanozolol is observed, then its presence is detected. The HPLC method would need to be modified to ensure that no other components in urine have the same retention time as stanozolol.

b. A calibration curve can be generated from the data in the table provided. Using the calibration curve and the area of the stanozolol peak in the urine sample, the concentration can be determined.

c. See figure at the bottom of the page*
Reading from the calibration curve, the urine sample gives a stanozolol concentration of 2.5 mg L^{-1}.

d. A concentration of 5.0 mg L^{-1} falls outside the range of the calibration curve.

e. The sample could be diluted so that its concentration results in a peak area within the range of the standards. Alternatively, standards of a higher concentration could be run to extend the range of the calibration curve.

10.6 Exam questions

1. a. Any one of the following:
   • Increasing the length of the column
   • Altering either the mobile or stationary phases
   • Altering the temperature of the column.

b. Measurement errors occur due to the overlap of the two peaks. If any of the stated factors in part a are altered this can result in improved separation/resolution of peaks in a HPLC. Improved resolution of peaks leads to a more reliable analysis of peak areas, and subsequently the concentrations of aspartame and DKP.

2. B

3. Chromatography can be used to separate a mixture of methyl esters because they have different strengths of attraction to the stationary and mobile phases
OR
the methyl esters all contain the ester functional group, so retention times will depend on molecular mass/size and extent of unsaturation.
Calibrate by running pure samples of each of the methyl esters through the HPLC under the same conditions to determine individual retention times.
The relative amounts and hence, percentage composition of the egg yolk can be determined from the relative areas under the peaks/peak heights at the retention times of the associate methyl esters.

4. B

5. C

10.7 Review

10.7 Review questions

1. D

2. C

3. B

4. a. $m/z = 86$

b. $m/z = 57$

c. The peak at $m/z = 57$ corresponds to the loss of $[CH_3CH_2]$.

d.
$$CH_3CH_2 - \overset{\displaystyle O}{\overset{\displaystyle \|}{C}} - CH_2CH_3$$

5. Ethanoic acid has a peak in the region 2500–3500 cm^{-1} due to the O–H (acid). Ethanoic anhydride has no peak in this region.

6. a. R–CH_3 (1.0 ppm); R_2–CH_2 (1.3 ppm); R_1–COOH (10.5 ppm)

b. Four

c.
$$H - \overset{\displaystyle H}{\underset{\displaystyle H}{C}} - \overset{\displaystyle H}{\underset{\displaystyle H}{C}} - C\overset{\displaystyle O}{\underset{\displaystyle O-H}{}}$$

*8. c.

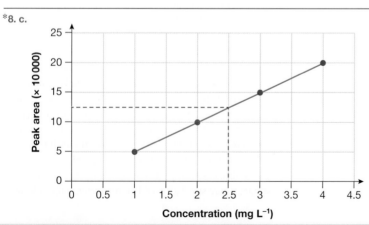

7. a.

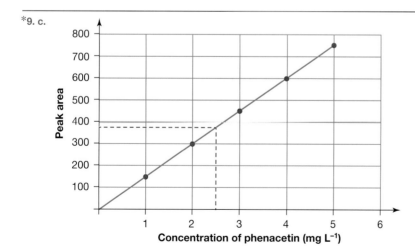

Propan-1-ol

H—C—C—C—H with H, H, H on top and H, O, H below, plus H below O

Propan-2-ol

b.

H—C—C—C—H (Propanone structure)

Propanone

The ^{1}H-NMR spectrum has indicated there is only one hydrogen environment. The oxidation of propan-1-ol to form propanoic acid would result in three hydrogen environments. The oxidation of propan-2-ol would produce propanone, which has two CH_3 groups in the same environment bonded to C=O.

c. The peak at $m/z = 43$ corresponds to $[CH_3CO]^+$, which can be formed by fragmentation of propanone.

d. Two

8. a. The standards would allow the dichlorobenzyl alcohol peak from the original chromatogram to be identified because the peak would have the same retention time as the standards.

b. A set of standards is used and then the area under their peaks is obtained to determine the relationship between concentration and peak area (a calibration curve), thus enabling the unknown concentration of other samples to be determined by interpolation.

c. The manufacturer's claim is true.

9. a. Peak 1

b. A pure sample of phenacetin could be analysed by HPLC under identical conditions to the APC sample. The retention time could then be matched to the retention times obtained for the APC sample.

c. See figure at the bottom of the page*

d. c(phenacetin) $= 2.5$ mg L^{-1}

e. % phenacetin $= 0.74$ %(m/m)

10. a. The empirical formula of X is $C_3H_6O_2$.

b. The molecular formula is $C_3H_6O_2$.

c. See figure at the bottom of the page**

d.

Wave number (cm^{-1})	Bond	Present or absent
≈ 3000	C–H	Present
≈ 1750	C=O (ester)	Present
2500–3500	O–H (acids)	Absent

e.

H—C—O—C—C—H (ester structure)

f. $m/z = 59$: $[CH_2COOH]^+$

$m/z = 45$: $[COOH]^+$

$m/z = 29$: $[CH_3CH_2]^+$ or $[CHO]^+$

$m/z = 15$: $[CH_3]^+$

10.7 Exam questions

Section A — Multiple choice questions

1. D

2. A

3. A

4. D

5. D

6. B

*9. c.

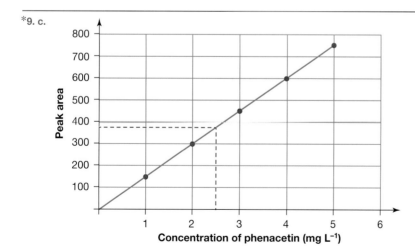

Peak area vs **Concentration of phenacetin (mg L^{-1})**

**10. c.

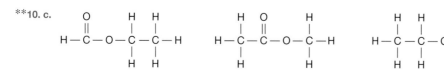

7. B

8. A

9. D

10. C

Section B — Short answer questions

11. a. $C_2H_4O_2$

   b. Two

   c.

   d. The broad peak at 2500–3300 (cm^{-1}) indicates an O–H (acid) bond.

   e. Ethanoic acid

12. a. 128 ppm (126–129 ppm is acceptable)

   b. Evidence may refer to:
   • no peak at the retention time of caffeine.
   • no peak at 96 seconds.

   c. i. The caffeine peak area is beyond the range of the calibration graph and extrapolation outside the range of the standard solutions may not be accurate.

   ii. The procedure should refer to diluting the espresso coffee sample either:
   • to bring its caffeine concentration within the range of the calibration curve
   • by a factor > 12.

13. a. Either of:
   • C=O
   • carbon, oxygen with double covalent bond.

   b. Three

   c. Three

   d. i. Two

   ii. Either of:

   CH_3CH_2

   Quartet   Triplet

   e. Either of:

14. a. i. $[CH_3CH_2]^+/CH_3CH_2{}^+/C_2H_5{}^+$

ii. Answers may include the following:
   • The relative intensities of the ^{79}Br and ^{81}Br isotopes are approximately equal.
   • The ^{79}Br isotope is slightly more abundant than the ^{81}Br isotope.

   Reasons may include the following:
   • The peaks at $m/z = 108$ — that is, $[C_2H_5{}^{79}Br]^+$ — and $m/z = 110$ — that is, $[C_2H_5{}^{81}Br]^+$ — are approximately the same height.
   • The peak at $m/z = 108$ — that is, $[C_2H_5{}^{79}Br]^+$ — is slightly higher than the peak at $m/z = 110$ — that is, $[C_2H_5{}^{81}Br]^+$.

b. i. Structure 1   Structure 2

ii. Structure 1

   Any two of the following:
   • The ^{1}H-NMR spectrum of CH_3CHBr_2 shows two sets of peaks/two signals.
   • The ^{1}H-NMR spectrum of CH_3CHBr_2 shows a doublet and a quartet.
   • This is consistent with the structure because there are the two different hydrogen environments — one for the three hydrogen atoms on CH_3 attached to CH_2Br, and one for the H on $CHBr_2$ attached to CH_3.
   • This is not consistent with the CH_2BrCH_2Br structure, in which all H atoms are in the same environment.
   • The splitting pattern is consistent with the different hydrogen environments on the CH_3CHBr_2 structure — a doublet for three hydrogen atoms on CH_3 attached to CH_2Br, and a quartet for H on $CHBr_2$ attached to CH_3.

15. a. i. Evidence from ^{13}C-NMR should include the following (or similar):
   • The number of different carbon environments (three)
   • Could be methylpropan-1-ol.
   Evidence from the reaction with $Cr_2O_7{}^{2-}/H^+$ should include either of the following:
   • The compound must be a primary alkanol.
   • Alcohol with –OH on C1.

ii.

   The systematic name should be any one of:
   • methylpropan-1-ol
   • methyl-1-propanol
   • 2-methyl-1-propanol.

b. i. Information about the structure of Z from IR data may include either of the following:
   • C=O is present.
   • Carbonyl group.

Information about the structure of Z from ^{1}H-NMR data may include the following:

- Two different hydrogen environments
- Splitting patterns are quartet and triplet → CH_3CH_2 is present
- 6 H in one environment, 4 H in a different environment.

ii. $CH_3CH_2COCH_2CH_3$

OR

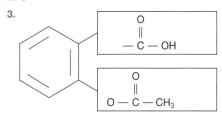

11 Medicinal chemistry

11.2 Structures and isolation of organic medicines

Practice problem 1

B

Practice problem 2

a. Ethyl ethanoate and methyl ethanoate. Due to the presence of the highly polar carboxyl group, propanoic acid and butanoic acid can form stronger interactions with water than ethyl ethanoate and methyl ethanoate, which contain an ester group. As such, ethyl ethanoate and methyl ethanoate are more likely to move to the non-polar hexane layer.

b. Methyl ethanoate. It is a smaller molecule than ethyl ethanoate, so will form weaker dispersion forces and have a lower boiling point.

11.2 Exercise

1. Chiral carbons have bonds that result in no plane of symmetry and the molecule mirror images are non-superimposable. Achiral molecules have symmetry and their mirror images can be superimposed.

2. Enantiomers are called optical isomers because they will rotate plane polarised light in opposite directions when it is passed through a sample of each one.

3.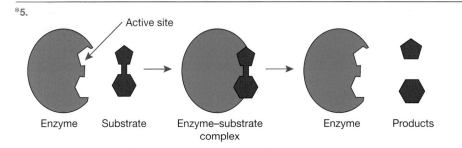

4. a. Anything polar — for example, water or ethanol — as shikimic acid is polar
   b. (Fractional) distillation

5. - Dissolve the mixture in a non-polar solvent.
   - Add an aqueous base/NaOH to convert A into the salt form (so it moves to the polar layer).
   - Separate the two layers.
   - (Add an acid/H$^+$ to convert A back into its acid form.)

11.2 Exam questions

1. C

2. C

3.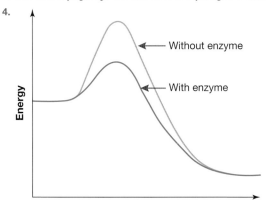

4. C

5. C

11.3 Enzymes and inhibitors

Practice problem 3

The fastest rate of reaction will be produced at 35 °C and at a pH of 2. These conditions are consistent with the conditions found in the stomach and represent the optimum pH and temperature for the functioning of this enzyme. These conditions allow the fastest rates of reaction to occur.

11.3 Exercise

1. Any two of:
   - arginine
   - lysine
   - histidine.

2. Hydrogen bonding

3. The carboxyl groups are able to form hydrogen bonds.

4.

5. See figure at the bottom of the page*

---

*5.

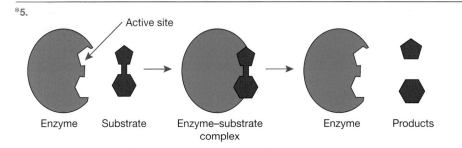

6. –CH_3 groups are non-polar, so the interaction would be dispersion forces with another non-polar group.

7. In order to have an optical isomer there must be four *different* groups attached to the central carbon (a chiral carbon). Glycine is the only amino acid to have two hydrogen atoms attached to the central carbon because the R group in glycine is just a hydrogen atom.

8.

$$CH_2 \text{---} COO^-$$
$$|$$
$$H_2N \text{---} CH \text{---} COO^-$$

9. a. The optimum temperature is the temperature at which an enzyme is most effective. In humans this is at about 37 °C. At lower temperatures the rate of reaction slows and at higher temperatures the enzyme can become denatured.

   b. The optimum temperature is 37 °C and the optimum pH is 8.

10. a. This will reduce enzyme activity.
    A higher concentration of the inhibitor will outcompete the substrate for active site binding.

    b. This will increase enzyme activity.
    A higher concentration of the substrate will outcompete the inhibitor for active site binding.

    c. This will reduce enzyme activity.
    Increasing similarity between the shape of the substrate and inhibitor will increase binding affinity of the inhibitor for the active site of the enzyme.

11.3 Exam questions

1. A
2. B
3. D
4. B
5. a.

   H   H   O   H   H   O   H   H   O
   | | ‖ | | ‖ | | ‖
 — N — C — C — N — C — C — N — C — C —
       | | |
       CH_2 H CH_2—〇—OH
       |
       OH

   or

   CH_2OH CH_2—〇—OH
   | |
 — HN — CH —[CONH]— CH_2 — CONH — CH — CO —

   b.

   H
    \\   H
     N — C — C = O
    / | |
   H   H   O⁻

11.4 Review

11.4 Review questions

1. D
2. All amino acids found in naturally occurring proteins have the following bonded to a central carbon atom: an amino group, a carboxyl group, a hydrogen atom and an R group (side chain), which determines what amino acid it is.

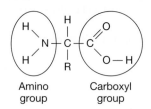

Amino   Carboxyl
group   group

3. Glutamine and glutamic acid have similar structures, but glutamine has the R group –CH_2–CH_2–$CONH_2$ (an amide group at the end), whereas glutamic acid has the R group –CH_2–CH_2–COOH (a carboxyl group at the end).

4. a. The intermolecular forces maintaining the secondary structure are hydrogen bonding between the peptide links on different parts of the protein chain.

   b. Any three of the following interactions between R groups:
   • Dispersion forces
   • Dipole–dipole interactions
   • Hydrogen bonding
   • Ionic bonding
   • Ion–dipole interactions
   • A covalent bond (disulfide bridge).

5. Any two of the following:
   • Enzymes can speed up reactions to a much greater degree than inorganic catalysts.
   • Enzymes operate under much milder conditions.
   • Enzymes operate within narrower temperature and pH ranges.
   • Enzymes are much more selective than inorganic catalysts as each enzyme catalyses a particular type of reaction. Inorganic catalysts can increase the rate of many different reactions.

6. a.

   CH_2—〇
   |
   H_3N^+ — CH — COOH

   b.

   CH_2—〇
   |
   H_2N — CH — COO⁻

   c.

   CH_2—〇
   |
   H_3N^+ — CH — COO⁻

7. a. An enzyme is said to fit together with a substrate like a lock and key. The substrate fits perfectly into the shape produced by the relevant enzyme. Bonds form between the substrate and the active site in the enzyme that weaken bonds within the substrate. The substrate is modified and the product separates from the enzyme.

b. A + B $\xrightarrow{\text{Catalyst}}$ C + D

See figure at the bottom of the page*

8. C

9. a. Carboxyl (group)/COOH

b. It has a similar structure to succinate. This allows malonate to compete with succinate for binding to the active site of succinate dehydrogenase.

10. a. Any one of the following:
- Hydroxyl
- Amide
- Ether (not in the VCE course).

b. A large proportion of the structure is non-polar. Non-polar substances dissolve in non-polar solvents.

c. Morphine would more readily dissolve in the ethanol because it has an additional hydroxyl group, so is more polar than codeine.

11.4 Exam questions

Section A — Multiple choice questions

1. D
2. A
3. B
4. B
5. B
6. C
7. D
8. C
9. A
10. B

Section B — Short answer questions

11. Primary structure: Covalent bonds between C and N in the peptide links between amino acids

Secondary structure: Hydrogen bonds between O (–C=O) and H (–N–H) on different peptide groups in the α-helix (β-sheet)

12. a.

b. Ethanoic anhydride

13. a. Any one of the following:
- Peptide bond/group
- Amide bond
- Covalent bond.

b.

c. Hydrogen bonding between the H atom on one peptide group and the O atom on a different peptide group in the same polymer chain

14. a. Any one of the following amino acids with a non-polar side chain:
- Alanine/Ala
- Glycine/Gly
- Isoleucine/Ile
- Leucine/Leu
- Methionine/Met
- Phenylalanine/Phe
- Proline/Pro
- Valine.

Either of the following amino acids with an acidic side chain:
- Aspartic acid/Asp
- Glutamic acid/Glu.

b. i. Cysteine/Cys

ii. Either of the following pairs:
- Asparagine and serine (or Asn and Ser)
- –CH$_2$CONH$_2$ and –CH$_2$OH.

iii. Either of the following pairs:
- Aspartic acid and lysine (or Asp and Lys)
- –CH$_2$COO– and –CH$_2$CH$_2$CH$_2$CH$_2$NH$_3$$^+$.

iv. Dispersion forces

***7. b.**

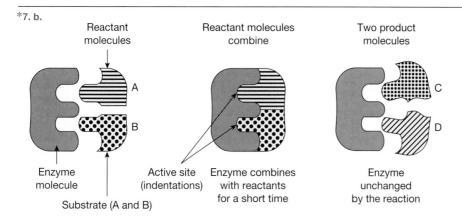

Reactant molecules | Reactant molecules combine | Two product molecules

Enzyme molecule

Active site (indentations)   Enzyme combines with reactants for a short time

Substrate (A and B)

Enzyme unchanged by the reaction

c. The tertiary structure/active site/side chain charges will be different at a lower pH.

15. a. Chemical bonding in tertiary structures:
The tertiary structure is made by the folding and twisting of the primary and secondary structures into shapes held in place by specific interaction between various R group residues. The actual bonding present will depend on the amino acids present in the primary structure.
Significance of the tertiary structure to the function of enzymes by maintaining the shape of the active site:
The shape of the active site, which is where the reaction occurs, is crucial to the function of the enzyme. It is shaped in such a way that the substrate 'fits' into the active site and attaches so that the reaction can occur. The tertiary structure of the enzyme is crucial for this.
Interaction between the substrate and enzyme:
According to the lock-and-key model, the substrate (key) and enzyme 'attach' at the active site (lock). The substrate interacts with the active site of the enzyme. The bonds that form can be ion–ion, dipole–dipole, hydrogen bonding or ion–dipole (or electrostatic).

b. Any one of the following factors:
- Temperature
- Inhibiting factors
- Substrate source/concentration.

An explanation consistent with the factor identified is required.

Unit 4 | Area of Study 2 Review

Section A — Multiple choice questions

1. B
2. D
3. C
4. B
5. A
6. B
7. C
8. D
9. B
10. C
11. D
12. B
13. A
14. C
15. D
16. B
17. B
18. C
19. D
20. A

Section B — Short answer questions

21. a. The iodide ions are in excess so that the amount of iodine produced is determined by the amount of hypochlorite present (hypochlorite is the limiting reagent).

b. $x = 2$

c. i. $n(S_2O_3^{2-}) = 0.00196 \text{ mol}$

ii. $n(OCl^-) = 0.00079 \text{ mol}$

iii. $c(OCl^-)_{undiluted} = 0.490 \text{ mol L}^{-1}$

22. a. i.   ii.

b. i. Two peaks

ii. Two peaks

c. i. One peak. There are no hydrogens on the adjacent neighbouring carbon. Using the $n + 1$ rule, this means the peak for CH_3 will be a singlet.

ii. Two peaks. There is one hydrogen on the adjacent neighbouring carbon atom. Using the $n + 1$ rule, this means the peak for CH_3 will be a doublet.

d. The IR spectrum for propanone will have a narrow peak at 1680–1850 cm^{-1}, whereas the IR spectrum for propan-2-ol will have a broad peak at 3200–3600 cm^{-1}.

e. $[CH_3CO]^+$

f. Compound 'P' is propan-2-ol.

23.

Any four *different* groups would be acceptable.

24. a. Any of the four identified below:

b. Gly, Leu, Lys, Asp, Gly

c. C=O (carbonyl) and N–H (amine) of the peptide links

d. Hydrogen bonding (due to –OH and –NH) and dispersion forces (due to non-polar Leu)

e. The kinetic energy would be transferred to the structure, disrupting the bonding in the quaternary, tertiary and secondary structure, causing a change to the active site, which denatures the enzyme.

GLOSSARY

α-helices refers to when hydrogen bonds are formed between an oxygen atom of a –C=O bond and a hydrogen atom of an –NH bond that is four amino acids away on the same chain

β-pleated sheets refers to when two sections of the peptide chain line up and are held together in a sheet-like structure by hydrogen bonds between one oxygen atom of a –C=O bond and a hydrogen atom of an –NH bond in the parallel or anti-parallel sheet

accuracy refers to how close an experimental measurement is to a known value

activation energy (E_a) the minimum energy required by reactants in order to react

addition reaction a reaction in which one molecule bonds covalently with another molecule without losing any other atoms

adsorption the adhesion of atoms, ions or molecules from a gas, liquid or dissolved solid to a surface

affinity the attraction of a component to a phase, either mobile or stationary

aim a statement outlining the purpose of an investigation, linking the dependent and independent variables

alcohols compounds in which a hydroxyl group (–OH) is the parent functional group

aliphatic describes organic compounds in which carbon atoms form open chains

aliquot the liquid from a pipette

alkaline fuel cell (AFC) a fuel cell that converts oxygen (from the air) and hydrogen (from a supply) into electrical energy and heat

alkanes the family of hydrocarbons containing only single carbon–carbon bonds

alkanols alkanes with a hydroxyl group replacing a hydrogen atom

alkenes the family of hydrocarbons that contain at least one carbon–carbon double bond

alkyl groups hydrocarbon branches joined to the parent hydrocarbon chain (e.g. CH_3 (methyl), CH_2CH_3 (ethyl))

alkynes the family of hydrocarbons with a carbon–carbon triple bond

allotropes the different physical forms in which an element can exist

amides organic compounds containing the amide functional group, –CONH–

amines organic compounds containing the amino functional group, $–NH_2$

amino acids molecules that contain an amino and a carboxyl group

anaerobic respiration the breakdown of glucose in the absence of oxygen

analyte the sample undergoing analysis

anode the electrode at which oxidation occurs; in a galvanic cell it is the negative electrode, since it is the source of negative electrons for the circuit; if the reducing agent is a metal, it is used as the electrode material; in an electrolytic cell it is the positive electrode, because it generates electrons that then travel back to the power supply

aquifer an underground rock layer that contains water; this groundwater can be extracted using a well

arenes aromatic, benzene-based hydrocarbons

aromatic describes a compound that contains at least one benzene ring and is characterised by the presence of alternating double and single bonds within the ring

assumptions ideas that are accepted as true without evidence in order to overcome limitations in experiments

atom economy a measurement of the efficiency of a reaction that considers the amount of waste produced, by calculating the percentage of the molar mass of the desired product compared to the molar mass of all reactants

backbone a peptide chain of covalently bonded nitrogen and carbon atoms

bar graph a graph in which data is represented by a series of bars; usually used when one variable is quantitative and the other is qualitative

base peak identifies the most abundant ion in a mass spectrum

benzene an aromatic hydrocarbon with the formula C_6H_6

bias an intentional or unintentional influence on a research investigation

biodiesel a fuel produced from vegetable oil or animal fats and combined with an alcohol, usually methanol

bioethanol ethanol produced from plants, such as sugarcane, and used as an alternative to petrol

biofuel a renewable, carbon-based energy source formed in a short period of time from living matter

biogas fuel produced from the fermentation of organic matter

biomimicry the act of copying and adapting processes that occur in nature

boiling point the temperature at which a substance changes between the liquid and gas states

bond energy the amount of energy required to break the bonds of a mole of molecules into its individual atoms

bond length the distance between two nuclei involved in covalent bonding, which depends on the size of the atoms

burette a graduated glass tube used for delivering known volumes of a liquid, especially in titrations

calibrate adjusting an instrument using standards of known measurements to ensure the instrument's accuracy

calibration curve a graph of concentration versus peak area; also known as a standard curve

calibration factor the amount of energy required to change the contents of a calorimeter by one degree, with units $J\,°C^{-1}$

calorimeter an apparatus used to measure heat changes during a chemical reaction or change of state

calorimetry a method used to determine the changes in energy of a system by measuring heat exchanges with the surroundings

carbohydrates the general name for a large group of organic compounds occurring in food and living tissues; includes sugars, starch and cellulose

carbon neutral no net release of carbon dioxide into the atmosphere

carboxylic acids the homologous series containing the —COOH functional group

catalyst a substance that increases the rate of a reaction without a change in its own concentration

cathode the electrode at which reduction occurs; in a galvanic cell it is the positive electrode, because the negative electrons are drawn towards it and then consumed by the oxidising agent, which is present in the electrolyte; in an electrolytic cell it is the negative electrode, because it receives electrons from the power supply

causation refers to when one factor or variable directly influences the results of another factor or variable

cell potential difference the difference between the reduction potentials of two half-cells

cellular respiration the process that occurs in cells to oxidise glucose in the presence of oxygen to carbon dioxide, water and energy

cellulose the most common carbohydrate and a condensation polymer of glucose; humans cannot hydrolyse cellulose, so it is not a source of energy

cellulosic fermentation the use of enzymes to obtain glucose from cellulose to make alcohol

chain isomers a type of structural isomer that involves branching

change in enthalpy the amount of energy released or absorbed in a chemical reaction

chemical calibration calibration of a calorimeter using a combustion reaction with a known ΔH

chemical energetics a branch of science that deals with the properties of energy and the way it is transformed in chemical reactions

chemical shift the horizontal scale on an NMR spectrum

chiral describes compounds that contain an asymmetric carbon atom or chiral centre; the molecule cannot be superimposed upon its mirror image

chiral centre an asymmetric carbon atom; a carbon atom bonded to four different groups of atoms

chlorophyll a series of green pigments that enable plants to capture sunlight for photosynthesis

chromatogram a chart that shows the results from analysis by chromatography

climate change changes in various measures of climate over a long period of time

closed system a system in which energy, but not matter, can be transferred to and from its surroundings; all reactants and products are contained

coagulation the process of turning a liquid into a solid or a thicker liquid

coal the world's most plentiful fossil fuel; it is formed from the combined effects of pressure, temperature, moisture and bacterial decay on vegetable matter over several hundred million years

Coldry Process a patented process that changes the naturally porous form of brown coal to produce a dry, dense pellet, via a process called 'brown coal densification'

combustion the rapid reaction of a compound with oxygen

competitive inhibitors molecules that bind to the active site of an enzyme and prevent the substrate from binding

concentration fraction essentially, the concentrations of the products divided by the concentrations of the reactants, including the coefficients of each component in the reaction

conclusion a section at the end of a scientific report that relates back to the question, sums up key findings and states whether the hypothesis was supported or rejected

concordant describes titres that are within a defined volume of each other, such as 0.10 mL

condensation reaction a reaction in which molecules react and link together by covalent bonding with the elimination of a small molecule, such as water or hydrogen chloride, from the bond that is formed

conformation the three-dimensional structure of a protein

continuous data quantitative data that can take any continuous value

control group a group that is not affected by the independent variable and is used as a baseline for comparison

controlled variable a variable that is kept constant across different experimental groups

correlation a measure of the relationship between two or more variables

covalent bonds bonds that involve the sharing of electron pairs between atoms

cyclic hydrocarbons also known as ring structures, because the carbon chain is a closed structure without open ends

Daniell cell one of the first electrochemical cells to produce a reliable source of electricity; it uses the redox reactions between zinc metal and copper ions to produce electricity

denaturation a change in the structure or function of a large molecule, such as a protein

dependent variable the variable that is influenced by the independent variable; the variable that is measured

desorption the removal of a substance from a surface; the opposite of adsorption

dietary fibre non-starch polysaccharides in both water-soluble and water-insoluble forms

dioxins highly toxic compounds formed from industrial processes and incomplete combustion of organics

dipeptide formed when two amino acids combine

dipole moment a way to describe the asymmetrical charge distribution in a polar molecule

direct methanol fuel cell (DMFC) a new technology that is powered by liquid methanol

disaccharide two sugar molecules (monosaccharides) bonded together

discrete data quantitative data that can only take on set values

discussion a detailed area of a scientific report in which results are discussed, analysed and evaluated; relationships to concepts are made; errors, limitations and uncertainties are assessed; and suggestions for future improvements are outlined

dissolution the process of solutes dissolving in solvents to form a solution

distillation a process for separating components in a mixture that is dependent on the differing boiling points of the components

Downs cell an electrolytic cell used for the commercial production of sodium and chlorine

dry cell an electrochemical cell in which the electrolyte is a paste, rather than a liquid; also called a Leclanché cell

efficiency (of energy conversion) the ratio between useful energy output and energy input

electrical calibration calibration of a calorimeter by supplying a known quantity of electricity

electrical potential the ability of a galvanic cell to produce an electric current

electrochemical cell a cell that generates electrical energy from chemical reactions

electrochemical series a series of chemical half-equations arranged in order of their standard electrode potentials

electrode a solid used to conduct electricity in a galvanic half-cell

electrolysis the process in which a non-spontaneous chemical reaction occurs by passing an electric current through a substance in solution or molten state

electrolytes liquids that can conduct electricity

electrolytic cell an electric cell in which a non-spontaneous redox reaction is made to occur by the application of an external potential difference across the electrodes; also known as an electrolysis cell

electron configuration the number of electrons and shells they occupy (e.g. 2,4 for a carbon atom)

electroplating the process of adding a thin metal coating by electrolysis

eluent a substance used as a solvent in separating materials; for example, the mobile phase in chromatography

enantiomers chiral molecules that are non-superimposable mirror images of one another

end point the point at which the indicator changes colour in a titration

endothermic describes a chemical reaction in which energy is absorbed from the surroundings

energy profile diagram a graph or diagram that shows the energy changes involved in a reaction from the reactants through the intermediate stages to the products

enhanced greenhouse effect the effect of increasing concentrations of greenhouse gases in the atmosphere as the result of human activity

enthalpy a thermodynamic quantity equivalent to the total heat content of a system

enzyme a protein that acts as a biological catalyst

equilibrium constant the value of the concentration fraction at equilibrium, which gives an indication of the extent to which reactants are converted into products; it is assigned the symbol K

equilibrium law the relationship between the concentrations of the products and the reactants, taking into account their stoichiometric values

equilibrium reaction a reaction in which both forward and reverse reactions are significant

equivalence point the point at which two reactants have reacted in their correct mole proportions in a titration

erroneous value an inaccurate value resulting from an error

esterification the process of ester formation

ethanol an alcohol with two carbons produced from fermentation of glucose by yeast

ethics principles of acceptable and moral conduct determining what is 'right' and what is 'wrong'

excited state refers to when electrons move to higher energy orbitals when energy is applied

exothermic describes a chemical reaction in which energy is released to the surroundings

experimental bias a type of influence on results in which an investigator either intentionally or unintentionally manipulates results to get a desired outcome

experimental group a test group that is exposed to the independent variable

external circuit a circuit composed of all the connected components within an electrolytic or a galvanic cell to achieve desired conditions

extrapolation an estimation of a value outside the range of data points tested

falsifiable able to be proven false using evidence

Faraday constant represents the amount of electric charge carried by 1 mole of electrons

Faraday's First Law states that the amount of current passed through an electrode is directly proportional to the amount of material released from it

Faraday's Second Law states that when the same quantity of electricity is passed through several electrolytes, the mass of the substances deposited are proportional to the stoichiometric coefficients in the balanced half-equations of the respective electrochemical reactions

fat a triglyceride formed from glycerol and three fatty acids

fatty acids long-chain carboxylic acids, usually containing an even number of 12–20 carbon atoms

feedstocks raw materials used to supply or fuel a machine or industrial process

fingerprint region the region of the infrared spectrum below 1500 cm^{-1} containing a pattern of peaks that is specific for an individual molecule

flashpoint the temperature at which a particular organic compound gives off sufficient vapour to ignite in air

fossil fuels fuels formed from once-living organisms

fracking the process of pumping a large amount of fluid, mainly water, under high pressure into a drilled hole, in order to break rock so that it will release gas or oil

fractional distillation the process of separating component fuels based on their different boiling points

fructose a pentose monosaccharide

fuel a substance that burns in air or oxygen to release useful energy

fuel cell an electrochemical cell that produces electrical energy directly from a fuel

fuel reformer a device or system that converts a fuel source — typically hydrocarbons or alcohols — into a hydrogen-rich gas mixture

functional group an atom or group of atoms that is attached to or part of a hydrocarbon chain, and influences the physical and chemical properties of the molecule

functional isomerism refers to when isomers contain different functional groups

global warming a gradual increase in the overall temperature of Earth's atmosphere

glucose a simple carbohydrate stored in the liver or muscles

glycerol an alcohol; it is a non-toxic, colourless, clear, odourless and viscous liquid that is sweet-tasting and has the semi-structural formula $CH_2OHCH(OH)CH_2OH$

glycogen the storage form of glucose in animals

green chemistry a relatively new branch of chemistry that emphasises reducing the amounts of wastes produced, the more efficient use of energy, and the use of renewable and recyclable resources

green hydrogen hydrogen that does not contribute to the enhanced greenhouse effect in its production

greenhouse effect a natural process that warms Earth's surface; when the Sun's energy reaches Earth's atmosphere, some of it is reflected back to space, and the rest is absorbed and re-radiated by greenhouse gases

greenhouse gases gases that contribute to the greenhouse effect by absorbing infrared radiation

half-cell one half of a galvanic cell containing an electrode immersed in an electrolyte that may be the oxidising agent or the reducing agent, depending on the oxidising strength of the other cell to which it is connected

half-equation an equation that gives one half of a redox reaction, showing the movement of electrons in either an oxidation or a reduction reaction

halogenation a reaction in which one or more halogen atoms are added

halogens elements in group 17 of the periodic table: F, Cl, Br, I and At

heat of reaction the heat evolved or absorbed during a chemical reaction taking place under conditions of constant temperature and of either constant volume or, more often, constant pressure

heterogeneous reaction a reaction in which some of the substances involved are in different phases

high-performance liquid chromatography (HPLC) a method used to separate the components of a mixture

histogram a graph in which data is sorted in intervals and frequency is examined; used when both pieces of data are quantitative

homogeneous reaction a reaction in which all of the substances involved are in the same phase

homologous series a series of organic compounds that have the same structure but in which the formula of each molecule differs from the next by a CH_2 group

hydrolytic reaction the chemical breakdown of a compound due to a reaction with water; also known as hydrolysis

hydrophilic describes molecules more likely to interact with water and other polar substances

hydrophobic describes non-polar molecules that repel water molecules

hygroscopic refers to when a substance has a tendency to absorb water vapour from the atmosphere

hypothesis a tentative, testable and falsifiable statement for an observed phenomenon that acts as a prediction for the investigation

ideal gas equation $PV = nRT$, where pressure is measured in kilopascals, volume is measured in litres and temperature is measured in kelvin

immiscible refers to liquids that do not form a homogeneous mixture when mixed with another liquid

independent variable the variable that is changed or manipulated by an investigator

indicator a chemical compound that changes colour and structure when exposed to certain conditions and is therefore useful for chemical tests

infrared (IR) spectroscopy describes spectroscopy that deals with the infrared region of the electromagnetic spectrum

internal circuit a circuit within a solution; anions flow to the anode and cations flow to the cathode

interpolation an estimation of a value within the range of data points tested

investigation question the focus of a scientific investigation in which experiments act to provide an answer

iodine number the mass of iodine that reacts with 100 g of a compound

isolated system a system in which neither matter nor energy is transferred to or from its surroundings

isomers molecules with the same formula but a different arrangement of atoms and different properties

isotope effect the generation of multiple peaks for fragments with the same formula due to the presence of isotopes of the constituent elements

kelvin the SI base unit of thermodynamic temperature, equal in magnitude to the degree Celsius

kerosene a mixture of hydrocarbons with molecules containing between 10 and 15 carbon atoms

kilojoule a unit of energy; one kilojoule (kJ) is equal to 1×10^3 joules (J)

kinetic energy energy associated with movement, in doing work

lactic acid an organic acid, $C_3H_6O_3$, present in muscle tissue as a by-product of anaerobic respiration

Le Chatelier's principle states that when a change is made to an equilibrium system, the system moves to counteract the imposed change and restore the system to equilibrium

lead–acid accumulator a battery with lead electrodes using dilute sulfuric acid as the electrolyte; each cell generates about 2 volts

Leclanché cell *see* **dry cell**

legumes plants that produce pods with a seed inside

limitations factors that affect the interpretation and/or collection of findings in a practical investigation

line graph a graph in which points of data are joined by a connecting line; used when both pieces of data are quantitative (numerical)

line of best fit a trend line added to a scatterplot to best express the data shown; these are straight lines, and are not required to pass through all points

lipids substances such as fats, oils and waxes that are insoluble in water

liquefied petroleum gas (LPG) a hydrocarbon fuel that consists mainly of propane and butane

lithium cells cells that use lithium anodes and can produce a high voltage

logbook a record containing all the details of progress through the steps of a scientific investigation

mass spectrometry the investigation and measurement of the masses of isotopes, molecules and molecular fragments by ionising samples and separating the fragments produced, using a combination of electric and magnetic fields

mass-to-charge ratio (*m/z*) the mass of a particle divided by its overall charge; when the charge is +1 the *m/z* and mass have the same numerical value

Maxwell–Boltzmann distribution curve a graph that plots the number of particles with a particular energy (vertical axis) against energy (horizontal axis)

measurement bias a type of influence on results in which an experimenter manipulates results to get a desired outcome; may be unintentional (i.e. through the placebo effect) or intentional

megajoule a unit of energy; one megajoule (MJ) is equal to 1×10^6 joules (J)

melting point the temperature at which a substance changes between the solid and liquid states

membrane cell an electrolytic cell used for the electrolysis of brine

metabolism the chemical processes that occur within a living organism to maintain life

methyl ester the product of a condensation reaction between a triglyceride and methanol

mistakes human errors or personal errors that can impact results, but should not be included in analysis; instead, the experiment should be repeated correctly

mobile phase the liquid or gas that flows through a chromatography system, moving the materials to be separated at different rates over the stationary phase

molar gas constant (*R*) the constant of the universal gas equation; $R = 8.31$ J mol^{-1} K^{-1} when pressure is measured in kPa, volume is measured in L, temperature is measured in K and the quantity of the gas is measured in moles (*n*)

molar gas volume the volume occupied by a mole of a substance at a given temperature and pressure; at SLC, 1 mole of gas occupies 24.8 L

molecular ion the positive ion produced by ionisation of a whole molecule

monosaccharide the simplest form of carbohydrate, consisting of one sugar molecule

_n_ + 1 rule a rule used for simple molecules; the number of peaks is one more than the number of equivalent hydrogen atoms on the neighbouring carbon atom(s)

natural gas a source of alkanes (mainly methane) of low molecular mass

NO$_x$ a term used for oxides of nitrogen, such as NO_2 and NO, that contribute to air pollution

nominal data qualitative data that has no logical sequence

non-renewable (with reference to energy sources) energy sources that are consumed faster than they are being formed

open system a system in which both energy and matter can be transferred to and from its surroundings; reactants and products are not contained

optical isomers *see* **enantiomers**

ordinal data qualitative data that can be ordered or ranked

outlier a result that is a long way from other results and seen as unusual

oxidation a loss of electrons; an increase in the oxidation number

oxidation numbers numbers used to find an oxidising agent and a reducing agent by a change in perceived valency

oxidising agents electron acceptors

particulates solid and liquid particles small enough to be suspended in the atmosphere

peptide link the link formed when a carboxyl group reacts with an amino group in a condensation reaction between two amino acids

percentage purity the percentage of a sample that is the desired substance

percentage yield a measurement of the efficiency of a reaction, found by calculating the percentage of the actual yield compared to the theoretical yield

petroleum a viscous, oily liquid composed of crude oil and natural gas that was formed by geological processes acting on marine organisms over millions of years; it is a mixture of hydrocarbons used to manufacture other fuels and many other chemicals

photodecomposition the use of light (photons) to break down molecules

photoelectrodes electrodes that achieve redox reactions utilising light as an energy source

photosynthesis in the presence of light, carbon dioxide + water → glucose + oxygen

pipette a piece of glassware used for transferring accurate volumes of liquid

polypeptide many amino acid residues bonded together

polysaccharide more than ten monosaccharides bonded together

positional isomers isomers in which the position of the functional group differentiates the compounds

potential energy energy that is stored, ready to do work

precision refers to how close multiple measurements of the same investigation are to each other; a measure of repeatability or reproducibility

pressure the force per unit area that one region of a gas, liquid or solid exerts on another

primary alcohol an alcohol in which the carbon atom that carries the –OH group is attached to only one other carbon atom

primary cell an electrolytic cell in which the cell reaction is not reversible

primary data direct or firsthand evidence about some phenomenon, obtained from investigations or observations

primary haloalkane a haloalkane in which the halogen atom is attached to a carbon that is only attached to one other carbon atom

primary source a document that is a record of direct or firsthand evidence about some phenomenon

primary standard a substance used in volumetric analysis that is of such high purity and stability that it can be used to prepare a solution of accurately known concentration

primary structure the order of amino acids in a protein molecule

proteins large molecules composed of one or more long chains of amino acids

proton exchange membrane fuel cell (PEMFC) a fuel cell being developed for transport applications, as well as for both stationary and portable fuel cell applications

qualitative analysis the determination of non-numerical information, such as the presence or absence of elements, ions, functional groups or molecules in a sample

qualitative data categorical data that examines the quality of something (e.g. colour or gender) rather than numerical values

quantitative analysis the determination of numerical information, such as the amount of a given element or compound in a known mass or volume of a sample

quantitative data numerical data that examines the quantity of something (e.g. length, time); also known as numerical data

quaternary structure the structure formed when individual protein molecules link together in a particular spatial arrangement

racemate a 50 : 50 mixture of two enantiomers; often occurs when optically active substances are synthesised in the laboratory

racemic mixture *see* **racemate**

random errors chance variations in measurements that result in a spread of readings

randomised assignment of individuals to an experiment or control group at random; not influenced by external factors

reaction quotient (Q) essentially, the concentrations of the products divided by the concentrations of the reactants, including the coefficients of each component in the reaction

rechargeable describes a battery that is an energy storage device; it can be charged again after being discharged by applying DC current to its terminals

recharging forcing electrons to travel in the reverse direction; because the discharge products are still in contact with the electrodes, the original reactions are reversed

redox reactions reactions that involve the transfer of one or more electrons between chemical species

reducing agents electron donors

reduction a gain of electrons; a decrease in the oxidation number

reduction potential a measure of the tendency of an oxidising agent to accept electrons and so undergo reduction

renewable (with reference to energy sources) energy sources that can be produced faster than they are used

repeatability refers to how close the results of successive measurements are to each other in the exact same conditions

reproducibility refers to how close results are when the same variable is being measured but under different conditions

residue what remains when two or more amino acids combine to form a peptide

resolution (with reference to chromatography) the degree of component separation; (with reference to scientific investigations) the smallest change of measurement that a particular piece of equipment can detect

response bias a type of influence on results in which only certain members of the target population respond to an invitation to participate in a scientific trial, resulting in an unrepresentative sample of the larger population

retention time the time taken for a component in a sample to travel from the injection port to the end of the column

risk assessment a document that examines the different hazards in an investigation and suggested safety precautions

salt bridge a component that provides a supply of mobile ions that carry the charge through the solution of a galvanic cell during a reaction

sample a substance to be analysed

sample size the number of trials in an investigation

sampling bias a type of influence on results in which participants chosen for a study are not representative of the target population

saturated describes hydrocarbons containing only single carbon–carbon bonds

scatterplot a graph in which two quantitative variables are plotted as a series of dots

scientific investigation methodology the principles of research based on the scientific method

scientific methodology the type of investigation being conducted to answer a question and resolve a hypothesis

secondary alcohol an alcohol in which the carbon atom that carries the –OH group is joined directly to two alkyl groups, which may be the same or different

secondary cell a cell that can be recharged once its production of electric current drops; often called a rechargeable battery

secondary data comments on, or summaries and interpretations of, primary data

secondary fuel a fuel that is produced from another energy source

secondary source a document that comments on, summarises or interprets primary data

secondary structure the structure formed from hydrogen bonding between carboxyl and amino groups in peptide links at different positions in a protein molecule

selection bias a type of influence on results in which test subjects are not equally and randomly assigned to experimental and control groups

serving size the recommended amount of food on a nutrition label for one serving

side chain an R group attached to an amino acid

solution calorimetry the process of using a calorimeter to measure heat changes in a solution; for example, heat of dissolution and neutralisation reactions

solvent extraction a technique used to separate solutes based on their relative solubility in two solvents with different polarity

specific heat capacity (*c*) the energy needed to change the temperature of 1 g of a substance by 1 °C

spectroscopy the investigation and measurement of spectra produced when matter interacts with or emits electromagnetic radiation

spontaneous reactions reactions that proceed on their own, without the need for any external supply of energy

standard cell potential difference the measured cell potential difference, under standard conditions, when the concentration of each species in solution is 1 M, the pressure of a gas (where applicable) is 100 kPa and the temperature is 25 °C (298 K)

standard electrode potential the voltage or potential difference due to the difference in charge on the electrode and electrolyte compared to the hydrogen half-cell

standard hydrogen half-cell a standard reference electrode; it is assigned 0.00 volts

standard laboratory conditions (SLC) 100 kPa and 25 °C

standard solution a solution that has an accurately known concentration

starch a condensation polymer of glucose

stationary phase a solid with a high surface area, or a finely divided solid coated with liquid; it shows different affinities for various components of a sample mixture when separating them by chromatography

stereoisomers two or more compounds differing only in the spatial arrangement of their atoms

stoichiometry calculating amounts of reactants and products using a balanced chemical equation

structural isomers molecules that have the same molecular formula but different structural formulas

substitution reaction a reaction in which one or more atoms of a molecule are replaced by different atoms

sustainable energy energy that meets present needs without compromising the ability of future generations to meet their own needs

systematic errors errors that affect the accuracy of a measurement and cannot be improved by repeating an experiment; usually due to equipment or system errors

tentative not fixed or certain; may be changed with new information

tertiary structure the structure formed in a protein molecule from side-group interactions, including hydrogen bonding, ionic bonding, dipole–dipole interactions and disulfide bridges

testable able to be supported or proven false through the use of observations and investigation

thermochemical equations balanced stoichiometric chemical equations that include the enthalpy change

thermochemical splitting refers to when very high temperatures are used to decompose molecules by breaking chemical bonds

thermochemistry the branch of chemistry concerned with the quantities of heat evolved or absorbed during chemical reactions

titration a type of volumetric analysis used to determine the concentration of a substance; a pipette is used to deliver one substance and a burette is used to deliver another substance until they have reacted exactly in the reaction equation mole ratios

titre the volume delivered by a burette

transesterification the conversion of one ester (triglyceride) into another ester (biodiesel)

triglycerides fats and oils formed by a condensation reaction between glycerol and three fatty acids

true value an accurate value

uncertainty a limit to the precision of data obtained; a range within which a measurement lies

United Nations Declaration on the Rights of Indigenous Peoples a universal framework of minimum standards for the survival, dignity and wellbeing of the indigenous peoples of the world

unsaturated describes hydrocarbons containing at least one double or triple carbon–carbon bond

urea a molecule synthesised in the liver to remove ammonia from the body

valence number the number of electrons occupying the orbitals in the outermost electron shell

validity describes how accurately an experiment investigates the claim it is intended to investigate

viscosity the resistance to flow of a liquid

volatility describes how readily a liquid substance will form a vapour

voltmeter a device used for measuring the potential difference between two points in a circuit

volumetric analysis determination of the concentration, by volume, of a substance in a solution, such as by titration

wave number the number of waves per centimetre

yeast a single-celled fungus

yield amount of product

INDEX

inert electrodes   285, 290
influenza   643
infrared radiation   540
infrared (IR) spectroscopy   540–50
  absorption and peak intensity
    544
  bond vibration energy   543–4
  definition   540
  infrared spectra   541–3
  principles   540–4
  vibration of covalent bonds
    540–1
inhibitors   627–50
inorganic catalysts   636
inorganic electrocatalysts   255
instrumental analysis
  chromatography *see*
    chromatography
  combining spectroscopic techniques
    566–79
  infrared spectroscopy   540–50
  mass spectrometry   529–39
  NMR spectroscopy   551–65
integration trace   554
intermolecular forces   393–6, 484
  and affinity   581
  dipole–dipole attractions   394
  dispersion forces   394
  in homologous series   396
  hydrogen bonding   395–6
  viscosity and   397
internal circuit   114
International Union of Pure and
  Applied Chemists (IUPAC)
  370
  naming of compounds   383–5
iodine number   360–1, 486–7
  definition   486
  for fats and oils   486
ionic compounds   276–7
ionic equations, for redox titration
  510
ionisation   529–31
ions
  in mass spectrum diagram   534
  notation   532
irreversible reactions   216
  rate and extent of   216–17
isolated system   188
isomers   389–93, 616–20
  definition   389
  introduction   389
  structural   389–91
    chain isomers   389
    drawing and naming   390
    functional isomers   390–1

positional isomers   390
  types   391
  summary   391
  types   616
isotope effect   534–6
IUPAC *see* International Union of
  Pure and Applied Chemists

K

kelvin   61
kerosene   10
ketones   379, 400
kilojoules   27
kinetic energy   25
knocking   11
KOHES method   108

L

lactic acid   44
Law of Conservation of Mass   181
lead–acid accumulator   304–7
  definition   304
  discharging process in   305
  recharging process in   305–7
Le Chatelier's principle   240–60
  applications   251
  changing temperature, effect of
    247–9
  changing volume, effect of   245–7
  equilibrium and   240–60
  mathematically   244–5
  yield of chemical reaction   240–1
Leclanché cell   131
legumes   41
limiting reactants   28
lipids   40, 442–3
  synthesis   450
liquid chromatography–mass
  spectrometry (LC–MS)   591–3
liquid fuel density   63–4
liquefied petroleum gas (LPG)   6, 10
lithium batteries   133–4
lithium cells   133
  definition   131
lithium-ion batteries   308–9
lock-and-key model   637–8
low-resolution proton NMR   554–5
LPG *see* liquefied petroleum gas

M

magnesium   102
magnesium metal   296
manganese dioxide cell   132–3
massive deforestation   19
mass–mass calculations   58–60
  in combustion of fuels   60

mass of substance   76
mass spectrometry   529–39
  carboxylic acid   532
  definition   529
  interpreting   531–6
    isotope effect   534–6
  ions in   534
  and IR and NMR spectroscopy
    567
  operation   530
  principles   529–31
    ionisation and fragmentation
    529–31
    technique overview   529
mass-to-charge ratio (*m/z*)   529
mass–volume calculations   60–1
  in combustion of fuels   61
Maxwell–Boltzmann distribution
  curve   185
medicinal chemistry
  competitive enzyme inhibitors as
    642–6
  enzymes and inhibitors   627–50
  organic medicines   615–27
megajoules   8
melting point   396–7
  determination   497–500
  laboratory, measuring in the
    498–9
membrane cells   298
metabolism   40, 434
metal ion–metal half-cells   115
methanal   378, 643
methanamide   401
methane gas   9
methanol   429–30
methyl ester   429
methyl halides   416
microbial electrolysis   142
mobile phase   580
modified electrochemical series   118
molar gas constant   61
molar gas volume   59
molecular ion   529
molten sodium chloride   276–7
  electrolysis of   295–6
monosaccharides   14, 41, 439–40
multiple carbon-to-carbon bonds
  359–60
multiplet   555

N

natural gas   9–10
Nernst equation   126
neutral carbon atom   357
neutralisation   78

APPENDIX
Periodic table of the elements

Key

1	2	← Atomic number
Hydrogen	Helium	← Name
H	He	← Symbol
1.0	4.0	← Relative atomic mass

Period 1

Group 1

| Period 1 | 1
Hydrogen
H
1.0 | | | | | | | | |

Group 2

| Period 2 | 3
Lithium
Li
6.9 | 4
Beryllium
Be
9.0 |
| Period 3 | 11
Sodium
Na
23.0 | 12
Magnesium
Mg
24.3 |

	Group 1	Group 2	Group 3	Group 4	Group 5	Group 6	Group 7	Group 8	Group 9
Period 4	19 Potassium **K** **39.1**	20 Calcium **Ca** **40.1**	21 Scandium **Sc** **45.0**	22 Titanium **Ti** **47.9**	23 Vanadium **V** **50.9**	24 Chromium **Cr** **52.0**	25 Manganese **Mn** **54.9**	26 Iron **Fe** **55.8**	27 Cobalt **Co** **58.9**
Period 5	37 Rubidium **Rb** **85.5**	38 Strontium **Sr** **87.6**	39 Yttrium **Y** **88.9**	40 Zirconium **Zr** **91.2**	41 Niobium **Nb** **92.9**	42 Molybdenum **Mo** **96.0**	43 Technetium **Tc** **(98)**	44 Ruthenium **Ru** **101.1**	45 Rhodium **Rh** **102.9**
Period 6	55 Caesium **Cs** **132.9**	56 Barium **Ba** **137.3**	57–71 Lanthanoids	72 Hafnium **Hf** **178.5**	73 Tantalum **Ta** **180.9**	74 Tungsten **W** **183.8**	75 Rhenium **Re** **186.2**	76 Osmium **Os** **190.2**	77 Iridium **Ir** **192.2**
Period 7	87 Francium **Fr** **(223)**	88 Radium **Ra** **(226)**	89–103 Actinoids	104 Rutherfordium **Rf** **(261)**	105 Dubnium **Db** **(262)**	106 Seaborgium **Sg** **(266)**	107 Bohrium **Bh** **(264)**	108 Hassium **Hs** **(267)**	109 Meitnerium **Mt** **(268)**

- Alkali metal
- Alkaline earth metal
- Transition metal
- Lanthanoids
- Actinoids
- Unknown chemical properties
- Post-transition metal
- Metalloid
- Reactive non-metal
- Halide
- Noble gas

Lanthanoids

57 Lanthanum **La** **138.9**	58 Cerium **Ce** **140.1**	59 Praseodymium **Pr** **140.9**	60 Neodymium **Nd** **144.2**	61 Promethium **Pm** **(145)**	62 Samarium **Sm** **150.4**	63 Europium **Eu** **152.0**

Actinoids

89 Actinium **Ac** **(227)**	90 Thorium **Th** **232.0**	91 Protactinium **Pa** **231.0**	92 Uranium **U** **238.0**	93 Neptunium **Np** **(237)**	94 Plutonium **Pu** **(244)**	95 Americium **Am** **(243)**

						Group 18
						2 Helium **He** 4.0

		Group 13	Group 14	Group 15	Group 16	Group 17	
		5 Boron **B** 10.8	6 Carbon **C** 12.0	7 Nitrogen **N** 14.0	8 Oxygen **O** 16.0	9 Fluorine **F** 19.0	10 Neon **Ne** 20.2

Group 10	Group 11	Group 12	13 Aluminium **Al** 27.0	14 Silicon **Si** 28.1	15 Phosphorus **P** 31.0	16 Sulfur **S** 32.1	17 Chlorine **Cl** 35.5	18 Argon **Ar** 39.9
28 Nickel **Ni** 58.7	29 Copper **Cu** 63.5	30 Zinc **Zn** 65.4	31 Gallium **Ga** 69.7	32 Germanium **Ge** 72.6	33 Arsenic **As** 74.9	34 Selenium **Se** 79.0	35 Bromine **Br** 79.9	36 Krypton **Kr** 83.8
46 Palladium **Pd** 106.4	47 Silver **Ag** 107.9	48 Cadmium **Cd** 112.4	49 Indium **In** 114.8	50 Tin **Sn** 118.7	51 Antimony **Sb** 121.8	52 Tellurium **Te** 127.6	53 Iodine **I** 126.9	54 Xenon **Xe** 131.3
78 Platinum **Pt** 195.1	79 Gold **Au** 197.0	80 Mercury **Hg** 200.6	81 Thallium **Tl** 204.4	82 Lead **Pb** 207.2	83 Bismuth **Bi** 209.0	84 Polonium **Po** (210)	85 Astatine **At** (210)	86 Radon **Rn** (222)
110 Darmstadtium **Ds** (271)	111 Roentgenium **Rg** (272)	112 Copernicium **Cn** (285)	113 Nihonium **Nh** (280)	114 Flerovium **Fl** (289)	115 Moscovium **Mc** (289)	116 Livermorium **Lv** (292)	117 Tennessine **Ts** (294)	118 Oganesson **Og** (294)

64 Gadolinium **Gd** 157.3	65 Terbium **Tb** 158.9	66 Dysprosium **Dy** 162.5	67 Holmium **Ho** 164.9	68 Erbium **Er** 167.3	69 Thulium **Tm** 168.9	70 Ytterbium **Yb** 173.1	71 Lutetium **Lu** 175.0
96 Curium **Cm** (247)	97 Berkelium **Bk** (247)	98 Californium **Cf** (251)	99 Einsteinium **Es** (252)	100 Fermium **Fm** (257)	101 Mendelevium **Md** (258)	102 Nobelium **No** (259)	103 Lawrencium **Lr** (262)